# *Contemporary Plant Systematics*

***Dennis W. Woodland***
*Andrews University*

*Prentice Hall, Englewood Cliffs, New Jersey 07632*

*This book is dedicated to all students with a sense of wonder.*

Signing Representative: Joyce Bast
Series Editor: Nancy Forsyth
Series Editorial Assistant: Christopher Rawlings
Production Administrator: Susan McIntyre
Text Designer: Anne Marie Fleming
Cover Administrator: Linda Dickinson
Cover Designer: Suzanne Harbison
Manufacturing Buyer: Megan Cochran

A Division of Simon & Schuster
Englewood Cliffs, New Jersey 07632



**Library of Congress Cataloging-in-Publication Data**

Woodland, Dennis W.
Contemporary plant systematics / by Dennis W. Woodland.
p. cm.
Includes bibliographical references and index.
ISBN 0-205-12182-9 :
1. Botany—Classification. I. Title.
QK95.W66 1991
581′.012—dc20 90-47770
CIP

Printed in the United States of America
10 9 8 7 6 5 4 3 2 1 95 94 93 92 91 90

# Contents

# *Preface*

Why another systematics textbook? This is a valid question, as there are several recent texts on the market. I will try to give a satisfactory answer.

For almost twenty-five years I have taught a basic course in systematic botany to undergraduate students. One of the exercises undertaken by the students each year toward the end of the course is to critique a published text in plant systematics. Some of the students' comments have been as follows: "Good text but too large for one term"; "Poorly illustrated"; "Most boring"; "Fine for the basic terminology, but doesn't go far enough"; "Only includes flowering plants"; "Gives basic information, but lacks contemporary material from the taxonomic field"; "Why not write a text yourself?"

Reflecting on the students' comments, I realized that perhaps they were right. Why not write a well-illustrated, broad-view, beginning text that would give students, wherever they may live in the world, sufficient botanical understanding of vascular plants?

## *Objectives*

*Contemporary Plant Systematics* has been written for the undergraduate student and serious amateur gardener-botanist who has taken at least a beginning biology course in high school or college.

The text has three objectives: (1) to teach basic botanical facts as applied to vascular plants, (2) to relate these facts to systematic principles, and (3) to show how systematic principles are important to contemporary botanical and environmental issues from a world perspective.

## *Features of the Text*

### *Organization*

The chapter sequence has been arranged to introduce students first to basic information and terminology; then to the nonflowering plants, followed by the terminology and descriptions of the flowering plants; then to historical aspects; and last, to contemporary approaches to and issues facing systematics. This systematic approach builds on the knowledge gained from the previous chapters. Terms encountered the first time are in **bold type** and are defined.

The last portion of the book attempts to show students the relevancy of systematic botany to modern society.

### *Illustrations*

Each chapter has been well illustrated using photographs and/or line drawings. Each of the plant families discussed is illustrated, emphasizing characteristics that help to distinguish the groups. The illustrations have all been sketched for this text by Ms. Anita Riess. They were prepared from actual specimens, photographs, and other drawings brought together to make a series of line drawings that is consistent in style and information. Certain parts of the drawings were styled to emphasize and clarify the characters.

### *Systematic Arrangement*

The classification of plant families follows, in general, well-published contemporary classification systems for each of the major vascular plant groups: ferns and fern allies, Pinophyta (gymnosperms), and Magnoliophyta (flowering plants). The use of these systems in no way reflects the author's preference of one system over another. The systems provide an organized, systematic sequence for study.

Many diverse families were chosen from all parts of the world and not just from families found predominantly in north temperate climates. Many families from tropical and Southern Hemisphere regions have been included to give students a more complete picture of family diversity. This was also done to make the text more usable outside the United States and to give students a world view.

### *Appendices*

Because references to floras are either out of print, not commonly found in most college or university libraries, or too expensive for most personal libraries, Appendix I gives an abridged bibliography of the published floras of the world's plants. This will be helpful to travelers who are interested in wild plants. Appendix II outlines the classification of the flowering plants according to a recent, well-known system.

## *Overview of the Content*

Chapter 1 introduces students to the field of systematics, the differences between identification and classification, and the purposes and significance of studying systematics. Chapter 2 reviews the rules of nomenclature that govern the naming of plant species. Chapter 3 focuses on important references and journals that are necessary for a student working in the taxonomic profession. Chapter 4 discusses how keys are constructed, how they are used, and the different types of keys. Chapter 5 describes how plants are collected, preserved, used, and housed; what techniques are followed; and why an herbarium is like a library of dried plants and a botanical resource center.

Chapter 6 explains reproduction and the terminology of ferns and fern allies, and briefly discusses some of the more common families, giving examples of distributions, relevant biological information, economic uses, and fossil records. Chapter 7 examines reproduction and the terminology in the naked-seed Pinophyta (gymnosperm) plants and briefly discusses the families, including examples of distributions, relevant biological information, economic uses, and fossil records. Chapter 8 introduces the terminology and reproduction of flowering plants, including vegetative and reproductive terms, and methods of pollination. Chapters 9 and 10 discuss the more common families of flowering plants found in the world (not just the North Temperate Region), using terminology learned in the previous chapter. The discussion includes the description, distribution, economic uses, classification views, and fossil record. Chapter 9 deals with the Magnoliopsida (dicots), while Chapter 10 deals with the Liliopsida (monocots).

Chapter 11 traces the historical development of botany from the earliest records to the present and how botanists' views on classifying plants have changed. Chapter 12 first reviews differing views held today on the origins of life and vascular plants, and then addresses current theories on the evolution of vascular plants. Chapter 13 shows how the field of systematics synthesizes information from anatomy (contributed by Nels Lersten), morphology (contributed by Rolf Sattler), palynology (contributed by Cliff Crompton), biochemistry (contributed by Loren Riesberg), and cytology and genetics to develop a more natural and complete taxonomy.

Chapter 14 explores the reasons why some plants are becoming extinct, what is being done about this problem, and why the work of a taxonomist is important. Chapter 15 traces the reasons why botanical gardens exist today, where and how they developed, and their functions. Representative gardens from all parts of the world are discussed and illustrated. The epilogue summarizes what has been studied and explains how the modern plant systematist fits into contemporary society.

## *Supplements*

Three supplementary materials are available to adopters of *Contemporary Plant Systematics:* a collection of 35-mm color slides of the plant families discussed in the text, a computer program for making labels for pressed specimens, and a 60-minute cassette tape, "Carl Linnaeus: The Second Adam," by W. T. Stearn. All items can be obtained from the author at cost.

## *Acknowledgments*

Many individuals have greatly influenced this book at various stages of its development. I am indebted to my students over some 25 years, who have sat in my classes and made many valuable comments on the relevancy of systematics.

To provide students with a sketch of a plant family instead of just "dry words," special credit goes to Ms. Anita Riess, a superb botanist, fine artist, and former student. Her keen eye for detail and quality is appreciated.

Special tribute goes to my friend Ms. Lynn E. Steil, who spent many hours proofreading, when the warm outdoors and lush putting greens beckoned.

Ms. Debbie Wasmer and Ms. Debbie Owen's computer expertise and typing skills were invaluable in the final preparation of the manuscript.

Many individuals contributed photographs and slides for the supplements. These included Dr. Ed Anderson, Dr. John Beaman, Dr. Felicity Coats, Ms. Helen Cohen-Reimer, Dr. David Delcher, Dr. Mike Dillon, Dr. Al Gentry, Dr. William Grant, Dr. Walter Lewis, Ms. Elizabeth Parnis, Dr. Barbara Parris, Dr. Asa Thoresen, and Ms. Marilyn Ward. Mr. Norbert Andrus helped with darkroom preparations.

Many of my colleagues and friends have provided encouragement and constructive criticism on all or parts of the book. These include:

Dr. John Beaman, Michigan State University, East Lansing, MI, USA
Dr. Arthur Coetzee, Andrews University, Berrien Springs, MI, USA
Dr. Peter Crane, Field Museum of Natural History, Chicago, IL, USA
Dr. Arthur Cronquist, New York Botanical Garden, New York, NY, USA
Dr. Mike Dillon, Field Museum of Natural History, Chicago, IL, USA
Ms. Suzanne Forget, Institute botanique, Universite Montréal, Montréal, Quebec, Canada
Dr. David Given, DSIR, Christchurch, New Zealand
Dr. Ron Hartman, University of Wyoming, Laramie, WY, USA
Dr. W. William Hughes, Andrews University, Berrien Springs, MI, USA
Dr. Walter Lewis, Washington University, St. Louis, MO, USA
Dr. John McNeill, Ontario Provencial Museum, Toronto, Canada
Dr. Gilbert Muth, Pacific Union College, Anguin, CA, USA
Dr. Barbara Parris, Royal Botanic Gardens, Kew, England
Dr. Richard Pippen, Western Michigan University, Kalamazoo, MI, USA
Ms. Kathy Pryer, National Museum of Natural History, Ottawa, Canada
Mr. Allen Radcliffe-Smith, Royal Botanic Gardens, Kew, England
Dr. Donald Rigby, Walla Walla College, College Place, WA, USA
Dr. Dan Skeen, Albion College, Albion, MI, USA
Dr. Roy Taylor, Chicago Botanic Garden, Chicago, IL, USA
Dr. John H. Thomas, Stanford University, Stanford, CA, USA
Dr. Robert Thorne, Rancho Santa Ana Botanical Garden, Claremont, CA, USA
Dr. Ed Voss, University of Michigan, Ann Arbor, MI, USA
Dr. Warren H. Wagner, Jr., University of Michigan, Ann Arbor, MI, USA
Ms. Marilyn Ward, Royal Botanic Gardens, Kew, England

Special thanks are also due to Dr. Pat Holmgren for providing up-to-date information on the world's herbaria.

Funding to help defray some of the costs for illustrations and general expenses came from a grant from the Office of Scholarly Research, Andrews University. The continued support of Dr. Arthur O. Coetzee, Vice President for Academic Administration, and Dr. Merlene Ogden, Dean, School of Arts and Sciences, has been greatly appreciated.

Lastly, I wish to thank my wife, Betty, and my daughters, Cherié and Heather, for their help and encouragement at many crucial times.

D.W.W.

# 1

# *The Significance of Systematics*

The study of systematics in the area of plant biology is a very old one. Its development has paralleled the age of world exploration. With the exploration of new lands, collections of specimens were sent back to the homelands.

To distinguish between the organisms observed or collected, humans have tried to recognize and organize them by using various characters or attributes that are common to a large number of individuals. By doing this, organisms are arranged (or "classified" as it were) into recognizable groups. It is often necessary for individuals to know the group to which an organism belongs, to be able to identify it, and to be able to communicate the information to others. These activities form the basis of what is known as systematics.

People of primeval cultures were interested in the numerous different plants that grew in their native environments. These plants were generally recognized and categorized according to their particular uses, such as for food, medicine, domestic uses, poisons, etc. Even in the late twentieth century, some ethnic peoples use a classification scheme for the plants they utilize in their day-to-day experience. An excellent example of this is reflected by the classification used by the Indians of Chiapas, Mexico (Berlin, Breedlove, and Raven, 1974).

In the libraries of the older botanical centers of the world, old handwritten or printed botanical works can be found. These systems of classification, expounded first by the Greeks from about 300 B.C. to the middle of the 1700s, were very crude systems developed by philosophers, botanists, and medical persons (herbalists) of their day.

From the 1400s to the late 1890s, many Western European voyages of discovery were sent to all parts of the world. The result was a great increase in the number of plants collected by botanists and brought back to Europe for study. These plants were in the form of seeds, fruits, living material, and pressed collections. The pressed collections were grouped together into a **herbarium** (pl. **herbaria**). Many of these collections, when brought together, became the starting point for some of the largest institutional collections in the world today.

In the mid to late 1800s, centers of botanical study developed in Europe and North America around botanical gardens and universities. As interest grew, various disciplines of botany began to emerge. Besides the traditional horticultural gardens of interest, the discipline of the **taxonomist** began to flourish. Individuals became interested in naming, describing, and classifying the world's plants. Emphasis was not on medicinal plants, as with the early herbal writers, or on new garden plants, as in the case of the gardener-horticulturist, but on basic botanical information and naming.

## Definitions

Before going further, some basic terms that are frequently encountered by beginning students will be defined.

### Taxonomy and Systematics

**Taxonomy** is generally considered as the study of classification, including its rules, theories, principles, and procedures. This old term, first used by A. P. de Candolle in 1813, was applied to the process of classification. It has been referred to by some as "alpha taxonomy." **Systematics,** on the other hand, has been applied to various kinds of organisms and to the diversity and relationships between them. Taxonomy, classification, naming, and identification are therefore encompassed under the broader term *systematics.* Despite these differences, others feel that the terms have little distinction between them and should be loosely treated as synonyms. Due to the introductory nature of this text, the two terms will be used interchangeably.

### Biosystematics

**Biosystematics** (or **biosystematy**) is a more recent descriptive term that was introduced by Camp and Gilly (1943) ". . . to delimit the natural biotic unities and to apply to these units a system of nomenclature adequate to the task of conveying precise information regarding defined limits, relationships, variability, and dynamic structure." Clausen et al. (1945) regarded genetics, comparative morphology, and ecology as supplying the necessary data that, taken and applied collectively to the study of speciation, make up biosystematics. More recently, botanists have sometimes referred to biosystematics as **beta taxonomy, experimental systematics,** or as the systematics of living organisms.

### Classification

**Classification** is the orderly arrangement of plants into a hierarchal system. This system of arrangement is derived from an accumulation of information about the individual plants, with the end result expressing an interrelationship.

There are three kinds of classification: artificial, natural, and phylogenetic. **Artificial systems** are based upon obvious or convenient fundamental items of information, called **characters,** for the purpose of categorizing or sorting, irrespective of any affinity. For example, the flowers of the grassland might be grouped according to their flower color or whether they are short or tall plants. This would not necessarily reflect any particular lineage or relationship. The early botanical works (until and including Linnaeus' *Species plantarum,* which emphasized the number of sexual parts of the flower) fall into this category. The **natural classification systems** of the past have been based upon large, morphological features that give a sense of having many characters of one plant correlating with those of another plant. This system has a predicting value to it, because the more attributes we include in it, the more natural it becomes. A morphologically-based system is a natural one only as far as the character information is concerned and is only a step in the direction of multiattribute utilization. The post-Darwin systems have been viewed as **phylogenetic.** These attempt to infer a particular lineage of diverse plants by the utilization of a wide variety of information. Ideally, a phylogeny should be recorded in the fossil record. However, the fossil record is rather fragmentary at best, and, therefore, modern classifications utilize a wide variety of information in constructing phylogenetic hypotheses.

### *Identification*

**Identification** is the assigning of an existing name or taxonomic group—usually a species name—to an unknown plant. The recognition is based on the comparison of a specimen with certain features of the root, stem, leaf, flower, fruit, habitat, or locality to a previously described plant. Because no two individual plants are exactly alike and are treated as being taxonomically identical, some botanists prefer to use **determination** to describe this process. In most beginning plant systematics courses, students are not concerned with classifying plants, but instead focus on identifying them. So, when a student says, "I am having difficulty classifying this plant," the incorrect word is being used. Instead, this should be phrased, "I am having difficulty identifying this plant," or "I am having difficulty determining the name of this plant," as the classification has already been done by someone else.

### *Keying*

**Keying** is the process that is used to identify an unknown plant. The scheme that is followed is called a taxonomic **key,** which usually uses a series of **couplets** (contrasting dichotomous choices); for example, leaves tomentose (with matted woolly hairs) on the undersurface versus leaf glabrous (without hair) on the undersurface. If the leaves of the specimen lack hairs (glabrous), then that choice in the key is followed, eliminating the other part or "leg" of the couplet. This process, if followed until all choices have been exhausted, will (hopefully) leave the correct answer remaining.

In any "natural classification," the basic unit is termed a **species,** which are units of populations of individuals distinguished on the basis of particular characters. Next, if similarities between different species are grouped together, they are termed **genera,** which may be grouped into successive, more inclusive categories called **families, orders, classes,** etc. Each of these categories may, on occasion, depending on the evidence, be divided into particular subdivisions (i.e., subspecies, subclass, etc.). Beginning students will

**TABLE 1.1 Important categories of taxonomic hierarchy as applied to the taxon *Solidago canadensis*, goldenrod.**

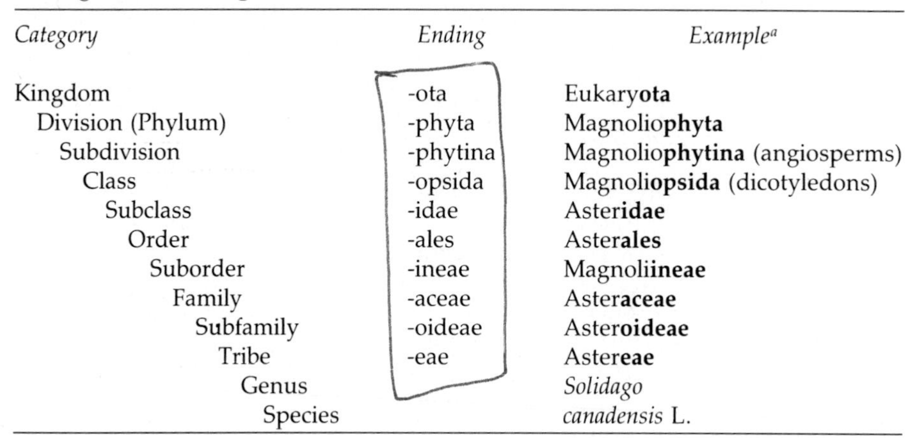

| *Category* | *Ending* | *Example*[a] |
|---|---|---|
| Kingdom | -ota | Eukary**ota** |
| Division (Phylum) | -phyta | Magnolio**phyta** |
| Subdivision | -phytina | Magnolio**phytina** (angiosperms) |
| Class | -opsida | Magnoli**opsida** (dicotyledons) |
| Subclass | -idae | Aster**idae** |
| Order | -ales | Aster**ales** |
| Suborder | -ineae | Magnoli**ineae** |
| Family | -aceae | Aster**aceae** |
| Subfamily | -oideae | Aster**oideae** |
| Tribe | -eae | Aster**eae** |
| Genus | | *Solidago* |
| Species | | *canadensis* L. |

[a]Note that the recommended endings are in boldface.

more commonly encounter categories below the species level. These are **subspecies, variety,** and **forma.** Clear distinctions are not necessarily made between subspecies and varieties, with some botanists using them synonymously. The forma name was used more in the past for minor character differences in a species (i.e., a white flower color), but is less used by contemporary botanists.

### *Nomenclature*

**Nomenclature** is the application of technical names to plants in accordance with an agreed set of rules. These rules give procedures and instructions for selecting the correct **taxon** (pl. **taxa**)—a term applied to any taxonomic group (i.e., species, genus, family, etc.) name—or the creation of a new one. A description of the characteristics of the taxon may be given. This taxon can then be incorporated into a list or description of the plants for a geographical region. This is called a **flora.** The preparation and collection of information for a flora is sometimes loosely referred to as **floristics.** A more in-depth discussion of taxonomic and nomenclature principles may be found in the next chapter.

## *Purposes and Significance of Systematics*

Today the main purposes of systematic botany are fourfold: (1) to compile an inventory of the world's plants, (2) to produce a classification system that reflects plant evolution, (3) to attempt to understand the great diversity within the botanical world, and (4) to assemble botanical knowledge about plants and helpful treatises for biological students, scientists, and the general public so that they can have a greater appreciation of the role of plants in the world we live in.

To reach these objectives is a very challenging goal for the taxonomist. The work of the taxonomist is the foundation of all the disciplines of botany, because it provides the names, lists, and classification of plants. The taxonomist's work provides the communication tools for sharing botanical information between cultures, languages, and people. Names provide a handle upon which to grasp each plant. Nameless plants lose their value because they cannot be understood until properly identified and described.

The field of systematic botany is very relevant in late twentieth century society. The potential use of wild plants for economic purposes has barely been explored, yet we must know the names of plants and which ones are related to one another. Wild forms of present-day cultivated species may have desirable characters, such as cold-hardiness, which, using modern experimental techniques, can be used for crop improvement. Many of the taxonomic works about various groups and floras of many parts of the world were written 100–150 years ago. Many new species have been described since that time, and much new information warrants new treatments and monographs of various groups and the writing of new floras. Some of these newer species may have horticultural value. This was recently illustrated when in England and Canada, a small, but very attractive little plant began to show up in flower shops without a name. After some plant "detective" work, it was determined to be *Pilea peperomioides* from the interior of China. It was previously known only from one preserved specimen in Edinburgh, Scotland. This plant's potential is limitless as a cultivated house plant.

Inventories of the tropical ecosystems, especially rain forests, are desperately needed before they are destroyed by lumbering, agricultural, mining, and human practices. Most of the ecosystems have not been studied, and the remaining ones are being cleared faster than they are being revegetated. Investigators are needed in countries such as the United States to inventory remaining natural areas due to new laws that require environmental impact studies before major alteration of the environment can take place.

The "green revolution" and the resulting environmental awareness have created a necessity for nature interpretation and information in local nature centers, and in local, state (provincial), and national parks. Taxonomists can find ample opportunity for instructing many civic, garden, and community groups in the local flora. Many times the energy and enthusiasm of groups such as these appear to be boundless.

You may ask, "How can I ever learn and remember all the names of plants and the terms used to describe them?" There are several ways to do this. One method is to ask a knowledgeable person and have him/her tell you. This is the easiest method, but also the quickest way to forget. The second method is to memorize terms and names, but after a brief time the words are usually forgotten. The third and best method is to learn by continual use and by relationship. When a famous person is observed briefly, his or her name does not become well established in the mind. The person's name is not forgotten when continually used. This same principle applies to plant names and terms. By continual use and application through the next four chapters, it is hoped that students will incorporate much of the information presented in their minds and will apply it throughout the remainder of this text.

## *SELECTED REFERENCES*

Berlin, B., D. E. Breedlove, and P. H. Raven. 1974. *Principles of Tzeltal Plant Classification, and Introduction to the Botanical Ethrography of a Mayan Speaking People of Highland Chiapas.* Academic Press, New York and London.

Camp, W. H., and C. L. Gilly. 1943. The Structure and Origin of Species. *Brittonia* 4:323–385.

Candolle, A. P. de. 1813. *Théorie élémentaire de la botanique.* Deterville, Paris.

Clausen, J., D. D. Keck, and W. M. Hiesey. 1945. *Experimental Studies on the Nature of a Species, II. Plant Evolution Through Amphidiploidy and Autoploidy, with Examples from the Madiinae.* Carnegie Inst., Washington D.C., publ. no. 564.

Davis, P. H., and V. H. Heywood. 1973. *Principles of Angiosperm Taxonomy.* Robert E. Krieger, Huntington, NY, 558 pp.

Gomez-Pompa, A., C. Vazques, and S. Guevara. 1972. The Tropical Rain Forest: A Nonrenewable Resource. *Science* 177:762–765.

Heywood, V. H., and D. M. Moore (eds.) 1984. *Current Concepts in Plant Taxonomy.* Sys. Assoc. Sp. Vol. No. 25. Academic Press, London, 450 pp.

Jones, S. B., and A. E. Luchsinger. 1986. *Plant Systematics,* 2nd ed. McGraw-Hill, New York, 512 pp.

Mason, H. L. 1950. Taxonomy, Systematic Botany, and Biosystematics. *Madroño* 10:193–208.

Mayr, E., E. G. Linsley, and R. L. Usinger. 1953. *Methods and Principles of Systematic Zoology.* McGraw-Hill, New York, 336 pp.

Naik, V. N. 1984. *Taxonomy of Angiosperms.* Tata McGraw-Hall, New Delhi, 304 pp.

Radcliffe-Smith, A. 1984. Plant Portrait. 5. *Pilea peperomioides. Kew Mag* 1(1):14–19.

Radford, A. E., W. C. Dickison, J. R. Massey, and C. R. Bell. 1974. *Vascular Plant Systematics.* Harper & Row, New York, 891 pp.

Raven, P. R., B. Berlin, and D. E. Breedlove. 1971. The Origins of Taxonomy. *Science* 174:1210–1213.

Ross, H. H. 1974. *Biological Systematics.* Addison-Wesley, Reading, MA, 345 pp.

Simpson, G. G. 1961. *Principles of Animal Taxonomy.* Columbia University Press, New York, 247 pp.

Stearn, W. T. 1961. Botanical Gardens and Botanical Literature in the Eighteenth Century. Catalogue of Botanical Books in the Collection of Rachel McMasters Miller Hunt 2:XIII–CXI.

Stuessy, T. F. 1975. The Importance of Revisionary Studies in Plant Systematics. *Sida* 6:104–113.

Turrill, W. B. 1938. The Expansion of Taxonomy. *Biol Rev* 13:342–373.

# 2

# *How Plants Get Their Names*

Many people are interested in learning about plants and their names. They may have various reasons for wanting to do so. These reasons may include:

Wanting to satisfy the desire to know "What plant is that?" Humans are by nature curious, and learning the name of an interesting wild or cultivated plant broadens our understanding of the living organisms around us.

Wanting to communicate about plants. In order to repeat either written or verbal plant information, there must be some commonly agreed way to express the information. This is usually done by names. Symbols or numbers could be used, but they are unsatisfying, because it is hard to relate to numbers.

Wanting to look up information about plants. Much information is available today about plants in general or about a specific plant in particular. This information includes information about the biology of the plant. Is it poisonous? Edible? How is it used? Is it a weed? Does the plant have an interesting history?

Satisfying these reasons has been accomplished by giving plants names. These are **common names** and **Latin names.** (Some persons use the term **scientific name** to apply to the Latin or botanical name, but there is really nothing scientific about the names. Botanical names are Latinized names applied to botanical organisms.) So-called common or vernacular names were used long before Latin names were given. It therefore seems best, at this time, to discuss the pros and cons for using either common or Latin names. There are advantages and disadvantages to both.

## Common, or Vernacular, Names

Common names are usually old names developed in the language of the culture and society where the plants are found. They are a means of communicating with the general public. There are reasons to use common names:

1. The names are usually simple and easy to remember (i.e., bluebell, may apple, redbud).
2. The names are familiar to the general population.
3. The names are often descriptive of the plant (i.e., old man's whiskers, scribble bark, pitcher plant).

There are also problems with the wide use of vernacular names:

1. Common names differ in different languages. This can be a hardship if one lives in parts of the world where different languages or dialects are spoken in the same geographic region, city, or state (i.e., Quebec, Canada [French and English]; Louisiana, USA [French and English]; Holland [Dutch, French, and German]; and Kenya [over 40 different languages and dialects].
2. The same species of plant may have many different common names depending on the locality and type of people. For example, the state flower of Utah is *Calochortus nuttallii* T. & G. in Beckwith, commonly called sego lily. The same plant in other regions of the western United States is referred to as mariposa lily. Which name is correct, or are they both correct?
3. Common names are many times misleading. For example, the pineapple is not a pine or an apple and is not related to either.
4. The same vernacular name may be applied to different, unrelated plants. For example, the word *lily* refers to many plants that look like lilies and *pine* and *oak* in Australia refer to different plants than those found as natives in the Northern Hemisphere.
5. Common names do not have any guiding rules or methods to follow that help to standardize vernacular name usage. There have been some feeble attempts to standardize common plant names, but most efforts have not succeeded. Two helpful sources of common names are the book *Standardized Plant Names* and the *Canadian Journal of Plant Science,* which publishes an ongoing series called "Biology of Canadian Weeds," in which the accepted common names of weeds are given.
6. If a plant is not overly widespread and encountered by the general public on a regular basis, it may not have a vernacular name. Therefore, an individual has difficulty communicating information about the plant in question. This may be very important, for example, if the plant is endangered or threatened in some way.
7. Some common names have been constructed by botanists writing books, by coining directly from the Latin name or from the name of the person who gave the plant its botanical name. For example, *Urtica*

*gracilis* (Aiton) Nuttall might be referred to as a slender nettle from the Latin *gracilis,* meaning slender, and *Cornus nuttallii* Nuttall would be called Nuttall's dogwood after Thomas Nuttall, who named the plant. These names would not be well-known by the general public and would be just as foreign as the botanical names. A few of these names, however, have been incorporated into general usage (i.e., *Cornus canadensis* L. is termed Canada dogwood, and *Pinus coulteri* D. Don. is called Coulter pine).

## *Latin Names*

Latin (also called botanical or technical) names were applied to plants at the time of history when Europeans were beginning to write about the medicinal and economic uses of plants. It was during the period, between the 1200s and 1700s, that many of the great exploratory voyages took place, and the language of educated communication was Latin. It, therefore, was only natural for writers to write a plant's name in Latin.

There are several good reasons to use Latin names:

1. A plant can have only one valid botanical name. Over the years, the plant may have obtained more than one name, but only one will be correct and the others are synonyms. On occasion this valid name may be used through error to apply to a different plant. We then refer to this name as a **homonym.**
2. Because a plant has only one valid Latin name, a widely distributed plant will have that same name everywhere in the world, and it will be written the same way, regardless of the country or language. Latin names stand out in articles written in such languages as Chinese, Japanese, Korean, and Russian, because the Latin name is printed in Roman letters and is in italics. For example, *Urtica dioica* L. can be read by all peoples irrespective of native language.
3. Latin names and original descriptions of plants are published following a definite agreed-upon system of regulations and rules. These rules, called the **International Code of Botanical Nomenclature,** are modified and updated every 6 years during the Nomenclature Section of the International Botanical Congress.
4. Latin names are often very descriptive of some aspect of the plant or have a particular meaning. For example, the name *Trifolium repens* L. means that the plant has creeping stems belonging to a group having leaves bearing three leaflets, and *Cornus canadensis* was possibly first described in and is distributed in Canada.

Not all things about Latin names are positive:

1. Botanical names are occasionally long, cumbersome, and difficult to pronounce and remember (i.e., *Parthenocissus quinquefolia*). However, some parts of the Latin name, especially the first or generic name, have become so well known and familiar as to be recognized by the

general public as common names (i.e., *Aster, Chrysanthemum, Cyclamen, Geranium, Rhododendron,* and *Zinnia.*)

2. The laws that govern botanical names were not always the same in all parts of the world. For many years botanists in North America followed the "American Code," while European botanists used the "Vienna Code." This led to different valid names for a large number of plants. Agreement to follow only one set of rules was not forthcoming until after 1930.
3. Older more familiar names may be changed to a less well-known name. This is due to one of the rules of the code, called the **rule of priority,** which in general states that the older of two or more conflicting Latin names is the correct name. This means that a well-known botanical name may have an older, recently rediscovered name. This earlier name has priority over the familiar name. More will be said about this later.
4. On occasion, improper meaning may be applied to the Latin name. This can best be understood by an illustration. The plant *Convallaria trifolia* L. has three leaves. E. L. Green felt the plant belonged in the genus *Unifolium* and therefore gave the combination name as *Unifolium trifolium* (L.) Green, a three-leaved species in a one-leafed genus.

On the practical side of the nomenclature issue, it may be best to learn both the common name as well as the Latin name of a plant to be able to communicate with all types of people. That is ultimately the goal of having names for plants.

### *Constructing the Latin Name*

Until the mid 1700s botanists who wrote about plants used too many Latin words to name a plant. This name became, in many cases, the Latin description of the plant, because of its length. These **polynomials** were most cumbersome.

In 1753 a Swedish naturalist-physician named Carl Linnaeus published a book called *Species plantarum,* in which he consistently gave all the plants two-word Latin names, or **binomial nomenclature.** Linnaeus was not the first person to use binomial names, for other authors in the 1500s and 1600s had used them occasionally. However, he was the first person to apply binomial nomenclature to all plants (and animals too) known at his time. The time was right for something better than the cumbersome polynomials, and the scientific community accepted Linnaeus's system. Linnaeus is therefore called the **Father of Modern Taxonomy.**

The first word of the name consists of the **genus** and is a noun. The second word of the Latin name is the **specific epithet** and is an adjective or a possessive noun. Many times the specific epithet is erroneously called the species. The **species** name is a two-word name consisting of the genus and the specific epithet (i.e., *Acer saccharum* Marsh, *Vitis vinifera* L.) The genus name is always capitalized, while the specific epithet is not normally capitalized. Because the species name is in Latin, it is italicized or underlined. The International Code of Botanical Nomenclature does allow the specific epithet to be capitalized if the name is the name of an individual, but the rule does recommend (recommendation 73F) that even persons' names used as specific epithets be written in lower case. For example, *Agropyron Smithii* Rydb. would be written as *Agropyron smithii* Rydb.

A part of the plant's Latin name that is frequently overlooked is the author's name. This is the name of the person or persons who were responsible for giving the plant its botanical name. The authority's name(s) follows after the specific epithet and is commonly abbreviated. For example, Carl Linnaeus is shortened to L. (i.e., *Trillium erectum* L.). Thomas Nuttall to Nutt. (i.e., *Urtica holosericea* Nutt.), and John Torrey and J. D. Hooker to Torrey & Hook. (i.e., *Carex oxylepis* Torrey & Hook.). Some taxa will have a second name(s), as in *Scirpus hudsonianus* (Michx.) Fern., a circumpolar bulrush. The first name in brackets means that the species epithet was first applied to this taxon by F. Michaux. Years later M. L. Fernald realized that the plant belonged in the genus *Scirpus* and therefore placed the species in that genus. Because the plant had been first called *Eriophorum hudsonianum* by F. Michaux, Michaux's name is placed in brackets first, followed by the individual who transferred and published the change (in this case, Fernald). The name supplied by Michaux is the basic name, or **basionym.**

The question is sometimes asked, why include the authority's name when writing the Latin name? This procedure is followed because different botanists in different places and times may, by accident, apply the same name to different plants. For example, *Urtica gracilis* Ait. was given to a plant species collected from Hudson's Bay in Canada. Unaware of this, the same name, *Urtica gracilis* Raf., was given later to a plant found in Louisiana that is today known as *Urtica chamaedryoides* Pursh.

Occasionally the authorities' names will be separated by an ampersand (&) or the preposition *ex* or *in*. When the ampersand is used it means the two authors described plants and worked together (i.e., *Marsilea vestita* Hook. & Grev.). Using "ex" means the first author proposed the name for the plant, but the second person published the name (i.e., *Picea engelmannii* Parry ex Engelm.). The use of "in" indicates that the first person described the plant in an article published or edited by the second (i.e., *Pinus flexilis* James in Rep.).

By consistently using Latin names followed by the accepted authorities in writing (especially in foreign languages), much confusion can be eliminated.

### *Infraspecific Names*

In many wide-ranging species there may be considerable variation in morphology. The morphological variants may become geographically isolated populations and therefore have sufficient differences to warrant being called by a subspecies, variety, or forma name.

Subspecies are more inclusive than varieties. However, variety names have been used in more plant groups than subspecies names. Unfortunately, the two taxonomic categories do not have clear distinctions between them and have been considered by some botanists to be used interchangeably. Whether a botanist uses either a subspecies or variety category may depend on personal preference, if past nomenclatural uses included one over the other, or if major name changes have to take place, depending on using one subspecific level over the other. In extremely plastic species, both subspecies and variety names may be used (i.e., *Sporobolus cryptandrus* (Torr.) Gray subsp. *cryptandrus* var. *occidentalis* Jones & Fassett). In the past, more than today, some botanists would recognize plants that were variable in only one or two characters. Such plants are called **formas** (i.e., *Aconitum columbianum* Nutt. in T. & G. forma *ochroleucum* (A. Nels.) St. John). The forma name is usually applied to plants that have one to a few characters that are different from the "normal" species and reflect simple genetic differences.

If a species is divided into at least one infraspecific taxon or if species are combined, the original species undergoes some name modification, with a readjustment of names taking place. The original species takes on the duplicate name of the specific epithet as its infraspecific name. The new infraspecific taxon is given a new name. For example, *Urtica dioica* L. subsp. *dioica* becomes the typical taxon and *U. dioica* L. subsp. *gracilis* (Ait.) Selander the new combination. It should be noted that the duplicate infraspecific name of the epithet does not have an author name following it.

## *Principles of the Code of Botanical Nomenclature*

The rules governing the use and application of Latin names are revised every 6 years during a special nomenclature session of the International Botanical Congress. The agreed-upon changes are then published in a book called the *International Code of Botanical Nomenclature.* The most recent guidelines were voted at a congress held in 1987 in West Berlin. Very few major changes were made in the "Berlin Code" as compared to the previous "Sydney Code." It seems best to understand the Code by briefly reviewing some of the main principles and articles that are sequentially given by it.

It should be emphasized that the best system of rules is a simple, specific nomenclature system that can be used by all botanists in all countries when dealing with taxonomic problems at all taxonomic levels.

**Principle 1. The Botanical Code is independent from the Zoological Code.** The Code applies to all plant groups except bacteria and blue-green algae. The principles of both codes are similar but differ in various details. For example, plants and animals can have similar names (i.e., *Corydalis* is a plant and a bird, the lark), and the oldest date to which the priority of one name applies over another is different. The starting point for plant names is 1 May 1753, C. Linnaeus' *Species plantarum,* while zoological names begin with *Systema naturae* in 1759 by the same author.

**Principle 2. The application of names of taxonomic groups is determined by nomenclatural types.** When a new taxon is described by a botanist, a particular specimen is designated as the **type specimen,** to which the name given is permanently attached. This individual specimen is designated the **holotype** and is the most important specimen of a taxon for determining the correct application of a plant name (Fig. 2.1). The holotype is carefully preserved in a dried collection of plants, an herbarium, where future botanists can refer to it if necessary. This specimen is the **nomenclatural type** that is permanently associated with the name applied. Likewise, the nomenclatural

**FIGURE 2.1 Example of a holotype specimen of *Urtica andicola* Wedd. from Peru, named by H. A. Weddell in 1852. The specimen is housed in the herbarium of the Muséum National d'Histoire Naturelle, Paris.**

type of a genus is the species upon which the genus name is based; the nomenclatural type of a family is the genus upon which the family is based; and the nomenclatural type of an order is a family upon which the order name is based. A duplicate specimen of the holotype is called an **isotype.** The isotype must be collected by the same individual, at the same location, and at the same time as the holotype (Fig. 2.2).

It is unfortunate that the type method is a relatively recent system and types were not designated in the earlier botanical literature. Most authors did, however, mention specimens seen that were considered to be the same taxon. These are known as **syntypes.** When a holotype has become lost or destroyed, a substitute must be chosen to replace the holotype. A **lectotype** is a specimen chosen from the original material studied by the original author (Fig. 2.3). A **neotype** is a specimen chosen as a nomenclatural type when all material studied by the original author is missing or destroyed. A **topotype** is a specimen collected at the same locality as the holotype but at a different time, regardless if collected by the same or a different collector. It should be pointed out that at no time does any of the above nomenclature types take precedence over a holotype or isotype.

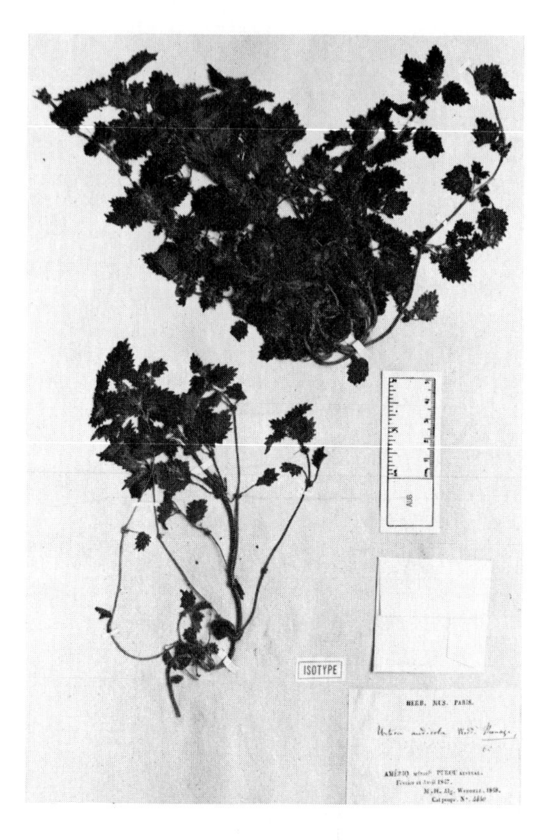


**FIGURE 2.2 Example of two isotypes of *Urtica andicola* Wedd. from Peru. The specimens are housed in the herbarium of the Muséum National d'Histoire Naturelle, Paris.**

**Principle 3. The nomenclature of a taxonomic group is based on priority of publication.** This is the so-called rule of priority that is so important to systematics (Articles 11–13). The principle means that the Latin name for a plant that is closest to an agreed-upon starting point for names is considered to be the correct name. The beginning point for binomial names is 1 May 1753, the date for the publication of Linnaeus' *Species plantarum.* Binomial names were occasionally used before this date, but they do not have validity as far as priority is concerned. It should be pointed out that Linnaeus published *Species plantarum* in two volumes, in May and August, 1753, respectively, but for nomenclature purposes the works are considered as having been published on 1 May 1753. The rule of priority does not apply to taxa above the level of family.

**Principle 4. Each taxonomic group with a particular circumscription, position, and rank can bear only one correct name, the earliest that is in accordance with the rules, except in specified cases.** The priority rule becomes important when it is remembered that only in recent years have botanists had access to almost all the world's literature and a wide variety of preserved specimens with which to study a taxon. The early botanists had few specimens to work with, little field observation, and many times had

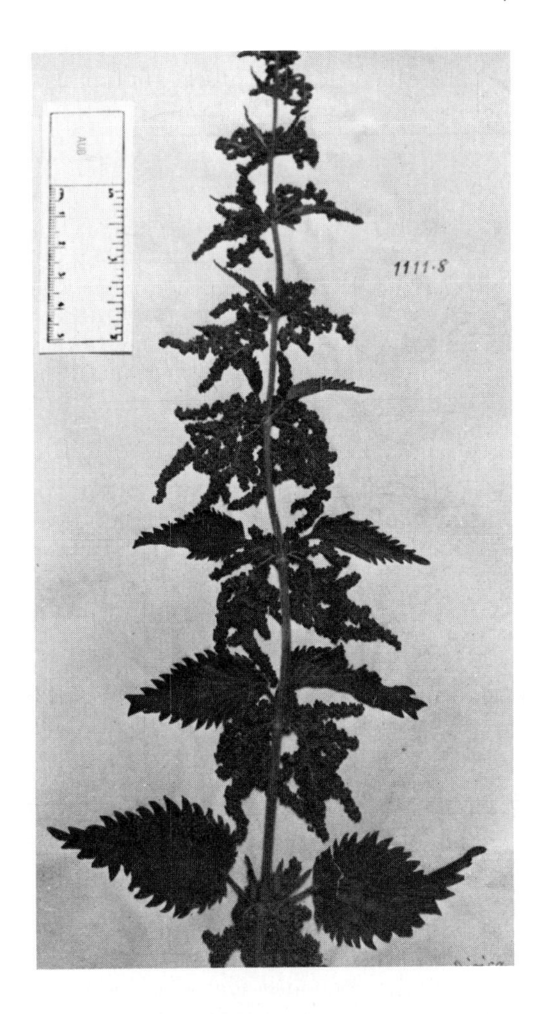

**FIGURE 2.3 A lectotype of *Urtica dioica* L., named by Carl Linnaeus in *Species plantarum* (p. 984). The specimen is housed in the herbarium of the Linnean Society of London.**

only small, poorly preserved fragments on hand. It was therefore inevitable that two or more names would be applied to the same taxon. For example, the following names have been given to plants of the genus *Penstemon* by different botanists at different times:

*Penstemon brachyanthus* Pennell 1941
*Penstemon formosus* A. Nels. 1904
*Penstemon micranthus* Nutt. 1834
*Penstemon procerus* Dougl. ex R. Grah. 1829
*Penstemon tolmiei* Hook. 1838

In the opinion of a botanist who studied the group, the above names apply to the same species, whose correct name should be *Penstemon procerus* Dougl. ex R. Grah., because it is the oldest validly published name. The remaining names are then treated as **taxonomic synonyms,** with each name based on different type specimens. If different epithets are given to the same type specimen, then these duplicate names are called **nomenclatural synonyms.** (The International Code of Nomenclature of Bacteria and the International Code of Zoological Nomenclature use the terms *subjective synonym* and *objective synonym,* respectively, for the taxonomic and nomenclatural synonyms of botany.)

It might be said at this point that at no time does a name or specific epithet have priority over another name at a different taxonomic level. In other words, an earlier epithet does not have precedence over a subspecies, variety, or another genus name; only at the species level does this priority exist. For example, *Penstemon procerus* Dougl. ex R. Grah., 1829, has priority over the names above but does not have priority over *Penstemon procerus* subsp. *tolmiei* Keck, 1957, and *P. procerus* subsp. *pulvereus* Pennell, 1920, because the latter two names are subspecies names.

The beginning point for most plant binary names is 1753. However, the beginning date for most mosses is 1 January 1801; that for fossil plants is 31 December 1820; and some fungi and algae start at various times, depending on the group. Generic names have their beginning date with Linnaeus's *Genera plantarum* ed. 5 (1754) and ed. 6 (1764). The rule of priority applies to all taxa of the rank of family and below, except for certain genus and family names that have been conserved *(nomina conservanda)* (i.e., *Buchloë, Brodiaea, Forsythia, Larrea, Leersia, Setaria, Wisteria*).

To conserve a later published, more well-known and used name takes a special voted decision at an International Botanical Congress to become valid after special proposals and committee approvals have taken place. For many years family and genus names were the only names that could be conserved. The most recent Code has a provision whereby, under restricted conditions, some species names (i.e., valuable economic plants, etc.) may be conserved. The conserved names are listed in the back of the most recently published Code.

Occasionally names do not conform to the guidelines given by the Code, and therefore these names are invalid and are termed **illegitimate names.** These names may be rejected because the type specimen was a monstrosity (that is, abnormal due to disease, contamination, a mixed type specimen of two different species, etc.), not validly published with a Latin description, or another earlier named plant was given the same name. Therefore, this later name, or homonym, is considered invalid and is rejected for use.

**Principle 5. Scientific names of taxonomic groups are treated as Latin regardless of their derivation.** As stated earlier in this chapter, botanical names of plants are given in Latin for the reasons mentioned. They are commonly Latin or Greek nouns or adjectives, or names from other languages written in Roman letters and Latinized. For example, when choosing genus names, the author should use words that are easily adapted to Latin, and the names should be short if possible and easy to pronounce. Names from different languages should not be combined, and the gender and endings of the word should be consistent. The species epithet may be a single word or hyphenated compound word. The gender of the epithet must agree with the gender of the genus (i.e., *Brassica nigra* L. Koch in Roehl., *Geranium maculatum*

L.). Trees are treated as being feminine (i.e., *Quercus alba* L.). When an epithet is the name of a person and ends in a consonant, except when the name ends in *er*, the letters *ii* are added (i.e., *Agropyron smithii* Rydb.). When the epithet is the name of a person and ends in a vowel, only one *i* is added (i.e., *Lilium grayi*).

**Principle 6. The rules of nomenclature are retroactive unless expressly limited.** The Code is a relatively recent set of rules that taxonomists did not have to guide them 100 or 200 years ago. Because early practices of botanists varied, the Code is amended at each International Botanical Congress. To apply the rules consistently, many of the regulations are made retroactive to a designated point in time. Some examples are as follows: since 1 January 1935, Latin descriptions for new taxa must be included in the publication. Since 1 January 1953, all new names published in newspapers, trade catalogues, and obscure nonbotanical journals and periodicals are *not* considered to be validly published, even with a Latin diagnosis. Since 1 January 1958, a holotype must be designated for all new taxa.

## *What Constitutes a Validly Published Name?*

Suppose that a recently collected plant from a tropical rain forest or unexplored region does not fit any description of similar known species for the geographical locality. There are a series of procedures that should be followed:

1. A comparison should be made between the proposed new species *(species nova)* and type specimens of similar taxa. It may be necessary to compare specimens at some large herbaria.
2. A detailed description in the author's language should be made.
3. A name should be chosen following the rules of the Code.
4. Holotype and isotype specimens should be designated.
5. A Latin diagnosis should be constructed. The diagnosis will include how the taxon differs from other taxa. It would also be advisable to show the diagnosis and documentation to a botanist who has had experience publishing similar articles. It is helpful to supply a detailed description and illustration along with the Latin diagnosis.
6. The information is published in a recognized professional journal or book that is available to and read by botanists (that is, **effectively published**). It is also advisable to deposit at least an isotype in a large herbarium for future reference if the author is employed at a smaller institution or does not have access to a well-maintained herbarium.

If the above procedure is followed, the new name will be **validly published** (that is, according to the proper format given in the Code), will be a **legitimate name** (one that has followed the rules of the Code), and will be correct. Whether the new name is accepted by the botanical community will depend on time and the strength of the author's arguments.

## *SELECTED REFERENCES*

Bailey, L. H., 1933. *How Plants Get Their Names.* Macmillan, New York.
Benson, L. 1962. *Plant Taxonomy.* Ronald Press, New York, 494 pp.
Berlin, B. 1973. Folk Systematics in Relation to Biological Classification and Nomenclature. *Ann. Rev. Ecol. Syst.* 4:259–271.
Brown, R. W. 1956. *Composition of Scientific Words.* Reese Press, Baltimore, MD.
Crozat, L. 1953. History and Nomenclature of the Higher Units of Classification. *Bull. Torrey Bot. Club* 72:52–75.
Gleason, H. A. 1932. The Pronunciation of Botanical Names. *Torreya* 32:53–58.
Greuter, W. and J. McNeill et al. (eds.). 1988. *International Code of Botanical Nomenclature. Regnum vegetabile Vol. 118.* Koeltz Scientific Books, Königstein, FRG, 328 pp.
Pesante, A. 1961. About the Use of Personal Names in Taxonomic Literature. *Taxon* 10:214–221.
Radford, A. E., et al. 1974. *Vascular Plant Systematics.* Harper & Row, New York, 891 pp.
Stern, W. T. 1983. *Botanical Latin,* 3rd ed. David & Charles, North Pomfret, VT, 566 pp.
Voss, E. G. et al. (eds.) 1983. *International Code of Botanical Nomenclature. Regnum vegetabile Vol. 111.* Bohn, Scheltema & Holkema, Utrecht, 472 pp.

# 3

# *The Literature of Systematics*

In the previous chapter, the way in which plants are named was described. Because the accepted plant literature dates back almost 250 years, a very complicated, old literature has developed. Innumerable books, journals, and pamphlets have been published, some with limited distribution and in many languages. Therefore, in order for taxonomists who write about plants to know the correct name for particular plants, they must have a good understanding of the taxonomic literature. The task would be almost impossible if it were not for certain bibliographic references and indices that make the task easier and less time-consuming. These aids are essential for conducting geographic, nomenclature, revision, and monographic studies; for locating literature dealing with bibliographic and biographic information; and for finding the economic uses of plants. Computerized data searches are also very helpful in gathering general information.

The following abridged reference compilations provide a starting point for information searches.

## *Indexes for Bibliographic, Illustrative, and Nomenclature Literature*

The following indexes are useful for working in plant systematics.

Christensen, C. F. A. et al. 1906–1965. *Index filicum.* Four supplements. Hagerup, Copenhagen.

This is an index dealing with ferns in the same manner that *Index Kewensis* deals with seed plants. R. E. G. Pichi-Sermoli authored Supplement 4 in *Regnum vegetabile,* 37, 1965.

Crabbe, J. A., A. C. Jermy, and J. M. Mickel. 1975. A New Arrangement for the Pteridophyte Herbarium. *Fern Gaz.* 11:141–162.

Due to the differences of opinion by botanists as to the limits of families and genera in ferns and fern allies, this work is very helpful in providing a quick reference to family and genus names in an arrangement scheme.

de Dalla Torre, C. G., and H. Harms. 1900–1907. *Genera siphonagamarum ad systema Englerianum conscripta.* Leipzig.

This reference is a numbered list of the families and genera according to the old Engler system of classification. The genera numbers are helpful for filing and finding specimens in herbaria that have arranged the collection following the numbers.

Engler, A. 1900–1968. *Das Pflanzenreich.* Nos. 1–107. Berlin.

This is a monographic series on the plant kingdom that is presently incomplete.

Engler, A. 1954–1964. *Sylabus der Pflanzenfamilien,* 12th ed., 2 vols. (ed. by H. Melchior and E. Werdermann). Gebrüder Borntraeger, Berlin.

This is the most recent summary of the Engler system of classification and gives descriptions of the higher taxonomic levels. Volume 1 covers the bacteria to the gymnosperms (Pinophyta), while volume 2 covers the flowering plants or angiosperms (Magnoliophyta).

Engler, A., and K. Prantl. 1887–1915. *Die natürlichen Pflanzenfamilien.* 23 vols. W. Engelmann, Leipzig.

This is the only detailed complete work dealing with the classification of plants from algae to flowering plants. It includes dichotomous keys, descriptions to all families and genera of plants, illustrations, and literature references. It is an enormous, valuable work. A second, incomplete edition is being produced.

Farr, E. R., J. A. Leussink, and F. A. Stafleu. 1979. *Index Nominum Genericorum (Plantarum). Regnum vegetabile* Vols. 100–102. Utrecht.

This is an index to genus names.

*Gray Herbarium Index.* 1896– . Gray Herbarium of Harvard University, Cambridge, MA.

This card index is continuously being updated and covers plants of the Western Hemisphere only. It is a very valuable reference used to locate the original literature source of generic, binomial, and infraspecific taxa. It now duplicates the new *Index Kewensis* and may be discontinued. Cards for 1896–1967 are reprinted in 10 volumes by G. K. Hall, Boston. A segment of a page is shown in Fig. 3.1.

*Index Kewensis plantarum phanerogamarum.* 1893–95. 2 vols. Oxford. 16 supplements, 1900–1988.

This work, compiled by the Royal Botanic Gardens, Kew, England, is an indispensible reference tool and the beginning point for all nomenclature searches. It is an alphabetical worldwide list of genus and species names of gymnosperms (Pinophyta) and flowering plants (Magnoliophyta). The information includes the author and the original reference to the genera and binomial names. The original two volumes (also some of the early supplements) gave some synonyms, country of origin for the name, and covered the botanical literature from 1753 to 1885. The supplements have been published at fairly regular intervals, with recent

**FIGURE 3.1 A segment of the *Gray Herbarium Index*.**

**Urtica dioica** subsp. **gracilis** var. **Lyallii** (S. Wats.) C. L. Hitchc.

Vasc. Pls. Pacif. NW. **2**:91. 1964.
*U. Lyallii* S. Wats.

**Urtica dioica,** var. **gracilis (Ait.) [incorrectly attributed to Wedd. by]** Seland.

Sv. Bot. Tidskr. **41**:271. 1947. [as *dioeca*].—In synon.
*U. gracilis* Ait.

**Urtica dioica mollis** (Steud.) Wedd.

Arch. Mus. Hist. Nat. Paris, **9**:78. 1856.
*U. mollis* Steud.

**Urtica dioica,** var. **mollis** Wedd.

DC. Prodr. **16**(1):51. 1869.—Chile & Mexico.

**Urtica dioica,** var. **occidentalis S. Wats.**

Bot. King's Exped. 321. 1871.—Nevada.

**Urtica dioica procera** (Willd.) Wedd.

Arch. Mus. Hist. Nat. Paris, **9**:78. 1856.
*U. procera* Willd.

**Urtica dioica,** var. **procera,** subvar. **duplicato-serrata** Wedd.

DC. Prodr. **16**(1):52. 1869.—Kentucky.

*SOURCE:* Used by permission of Gray Herbarium, Harvard University, Cambridge, Massachusetts.

volumes including infraspecific taxa and references to illustrations and excluding synonyms. B. D. Jackson compiled the original volumes, following the direction of J. D. Hooker. A segment of a page is shown in Fig. 3.2.

*Index Londinensis to Illustrations of Flowering Plants, Ferns and Fern Allies.* 1926–1931. 6 vols. Oxford. One 2-volume supplement, 1941.

This is an alphabetical genus and species index to vascular plant illustrations from 1753 to 1935. (After 1935 the information given was published in *Index Kewensis.*) This is especially helpful in locating illustrations

**FIGURE 3.2 A segment of the *Index Kewensis plantarum phanerogamarum* for the genus *Urtica*.**

**URTICA**, [Tourn.] Linn. Syst. ed. I (1735). *URTICACEAE*, Benth. & Hook. f. iii. 381.
ADICEA, Rafin. Cat. 13 (1824).
ILDEFONSIA, Mart. ex Steud. Nom. ed. II. i. 802 (1840).
RUTICA, Neck. Elem. ii. 202 (1790).
SELEPSION, Rafin. Fl. Tellur. iii. 48 (1836).
*acerifolia*, Zenker, Pl. Ind. Dec. i. tt. 3, 4 = Girardinia palmata.
*acuminata*, Poir. Encyc. Suppl. iv. 224 = Urera acuminata.
*adoënsis*, Hochst. in Flora, xxiv. (1841) I. Intell. 21 = Girardinia condensata.
*aestuans*, Linn. Sp. Pl. ed. II. 1396 = Fleurya aestuans.
*aestuans*, Sieber, ex Steud. Nom. ed. II. ii. 734 = Urera alceaefolia.
*affinis*, Hook. & Arn. Bot. Beech. Voy. 69 = Fleurya interrupta.
*alata*, J. F. Gmel. Syst. 268, sphalm. = elata.
*alba*, Blume, Bijdr. 499 = Leucosyke alba.
*alba*, Korth. ex Blume, Mus. Bot. Lugd. Bat. ii. 165 = Leucosyke sumatrana.
*alba*, Rafin. Fl. Ludov. 114 (Quid ?).—Am. bor.
*alba*, Reinw. ex Blume, Mus. Bot. Lugd. Bat. ii. 165 = Leucosyke celebica.
*alba*, Zipp. ex Blume, l. c. = Leucosyke ochroneura.
*albido-punctata*, Steud. Nom. ed. II. ii. 734 = Pipturus argenteus.
*alceaefolia*, Poir. Encyc. Suppl. iv. 227 = Urera alceaefolia.
*alienata*, Jacq. ex Blume, Mus. Bot. Lugd. Bat. ii. 231 = Pouzolzia confinis, indica.
*alienata*, Linn. Syst. ed. XII. 622 = Pouzolzia indica.

*SOURCE:* From *Index Kewensis* . . . , 1977 reprint ed., vol. II, p. 1153, by permission of Oxford University Press.

for horticultural plant names. It was compiled by O. Stapf under sponsorship of the Royal Horticultural Society of London, England.

*Index Nominum Genericorum*. 1955– .

This reference tool was begun initially as a card index to all the world's validly published plant genera. On each card is found the genus name, author, date, and name of publication; type species, and, if necessary, the basionym; a serial number; and the name of the person who prepared the card via a code number. The card file was discontinued in 1972 but is now computerized and is produced in volumes.

Jackson, B. D. 1881. *Guide to the Literature of Botany.* Longmans, Green, London. (Reprint 1974, Otto Koeltz Antiquariat, Koenigstein, FRG).

This is an older botanical literature guide with about 6000 references not given in Pritzel's work.

*Kew Record of Taxonomic Literature Relating to Vascular Plants.* 1971– . Royal Botanic Gardens, Kew, England.

This relatively new publication lists all of the world's taxonomic literature. The citations are in systematic groups. It includes all levels of taxonomic rearrangement and now includes the information included in *Index Kewensis.*

Mabberley, D. J. 1987. *The Plant Book. A Portable Dictionary of the Higher Plants.* Cambridge Univ. Press, Cambridge, England.

This work is an alphabetical listing of families and genera of vascular plants. It follows the Cronquist system of flowering plants and has a layout much like Willis's *A Dictionary of Flowering Plants.* It includes detailed information on family descriptions, distributions, number of species, economic uses, and examples, as well as less detailed information on synonymy, tribal, and subfamily levels. Pertinent recent literature is also included. This is a very handy reference.

*National List of Scientific Plant Names.* 1982. 2 vols. U.S.D.A., Soil Conservation Service, Washington, D.C.

A list of plant names for North America, Hawaii, and the Caribbean area that includes families, genera, species, and infraspecific categories, authors of the names, distribution, habit, and manuals.

Pritzel, G. A. 1872–1877. *Thesaurus literaturae botanicae,* rev. ed. (Reprint 1972. Otto Koeltz Antiquariat, Koenigstein, FRG).

This is a guide used in conjunction with B. D. Jackson's *Guide to the Literature of Botany* for early botanical literature up to 1851.

Rouleau, E. 1981. *Guide to the Generic Names Appearing in the Index Kewensis and Its Fifteen Supplements.* Chatelain, Lac de Brome, Quebec.

This is an alphabetical list to the genera and volumes published in *Index Kewensis* and its supplements. This is a time-saving device for use with *Index Kewensis.*

Stafleu, F. A. 1967. *Taxonomic Literature. Regnum vegetabile.* Vol. 52. Utrecht.

This is an abridged guide to bibliographies and publications of important early taxonomists. It is an important tool in the study of the priority of names.

Stafleu, F. A. and R. S. Cowan. 1976– . *Taxonomic Literature,* 2nd ed. *Regnum vegetabile.* Vols. 94, 98, 105, 110, 112, 115, and 116.

This reference is a greatly expanded edition of the 1967 version. It is an indispensable tool when validating dates of old literature, searching for type material, studying the priority of names, and looking for bibliographic information.

Tralau, H. (ed.). 1969– . *Index Holmensis, A World Phytogeographic Index.* 5 vols. Scientific Publishers, Zurich.

This is an alphabetical list to distribution maps found in the taxonomic literature of vascular plants. Volume 6 was edited by J. Lundquist and B. Nordenstram in 1988 and published by Koeltz Scientific Books, Koenigstein, W. Germany.

Willis, J. C. 1973. *A Dictionary of the Flowering Plants and Ferns*, 8th ed., revised by H. D. A. Shaw. Cambridge, England.

Affectionately referred to as "Willis's dictionary" by many taxonomists, this work is an alphabetical list of vascular generic and family names published since 1753 and 1789, respectively. The information included is author, distribution, family, and number of species in a genus. It proves very helpful in the herbarium for looking up obscure genera and family names. *The Plant Book* by Mabberley covers some of the same information, but is more recent in scope.

## Terminology and Dictionaries

From time to time, when dealing with old references or translating literature, it is advisable to have written definitions to standardize terms. The following can be most helpful:

Featherly, H. I. 1954. *Taxonomic Terminology of the Higher Plants.* Iowa State University Press, Ames, IA (reprint 1965. Hafner, New York), 166 pp.

Harrington, H. D. and L. W. Durrell. 1957. *How to Identify Plants.* Swallow Press, Chicago, IL, 203 pp.

Jackson, B. D. 1928. *A Glossary of Botanic Terms, with Their Derivation and Accent.* Duckworth, London (reprint 1960. Hafner, New York), 481 pp.

Lawrence, G. H. M. 1951. *Taxonomy of Vascular Plants.* Macmillan, New York, 823 pp.

Lincoln, R. T., G. A. Boxshall, and P. F. Clark. 1982. *A Dictionary of Ecology, Evolution and Systematics.* Cambridge Univ. Press, 228 pp.

Lindley, J. 1964. *Excerpt from Illustrated Dictionary of Botanical Terms.* School of Earth Science, Stanford Univ., Stanford, CA (reprint of 1848. *Lindley's Illustrated Dictionary of Botanical Terms*).

Little, R. J., and C. E. Jones. 1980. *A Dictionary of Botany.* Van Nostrand Reinhold, New York, 400 pp.

Redford, A. E. et al. 1974. *Vascular Plant Systematics.* Harper and Row, New York, 891 pp.

Stern, W. T. 1983. *Botanical Latin,* 3rd ed. David and Charles, North Pomfret, VT, 566 pp.

Usher, G. 1966. *A Dictionary of Botany.* Constable, London, 408 pp.

Woods, R. S. 1966. *An English-Classical Dictionary for the Use of Taxonomists.* Pomona College, Pomona, CA.

## Specific and Comprehensive Guides

Baillon, H. 1867–1895. *Histoire des plantes.* 13 vols. Hachette, Paris.

This is a well-illustrated work discussing families and genera, and providing an extensive reference list for the day.

Bentham, G. and J. D. Hooker. 1862–1883. *Genera plantarum.* 3 vols. Reeve, London.

Written in Latin, the work provides generic descriptions for gymnosperms and angiosperms, and was an important early reference dealing with genera through family levels.

Blake, S. F. and A. C. Atwood. 1942–1961. *Geographical Guide to Floras of the World*. U.S.D.A. Misc. Publ. No. 401 and 797, Washington D.C.

This is an older index to floras that has now been superceded by Frodin (1984).

Chaudhri, M. M. 1980. *Draft Index of Author Abbreviations*. H. M. Stationery Office, London.

This is an alphabetical list of the authors who named plants and their abbreviations.

Chaudhri, M. M., I. H. Vegter, and C. del Wall. 1972. *Index Herbariorum. Part 2, Index to Collectors. (3) I–L. Regnum vegetabile* Vol. 86. Urtrecht.

Cronquist, A. 1981. *An Integrated System of Classification of Flowering Plants*. Columbia Univ. Press, New York, 1262 pp.

This is a most comprehensive study of flowering plant families, with detailed description, examples, illustrations, keys to the families, and fossil record information, all of which are based around Cronquist's classification scheme.

Cronquist, A. 1988. *The Evolution and Classification of Flowering Plants,* 2nd ed. New York Botanical Garden, Bronx, New York, 555 pp.

Dahlgren, R. M. T., H. T. Clifford, and P. F. Yeo. 1985. *The Families of Monocotyledons*. Springer Verlag, Berlin, 520 pp.

This is a good descriptive work that puts Dahlgren's views on monocot families in perspective.

Dandy, J. E. 1967. *Index to Generic Names of Vascular Plants* 1753–1774. *Regnum vegetabile* Vol. 51. Utrecht.

This work arranges the genera according to Dalla Torre & Harms, *Genera Siphonogamarum* (1900–1907) numbers. In addition to the genus name, the author's name, synonyms, dates, and references are given.

de Candolle, A. P., et al. 1824–1873. *Prodromus systematis naturalis regni vegetabilis*. 17 vols. and 4 index vols. Paris, France.

This was an early attempt to give descriptions to all plant genera, but described only dicots. It is still a main reference to some genera.

Federov, A. A. (ed.). 1969. *Chromosome Numbers of Flowering Plants*. Academy of Sciences of the USSR, V.L. Komarov Botanical Institute.

This is the most complete listing for published plant chromosome numbers through 1966, along with literature citations.

Frodin, D. G. 1984. *Guide to Standard Floras of the World*. Cambridge Univ. Press, Cambridge, England, 619 pp.

With regard to world floras, this is a most complete bibliographic guide to the subject. It is filled with history, development, information on styles, and philosophy of floras.

Goldblatt, P. (ed.). 1981, 1983, 1985, 1988. *Index to Plant Chromosome Numbers 1975–1978; 1979–1981; 1982–1983; 1984–1985*. Monogr. Syst. Bot. Vols. 6, 8, 12, 23.

This index covers the published plant chromosome numbers for 1975–1985.

Henderson, D. M. (Comp.). 1983. *International Directory of Botanical Gardens,* 4th ed. Koeltz Scientific Books, Koenigstein, FRG.

This is a guide to the world's botanical gardens arranged by cities and with pertinent general information about each.

Holmgren, P. K., N. H. Holmgren, and L. C. Barnett. 1990. *Index Herbariorum. Part I. The Herbaria of the World,* 8th ed. New York Botanic Gardens, Bronx, New York, 693 pp.

This is a most valuable guide to the world's herbaria arranged by cities and with pertinent general information about each herbarium, including a standard acronym.

Hutchinson, J. 1964–1967. *Genera of Flowering Plants.* 2 vols. Clarendon Press, Oxford, England.

This publication gives Hutchinson's classification scheme, along with family and genus descriptions.

Hutchinson, J. 1973. *The Families of Flowering Plants. Vol. I. Dicotyledons, Vol. II. Monocotyledons,* 3rd ed. Clarendon Press, Oxford, England.

This publication also gives Hutchinson's classification scheme, along with family descriptions and illustrations.

Lawrence, G. H. M., et al. (eds.) 1968. *B-P-H. Botanico-Periodicum-Huntianum.* Hunt Botanical Library, Pittsburgh, PA.

This reference provides standardized abbreviations to over 12,000 botanical journals and is especially helpful in identifying older reference citations. It is used by many current botanical journals as a guide for abbreviating literature citations.

Lanjouw, J. and F. A. Stafleu. 1954, 1957. *Index Herbariorum, Part 2. Index to Collectors. (1) A-D; (2) E-H. Regnum vegetabile* Vols. 2, 9. Utrecht.

Moore, R. J. (ed.) 1973. *Index to Plant Chromosome Numbers 1967–1971. Regnum vegetabile* Vol. 90. Utrecht.

This is an index to chromosome numbers and references for the time period.

*Regnum vegetabile.* International Association for Plant Taxonomy, Utrecht.

This is a numerical series of indexes and taxonomic references published under the sponsorship of the International Association of Plant Taxonomists. At present, well over 100 volumes have been issued.

Rendle, A. B. 1925–1930. *The Classification of Flowering Plants, Vol. 1. Gymnosperms and Monocotyledons; Vol. 2. Dicotyledons.* 2nd ed. Cambridge Univ. Press, Cambridge, England.

This reference includes very good descriptions of families organized according to the Engler system of classification.

Solbrig, O. and T. W. J. Gadella. 1970. *Biosystematic Literature: Contributions to a Biosystematic Literature Index (1945–1964). Regnum vegetabile* Vol. 69. Utrecht.

This reference provides selected experimental taxonomy literature under systematic headings. It provides a very good reference for beginning a literature search.

Takhtajan, A. 1959. *Die Evolution der Angiospermen* (transl. to German by W. Hoppner). Gustav Fischer Verlag, Jena.

Takhtajan, A. 1969. *Flowering Plants–Origin and Dispersal* (transl. from Russian by C. Jeffrey). Smithsonian Institution Press, Washington, D.C.

Takhtajan, A. 1987. *Systema Magnoliophytorum.* Soviet Sciences Press, Leningrad.

This and the two preceding references give the basis for Takhtajan's system for classification.

Thorne, R. F. 1983. *Proposed New Realignments in the Angiosperms. Nordic J Bot* 3:85–117.

This reference gives the most recent basis and organization for Thorne's classification system.

Vegter, I. H. 1976, 1983. *Index Herbariorum, Part 2. Index to Collectors. (4) M; (5) N-R. Regnum vegetabile.* Vols. 93, 109. Utrecht.

These references, like the others in the series, are most helpful in locating type specimens. They are alphabetical lists of collectors and of where their specimens are deposited.

## *Selected Botanical Journals*

There have been many botanical journals published in the world. Some were privately published by botanists and had a very short lifespan (i.e., Pittonia 1887–1905; Silva 1908–1912; Zoe 1890–1900/08), while others were or have been in existence longer (i.e., *Leaflets of Western Botany,* 1932–1966; *Phytologia,* 1933– ; *Sida,* 1962– ). Some are published by professional societies (i.e., *American Journal of Botany,* the Botanical Society of America; *Botanical Journal of the Linnean Society, London*), others from herbaria and museums (i.e., *Fieldiana: Botany by the Field Museum of Natural History, Chicago; Contributions from the Gray Herbarium of Harvard University*), some from botanical gardens (i.e., *Kew Bulletin; Journal of the Arnold Arboretum*), universities (i.e., *University of California Publications in Botany; Journal of the Faculty of Science, University of Tokyo, Botany*), and governments (i.e., *Canadian Journal of Botany; Iowa Agricultural Experiment Station Research Bulletin*). Some journals are very experimental and deal with botanical subjects that taxonomists do not normally refer to for taxonomic information.

The abridged periodical and review list below deals with current taxonomic research topics. Those with an asterisk are published in North America.

*Acta Botanica Neerlandica*
*Acta botanica sinica*
*Adansonia*
**Aliso*
**American Fern Journal*
**American Journal of Botany*
**American Midland Naturalist*
*Annales des sciences naturelles* (Paris)
**Annals of Botany*
**Annals of the Missouri Botanical Garden*
**Annual Review of Ecology and Systematics*
*Australian Journal of Botany*
**Baileya*
**Bartonia*
*Bauhinia*
*Beiträge zur Biologie der Pflanzen*
*Biochemical Systematics*
*Biota*
*Biotropica*
*Blumea*
**Boletin de la Sociedad Botánica de Mexico*
*Botanical Journal of the Linnean Society, London*
**Botanical Museum Leaflets Harvard University*
**Botanical Review*
*Botaniska notiser*
*Botanisk tidsskrift*
*British Fern Gazette*
**Brittonia*
*Bulletin on the Botanical Survey of India*

*Bulletin on the British Museum (Natural History) (Series E) Botany*
*Bulletin du muséum d'histoire naturelle*
**Bulletin Pacific Tropical Garden*
**Bulletin of the Torrey Botanical Club*
**Canadian Field Naturalist*
**Canadian Journal of Botany*
*Candollea*
**Castanea*
*Cladistics*
**Contributions from the Gray Herbarium of Harvard University*
**Contributions from the United States National Herbarium*
*Darwiniana*
**Economic Botany*
**Evolution*
*Fern Gazette*
**Fieldiana (Botany)*
*Israel Journal of Botany*
**Journal of the Arnold Arboretum*
*Journal of Biogeography*
*Journal of Phytogeography and Taxonomy*
*Kew Bulletin*
*Korean Journal of Botany*
**Madroño*
*Mémoires du muséum national d'histoire naturelle*
**Memoirs of the New York Botanical Garden*
**Memoirs of the Torrey Botanical Club*
**Michigan Botanist*
**Naturaliste Canadien*
*New Phytologist*
*New Zealand Journal of Botany*
*Oikos*
**Phytologia*
**Proceedings of the Academy of Natural Sciences of Philadelphia*
**Proceedings of the California Academy of Sciences*
*Reinwardtia*
**Rhodora*
**Sida*
**Smithsonian Contributions to Botany*
*South African Journal of Botany*
*Svensk botanisk tidskrift*
**Systematic Botany*
*Taiwania*
*Taxon*
*Telopea*
*Vegetatio*
*Watsonia*
*Webbia*
*Willdenowia*

The above list of periodicals may seem rather long, but is very short when compared to the over 12,000 journal titles (past and present) listed for the world literature by Lawrence et al. (1968). In reality, modern botanists cannot work in the isolation of their own languages, cultures, or countries. Taxonomy is a broad, ever-expanding field of science that bridges many barriers.

## Floras and Manuals

A **flora** is a taxonomic treatment of all plants occurring in a particular geographical area. It can also refer to the plants living in a region. This area can be small, such as a nature preserve of a few hectares (or acres), or more extensive such as a regional, state, or country flora. Each written flora is developed following predetermined guidelines on the type of coverage to be included. These guidelines might involve all plants or only vascular plants; only a listing of names; descriptions, keys, and distributions, and ecological, geological, or soil information; how it will be used; etc. The book including all or part of these things is called a **manual.** However, you may ask, why a manual to a flora in the first place, especially in the age of recombinant DNA, genetic engineering in plants, and "star wars"? This is a valid twenty-first century question!

Human nature is basically inquisitive. As pointed out earlier, there are early historical records of various cultures (i.e., Chinese, Hebrew, Indian, New World, etc.) giving names to plants and classifying plants in an attempt to preserve a "mental" list of a geographical region for recognition. This is done today for the curious individual who asks, "What plants live there?" Also, with today's ease of travel to diverse regions; with greater environmental awareness; with many local, national, and international laws governing and protecting plants and animals (see Chapter 14); and with greater emphasis being placed on discovering new plants for medicine, food, and as a genetic gene pool for future research, there are good reasons for knowing what plant lives where.

## *Historical Background*

Following the invention of the printing press, various botanical papers and plant lists were published. It did not take long until a considerable wealth of information was available, necessitating the need for a guide to the floristic literature.

The first worldwide index or list was published by Carl Linnaeus himself in 1736 as *Bibliotheca botanica* (2nd ed. 1751). In this guide he arranged the floras on a geographical, hierarchical basis and included a commentary. Linneus's bibliography was the only guide for 50 years until J. Dryander (1798) published the next guide to floras and related publications in volume III (botany) of *Catalogus bibliothecae historico-naturalis* Joseph Banks, which was one of five volumes printed between 1797 and 1800. This scholarly work quickly became the standard bibliographic reference in biology for another 50 years.

In the mid-nineteenth century, G. A. Pritzel published his famous *Thesaurus Literaturae* (1847–52; 2nd ed. 1871–77), which, with its author index, arrangements, biographical notes, and classes of floristic literature, became the classic of botanical bibliographies. At about the same time as the second edition was published, B. D. Jackson published *Guide to the Literature of Botany* (1881). It was touted as a companion to Pritzel's thesaurus but was independent from it and included greater geographical subdivisions, especially outside Europe and in North America.

During the intervening 60+ years, various attempts were made to provide some type of bibliographic index to floras. However, it was not until Blake and Atwood published *Geographical Guide to the Floras of the World* (Part I, 1942; Part II, 1961) that a wide-ranging guide was produced. This guide covered continents, continental subdivisions, countries, checklists, serial literature, and works on applied botany. Although this guide was by far the most comprehensive produced up to that time, it lacked the continent of Asia and eastern Europe, due to the death of Blake before its completion.

Frodin (1984) has most recently published the most comprehensive *Guide to Standard Floras of the World.* It is filled with much information on history and development, styles, and philosophy, and a detailed geographical breakdown of floras. It describes and lists the "standard floras" for large areas only and generally includes literature dating after 1840. It judiciously selects only specific groups (i.e., woody plants, pterdophytes, etc.) and does not include works on applied botany, nonvascular plants, or popular publications.

Most recently, Davis et al. (1986) and the International Union for Conservation of Nature and Natural Resources have attempted to document the published information on the world's endangered plants. This work, called *Plants in Danger. What Do We Know?*, discusses the state of the native flora of each country of the world. It summarizes what is known about the country relating to floristics, vegetation, published checklists and floras, information on threatened plants, laws pertaining to plants of the region, botanical gardens, plant societies of the country, and pertinent references.

Today there is a need for amateur and professional botanists to study the flora of many regions, especially regions under pressure from development and habitat modification. Specimens collected and preserved in a recognized herbarium help support many aspects about plants.

An abridged list of floras and manuals is found in Appendix I. It is a world list, with emphasis on North America, and should provide help to travelers.

One aspect of floras that should also be mentioned is the increasing number of popular "flower books" or "picture books" on the market for the general public. Many of these have very good color photographs or illustrations, so that a person can become very knowledgeable about plants of an area. These nontechnical works often include information that professional floras lack (i.e., folklore, uses, etc.), and therefore they often make welcome gifts and additions to any botanical library. A few words of caution about these books: (1) Be sure that the figures are clear and distinctive, showing the plant's features. (2) Stay away from books that only show common roadside plants, as these are mostly introduced aliens and many interesting native plants are omitted. (3) Be certain that the book contains a large enough number of species within it to make the cost worthwhile.

## *Detailed Taxonomic Studies*

The backbone, as it were, of contemporary systematic botany is made up of the monographic study and the revision. A **monograph** is a comprehensive study of a family, genus, or distinct group of species. This study involves an extensive library search of all literature pertaining to the plant group and a presentation of original research conducted by the investigator. This original research may include a literature survey; taxonomic names, including type specimens, synonymns, authors, and common names; and information gathered from experimentation, such as anatomy, chemistry, cytology, distribution, ecology, hybridization with other species, morphological variation, transplant garden studies, etc. In other words, as much information as possible about the biology and evolution of plants and how these data help develop a workable taxonomy is included.

A taxonomic **revision** is similar to a monograph but is usually less involved in completeness and scope. The revision may bring together and synthesize predominantly other botanists' research and is much less detailed than a monograph.

Beginning students reading almost any of the journals mentioned earlier will observe that the great majority of systematic research papers are monographs and revisions.

When writing a flora of a region, an author will refer to the information presented in the monographs and revisions, look at the authors' evidence, agree or disagree with it, and then incorporate the pertinent usable data, names, etc. into the floristic study. The information is then passed on to the reader and is disseminated for the benefit of all who use it.

## *Basic Systematic Texts*

Students may wish to search for another basic text to help better explain a concept or to gather further information. The following list includes those books that are currently in print or are accessible in most university libraries.

Benson, L. 1979. *Plant Classification,* 2nd ed. D. C. Heath, Lexington, MA, 901 pp.

Harrington, H.D. 1957. *How to Identify Plants.* Swallow Press, Chicago, 203 pp.

Heywood, V. H. (ed.) 1985. *Flowering Plants of the World.* Prentice Hall, Englewood Cliffs, NJ, 335 pp.

Jones, S. B., Jr. and A. E. Luchsinger, 1986. *Plant Systematics,* 2nd ed. McGraw-Hill, New York, 512 pp.

Lawrence, G. H. M. 1951. *Taxonomy of Vascular Plants.* Macmillan, New York, 823 pp.

Naik, V. N. 1984. *Taxonomy of Angiosperms.* Tata McGraw Hill, New Delhi, 304 pp.

Porter, C. L. 1967. *Taxonomy of Flowering Plants,* 2nd ed. W. H. Freeman, San Francisco, 472 pp.

Radford, A. E. 1986. *Fundamentals of Plant Systematics.* Harper & Row, New York, 498 pp.

Skukla, P. and S. P. Misra. 1979. *An Introduction to Taxonomy of Angiosperms.* Vikas Publ. House, New Delhi, 546 pp.

Smith, J. P. 1977. *Vascular Plant Families.* Mad River Press, Eureka, CA, 320 pp.

Walters, D. R. and D. J. Keil. 1988. *Vascular Plant Taxonomy,* 3rd ed. Kendall/Hunt, Dubuque, IA, 488 pp.

## *SELECTED REFERENCES*

Benson, L. 1962. *Plant Taxonomy.* Ronald Press, New York, 494 pp.
Davis, P. H. and V. H. Heywood. 1973. *Principles of Angiosperm Taxonomy.* Robert E. Krieger, Huntington, NY, 558 pp.
Davis, S. D. et al. 1986. *Plants in Danger. What Do We Know?* IUCN, Gland, Switzerland, 461 pp.
Frodin, D. G. 1984. *Guide to Standard Floras of the World.* Cambridge Univ. Press, Cambridge, England, 619 pp.
Jones, S. B., Jr. and A. E. Luchsinger. 1986. *Plant Systematics,* 2nd ed. McGraw-Hill, New York, 512 pp.
Lawrence, G. J. M. 1951. *Taxonomy of Vascular Plants.* Macmillan, New York, 823 pp.
Radford, A. E. 1986. *Fundamentals of Plant Systematics.* Harper & Row, New York, 498 pp.
Radford, A. E. et al. 1974. *Vascular Plant Systematics.* Harper & Row, New York, 891 pp.
Shukla, P. and S. P. Misra. 1979. *An Introduction to Taxonomy of Angiosperms.* Vikas Publ. House, New Delhi, 546 pp.

# 4

# *How Plants Are Identified*

The process of learning the names of the plants of a particular location or region, or just the name of a particular plant, can be accomplished in various ways. As previously stated, the easiest and probably the most used method is to ask someone who already knows. In this way, the knowledge of another individual is utilized. Probably the second most used method of learning plant names is to compare an unknown plant with a photograph or picture of a similar plant in a book. Many picture books about plants have examples of the more common plants, but most species cannot be identified in this fashion. The third way is to identify the plant oneself. To have the skill to identify an unknown plant oneself is a most valuable asset and a valuable part of the study of systematics.

Identification of unknown specimens is usually made by using a **key,** that is, a device whereby successive choices between contrasting statements are followed until the correct name is found by the process of elimination. Keys play an integral part in a flora, allowing for the proper identification of families, species, and infraspecific taxa.

Most modern keys are constructed of paired choices. Each half of the paired choice makes a statement or several statements that are either true or false. These "two-forked", "couplet," or "paired-choice" type of keys are called **dichotomous** (meaning forking) **keys.** The first of this type was developed by Jean de Lamarck, the famous French botanist, who in 1778 as a short-cut method to reading down the long lists and descriptions of plants found in the botanical literature of the day, provided an artificial key as a means to identify the plants of France *(Flore Françoise).*

The use of a dichotomous key is like following directions given by a friend to a distant destination. As one travels down a road or highway, decisions are made at road junctions as to which way to turn. If the roads are properly marked, and if the directions are followed carefully, the destination should be reached. The same is true with following the paired choices of a dichotomous key.

**FIGURE 4.1 An example of an indented key.**

| | |
|---|---|
| **1.** Object with curved sides | a sphere |
|   **2.** Sphere solid white in the middle | ○ |
|   **2.** Sphere black or partially black | |
|     **3.** Sphere all black | ● |
|     **3.** Sphere one-half white and one-half black | ◑ |
| **1.** Object with straight sides | |
|   **4.** Object with four equal sides | a square |
|     **5.** Square white | □ |
|     **5.** Square one-half black and one-half white | ◧ |
|   **4.** Object with three equal sides | a triangle |
|     **6.** Triangle with spines at the points | △ |
|     **6.** Triangle lacking spines | △ |

There are two main types of botanical keys: the **indented key** and the **bracketed key.** Figures 4.1 and 4.2 illustrate how both these types of keys function for seven objects. In both keys, the objects are divided into subgroups, and by the process of elimination the unknown can be identified. All objects must be accounted for, and none may be left out.

In an indented key, the paired-couplet choices are identified in the same way and given the same number. This is important because in large keys to

**FIGURE 4.2 An example of a bracketed key.**

| | |
|---|---|
| **1.** Object with curved sides; a sphere | 2 |
| **1.** Object with straight sides | 4 |
|   **2.** Sphere solid white in middle | ○ |
|   **2.** Sphere with black | 3 |
| **3.** Sphere all black | ● |
| **3.** Sphere one-half white and one-half black | ◑ |
|   **4.** Object with four equal sides; square | 5 |
|   **4.** Object with three equal sides; triangle | 6 |
| **5.** Object white | □ |
| **5.** Object one-half white and one-half black | ◧ |
|   **6.** Triangle with spines at the points | △ |
|   **6.** Triangle lacking spines | △ |

many species, the halves of the couplet may be separated by some distance from one another and may even be on different pages. Some authors will use letters (i.e., a, b, c, . . . or aa, bb, cc, etc.) or symbols (i.e., 1, 1′, 2, 2′, . . . etc.) in place of numbers to keep from confusing couplets.

A bracketed key follows the same basic principles of contrasting choices and choices given the same number, letter, or symbol as an indented key. The choices, however, are always placed on adjacent lines, thereby requiring less room on a page and keeping the couplets together. Relationships between taxa are not as easily observed in bracket keys, and when mistakes in keying are made, backtracking is more difficult. Bracketed keys are commonly used in beginning, student-type keys.

## *Some Basic Rules for Constructing and Using Dichotomous Keys*

There are some basic rules that are common to all types of keys that should be followed when constructing and using dichotomous keys. These will be discussed below.

1. All parts of the key should be constructed in a dichotomous fashion. One will sometimes encounter keys with many branches in certain disciplines of biology, or in older botanical manuals. This can be very confusing, time consuming, and can lead to incorrect choices being made.
2. No character of the plant should be used alone in the key without using its contrasting alternative condition in the other half of the couplet. This means that if the character is a leaf character, the alternating leaf character must also deal with the same leaf feature. For example,

   1\. Leaves attached alternate
   1\. Leaves attached opposite

   Using as the alternating choice "leaves heart-shaped" would not be a correct contrasting choice, even though it is a choice dealing with leaves.
3. Construct the leads parallel within each couplet. This means that the beginning word of each alternating choice should be the same. For example, if the word of the first lead is *petals,* then the first word of the contrasting couplet should also be *petals.*
4. Describe the characters of the plant in a positive manner. The person using the key should be able to get a visual image of the condition of each character used.
5. The use of vague, unclear, overlapping measurements and general broad terms should be avoided, for example, large versus small, dark versus light colored, 5.0—10.0 mm versus 8.0—12.0 mm long.
6. The season of the year during which the key is to be used should not be changed to another season within the same key, that is,

if the key is to winter twig features, they should not be mixed with the characters of leaves. An easy way to accomplish this and to make the key more seasonably usable is to include flower, fruit, and vegetative characters together in the key.

7. Taxonomic names should not be used as key characters. The key may, however, be designed to key out taxonomic categories above the rank of species, as well as at the species level or below.
8. The characters chosen to be used in a key should be the most reliable, least variable characters available, but should still be easily observable.
9. Highly technical and obscure characters should not be used as key characters, for example, chromosome morphology or number, detailed anatomical features, the presence or absence of compounds, and localities. The person using the key will not normally go to great lengths to get the information requested by a couplet.
10. In species that are dioecious (reproductive structures on different plants), the key should consider both sexes, since the specimen to be identified may have flowers of only one sex.
11. Try to avoid difficult or irregular arrangements or identification in the couplets. The purpose of the key is to provide an easy method to identify an unknown plant specimen. Therefore, the design of the keys should be for convenient and self-evident use.
12. Remember that keys are not absolutely reliable and without mistakes. The key is only as good as the material used in its construction. Plant specimens that exhibit extreme variation and have not been observed by the writer of the key may not key properly. Also, interpretation of characters may vary, and therefore, the student should understand the author's use of a term.

It should be remembered that a key is another person's way of interpreting how a group of plants should be identified. The clarity and ease of use of a key will depend on many factors; therefore, it should be remembered that a key is not magic, that to master a key may take some time before one can identify unknown plants without making errors. Keying specimens takes skill and considerable practice. One should not guess when keying, if at all possible. Guessing will almost always bring a wrong determination. Correctly learning and applying terminology is the best way to key plants.

## *Other Identification Methods*

The keys we have been discussing provide the traditional tools that botanists use to identify plants. In recent years, however, other techniques have developed that do not follow the very structured, single beginning point of the dichotomous key. These are polyclaves and computer methods.

**Polyclave,** or **synoptical, keys** are multientry, any-order keys that use cards stacked in any arrangement on top of one another, with holes or edges punched in such a way as to allow cards with the desired taxa to be retained or eliminated until the card with the desired taxon listed on it is the only card remaining. The holes or punched edges correspond to characters chosen by the investigator.

During the 1960s, computer technology and its associated programs developed greatly. Programs began to be developed that would provide for automated identification of specimens by the computer, computer-developed multientry keys, and computer-stored information. The late 1970s and 1980s have seen the rapid development and use of the microcomputer. These machines and the programs perfected for them have expanded the potential for their use, especially for smaller, more specific groups. As the technology and programs are further perfected, future use of them will certainly increase.

## SELECTED REFERENCES

Abbott, L. A., F. A. Bisby, and D. J. Rogers. 1985. *Taxonomic Analysis in Biology. Computers, Models and Databases.* Columbia University Press, New York.

Duncan, T. and C. A. Meacham. 1986. Multiple-Entry Keys for the Identification of Angiosperm Families Using A Microcomputer. *Taxon* 35:492–494.

Geesink, R. 1987. Structure for Keys for Identification. In de Vogel, E. F. (ed.). *Manual of Herbarium Taxonomy. Theory and Practice.* United Nations Education, Scientific and Cultural Organization, Regional Office for Science and Technology, Jakarta, Indonesia, pp. 77–90.

Jones, S. B., Jr. and A. E. Luchsinger. 1986. *Plant Systematics,* 2nd ed. McGraw-Hill, New York, 512 pp.

Leenhouts, P. W. 1966. *Keys in Biology: A Survey and A Proposal of A New Kind.* Proc. Koninklijke Nederlandse Acad. Van Wetenschappen 69 (Ser. C): 571–596.

Morse, L. E. 1971. Specimen Identification and Key Construction with Time-Sharing Computer. *Taxon* 20:269–282.

Pankhurst, R. J. (ed.). 1975. *Biological Identification with Computers.* Academic Press, London.

Pankhurst, R. J. 1978. *Biological Identification.* Edward Arnold, London, 104 pp.

Radford, A. E. et al. 1974. *Vascular Plant Systematics.* Harper and Row, New York, 891 pp.

Shetler, S. G. 1974. Demythologizing Biological Data Banking. *Taxon* 23:71–100.

Voss, E. G. 1952. The History of Keys and Phylogenetic Trees in Systematic Biology. *J. Sci. Lab. Denison Univ.* 43:1–25.

Walters, D. R. and K. J. Keil. 1988. *Vascular Plant Taxonomy,* 3rd ed. Kendell/Hunt, Dubuque, IA, 488 pp.

# 5

# *Collecting, Handling, and Preserving Specimens*

From time to time in all areas of botany and plant taxonomy, it will become necessary to preserve the plant(s) being studied. Over the years many methods to preserve plant specimens have been tried, with the easiest and cheapest method involving pressing and drying the desired plant material. When treated in this way, plants may lose their aesthetic appeal but do not lose their scientific value if the material is treated properly and proper procedures are followed.

The specimens collected may have various uses. First, a specimen documents the presence of a species at a particular location. Second, the specimen records the appearance of the plant at the time of collection. This appearance can be used to compare with other plants growing at the same or at different times of the year. Third, living wild plants can be studied only during the growing season or when an individual is in the field. A preserved, pressed specimen can be studied any time. Fourth, a preserved specimen can provide a wide range of data, depending on the information included on the label with the specimen. This can be very helpful in historical and phytogeographic studies, and monographic and revision research. Fifth, a collection of dried and pressed botanical specimens arranged for reference is called a **herbarium.** These specimens can be borrowed, loaned, and returned like library books. A large number of specimens from a geographical region or of a particular taxon may be necessary for preparing local or regional floras, or for studying the taxonomy or biology of a group. Sixth, a specimen is a permanent document of the name given to a taxon (discussed in Chapter 2).

Because of these uses, one should make every effort to prepare good quality specimens that can provide the maximum amount of information preserved in the specimen for future time.

## *Materials Needed for Collecting Plants*

The basic supplies needed for preparing good-quality pressed specimens are summarized in Table 5.1 and are shown in Fig. 5.1. We will discuss each in detail. The amount and kind of equipment needed will depend on the objectives, space, transportation, and weight restrictions.

### *Plant Press*

The plant press is a simple, inexpensive device constructed to not only press, but also to remove the moisture from the plant. It is usually constructed of a sturdy metal, plywood, or wooden grid frame 30 × 46 cm (12 × 18 in.). This frame provides rigidity to the press. The insides are made up alternately of corrugated cardboard ventilators (also called *corrugates*) with the air spaces running crosswise and blotters (some use folded newspaper), and all measure 30 × 46 cm (12 × 18 in.). The cardboard provides air circulation, and the blotters hasten drying by absorbing moisture. Specimens to be pressed are placed in a folded sheet of newspaper or blank newsprint paper that, when folded, is approximately 28 × 40 cm (11 × 16 in.). This in turn is placed between two blotter sheets, followed by a ventilator. The process is then repeated for each specimen. The press is kept closed and pressure is applied to the specimens by two belts, ropes, or straps (twine) around the outside.

**TABLE 5.1 Equipment helpful for collecting plants.**

| *Essential Equipment* | *Use* |
|---|---|
| Plant press | Pressing specimens |
| Field press | Pressing specimens |
| Old newspaper | Pressing specimens |
| Field notebook | Recording field data |
| Various digging tools: dandelion digger, geology pick, large screwdriver, trowel | Extracting underground parts and soil samples |
| Clippers and/or knife | Collecting woody material |
| Plastic bags | Collecting and holding specimens |
| Flexostat | Holding plant parts together at bends |
| Magnifying lens | Observing plant micro-characters |
| Insect repellent | Protection against biting bugs |
| *Additional Equipment* | *Use* |
| Small paper envelopes | Holding fruits and seeds |
| Vials with preserving fluid | Holding anatomical, cytological, and morphological specimens |
| Boots | Collecting in wet habitats |
| Darkroom tongs | Handling toxic plants |

**FIGURE 5.1 Equipment helpful for collecting plants. Top (from left to right): newspaper, plant press, plastic (poly) bags; middle: small paper envelopes, trowel, snippers, field press, field notebook; bottom: magnifier, vials, pencil, flexostats, dandelion digger, pick, spade, and file cards.**

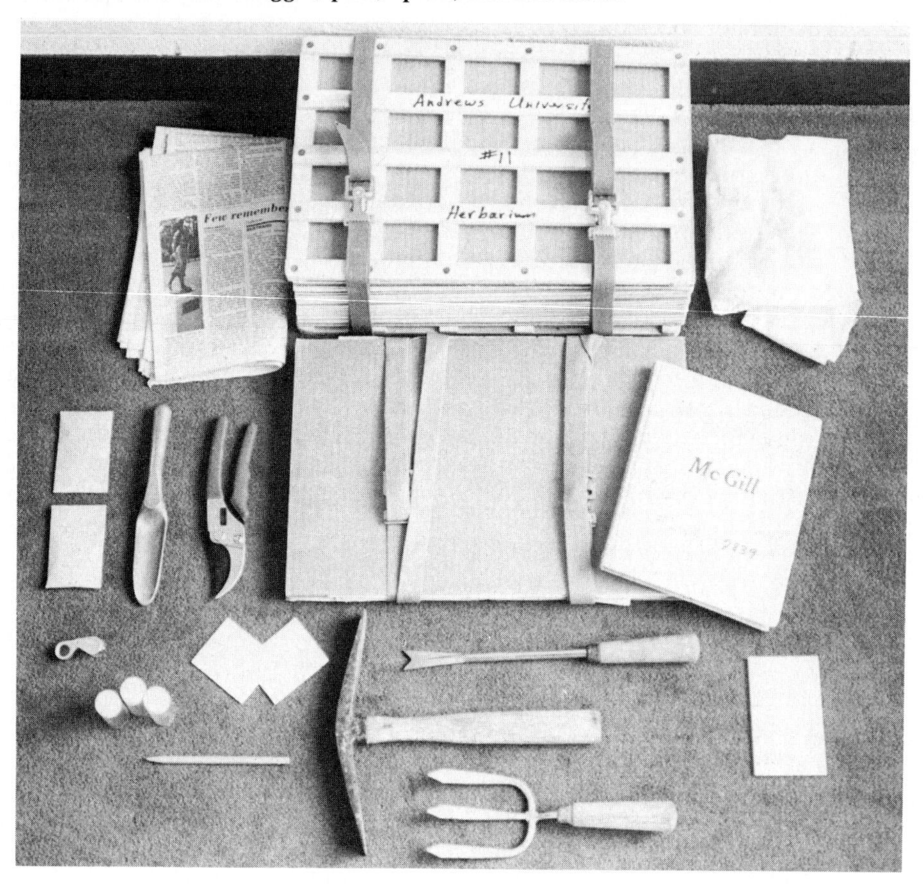


The press can be inexpensively made from wood scraps, old cardboard boxes, old newspaper, and a small rope, or it can be purchased from a biological supply company.

### *Field Press (or Satchel)*

A field press is a collecting device developed in North America to collect and keep relatively flat a series of specimens. The construction varies in size and material, but is relatively lightweight and is made of a series of folded newspapers, blotters, and heavy cardboard, press board, or similar material. Specimens in newspaper are transferred to a plant press for final pressing and drying.

### *Newspaper (or Blank Newsprint)*

The specimens are pressed in newspaper. The paper provides a surface for writing information about the plant and a way, when stacked in bundles, to handle, keep separate, and store specimens temporarily.

### *Field Notebook*

This is a small multileaved book used to record field information, notes, names, etc. that will be necessary when specimen labels are prepared. Information should be recorded in pencil or waterproof India ink. Many a botanist has gotten his or her notebook wet for one reason or another, with nonreplaceable information being destroyed as a result. Extra pencils should be included.

### *Digging Tools*

A completely preserved herbaceous specimen may need an underground segment for proper identification. A bricklayer's hammer, "dandelion digger," geologist's hammer, large screwdriver, trowel, or small shovel may be most helpful. A digger that has a solid steel shank from the handle to the blade is best. Poorly constructed tools are cheaper to buy but are ultimately more costly in money and frustration due to bending and breakage.

### *Clippers and/or Knife*

These tools are useful for removing specimens from woody or large herbaceous plants, and for pruning a specimen. Painting the handles of clippers and diggers bright colors helps to avoid losing them in the field.

### *Plastic Bags*

The plastic poly bag originally designed for holding household waste or for carrying goods home from the market is ideal for use in the field. Collected specimens are placed in the bag, the bag is sealed, and it is taken to where pressing and identification takes place. Having a few smaller poly bags for delicate or tiny plants helps keep the specimens in good quality. The plastic bag has replaced the **vasculum,** a metal or hard-plastic cylindrical container with a hinged door on one side and a carrying strap or handle.

### *Flexostat*

The flexostat is really a small, stiff piece of paper with a 2.5-cm (1-in.) slit made in the middle. The flexostat is used by slipping the folded end of a large plant through the slit when pressing the specimen. The flexostat holds the bent plant and its associated parts in place while drying, making a neater specimen. The flexostat is easily constructed by cutting a standard 76 × 127 mm index card (3 × 5 in.) in half crossways and cutting a 2.5-cm (1-in.) slit in the middle with a sharp blade.

### *Magnifying Lens*

A small 10X to 20X folding magnifier *(hand lens)* is essential for proper plant identification. This lens is also most helpful when observing microcharacters in the field, especially when collecting similar species.

### *Insect Repellent*

Depending upon the location, season, habitat, etc., biting insects can be a problem. A small bottle of proven insect repellent should always be in the collecting bag.

## *Procedure for Pressing, Drying, and Labeling Specimens*

Remember that a major purpose of collecting and preserving a plant specimen is to preserve a plant for posterity in as lifelike a condition as possible. In order to accomplish this, the following procedure is given as a general outline.

**1.** Choose the plant that appears most typical of those in the population. Good specimens will include flowers or fruits (or both), leaves, and underground parts. You must ultimately identify the specimen, and correct identification may depend on some or all of these features. For example, the families Brassicaceae (mustard), Fagaceae (oak), and Juglandaceae (walnut) need mature fruits for proper identification; bulb characters are important in identifying Liliaceae (lily); and knowing the type of host plant helps in identifying Loranthaceae (mistletoe).

The specimen should be smaller than the size of the single sheet of newspaper the plant is pressed in, smaller than the plant press size of 30 × 46 cm (12 × 18 in.), and smaller than the standard-size mounting sheet, which is 29 × 41.5 cm (11.5 × 16.5 in.). Larger plants may be bent into a V shape, N shape, or W shape so as to fit onto a mounting sheet. A flexostat slipped over the bend will help to hold smaller plant parts in place (especially useful with grasses, rushes, and sedges). The flexostat is removed after the plant has dried. Care should be maintained in displaying the necessary parts.

Some specimens will need special care and handling. Extra large plants may be trimmed or cut, and made into multiple sheets. Plants that are bulky or succulent may have to be sliced and/or sprinkled with table salt to help draw off the plant's moisture.

Aquatic plants are a special problem because many species have finely dissected plant parts. The best results are obtained by placing a sheet of high-quality mounting paper under a floating or submerged plant, then slowly raising the paper until the specimen is lying on the paper and out of the water.

Plastic bags containing plants can be stored in a cool place or refrigerator, as can a field press. The plants keep best with a small amount of moisture or moist moss in the bag. Plants that might become easily damaged or lost among larger plants from the same collection site can be placed in small bags within the larger bag. Notes with specific information can be placed on a piece of paper in each bag.

It is best to press the specimen as soon as possible within a single sheet of old newspaper. This can be done in the field using the plant press or field press (Fig. 5.2a). It is at this time that cleaning, pruning, removal of dead

**FIGURE 5.2** Steps in the handling of plant specimens. (a) Arranging a plant in a plant press. (b) Positioning filled plant presses on a light bulb dryer. (c) Example of entries in a field notebook: the four-digit number is the collection number; the circled numbers are the numbers of duplicates collected; and $n = 13$ is the chromosome number determined from buds collected at the same time as the specimens. (d) Placing specimens in a fumigator for insect control.

(a)

(b)

2726 *U. dioica* L. var. *angustifolia*

Puebla, Mexico 17/XII/80
⑤ 8 Km N. of highway 150D and Puebla in the village of San Miguel Canoa. 3000 m. Along hedges and waste area 0.2 Km S. of center of village
$n = 13$ 19°08′N; 98°05′W.

2727 *U. subincisa*

Same as 2726
⑤ 1.5 Km N. of center of village. 3,100 m.

2728 *U. subincisa*

Puebla, Mexico 18/XII/80
⑥ 0.2 Km east of center of Tlachichuca, along edge of road at base of old adobe wall. Scattered. Annual weed. 0.2–0.8 m. In flower and fruit. Called "chichicastle." 22 Km E of highway 125. 2300 m. 19°10′N; 97°25′W.

2729 *Urtica urens*

Puebla, Mexico 18/XII/80
⑤ In pasture and disturbed area in the village of San Miguel Zoapan; 10 Km SE of Tlachichuca. 3000 m. annual. scattered; in flower and fruit. NE slope of Mont Orizaba. *Pinus*, *Abies* zone.
$2n = 26$ 19°05′N; 97°21′W.

(c)

(d)

leaves, and the arrangement of the plant should be undertaken. Dirt should be washed from the underground plant parts. The specimen is arranged in the manner that it should appear when dried.

The specimen in the newspaper is usually arranged in the press in the following order: ventilator - blotter - specimen - blotter - ventilator - blotter - specimen, etc. Some collectors prefer to use only one blotter, others prefer to use no blotters, while still other collectors use ventilator - specimen - blotter - specimen - ventilator. Still others use no blotters, using only ventilators and newspapers. Only one species should be placed in each sheet of folded newspaper, but several small specimens of the same species can be pressed in the same sheet. Identifying information about the species (collector's number, Latin name, location, etc.) should be written on the outside of the newspaper in an easily accessible place for future reference.

**2.** After the press is filled or all the specimens have been pressed, the plant press is closed and pressure is applied by means of tightening the belts, ropes, or straps. As the plants dry, excess slack may appear and the belts are tightened again. The press can be placed in the warm sunshine or dried on the top of a car or trailer as one travels. Best results are obtained by placing the press over an artificial dryer. This dryer can be as simple as a metal 20 L (5 gal) empty container with a light bulb in the bottom. Another dryer is a box in which the presses are suspended over moderate heat provided by a row of light bulbs or electric heaters (Fig. 5.2b). For field work, a portable dryer can be constructed from a collapsible aluminum frame and a removable fire-retardant canvas or fiberglass skirt. The heat source is a catalytic heater, gas lantern used in camping, or electric bar heater when in a building and near electricity. **Caution! The use of gas heat of any type increases the risk of fire.** Also, proper ventilation must be provided.

How long specimens take to dry depends on many variables. The usual time for drying small specimens is 24–48 hours. Thick, fleshy specimens may take 3–6 days; small, delicate plants may dry in 8–12 hours. Other factors affecting drying time are the number of presses on the dryer, humidity, the type of heat source, climate, and temperature. Too high a temperature for too long a time period will cause a specimen to become brittle and discolored. Too short a drying period or too low a temperature will keep the specimen from drying, make it moist to the touch, and possibly cause mildew. Specimens will need to be checked regularly until they are dry. A specimen is not dry if it is still limp when picked up and cool and moist to the touch.

**3.** At the time a collection is made and the specimen is pressed, field data about that plant should be entered in pencil into a field notebook as a permanent record (Fig. 5.2c). The information should be complete and should include sufficient data so that a permanent label can be prepared from it. The information should include: detailed location (i.e., state or province, county, latitude, and longitude, or township, range, and section, distance from definitive point, etc.) habitat, plant information, location, unique features (i.e., soil type, pH, number of duplicate specimens collected, etc.), date, and **collection number.** The collection number is a *nonrepeatable* number given by the collector to different collections made throughout a lifetime. These numbers usually begin with 1, 2, 3, etc. and continue indefinitely. Some collectors use a modified system beginning again each new year (i.e., 89-1, 89-2, . . . 90-1, 90-2, etc.). This number can be used as a **specimen tag** put on the newspaper the specimen was pressed in or on duplicate newspapers in which duplicate specimens of the same species were collected at the same time and location. It should not be used as a "species number" whenever the same species is collected during a lifetime. This was done by some nineteenth century collectors and can lead to confusion later when doing monographic studies. Later, when the correct Latin name is determined, the complete botanical name, including the author, is added to the field notes and to the outside of the newspaper. Most professional taxonomists do not put the family in the field notes, but it is advisable for students to do so because it helps them to distinguish the characters of families and, if added to labels, facilitates filing specimens in a collection.

Finally, *always* record the field information promptly. Memory cannot be trusted.

**4.** After the specimens have been dried, the next step is treatment for pests. Much care must be taken to protect permanent specimens from troublesome pests, such as silverfish and especially the **dermestid beetle.** This is the larval stage of various kinds of insects that attack not only plant material, but carpet, cloth, and tapestry. Damage by these pests can occur when the specimens are in unattended temporary or permanent storage. How specimens are treated varies considerably from institution to institution. Methods of treatment include: a) freezing at −20 to −60°C; b) heating at high temperatures to 60°C for 4–8 hours; c) placing dried specimens in a microwave oven; d) treating the specimen with an insect retardant, such as naphthalene, paradichlorobenzene (PDB), or formaldehyde; and e) fumigation in a special airtight exhaustible cabinet, chamber, or room (Fig. 5.2d). Because many of the fumigants used are extremely toxic to human health, extreme attention and careful procedure must be followed. The use of some fumigants is under various governmental controls, and therefore individuals must comply with the applicable laws to ensure human safety.

**5.** Following the treatment of the specimens for pests, they are "clean" and can be prepared for temporary storage and identification. It is usually not possible to identify many specimens when collected fresh, because time is generally a priority. Therefore, most professional botanists identify specimens that have been pressed and dried. The specimens are grouped into bundles according to a system (i.e., taxonomic lines, collector, collection location, etc.). These bundles are properly tagged or labeled, and placed in the priority sequence to be studied or mounted.

**6.** Beginning students start to identify plants using fresh specimens. As experience is gathered, most identification will be with dried or pressed material. The plant part (i.e., flower, etc.) needed can be made more lifelike by boiling or softening with a wetting agent. The simplest wetting agent is drops of laundry detergent in water applied to the desired plant part. A better type of agent is **Pohlstöeffe,** which comprises distilled water, 74%; methanol, 25%;

and aerosol OT, 1% (a commercial name for dioctyl sodium sulfosuccinate). (The name *Pohlstöeffe* was given by graduate students to honor Dr. R. Pohl of Iowa State University, who developed the formula, and is not a commercial name for the formula.) Pohlstöeffe applied directly to a specimen softens the plant part in usually less than one minute. After study, the plant specimen can be repressed without damage.

**7.** After completing the identification of a specimen, a permanent **specimen label** is prepared. The recorded information in the field notebook is now transferred to this label. The label is made of nonacid, high-quality, rag content paper. The size and shape may vary slightly but it will usually be a rectangle and range under 10 × 15 cm (4 × 6 in.). Labels should be typed or carefully handwritten in waterproof India ink. Labels today are commonly printed by personal computer with exact duplicates easily generated. A label should *never* be written in ballpoint pen! The ink of most ballpoint pens is not permanent and may diffuse into the fibers of the paper through time; some glues also blur the ink. An example of a good-quality label with pertinent information is given in Fig. 5.3. A good "rule of thumb" to follow is to give sufficiently clear location and habitat information so that another person in the future can locate the exact population, providing the original population has not become destroyed or the location greatly modified by human activities. Some labels even have maps printed on them.

Another label that may be placed with the specimen is a **voucher label.** This label establishes the authenticity of the plant. An **annotation label** is added when an authority checks the identification of the specimen and either verifies or changes the name. The new correct name is placed on the annotation label, *never* on the original label.

**8.** Mounting a specimen means attaching the plant permanently to a standard sheet of special paper (Fig. 5.4). In North America, standard herbarium mounting sheets are 29 × 41.5 cm (11.5 × 16.5 in.). In Europe,

ANDREWS UNIVERSITY HERBARIUM
Flora of "Ontario Road Prairie"

Juglandaceae

*Carya ovata* (Mill.) K. Koch.

BERRIEN COUNTY, MICHIGAN, U.S.A.
Niles Township, south of Niles, 1.5 km east of U.S. Highway 33. Lat. 41° 46'N; Long. 86° 15'W. R17W; T8S; Sec. 24.
Located between Ontario Road and State Line Road.

Habitat type: Oak Forest
Mature tree, 3.0 dm. diameter, many fruits.

James Ng — No. 729
September 20, 1981 — Dup. 3

**FIGURE 5.3** **A good example of a typed label with pertinent information.**

**FIGURE 5.4 A mounted specimen ready to file.**

herbarium mounting sheets are usually a little narrower and longer. The mounting sheet is a heavyweight, high rag content (100% preferred) paper with its fibers running lengthwise. It can be obtained from various biological supply and specialty paper companies. The paper should be sturdy enough to support the specimen.

The specimen may be attached by various methods (Fig. 5.5). A common method involves smearing a glass plate or plastic sheet with a water-soluble paste, placing the specimen on the paste, and then transferring the glued plant to the mounting sheet. The stem and bulky parts may be strapped with strips of glued linen tape, or special plastic glue, sewed to the sheet and tied on the back, or held to the sheet by glued paper strips. Some older European herbaria glue the specimens to small sheets of paper and then these sheets are pinned to larger sheets. The label is attached to the lower right-hand corner. To keep this corner from curling, glue is added only to the edges and center of the label. Weights made of lead, scrap metal, large washers, or flat stones are used to hold the specimen to the herbarium mounting sheet until

**FIGURE 5.5 Mounting herbarium specimens with a nontoxic glue dispensed from a plastic squeeze bottle. Metal washers help hold the plant to the sheet until the glue hardens.**

the glue dries. Small paper envelopes called **fragment packets** are attached to the sheet to hold seeds, loose plant parts, or anything pertinent to the specimen.

After the specimen has been mounted, the sheet is stamped with the name of the institution and is usually given an accession number. The specimen may be fumigated a second time if considerable time has elapsed since the first pest-control treatment, if the building is not climate controlled. Following this last step, the specimen is ready to be added into the collection.

## *The Herbarium*

As stated at the beginning of this chapter, the classic **herbarium** (pl. **herbaria**) is a collection of dried, pressed, and preserved botanical specimens arranged for reference. The specimens can be the flat and mounted ones, as discussed in the previous section, or preserved within jars and vials in fluid. Associated with these specimens may be found wood, bark, fruits, objects constructed of plant material, botanical illustrations, and reference works dealing with plants. In short, the contemporary herbarium has become a library of preserved plant specimens and an associated reference and botanical resource center.

Herbaria have been organized for different reasons. Small collections may house specimens collected from a specific locality (i.e., research study area, botanical garden, park, etc.) and gathered by one to a few individuals. Larger herbaria are made up of multiple collections and represent regional, state or province, country, or world areas.

The history of making pressed specimens and keeping collections can be traced to Italy. Luca Ghini, who lived during the sixteenth century, has been attributed as being the first person to preserve plants by pressing and sewing them to sheets of paper. This technique quickly spread across Europe, with the mounted specimens being bound into book volumes. The current method of housing specimens loosely was emphasized by the Swedish botanist Carl Linnaeus and perpetuated by his many students.

The modern herbarium houses specimens in specially constructed steel cabinets that are airtight and free from dust, insects, and light. Each cabinet has a series of shelves upon which the specimens are placed (Fig. 5.6). Some cabinets are constructed with small sliding shelves upon which specimens can be rested when being studied or filed.

In growing herbaria, the lack of space within the herbarium cabinet and/or herbarium room(s) can become a problem. To reduce the space dilemma, **curators** of herbaria are using compactor cases. These movable banks of shelves ride on tracks and can open or close, thus eliminating the space lost by aisles between rows of traditional herbarium cabinets (Fig. 5.7).

The specimens of the herbarium are arranged according to a system that allows for ease of filing and retrieval of material. The simplest system is to file the specimens alphabetically according to family, genus, and species. Many small herbaria do this. However, some large herbaria like the Field Museum, Chicago; Jardin Botanique, Genève; Iowa State University, Ames, Iowa; Rocky Mountain Herbarium, Laramie, Wyoming; and McGill Univer-

**FIGURE 5.6 The standard herbarium case with specimens in manila folders on the shelves.**

**FIGURE 5.7 Modern compactor herbarium cases that move on tracks and open electrically to provide access aisles (Missouri Botanical Garden, St. Louis, MO).**

sity, Montreal, have the families, genera, and species arranged alphabetically within large taxonomic groups (i.e., pteridophytes, gymnosperms, monocots, dicots). Some herbaria have each family and genus filed according to the Dalla-Torre and Harms (1900–1907) number code, a sequence based on the Engler classification system. This system has been followed a great deal in Europe and North America. Others, such as the Royal Botanic Gardens, Kew, England, file their specimens after the Bentham and Hooker family arrangement. This system is most common within British and British Commonwealth herbaria. Today some curators of herbaria are deciding to break with tradition and to arrange the herbarium specimens according to a contemporary scheme (e.g., Michigan State University, after the A. Cronquist system).

The specimens are filed in large folders that completely enclose the specimens. These folders may be color coded to fit a geographical scheme that is distinctive for the herbarium. Color coding geographically reduces the damage to specimens and helps in locating specimens. Each folder may include a genus or species, or in small herbaria, a family. Unknown specimens identified only to family or genus are placed at either the beginning or end of the family or genus. This will allow experts viewing the collection to identify these plants together.

Type specimens (usually holotypes and isotypes) are kept separate from the main collection or filed at the beginning of a taxonomic group in specially marked folders.

The large institutional herbaria are not the product of one individual's collecting but the joining of many individual collections. Tables 5.2 and 5.3 give the approximate size of the world's and North America's largest collections, respectively. Each of these herbaria are given an abbreviated code acronym recorded in the *Index Herbariorum* (see Chapter 3). Two of the fastest growing collections in the world today are found at the Royal Botanic Gardens, Kew, and the Missouri Botanical Garden, St. Louis, MO.

The personal computer has now become a most important piece of equipment in the modern herbarium. These electronic wonders, combined with some of the various programs available, can be used to accomplish many tasks. The tasks, to name a few, include the preparation and printing of labels (including as many exact replicas as needed), all types of data storage and information retrieval, statistical analysis, mapping of specimens, collection management, specimen determination, and manuscript preparation.

In spite of all the wondrous things accomplished with the computer, it should be pointed out that the work and activity of the herbarium will not become simpler. In fact, it may become even more complex because of the many more options open to the user of herbarium specimens. This is not all bad, for as society prepares to enter the twenty-first century, old, dry specimens will be given new life and value heretofore unknown in the past.

## *Ethics of Collecting and Handling Specimens*

Today there is concern about the conservation of flora and fauna. Many plant species are becoming rare due to habitat modification, the introduction of alien species, pollution, overcollecting, etc. (see Chapter 14). It therefore

**TABLE 5.2 Largest herbaria in the world based on information from *Index Herbariorum* I, ed. 8 (1990). Only those herbaria with over 1,500,000 specimens have been included. The number of specimens has been rounded to the nearest 5000.**

| *Rank* | *Location* | *Abbreviation* | *Number of Specimens* |
|---|---|---|---|
| 1 | Paris, FRANCE | P, PC | 8,880,000 |
| 2 | Kew, UNITED KINGDOM | K | 6,000,000 |
| 3 | Leningrad, U.S.S.R. | LE | 5,770,000 |
| 4 | Stockholm, SWEDEN | S | 5,600,000 |
| 5 | New York, NY U.S.A. | NY | 5,300,000 |
| 6 | London, UNITED KINGDOM | BM | 5,200,000 |
| 7 | Genève, SWITZERLAND | G | 5,000,000 |
| 8 | Cambridge, MA U.S.A. (Harvard U. Herbaria) | A, AMES, ECON FH, GH, NEBC | 4,860,000 |
| 9 | Washington, D.C. U.S.A. | US, USNC | 4,370,000 |
| 10 | Montpellier, FRANCE | MPU | 4,000,000 |
| 11 | Villeurbanne, FRANCE | LY | 3,800,000 |
| 12 | Wien, AUSTRIA | W | 3,750,000 |
| 13 | St. Louis, MO U.S.A. | MO | 3,700,000 |
| 14 | Firenze, ITALY | FI | 3,600,000 |
| 15 | Jena, FRG (FDR) | JE | 3,000,000 |
| | Leiden, NETHERLANDS | L | 3,000,000 |
| 16 | Helsinki, FINLAND | H | 2,720,000 |
| 17 | Berlin, FRG (FDR) | B | 2,500,000 |
| | Uppsala, SWEDEN | UPS | 2,500,000 |
| 18 | Chicago, IL U.S.A. | F | 2,415,000 |
| 19 | Lund, SWEDEN | LD | 2,400,000 |
| 20 | München, FRG (FDR) | M | 2,300,000 |
| 21 | Copenhagen, DENMARK | C | 2,225,000 |
| 22 | Meise, BELGIUM | BR | 2,040,000 |
| 23 | Praha, CZECHOSLOVAKIA | PR | 2,000,000 |
| | Praha, CZECHOSLOVAKIA | PRC | 2,000,000 |
| | Zürich, SWITZERLAND | ZT | 2,000,000 |
| | Edinburgh, UNITED KINGDOM | E | 2,000,000 |
| 24 | Beijing (Peking), P.R. CHINA | PE | 1,800,000 |
| 25 | Berkeley, CA U.S.A. | JEPS, U.C. | 1,725,000 |
| 26 | Budapest, HUNGARY | BP | 1,620,000 |
| 27 | Ann Arbor, MI U.S.A. | MICH | 1,615,000 |
| 28 | Bogor, INDONESIA | BO | 1,600,000 |
| | San Francisco, CA U.S.A. | CAS, DS | 1,600,000 |
| 29 | Philadelphia, PA U.S.A. | PH, ANSP, PENN | 1,590,000 |
| 30 | Zürich, SWITZERLAND | Z | 1,500,000 |

*SOURCE:* Courtesy of P. Holmgren.

**TABLE 5.3 Largest herbaria in North America based on information from *Index Herbariorum* I, ed. 8 (1990). Only those herbaria with over 500,000 specimens have been included. The number of specimens has been rounded to the nearest 5000.**

| *Rank* | *Location* | *Abbreviation* | *Number of Specimens* |
|---|---|---|---|
| 1 | New York, NY | NY | 5,300,000 |
| 2 | Cambridge, MA (Harvard U. Herbaria) | A, AMES, ECON FH, GH, NEBC | 4,860,000 |
| 3 | Washington, D.C. | US, USNC | 4,370,000 |
| 4 | St. Louis, MO | MO | 3,700,000 |
| 5 | Chicago, IL | F | 2,415,000 |
| 6 | Berkeley, CA | JEPS, UC | 1,725,000 |
| 7 | Ann Arbor, MI | MICH | 1,615,000 |
| 8 | San Francisco, CA | CAS, DS | 1,600,000 |
| 9 | Philadelphia, PA | PH, ANSP, PENN | 1,590,000 |
| 10 | Ithaca, NY | BH, CU, CUP | 1,095,000 |
| 11 | Ottawa, ONT | DAO, DAOM | 1,050,000 |
| 12 | Beltsville, MD | BPI, BARC | 1,010,000 |
| 13 | Claremont, CA | POM, RSA | 960,000 |
| 14 | Austin, TX | TEX, LL | 950,000 |
| 15 | Ottawa, ONT | CAN, CANA, CANL, CANM | 905,000 |
| 16 | Madison, WI | WIS | 900,000 |
| 17 | St. Paul, MN | MIN, MPPD | 845,000 |
| 18 | Champaign-Urbana, IL | ILL, ILLS, CEL | 795,000 |
| 19 | Laramie, WY | RM, RMS, WYAC, USFS | 655,000 |
| 20 | Durham, NC | DUKE | 635,000 |
| 21 | Montréal, QUE | MT, MTJB | 630,000 |
| | Chapel Hill, NC | NCU | 630,000 |
| 22 | Washington, D.C. | NA | 600,000 |
| 23 | Pittsburgh, PA | CM | 575,000 |
| 24 | East Lansing, MI | MSC | 520,000 |
| 25 | Seattle, WA | WTU | 515,000 |
| 26 | Columbus, OH | OS | 500,000 |

*SOURCE:* Courtesy of P. Holmgren.

behooves all students of botany to be aware of the consequences of their behavior and how one's actions are perceived by others. Following are a few guidelines with regard to collecting plants:

1. Get permission to collect from the owners of private property.
2. Obtain the necessary permit beforehand when wishing to collect in parks, preserves, and areas owned by private, local, state, and governmental agencies.
3. Collect the necessary specimens away from areas frequented by the public.

4. Refrain, if at all possible, from mass collecting rare species when a single aboveground specimen or color photograph will do as a voucher.
5. Be aware of laws that restrict or regulate the collection of certain plants.
6. Do not leave excess plant material, large holes in the ground, and the collected area a mess when finished. Be neat.
7. If there is any question about whether to collect or not, don't.

Dried herbarium specimens are brittle and easily damaged or destroyed and type specimens and vouchers cannot be replaced, therefore, much care must be practiced when using them. The following rules will cover common difficulties:

1. Do not bend sheets, but keep them flat and supported when carrying.
2. Do not put objects on top of specimens.
3. Keep food and drink away from specimens.
4. Do not turn specimens like pages in a book while in the folders, but lift off each specimen individually while supporting the folder on a firm surface.
5. Follow the instructions provided by the herbarium curator.
6. Remove folders straight in and out of the cabinets, not at an angle over the edge of the shelf. This is especially a problem with high shelves, as the bottom specimens are easily broken.
7. Carry out detailed study of the specimens on a flat surface, using a long-armed microscope that permits the specimens to be kept flat.
8. Remember, if specimens are cared for properly, they can last almost indefinitely. Students should care for specimens as though a project's success or failure depends on the survival of that specimen. It just may.

## SELECTED REFERENCES

Archer, W. A. 1950. New Plastic Aid in Mounting Herbarium Sheets. *Rhodora* 52:298–299.

Benson, L. 1979. *Plant Classification*, 2nd ed. D. C. Heath, Lexington, MA, 901 pp.

Brayshaw, T. C. 1973. *Plant Collecting for the Amateur.* Museum Methods Manual 1. British Columbia Provincial Museum, Vancouver, BC, 15 pp.

*Computer Use in Botanical Systematics.* 1988. Report of the Systematics Collections Committee, American Society of Plant Taxonomists.

Croat, T. B. 1978. Survey of Herbarium Problems. *Taxon* 27:203–218.

Davis, P. H. and V. H. Heywood. 1973. *Principles of Angiosperm Taxonomy.* Robert E. Krieger, Huntington, NY, 558 pp.

de Vogel, E. F. (ed.). 1987. *Manual of Herbarium Taxonomy. Theory and Practice.* UNESCO, Jakarta, Indonesia, 164 pp.

De Wolf, G. P., Jr. 1968. Notes on Making an Herbarium. *Arnoldia* 28:69–111.

Edwards, S. R., B. M. Bell, and M. E. King (eds.). 1981. *Pest Control in Museums: A Status Report.* Association of Systematics Collections. Lawrence, KS.

Fosberg, F. R. and M. H. Sachet. 1965. *Manual for Tropical Herbaria. Regnum vegetabile* 39, Utrecht.

Franks, J. W. 1965. *A Guide to Herbarium Practice.* Handbook for Museum Curators, Part E, Section 3. The Museum Association, London.

*Herbarium News.* (A monthly newsletter published by the Missouri Botanical Garden, St. Louis, MO USA).

Hill, S. R. 1983. Microwave and the Herbarium Specimen: Potential Dangers. *Taxon* 32:614–615.

Lawrence, G. H. M. 1951. *Taxonomy of Vascular Plants.* Macmillan, New York, 823 pp.

Lee, W. L., B. M. Bell, and J. F. Sutton (eds.) 1982. *Guidelines for Acquisition and Management of Biological Specimens.* Association of Systematic Collections, Lawrence, KS.

Leenhouts, P. W. 1968. *A Guide to the Practice of Herbarium Taxonomy. Regnum vegetabile* Vol. 58. IAPT, Utrecht, 60 pp.

MacFarlane, R. B. A. 1985. *Collecting and Preserving Plants for Science and Pleasure.* Arco, New York, 184 pp.

McNeill, J. 1968. Regional and Local Herbaria. In: V. H. Heywood (ed.) *Modern Methods in Plant Taxonomy.* Academic Press, London, pp. 33–44.

Nevling, L. I. Jr. 1973. Report of the Committee for Recommendations in Desirable Procedures in Herbarium Practice and Ethics, II. *Brittonia* 25:307–310.

Perkins, K. D. 1990. Should Herbarium Specimens Be Mounted on Non-buffered Paper? *Herbarium News* 10(7/8):45, 46.

Pohl, R. W. 1965. Dissecting Equipment and Materials for the Study of Minute Plant Structures. *Rhodora* 67:96–95, 96.

Radford, A. E. et al. 1974. *Vascular Plant Systematics.* Harper & Row, New York, 891 pp.

Regalado, J. C., R. K. Rabeler, and J. H. Beaman. 1986. *Label S3 User's Manual.* Beal-Darlington Herbarium, Dept. of Botany and Plant Pathology, Michigan State Univ., East Lansing, MI, 102 pp.

Robertson, K. R. 1980. *Observing, Photographing, and Collecting Plants.* Illinois Natural History Survey Circular 55, Urbana, 62 pp.

Savile, D. B. O. 1962. *Collection and Care of Botanical Specimens.* Publ. 1113. Research Branch Canada Dept. of Agriculture, Ottawa, Canada, 124 pp.

Smith, C. E., Jr. 1971. *Preparing Herbarium Specimens of Vascular Plants.* Agricultural Information Bulletin. No. 348. U.S.D.A., Washington, D.C., 29 pp.

von Reis Altschul, S. 1977. Exploring the Herbarium. *Sci Amer* 236(5):96–104.

Waddington, J. and D. M. Rudkin. 1986. *Proceedings of the 1985 Workshop on Care and Maintenance of Natural History Collections.* Royal Ontario Museum, Life Sciences Miscellaneous Publications, Toronto, Ontario, 121 pp.

Womersley, J. S. 1981. *Plant Collecting and Herbarium Development.* FAO Plant Production and Protection Paper, No. 33, FAO.

Woodland, D. W. 1974. The McGill University Herbarium—A Library of Plants. *Macdonald J.* 35(6):7–9, 20.

# 6

# Families of Ferns and Their Allies (Pteridophytes)

Many plant taxonomy books emphasize only the flowering plants (angiosperms) and spend little or no time on the other groups of vascular plants. Because other vascular plants, such as ferns, that reproduce by spores and the nonflowering seed plants, such as pines, spruces, firs, and cycads (gymnosperms), are so prevalent in many parts of the world, it seems only natural that a discussion of systematic principles should also include these groups.

In this chapter we shall consider the four major divisions of "seedless" vascular plants (pteridophytes) that have living representatives: Psilotophyta (whisk ferns), Microphyllophyta (club mosses, quillworts, spike moss), Arthrophyta (horsetails), and Polypodiophyta (true ferns). The division names follow Bold, Alexopoulos, and Delevoryas (1987), and the family names are as given in Crabbe, Jermy, and Mickle (1975). Different but similar division names may be used in other books.

The pteridophytes were one of the most common groups of plants during geological periods of the past, especially during the Carboniferous. They are represented today by approximately 9085 species worldwide. There are about 10 species of whisk fern, 300 species of club mosses, 700 species of spike mosses, 150 species of quillworts, 25 species of horsetails, and 7900 species of ferns. These numbers may actually be on the conservative side. Many botanists feel that there are almost 12,000 fern species alone, thereby increasing the total number of pteridophyte species to almost 13,000. Most pteridophytes are generally **perennials** (live 3 or more years), found in moist, shady terrestrial habitats. Some are **epiphytes** (live attached to other plants but do not obtain nourishment from them), others are aquatic, while a few are found in arid to semiarid (xeric) conditions. The ferns genus *Anogramma* is unique among ferns in being annual. About two thirds of the living species are found in tropical regions, with about 75% of these species occurring in two regions: the New World tropics with 2250 species (including southern Mexico, Central America, and Andes of western

Venezuela, south to Bolivia and the Greater Antilles), and Malaysia and southeastern Asia, with 4500 species.

The morphology of pteridophytes is variable and at times quite "unfernlike," such as *Salvinia* (salvinia) and *Isoetes* (quillwort), two aquatic genera that have floating, small overlapping leaves about 2 cm long and submerged grass-like leaves, respectively. At the other extreme in size are tree ferns of the genus *Cyathea*, which may reach over 25 m in height with **fronds** (leaves) of 7 m in length. Most pteridophytes have roots, stems, and leaves, even though the stem is only a modified one, termed a **rhizome,** from which adventitious roots arise. The most distinctive part of the plant is the frond, which is comprised of a **lamina,** which at times may be many times divided, and a **stipe,** which attaches the frond to the rhizome (Fig. 6.1).

A divided lamina is called a **compound frond,** with the individual divided portions called **pinna** if they are divided once and **pinnules** if divided again. The pinna are attached to the **rachis** or main axis, which is an extension of the stipe. In nearly all ferns, the young leaves are coiled and characteristically unroll during their development, called **circinate vernation,** producing a fern **crozier,** or **fiddlehead.**

## *Reproductive Cycle of the Pteridophytes*

The overall reproductive cycle of the pteridophytes shows less variability as a whole when compared to other plant groups, and this is one of its unifying factors. The life cycle alternates between two distinct morphological generations of unlike plants. For example, in the common Christmas fern of North America, *Polystichum acrostichoides,* the familiar frond and rhizome represent the **sporophyte** or **diploid** phase of generation (Fig. 6.2). As the frond ages, **sporangia** (sing. **sporangium**), which are spore-producing structures, begin to appear on the undersurface in clusters called **sori** (sing. **sorus**). In many genera the sori are arranged in clustered patterns, which are helpful in proper identification. Some sori are covered by specialized outgrowths of the leaf, the **indusia** (sing. **indusium**) or by a rolling of the leaf margin itself, called a **false indusium.**

At this time, mature **spores,** the result of meiosis of the **spore mother cells** in the sporangia, are released from each individual sporangium. A row of unevenly thick-walled cells on the sporangium, the **annulus,** facilitates spore release. Contraction of the annulus due to drying causes the **lip cells** to tear. A sudden expansion of the annulus followed by contraction catapults the spores into the air, where they are dispersed by air currents.

The spore germinates after landing on a suitable moist substrate and forms an independent **haploid** generation structure, the **gametophyte** or **prothallus.** The resulting gametophyte is bisexual with both male sex organs, **antheridia** (sing. **antheridium**), and female sex organs, **archegonia** (sing. **archegonium**). The shape and size of the prothallus varies somewhat from a dumbbell shape (2–3 mm in length) in *Psilotum;* a more cylindrical, thallus shape (2–3 cm long) in *Lycopodium;* to a flat, heart-shaped membranous structure, 1 cm across, in some ferns. Motile, flagellated **sperm** produced in the

**FIGURE 6.1 Terminology and structure of a fern.**

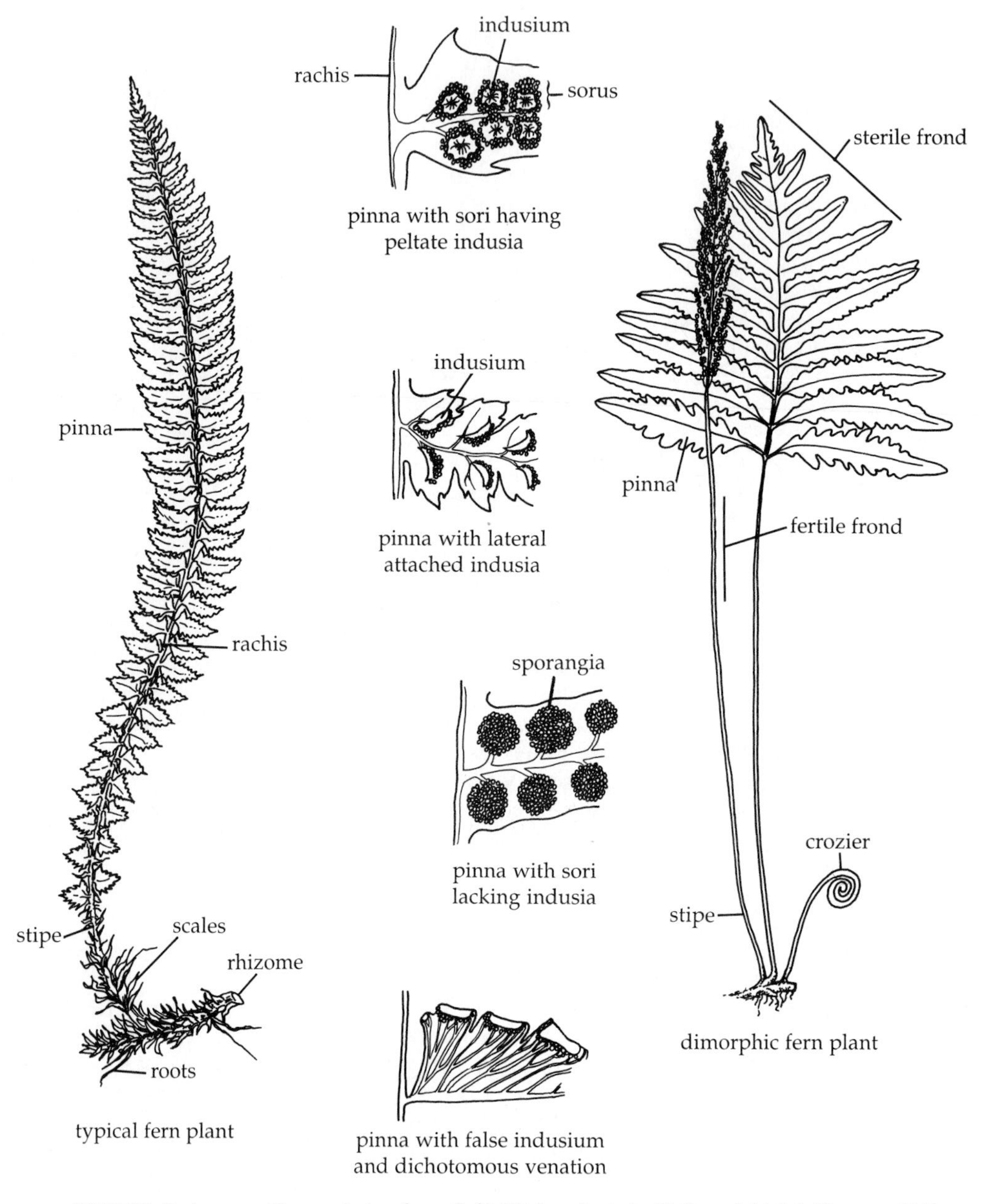


*SOURCE:* Redrawn with permission from (left) Hitchcock et al., 1969, and (right) Gleason, 1963.

**FIGURE 6.2 Typical reproductive cycle of a fern.**

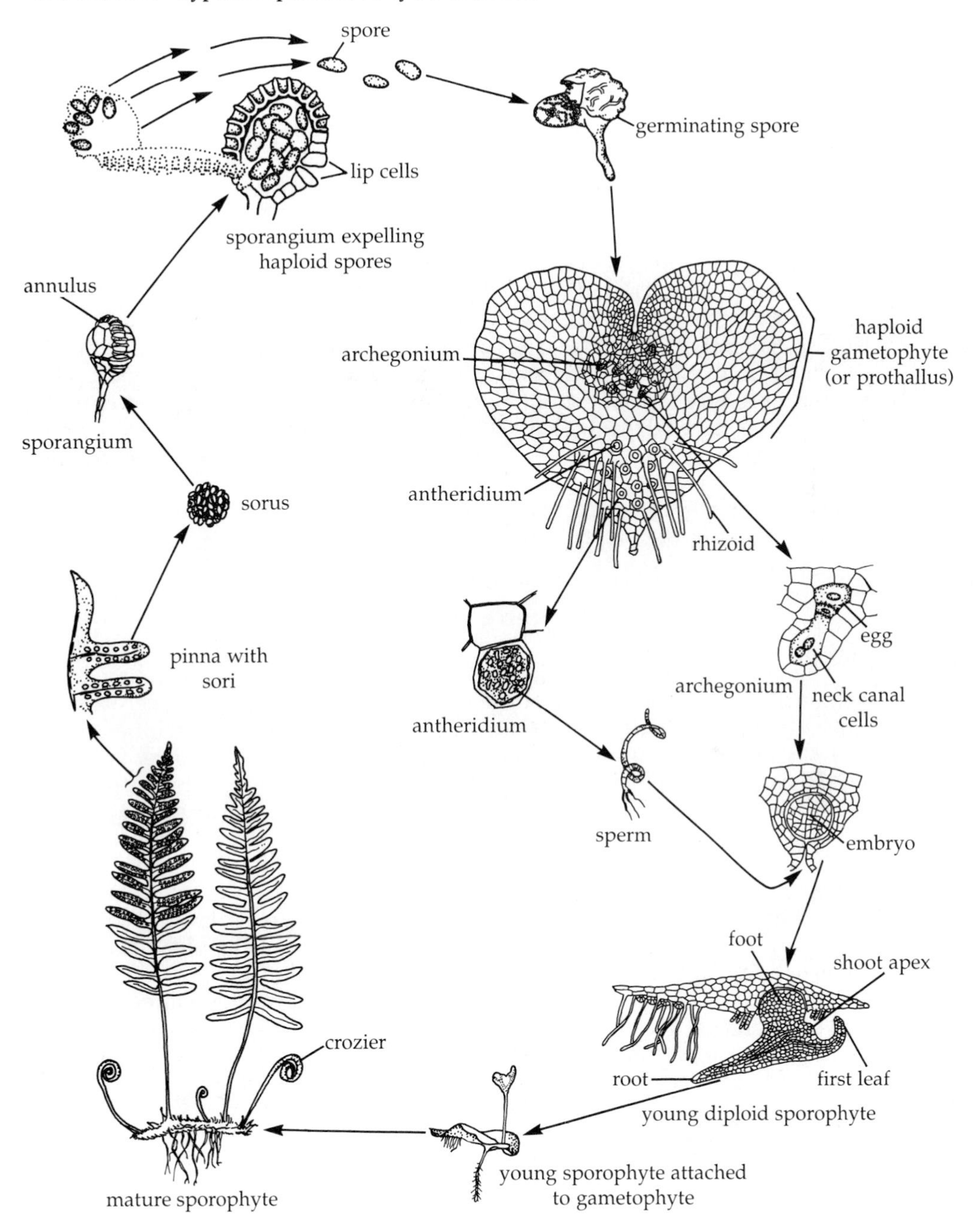

antheridia fertilize an egg in the archegonium. The resulting diploid **zygote** grows into a young sporophyte plant that completes the alternating phases of the life cycle. The development is rapid, with the sporophyte becoming an independent plant. The gametophyte soon disintegrates.

In the example of the Christmas fern given above, only one type of spore is produced by the sporophyte. These types of plants are called **homosporous.** A few living pteridophytes are **heterosporous,** that is, they produce large spores called **megaspores** that develop into a separate gametophyte, which produces only archegonia, and smaller spores, termed **microspores,** that form gametophytes that produce antheridia. Examples of pteridophytes that reproduce in this way are *Azolla, Isöetes, Salvinia,* and *Selaginella.*

As botanists have studied the development of pteridophytes, it has been learned that there are two types of sporangia formed: **eusporangiate** and **leptosporangiate.** The eusporangiate are large in size, develop from more than one initial cell on the stem or leaf, and have many hundreds to thousands of spores produced in each sporangium. The leptosporangiate-type sporangia are usually smaller than the eusporangiate type, are on a narrow stalk, and produce fewer spores, usually 64–128 for each sporangium. Each sporangium develops from a single initial cell and has an annulus that helps in spore dispersal. Modern pteridophyte classification uses these two development types in dividing the plants into two groups.

## *Selected Families of Pteridophytes*

The divisions of the pteridophytes represent distinct groups and are well recognized. However, the classification of the ferns has undergone considerable change during the past 100 years, especially at the inter- and intrafamily levels. Unlike most families of flowering plants, the fern families are not always easily distinguished from each other. This is due to different philosophies of fern specialists (pteridologists), the lack of understanding of the evolutionary relationships, and the use of small characters that do give distinct taxonomic recognition. This can be easily seen by comparing some North American floras, which group many genera under the family Polypodiaceae, while most European written floras will have the genera in several families. A helpful guide to fern genera, generic synonyms, and families has been published by Crabbe, Jermy, and Mickel (1975). Table 6.1 is a summary of the classification of ferns and their allies gathered from various sources, using the family names given by Crabbe, Jermy, and Mickel. This system is followed for the convenience of instruction purposes only.

### DIVISION PSILOTOPHYTA

**Family Psilotaceae (Whisk Fern)** Terrestrial or epiphytic, perennial, rhizomes without roots but with rhizoids and mycorrhizal fungi present, stems erect or pendulent with small scale-like outgrowths, and dichotomously branched, as in *Psilotum.* The other genus in the family, *Tmesipteris,* has leaf-like appendages that are much larger than the scale-like outgrowths of

**TABLE 6.1 Classification of the living ferns and their allies (pteridophytes) as followed in this text.**

Division Psilotophyta
  Class I. Rhyniopsida
    Subclass A. Rhyniidae
      *Family 1. Psilotaceae (2 genera, 4–8 species)

Division Microphyllophyta
  Class II. Lycopodiopsida
    Subclass A. Lycopodiidae
      *Family 1. Lycopodiaceae (2–5 genera, 450 species)
    Subclass B. Selaginellidae
      *Family 1. Selaginellaceae (1 genus, 700 species)
    Subclass C. Isöetidae
      *Family 1. Isöetaceae (2 genera, 77 species)

Division Arthrophyta
  Class III. Equisetopsida
    Subclass A. Equisetidae
      *Family 1. Equisetaceae (1 genus, 29 species)

Division Polypodiophyta
  Class IV. Polypodiopsida
    Subclass A. Ophioglossidae
      *Family 1. Ophioglossaceae (4 genera, 65 species)
    Subclass B. Marattiidae
      *Family 1. Marattiaceae (7 genera, 100 species)
    Subclass C. Osmundidae
      *Family 1. Osmundaceae (3 genera, 18 species)
    Subclass D. Plagiogyriidae
      Family 1. Plagiogyriaceae (1 genus, 37 species)
    Subclass E. Schizaeidae
      *Family 1. Schizaeaceae (4 genera, 150 species)
      Family 2. Parkeriaceae (1 genus, 4 species)
      Family 3. Platyzomataceae (1 genus, 1 species)
      *Family 4. Adiantaceae (47–56 genera, 1100 species)
    Subclass F. Gleicheniidae
      *Family 1. Gleicheniaceae (4 genera, 140 species)
      Family 2. Matoniaceae (2 genera, 4 species)
      Family 3. Cheiropleuriaceae (1 genus, 1 species)
      Family 4. Dipteridaceae (1 genus, 8 species)
      *Family 5. Polypodiaceae (40–52 genera, 550 species)
      Family 6. Grammitidaceae (11–13 genera, 500 species)
    Subclass G. Hymenophyllidae
      Family 1. Loxsomaceae (2 genera, 4 species)
      *Family 2. Hymenophyllaceae (6–33 genera, 465 species)
      Family 3. Hymenophyllopsidaceae (1 genus, 2 species)
      Family 4. Stromatopteridaceae (1 genus, 1 species)

*(continued)*

*TABLE 6.1 (continued)*

Family 5. Metaxyaceae (1 genus, 1 species)
Family 6. Lophosoriaceae (1 genus, 1 species)
*Family 7. Cyatheaceae (2–4 genera, 625–651 species)
Family 8. Thyrsopteridaceae (3 genera, 20 species)
*Family 9. Dennstaedtiaceae (24 genera, 410 species)
*Family 10. Thelypteridaceae (1–30 genera, 900 species)
*Family 11. Aspleniaceae (73–78 genera, 2700 species)
*Family 12. Davalliaceae (13 genera, 220 species)
*Family 13. Blechnaceae (10 genera, 260 species)
Subclass H. Marsileidae
*Family 1. Marsileaceae (3 genera, 72 species)
Subclass I. Salviniidae
*Family 1. Salviniaceae (1 genus, 12 species)
*Family 2. Azollaceae (1 genus, 6 species)

Total: 35 families; 278–367 genera, 9637–9667 species

*NOTE:* Division names follow Bold, Alexopoulos, and Delevoryas (1987); class and subclass names follow Lellinger (1985); family names follow Crabbe, Jermy, and Mickel (1975); and the number of taxa for each taxonomic level follows Mabberley (1987). Families with an asterisk are discussed and illustrated in the text.

*Psilotum,* which are considered by some botanists to be flattened branches, although others disagree. Sporangia are two or three fused as a thick-walled **synangium.** There are homosporous spores without chlorophyll. Gametophyte is dumbbell-shaped and subterranean with a symbiotic fungus, nongreen (Fig. 6.3).

The family consists of 2 genera and 4–8 species. *Psilotum* is pantropical in distribution, occurring in North America in the southern states, while *Tmesipteris* occurs in some Polynesian islands, New Zealand, Australia, New Guinea, and the Philippines.

The family lacks a fossil record but has traditionally been considered to be similar to the fossil genera *Cooksonia, Psilophyton,* and *Rhynia,* the first known vascular plants from the Silurian and Devonian periods.

## DIVISION MICROPHYLLOPHYTA

**Family Lycopodiaceae (Clubmoss, Ground Pine)** Terrestrial or epiphytic, perennial, roots present, dichotomously branched, protostelic, slender; stems erect, pendant, or creeping-prostrate. Leaves (microphylls) small, simple, borne spirally around the stem, with one vein. Sporangia large, sessile, or short stalked on the upper (adaxil) side of the leaf or on the stem, associated among leaves or with well-differentiated **sporophylls** (leaf or leaf-like organs that bear sporangia) into a definite **strobilus** or cone at the end of arial branches. Homosporous. Subterranean gametophytes are nongreen or green, irregular lobed structures, may live as long as 25 years; development of mature sex organs may take 6–15 years (Fig. 6.4, p. 64).

The family consists of 2 (possibly more) genera and 450 species. *Lyco-*

**FIGURE 6.3 Family Psilotaceae. *Psilotum:* (a) mature plant; (b) three-parted synangium. *Tmesipteris:* (c) frond showing sterile and fertile appendages; (d) two-parted synangium between fertile appendages.**

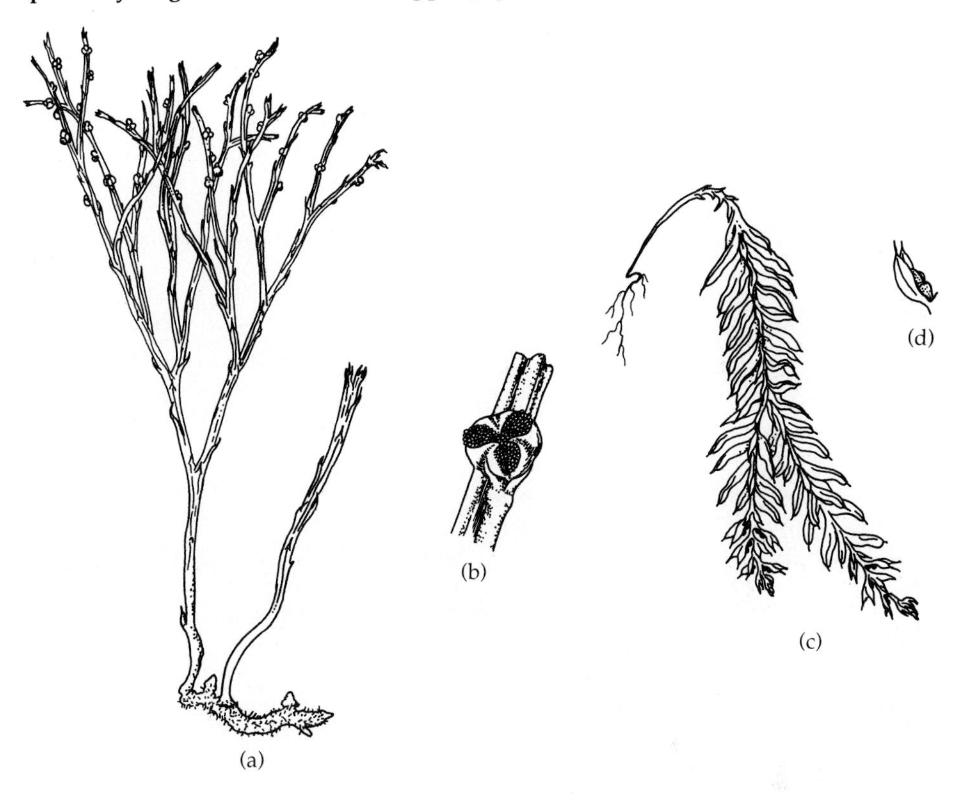


*podium* is the most widely distributed genus of pteridophytes and is found on all continents from north of the Arctic Circle south to South America, Africa, Tasmania, and New Zealand, and in the Pacific and South Atlantic islands as well. *Phylloglossum* of New Zealand, Australia, and Tasmania has a single species, *P. drummondii.*

Whether *Lycopodium* should be divided into various smaller genera is presently being debated by pteridologists. Some classifications have included other genera, such as *Diphasiastrum, Lycopodiastrum,* and *Urostachys,* but most taxonomists at present consider these as synonyms of *Lycopodium.*

The spores of *Lycopodium* have a highly volatile oil that made them useful in early Chinese fireworks. They are produced in great abundance and can be collected in North America in mid to late autumn. Pouring them over a flame produces a sparkling display. Caution must be taken when doing this, so that excess nonburnt spores are not inhaled. Some individuals are highly allergic to them.

The fossil record of *Lycopodium* is not well known due to the lack of sufficiently preserved material that is distinctly *Lycopodium* beyond the Cretaceous. However, the fossil forests of the Carboniferous period were predominantly lycopod trees.

**FIGURE 6.4 Family Lycopodiaceae.** ***Phylloglossum:*** **(a) plant showing "grass-like leaves" and tuberous underground parts.** ***Lycopodium:*** **(b) microphyllous leaf from stem; (c) sporophyll from strobilus; (d) plant with strobilis; (e) plant lacking strobilis, sporongia among upper stem leaves.**

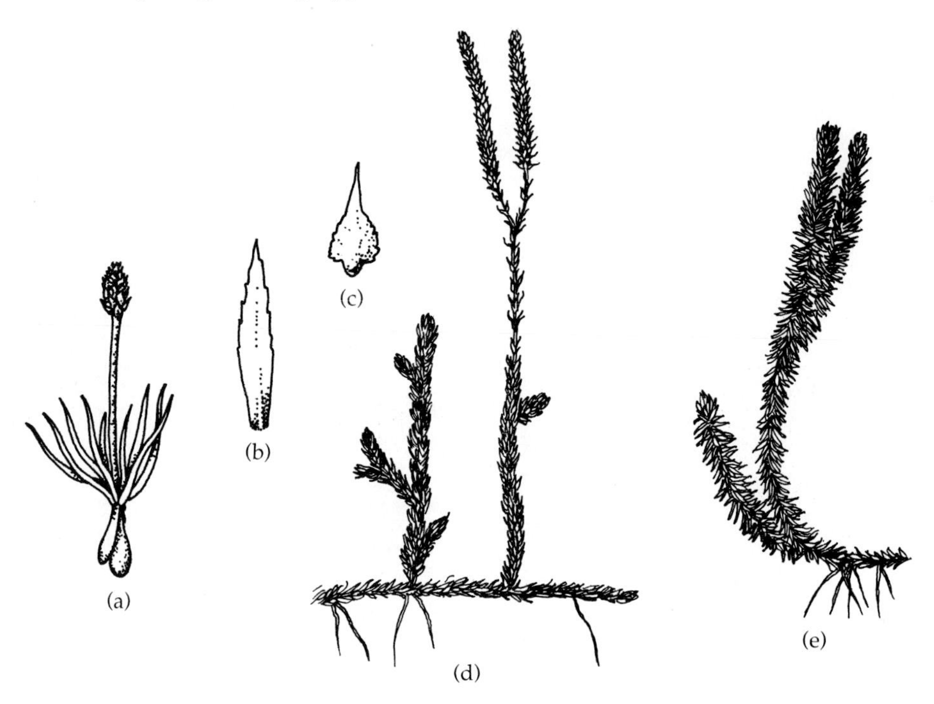


**Family Selaginellaceae (Spike Moss)** Terrestrial, moist habitats with few in arid to semiarid areas, or rarely epiphytic, perennials; root and stem protostelic with vessels, prostrate, creeping to erect. Small scale-like outgrowth called a **ligule** present on the upper surface of the leaf (microphyll) toward the base, leaves spirally arranged. Heterosporous (a most significant difference between *Lycopodium* and *Selaginella*). **Megasporangia** are borne on **megasporophylls** toward the apex of the strobilus, while **microsporangia** are borne on **microsporophylls** below the megasporophylls; the unisexual gametophytes are formed within the various spore walls (Fig. 6.5).

The family consists of 1 genus and 700 species. *Selaginella* is nearly worldwide in distribution, from northern Alaska and Greenland to South America, Africa, Australia, and the South Pacific, most common in the tropics. *Selaginella lepidophylla*, the so-called resurrection plant, inhabits the deserts of Mexico north to New Mexico and Texas. The smallest chromosomes within the pteridophytes are found in *Selaginella.*

A genus *Selaginellites,* a Carboniferous genus, includes some species that have been put in *Selaginella.* Some lycopod fossils put in *Lycopodites* of pre-Cretaceous deposits may actually be *Selaginella.*

FIGURE 6.5 **Family Selaginellaceae.** ***Selaginella:*** **(a) creeping, leafy habit; (b) shoot apex, showing difference of microphyllous leaves and sporophylls; (c) microphyllous leaf; (d) sporophyll; (e) habit of prostrate plant, showing roots.**

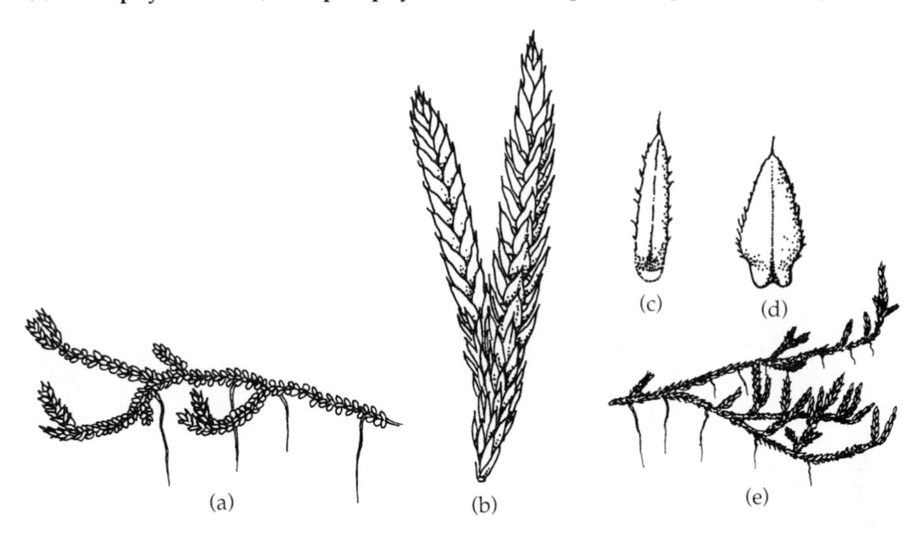


**Family Isoetaceae (Quillwort)** Terrestrial or aquatic, stems protostelic, with cambium horizontal or erect and corm-like, with many long, firm, usually dichotomously branched roots attached. Leaves (microphylls) long, quill- or grass-like, with a single vein. Sporangia large and single, buried in the spoon-shaped base of the leaves on the abaxal side. Heterosporous. Microsporangia borne toward center of plant, while megasporangia borne toward outside of plant. Gametophytes form within the spore wall, as in *Selaginella* (Fig. 6.6, p. 66).

The family consists of 2 genera and 77 species. *Isöetes* is widely distributed in the Americas, Europe, Asia, Africa, Australia, and New Zealand; absent from oceanic islands. It grows in shallow water at least part of the year and is most frequent in clear, sandy-bottomed lakes, ponds, and rivers that are high in oxygen. It is easily overlooked because it appears to be grass-like. The chromosomes are easily fragmented, and this may account for the many chromosome number reports with extra chromosomes.

*Isöetes*-like fossils have been collected from Upper Triassic deposits.

## DIVISION ARTHROPHYTA

**Family Equisetaceae (Horsetail, Scouring Rushes)** Terrestrial or aquatic, annual or perennials, sometimes evergreen, stems jointed, subterranean ones horizontal (rhizomes) with numerous slender roots; aboveground stems generally green, jointed, longitudinally ribbed, 10 cm to 8 m long, unbranched or with whorled branches from the nodes that are easily mistaken as leaves; anatomy complex, with various canals and air spaces. Leaves greatly

**FIGURE 6.6 Family Isoetaceae. *Isoetes:* (a) habit of plant, showing cormlike base, roots, and grass-like sporophylls; (b) spoon-shaped base of a megasporophyll, showing megasporangium and ligule; (c) Y-shaped ridge and spines on surface of a megaspore.**

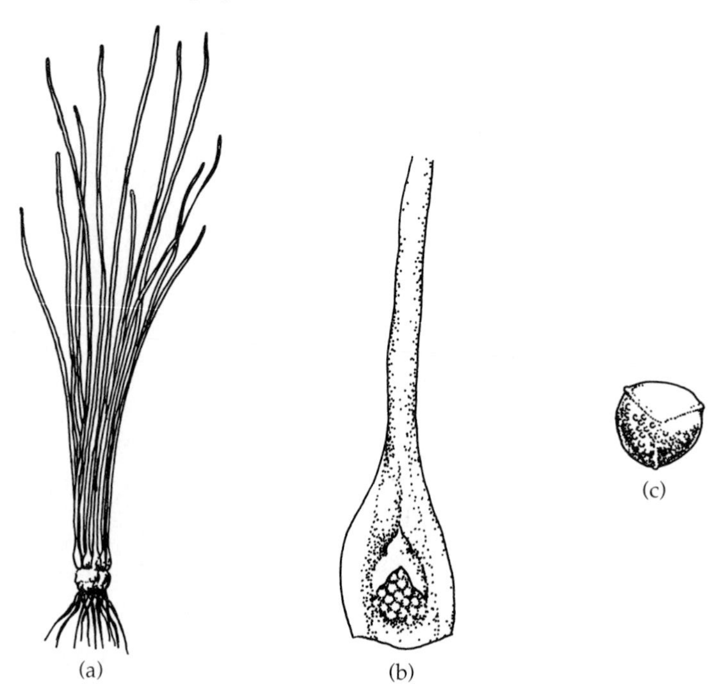


reduced with a single vein, borne as a whorl, fused as a sheath. Sporangia large and attached to peltate (umbrella-like) structures called **sporangiophores** into a terminal strobilus or "cone." Spores homosporous, green with chlorophyll, **elaters** (arm-like structures) on the outer spore wall coil when dry and elongate when wet, presumably enhancing spore dispersal. Pinhead-sized thallus-shaped green gametophytes develop near the soil surface, some functioning unisexually with antheridia only (Fig. 6.7).

The family consists of the genus *Equisetum* and 29 species. *Equisetum* is distributed almost worldwide in temperate and tropical climates but is absent from the Amazon basin and Australia, though recently naturalized in New Zealand. The rough silica-stemmed species, such as *Equisetum hyemale,* were used by early North American settlers to scour pots and pans, hence the common name, scouring rush. The consistency of chromosome number in *Equisetum* is unique among pteridophytes and probably represents a derived level, followed by extinctions of lower numbers.

Horsetails, like the lycopods, have been found in Devonian deposits and are most abundant in the Paleozoic era. During the Carboniferous period, *Calamites,* trees over 2 dm in diameter and 15 m tall, were represented.

**FIGURE 6.7 Family Equisetaceae.** ***Equisetum:*** **(a) node of stem showing whorled branches; (b) strobilus, or "cone," and stem node with fused whorled leaves; (c) sterile stem with whorled branches; (d) cluster of unbranched stems.**

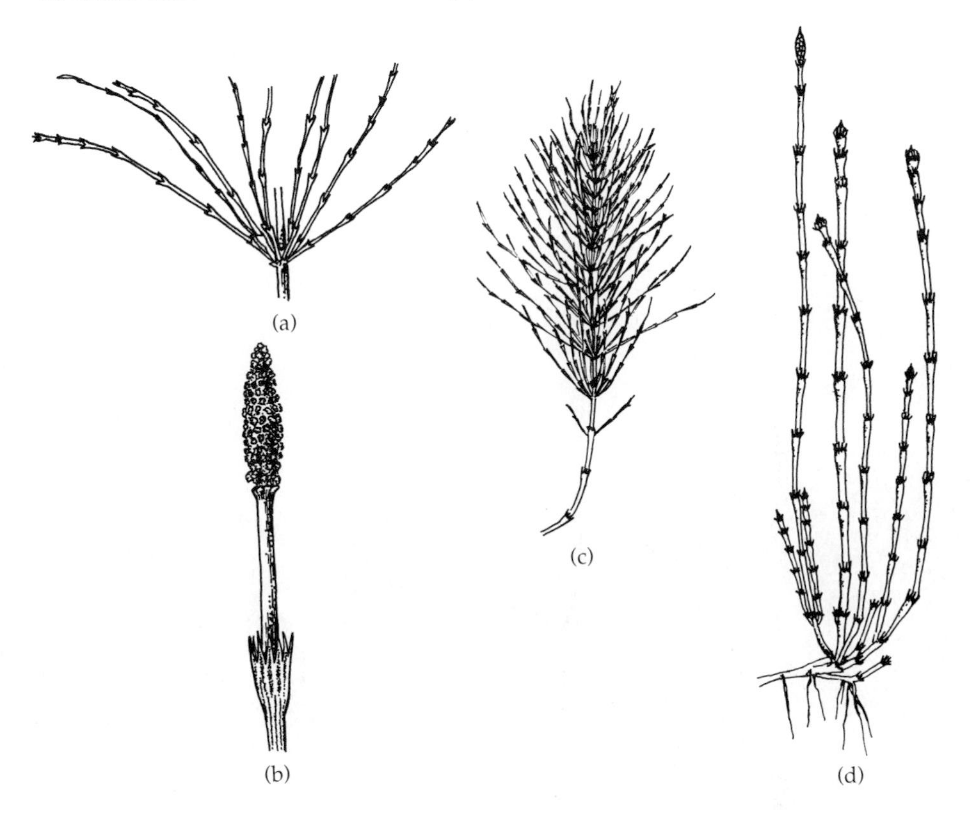


## DIVISION POLYPODIOPHYTA

**Family Ophioglossaceae (Adder's Tongue, Grape Fern, Moonwort)** Terrestrial or rarely epiphytic, perennial from tubers or rhizomes. Leaves simple or dissected, not circinate in *Ophioglossum,* but circinate in *Botrychium.* From the surface of the leaf develops a simple or branched axis bearing large spherical sporangia of the eusporangiate type. Homosporous. Gametophytes are subterranean. Cytologically the genus holds the world record for the most chromosomes for any living organism at $n = 621 + 10$ fragments (Fig. 6.8, p. 68).

The family consists of 4 genera and 65 species with a cosmopolitan distribution in temperate and tropical areas.

Spores ascribed to *Ophioglossum* have been found from Jurassic deposits.

**Family Marattiaceae (Marattia)** Terestrial, perennial, stems erect often forming large, globose shapes due to persistent frond bases, scales, or thick-

**FIGURE 6.8 Family Ophioglossaceae. *Ophioglossum:* (a) large eusporangiate sporangia; (b) mature plant. *Botyrichium:* (c) mature plant; (d) compound sterile leaf.**

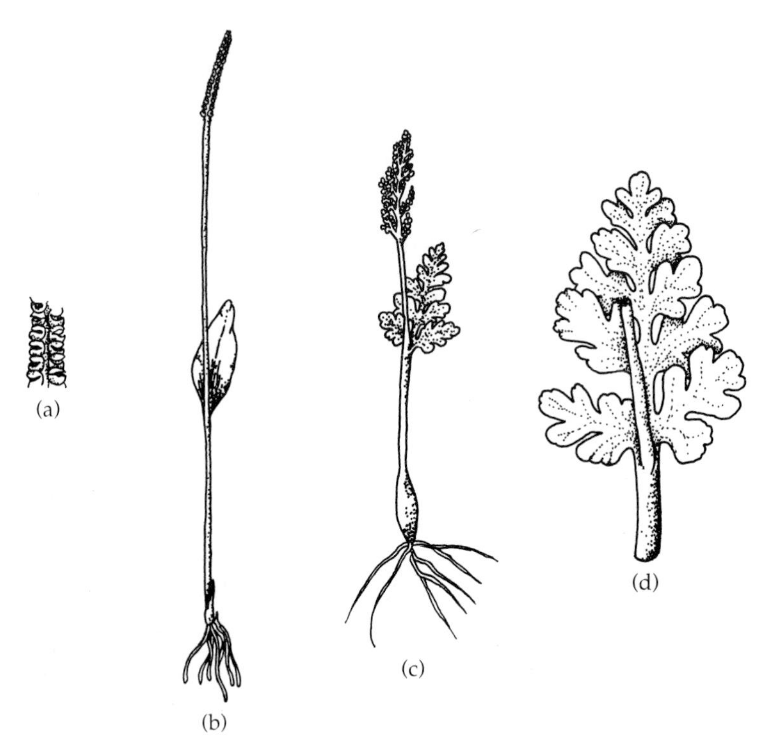


ened roots. Lamina 2–4 pinnate, the frond 2–3.5 m long, rachis often with a swollen node **(pulvinus)** at the base of each pinna. Sporangia of sori fused into an elongate synangium (Fig. 6.9). The family consists of 7 genera and 100 species distributed pantropically. The spores are among the smallest. Fossils of the family were first recorded from Carboniferous deposits.

**Family Osmundaceae (Royal Fern)** Terrestrial, erect to decumbent stems, dictyostelic, often massive with persistent stipe bases, having wiry roots. Fronds wholly or partially **dimorphic** (sterile green frond or pinnae different from the fertile brown ones), pinnately compound. Sporangia large with many spores, making it intermediate between the eusporangiate to leptosporangiate condition, annulus of sporangium a shield-like plate or broad and horizontal band forming around the sporangium. Gametophyte is green and develops on the soil surface (Fig. 6.10).

The family is comprised of 3 genera and 18 species. *Osmunda* is distributed almost worldwide, except in cold or arid climates, Australia and New Zealand (recently introduced into New Zealand), while *Todea* is found in South Africa, New Guinea, Australia, and New Zealand. *Leptopteris* is generally an

**FIGURE 6.9 Family Marattiaceae.** ***Marattia:*** **(a) pinna of a frond with pulvinus at base; (b) mature plant; (c) undersurface of pinna showing fused synangia.**

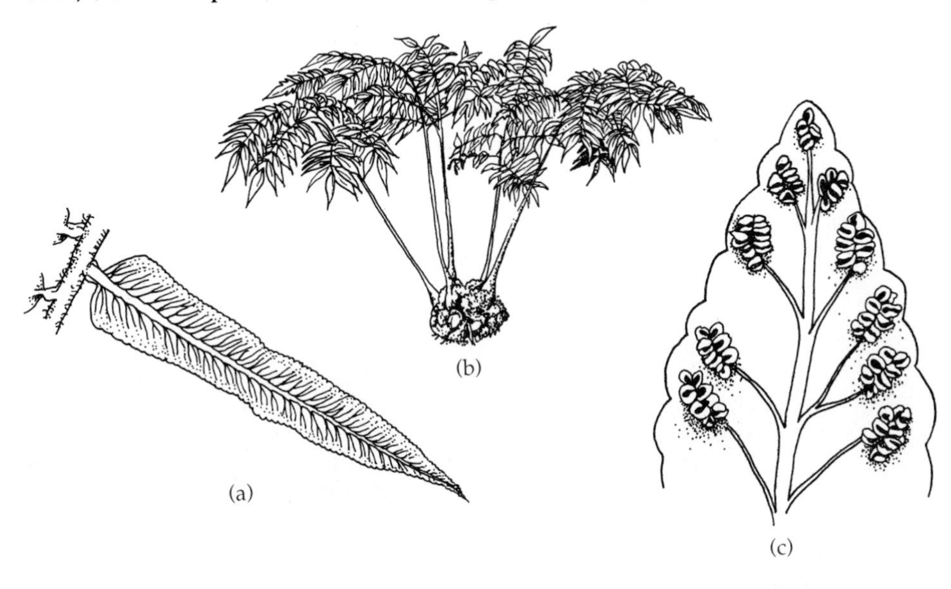


**FIGURE 6.10 Family Osmundaceae.** ***Osmunda:*** **(a) frond showing fertile and sterile pinna; (b) sporangium with annulus; (c) stout rhizome and stipes.**

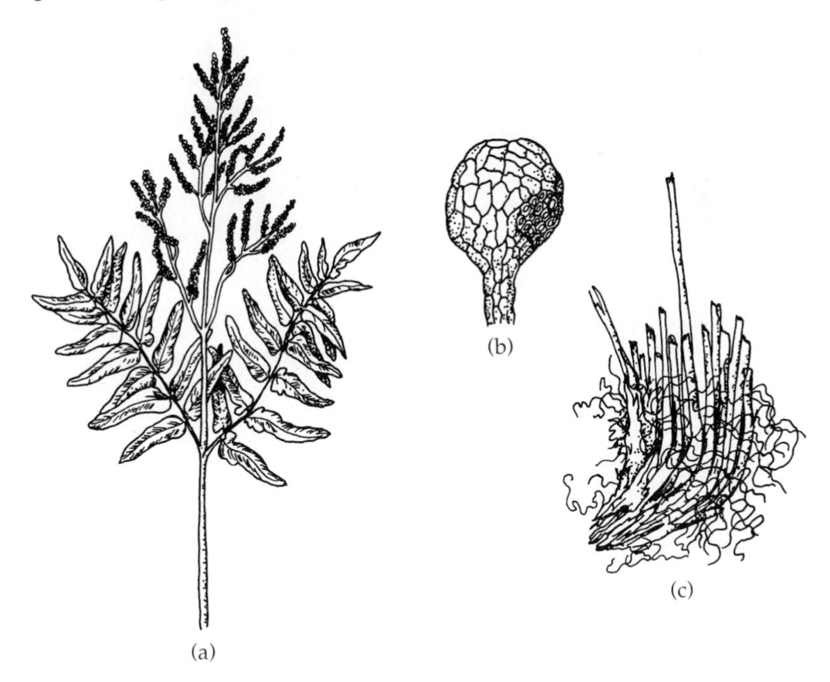


*SOURCE:* Redrawn and modified from Gleason, 1963.

arborescent genus and is restricted to the southwestern Pacific Ocean area. All species have the same chromosome number, $2n = 44$.

Fossils attributed to the family have been collected from Carboniferous deposits. Some fossils similar to the living genus *Osmunda* have been found in Cretaceous and Paleocene strata.

**Family Schizaeaceae (Climbing Fern)** Terrestrial, perennial, stems grass-like. Leaflets of two morphological types: one produces sporangia, the other lacking sporangia. Sporangia-bearing leaves with leaflets generally smaller. Sporangium solitary with a horizontal annulus, cap-like on the apex (Fig. 6.11).

The family consists of 4 genera and 150 species. *Anemia* is primarily a genus of the American tropics but gets into Texas and Florida, and also into Africa, Madagascar, and southern India; *Lygodium,* a pantropical genus, is also found in the eastern United States.

*Schizaea* is a pantropical genus with a few species in southern Florida and with *S. pusilla* found in eastern North America. *Mohria* is a genus of Africa and Madagascar. The varying chromosome numbers found in the family indicate both polyploidy and aneuploidy, and possibly similar genetic mechanisms at work as those present in *Ophioglossum.*

Fossil representatives have been found in Jurassic deposits to the present.

**FIGURE 6.11 Family Schizaeaceae. *Lygodium:* (a) venation of sterile leaf; (b) mature vine-like plant having fertile and sterile pinna; (c) fertile leaflets with sporangia on underside; (d) cap-like annulus on sporanga. *Schizaea:* (e) dichotomously branched frond; (f) sporangia.**

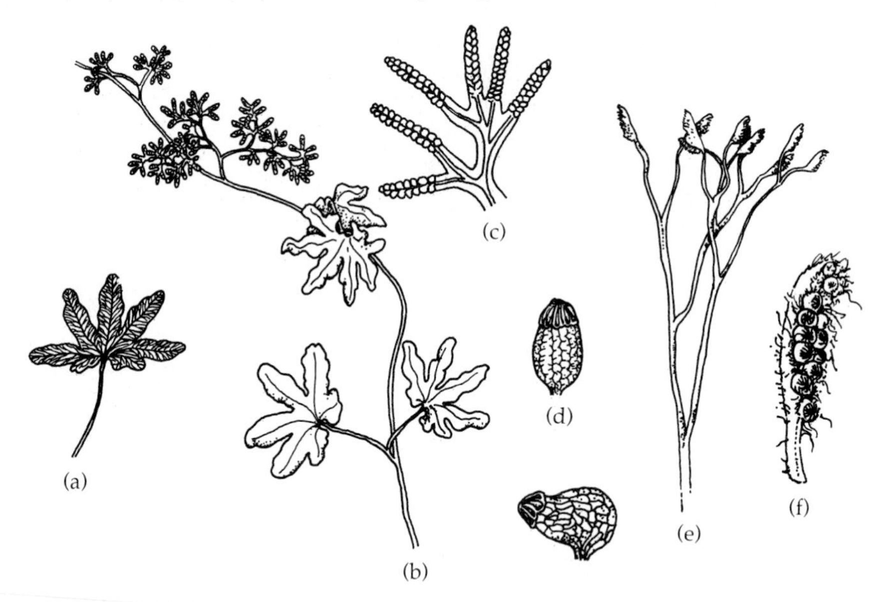

**Family Adiantaceae (Maidenhair Fern)** Terrestrial or epiphytic, annual or perennial with long fibrous roots bearing scales; generally small ferns of unique habitats, stems erect or decumbent, small and compact to long and creeping. Fronds monomorphic or dimorphic, veins of pinnae dichotomously branched. Sori along curved margins of pinnae (Fig. 6.12).

The family consists of 47–56 genera and a varied number of species. The family distribution is pantropical and temperate. Some of the common genera include: *Adiantum,* maidenhair fern; *Cheilanthes,* rock ferns; *Cryptogramma,* rock brake fern; *Pellaea,* cliffbrake fern; and *Pteris,* wall fern. *Anogramma,* annual fern, is the only known annual fern.

The delineation of this family is very much in debate and needs a broad-scope study. The cytology of the genera is fairly consistent, with haploid numbers being $n = 29$, 30, or multiples of these. The fossil record of this family is only known definitively from the Eocene.

**Family Hymenophyllaceae (Filmy Fern)** Terrestrial or epiphytic, perennial, small ferns, found in wet habitats; stem erect or decumbent, usually

**FIGURE 6.12 Family Adiantaceae. *Pellaea:* (a) mature plant; (b) undersurface of pinna showing sporangia under the false indusia. *Adiantum:* (c) pinnae showing dichotomous branched veins and false indusia; (d) dichotomously branched frond; (e) close-up of false indusium.**

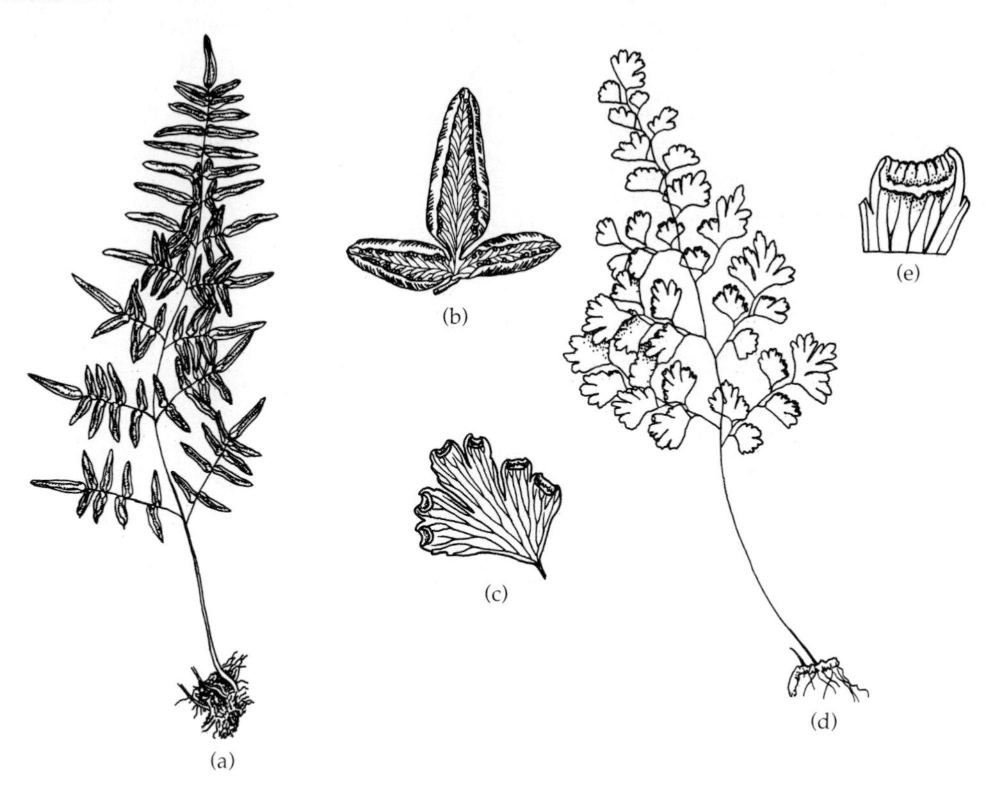

long and creeping. Leaves small to minute, thin, one cell thick, except for the veins, which makes some fronds semitransparent. Sporangia in sori along margin of pinnae and enclosed by a tubular or two-lipped indusium; the annulus is oblique on the sporangium, causing it to open horizontally (Fig. 6.13).

The family consists of 6–33 genera and 465 species. Both the genera *Hymenophyllum* and *Trichomanes* are common worldwide in tropical regions and are less common in wet temperate areas such as the southeastern United States. It should be noted that the generic concepts in this family are much undecided at present, and the use of 2 genera in this book is purely for convenience.

Fossil representatives have been found from Jurassic deposits of Queensland, Australia to the present.

**Family Gleicheniaceae (Gleichenia)** Terrestrial in wet areas, forming dense stands in open habitats; rhizome creeping, slender, protostelic. Fronds branched, with the main rachis periodically stopping growth, allowing successive pairs of branches to grow, giving a **bipinnatifid** (twice-branched) or **pinnatifid** (once-branched) lamina. Sporangia pear-shaped in sori, which are

**FIGURE 6.13 Family Hymenophyllaceae.** ***Hymenophyllum:*** **(a) mature plant with sporangia.** ***Trichomanes:*** **(b) oblique annulus of sporangium; (c) sporangia of sori in cup-shaped indusium; (d) mature frond showing sori along margin.**

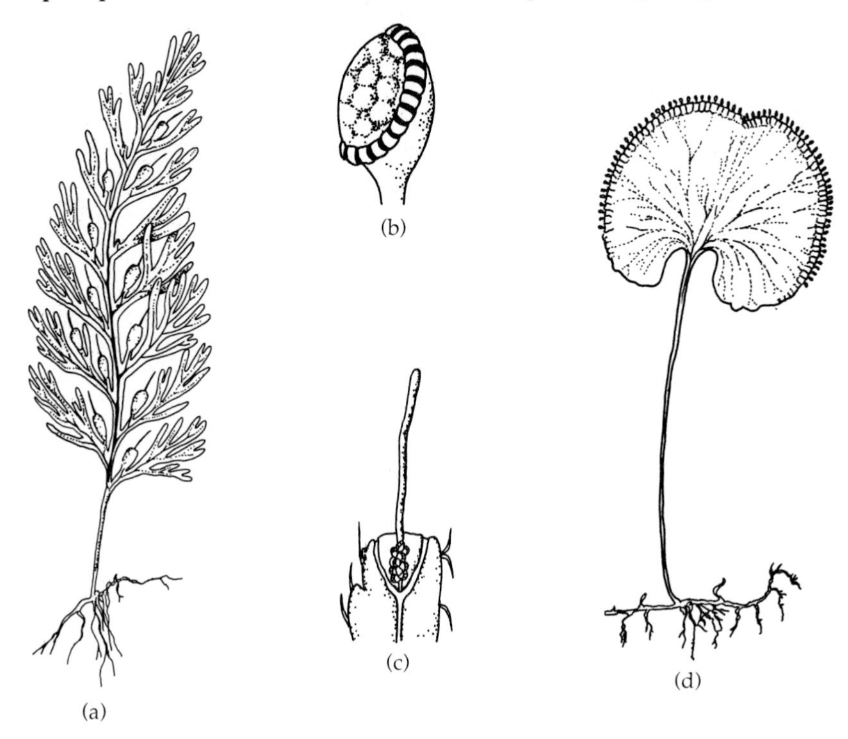

in a small receptacle, splitting open vertically by a paired oblique annulus (Fig. 6.14).

The family is comprised of 2–4 genera and 120 species distributed pantropically to subtropically in wet areas, forming dense stands in open habitats. Two genera are *Gleichenia* with 110 species and *Dicranopteris* with 10 species.

Fossils have been observed from Carboniferous deposits, with Cretaceous fossils showing characteristic branched fronds that are similar to living species.

**Family Polypodiaceae (Polypody)** Mostly small to large, epiphytic with a few terrestrial, rhizomes creeping, covered with scales, solenostele, or dictyostele. Leaves simple, lobed to tripinnate. Sporangia in round, separate, raised sori or over the whole fertile frond surface; indusium absent; annulus an incomplete vertical crest not completely encircling the sporangium (Fig. 6.15, p. 74).

The family consists of 40–52 genera and 550 species distributed chiefly

**FIGURE 6.14 Family Gleicheniaceae. *Gleichenia:* (a) habit of fertile frond; (b) sterile frond showing branching pattern; (c) crozier of terminal pinna and young frond.**

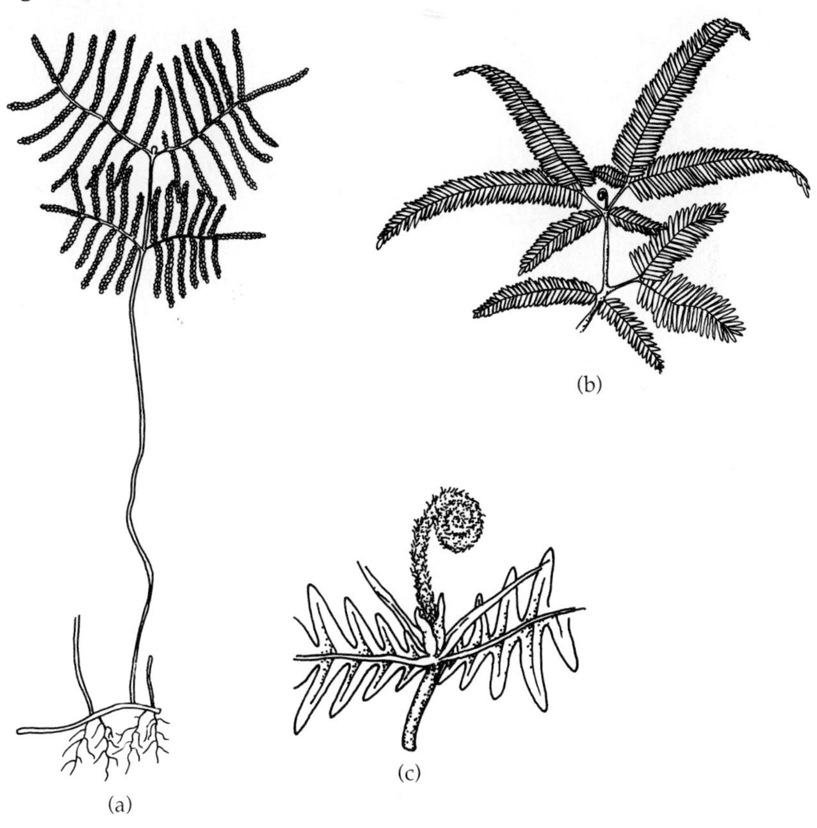

**FIGURE 6.15 Family Polypodiaceae.** ***Polypodium:*** **(a) rhizome with fronds; (b) pinna undersurface showing sori; (c) sori lacking indusia.**

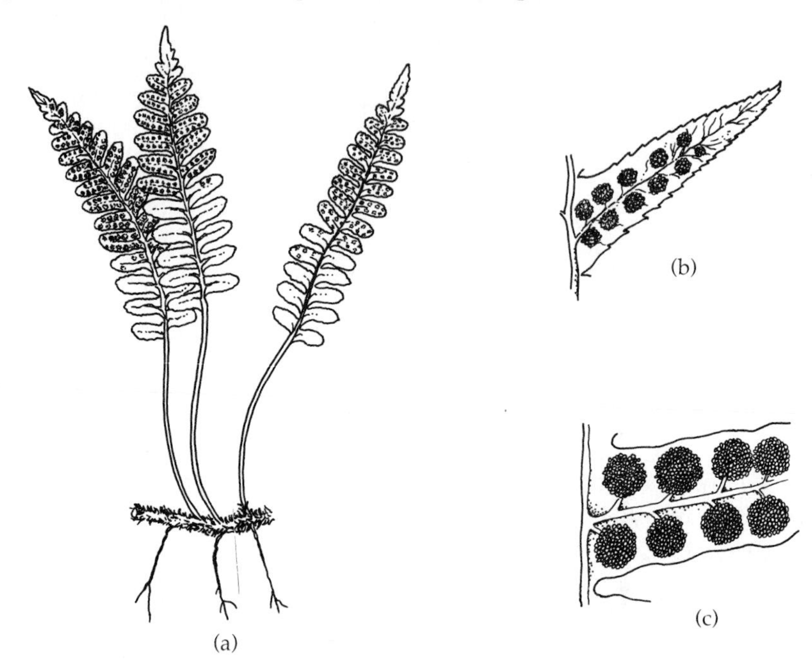


tropically with some temperate representatives, such as *Polypodium vulgare* of Europe and South Africa and *P. virginianum* of North America. The tropical staghorn fern, *Platycerium,* is commonly cultivated in greenhouses. The number of genera included in this family differs greatly. The family is sometimes used in North American floras to include most genera. A major problem in the classification is agreement among pteridologists on the affinity of many small characters. The chromosome ploidy levels within the family are low, being mainly tetraploid with a few hexaploids.

Some fossils from Pliocene deposits of Russia have been reported.

**Family Cyatheaceae (Tree Fern)** Palm-like stem, generally growing above ground (some species have decumbent, creeping stems), usually massive (may reach 25 m in height), with a dictyostele. Fronds usually large, 2–7 m long. Sporangia in sori, with or without indusia, sometimes completely enclosing the sorus; annulus oblique, allowing the sporangium to open horizontally (Fig. 6.16).

The family consists of 2–6 genera and 625 species distributed largely subtropically to tropically. Common genera include *Alsophila, Cyathea, Nephelea,* and *Sphaeropteris.* A uniform chromosome number of $2n = 138$ throughout the family makes it distinctive among ferns.

The earliest fossil records have been a stem of *Cyathea* from Jurassic deposits and fertile fronds from Upper Jurassic.

**FIGURE 6.16 Family Cyatheaceae. *Cyathea:* (a) habit of a large tree fern; (b) sori of pinna in cup-like indusia; (c) oblique annulus of sporangium.**

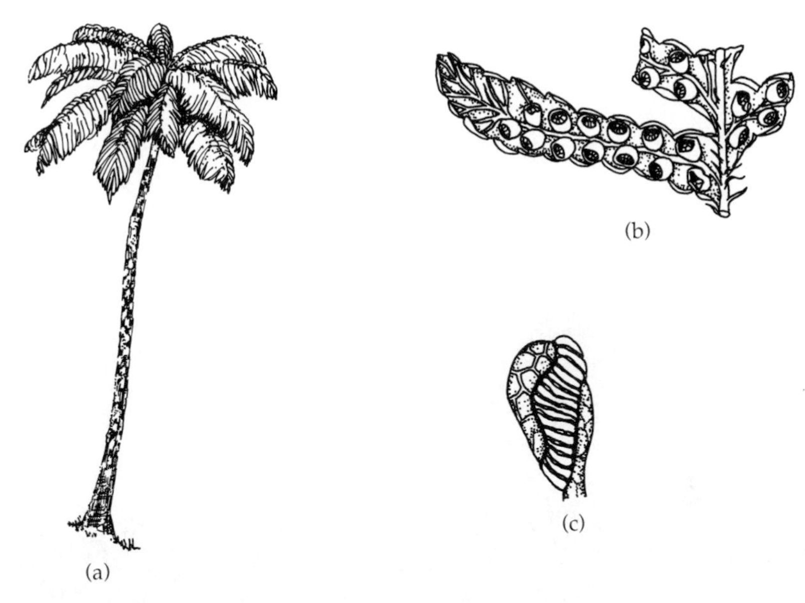


*SOURCE:* After Scagel et al., 1984.

**Family Dennstaedtiaceae (Bracken, Hayscented Fern)** Terrestrial or rarely climbing, perennial, rhizome erect or long-creeping with scales. Leaves large and pinnately divided. Sporangia marginal, indusium cup- or purse-shaped, or with a **false indusium** (recurved margin); annulus vertical or slightly oblique (Fig. 6.17, p. 76).

The family is comprised of 24 genera and 410 species distributed mostly pantropically, with a few species found in temperate climates. *Pteridium aquilinum,* a weedy genus with 6 subspecies and 6 varieties, is found in extremely cold, dry to wet tropical climates. In some tropical areas the genus forms almost pure stands, possibly through allelopathy (Gliessman & Muller, 1978), and invades cleared lands and pastures. It is long lived and its capacity to survive is evidenced by regeneration through several meters of volcanic ash on Mt. St. Helens in Washington state in North America within 1 to 2 years of the volcano's eruption. *Dennstaedtia punctilobula* has the aroma of dried hay, hence the name *hayscented fern.*

The fossil record of the family is sparse, being confined to Paleocene to Eocene deposits.

**Family Thelypteridaceae (Maiden Fern, New York Fern)** Terrestrial, perennial, rhizome slender erect or creeping, bearing many fibrous roots and scales. Frond pinnate with two vascular strands in the petiole (stipe). Sporangia in round sori covered by a kidney-shaped indusium. Spores are bilateral shaped instead of tetrahedral (Fig. 6.18, p. 77). The family consists of

**FIGURE 6.17 Family Dennstaedtiaceae. *Pteridium:* (a) portion of frond; (b) sporangium and annulus; (c) undersurface of pinnule showing sporangia and false indusium.**

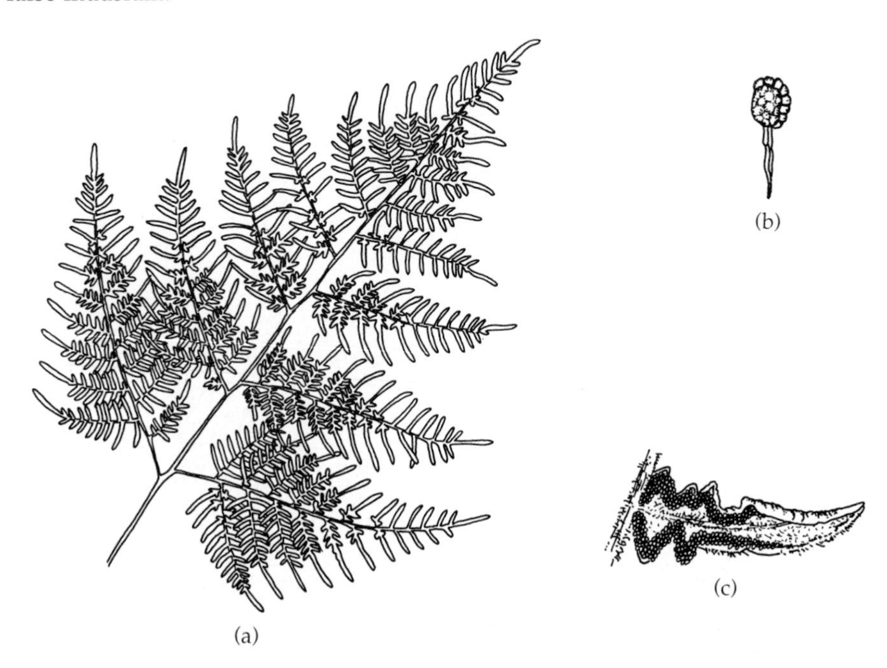


1–30 genera *(Thelypteris)* and 900 species distributed worldwide, mostly in wet forested tropical regions, with a few species boreal or temperate. Some species have been shown to have allelopathic abilities, while others are weedy. The generic classification is unsettled in this family, with some authors recognizing one monotypic genus, *Thelypteris,* others 3 genera, while still others as many as 32 genera. Many of them until recently were considered under *Dryopteris.* This book will follow Tryon and Tryon (1982) and consider only *Thelypteris.* A suspected thelypteroid fern from Jurassic strata has been reported.

**Family Aspleniaceae (Spleenwort, Wood Fern)** Terrestrial or epiphytic, perennial, rhizome erect or creeping, dictyostelic, with scales. Frond simple to pinnately compound, membrane-like, leathery to touch. Sori round, kidney-shaped to elongate with similarly shaped indusia. Spores bilateral (instead of tetrahedral) and surrounded by a wrinkled sheath-like covering called the **perispore** (Fig. 6.19, p. 78).

The family consists of 78 genera and 2700 species, both temperate and tropical, in moist, shady forests. This is one of the most abundant fern groups. Some important genera are *Asplenium,* spleenwort; *Athyrium,* lady fern; *Cystopteris,* brittle fern; *Dryopteris,* wood fern; *Polystichum,* Christmas fern; and *Woodsia,* cliff fern. *Camptosorus rhizophyllus,* walking fern of North America, forms large populations mostly on damp, shady, calcareous rocks, by rooting at the end of the frond, forming new asexual plantlets. The ostrich fern,

FIGURE 6.18 **Family Thelypteridaceae. *Thelypteris:* (a) mature plant; (b) cluster of fronds; (c) sori and kidney-shaped indusia.**

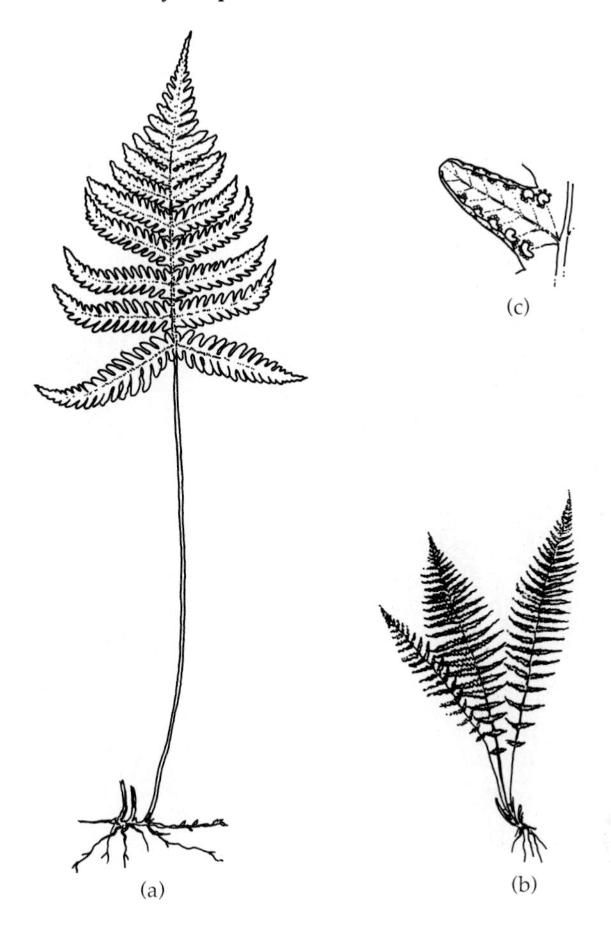


*Matteuccia struthiopteris,* is collected in the spring in eastern North America for its delicious fiddleheads. Like some other families of ferns, the Aspleniaceae is open to question as to which genera should be included, hence the number of genera and species varies greatly. The fossil record of this family is very poor, with only a record of *Onoclea* from Upper Cretaceous deposits of British Columbia, Canada.

**Family Davalliaceae (Rabbit's Foot and Boston Fern)** Mostly epiphytic, perennials, rhizomes erect or creeping, moderately stout, dictyostelic bearing scales. Sori roundish to elongate with round, kidney-shaped, or elongate indusia; annulus interrupted by the sporangium stalk. Spores bilateral, lacking chlorophyll and with a perispore (a wrinkled envelope covering the spore) (Fig. 6.20, p. 79).

The family consists of 13 genera and 22 species distributed mostly trop-

**FIGURE 6.19 Family Aspleniaceae.** ***Asplenium:*** **(a) mature plant; (b) sori and elongate, laterally attached indusia.** ***Dryopteris:*** **(c) mature plant.** ***Polystichum:*** **(d) sori arrangement; (e) sorus with umbrella-shaped indusium.**

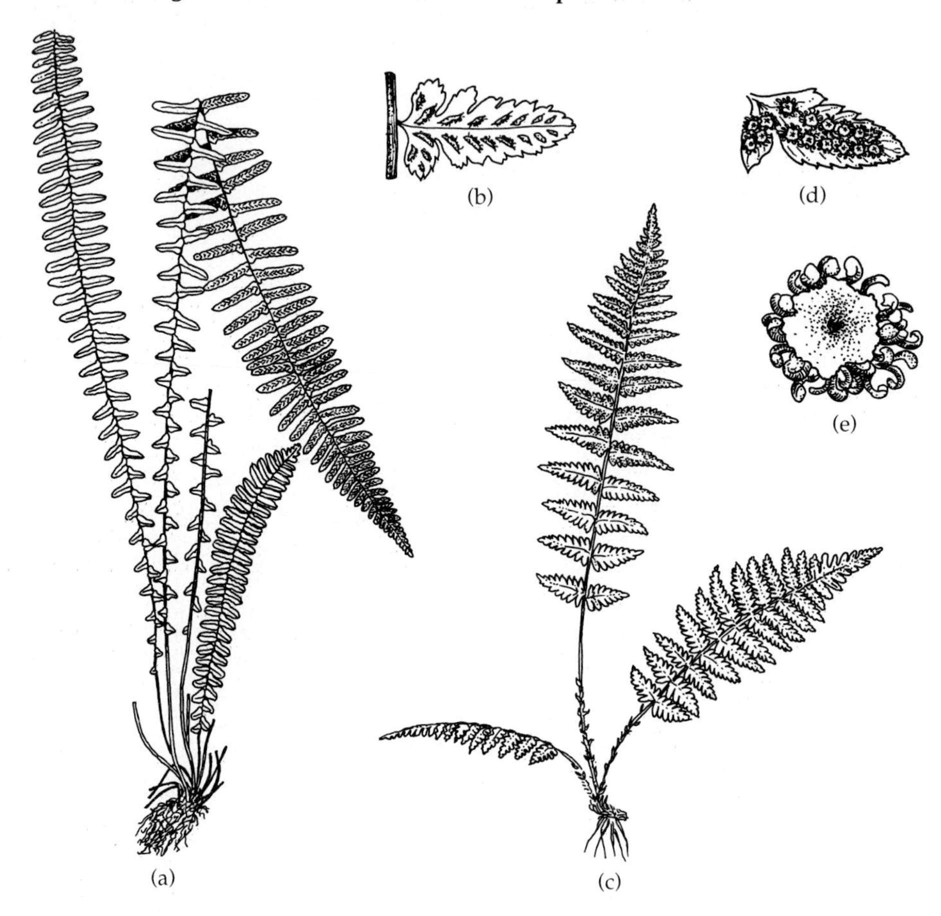


ically in the Old World. *Davallia feejeensis* (Fijian rabbit's foot fern) is a common hanging basket house plant. Many cultivars of the Boston fern, *Nephrolepis exaltata,* are also popularly grown indoors.

There is no known fossil record of this family.

**Family Blechnaceae (Deer Fern)** Mostly terrestrial, perennial, rhizome various with scales, dictyostelic. Fronds usually pinnate to pinnatifid. Sori elongate and covered with an indusium that opens toward the rachis axis. Sporangium with annulus interrupted by the stalk. Spores with or without chlorophyll (Fig. 6.21, p. 80).

The family consists of 10 genera and 260 species distributed mostly tropically, but some species are temperate. *Blechnum spicant* (deer fern) is a dominant understory species in the Pacific Coast temperate rain forest of

**FIGURE 6.20 Family Davalliaceae.** ***Davallia:*** **(a) fronds from rhizome; (b) pinnule of frond; (c) sori with cup-shaped indusia along margin of pinnule; (d) very scaly rhizome.**

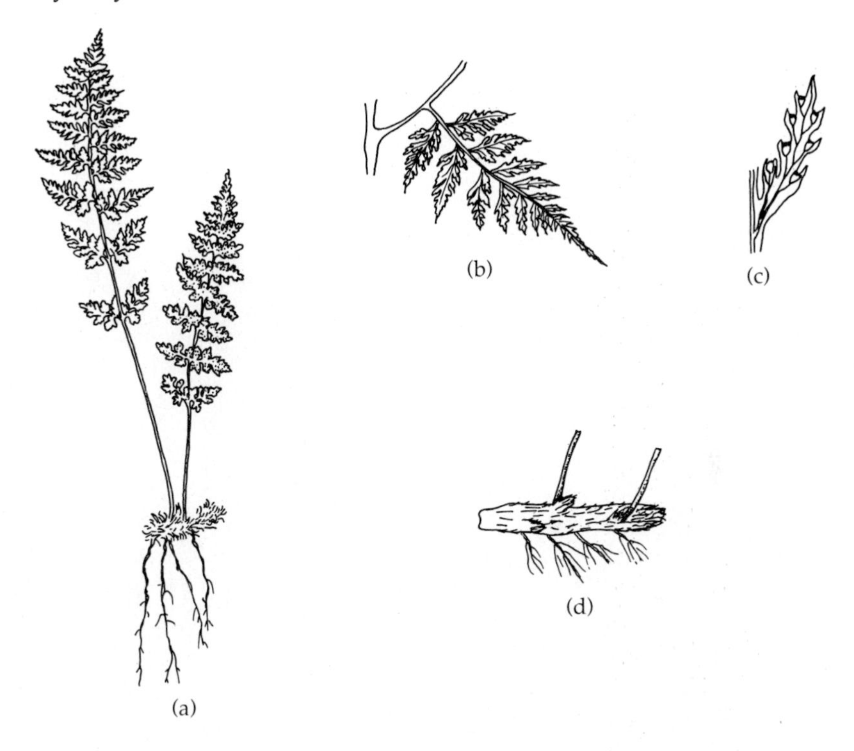


North America. *Woodwardia* is a genus of eastern North American swamps. The correlation of chromosome number and geography has been helpful in understanding the systematics of *Blechnum.*

Fossils of *Woodwardia* have been found from Paleocene and Pleistocene deposits of Wyoming and Oregon, respectively, in North America.

**Family Marsileaceae (Clover-leaf Fern, Pillwort)** Aquatic or rooted in mud, perennial, stem short or long, slender creeping, branched and solenostelic. Leaves resemble narrow blades of grass or a two- or four-leaved clover at end of a long petiole. Sori are enclosed in a hard structure, termed a **sporocarp,** which is formed from a specialized leaf at the base of the frond. Sporangia heterosporous with microsporangia and megasporangia attached to a gelatinous ring-like structure **(sorophore)** that swells with water (Fig. 6.22, p. 81).

The family consists of 3 genera and 72 species distributed in temperate and tropical climates. *Marsilea* looks like a four-leaved clover and is sometimes used as such in plastic charms. *Regnellidium,* a monotypic genus of Brazil and adjacent Argentina, looks like a two-leaved clover, while the 5 species of *Pilularia* look like narrow blades of grass attached to a slender rhizome.

**FIGURE 6.21 Family Blechnaceae. *Blechnum:* (a) habit of fertile dimorphic frond plant; (b) central section of fertile pinna showing elongate sporangia and indusia; (c) sporangium; (d) fertile frond; (e) sterile frond.**

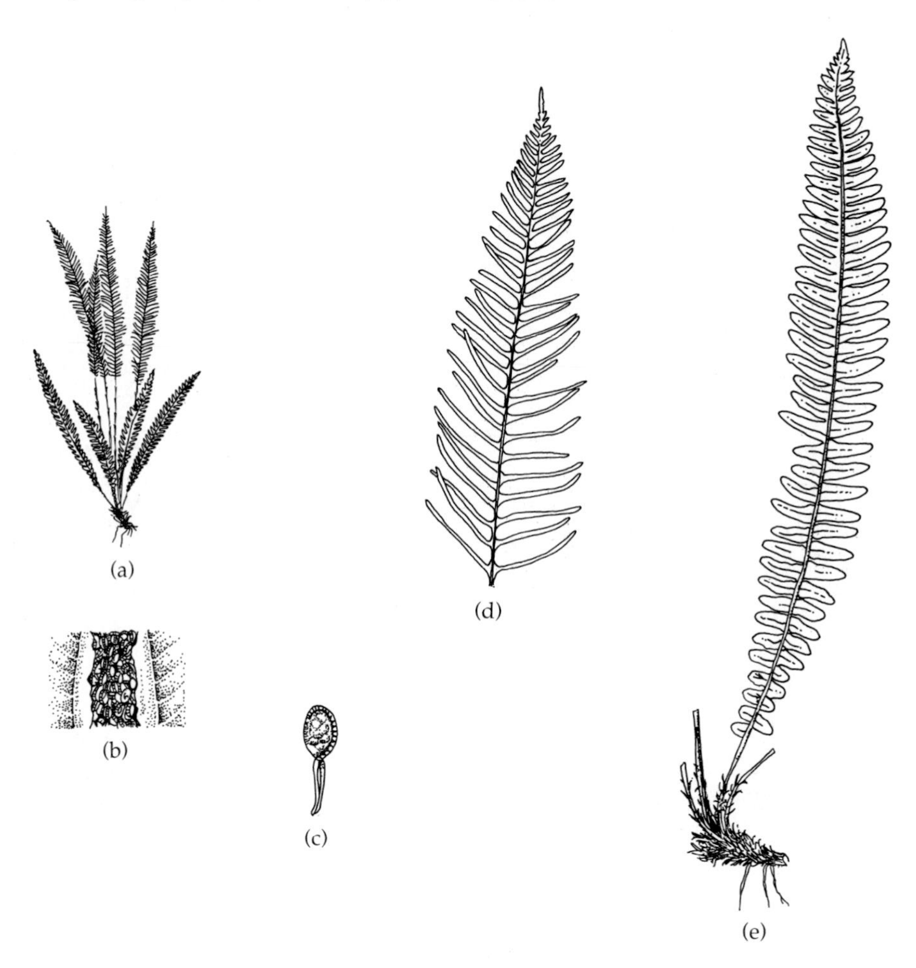


Fossil *Marsilea* have been reported from the Cretaceous to more recent deposits.

**Family Salviniaceae (Salvinia)** Floating aquatics, stems zig-zag. Leaves in threes, with two floating and covered with hairs, while the third leaf is submerged, finely dissected, resembling a branched root system. Sori enclosed in a soft sporocarp among the root-like leaves. Sporangia heterosporous, with microsporangia and megasporangia in different sporocarps (Fig. 6.23). The family consists of the genus *Salvinia* and 12 species distributed widely in the tropics. *Salvinia* is frequently sold as an aquarium plant. It can form dense floating mats in still tropical waters. The genus is of interest for

**FIGURE 6.22 Family Marsileaceae.** ***Marsilea:*** **(a) mature plant bearing four-lobed fronds and sporocarps; (b) sporocarp.** ***Pilularia:*** **(c) plant with hairy sporocarps.**

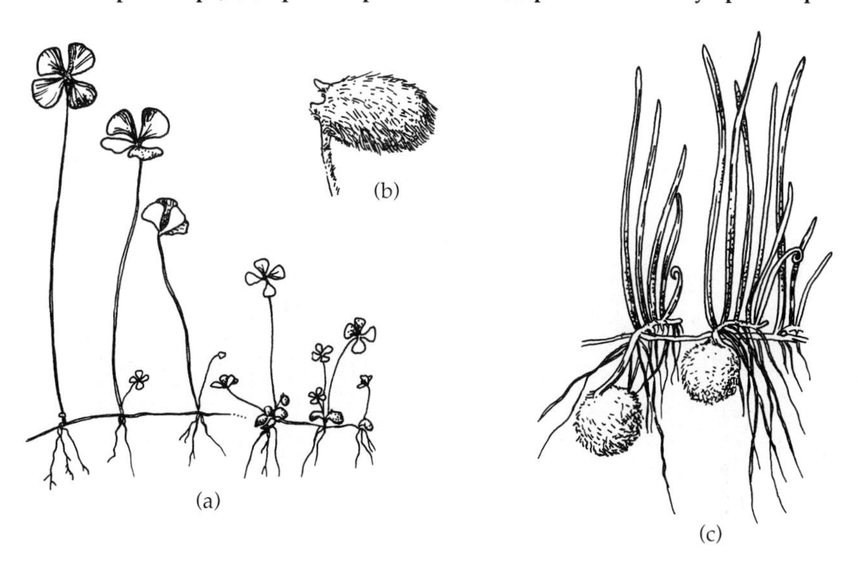


**FIGURE 6.23 Family Salviniaceae.** ***Salvinia:*** **(a) single floating plant segment showing "branched roots" and sporocarps; (b) rhizome; (c) sporocarps.**

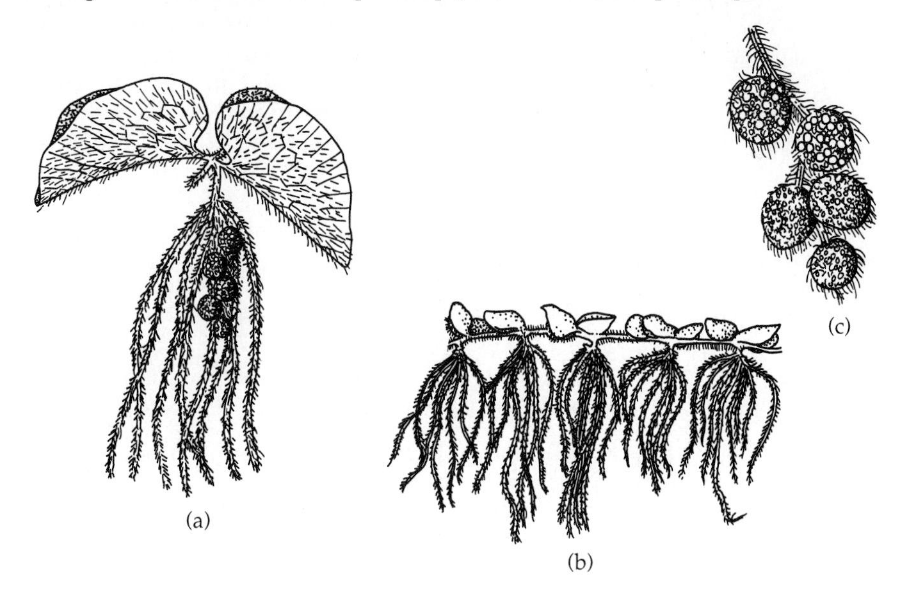


*SOURCE:* Redrawn with permission from L. Benson, *Plant Classification*, 2nd ed. Lexington, MA: D.C. Heath, 1979.

**FIGURE 6.24 Family Azollaceae.** ***Azolla:*** **(a) floating plant; (b) sporocarp; (c) overlapping leaves.**

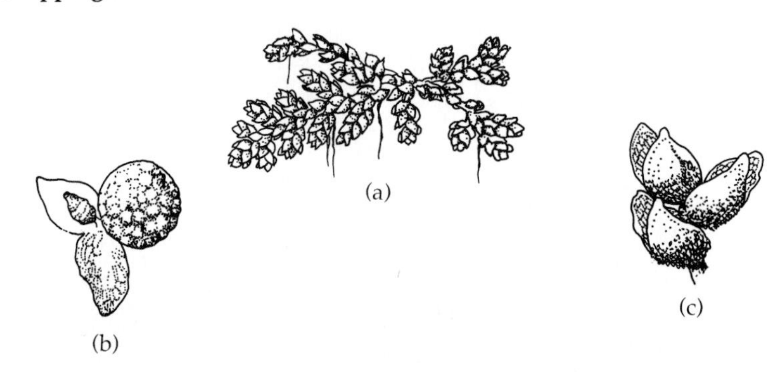


its low chromosome number and its large (one of the largest) morphologically different chromosomes.

Fossils attributed to *Salvinia* have been known from Upper Cretaceous deposits.

**Family Azollaceae (Mosquito Fern)** Small feather-like floating aquatic plants, stems zigzag, horizontal with short roots hanging downward. Leaves **imbricate** (overlapping) in two rows, harboring nitrogen-fixing blue-green algae *(Anabaena azollae)* in leaf cavities. Sporangia enclosed in an indusium with either one megasporangium or several microsporangia (Fig. 6.24).

The family consists of the genus *Azolla* and 6 species widely distributed in the Americas, Africa, Madagascar east to New Zealand, Australia, New Guinea, New Caledonia, and Japan; introduced into Europe, South Africa, New Zealand, and Hawaii. It is usually found floating on quiet streams, ponds, and lakes, often stagnant water, forming velvet-like green to red mats. It is easily recognized by the red or reddish-green color of the leaves, especially in autumn. It is important in rice culture in the Orient and as a green manure.

Like *Salvinia,* fossils attributed to *Azolla* have been found in Upper Cretaceous deposits. *Azolla* is sometimes placed with *Salvinia* in the same family.

## *SELECTED REFERENCES*

Bierhorst, D. W. 1971. *Morphology of Vascular Plants.* Macmillan, New York, 560 pp.

Bierhorst, D. W. 1977. The Systematic Position of *Psilotum* and *Tmesipteris. Brittonia* 29:3–13.

Bold, H. D., C. J. Alexopoulos, and T. Delevoryas. 1987. *Morphology of Plants,* 5th ed. Harper & Row, New York, 912 pp.

Bower, F. O. 1923–1928. *The Ferns.* Vols. 1–3. Cambridge Univ. Press. Cambridge.

Brownsey, P. J., D. R. Given, and J. D. Lovis. 1985. A Revised Classification of New Zealand Pteridophytes with a Synonymic Checklist of Species. *New Zealand J. Bot.* 43:431–489.

Bruce, J. G. 1979. Gametophyte and Young Sporophyte of *Lycopodium carolinianum. Amer. J. Bot.* 66:1156–1163.

Christensen, C. 1928. Filicinae. In: F. Verdoorn (ed.) *Manual of Pteridology.* Nijhoff, The Hague, pp. 522–550.
Copeland, E. B. 1947. *Genera Filicum, The Genera of Ferns.* Chronica Botanica, Waltham, MA, 247 pp.
Crabbe, J. A., A. C. Jermy, and J. M. Mickel. 1975. A New Arrangement for the Pteridophyte Herbarium. *Fern Gaz.* 11:141–162.
Cronquist, A., A. Takhtajan, and W. Zimmerman. 1966. On the Higher Taxa of Embryobionta. *Taxon* 15:129–134.
Datta, S. C. 1965. *A Handbook of Systematic Botany.* Asia Publ House, London, 435 pp.
Davis, P. H. and V. H. Heywood. 1973. *Principles of Angiosperm Taxonomy.* Robert E. Krieger, Huntington, NY, 558 pp.
Delevoryas, T. (ed.). 1964. Origin and Evolution of Ferns. A Symposium. *Mem. Torrey Bot. Club.* 21:1–95.
Dixit, R. D. 1984. *A Census of Indian Pteridophytes.* Flora of India Ser. IV. Botanical Survey of India. Dept. of Environment, New Delhi, 177 pp.
Dyer, A. F. (ed.). 1979. *The Experimental Biology of Ferns.* Academic Press, London, 657 pp.
Foster, F. G. 1971. *Ferns to Know and Grow,* rev. ed. Hawthorn Books, New York, 258 pp.
Gleason, H. A. 1963. *New Britton and Brown Illustrated Flora of the Northeastern United States and Adjacent Canada.* 3 vols. Hafner, New York.
Gliessman, S. R. and C. H. Muller. 1978. The Allelopathic Mechanisms of Dominance in Bracken *(Pteridium aquilinum)* in Southern California. *J. Chem. Ecol.* 4:337–362.
Hitchcock, C. L. et al. 1955–1969. *Vascular Plants of the Pacific Northwest.* 5 parts. Univ. of Washington Press, Seattle, WA.
Holttum, R. E. 1949. The Classification of Ferns. *Biol. Rev.* 24:267–296.
Holttum, R. E. 1971. The Family Names of Ferns. *Taxon* 20:527–531.
Holttum, R. E. 1973. Introduction to Pteridophyta (and) Family Names of Pteridophyta. In: J. C. Willis (ed.). *A Dictionary of the Flowering Plants and Ferns,* 8th ed. Revised by H. K. A. Shaw. Cambridge Univ Press, Cambridge.
Jermy, A. C., J. A. Crabbe, and B. A. Thomas (eds.). 1973. The Phylogeny and Classification of the Ferns. *J. Linn. Soc., Bot.* 67 (Suppl 1):1–284.
Johnson, A. M. 1931. *Taxonomy of the Flowering Plants.* The Century Co., New York, 864 pp.
Johnson, D. M. 1986. *Systematics of the New World Species of Marsilea (Marsileaceae).* The Amer. Soc. Pl. Taxonomy. Sys. Bot. Monogr. 11. 1–87.
Jones, D. L. 1987. *Encyclopedia of Ferns.* Timber Press, Portland, OR, 433 pp.
Klekowski, E. J. 1972. Genetical Features of Ferns as Contrasted to Seed Plants. *Ann. Missouri Bot. Gard.* 59:138–151.
Krassilov, V. 1978. Mesozoic Lycopods and Ferns from the Bureja Basin. *Palaeontographica* 166, Abt. B:16–29.
Kremp, G. O. W. and T. Kawasaki. 1972. *The Spores of the Pteridophytes.* Hirokawa, Tokyo, 398 pp.
Lellinger, D. B. 1985. *A Field Manual of the Ferns & Fern-Allies of the United States & Canada.* Smithsonian Inst. Press, Wash. D.C., 389 pp.
Love, A., D. Love, and R. E. G. Pichi-Sermolli. 1977. *Cytotaxonomical Atlas of the Pteridophyta.* J. Cramer, Vaduz, 398 pp.
Lumpkin, T. A. and D. L. Plucknett. 1980. *Azolla:* Botany, Physiology and Use as a Green Manure. *Econ. Bot.* 34:111–153.
Mickel, J. T. 1974. Phyletic Lines in the Modern Ferns. *Ann. Missouri Bot. Gard.* 61:474–482.
Mickel, J. T. 1979. *How to Know the Ferns and Fern Allies.* Wm. C. Brown, Dubuque, IA, 229 pp.

Morton, C. V. 1968. The Genera, Subgenera and Sections of the Hymenophyllaceae. *Contrib. U.S. Natl. Herb.* 38:153–214.

Naik, V. N. 1984. *Taxonomy of Angiosperms.* Tata McGraw-Hill, New Delhi, 304 pp.

Ogura, Y. 1972. *Comparative Anatomy of Vegetative Organs of the Pteridophytes,* 2nd ed. Borntraeger, Berlin, 502 pp.

Parihar, N. S. 1977. *The Biology and Morphology of the Pteridophytes.* Central Book Depot, Allahabad, India, 660 pp.

Pichi-Sermolli, R. E. G. 1959. Pteridophyta. In: Turrill, W.B. (ed.) *Vistas in Botany.* Pergamon Press, London, pp. 421–493.

Pichi-Sermolli, R. E. G. 1970. A Provisional Catalogue of the Family Names of Living Pteridophytes. *Webbia* 25:219–297.

Pichi-Sermolli, R. E. G. 1977. Tentamen Pteridophytorum in Taxonomicum Ordinem Redigendi. *Webbia* 31:313–512.

Porter, C. L. 1967. *Taxonomy of Flowering Plants,* 2nd ed. W. H. Freeman, San Francisco, 472 pp.

Scagel, R. F. et al. 1984. *Plants: An Evolutionary Survey.* Wadsworth, Belmont, CA, 757 pp.

Shukla, P. and S. P. Misra. 1982. *An Introduction to Taxonomy of Angiosperms,* 3rd ed. Vikas Publ. House, New Delhi, 556 pp.

Smith, G. M. 1955. *Cryptogamic Botany. Vol. II. Bryophytes and Pteridophytes,* 2nd ed. McGraw-Hill, New York.

Sporne, K. R. 1975. *The Morphology of Pteridophytes,* 4th ed. Hutchinson & Co., London, 191 pp.

Taylor, T. N. and J. T. Mickel (eds.) 1974. Evolution of Systematic Characters in the Ferns. (A symposium of papers presented at Amherst, Mass., June 1973). *Ann. Missouri Bot. Gard.* 61:307–482.

Tryon, R. 1970. Development and Evolution of Fern Floras of Oceanic Islands. *Biotropica* 4:121–131.

Tryon, R. M. and A. F. Tryon. 1982. *Ferns and Allied Plants.* Springer-Verlag, New York, 857 pp.

Verdoorn, Fr. ed. 1938. *Manual of Pteridology.* Martinus Nijhoff, The Hague, 640 pp.

Wagner, W. H. 1954. Reticulate Evolution in the Appalachian Aspleniums. *Evolution* 8:103–118.

Wagner, W. H. 1977. Systematic Implications of the Psilotaceae. *Brittonia* 29:54–63.

Walker, T. G. 1966. A Cytotaxonomic Survey of the Pteridophytes of Jamaica. *Trans. Roy. Soc. Edinburgh* 66:169–237.

White, R. A. et al. 1977. Taxonomic and Morphological Relationships of the Psilotaceae. *Brittonia* 29:1–68.

# 7

# Families of the Pinophyta (Gymnosperms)

The two divisions Pinophyta (gymnosperms) and Magnoliophyta (flowering plants, angiosperms) are different from all other living plants in that they produce seeds.

The gymnosperms living today include three subdivisions: Cycadicae (cycads), Pinicae (conifers), and Gneticae (vessel-containing gymnosperms). The name **gymnosperm** comes from **gymnos** (Gr.), naked, and **sperma** (Gr.), seed, and emphasizes the fact that **ovules**—the structures that develop into seeds—are not enclosed in any surrounding structure. Instead, they are exposed on the surface of sporophylls or some other structure. On the other hand, the ovules and seeds of the flowering plants are enclosed within a structure called an **ovary** and are therefore called **angiosperms,** a word derived from **angeion** (Gr.), enclosing vessel, and **sperma** (Gr.), seed.

Living gymnosperms consist of 11 living families, 68 genera, and approximately 900 species (Table 7.1). They are distributed throughout the world, making up extensive forests in North America, Europe, and Asia. These forests are exploited economically for the wood, which is important for lumber, furniture, pulp for paper, chemicals, and as ornamentals. Ecologically they are the dominant vegetation; they provide habitat for big game (deer, elk, black and grizzly bear), control erosion, and are some of the earliest trees in secondary succession. They include the largest, tallest, and oldest living organisms in the world.

## *Reproduction of Gymnosperms*

To gain an understanding of the Pinophyta, it seems best to review the way in which gymnosperms reproduce (Fig. 7.1, p. 88). This is necessary because many characters of the cones and reproductive processes are used in classification and identification. In gymnosperms and flowering plants an alternation of generation of two independent life forms is present. The obvious large plant (i.e., cycad, pine tree, etc.)

**TABLE 7.1 Classification of the Pinophyta.**

Division Pinophyta
  Subdivision Cycadicae
    *Class I. Lyginopteridatae
    Class II. Cycadatae (3 families, 10 genera, 107–119 species)
      Family 1. Cycadaceae (1 genus, 20 species)
        *Cycas*, 20 species, Polynesia to Madagascar, North to Japan
      Family 2. Zamiaceae (8 genera, 86–98 species)
        *Bowenia*, 2 species, northern Australia
        *Ceratozamia*, 4 species, Mexico
        *Dioon*, 3–5 species, Mexico and Central America
        *Encephalartos*, 30 species, tropical and southern Africa
        *Lepidozamia*, 2 species, eastern Australia
        *Macrozamia*, 14 species, temperate Australia
        *Microcycas*, 1 species, Cuba
        *Zamia*, 30–40 species, tropical America, West Indies
      Family 3. Stangeriaceae (1 genus, 1 species)
        *Stangeria*, 1 species, s.e. Africa
    *Class III. Bennettitatae
  Subdivision Pinicae
    Class I. Ginkgoatae (monotypic)
      Family Ginkgoaceae (1 genus, 1 species)
        *Ginkgo*, 1 species, China
    Class II. Pinatae (7 families, 53 genera, 520–622 species)
      *Subclass A. Cordaitidae
      Subclass B. Pinidae
      Family 1. Araucariaceae (2 genera, 34 species)
        *Agathis*, 16 species, Indochina, Western Malaysia, northern Australia to New Zealand
        *Araucaria*, 18 species, Brazil, Chile, New Caledonia, Australia and New Zealand
      Family 2. Cephalotaxaceae (1 genus, 4–8 species)
        *Cephalotaxus*, 4–8 species, eastern Asia
      Family 3. Cupressaceae (18 genera, 128–133 species)
        *Actinostrobus*, 3 species, southwest Australia
        *Austrocedrus*, 1 species, temperate South America
        *Callitris*, 16 species, Australia, New Caledonia
        *Calocedrus*, 3 species, southeastern Asia, Pacific North America
        *Chamaecyparis*, 7 species, eastern Asia, North America
        *Cupressus*, 15 species, Mediterranean, Asia, western North America
        *Diselma*, 1 species, Tasmania
        *Fitzroya*, 1 species, Chile
        *Fokienia*, 1 species, China, Indochina
        *Juniperus*, 60 species, Northern Hemisphere
        *Libocedrus*, 5 species, New Caledonia, New Zealand
        *Neocallitropsis*, 1 species, New Caledonia
        *Papuacedrus*, 3 species, New Guinea, Moluccas
        *Pilgerodendron*, 1 species, southern Chile
        *Tetraclinis*, 1 species, Spain to Tunis, Malta
        *Thuja*, 5 species, China, Japan, North America
        *Thujopsis*, 1 species, Japan
        *Widdringtonia*, 3 species, tropics and South Africa

Family 4. Pinaceae (10 genera, 189–269 species)
*Abies*, 40 species, North Temperate, Central America
*Cathaya*, 1 species, China
*Cedrus*, 4 species, Middle East, Algeria, Cyprus, Himalayas
*Keteleeria*, 4–8 species, eastern Asia, Indochina
*Larix*, 10–12 species, North Temperate
*Picea*, 36–80 species, North Temperate
*Pinus*, 70–100 species, North Temperate, mountains in tropics
*Pseudolarix*, 1 species, China
*Pseudotsuga*, 5 species, eastern Asia, western North America
*Tsuga*, 18 species, eastern Asia, North America
Family 5. Podocarpaceae (7 genera, 134–139 species)
*Acmopyle*, 3 species, New Caledonia, Fiji
*Dacrydium*, 20–25 species, New Zealand to Indo-Malaysia, Chile
*Microcachrys*, 1 species, Tasmania
*Microstrobus*, 2 species, Australia, Tasmania
*Phyllocladus*, 7 species, New Zealand, Tasmania to Philippines
*Podocarpus*, 100 species, New Zealand, tropics north to Himalayas and Japan
*Saxegothaea*, 1 species, Andes of Argentina
Family 6. Taxaceae (5 genera, 17–22 species)
*Amentotaxus*, 1–4 species, Assam, western China
*Austrotaxus*, 1 species, New Caledonia
*Pseudotaxus*, 1 species, China
*Taxus*, 8–10 species, North Temperate
*Torreya*, 6 species, eastern Asia, California, Florida
Family 7. Taxodiaceae (10 genera, 14–17 species)
*Athrotaxis*, 3 species, Australia and Tasmania
*Cryptomeria*, 1 species, Japan, China
*Cunninghamia*, 2–3 species, China, Formosa
*Glyptostrobus*, 1 species, China
*Metasequoia*, 1 species, China
*Sciadopitys*, 1 species, Japan
*Sequoia*, 1 species, California, s.w. Oregon
*Sequoiadendron*, 1 species, California
*Taiwania*, 1–2 species, China, Formosa
*Taxodium*, 2–3 species, southeastern United States and Mexico

Subdivision Gneticae
Class I. Gnetatae (3 families, 3 genera, 83 species)
Subclass A. Ephedridae
Family 1. Ephedraceae, 1 genus, 42 species
*Ephedra*, 42 species, New World, Mediterranean to China
Subclass B. Gnetidae
Family 1. Gnetaceae, 1 genus, 40 species
*Gnetum*, 40 species, Amazon, West Africa, India, southern China, and Malaysia
Subclass C. Welwitschiidae
Family 1. Welwitschiaceae, 1 genus, 1 species
*Welwitschia*, 1 species, Southwest Africa

*SOURCE:* After Cronquist, Takhtajan, and Zimmerman, 1966; Bierhorst, 1971; Krüssmann, 1985.
*NOTE:* An asterisk means fossil groups only.

**FIGURE 7.1 The reproductive cycle of a pine.**

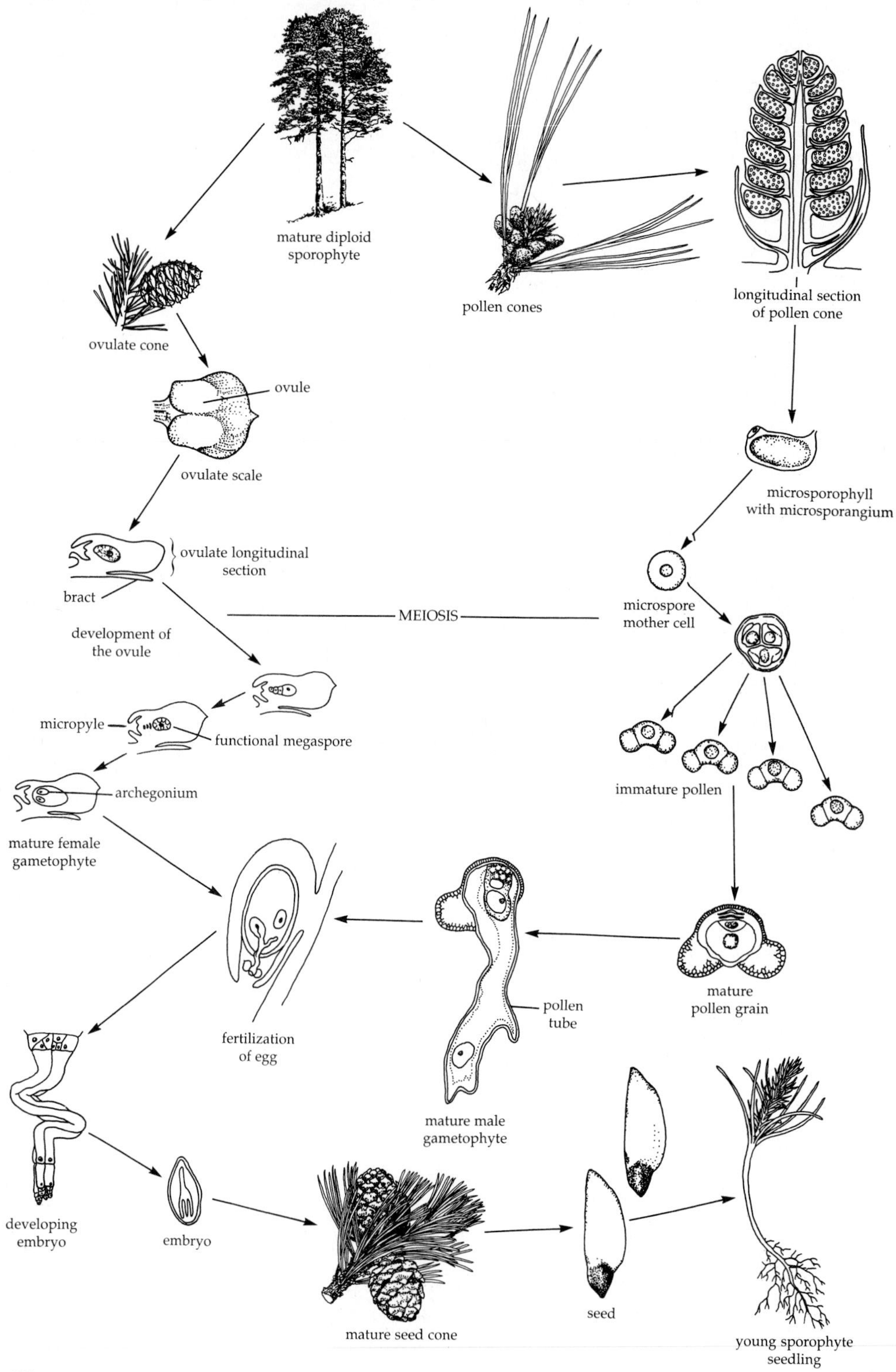

is the **sporophyte** of the **diploid** phase, while the **gametophyte** or **haploid** phase is very inconspicuous. All seed plants are heterosporous and produce spores by meiosis, like ferns, but these spores are not released. In most gymnosperms reproduction involves **cones**. A male cone consists of a collection of **microsporophylls** with **microsporangia** attached. With these, **microspore (pollen) mother cells** undergo meiosis and produce four haploid **microspores;** the nucleus of each divides several times, producing a small immature male gametophyte, which forms a wall and becomes **pollen.**

The female cones consist of a group of **megasporophylls** on which **megasporangia** develop. The **megaspore mother cell** in the megasporangium is surrounded by two sporophyte tissue layers called the **nucellus** and **integument,** respectively. An opening, the **micropyle,** develops at the end of the integument. Meiosis in the megaspore mother cell forms four haploid **megaspores,** three of which abort, leaving one **functional megaspore.** More rapid nuclear division followed by more slowly forming cell walls produce a multinucleated **female gametophyte** with several **archegonia,** consisting of an enlarged **egg** cell and two or four **neck cells.**

Pollen carried by wind to the female cone filters between the layers of megasporophylls to the micropyle. The pollen adheres frequently to a sticky fluid that has formed there, and as the fluid dries it pulls the pollen to the surface of the nucellus, resulting in **pollination.** The pollen germinates, producing a **pollen tube** that grows slowly through the nucellus. During pollen tube development, the cells within the tube divide and form two **sperm.** Some gymnosperms produce motile, multiflagellated sperm, while others produce nonmotile ones. In some Pinicae the sperm are discharged into a small cavity that forms just above the necks of the archegonia. The sperm swim to the eggs and fuse with them, forming a **zygote,** the end product of **fertilization.** As much time as 1 year (i.e., *Pinus)* can elapse between pollination and fertilization. At first, more than one **embryo** usually begins to develop, but only one survives in the mature **seed;** the remaining embryos abort.

The gymnosperm seed is comprised of three layers, with the middle layer making up the hard, outer **seed coat.** Inside is found the young sporophyte embryo and the remains of the female gametophyte tissue. Germination of the seed produces a young sporophyte seedling. It usually takes many years before the plant is mature enough to produce cones.

Gymnosperms are either **dioecious,** producing male and female reproductive structures on separate plants, or **monoecious,** bearing female and male reproductive structures on the same plant, so both types of plants and/or their structures may be needed for study.

## *Classification of Gymnosperms*

The approximately 900 species of modern living gymnosperms are a very diverse group of plants. They range from the dioecious cycads with large frond-like leaves and attractive cones, to some conifers with scale-like leaves and berry-like reproductive structures, to broad, flat-leaved *Gnetum* and its cherry-like seeds in "flower-like" clusters.

The group is even more diverse in the fossil record from the Carboniferous, Permian, and the Mesozoic eras until the Lower Cretaceous. Charles Beck (1962) of the University of Michigan has described "progymnosperm" fossil plants from the Devonian period. These fossils have elaborate branching systems that are fern-like, three-dimensionally branched, and have a unique vascular system. Each family of conifers (i.e, pines, redwoods, etc.) has a fossil record extending to the Mesozoic. *Ginkgo* (maidenhair tree) leaves are found as far back as the Permian.

An exclusive fossil group of plants with frond-like leaves but with seeds are known as the **seed ferns** or **pteridosperms** (class Lyginopteridatae). They are found in Lower Carboniferous to Jurassic rocks, being most abundant in the Carboniferous, with many transformed into coal. The seed ferns resemble modern cycads more than ferns, in having firmer leaves with a thicker cuticle and bearing seeds along the frond margin. The seed ferns were so common in the Carboniferous that this period has sometimes been referred to as the Age of the Seed Ferns.

The second class of exclusive fossil gymnosperms is the Bennettitatae. Some resemble cycads in their short, stout stems (trunks) and similar pinnate leaves but differ in the presence of "flower-like" bisporangiate cones with seeds. They are commonly from Triassic to Cretaceous rocks. It is this group that has caused the Mesozoic era to be called the Age of the Cycads.

As a result of this diversity, taxonomic interpretations have at times been controversial. Some classification schemes assign each of the groups of gymnosperms to separate divisions: Cycadophyta, Ginkgophyta, Coniferophyta, and Gnetophyta (Arnold, 1948). This is based on the interpretations that diverse genera and paleobotanical differences are more important than many gymnosperm similarities. More recently (Cronquist, Takhtajan, and Zimmerman, 1966), the naked seed plants have been grouped into one division, Pinophyta, three subdivisions, and four classes of living gymnosperms and three classes known only from the fossil record. The final classification of the Pinophyta is yet to be written. This book will follow the most recent system mentioned above and that in Table 7.1.

## DIVISION PINOPHYTA

### Subdivision Cycadicae; Class Cycadatae (Cycads)

**Family Cycadaceae (Cycad)** Stems palm-like and rough, usually not branched. Leaves fern-like, pinnately compound, thick, and leathery; attached spirally at the stem apex, young pinnae circinate, leaf bases remaining after the leaves drop. Dioecious. Whorls of woolly-covered micro- and megasporophylls alternate with whorls of scales and foliage leaves at the stem apex. Ovules borne along the sporophyll margins. Seeds almond or plumlike (Fig. 7.2).

The family consists of the genus *Cycas* and 20 species distributed in the tropics and subtropics from Africa, Madagascar to Australia, Polynesia, Philippines, and Japan.

*Cycas* has limited economic value, except as ornamental plants, especially *C. revoluta*. They are commonly grown in the southern United States. A starch,

**FIGURE 7.2 Family Cycadaceae. *Cycas:* (a) sporophyll with two developing ovules and two aborted ovules; (b) mature cycad plant; (c) cross section of leaf rachis showing two circinate pinna; (d) circinate vernation of rows of pinna of a young leaf.**

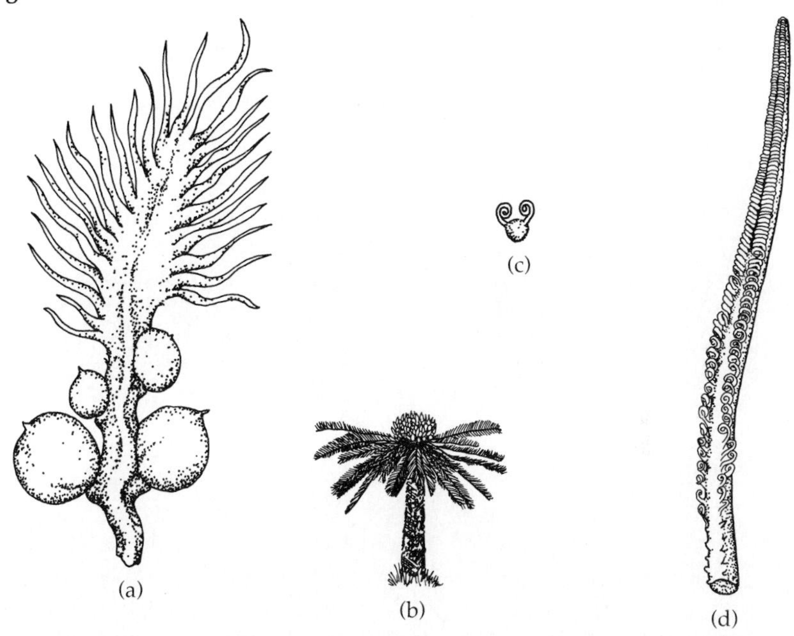


*SOURCE:* Parts (c) and (d) redrawn with permission from H. C. Bold, C. J. Alexopoulos, and T. Delevoryas, *Morphology of Plants and Fungi,* 5th ed. © 1987 by Harper & Row Publishers Inc.

sago starch, is extracted from the pith of *C. circinalis* in the Philippines and Indonesia. Sometimes *Cycas* is put by authors in the family Zamiaceae.

Fossils attributed to the family have been found in Mesozoic strata.

**Family Zamiaceae (Cycad)** Stems palm-like and rough, usually not branched. Leaves pinnately compound, thick, and leathery, with an additional protective layer beneath the upper epidermis, called a **hypodermis;** some may reach 3–4 m in length, attached spirally at the stem apex; the young pinnae or leaves of some species have circinate vernation, others are straight. Dioecious. Microsporangia and megasporangia produced in strobili (cones), which vary greatly in size, up to 7 dm in length. Seeds are cherry-like and may be brightly colored (Fig. 7.3).

The family consists of 8 genera and 86–98 species distributed in tropical and subtropical regions in Mexico and Central America, the West Indies, South America, South Africa, and Australia.

*Zamia,* the largest genus, found in tropical America, extends into Florida in the form of *Z. pumila.* Its starchy underground stem was used by the Seminole Indians and early settlers as a flour. The seeds of *Dioon edule* are

**FIGURE 7.3 Family Zamiaceae.** ***Zamia:*** **(a) male pollen cone; (b) many microsporangia on underside of microsporophyll; (c) mature female plant showing stout tap root; (d) two ovules; (e) mature ovulate cone.**

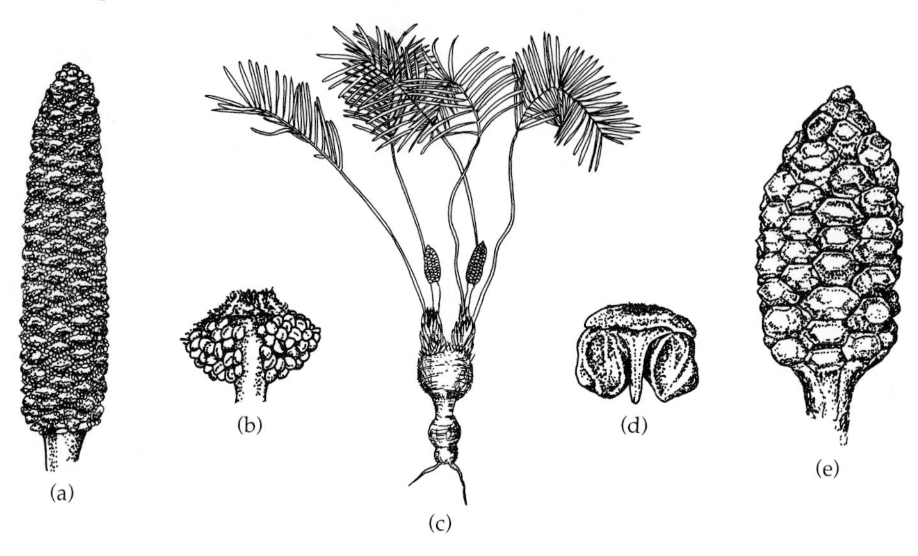


ground into a meal and eaten for food. In spite of this, some species contain very potent toxins. The longest cones are produced by *Dioon.* Some genera have very restricted distributions, such as *Microcycas* in Cuba and *Bowenia* in extreme North Queensland, Australia. It is feared by conservationists that many cycads may soon become extinct in their native habitat. The South African National Botanic Garden (Kirstenbosch), Cape Province, and the Fairchild Tropical Gardens, Coral Gables, Florida, are reported to have most of the living cycad genera under cultivation. The motile, flagellated sperm cells of cycads and the maidenhair tree are some of the largest sperm in the plant kingdom.

Fossil cycads are known from upper Triassic deposits and therefore are sometimes called (along with two other gymnosperms, the maidenhair tree and dawn redwood) "living fossils."

## Subdivision Pinicae; Class Ginkgoatae (Ginkgo)

**Family Ginkgoaceae (Maidenhair tree, Ginkgo)** Tree, tall, stately with curving branches attached to a stout trunk. Leaves fan-shaped, **deciduous** (shed each year), attached in whorls to the end of "short shoots" growing from the longer branches ("long shoots"); veins of the leaves dichotomously branched. Dioecious. Paired ovules at the end of a stalk and naked, hanging like cherries. Seeds enclosed in a fleshy whitish-pink covering; when crushed smell like rancid butter or human vomit (Fig. 7.4); originally cultivated widely in China and Japan for its edible seeds, which when roasted are eaten like nuts. They are sold commercially as ginkgo or silver nuts.

*Ginkgo biloba* is a desired tree in cities as far north as southern Canada and northern Europe because it is somewhat disease resistant, survives in

**FIGURE 7.4 Family Ginkgoaceae.** ***Ginkgo:*** **(a) spur shoot with microstrobil; (b) spur shoot with ovules; (c) pair of ovules.**

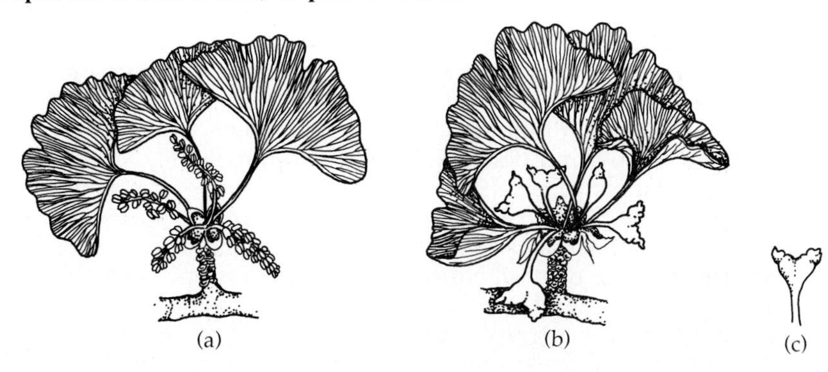


regions of high air pollution, and its leaves become golden-yellow in autumn. It is desirable to plant male trees to keep one's neighbors happy for the obvious reasons mentioned above. *Ginkgo* is interesting genetically because of the presence of X and Y sex chromosomes. *Ginkgo biloba* is known from fossils in rocks from Triassic and Jurassic periods and is referred to as a "living fossil."

## Subdivision Pinicae; Class Pinatae (Conifers)

This group of gymnosperms has the largest number of living representatives (7 families, 54 genera, 700 species). It forms vast forests of pine, spruce, fir, Douglas fir, cedar, and other conifers across western and northern North America, northern Europe and Asia, China, areas of Australia and New Zealand, and temperate Central and South America. They are widely used as ornamentals. Some types have been greatly prized or honored, such as individual trees used as shrines in China or young pine, Douglas fir, spruce, or fir trees used as Christmas trees. Despite relatively few species, the conifers are extremely important as a building material, as wood pulp, and for ecosystem management.

In size they range from small prostrate shrubs to the giant redwoods, which are over 110 m tall and 4500–5000 years old.

The conifers are shrubs or trees with **excurrent branching** (with a central axis and smaller lateral branches). The woody stem is comprised of tracheids and generally lacks vessel-conducting cells. The stem also has narrow rays radiating out through the xylem. Resin canals occur in the cortex and throughout the xylem, as well as in the leaves and roots.

The leaves are all simple, with most being needle-like or scale-like in form. In some the leaves may be very broad, tough, and leathery.

The reproductive structures are generally cones, but some may be modified into a fleshy berry-like structure, such as "juniper berries," or naked seeds on a fleshy stalk, as in *Podocarpus* and *Taxus*. Most species are monoecious, with male and female cones on the same plant.

The Pinicae is well known in the fossil record. The completely fossil subclass Cordaitidae is found in Upper Devonian, Carboniferous, and Permian deposits. The most notable was *Cordaites,* trees up to 30 m tall with long, narrow strap-shaped leaves up to 1 m long. The seeds were in cones that were different from other gymnosperm cones.

**Family Araucariaceae (Monkey-Puzzle Tree)** Large trees with buds naked. Leaves needle-like (a few species) to broad, lanceolate; opposite or spirally arranged on branches; evergreen but small branches (branchlets) and leaves deciduous, dropping in time. Female cones large, dry, or fleshy; ovules one to a scale. Wind pollinated, sperm nonmotile (Fig. 7.5).

The family consists of 2 genera and 30 species distributed almost completely in the Southern Hemisphere. Many of the species are called *pines,* such as *Araucaria heterophylla,* Norfolk island pine, which is a common house and greenhouse plant in North America and Europe. The Kauri, *Agathis australis* of New Zealand, is one of the largest species in the family, with a height of over 45 m and a trunk diameter of over 3 m.

Several species of *Araucaria* are cultivated in milder European and North American climates with *A. araucana,* the monkey-puzzle tree from Chile, being rather spectacular with its whorled, gently arching branches covered with stiff, broadly lanceolate leaves.

The family is known from fossils found in Triassic Period deposits of both hemispheres to the present.

**FIGURE 7.5 Family Araucariaceae.** ***Agathis:*** **(a) leafy branch bearing mature female cone; (b) young male cone.** ***Araucaria cunninghamii:*** **(c) male pollen cone; (d) habit of young tree.**

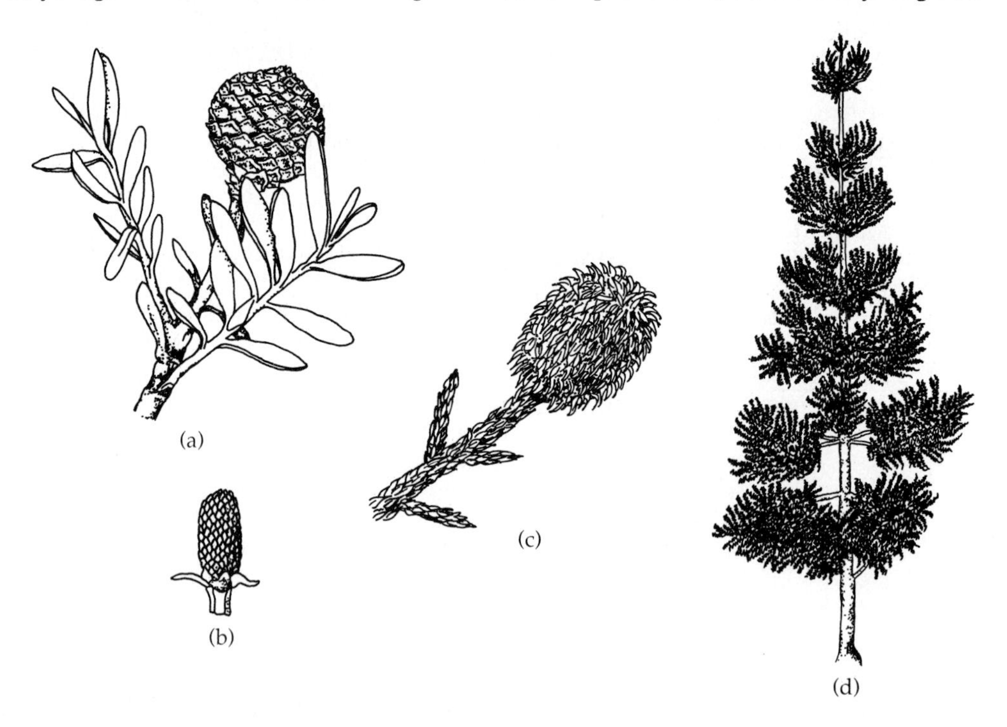

**Family Cephalotaxaceae (Plum Yew)** Shrubs to small trees, opposite or whorled branches. Leaves evergreen, linear, dark green, pointed, and spirally arranged. Dioecious. Microsporophylls with 3–7 pollen sacs. Ovules (subtended by bracts) in pairs on naked stalks that become enlarged and fleshy at maturity, to 2.5 cm long and plum-like (Fig. 7.6).

The family consists of the genus *Cephalotaxus* and 4–7 species distributed in eastern China.

Several species, such as *C. fortunei,* are cultivated in North America.

The group is recorded in the fossil record from western North America and Europe.

**Family Cupressaceae (Cypress)** Prostrate shrubs to tall trees; branches **diffuse** (spreading widely in all directions), buds without bud scales. Leaves small, scale-like, or **awl-shaped** (flat, tapering gradually to a sharp point), often closely appressed to the branches; the parts are opposite or whorled. Dioecious or monoecious. Male cones small, inconspicuous; female cones terminal or lateral on short branches, with the ovule scales and bract scales fused; ovules several to many. Fruits small, woody or fleshy, and berry-like (Fig. 7.7, p. 96).

The family is comprised of 18 genera and 130–136 species with a cosmopolitan distribution. The most common genus is *Juniperus,* the juniper, which is used extensively as an ornamental shrub or small tree. The wood of *J. virginiana,* eastern red cedar of eastern North America, is used in cedar chests, to line closets, for pencils, and for shingles. An oil used to flavor gin is extracted from the berry-like fruit of *J. communis.* Other species of juniper in western North America are used for fence posts. Several species of *Cupressus,* true cypress, occur in southwestern North America and Mexico. *Cupressus macrocarpa,* Monterey cypress from a small area in California, is extensively planted in Australia and New Zealand as a wind break or hedgerow tree. *Cupressus sempervirens* is planted as a cemetery tree in the Mediterranean region. *Thuja* occurs in eastern Asia and North America. *Thuja plicata,*

**FIGURE 7.6 Family Cephalotaxaceae.** ***Cephalotaxus:*** **(a) branch with mature female fruits; (b) longitudinal section of ovule and bract scale; (c) end of young female shoot.**

**FIGURE 7.7 Family Cupressaceae.** ***Juniperus:*** **(a) branch; (b) imbricate, scale-like leaves; (c) berry-like fruits.** ***Cupressus:*** **(d) cones with peltate scales.**

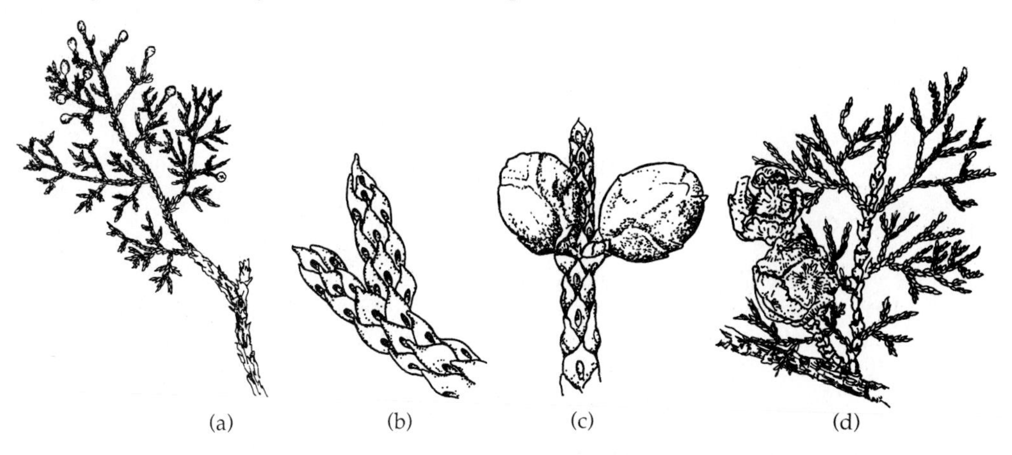


Western red cedar, is a large tree reaching 75 m tall. Its rot-resistant wood makes it a valuable lumber tree. *Thuja occidentalis,* the arbor vitae of eastern North America, and *T. orientalis* are valuable ornamentals. *Callitris,* cypress pine, is common in parts of Australia. The genus *Chamaecyparis* occurs in North America and eastern Asia; *C. nutkatensis,* yellow cedar, occurs from Alaska to Oregon; *C. lawsoniana,* Port Orford cedar, occurs in Oregon and California; and *C. thyoides,* white cedar, occurs along the east coast of North America. All three cedars are used extensively as ornamentals. Most species of the family have rot-resistant, weather-resistant, straight-grained wood.

Fossils attributed to the family extend from Upper Cretaceous deposits to the present.

**Family Pinaceae (Pine)** Trees (rarely shrubs), small or large, trunk elongate with whorled branches. Leaves evergreen, less commonly deciduous, needle-like, single, spirally attached to the branch or in sheathed **fascicles,** clusters of generally 2–5; buds enclosed in bud scales. Monoecious. Male cones small, papery in clusters; female cones woody with persistent scales (except *Abies, Cedrus, Pseudolarix),* spirally attached to an axis with two ovules on the upper surface of each scale, scale in axil of a **bract** (modified leaf); mature female cones take 1 or 2 years to develop to maturity. Seeds dropping following elongation of cone axis allowing scales to open (Fig. 7.8).

The family consists of 10 genera and 214–250 species distributed extensively in the Northern Hemisphere.

The family is very important economically for lumber, pulpwood, furniture, and ornamentals, as well as chemically and ecologically. It makes up the bulk of the boreal or Northern Coniferous forest.

The genera *Abies,* firs; *Cedrus,* cedar; and *Pseudolarix* have the scales and seeds falling free from the central axis and cones being upright on upper branches. *Abies procera* produces large attractive cones with exserted bracts. *Abies balsamea,* balsam fir, is an important component of the Canadian boreal forest and for the production of balsam mordant. Other important firs are *A.*

**FIGURE 7.8 Family Pinaceae.** ***Abies:*** **(a) upright, deciduous cone; (b) branch with needles.** ***Pinus:*** **(c) branch with needles in fasicles; (d) mature female cone; (e) cluster of male pollen cones at end of branch.**

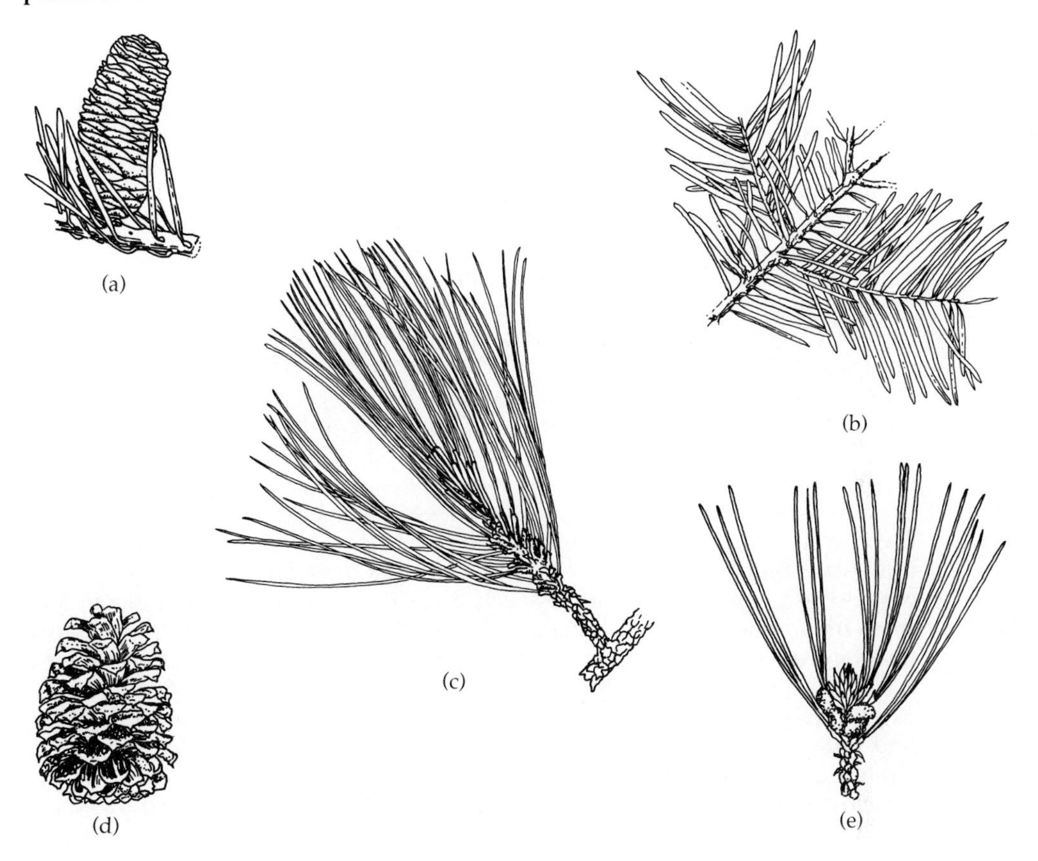


*concolor*, white fir; *A. grandis*, grand fir, from western North America; and *A. fraseri*, balsam fir, from the Appalachian Mountains of southeastern North America.

*Cedrus*, true cedar, with its aromatic wood, is found from the Mediterranean to Tibet and includes *C. libani*, cedar of Lebanon, of the Bible. *Cedrus atlantica*, Atlas cedar, and *C. deodara* are commonly grown as ornamental trees in warmer parts of North America and Europe.

The tamarack or larch, *Larix*, is the largest genus of deciduous conifers losing its needles each autumn. *Larix decidua*, European larch, is commonly planted as an ornamental. *Larix dahurica*, Dahurian larch, comprises the most northern forest in the world to above 72°N latitude in Siberia. Here the mean January temperature is below −30°C. Two common larches in North America are *L. laricina* in the boreal or Northern Coniferous forest and *L. occidentalis* in the northern Rocky Mountains.

Spruces, *Picea,* predominate in the boreal forests and high mountains. The genus has become important for lumber, wood pulp, and ornamentals. The single needles are generally four sided (one species has flat needles), sharp pointed, and attached to the branches on short "woody stumps," which give the young twigs a very rough appearance when the needles are missing. Important species are *P. glauca,* white spruce, and *P. mariana,* black spruce, of the boreal or the Northern Coniferous forest, *P. engelmannii,* Engelmann's spruce, and *P. sitchensis,* Sitka spruce, of western North America. The wood of mature Sitka spruce is used today for the sound boards of the finest harpsichords, for grandfather clocks, and for the top plate (belly) and sound posts of fine violins. *Picea abies,* Norway spruce, native of Europe, and *P. pungens,* Colorado blue spruce, are planted as ornamentals.

The genus *Pinus* is the largest and economically the most important genus of the family. It is characterized by having needles in fascicles of 2–5 (rarely 1 or 8) on short sheathing shoots. The single-leaf pinion pine, *P. monophylla* of the Great Basin of North America, has only one round needle, while Parry pine, *P. quadrifolia,* of California may have four needles. The cones are small (1–3 cm) to large (over 6 dm) in length in *P. lambertiana,* sugar pine, and massive (reported to be over 2 kg in *P. coulteri* of California). Some other common pines are *P. elliotti,* slash pine; *P. palustris,* longleaf pine; *P. ponderosa,* western yellow pine; *P. strobus,* eastern white pine; and *P. sylvestris,* Scotch pine. The cones of some species such as *P. contorta,* lodgepole pine, in western North America and *P. banksiana,* jack pine, its eastern counterpart, commonly remain closed on the tree for many years. Normally they open in mass only after a forest fire has provided heat; seeds are released, resulting in an even-aged stand. These types of pines are sometimes loosely termed *fire* species. *Pinus radiata,* Monterey pine, from a small area in California, is extensively used in reforestation in Africa, Australia, Chile, and New Zealand, and today covers over 1 million hectares in Australia alone. *Pinus longaeva,* Great Basin bristle cone pine of Nevada in the American southwest, is the oldest living organism in the world, with some recently discovered members reaching approximately 5000 years of age.

*Pseudotsuga menziesii,* Douglas fir of western North America, is one of the most important lumber trees in the world. Some trunks are 3–4 m in diameter and over 90 m tall. The wood is strong, durable, and when the wood is dry, nails previously driven into it are very difficult to remove.

The hemlocks, *Tsuga,* are trees with droopy tops and flat, petioled, blunt needles found in North America and eastern Asia. *Tsuga heterophylla,* western hemlock, and its eastern North American counterpart *T. canadensis,* eastern hemlock, are used for lumber and as ornamentals.

The genera *Abies, Pinus,* and *Peudotsuga* are used extensively for Christmas trees. Some vendors will try to sell cut *Picea* and *Tsuga,* but the buyer may be disappointed because both genera tend to drop their needles on drying after cutting.

Fossils of the family are found in Lower Cretaceous rocks to the present.

**Family Podocarpaceae (Podocarps)** Shrubs to large trees. Leaves evergreen thick, alternate, or opposite, needle-like or lanceolate to broadly oblong. Monoecious or dioecious. Two pollen chambers in each microsporophyll.

Ovules at maturity generally solitary or one per fleshy cone scale, which at maturity subtends and becomes a fleshy stalk or **aril** (Fig. 7.9).

The family consists of 7 genera and 134–139 species distributed largely in the Southern Hemisphere and Orient, but extending northward to Central America and the West Indies. It is the most important conifer south of the Equator.

The genus *Podocarpus* is widely cultivated as a shrub in Europe and in the southern United States and is used for timber in Australasia. Large trees of *P. dacrydioides,* kahikatea, and *P. totara,* totara, of over 60 m tall are reported from New Zealand. At the time of European settlement in New Zealand, 100–150 years ago, various podocarps formed extensive stands in the lowland, mixed hardwood forests. *Dacrydium cupressinum,* rimu; *D. biforme,* yellow pine; and *Phyllocladus,* which replaces the linear leaves with broad, flat **phylloclads** (a flat branch, looking like and functioning as a leaf), are some common examples. In the past the family was thought by some to be similar to the *Taxaceae* family.

**Family Taxaceae (Yew)** Shrubs or trees, much branched. Leaves dark green, evergreen, alternate, generally with abruptly tapering points, bases with **decurrent** petiole (extends down from point of attachment). Dioecious. Female ovule solitary, a fleshy aril at the base, sometimes becoming bright red at maturity and surrounding seeds (Fig. 7.10).

The family consists of 5 genera and 19–22 species distributed throughout the Northern Hemisphere, south to Central America, North Africa, and New

**FIGURE 7.9 Family Podocarpaceae.** ***Podocarpus:*** **(a) branch with mature ovules having a fleshy stalk, or aril; (b) branch with broad leaves and large aril and seed; (c) branch with male pollen cones.** ***Phyllocladus:*** **(d) flat branch, or phylloclad; (e) phylloclads and ovules.**

**FIGURE 7.10 Family Taxaceae.** ***Taxus:*** **(a) branch with ovule surrounded by aril; (b) male pollen cone; (c) sterile branch.** ***Torreya:*** **(d) branch showing long needles and fleshy fruits.**

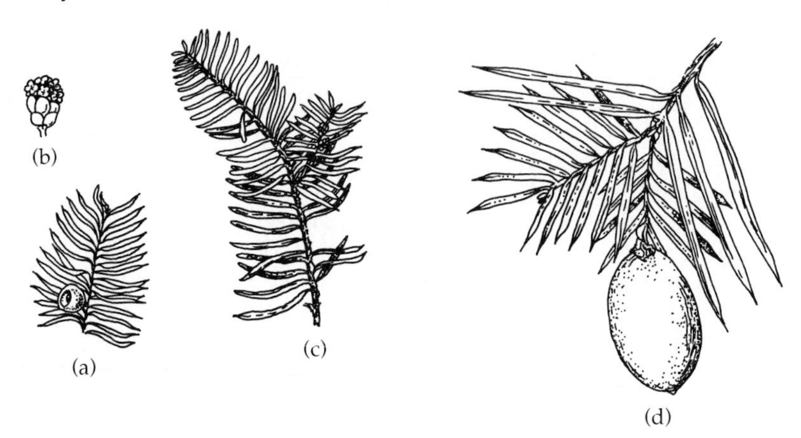


Caledonia. *Taxus,* yew, is the largest and most well-known genus. Various cultivars of *T. baccata,* English yew, and *T. cuspidata,* oriental yew, are commonly planted as ornamentals. *Taxus brevifolia,* Pacific yew, and *T. canadensis* are two native species in North America.

One species of *Torreya, T. californica,* California nutmeg, has very sharp leaf tips, while *T. taxifolia,* stinking cedar of Florida, is less sharply pointed. Three additional species are found in China and Japan.

The wood of the yew family has been popular since the Middle Ages for making bows. This is due to extra spiral thickenings on the xylem cells.

**Family Taxodiaceae (Taxodium, Bald Cypress)** Trees, large, often with massive trunks; young twigs and leaves deciduous or persistent. Leaves scale-like or needle-like, sometimes **dimorphic** (two different types on same plant). Monoecious. Male cones small and hanging together in clusters; female cones woody, terminal with flat or **peltate** (umbrella-shaped) scales, bracts, fused, with two to several ovules for each scale (Fig. 7.11).

The family consists of 10 genera and 18 species distributed in North America, Mexico, eastern Asia, and Tasmania.

*Metasequoia glyptostrobioides,* dawn redwood of central China, is called a "living fossil" because it was described from fossil material before it was observed living and brought to the western world in the 1940s. It is now being planted in milder climates in the United States.

*Sequoia sempervirens,* coastal redwood of extreme southwestern Oregon and coastal Northern California, is the tallest tree on earth, reaching over 110 m in height, just slightly more than the big tree, *Sequoiadendron giganteum,* of the California Sierra Nevada, at 96 m in height. The big tree has a greater diameter, however. Next to the bristle cone pine, the big tree may be the oldest living plant, with estimates to over 4000 years old. Redwood has been used for many years in reforestation in New Zealand and as a cultivated tree

**FIGURE 7.11 Family Taxodiaceae.** ***Taxodium:*** **(a) branches and needles; (b) panicle of male pollen cones; (c) female cone.** ***Sequoia sempervirens:*** **(d) female cone and leafy branch.** ***Sequoiadendron giganteum:*** **(e) female cone and leafy branch.**

in England. Several almost 200-year-old big trees can be seen in the botanical garden in Geneva, Switzerland.

The bald cypress, *Taxodium distichum*, of the southeastern United States, is characteristic in swamps and wet woods, with its broad, tapering trunk base and "knees" sticking out of the water. The cypress of Mexico, *T. mucronatum*, grows to enormous trunk size (over 42 m in circumference), as seen by the tule tree in the small village of Santa Maria just south of the city of Oaxaca, Oaxaca, Mexico. The same species is planted extensively in Australia and New Zealand as wind breaks and along fence rows.

The family is highly prized for its rot- and weather-resistant wood, which is used in furniture and building materials. The small genera *Cryptomeria, Cunninghamia,* and *Sciadopitys* from eastern Asia are grown as ornamentals in North America and Europe.

The monotypic *Sciadopitys verticillata* is sometimes put into its own family. Some botanists are presently advocating placing the Taxodiaceae in the Cupressaceae.

Fossils from many parts of North America and Eurasia of *Metasequoia, Sequoia, Sequoiadendron,* and *Taxodium* from rocks of the Cretaceous to the present indicate that this family was much more common and more widespread in the past than today.

## Subdivision Gneticae; Class Gnetatae

**Family Ephedraceae (Ephedra, Ma Huang, Mormon Tea)** Shrubby or trailing, stems scraggly, diffusely branched, jointed, green and photosynthetic, "horsetail-like." Leaves **decussate,** opposite, or whorled reduced to dry brown-tan scales. Monoecious and dioecious. Male pollen cone with compound, stalked microsporophylls; surrounded at the base by paired bracts. Female ovulate cone reduced, 1–4 at a node. Ovules single or in pairs, surrounded by a fleshy cup attached at the base, the micropyle opening within an elongated extended tube. Strobili at maturity become dark and leather-like covered seeds, colored scarlet (Fig. 7.12).

The family consists of the genus *Ephedra* and 42 species distributed in arid regions of North America, Mexico, and Eurasia.

Various species of *Ephedra* in the American southwest are called joint-fir or Mormon tea for its beverage use by early Mormon settlers. The decongestant alkaloid drug ephedrine is prepared from the Chinese species, *E. sinica,* ma huang, and has been used in China since before 2500 B.C. *Ephedra antisyphilitica* and *E. nevadensis* are common species in Texas and the intermountain United States, respectively. Xylem vessels are generally absent in most gymnosperms, but are present in *Ephedra* and *Gnetum.* Pollen microfossils attributed to *Ephedra* have been found from the Upper Cretaceous and Tertiary strata.

**FIGURE 7.12 Family Ephedraceae.** ***Ephedra:*** **(a) whorled branches with ovules; (b) stalked male pollen cones subtended by bracts; (c) paired ovules of female cone.**

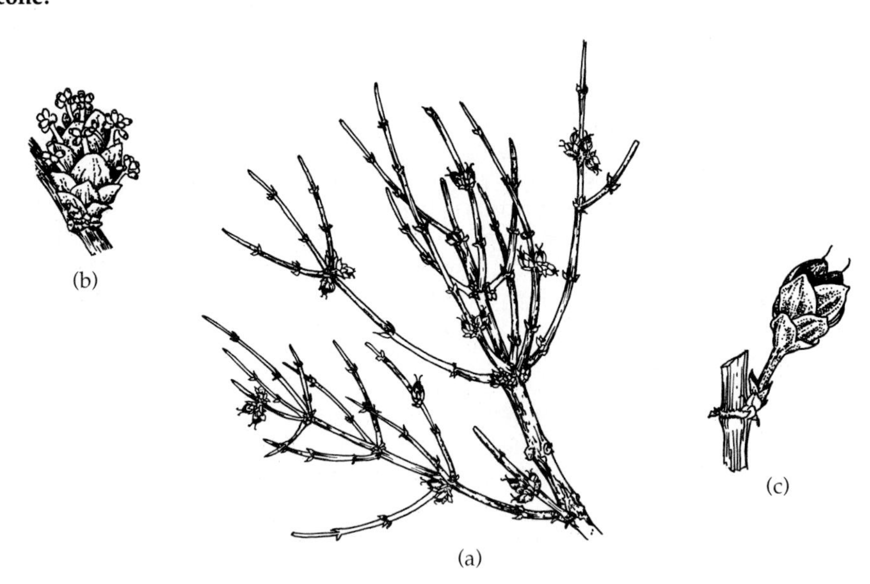

**Family Gnetaceae (Gnetum)** Shrubs, woody vines, rarely trees. Leaves opposite, simple, thick, ovate to oblong with netted veins and resembling the leaves of flowering plants. Dioecious. Male strobilus slender elongate with microsporophylls arising in the axis of bract-like leaves, "inflorescence-like." Female strobilus with ovules in 5–8 separated whorls. Seeds large, fleshy (Fig. 7.13).

The family consists of the genus *Gnetum* and 30 species distributed tropically from Brazil, tropical West Africa, India, and Southeast Asia. *Gnetum gnemon,* a small tree in Malaya, is cultivated in Java for food (leaves and reproductive strobili are cooked in coconut milk) and fiber for making rope.

Xylem vessels are generally absent from most gymnosperms, but are present in *Gnetum* and *Ephedra*. *Gnetum* is like *Welwitschia* in having a female gametophyte without organized archegonia and only free nuclear development.

**Family Welwitschiaceae *(Welwitschia)*** Woody, fleshy inverted conical stem, protruding slightly above ground level, may be over 1 m in diameter at ground level. Leaves two, opposite, leathery ribbon-like, continual basal growth persisting throughout the plant's life, but becoming tattered and ripped by wind into several to many ribbons. Dioecious with strobili on compound branched axes. Microsporophylls of male cone surround a sterile ovule (like stamen around a pistil) enclosed by bracts. Fertile ovule of female cone enclosed by bracts (Fig. 7.14).

The family consists of only the genus *Welwitschia* and the species *W. bainesii,* endemic to the coastal desert of Angola and Southwest Africa.

*Welwitschia bainesii* lives in a semiarid region, where annual rainfall ap-

**FIGURE 7.13 Family Gnetaceae. *Gnetum:* (a) axis with whorls of male cones; (b) male cone with a ring of pollen sacs; (c) leaves and axis of male cones.**

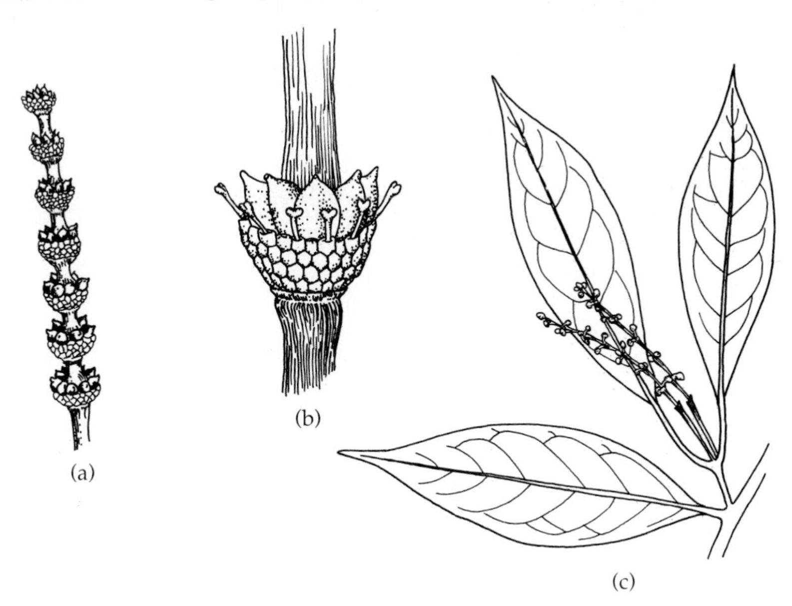

**FIGURE 7.14 Family Welwitschiaceae. *Welwitschia:* (a) cluster of male pollen cones; (b) cluster of female cones; (c) young plant showing paired leaves and large taproot; (d) male cone with exserted pollen sacs and overlapping bracts.**

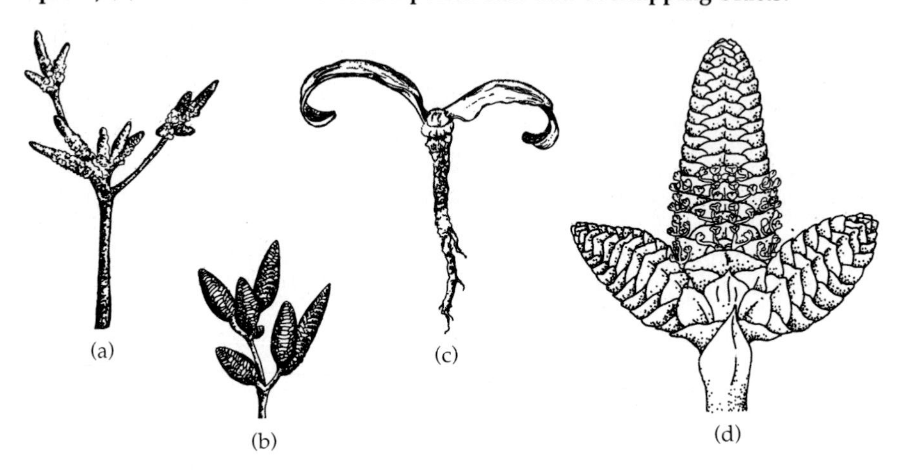


proximates 20–25 cm per year. *Welwitschia* is unique from most gymnosperms in being like *Gnetum* in 1) having a female gametophyte without organized archegonia, which is only free nuclear in development, and 2) having tubular outgrowths from the egg cells that grow toward the pollen tubes. Fertilization occurs in the united tubes.

Plants of *Welwitschia* can be grown in greenhouses, provided the plants are given little water and the underground stem and roots are not disturbed and repotted. It is best to plant in large drainage tile surrounded by sand and washed gravel to allow for the expanding stem.

## *SELECTED REFERENCES*

Andrews, H. A. (ed.). 1948. Evolution and Classification of Gymnosperms. A Symposium. *Bot. Gaz. (Crawfordsville)* 110:1–103.

Arnold, C. A. 1948. Classification of the Gymnosperms from the Viewpoint of Paleobotany. *Bot. Gaz. (Crawfordsville)* 110:2–12.

Beck, C. B. 1962. Reconstruction of *Archaeopteris* and Further Consideration of its Phylogenetic Position. *Amer. J. Bot.* 49:372–382.

Beck, C. B. 1988. *Origins and Evolution of Gymnosperms.* Columbia University Press, New York, 504 pp.

Benson, L. 1970. *Welwitschia mirabilis* in the Namib Desert, South West Africa. *Cact. Succ. J. (Los Angeles)* 42:195–290.

Bierhorst, D. W. 1971. *Morphology of Vascular Plants.* Macmillan, New York, 560 pp.

Bold, H., C. J. Alexopoulos, and T. Delevoryas. 1987. *Morphology of Plants and Fungi,* 5th ed. Harper & Row, New York, 912 pp.

Bornman, C. H. 1978. *Welwitschia.* Paradox of a Parched Paradise. C. Struik, Capetown, South Africa, 71 pp.

Chamberlain, C. J. 1919. *The Living Cycads.* Univ Chicago Press, Chicago, IL, 172 pp.

Chamberlain, C. J. 1935. *Gymnosperms: Structure and Evolution.* Univ Chicago Press, Chicago, IL, 484 pp.

Coulter, J. M. and C. J. Chamberlain. 1917. *Morphology of Gymnosperms,* rev. ed. Univ Chicago Press, Chicago, IL, 466 pp.
Cronquist, A., A. Takhtajan, and W. Zimmerman. 1966. On the Higher Taxa of Embryobionta. *Taxon* 15:129–134.
Datta, S. C. 1966. *An Introduction to Gymnosperms.* Asia Publishing House, London, 168 pp.
Debazac. E. F. 1964. *Manual des Coniferes.* Imprimerie Louis-Jean, Gap, France, 172 pp.
Eckenwalder, J. E. 1980. Taxonomy of the West Indian Cycads. *J. Arnold Arbor.* 61(4):701–722.
Farjon, A. 1984. *Pines.* E. J. Brill/Dr. W. Backhuys, Leiden.
Giddy, C. 1974. *Cycads of South Africa.* Purnell, Capetown, South Africa, 122 pp.
Greguss, P. 1972. *Xylotomy of the Living Conifers.* Akademiai Kiado, Budapest, Hungary, 329 pp.
Hu, S. 1969. *Ephedra* (Ma-Huang) in the New Chinese Materia Medica. *Econ. Bot.* 23:346–351.
Hutchinson, J. 1924. Contributions Towards a Phylogenetic Classification of Flowering Plants. III. The Genera of Gymnosperms. *Kew Bull.* 2:49–66.
International Union of Forestry Research Organizations (IUFRO). 1980. *Forestry Problems of the Genus* Araucaria. Fundacao de Pesquisas Florestais do Parana (FUPEF), Curitiba, Brasil, 382 pp.
Johnson, L. A. S. 1959. The Families of Cycads and the Zamiaceae of Australia. *Proc. Linn. Soc. New South Wales* 84:64–117.
Krüssmann, G. 1985. *Manual of Cultivated Conifers.* Timber Press, Portland, OR, 361 pp.
Maheshwari, P. and C. Biswas. 1970. *Cedrus.* Botanical Monogr. No. 5. Council of Scientific & Industrial Research, New Delhi, 115 pp.
Maheshwari, P. and R. N. Konar. 1971. *Pinus.* Botanical Monogr. No. 7. Council of Scientific & Industrial Research, New Delhi, 130 pp.
Maheshwari, P. and V. Vasil. 1961. *Gnetum.* Botanical Monogr. No. 1. Council of Scientific & Industrial Research, New Delhi, 142 pp.
Martens, P. 1971. *Les Gnetophytes.* Gebruder Borntrager, Berlin, 295 pp.
Martinex, M. 1963. *Las Pinaceas Mexicanas,* 3rd ed. Universidad Nacional Autonoma de Mexico, Mexico D.F., 400 pp.
Mirov, N. T. 1967. *The Genus Pinus.* Ronald Press, New York, 602 pp.
Ouden, P. D. and B. K. Boum. 1978. *Manual of Cultivated Conifers.* (Reprint of 1965 ed.) Martinus Nijhoff, The Hague, 526 pp.
Pant, D. D. and B. Mehra. 1962. *Studies in Gymnospermous Plants. Cycas.* Central Book Depot, Allahabad, India, 179 pp.
Raven, P. H., R. F. Evert, and H. Curtis. 1981. *Biology of Plants,* 3rd ed. Worth, New York, 686 pp.
Roden, R. J. 1953. The Distribution of *Welwitschia mirabilis. Amer. J. Bot.* 40:280–285.
Salmon, J. T. 1980. *The Native Trees of New Zealand.* A. H. & A. W. Reed, Wellington NZ, 384 pp.
Silba, J. 1986. *Encyclopaedia Coniferae.* Phytologia Memoirs, VIII, 217 pp.
Sporne, K. R. 1971. *The Morphology of Gymnosperms,* revised ed. Hutchinson & Co. Univ. Library, London.
Sprecher, A. 1907. *Le Ginkgo biloba* L. Imprimerie Atar, Genève, 207 pp.
Trivedi, B. S. and K. Singh. 1966. *Structure and Reproduction of the Gymnosperms.* Shashidhar Malaviya Prakashan, Lucknow, India, 185 pp.
Turrill, W. B. 1959. Gymnospermae. In: W.B. Turril (ed.) *Vistas in Botany, I.,* Pergamon Press, London, pp. 494–518.

# 8

# Terminology of Flowering Plants (Angiosperms)

Most people may look out over a grassland, garden, mountain meadow, or tropical forest and exclaim: "Look at the beautiful flowers." And rightly so, because the flowering plants or Division Magnoliophyta (in older systems of classifications, called **angiosperms**) are generally the plants that provide much of the bright colors observed in a sea of green. These colors of red, orange, yellow, blue, and violet, and their various hues, are confined mostly to the flowers, fruits, and leaves of the flowering plants. Only several gymnosperms, such as *Ginkgo* (maidenhair tree) and *Larix* (larch), change their green leaves to yellow in the autumn.

The flowering plants not only dominate most world floras but also appeal to our senses. The color and aromas of the flowers and fruits attract pollinators and appeal to our senses of sight and smell. The surface features of hairs, sticky glands, hooks, and spines on fruits and seeds provide mechanisms by which plants are dispersed and are discernible to the touch.

The flowering plants, Division Magnoliophyta, is made up of approximately 235,000 species and is the largest group of plants. The division ranges in size from giant *Eucalyptus* (gum) trees of southwestern Australia, which are over 100 m tall and almost 20 m in circumference, to the aquatic plant *Wolffia* (duckweed), which is barely 1 mm in size. There are herbs, shrubs, vines, and trees. Some species are adapted to living in very wet tropical rain forests, while others live in the driest arid regions of the world. Still others live in the extreme cold of high alpine and arctic regions.

Most flowering plants are independent of one another, but some are **parasites** (a live plant attached to another plant to the detriment of that plant) and others are **epiphytes** (a live plant attached to another plant for support only, without doing harm to the host plant) and may live in the upper regions of a tropical rain forest attached to tree bark. Some are **saprophytes** and depend upon decaying organic matter for necessary nutrients. Many parasites and saprophytes lack chlorophyll and are white, yellow, or red. A few plants are called **semiparasites** because they are green with chlorophyll but need necessary substances from a host plant for particular biological processes like flowering and seed set. The genus *Castilleja* (Indian paintbrush) of western North America is an example of a semiparasite.

Traditionally flowering plants have been divided into two groups: monocots (Monocotyledoneae) and dicots (Dicotyledoneae). In a more recent classification (Cronquist, Takhtajan, and Zimmerman, 1966), they are termed Class Liliopsida and Magnoliopsida, respectively. These two groups are distinguished by the characters summarized in Table 8.1.

In this chapter we will discuss and illustrate the general features and terms that apply to the plant body and the flower, and review the reproductive process in flowering plants. This will be followed in Chapters 9 and 10 by a discussion and illustration of selected orders and families of flowering plants. The above sequence of studying flowering plants is best followed because the common terminology needed for using identifying manuals must be learned. This can be done by learning them while studying selected plant material.

## *Vegetative Organs*

### *Roots*

The first structure to emerge from a swollen germinating seed is the **radicle** (Fig. 8.1). This young root tip is forced outside the seed coat by the elongating **hypocotyl** and becomes the **primary root.** This root usually becomes elongate and tapering, with smaller secondary roots arising from it. It is termed a **taproot** (Fig. 8.2). A root may be long and slender, or fleshy and enlarged considerably, as in carrots, beets, or sweet potatoes, with internal storage tissue. This type is most common in the "dicot" group, while most

**TABLE 8.1 Characteristics That Distinguish Magnoliopsida (Dicots) from Liliopsida (Monocots).**

| *Magnoliopsida* | *Liliopsida* |
|---|---|
| 1. Seed embryo with two cotyledons | 1. Seed embryo with one cotyledon or seed leaf |
| 2. Flower parts usually in fours or fives or multiples thereof | 2. Flower parts usually in threes or multiples thereof |
| 3. Veins of leaf usually netted[a] | 3. Veins of leaf usually parallel |
| 4. Primary vascular bundles of stem in a ring[b] | 4. Primary vascular bundles of stem scattered |
| 5. Vascular cambium for secondary growth present | 5. Vascular cambium for secondary growth absent |
| 6. Root system characterized by a large taproot with branch roots growing from it | 6. Root system characterized by slender fibrous roots of equal size |
| 7. Pollen usually having three or more pores or furrows | 7. Pollen usually having only one pore or furrow |

[a]Character with the most exceptions to it.
[b]Character with few exceptions; determine by cutting stem internode and observing bundles.

**FIGURE 8.1 Structure of the flowering plant seed and seedling. Magnoliopsida (dicot): (a) seed; (b) longitudinal section through seed; (c) young seedling; (d) longitudinal section through a seed showing remaining endosperm. Liliopsida (monocot): (e) longitudinal section through a grass seed; (f) young seedling.**

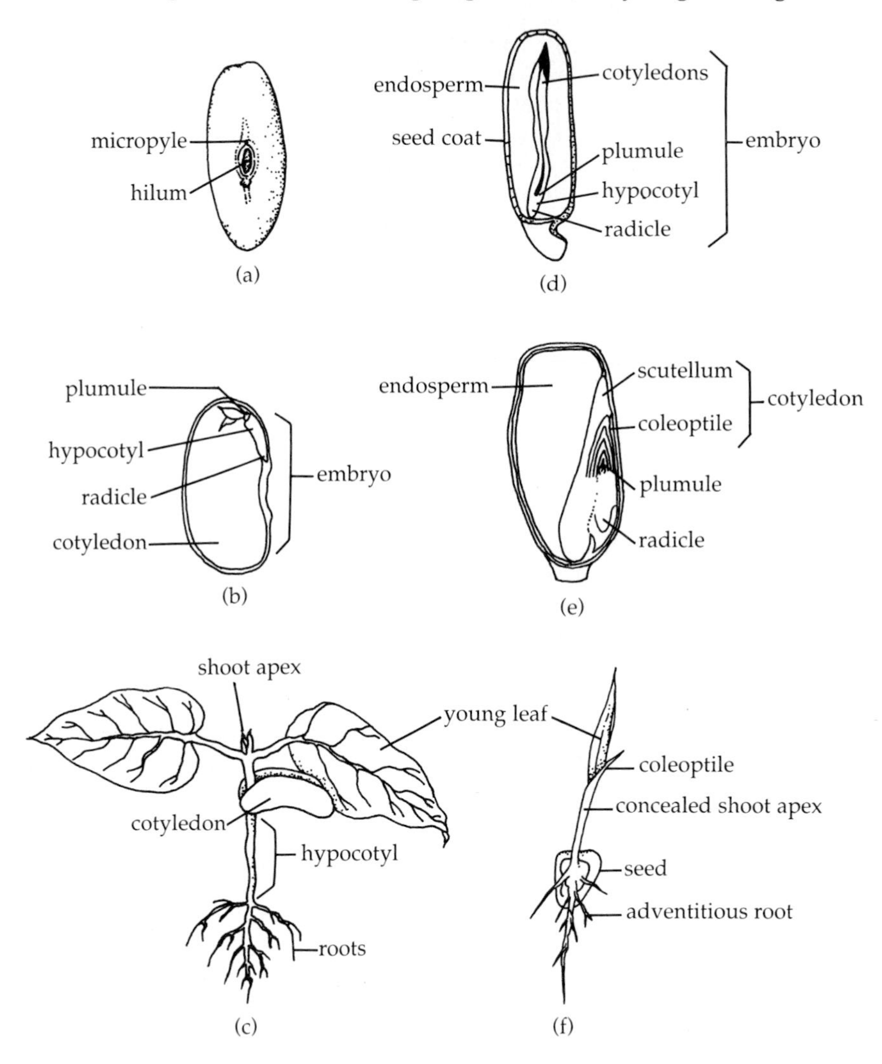


"monocots" have several to many slender roots of similar size that are present in the single taproot. These are known as **fibrous roots.** If roots are formed in some other way than from the primary root, they are termed **adventitious roots.** Some of the best known types of adventitious roots arising from some part of the stem are **aerial roots, climbing** or **holdfast roots** of various vines, or **prop roots** at the base of some stems, as in maize.

The main functions of the roots are for anchorage, water and mineral

**FIGURE 8.2** **Types of roots in flowering plants: (a) taproot; (b) fibrous roots; (c) prop roots; (d) aerial roots; (e) buttress roots; (f) storage taproots; (g) storage taproots; (h) holdfast roots.**

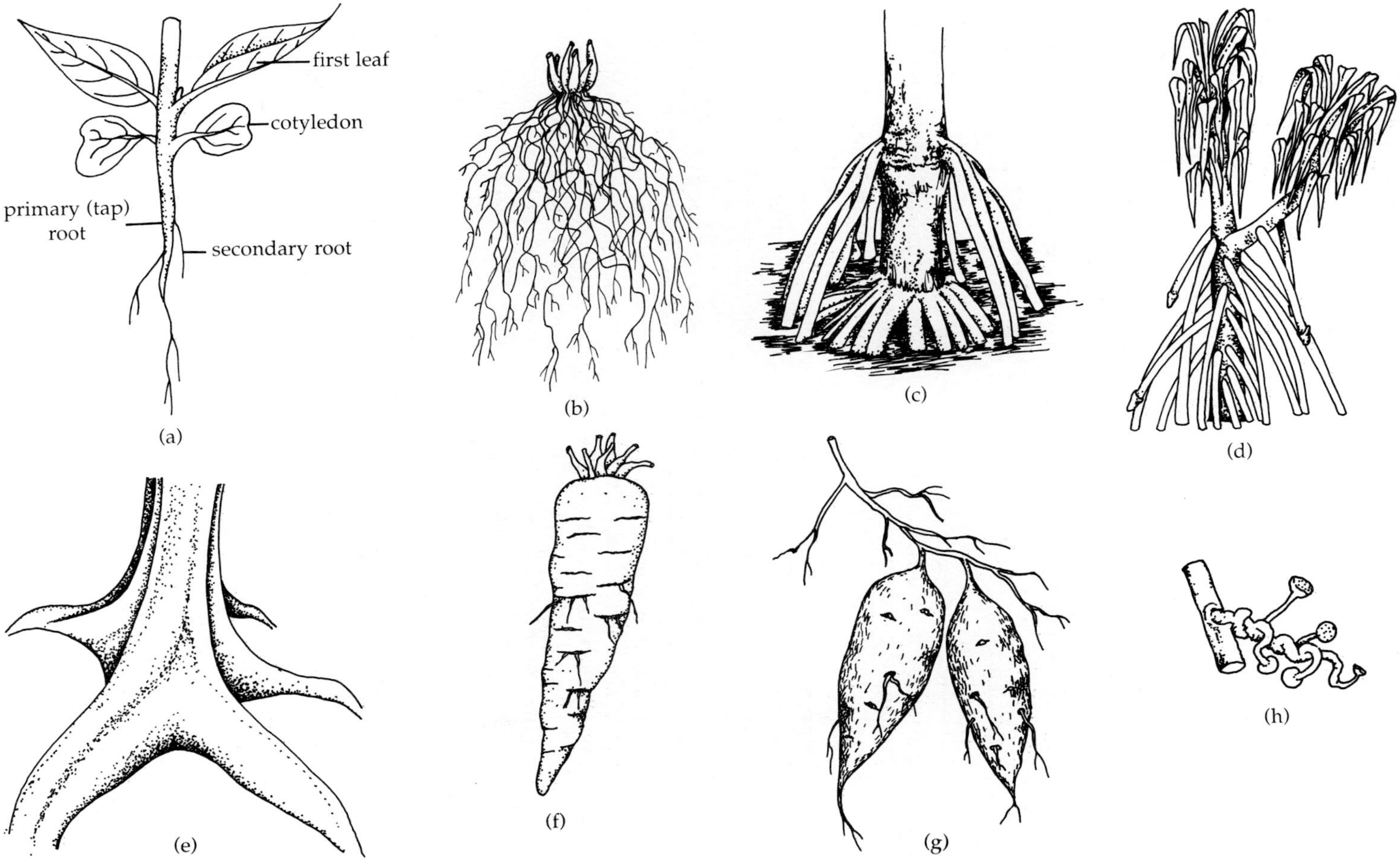

absorption, and the storage of various manufactured metabolites of photosynthesis such as starch. Some plants, such as members of the heath family and certain trees, have root systems that are symbiotically associated with a fungus, which by its slender **hyphae** is able to obtain nourishment from roots. The fungus in turn provides water and nutrients. This is termed a **mycorrhizal association.**

### *Stems*

The main axis of the plant is termed the **stem** (Fig. 8.3). Stems are recognized as generally being above ground, having **nodes** where leaves and buds in the leaf axil are attached, and having **internodes** between each node.

Stems can vary considerably in size, position above or below the ground, angle, and the direction from which they grow, as well as their longevity or duration.

Since the time of the Greek philosopher Theophrastus, plants have been classed by their duration as herbs, shrubs, or trees. An **herb** is a plant containing very little woody tissue and dying back to the ground each year. Its duration may be **annual** if it lives for only one year; **biennial** if the plant is vegetative the first year and blooming the second; or **perennial** if the plant lives for an indefinite number of years from the same root stock or underground system. To distinguish to which of the three duration categories a plant belongs is not always easy, but the characters listed in Table 8.2 are helpful.

A **shrubby** (or **fruticose**) plant is generally woody and lacks a main stem or trunk; instead there are usually several to many branches arising at ground level that are less than 3–4 m tall. Some botanists may refer to extremely large herbaceous perennial plants, such as *Polygonum cuspidatum* (Mexican bamboo), which may reach 3 m tall in one summer, as an "herbaceous shrub."

A **tree** has a single woody main trunk with radiating branches, usually on the upper portion of the plant. Trees are generally larger than shrubs, but when in a young **sapling** stage they may be hard to distinguish from shrubs.

Many genera and species of temperate trees and shrubs can be identified

**TABLE 8.2 Characteristics Helpful for Separating Annual, Biennial, and Perennial Plants**

1. Annual plants usually have a small, slender taproot.
2. All plants with woody stems and/or roots with woody crowns are perennial.
3. A plant with only a rosette of leaves late in the growing season may be a biennial.
4. Annual plants usually have root and stem joining, without any constrictions or scars.
5. Plants with large roots are usually perennial.
6. Plants with modified underground structures for food storage (i.e., bulbs, corms, rhizomes, tubers, etc.) are perennial.
7. Remnants of last year's leaves or stems attached to the root crown usually indicates the plant as being perennial or biennial.

FIGURE 8.3 Terminology relating to duration of stems: (a) annual; (b) perennial; (c) biennial, first year; (d) biennial, second year.

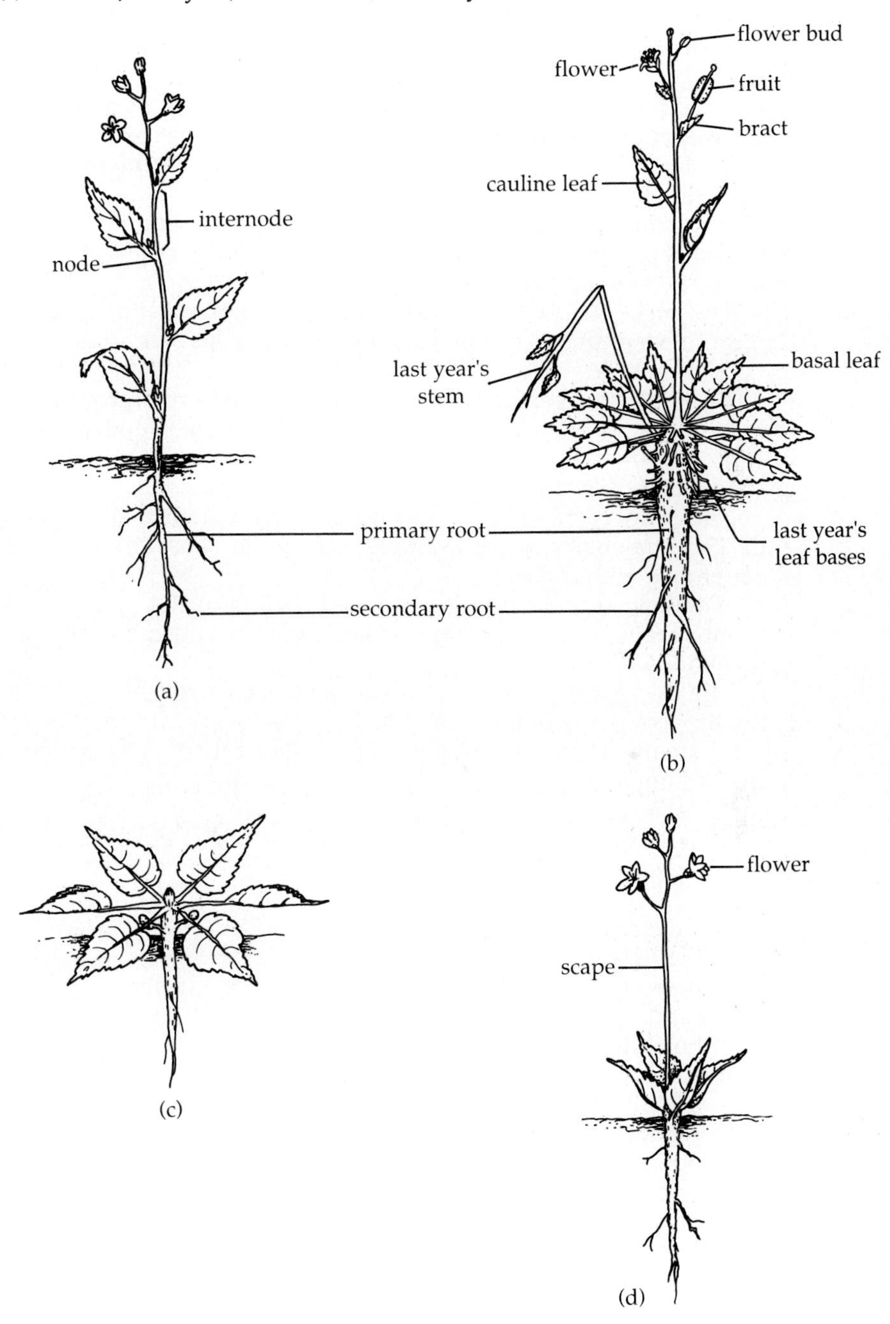

in winter or dormant periods by using stem characters. Figure 8.4 gives some features of a woody stem or twig in the winter condition.

***Modified Special Stems*** Several types of modified stems are described below.

A **bulb** is a subterranean thickened stem bearing many fleshy or scale-like leaves surrounding a fleshy bud, and with fibrous roots coming from the bottom (Fig. 8.5). *Allium* (onion) is a good example. **Bulblets** are small bulbs produced below or above ground.

A **caudex** is a slow-growing, woody upright underground base of a herbaceous perennial that each year gives rise to leaves and flowering stems; many arctic and alpine species have caudices.

The **corm** is a vertical, usually broader than tall, thickened fleshy underground stem that is covered with papery dry leaves. *Crocus* and *Gladiolus* are good examples.

**Rhizomes** are generally horizontal elongated underground or prostrate stems rooting at the nodes, covered with scale-like modified leaves, and upturned at the apex. Buds in the axils of the scales may grow into aboveground stems. *Iris* and Bermuda grass are examples.

The **runner** is a very slender aboveground stem that roots only at the end. The strawberry is an example. *Runner* is considered by some to be synonymous with *stolon.*

A trailing stem on the ground and rooting at the nodes is a **stolon.** It may intergrade with a prostrate stem. *Stolon* is considered by some to be synonymous with *runner.*

The thickened fleshy tip of a subterranean stem (rhizome) is a **tuber.** It is usually covered with numerous buds (''eyes''), which can grow into aboveground stems. Thick, fleshy taproots are not tubers. The potato is a well-known example of a tuber.

***General Stem Terminology*** The following list summarizes important terminology concerning stems.

**Acaulescent** A naked stem, usually with leaves clustered at or near the base.

**Arborescent** Tree-like, usually with a single trunk.

**Ascending** Stem growing upward at about 45–60° angle from the horizontal.

**Bud** An embryonic shoot of a plant (see Fig. 8.4). Buds can be very useful in identifying woody plants in the dormant or winter condition. **Adventitious buds** are buds that develop at places other than at the nodes; **axillary buds** are found in leaf axils; **flower buds** contain only embryonic flowers, while **mixed buds** containing both leaves and flowers are produced on the side of the stem, in contrast to **terminal buds,** which are located at the stem apex.

**Caespitose** (or **cespitose**) Stems growing in a clump or tuft.

**Caulescent** Having a leafy stem above ground.

**Cladophyll** A flattened, green stem that is leaf-like; also referred to as a **phylloclad** (i.e., *Ruscus, Asparagus*).

**FIGURE 8.4 Terminology relating to woody stems: (a) alternate-leaved stem; (b) opposite-leaved stem; (c) solid pith; (d) diaphragmed pith; (e) chambered pith.**

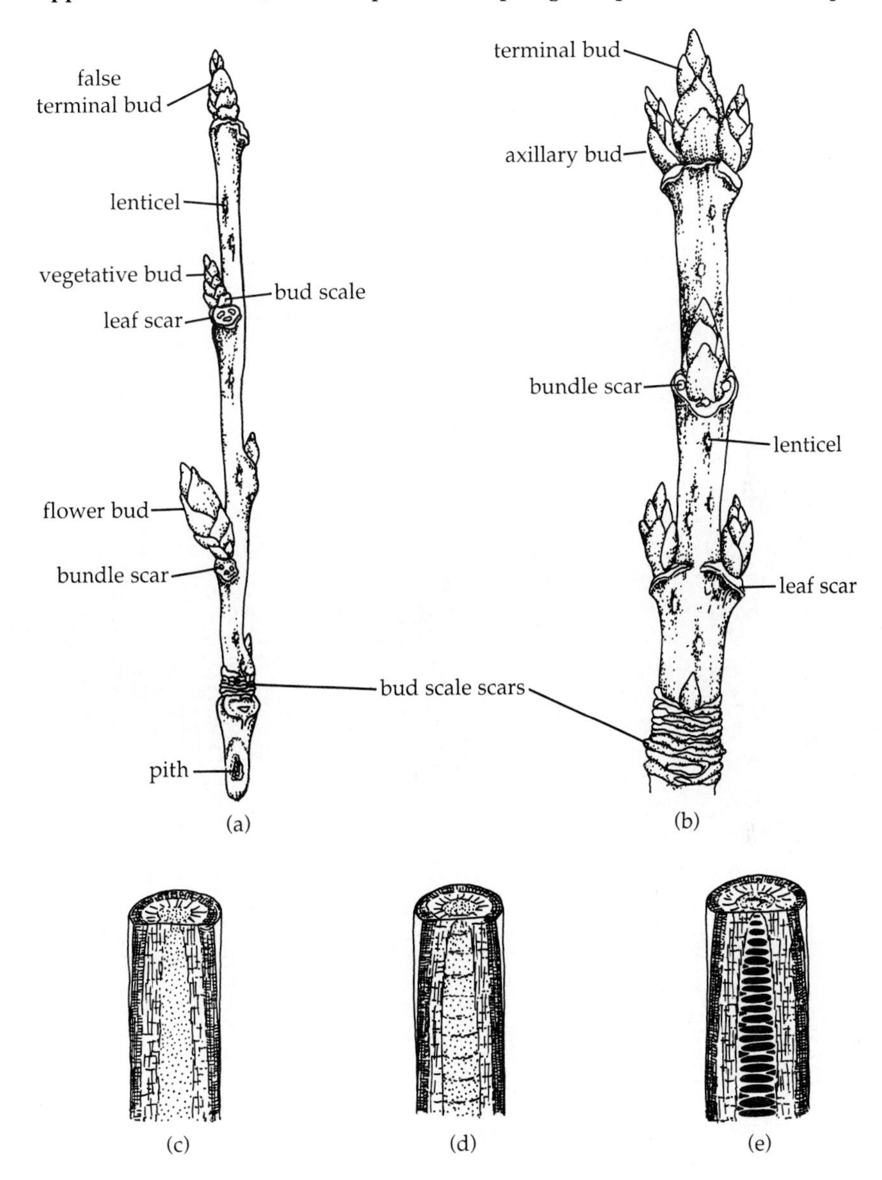


*SOURCE:* Parts (a) through (e) redrawn with permission from B. V. Barnes and W. H. Wagner, Jr., *Michigan Trees*. Ann Arbor, MI: University of Michigan Press, 1981.

**FIGURE 8.5 Terminology of modified stems: (a) bulb; (b) caudex; (c) cladophyll; (d) corm; (e) rhizome; (f) runner; (g) stolon; (h) tuber.**

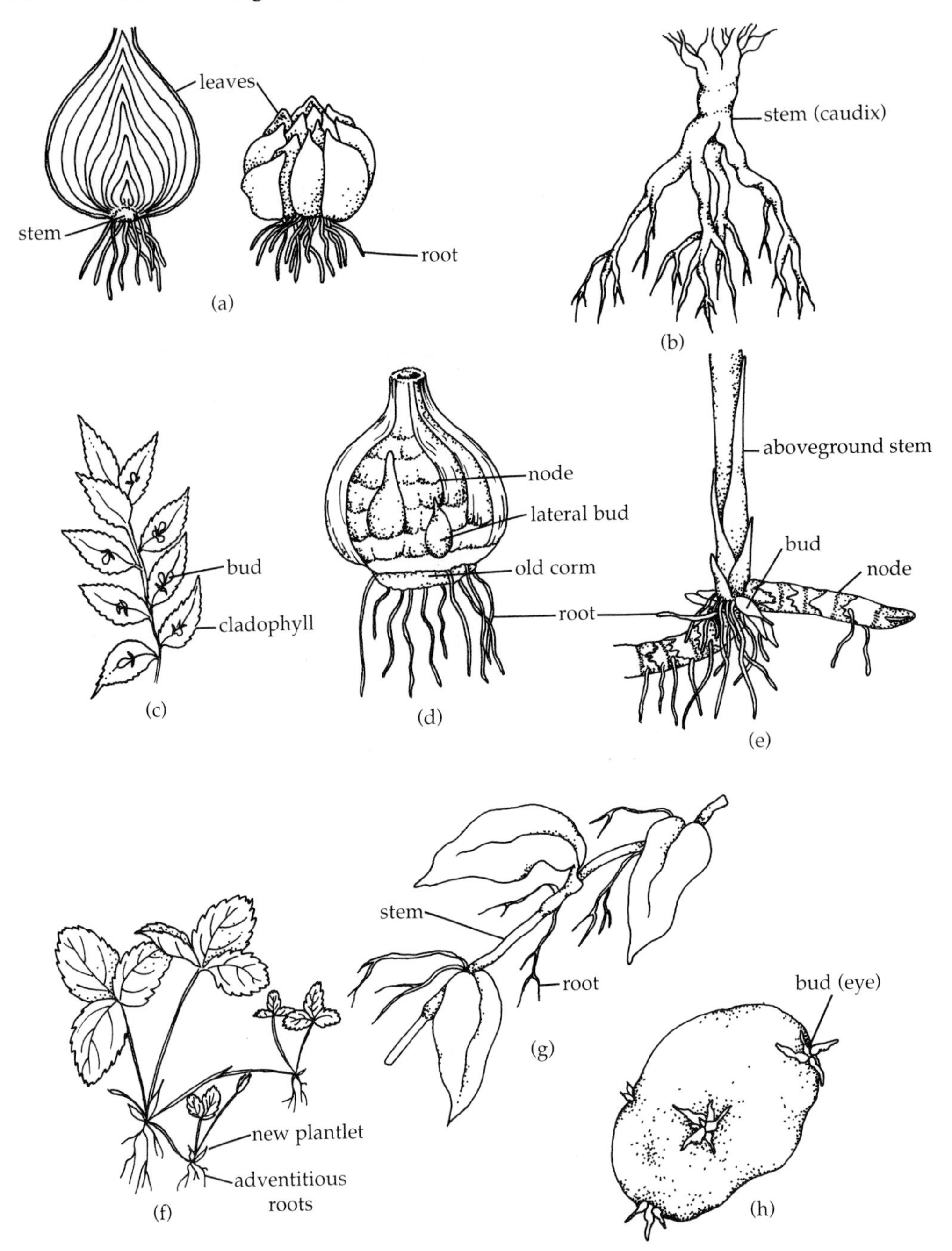

**Climbing** Clinging or twining to other objects, **scandent.**
**Decumbent** Stems flat on the ground with upturned ends.
**Erect** Stems growing upright; sometimes called **strict.**
**Lenticels** Spongy tissue on the stem surface that allows for gas exchange to and from the stem interior.
**Pith** The spongy center of a stem (see Fig. 8.4). The pith may be solid throughout and **continuous** or **chambered,** with solid tissue forming cross-partitions only and separated by air spaces.
**Prickle** Sharp outgrowth of the stem epidermis or cortex that lacks any conductive tissue; roses have prickles.
**Procumbent** Lying on the ground but not rooting at the nodes.
**Prostrate** Lying flat on the ground; prostrate stems may or may not root at the nodes.
**Rootstock** Loosely used term applied to underground specialized stems such as the rhizome or caudex.
**Scape** Naked flowering stem arising from the ground without leaves.
**Scapose** Bearing a scape.
**Scar** Remains of a point of attachment (see Fig. 8.4). Various kinds are **leaf scars,** where leaves were attached; **stipule scars,** where stipules were attached; and **bud scale scars,** where bud scales have dropped off. The age of young branches of temperate woody plants may be determined by counting the number of clustered rings of **terminal bud scale scars.**
**Spine** Sharp-pointed, stiff outgrowth of a stem, usually lacking a vascular supply; considered by some to be synonymous with **thorn.**
**Tendril** A slender twisting appendage that attaches to other plants or structures; an organ for support. It may be a modified stem or branch tip, leaf, leaflet, or petiole.
**Thorn** Sharp-pointed, stiff outgrowth of a stem, usually with a vascular supply; considered by some botanists to be synonymous with **spine.**
**Woody** Hard textured with wood.

### *Leaves*

Leaves of flowering plants are generally broad, flattened organs that are attached at the stem nodes. They are the primary food-producing organ by the process of photosynthesis. Characters of the leaf and its surrounding attachment area are many times most helpful in the identification and classification of some plants. Very few magnoliophytes lack leaves completely, but in temperate regions many species are **deciduous** and drop their leaves at the end of the growing season or autumn. Other plants have leaves that persist for longer than one growing season and are termed **evergreen.** The duration of leaves on the plant may be regulated by drought, lowering of temperature, or a change in day versus night length.

*Parts of a Leaf* A complete leaf is composed of a flat **blade** or **lamina** attached to a slender stalk or **petiole,** which fastens to the stem at the node by a thickened **leaf base.** In some plants, small **stipules** may be produced in pairs on each side of the leaf base. Generally stipules are small, thin brownish and insignificant, but in some plants they can be large and leaf-like (i.e.,

*Lathyrus*), become **stipular spines** (sharp pointed, stiff outgrowths), **tendrils** (slender, twisting appendages, which attach to other plants or structures for support), or **glands.** Sometimes they form a thin paper-like sheath around the stem called an **ocrea.** A plant having stipules is termed **stipulate,** while plants lacking stipules are said to be **exstipulate.** Some plants, like grasses, have the lower part of the leaf wrapping around the stem forming a **sheath** with a **ligule** or a small extension of tissue where the sheath joins to the blade.

Any of these parts may be absent. If the petiole is missing with the blade attached directly to the stem, the leaves are said to be **sessile;** if the blade is absent, the leaf is then **bladeless;** if a broadened petiole takes the place of a blade, the resulting leaves may be termed **phyllodes.** Phyllodes are observed in various species of *Acacia.*

***Leaf Venation*** The leaf blade consists of **veins** (vascular bundles) and the softer tissue between. The position of the veins is the leaf's **venation** (Fig. 8.6). Most leaves have a central large vein, the **midrib** with many smaller veins radiating and branching from it, forming a **reticulate** or **netted** venation, common in magnoliophytes (dicots). In most Liliopsida (monocots) the main veins often lie **parallel** (or **nerved**) to each other. In netted venation, if the main veins radiate from one point at the end of the petiole it is termed **palmate,** and if veins originate along both sides of the midrib like a feather it is called **pinnate.**

**Dichotomous** venation is when the veins are continually forking twice into equal-sized branches. Sometimes various combinations of venation patterns exist, as in some species of *Myriocarpa* or *Rhamnus.*

***Leaf Arrangement*** The arrangement of leaves at the nodes is sometimes referred to as **phyllotaxy** (Fig. 8.6). One leaf at a node with the next leaf spirally attached at the next node is called **alternate** arrangement. When two leaves are at a node and attached across from each other, it is termed **opposite.** If three or more are present, it is termed **whorled** (or **verticillate**). **Equitant** leaves are alternate but have their leaf bases overlapping and flattened lengthwise, as in *Iris.* In **acaulescent** (without leafy stems) species, leaves may arise from near the junction of the root and stem, forming a **rosette** (or **radical**). If most leaves are crowded toward the base of the stem, they are called **basal.**

Most plants have a consistent type of arrangement, but in some species there may be both an alternate and an opposite arrangement, or opposite and whorled, or all three types.

***Leaf Type*** A leaf made up of a single blade is a **simple leaf** (Fig. 8.7). A leaf with the blade divided into smaller blade-like segments is called a **compound leaf,** with its segments called **leaflets** (or **pinnae**). A compound leaf with all the smaller leaflets attached from one point and radiating outward like fingers on a hand is termed **palmately compound,** while one with leaflets attached to a main axis or **rachis** is **once-pinnately compound.** In pinnately compound leaves, if there is an odd number of leaflets (easy to tell by the presence of a leaflet at the end of rachis) the leaf is **odd-pinnately compound,** and if there is an even number of leaflets (lacking a leaflet at the end of rachis) it is **even-pinnately compound.**

**FIGURE 8.6 Terminology related to the venation and arrangement of leaves. Venation: (a) simple leaf, netted venation; (b) simple leaf, dichotomous venation; (c) simple leaf, netted venation; (d) simple leaf, palmate venation; (e) simple leaf, parallel venation. Arrangement: (f) alternate; (g) opposite; (h) whorled.**

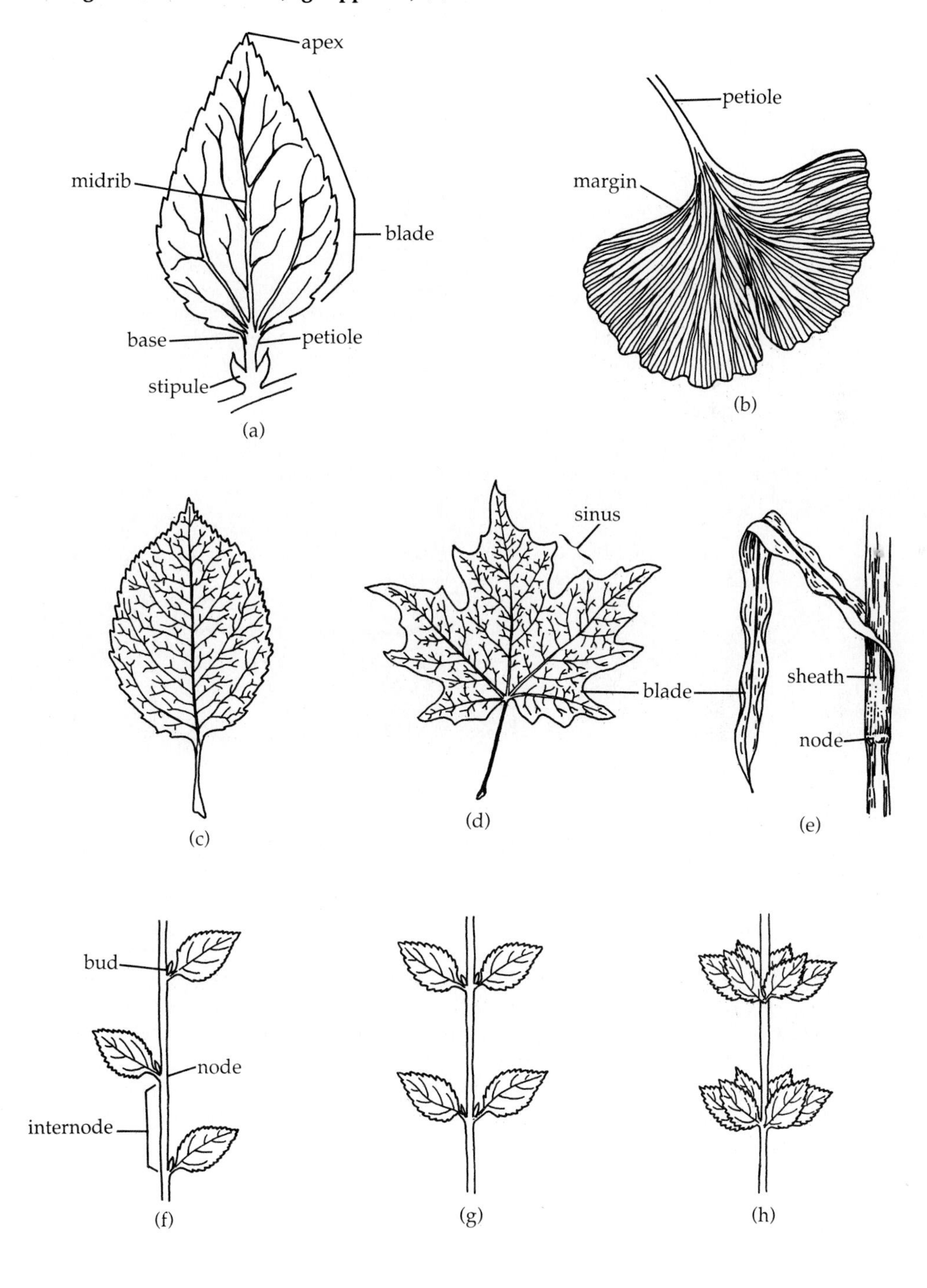

**FIGURE 8.7 Types of leaves: (a) simple. Compound: (b) odd pinnate; (c) even pinnate; (d) pinnately trifoliate; (e) palmately compound; (f) ternate (palmately trifoliate).**

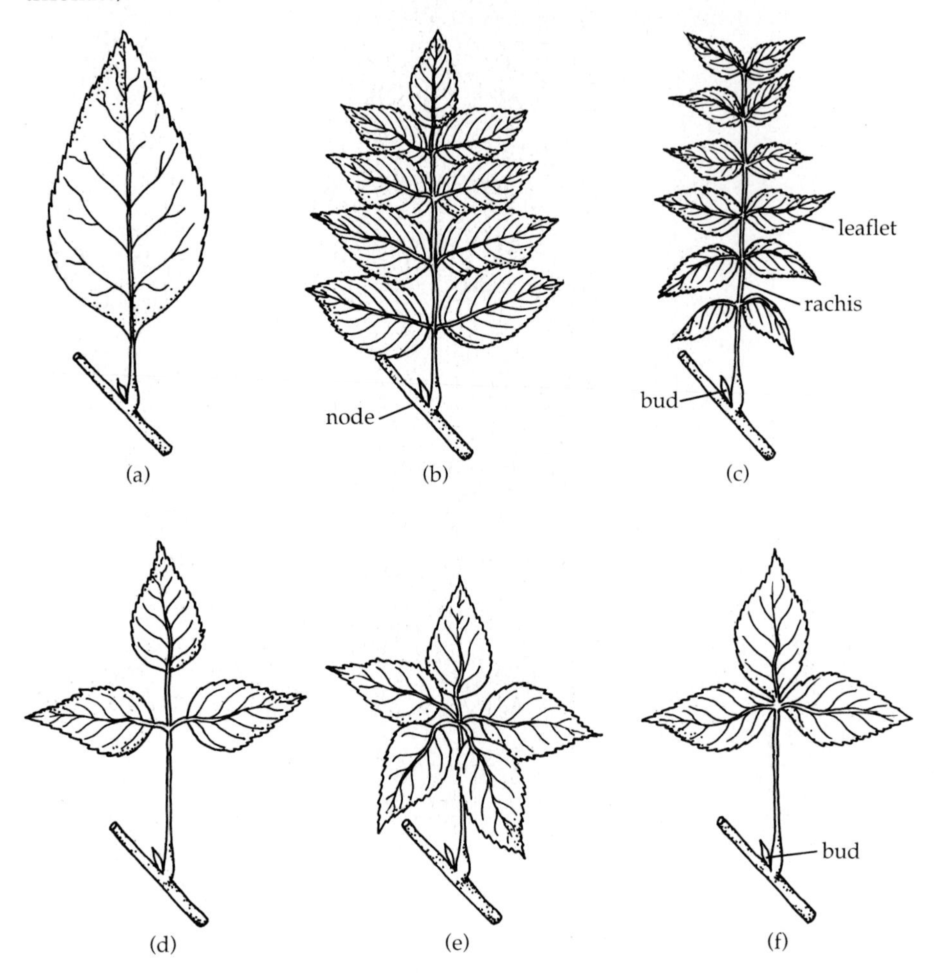


Pinnate leaves may be further divided. If the leaflets are divided again into secondary leaflets, the leaves are called **twice-pinnately** (or **bipinnately**) **compound;** if the leaflets are twice divided, **tri-pinnately compound,** etc. The axis of a pinnate leaflet is the **rachilla.** When the leaflets are in threes, as in wood soral or clover leaves, they are called **trifoliate** (or **ternate**). Trifoliate leaves can be either pinnate or palmate.

It may at times be difficult to distinguish between a compound leaf and leaflets. **Remember, buds occur only in the axils of leaves and not in the axils of leaflets of compound leaves.** One should also bear in mind that the bud may be very difficult to find or may have already developed into a branch, flower, or cluster of flowers.

*Leaf Shape* The shape of a leaf is determined by the general outline of the blade or of all the leaflets of a compound leaf minus the petiole. Sometimes the shape of leaflets is more desirable for identification than that of the leaf. Some of the more common shapes are given in Figure 8.8. If the prefix **ob** is used, it means that the shape has been inverted from the typical shape (i.e., obovate versus ovate). Some common shapes are as follows:

**Awl-shaped** (or **subulate**) Small, sharp pointed, narrowly triangular shaped leaf, as found in some junipers.
**Cordate** Broadly valentine heart-shaped, with the petiole attached at the broad end.
**Deltoid** Triangular-shaped.
**Elliptic** (or **elliptical**) Broadest at the middle, with the length usually more than twice the width.
**Falcate** Curved sideways and tapering upward; asymmetric.
**Hastate** Arrowhead-shaped, with the basal lobes flared outward.
**Lanceolate** Long tapering leaf, widest toward the base.
**Linear** Narrow leaf with parallel sides and the length usually over four times the width.
**Needle-shaped** (or **acicular**) Very long and narrow like the leaves of *Pinus* (pine) or *Abies* (fir).
**Obcordate** Broadly heart-shaped, with the petiole attached at the narrow tapering end.
**Oblanceolate** Lanceolate but with the petiole attached at the narrow end.
**Oblong** The sides generally parallel, with the ends rounded and two to three times longer than broad.
**Obovate** Egg-shaped or ovate but connected at the narrow end.
**Oval** A loose term meaning broadest at the middle (elliptic) but usually rounded at the ends; the width over one-half the length.
**Ovate** Egg-shaped and connected at the broad end.
**Peltate** Umbrella-shaped leaf, with the petiole attached to the lower surface of the blade, usually away from the margin.
**Perfoliate** Stem apparently passing through the leaf blade or opposite leaf bases seemingly unite around the stem. Considered by some authors as a leaf base type.
**Reniform** Kidney-shaped and broader than long.
**Sagittate** Arrowhead-shaped but with the basal lobes turned inward.
**Spatulate** (or **spathulate**) Resembling a spatula, rounded, and broad at the apex and elongate tapering toward the base.

*Leaf Margins* The edge of the leaf blade is its **margin.** Some common margins are given in Figure 8.9. There is considerable variation in the types of margins of leaves, but, in spite of this, margin characters have been used extensively in taxonomic descriptions of many plants.

**Cleft** Generally applies to margin segments and sinuses that are sharp and not cut over one-half way to the midrib.
**Crenate** Low, broad rounded teeth.

**FIGURE 8.8 Shapes of leaves: (a) awl (subulate); (b) cordate; (c) deltoid; (d) elliptic; (e) falcate; (f) hastate; (g) lanceolate; (h) linear; (i) needle-shaped (acicular); (j) obcordate; (k) oblanceolate; (l) oblong; (m) obovate; (n) oval; (o) ovate; (p) peltate; (q) perfoliate; (r) reniform; (s) sagittate; (t) spatulate.**

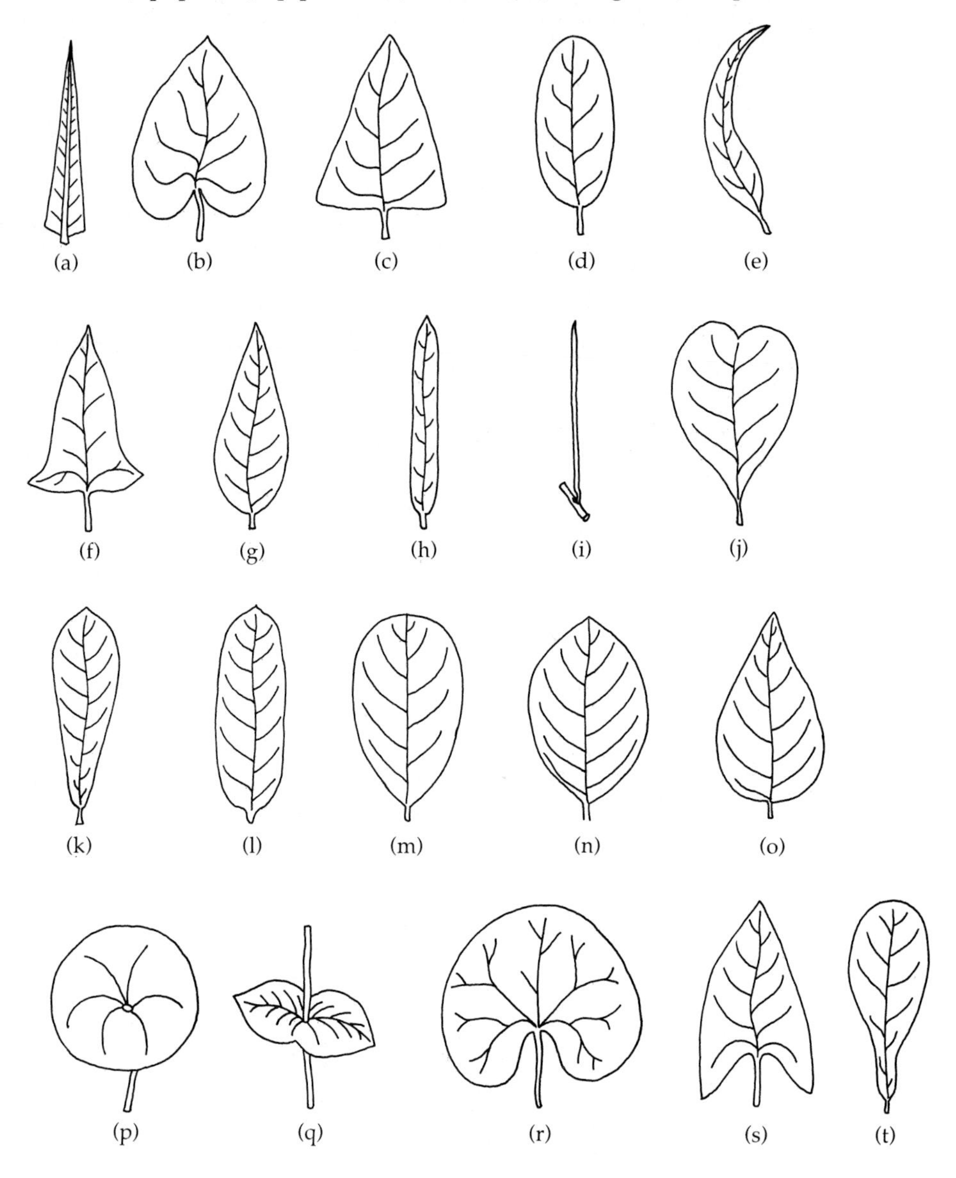

**FIGURE 8.9 Types of leaf margins: (a) ciliate; (b) crenate; (c) crenulate; (d) dentate; (e) denticulate; (f) double serrate; (g) entire; (h) revolute; (i) serrate; (j) serrulate; (k) sinuate; (l) unulate; (m) pinnately cleft; (n) palmately cleft; (o) pinnately incised; (p) palmately incised; (q) pinnately lobed; (r) palmately lobed; (s) pinnately parted; (t) palmately parted; (u) pinnately divided; (v) palmately divided.**

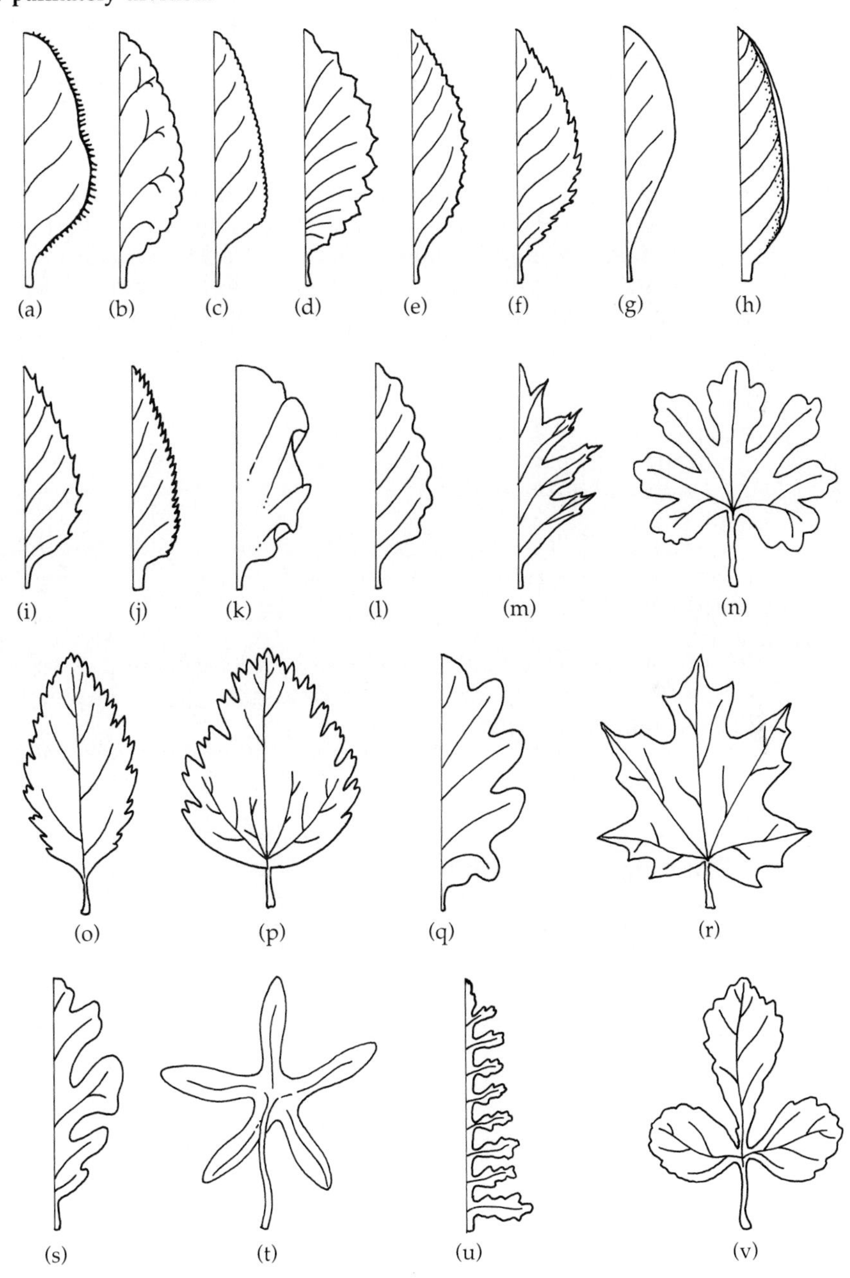

**Crenulate** Very small rounded teeth.
**Dentate** Sharp marginal teeth projecting at right angles to the margin.
**Denticulate** Very small, sharp marginal teeth projecting at right angles to the margin.
**Dissected** Cut into many fine segments.
**Divided** Margin segments cut to the base or midrib.
**Doubly serrate** Larger sharp, forward pointing marginal teeth; these in turn have small serrations.
**Entire** Smooth margin lacking any teeth or indentations.
**Incised** Sharp, irregularly cut segments and sinuses cut not more than one-third the way to the midrib.
**Lobed** Loosely used term that technically applies to round segments and sinuses not cut over one-half the way to the midrib.
**Parted** Margin segments cut one-half to three-fourths the way to the leaf base or midrib; sinuses may be rounded or sharp.
**Pinnatifid** A pinnately cleft, lobed, or parted leaf margin divided one-half to three-fourths the way to the midrib.
**Revolute** Margin rolling under toward the leaf underside.
**Serrate** Sharp teeth of the leaf margin directed forward toward the apex.
**Serrulate** Very small serrate teeth.
**Sinuate** A pronounced wavy margin.
**Undulate** (or **repand**) A slightly wavy margin.

*Leaf Apex* The apex of the leaf blade refers to the general area of the leaf tip. Some commonly used terms are illustrated in Figure 8.10 and are defined below.

**Acuminate** Gradually tapering to a prolonged point, with the two margins pinched slightly before reaching the tip. The tip may be short or long, and narrow or broad.
**Acute** Tapering to a straight point, with the two margins straight and not pinched and the angle less than 90°. The tip may be narrow or broad.
**Apiculate** Ending with a slender, not stiff tip.
**Aristate** With a stiff bristle tip.
**Cuspidate** With an abrupt, short, sharp rigid tip.
**Emarginate** (or **retuse**) Having a shallow notch at a broad apex.
**Mucronate** With an abrupt, short soft tip.
**Obtuse** Having a nonpointed but rounded apex.
**Rounded** With a broadly rounded apex.
**Truncate** Tip straight across the apex.

*Leaf Base* The base of the blade is where the petiole is attached. Common terms applied to leaf bases are given below and are illustrated in Fig. 8.10.

**Acute** Tapering to a point with two straight sides and with the angle less than 90°, wedge-shaped. Sometimes referred to as **cuneate,** which means wedge-shaped.

**FIGURE 8.10 Types of leaf apices and bases: (a) acuminate; (b) broadly acute; (c) narrowly acute; (d) apiculate; (e) aristate; (f) cuspidate; (g) emarginate; (h) mucronate; (i) obcordate; (j) obtuse; (k) rounded; (l) truncate; (m) acute (cuneate); (n) attenuate; (o) auriculate; (p) clasping; (q) cordate; (r) decurrent; (s) hastate; (t) oblique; (u) perfoliate; (v) rounded; (w) sagittate; (x) truncate.**

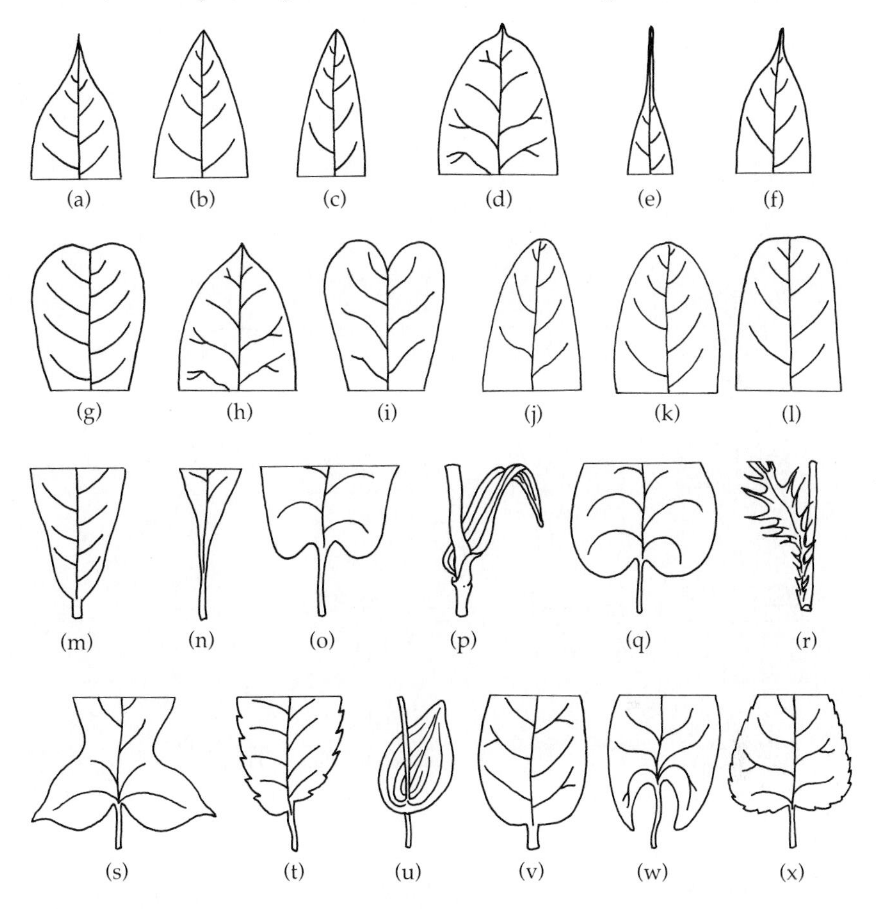


**Auriculate** Having ear-like lobed appendages from the base of the leaf blade.
**Cordate** Base broadly lobed, like that of a valentine heart.
**Hastate** The base lobed with the lobes flaring outward.
**Oblique** Both sides of the base are unequal and asymmetrical. Some are rounded on one side and acute on the other.
**Rounded** The sides of the blade rounded into the petiole.
**Sagittate** The base arrowhead-shaped with the lobes turned inward.
**Truncate** Base cut off squarely, as if cut by a blade.

### *Surface Features*

Some very distinctive features of the surface of the epidermis are often very helpful in distinguishing taxa (Fig. 8.11). These terms are most commonly

**FIGURE 8.11 Common surface features: (a) arachnoid; (b) barbellate; (c) canescent; (d) ciliate; (e) comose; (f) floccose; (g) glandular; (h) glandular hairs; (i) glochidiate; (j) hirsute; (k) hirsutuous; (l) hispid; (m) lanate; (n) pilose; (o) puberulent; (p) pubescent; (q) scabrous; (r) scurfy; (s) sericeous; (t) stellate; (u) strigose; (v) tomentose; (w) uncinate; (x) velutinous; (y) villous.**

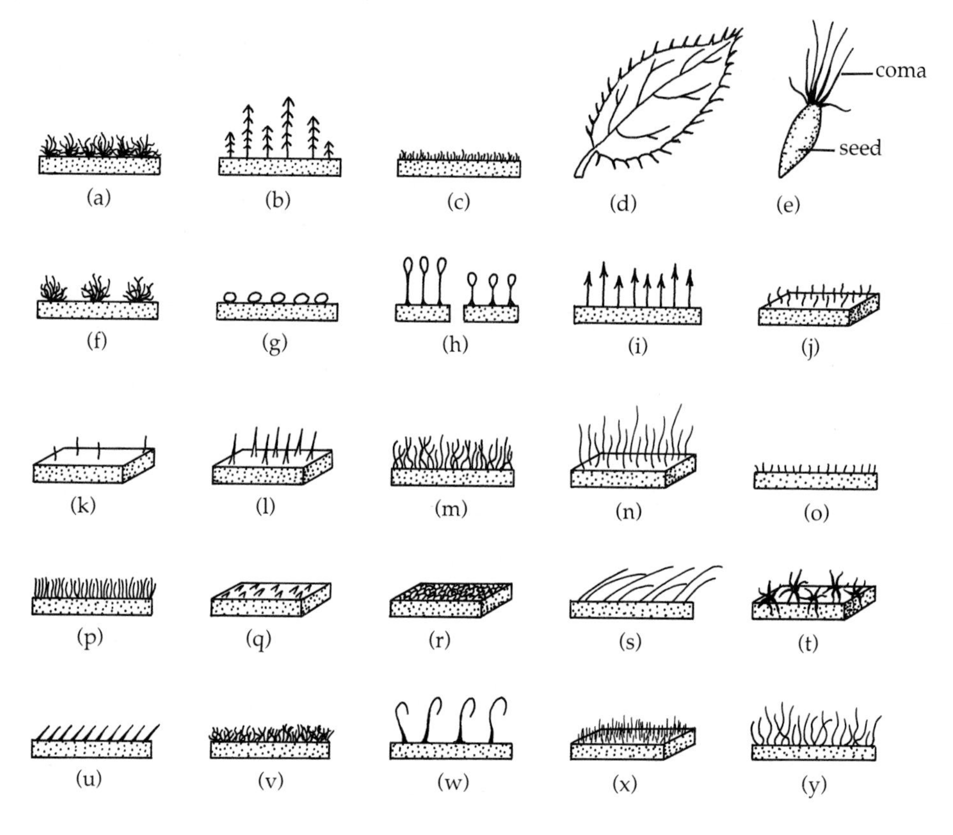


used with leaf surfaces, but can also be applied to the surfaces of other plant organs (i.e., stems, etc.). It should be mentioned that these terms are some of the most difficult for beginning students to master. This is due to the subjective nature of the descriptions. For example, how long is *long* or how short is *short*, or how "short" does hair have to be before it is no longer "long" hair? The illustrations should be consulted frequently, along with comments from an expert, until the student is well versed in each term. It should also be noted that the illustrations are only approximations and are relative.

*Nonhairy Surface Features* Terms used to describe nonhairy surface features are as follows (see Fig. 8.11).

**Glabrate** (or **glabrescent**) Nearly glabrous or becoming glabrous with age.

**Glabrous** No hairs present; surface smooth and free of hairs.
**Glandular** Having secretory structures or glands.
**Glaucous** (or **bloom**) Covered with a waxy covering giving a "whitish" appearance.
**Punctate** Having pits or dots formed by glands or waxy spots.
**Rugose** Being wrinkled.
**Scabrous** A rough rasp-like surface to the touch.
**Scarious** Thin, dry, membranous, more or less translucent.
**Scurfy** Covered with overlapping scales; give the appearance of corn-meal or "dandruff" on the surface.
**Verrucose** Covered with wart-like structures.
**Viscid** Surface sticky, as if covered with honey.

***Hairy Surface Features*** The following terms are used for hairy surface features.

**Arachnoid** With tangled cobweb-like hairs.
**Barbellate** Stiff hairs with barbs down the sides.
**Canescent** Densely covered with white or gray short hairs that give color to the surface.
**Ciliate** With a marginal fringe of small hairs.
**Comose** A clump or tuft of long hairs attached at the apex (i.e., apex of a seed).
**Coriaceous** Thick, tough, and leather-like.
**Fimbriate** Cut into fine fringe.
**Floccose** Scattered patches of interwoven hairs.
**Glandular hair** Enlarged gland or secretory structure at the apex of the hair.
**Glandular punctate** Glands dotting the surface.
**Glochidiate** Stiff hairs barbed at the apex.
**Hirsute** Long, shaggy stiff hairs. A less pronounced hirsute condition is sometimes termed **hirsutulous.**
**Hispid** Very long, sharp, stiff hairs. A less pronounced hispid condition is sometimes called **hispidulous.**
**Lanate** Long, interwoven wooly hairs.
**Pilose** Scattered, long, soft, nearly straight hairs.
**Puberulent** Very short hairs.
**Pubescent** A general term applied to hairs of any type. Some apply the term to soft, short- to medium-length hairs.
**Sericeous** Surface covered with long, soft, appressed hairs, mostly all pointed in the same direction and giving a silky appearance.
**Stellate** Star-shaped hairs with the segments radiating from a central point.
**Strigose** With short, stiff appressed hairs pointed in one direction.
**Tomentose** Densely interwoven (wooly) soft hairs forming a covering that can conceal the true surface.
**Uncinate** Stiff hairs or spines with a hook at the end.
**Velutinous** Velvety.
**Villous** (or **villose**) With long, soft, wavy hairs.

## *Reproductive Structures*

Many of the important characters used in the identification and classification of flowering plants are centered upon the flower. A **flower** is a very specialized shoot that has specialized "leaves," some of which produce reproductive structures. The flower is attached at the tip of a stem, called a **receptacle.** It may develop in the **axil** (angle formed by the petiole of a leaf and the internode of the stem) of a leaf or a reduced leaf, called a **bract.**

### *Parts of the Flower*

The four basic parts of the flower are **sepal, petal, stamen (androecium),** and **pistil (gynoecium)** (see Fig. 8.12).

*Perianth* The **sepals** are the outermost set of "floral leaves" and are collectively termed the **calyx.** They are commonly leaf-like and green, but in some plants (i.e., lilies or tulips) they are brightly colored and appear as petals **(petaloid).** Some botanists refer to these as **tepals.** The calyx encloses the flower in bud, protecting the other flower parts from injury, and may or may not persist for the life of the flower. The sepals may be individual, fused toward the base forming a **calyx tube,** or cup, or modified into hairs or scales. The most common number of sepals is five in dicots and three in monocots, even though other numbers are found. Some plants may have very small leaves or leaf-life structures at the base of the flower stalk or outside of the calyx. These are called **bracts,** or if very small, **braceoles** (or **bractlets**).

**Petals,** collectively called the **corolla,** form the second set of delicate "floral leaves" between the sepals and the stamens. The petals are typically colored or white and are important for attracting animal pollinators. The petals are usually larger than the sepals and may drop soon after the flower opens. The sepals and petals collectively are termed the **perianth.** This term is helpful when there is no distinction between the calyx and corolla. Sometimes the petals are fused together toward the base, forming a **corolla tube,** or cup. The corolla is then termed **sympetalous** (or **gamopetalous**). If the petals are not united at all to each other, the corolla is **polypetalous.** One might expect this word to mean "having many petals," but this is not the case. On some occasions the calyx, corolla, and stamens may be fused together at their base to form a **hypanthium** or cup. This may be joined to the ovary or form a cup around it. The number of petals is usually the same number as the sepals, but exceptions do exist. When the petals are all alike in shape and have radial symmetry, they are termed **regular** (or *actinomorphic*), but when they are different from one another they are called **irregular** (or *zygomorphic*). Some botanists use the word *irregular* for a flower that cannot be divided into matching halves.

*Stamen* The **stamen** is the male sex organ of the flower and collectively the stamens of a flower are called the **androecium** (Fig. 8.12). They are positioned just inside the perianth. Each stamen usually consists of a slender stalk or **filament** and an enlarged terminal **anther** that contains the **pollen.** The anther usually consists of two mature **pollen sacs** from which the mature pollen is shed. In some plants one or more of the stamens may be modified

FIGURE 8.12 **Parts and union and symmetry of a flower: (a) flower parts; (b) regular (actinomorphic) flower; (c) irregular (zygomorphic) flower; (d) superior ovary, hypogynous insertion of parts; (e) inferior ovary, epigynous insertion of parts; (f) superior ovary, perigynous insertion of parts.**

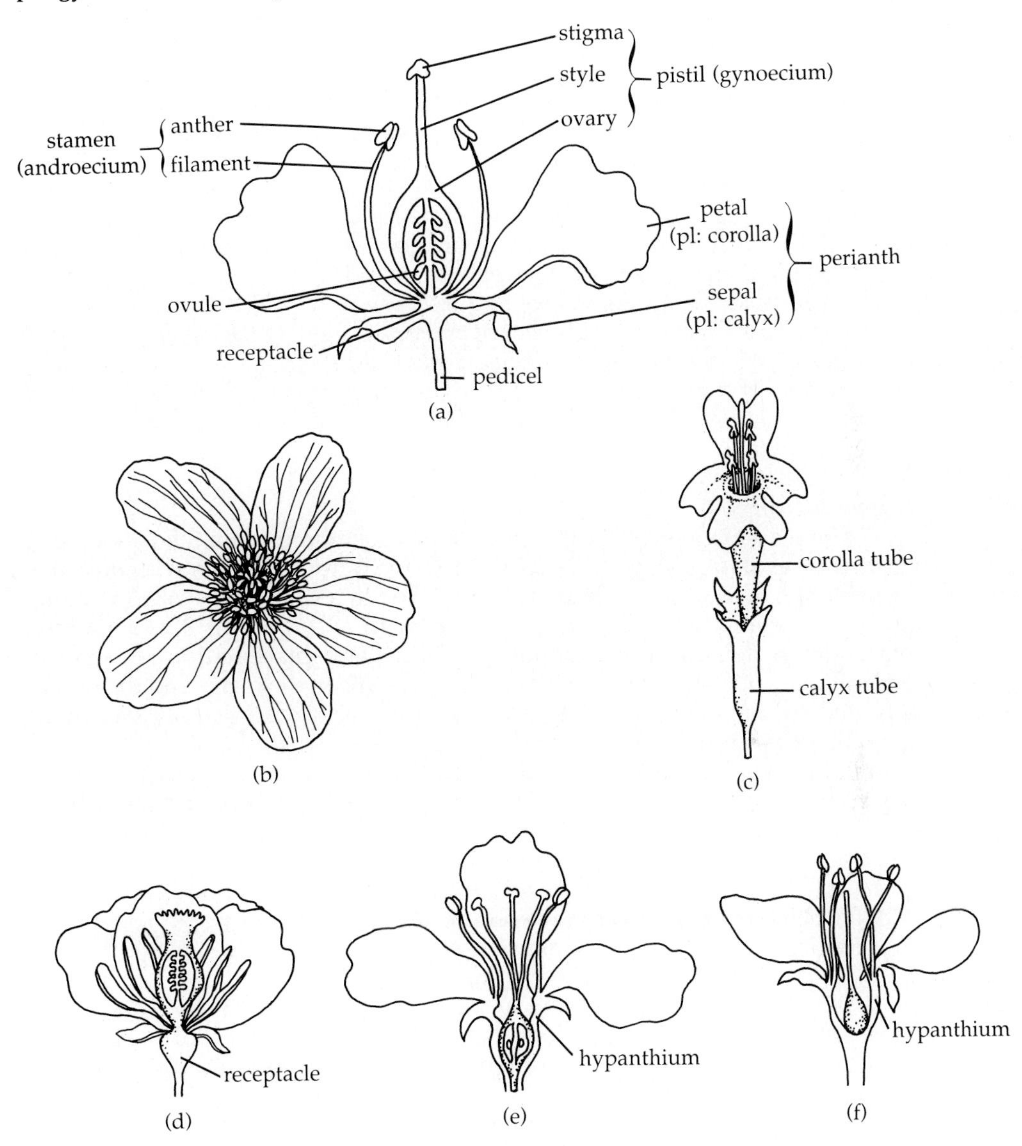


to become nonfunctional. This modified stamen is termed a **staminode.** An example of a flower with a staminode is *Penstemon,* the beardtongue.

***Pistil*** **Pistils** are the female sex organs of the flower and collectively the pistils of a flower are called the **gynoecium** (Fig. 8.12). They are positioned in the center of the flower. The gynoecium may consist of only one pistil (i.e., bean) or it may consist of two or more separate pistils (i.e., magnolia). The pistil generally comprises three parts: the **stigma** or pollen receptive region of the pistil is elevated above the enlarged basal region, the **ovary,** by an elongated stalk, the **style.** The stigma may be single lobed or branched; the style may also be single, cleft, two or more, or absent altogether. The ovary contains one or more immature seeds or **ovules.**

The basic constructive unit of the pistil is the **carpel.** It is a modified "seed-bearing leaf" or **megasporophyll,** with rows of **ovules** along each leaf edge and the edges fused so that the ovules appear to be in pairs. The line of ovule attachment is the **placenta.** A pistil may consist of a single carpel with one placenta region (i.e., bean, legumes) and is therefore termed a **simple pistil.** If the pistil is comprised of two or more united carpels, then the pistil is a **compound pistil** (i.e., two carpels in snapdragons; three carpels in orchids). *The number of carpels making up a pistil can be easily determined by carefully sectioning across the middle of an enlarged ovary and counting the number of placenta where ovules are attached, the partitions* ***(septa),*** *the spaces in the ovary* ***(locules),*** *or the number of stigmas and styles. The largest number obtained will be the number of carpels.* This procedure will usually be necessary to identify an unknown flowering plant.

### *Placentation*

The placenta may be located in different positions within the gynoecium, termed **placentation** (Fig. 8.13). In a simple or compound pistil with only one locule in the ovary, the ovules may be attached in a vertical row or rows along the ovary wall. It is **marginal placentation** in a simple pistil and **parietal placentation** if found in a compound pistil. **Basal placentation** is found in both simple and compound pistils, with one locule in the ovary and the ovules attached to the locule floor. As a rule of thumb, if there is only one ovule present, regardless of where it is attached, it is usually basal. In **axile placentation,** the ovules are attached where the septa of a compound pistil are united in the middle of the ovary. In some ovaries of a compound pistil, there may be a single locule with a column of tissue sticking up in the center of the ovary bearing the ovules. This is **free-central placentation.** There may be some difficulty in determining the proper placentation type if only a cross section is made of the ovary. Therefore, to distinguish between some basal and free-central types, a longitudinal section of a mature ovary will have to be made.

When flowers consist of two or more separate carpels (i.e., strawberry), they are said to have **apocarpous** flowers. Other flowers that have single pistils are made up of two or more joined carpels (compound pistil) and are termed **syncarpous** flowers.

Most flowers have both stamens and pistil(s). Flowers with both sexual structures within are **bisexual** (or **hermaphroditic**) and are **perfect** flowers. **Unisexual** flowers with only stamens or only pistil(s) are called **imperfect** flowers.

If a flower has all four of the floral parts—sepals, petals, stamens, and pistils—it is referred to as a **complete** flower, while an **incomplete** flower

**FIGURE 8.13 Placentation types: (a) axile; (b) basal (longitudinal section); (c) free central; (d) free central (longitudinal section); (e) marginal (cross section and longitudinal section); (f) parietal; (g) superficial.**

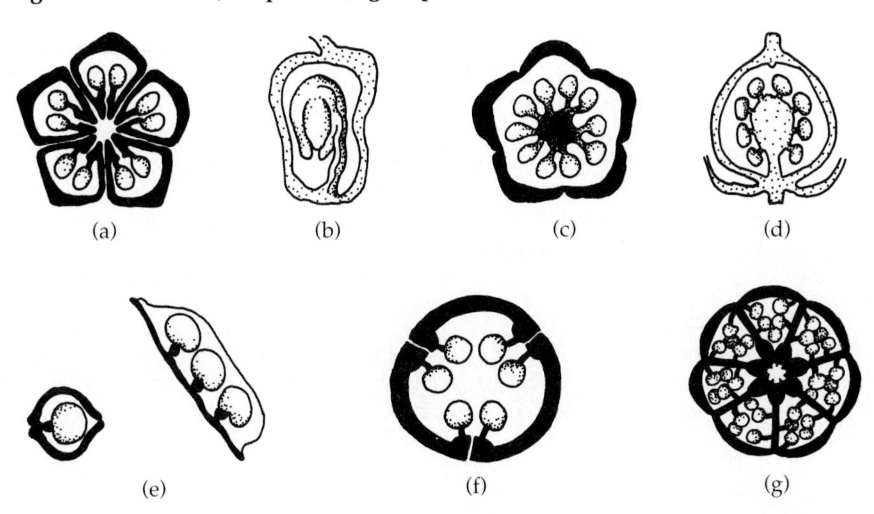


lacks one or more of the parts. Therefore, a perfect flower can be complete or incomplete, and an imperfect flower is always incomplete.

### *Insertion of Parts and Position of the Ovary*

The four floral parts of the flower may be arranged or attached to one another in different ways. This is termed **inserted** or the **insertion of floral parts.** It implies that one or more parts may grow out of or be fused to another part. If like parts are united **(connate)**, the parts are said to be **coalescent** (i.e., sepals united), but if unlike parts (i.e., calyx and corolla) are united, then the parts are said to be **adnate.**

All the floral parts are attached to the receptacle. The relative position of the ovary to the other parts is an important diagnostic character. If the floral parts are attached to the receptacle below the ovary, the position of the ovary is **superior** (see Fig. 8.12). If the perianth and androecium are attached to the top or the wall of the ovary, the ovary position is **inferior.**

In some situations the perianth and androecium fusion to the pistil may extend only part way up the ovary wall. Some botanists call this **partly (half) inferior.**

Flowers may also be grouped according to the positions of the calyx, corolla, and androecium to the pistil. A flower with the perianth and androecium attached under the ovary is said to have an insertion of parts that is **hypogynous.** The stamens may be adnate to the corolla with the calyx and corolla separate. A hypogynous flower always has a **superior** ovary. An **epigynous** flower has the floral parts adnate to the ovary wall, with the ovary appearing sunken in tissues below the flower. An epigynous flower is always inferior. The **perigynous** flower has the calyx and corolla fused into a "floral cup" or **hypanthium,** which arises from around the pistil and is attached below the ovary. The position of the ovary is always superior.

To determine the proper position of the ovary or insertion of parts, a longitudinal section of the flower will have to be made.

Unisexual flowers having only stamens are called **staminate** flowers, and flowers with only pistils are termed **pistillate** flowers. When both staminate and pistillate flowers occur on different plants (i.e., cottonwoods and willows) the plants are said to be **dioecious.** Other times both staminate and pistillate flowers are on the same plant (i.e., birches, walnuts, and some nettles) and the plants are said to be **monoecious.** Sometimes plants will have both perfect and unisexual flowers on the same plant (i.e., rhubarb) and the plants are known as **polygamous.**

## *Inflorescences*

Plants often have their flowers grouped in various arrangements or clusters, termed **inflorescences.** These arrangements may be very **simple** and easy to determine, while others may be very complicated and **compound** and made up of two or more simple inflorescences.

The main stalk supporting a single flower is the **pedicel,** whereas the stalk supporting an entire inflorescence is called a **peduncle.** The central axis of the inflorescence is the **rachis.** An inflorescence that has the oldest flower terminating the rachis with the blooming pattern being outward and downward is called a **determinate** type. Conversely, if the youngest flower is central or terminal and the blooming pattern is progressively inward and upward, the inflorescence is called **indeterminate.**

Some of the common inflorescence types are illustrated in Fig. 8.14 and are described below.

**Ament** (or **catkin**) A deciduous, pendent, or erect spike-like inflorescence comprised of unisexual, apetalous flowers.

**Corymb** A flat-topped indeterminate inflorescence in which the lower pedicels become progressively elongate and the rachis shortened. A corymb can be **compound** with several corymb-like clusters to each pedicel, giving a larger collective corymb.

**Cyme** A determinate inflorescence in which the terminal flower is older than the subtending lateral flowers. A cyme can be compound.

**Head** (or **capitulum**) An indeterminate dense cluster of sessile flowers.

**Glomerule** A general term applied to a dense cluster of flowers.

**Panicle** An indeterminate inflorescence comprised of two or more flowers on each pedicel; may be compound.

**Raceme** An indeterminate inflorescence with single flowers on pedicels arranged along the rachis.

**Scorpioid cyme** A single axis inflorescence with one-sided flowers appearing to coil like a scorpion's tail.

**Solitary** A single flower at the end of the peduncle.

**Spadix** A thick, fleshy, spike-like inflorescence, usually of imperfect flowers and subtended by a large bract called a **spathe.** Sometimes the spathe is showy and brightly colored.

**FIGURE 8.14 Inflorescence types (large circle = older flower, smaller circle = younger flower): (a) ament (catkin); (b) corymb (simple); (c) corymb (compound); (d) cyme (simple), or dichasium; (e) cyme (compound), or dichasium; (f) cyme (scorpioid); (g) head; (h) panicle; (i) raceme; (j) solitary; (k) spadix (with spathe); (l) spike; (m) umbel (simple); (n) umbel (compound); (o) verticil.**

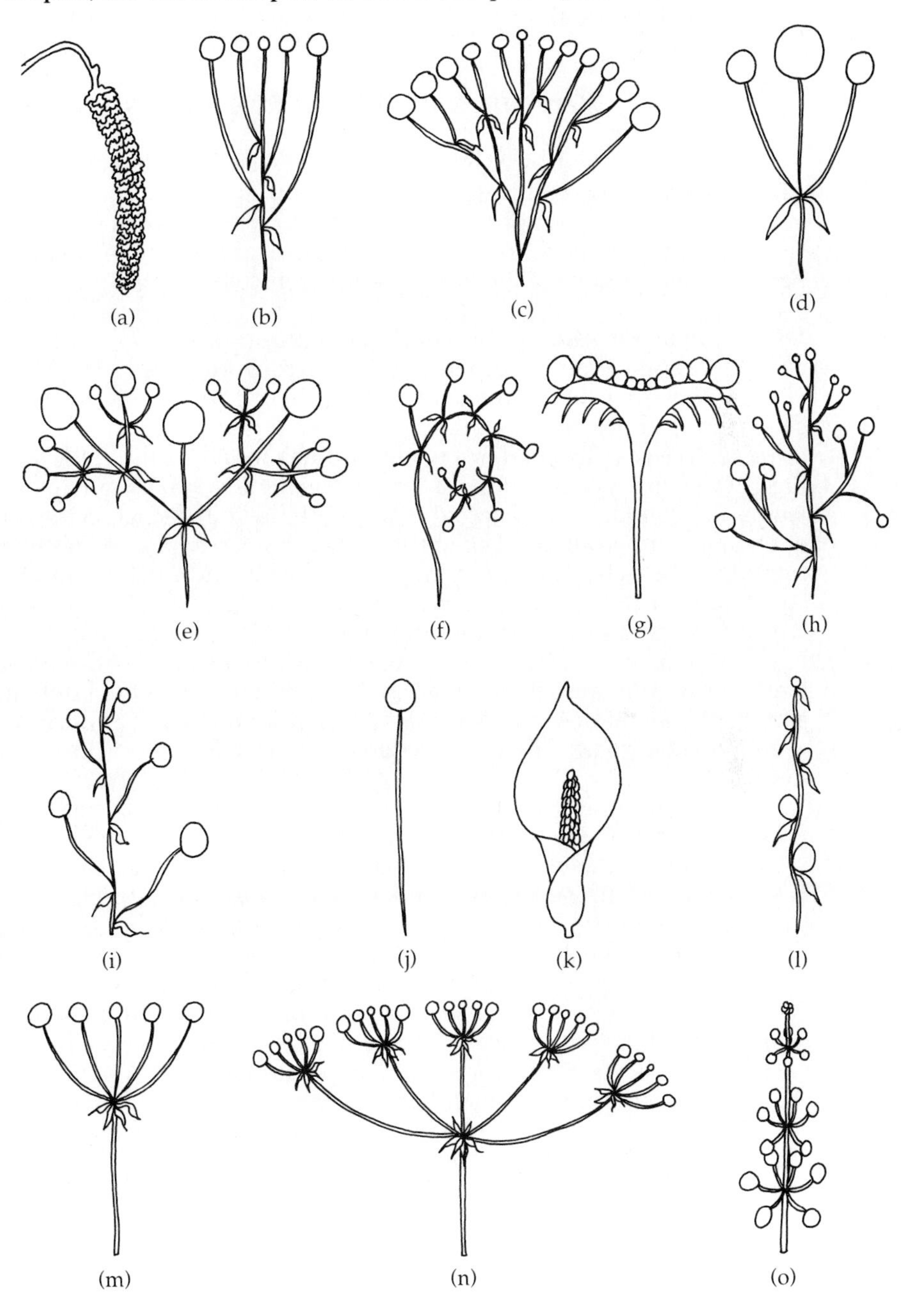

**Spike** An indeterminate inflorescence with sessile flowers along the rachis.

**Umbel** An indeterminate, generally flat-topped or orbicular inflorescence with equal-length pedicels arising from a single point at the end of the rachis. An umbel may be compound.

**Verticil** An inflorescence with the flowers in whorls at the nodes.

## *Reproduction of Flowering Plants*

As we observed in previous discussions of ferns and pinophytes, the gametophytes were relatively large in size. In the flowering plants they are very much reduced, with the male gametophyte **(microgametophyte)** consisting of only three cells, while the female gametophyte (mature **megagametophyte**) is generally comprised of seven cells (Fig. 8.15).

### *Development of Male and Female Gametophytes*

The **stamens** are considered to be **microsporophylls** and are made up of a slender stalk or **filament** and an **anther.** The anther consists of four chambers or **anther sacs,** with each chamber at immaturity containing a mass of diploid cells called **pollen** (or **microspore**) **mother cells** (PMCs). Each of the PMCs undergoes meiosis to produce generally spherical haploid **microspores** in a tetrad. These haploid *(n)* cells are the initial stage in the formation of the male gametophyte. The nucleus of each microspore divides by mitosis, forming ultimately a **pollen grain** with two cells and nuclei: a **tube cell** and a **generative cell.** This latter generative cell will later divide into two **sperm,** which function as male gametes in the fertilization process. When the pollen is shed from the anther, it will have a thick **exine** wall with characteristic markings on the surface. The shed pollen might be called the **immature male gametophyte** and does not become mature until following pollination, when a **pollen tube** grows from the pollen grain and the two sperm nuclei are formed.

The **pistils** or female parts of the flower are thought to be highly modified **megasporophylls.** They often include three regions. The **stigma,** which receives pollen, is the elevated terminal portion of the **style.** The **ovary** is the basal portion of the pistil and contains the **ovules,** which develop into seeds. Depending on the type of plant, the ovary may ultimately contain from one to several thousand seeds.

Each ovule begins as a small swelling along the inner wall of the ovary. As the swelling grows, it becomes a dome-like mass called the **megasporangium** or **nucellus,** with one or two ring-like layers of tissue, the **integuments,** surrounding it. As time progresses, the integuments will completely cover the nucellus, except for a small opening at the tip, called the **micropyle.**

A cell deep within the nucellus will ultimately enlarge and undergo meiosis. This **megaspore mother cell,** which represents the last stage of the sporophyte generation, forms a linear tetrad of haploid **megaspores.** Typically three of the four megaspores ultimately disintegrate and disappear, leaving one **functional megaspore** furthest from the micropyle. This megaspore will give rise to the **female gametophyte** or **embryo sac** by typically undergoing

**FIGURE 8.15 Life cycle of a flowering plant.**

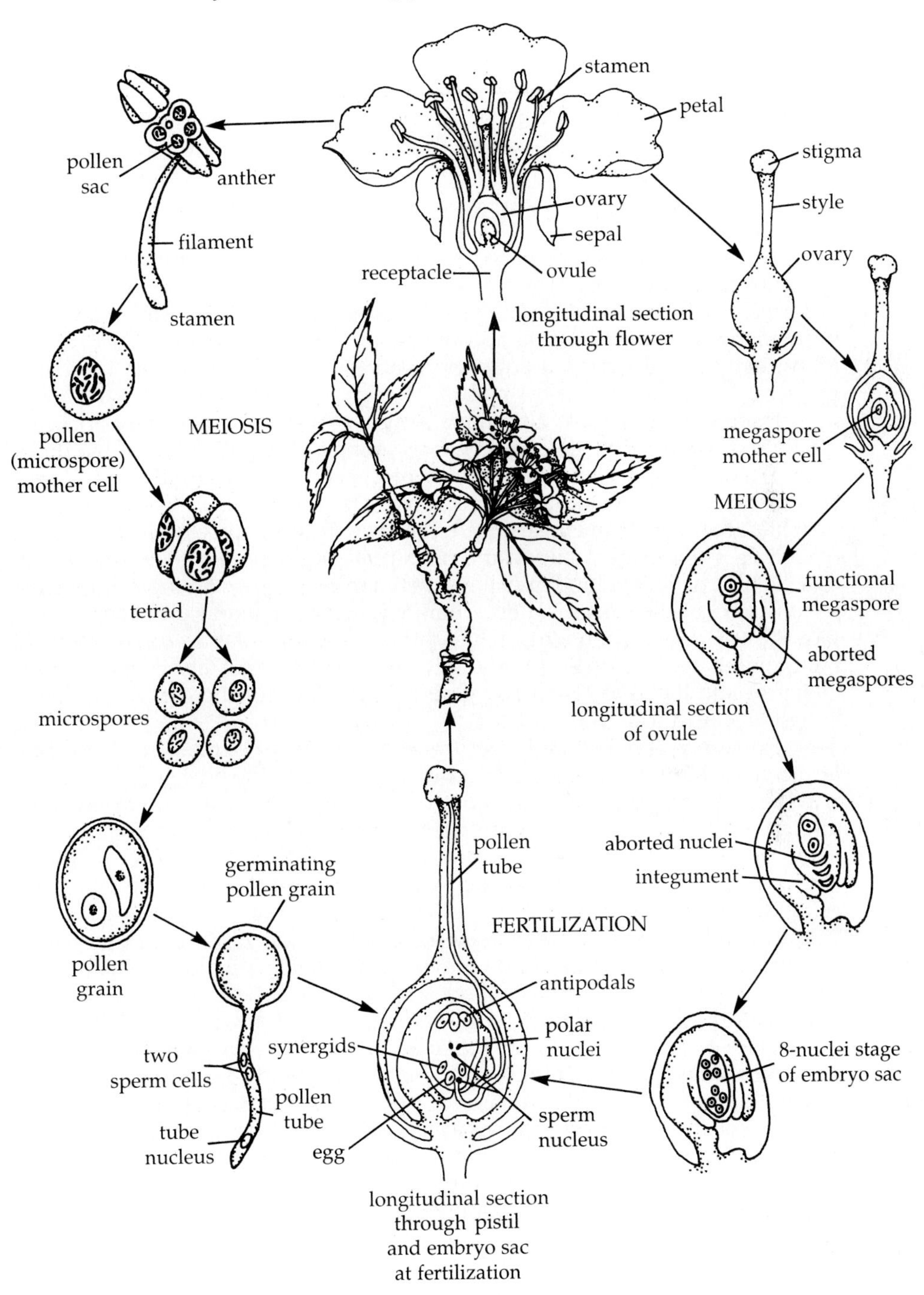

three separate mitotic divisions, giving eight haploid nuclei lying in the cytoplasm free of the cell walls. Some species have as few as four nuclei, while as many as 128 nuclei are known. As the female gametophyte matures, the nuclei become oriented into three areas: three nuclei at the pole furthest from the micropyle are the **antipodal nuclei** and have no known function; one of the three nuclei at the micropyle end becomes the **egg,** with the other two **synergids** assisting fertilization; and two nuclei in the center are called the **polar nuclei.**

### *Pollination*

The anthers open and disperse the pollen grains at about the time that the embryo sac is mature. The transfer of pollen from the anther to the stigma is termed **pollination.** It should be noted that most flowering plants are normally cross-pollinated, but it is not uncommon for self-pollination to occur. Many plant species have unique morphologies and mechanisms to ensure proper cross pollination takes place.

Insects are the most common carriers of pollen for cross-pollination. Flowers that attract beetles may be large, single flowers, such as magnolia, wild rose, or California poppy; others are small but clustered into inflorescences, such as dogwood and elderberry. The flowers that are dull colored or white frequently have a strong fermentive, foul or spicy odor.

Bees are the most important cross-pollinating agents. Bee flowers are usually blue or yellow, and may have distinctive markings on them to indicate the position of the nectar. These "honey guides" as found on *Digitalis* (foxglove), are very prominent, while other flowers such as *Caltha* (marsh marigold) have markings visible only under ultraviolet light. Bee flowers characteristically have the **nectar glands** at the base of the corolla and have a "landing platform" for the bee.

Flowers pollinated by butterflies and moths are similar to those that attract bees, but may be orange or reddish along with the blue or yellow ones, and lack landing platforms. Many of the flowers pollinated by moths are open and give off strong penetrating odors at night or have color displays that stand out against a dark background. The nectar glands are usually found at the base of a long corolla tube, which accommodates the long sucking tongues of moths and butterflies.

Bird-pollinated flowers usually have little or no odor, and a thin nectar that may even drip from the flower when the pollen is mature. This may be correlated to the fact that birds have a poorly developed sense of smell but keen eyesight to see colorful yellow, orange, and red flowers. Bird-pollinated flowers are generally large or are part of large inflorescences and contain large quantities of nectar. Hummingbirds in North and South America are the main bird pollinators, but in other regions of the world other specialized bird families visit the flowers. Flower examples include columbine, *Fuchsia, Eucalyptus,* and *Hibiscus.*

Flowers visited by bats are similar in many respects to bird-pollinated flowers, in being large nectar producers and large and stout. Since bats feed at night, the flowers are usually dull colored, open at night only, and have strong fermenting or fruity odors. Most of the bats that visit flowers are found in the Old and New World tropics.

In recent years research has shown that many kinds of animals are involved in cross-pollination. These include mosquitoes, midges, ants, flies, wasps, and small rodents. It should be pointed out that the animal does not intentionally set out to pollinate the flower, but instead does so in its normal feeding behavior. This is certainly an interesting coadaptation between plant and animal.

Wind-pollinated **(anemophilous)** plants are easily recognized from animal-pollinated plants. Wind pollination is very wasteful because large amounts of light, smooth-walled pollen is produced from very large exposed anthers. Unfortunately, most of the pollen ultimately lands on the ground or in the air, except where plants are close together. Most wind-pollinated plants have small, inconspicuous green or brown flowers, without petals; many have brush-like or feather-like stigmas or are monoecious or dioecious. Most wind-pollinated flowers have ovaries with only one ovule and single-seeded fruits. Some examples of wind-pollinated plants are grasses, cottonwoods, nettles, oaks, and ragweed. It is therefore not surprising that many people show signs of "hayfever" or allergic reactions to many of these plants. It might be interesting to note that in the tropical rain forest, where individuals of a species are widely separated, wind pollination is not very common.

### *Fertilization*

On the surface of the stigma is found a sticky, sugary fluid, which provides the ideal medium for the pollen to germinate. Through openings in the exine wall, called **pores,** the pollen tube emerges and penetrates the stigma surface and grows down through the style tissue. As mentioned earlier, the pollen grain has two haploid cells, the **generative cell** and the **tube cell.** The nucleus of the tube cell stays near the end of the growing pollen tube and appears to influence the tube's growth. The nucleus of the generative cell flows with cytoplasm into the tube, where, following a mitotic division, two **sperm** are produced. Therefore, when the pollen tube reaches the micropyle region of the ovule, the mature male gametophyte has three haploid nuclei.

In most species the embryo sac is reached through the side of the ovule. Passing through the nucellus, the tube discharges its contents into the embryo sac. The tube nucleus begins to disintegrate. One of the male gametes or sperm unites with the **egg cell** forming a diploid *(2n)* **zygote,** the first stage of sporophyte generation. The other sperm unites with the two **polar nuclei** in the middle of the embryo sac, generally resulting in a triploid *(3n)* **triple fusion nucleus** or **primary endosperm nucleus.** This process, which has involved two separate sperm in separate nuclear fusions, is called **double fertilization.** This feature is unique only to the flowering plants and is not present in the same way in any other group of plants.

The primary endosperm nucleus undergoes a series of mitotic divisions and ultimately forms **endosperm.** The endosperm can form in various ways, but its prime function is to provide nutrients for the developing **embryo** or young seedling following germination. This differs from what was observed in pinophytes, where the stored nutritive tissue was provided by the female gametophyte.

The early developmental stages of the Liliidae (monocots) and Magnoliidae (dicots) embryo are basically the same. The zygote undergoes a series of divisions, resulting in a large **basal cell,** a **suspensor,** which functions by pushing the spherical-shaped embryo into the endosperm tissue. The **embryo** soon loses its roundish shape, first becoming heart-shaped and then followed by the development of **cotyledons:** two in the Magnoliidae and one in the Liliidae. The embryo is composed of the **plumule,** which develops into the future shoot, and the **radicle,** the root-growing region, which is located at the basal end of the major embryo axis, called the **hypocotyl** (see Fig. 8.1).

The mature ovule is termed the **seed.** It has two basic parts, the **seed coat,** which developed from the integuments of the ovule, and the **embryo,** which developed from the fertilized egg. Often the embryo may be embedded in or associated with endosperm, which still remains following embryo development. The seed when mature often undergoes a period of dormancy before the embryo undergoes growth and **germination** occurs. The resulting new plant is called a **seedling.**

### *Fruits*

Seeds are born in mature ovaries called **fruits.** The mature ovary wall of the fruit is called the **pericarp,** and sometimes three layers can be distinguished: the inner, **endocarp;** the middle, **mesocarp;** and the outer, **exocarp.** The fruits may be **dry** or **fleshy.** Dry fruits may open up at maturity and are termed **dehiscent. Indehiscent** means that the fruit does not open to release the seeds. One of the segments of a dehiscent fruit after opening is a **valve,** and a **suture** is the line of dehiscence. A partition between the **locule** (cell) of the fruit or ovary is the **septum.** A **simple fruit** forms from an individual pistil, regardless of whether the pistil is simple or compound.

Some common types of fruits are listed below.

*Fleshy Fruits* The following terms are used for fleshy fruits (see Fig. 8.16).

**Accessory** A fruit made up of a succulent receptacle covered with several to many pistils, each forming a dry achene-like fruit (i.e., strawberry).

**Aggregate** A fruit with a receptacle that is not especially fleshy and has several to many pistils, these each becoming fleshy drupes (i.e., raspberry, blackberry).

**Berry** A fruit formed from one compound ovary, with few to many seeds (i.e., blueberry, currant, grape, tomato, kiwi).

**Drupe** A one-seeded indehiscent fruit with a stony endocarp (i.e., apricot, cherry, peach, plum).

**Hesperidium** A berry-like fruit covered with a thick leathery rind and with locules filled with fleshy hairs (i.e., lemons, oranges, limes).

**Hip** A cluster of achenes surrounded by a hypanthium or cup-shaped receptacle (i.e., rose).

**Multiple** A fruit made up of more than one flower, usually with superior ovaries (i.e., mulberry, pineapple, custard apple).

**Pepo** A fruit derived from a compound inferior ovary in which the outer wall (exocarp) becomes hard and tough (i.e., cucumber, gourd, watermelon).

**FIGURE 8.16 Types of fleshy fruits: (a) accessory; (b) accessory (longitudinal section); (c) aggregate; (d) aggregate (longitudinal section); (e) berry; (f) berry (cross section); (g) drupe; (h) drupe (longitudinal section); (i) hesperidium (cross section); (j) hip (longitudinal section); (k) multiple; (l) pepo; (m) pepo (cross section); (n) pome; (o) pome (longitudinal section); (p) synconium (longitudinal section).**

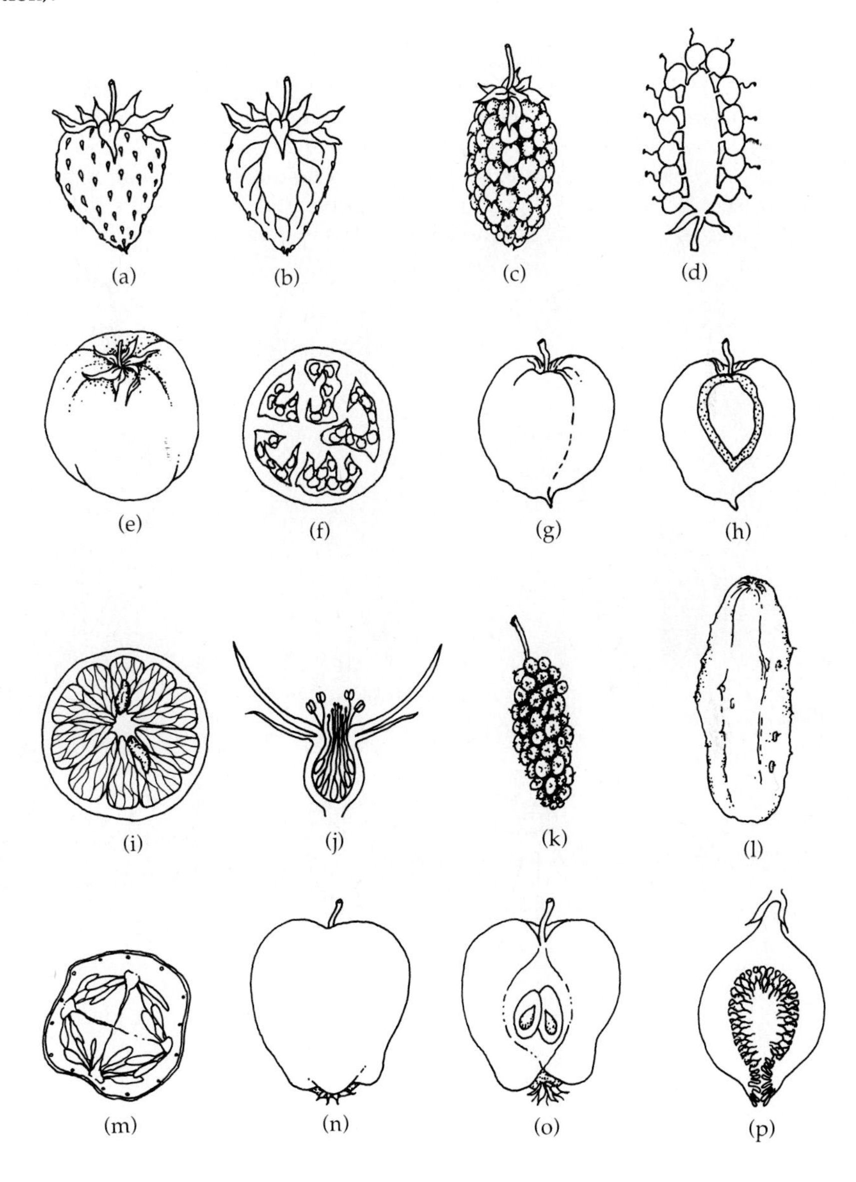

**Pome** A fruit formed from a compound inferior ovary, in which the receptacle (or calyx tube) becomes thick and fleshy (i.e., apple, pear).

**Synconium** A hollow receptacle or peduncle that houses many small achenes inside; a type of multiple fruit (i.e., fig).

**FIGURE 8.17 Types of dry fruits: (a) achene (longitudinal section and side view); (b) acorn; (c) caryopsis (grain) (front view and longitudinal section); (d) nut (with part of wall removed); (e) samara; (f) schizocarp (cross section and dehisced fruit); (g) circumscissile capsule; (h) loculicidal capsule; (i) poricidal capsule; (j) septicidal capsule; (k) follicle; (l) legume; (m) loment; (n) silicle (dorsal view and cross section); (o) silique (side view and cross section).**

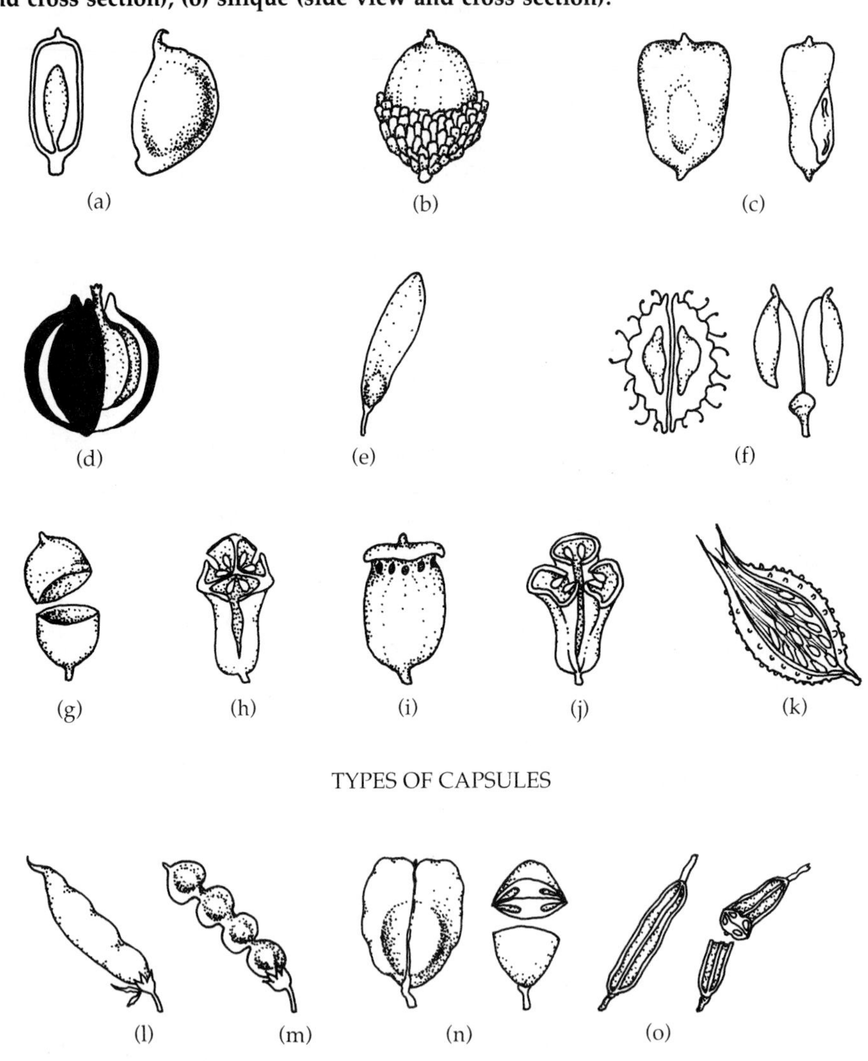

*Dry Indehiscent Fruits* The following terms are used for dry indehiscent fruits (see Fig. 8.17).

**Achene** A one-seeded fruit with the seed connected to the pericarp at only one point (i.e., sunflower seed).
**Acorn** A one-seeded fruit with a hard coat and surrounded by a "cap" of dried bracts (i.e., oak).
**Caryopsis** (or **grain**) A one-seeded fruit with the seed connected to the pericarp by all sides (i.e., maize, rye, wheat).
**Nut** A loosely used term applied to a one-seeded fruit with a hard coat (i.e., hickory, hazelnut (filbert), pecan).
**Samara** A winged achene (i.e., ash).
**Schizocarp** A fruit made up of two or more one-seeded carpels that separate from each other, leaving a connection between them (i.e., parsley).

*Dry Dehiscent Fruits* The following terms are used for dry dehiscent fruits (see Fig. 8.17).

**Capsule** A more than one-carpeled fruit with two or more placentae. There are several types of capsules: **circumscissile,** opens by a lid formed along a horizontal circular suture (i.e., plantain); **loculicidal,** splits open along the middle of the locule (i.e., iris); **poricidal,** opens by pores at the top (i.e., poppy); and **septicidal,** splits open along the septa splitting it in half (i.e., agave).
**Follicle** A one-celled, one carpellate fruit splitting down only one side (i.e., milkweed, dogbane).
**Legume** A one-celled, one-carpellate fruit splitting down two sides; loosely called a **pod** (i.e., bean, pea).
**Loment** A legume constricted between the seeds so that the fruit falls apart in one-seeded segments (i.e., *Desmodium).*
**Silicle** A short, two-locular fruit splitting with each half (valve) separating from one another, leaving a thin septum remaining (i.e., *Capsella, Lepidium*).
**Silique** A long (length more than twice the width) two-locular fruit splitting with each half (valve) separating from one another, leaving a thin septum remaining (i.e., *Brassica,* mustard).

## SELECTED REFERENCES

Bedevian, A. K. 1936. *Illustrated Polyglottic Dictionary of Plant Names.* Argus & Papazian Presses, Cairo, part I, 644 pp; part II, 454 pp.
Borror, D. J. 1960. *Dictionary of Word Roots and Combining Forms.* Mayfield, Palo Alto, CA.
Cronquist, A. 1981. *An Integrated System of Classification of Flowering Plants.* Columbia Univ. Press, New York, 1262 pp.
Cronquist, A., A. Takhtajan, and W. Zimmerman. 1966. *On the Higher Taxa of Embryobionta. Taxon* 15:129–134.
Dayton, W. A. 1950. *Glossary of Botanical Terms Commonly Used in Range Research.* (rev. ed.) U.S.D.A. Misc. Publ. No. 110. Supt. of Doc., Washington D.C., 41 pp.

Featherly H. I. 1954. *Taxonomic Terminology of Higher Plants.* (Reprinted 1965. Hafner, New York) Iowa State College Press, Ames, IA, 166 pp.

Ferri, M. G., N. L. de Menezes, and W. R. Monteiro-Scanavacca. 1969. *Glossario de Termos Botanicos.* Editora Edgard Blucher, Sao Paulo, Brasil, 199 pp.

Gledhill, D. 1985. *The Names of Plants.* Cambridge Univ. Press, Cambridge, 159 pp.

Harrington, H. D. 1957. *How to Identify Plants.* The Swallow Press, Chicago, IL, 203 pp.

Henderson, I. F. and W. D. Henderson. 1957. *A Dictionary of Scientific Terms,* 6th ed. (rev. by J. H. Kenneth) Oliver and Boyd, Edinburgh.

Horticultural Colour Chart. Vol. I 1938, Vol. II 1941. The British Colour Council and the Royal Horticultural Society, Henry Stone and Son, Banbury, England.

Instituta Botanica Academea Senica (ed.) 1979. *Advanced Chinese Botany—A Simplified Dictionary* (in Chinese). Science Publ. House, Beijing, China, 733 pp.

Jackson, B. D. 1971. *A Glossary of Botanic Terms.* (Reprint 1928, 4th ed.) Duckworth & Co., London, 481 pp.

Jaeger, E. C. 1950. *A Source-book of Biological Names and Terms,* 2nd ed. Charles C. Thomas, Springfield, IL.

Lecoq, H. and J. Juillet. 1831. *Dictionnaire Raisonne des Termes de Botanique et des Familes Naturelles.* Clermont-Ferrand, Imprimerie de Thibaud-Landriot, Paris, 719 pp.

Lindley, J. 1848. *A Glossary of Technical Terms Used in Botany.* Bradbury and Evans, London, 100 pp.

Lindley, J. 1848. *Illustrated Dictionary of Botanical Terms.* (Reprinted 1964) School of Earth Sciences, Stanford Univ., Stanford, CA.

Little, R. J. and C. E. Jones. 1980. *A Dictionary of Botany.* Van Nostrand Reinhold, New York, 400 pp.

Lloyd, G. N. 1826. *Botanical Terminology, or Dictionary Explaining the Terms Most Generally Employed in Systematic Botany.* Bell & Bradfute, Edinburgh, 228 pp.

Moreno, N. P. 1982. *Glosario de Terminos Botanicos para Plantas Superiores.* Instituto Nacional de Investigaciones Sobre Recursos Bioticos, Xalapa, Veracruz, Mexico, 158 pp.

Nayar, M. P. 1985. *Meaning of Indian Flowering Plant Names.* Bishen Singh Mahendra Pal Singh, Dehra Dun, India, 409 pp.

Plowden, C. C. 1968. *A Manual of Plant Names.* George Allen and Unwin, London, 260 pp.

Radford, A. E., W. C. Dickison, J. R. Massey, and C. R. Bell. 1974. *Vascular Plant Systematics.* Harper & Row, New York, 891 pp.

Sanchez-Monge, E. 1980. *Diccionario de Plantas Agricolas.* Premio Nacional de Publicaciones, Madrid, 466 pp.

Schubert, R. and G. Wagner. 1971. *Pflanzennamen und Botanische Fachworter.* Neumann Verlag, Leipzig, 428 pp.

Sporn, K. R. 1974. *The Morphology of Angiosperms.* Hutchinson & Co., London, 207 pp.

Swartz, D. 1971. *Collegiate Dictionary of Botany.* Ronald Press, New York, 520 pp.

Tierno, J. C. 1958. *Dictionario Botanico.* Joao Francisco Lopes, Lisboa, 1299 pp.

Usher, G. 1966. *A Dictionary of Botany.* Constable, London, 408 pp.

Van Wijk, H. L. G. 1911. *A Dictionary of Plant Names.* Vols. I–IV. Dutch Soc. of Sci. at Haarlem, The Hague, I–II, 1444 pp; III–IV, 1696 pp.

Walters, D. R. and K. J. Keill. 1988. *Vascular Plant Taxonomy,* 3rd ed. Kendall/Hunt, Dubuque, IA, 488 pp.

Woods, R. S. 1944. *The Naturalist's Lexicon.* Abby Garden Press, Pasadena, CA.

# 9

# *Families of Flowering Plants I. Magnoliopsida (Dicots)*

Within recent years there has been much discussion as to the best way to classify the biological world. The flowering plants have been scrutinized in great detail as new paleobotanical and experimental data have been published. A number of individuals have proposed various classifications, with four of them, A. Cronquist (New York Botanical Garden, U.S.A.), A. Takhtajan (Botanical Institute of the Academy of Sciences of the U.S.S.R.), R. Thorne (Rancho Santa Ana Botanical Garden, U.S.A.), and the late R. Dahlgren (Copenhagen) proposing systems that are relatively similar and comparable. The late J. Hutchinson's (Royal Botanic Garden, Kew, U.K.) classification is significantly different from the other four and has received less acceptance by botanists. In the following arrangement, the extensive works by Cronquist (1981, 1988) will be followed for the most part. This classification system is chosen primarily because students will have a detailed reference to refer to for further information. It also provides an organized sequence to follow, whether or not one agrees with the classification. Within each order, the families follow Cronquist (1981, 1988), with their position based on the most recent evidence, though speculative in many instances.

## DIVISION MAGNOLIOPHYTA (ANGIOSPERMS)

The division Magnoliophyta (class Angiospermae of older classifications) is comprised of 83 orders, 383 families, and approximately 215,000 species. It consists of two unequal-sized classes: the Magnoliopsida (dicotyledons) made up of 64 orders, 318 families, and 165,000 species and the Liliopsida (monocotyledons) with 19 orders, 65 families, and about 50,000 species.

## Class Magnoliopsida (Dicotyledons)

The Magnoliopsida are herbaceous or woody plants, with the woody forms and many of the herbaceous forms with secondary tissues derived from a vascular cambium. The vascular bundles in most herbaceous species are in a generalized ring around a central pith. The leaves are generally reticulate veined, commonly with a petiole and expanded blade. In young plants taproots are most common, rarely fibrous roots. The flower's perianth parts are generally in sets of fours, fives, or multiples of these. The pollen grains are most often with 3 pores (triaperturate) or derived therefrom, except some 1-pore (uniaperturate) ones. The cotyledons are 2 in number (rarely 1, 3 or 5).

**Subclass I. Magnoliidae** The Magnoliidae consist of 8 orders, 39 families, and approximately 12,000 species, with three of the orders—Laurales, Magnoliales, and Ranunculales—having over two thirds of the species. (See Table 9.1.) There are no exclusive characters setting this group apart. The flowers typically are showy, polypetalous with well-developed perianths. The pollen has 1 or 3 pores (uniaperturate or triaperturate), and is produced by numerous spirally arranged stamens. The pistil is apocarpous. Some members have a unique isoquinoline alkaloid and related compounds or unique volatile oils found in special cells. The delineation of some families is debatable.

**TABLE 9.1 A list of orders and family names of flowering plants found in the subclass Magnoliidae.**

Division Magnoliophyta (Flowering Plants)
Class Magnoliopsida (Dicots)

| | |
|---|---|
| Subclass I. Magnoliidae | |
| Order A. Magnoliales | Order D. Aristolochiales |
| Family 1. Winteraceae* | Family 1. Aristolochiaceae* |
| 2. Degeneriaceae | Order E. Illiciales |
| 3. Himantandraceae | Family 1. Illiciaceae* |
| 4. Eupomatiaceae | 2. Schisandraceae |
| 5. Austrobaileyaceae | Order F. Nymphaeales |
| 6. Magnoliaceae* | Family 1. Nelumbonaceae* |
| 7. Lactoridaceae | 2. Nymphaeaceae* |
| 8. Annonaceae* | 3. Barclayaceae |
| 9. Myristicaceae* | 4. Cabombaceae |
| 10. Canellaceae | 5. Ceratophyllaceae* |
| Order B. Laurales | Order G. Ranunculales |
| Family 1. Amborellaceae | Family 1. Ranunculaceae* |
| 2. Trimeniaceae | 2. Circaeasteraceae |
| 3. Monimiaceae | 3. Berberidaceae* |
| 4. Gomortegaceae | 4. Sargentodoxaceae |
| 5. Calycanthaceae* | 5. Lardizabalaceae |
| 6. Idiospermaceae | 6. Menispermaceae* |
| 7. Lauraceae* | 7. Coriariaceae |
| 8. Hernanidiaceae | 8. Sabiaceae |
| Order C. Piperales | Order H. Papaverales |
| Family 1. Chloranthaceae | Family 1. Papaveraceae* |
| 2. Saururaceae* | 2. Fumariaceae* |
| 3. Piperaceae* | |

*NOTE:* Family names followed by an asterisk are discussed and illustrated in the text.

## ORDER MAGNOLIALES
### Family Winteraceae (Wintera)

Figure 9.1 Family Winteraceae. *Drimys:* (a) leafy branch with flowers; (b) flower lacking petals and stamens and showing spirally arranged stamen scars and distinct carpels; (c) individual stamen; (d) front view of flower. (*SOURCE:* Redrawn with permission from Cronquist, 1981.)

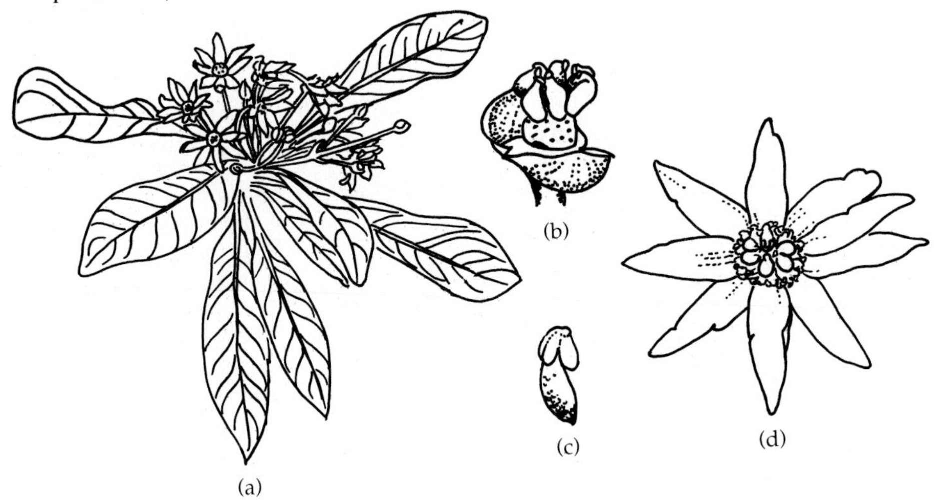


Shrubs or trees, aromatic, glabrous ethereal oil cells present in the parenchyma tissue, but calcium oxalate crystals rarely present. Leaves alternate, simple, entire; dotted with translucent glands and an unorganized venation; stipules lacking. Flowers small, regular to slightly irregular, hypogynous, perfect and unisexual; inflorescence of solitary flowers or flowers borne in cymes. Sepals 2–6, distinct or connate at the base. Petals 5–many, generally in two whorls, distinct. Stamens many, filaments strap-shaped; insect or wind pollinated. Pistil simple of few to numerous individual carpels, generally in only one whorl; distinct but may be slightly connate and unsealed in fruit; ovary superior. Fruit berry-like or a follicle. Seeds small, oily, endosperm present (Fig. 9.1).

The family Winteraceae consists of 9 genera and 100 species distributed in the montane subtropics and tropics of Mexico, Central and South America, most diverse in southeastern Australiasia, and absent from Africa, except Madagascar. The largest genera are *Tasmannia* (40 species) and *Bubbia* (30 species).

Economically the group is of very little value, except *Drimys winteri,* winter's bark from South America, which is used as a tonic locally.

Along with the Magnoliaceae, the Winteraceae is considered by most modern classifications to be one of the oldest known flowering plant families. Pollen remains attributed to the Winteraceae come from the Upper Cretaceous deposits, with other plant parts from Oligocene formations.

## Family Magnoliaceae (Magnolia)

Figure 9.2 Family Magnoliaceae. *Liriodendron:* (a) flower and leaves; (b) winter twig. *Magnolia:* (c) flower and leaves; (d) follicular fruits with suspended seeds.

Shrubs or trees. Leaves deciduous or evergreen, alternate, simple; stipules large deciduous, enclosing the terminal bud, but often forming an ocrea. Flowers large and showy, regular, perfect, hypogynous; inflorescence of a solitary flower; insect pollinated. Perianth free and not always differentiated; sepals generally three; petals 6–many. Stamens many, filaments distinct and spirally arranged on an elongate receptacle. Pistils simple of many individual carpels (rarely 1–3) inserted spirally on the receptacle; locule usually 1; ovules 1–5 and borne on parietal placentas; ovary superior; style 1; stigma terminal. Fruit a follicle, samara, or rarely a berry. Seed usually large, suspended on an elongate funiculus in the dehiscent fruits; embryo with prominent suspensor, oily endosperm (Fig. 9.2).

The family Magnoliaceae consists of 12 genera and about 220 species, widespread in warm temperate regions of the world, with diversity centers in the southeast United States and eastern Asia. The largest genus is *Magnolia,* the magnolias (80 species).

Economically the family is of some importance. *Magnolia,* magnolia; cucumber tree species are planted as ornamentals. *Liriodendron tulipifera,* tulip tree, is a valuable lumber tree in the eastern United States.

The Magnoliaceae is considered by most modern classifications to be one of the oldest known flowering plant families.

Fossil wood considered to be Magnoliaceae is known from Upper Cretaceous and Tertiary deposits.

## Family Annonaceae (Custard Apple)

Figure 9.3 Family Annonaceae. *Asimina:* (a) winter twig with fuzzy buds; (b) front view of flower; (c) banana-like fruit; (d) obovate-shaped leaf. (*SOURCE:* Redrawn with permission from B. V. Barnes and W. H. Wagner, Jr., *Michigan Trees*. Ann Arbor, MI: University of Michigan Press, 1981.

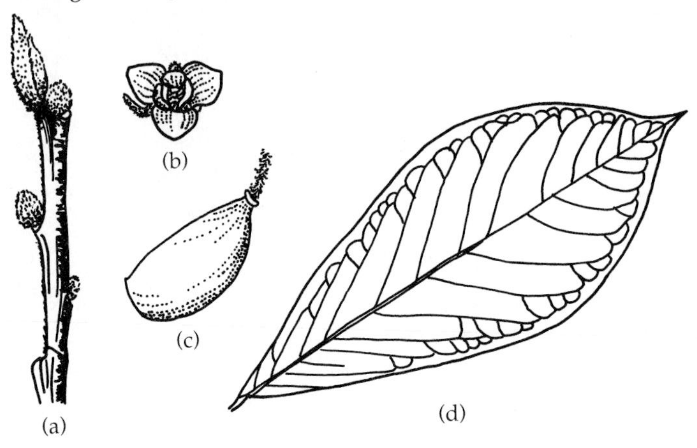


Shrubs, trees, or lianas; aromatic, scattered round ethereal oil cells in the parenchyma tissue. Leaves alternate, simple, entire; stipules lacking. Flowers regular, perfect (rarely unisexual); hypogynous inflorescence of solitary flowers or flowers borne in cymes. Sepals 3 (rarely 2 or 4). Petals often 6 in 2 rows (rarely 3 or 4), imbricate. Stamens many and spirally arranged, filaments generally distinct, may appear as peltate-shaped; anthers opening by longitudinal slits. Pistil simple of 1–many carpels attached to a cone-shaped receptacle; locules 1; ovules 1–many and borne on parietal placentas; ovary superior; styles 3, short and thick; stigmas sessile. Fruit dry or with fleshy berry-like carpels or fleshy carpels fused to form an aggregate. Seeds small, embryo small, endosperm present, often aril-like (Fig. 9.3).

The Annonaceae is the largest family in the Magnoliales and consists of 130 genera and 2300 species distributed in subtropical to tropical regions; an exception to this is *Asimina*, pawpaw, found in the eastern United States. The largest genera are *Guatteria* (250 species), *Uvaria* (175 species), *Xylopia* (160 species), *Polyalthia* (150 species), and *Annona* (120 species).

Economically the family is of minor importance. Several species of *Annona* are cultivated for their edible fruit: *A. muricata*, sour sop; *A. squamosa*, sweet sop; and *A. reticulata*, custard apple.

The family is fairly distinctive and is considered by taxonomists to be part of the group of woody families having many spirally arranged stamens and floral parts, here listed under the order Magnoliales.

Fossils attributed to the family are known from Eocene deposits with pollen records from the Oligocene.

## Family Myristicaceae (Nutmeg)

Figure 9.4 Family Myristicaceae. *Myristica:* (a) leafy branch with flowers and mature fruit; (b) longitudinal section of male flower; (c) longitudinal section of fruit; (d) longitudinal section of female flower showing 3-lobed calyx.

Trees (rarely shrubs), very aromatic. Leaves alternate, simple, entire; ethereal oil cells and calcium oxalate crystals commonly present in the parenchyma tissue; stipules lacking. Flowers small, apetalous, hypogynous, unisexual and commonly dioecious; inflorescence of axillary or terminal cymes or racemes. Calyx cup-shaped, 3-lobed. Petals absent. Stamens of male flower 2–many, filaments united into a column, rudimentary pistil absent; anthers opening by 1 longitudinal slit. Female flower lacks staminodes, with a single unsealed carpel; locule 1; ovule 1 and borne on a basal placenta; ovary superior; style long, stigma various. Fruit more or less a fleshy berry dehiscing along two sutures. Seed often with an aril; embryo small, endosperm present (Fig. 9.4).

The family Myristicaceae consists of 15 genera and over 300 species with a widespread distribution in the lowland tropics, especially the Amazonian basin and New Guinea. The largest genus is *Myristica* (100 species).

Economically the family is of some importance. Nutmeg is obtained from *Myristica fragrans,* hallucinogenic snuff from the bark of *Virola* in Amazonia, and timber for plywood from *V. surinamensis.*

The family has many features that are similar to other families of the order Magnoliales but is distinct enough to be treated as a family.

The family has no known fossil record.

## ORDER LAURALES
## Family Calycanthaceae (Strawberry Shrub)

Figure 9.5 Family Calycanthaceae. *Calycanthus:* (a) branch with flowers and opposite leaves; (b) front view of flower; (c) stamen; (d) longitudinal section of flower showing cup-shaped hypanthium. (*SOURCE:* Redrawn with permission from Cronquist, 1981.)

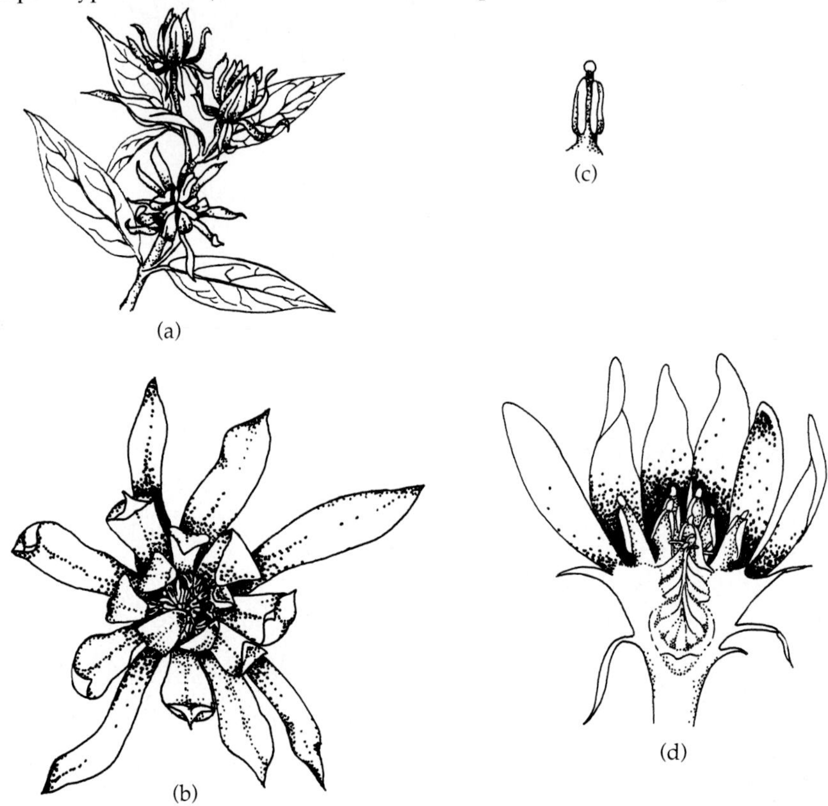


Shrubs, very aromatic bark with ethereal oil glands in the parenchymous tissue, cells around epidermal hairs with silica in walls. Leaves simple, opposite, stipules lacking. Flowers relatively large, solitary, fragrant, perigynous with a cup-shaped hypanthium. Perianth parts many, distinct, attached to the outside of the hypanthium. Stamens 5–many, filaments distinct, adnate to the top of the receptacle, the inner ones sterile; anthers opening by longitudinal slits. Pistil simple of 5–many individual spirally arranged carpels attached within the hypanthium; ovules 2; ovary superior. Fruit of many achenes attached inside the fleshy, oily hypanthium. Seed with large embryo and cotyledons; endosperm lacking, poisonous (Fig. 9.5).

The family Calycanthaceae consists of 3 genera and 5–8 species, confined mostly to North America and China.

Economically the family is useful as aromatic cultivated shrubs and for *Calycanthus floridus*, Carolina allspice. Extracts of *C. fertilis* from eastern United States are used for medicinal purposes.

In older classifications the family was placed in the order Rosales. It does have some features similar to the families Annonaceae or Magnoliaceae, in having the perianth parts spirally arranged and many stamens and individual carpels.

The family has no known fossil record.

## Family Lauraceae (Laurel)

Figure 9.6 Family Lauraceae. *Sassafras:* (a) male flower; (b) examples of leaf lobing; (c) female flower; (d) twig with female flowers and young leaves.

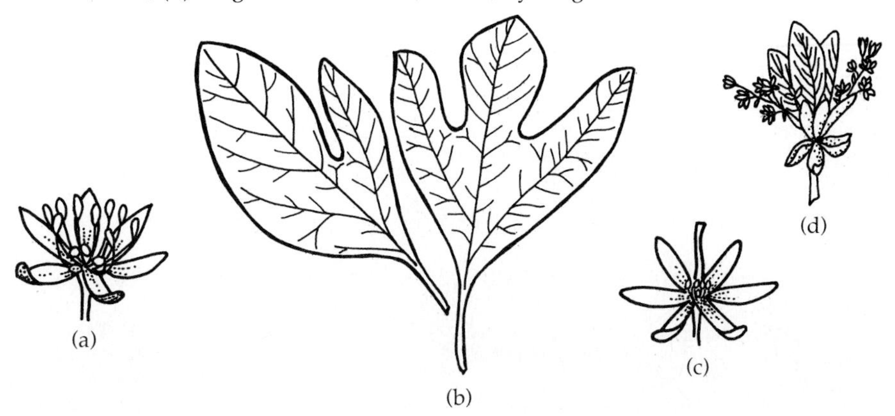


Shrubs or trees (rarely parasitic herb); very aromatic. Leaves evergreen or less commonly deciduous, alternate, simple (rarely opposite, whorled scale-like, or lacking); entire or lobed with various venation; stipules lacking. Flowers small, yellow or greenish, regular, perfect or unisexual, perigynous or rarely epigynous; inflorescence of cymes, heads, panicles, racemes, or spikes. Perianth usually 6 in one whorl or in 2 whorls of 3. Stamens 3–12, filaments distinct, free or adnate to the perianth forming a hypanthium, the inner ones commonly as staminodes; anthers opening by slits from the base of the anther upward by flaps of tissue. Pistil simple of a single carpel, locule 1, ovule 1 and borne on an apical placenta; ovary superior or inferior; style 1, stigma capitate or lobed. Fruit a berry or drupe. Seed with a large oily embryo, endosperm lacking (Fig. 9.6).

The family Lauraceae consists of 30–50 genera and over 2000 species distributed widely in subtropical and tropical regions, with its greatest diversity found in Southeast Asia and Brazil. The largest genera are *Ocotea* (400 species) and *Litsea* (250 species).

Economically the family has many useful trees. These include the bark of *Cinnamomum zeylanicum,* cinnamon; the leaves of *C. camphora,* camphor, and *Laurus nobilis,* laurel; the fruit of *Persea americana,* avocado; and oil from *Sassafras albidum,* sassafras. The latter has been shown to be carcinogenic.

The family is fairly distinctive from other families in the manner in which the anthers open and its single carpel.

Fossil wood attributed to the Lauraceae has been found in Upper Cretaceous and Eocene deposits in California and in Yellowstone Park, Wyoming.

## ORDER PIPERALES
## Family Saururaceae (Lizard Tail)

Figure 9.7 Family Saururaceae. *Saururus:* (a) plant with characteristic leaves and inflorescence; (b) flower. (*SOURCE:* Redrawn with permission from Cronquist, 1981.)

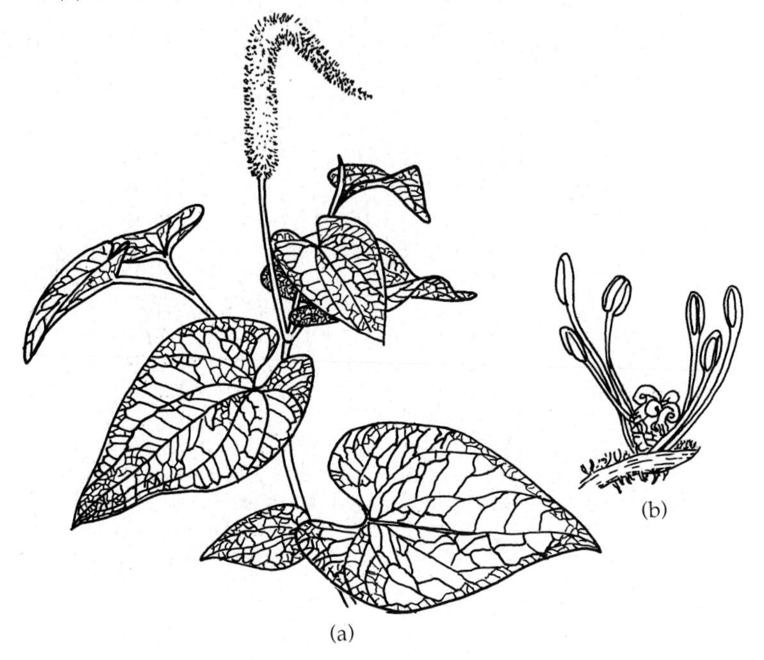


Herbs with spherical ethereal cells present in the parenchymous tissue, perennial, aromatic. Leaves alternate; simple stipules adnate to the petiole. Flowers small, naked, perfect, hypogynous or epigynous; inflorescence in dense terminal racemes or spikes. Perianth lacking. Stamens 3, 6, or 8 in two alternating whorls, filaments distinct, long, and free or adnate to the ovary; anthers large, opening by longitudinal slits. Pistil simple or compound of 3–5 carpels, distinct above the base or united to be compound; locules 1; ovules 2–10 and borne on parietal placentas; ovaries superior or inferior; style 1, stigma decurrent. Fruit somewhat fleshy or a capsule. Seeds with small embryo, little endosperm (Fig. 9.7).

The family Saururaceae consists of 5 genera and 7 species distributed on both coasts of North America and in eastern Asia. No genus has more than 2 species.

Economically this small family is of little value but is used as a garden ornamental, a ground cover *(Houttuynia cordata),* and in various folk medicines.

In many respects the family is similar to the Piperaceae.

The family has no known fossil record.

## Family Piperaceae (Pepper)

Figure 9.8 Family Piperaceae. *Peperomia:* (a) leafy plant with terminal spikes; (b) flower with its bract. *Piper:* (c) leafy branch with spikes; (d) segment of spike showing flower buds; (e) flower comprised of single pistil with tufted stigma, two stamens, and a peltate (umbrella-shaped) fleshy bract. (*SOURCE:* Part (b) redrawn with permission from Cronquist, 1981; parts (c), (d), and (e) adapted from Degener, 1946.)

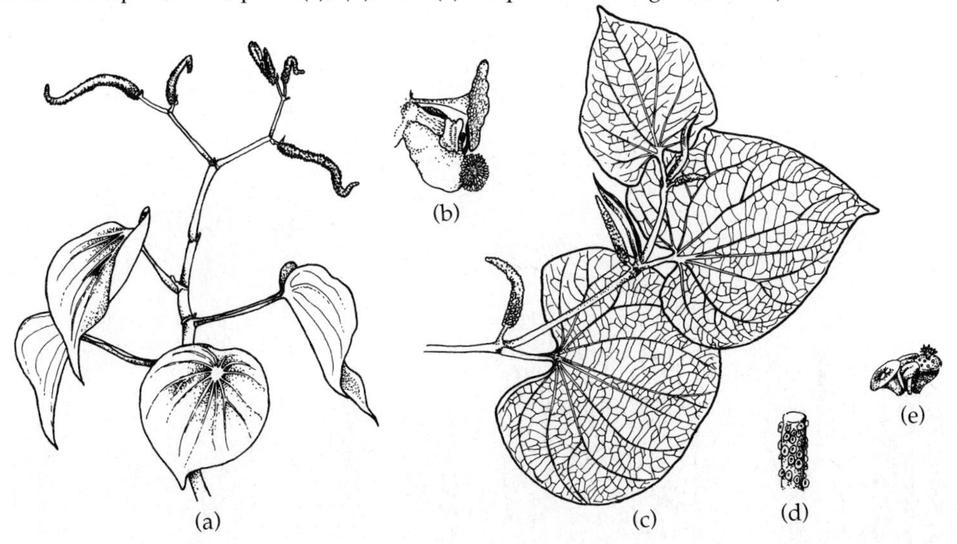


Herbs, shrubs, trees, vines, or epiphytes; aromatic ball-shaped ethereal oil cells generally present in the parenchyma tissue; vascular bundles scattered (like monocots) or in two rings. Leaves alternate, opposite or whorled; simple with adnate stipules to the petiole or lacking. Flowers small, unisexual or perfect, each with a peltate bract; inflorescence in dense fleshy spikes. Perianth lacking. Stamens 1–10, filaments distinct; anthers opening by longitudinal slits. Pistil compound of 1 or 5 united carpels; locule 1; ovule 1, attached on a basal placenta; ovary superior; stigma 1–4, tufted and brush-like. Fruit fleshy and drupe-like. Seed small; embryo small, endosperm scanty (Fig. 9.8).

The family Piperaceae is large and consists of about 10 genera and 1500–2000 species distributed widely in shady, mesic habitats in New and Old World Tropics. Most of the species belong to 2 genera, *Peperomia* and *Piper*, each with about 1000 species.

Economically, the family has some value. *Piper nigrum*, pepper, a condiment, is the most used species, while other *Piper* species are important in local medicinal practice. Some species of *Peperomia* are cultivated as house plants and eaten uncooked *(P. vividispica)* in Central and South America.

Anatomically the family is interesting because the vascular bundles are commonly scattered, as in "monocots." Some botanists place the genus *Peperomia* in its own family, Peperomiaceae.

The family has no fossil record.

## ORDER ARISTOLOCHIALES
### Family Aristolochiaceae (Birthwort)

Figure 9.9 Family Aristolochiaceae. *Aristolochia:* (a) vine segment showing heart-shaped leaves and flowers with the showy calyx; (b) flower with showy calyx; (c) longitudinal section through flower. (*SOURCE:* Redrawn with permission from Cronquist, 1981.)

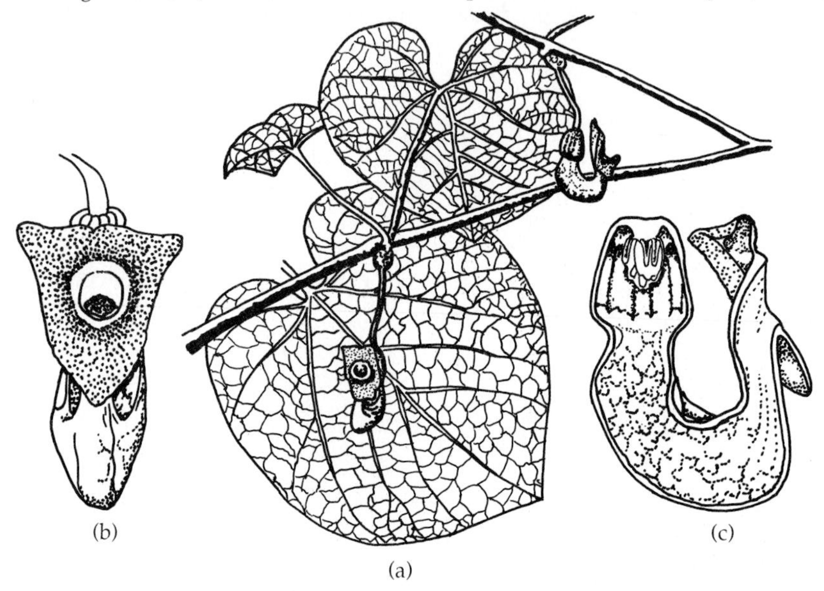


Herbs, shrubs, or most commonly woody lianas; aromatic with spherical oil cells and silica bodies commonly present in the parenchyma tissues along with aristolochic acid, a yellow, bitter nitrogenous alkaloid. Leaves alternate, simple, entire (rarely lobed), sometimes punctuate; stipules lacking. Flowers regular or irregular, perfect, epigynous or perigynous, often with a rotting meat smell; inflorescence of a solitary flower or flowers in cymes or racemes. Calyx well developed, fused and tubular, sometimes corolla-like. Corolla lacking or greatly reduced. Stamens 4–many, filaments distinct or adnate to the style; nectaries present. Pistil compound of 3–6 united carpels, locules 3–6; ovules many and borne on axile or parietal placentas; style 1, stigmas 3–6. Fruit usually a capsule. Seeds with small embryos, endosperm oily (Fig. 9.9).

The family Aristolochiaceae consists of about 10 genera and 600 species distributed in the tropical regions with a few species being temperate. The largest genera are *Aristolochia* (about 300 species) and *Asarum* (70 species).

Economically the group is of little value except for cultivated "porch vines."

How the family Aristolochiaceae should be classified is debated among botanists. Characters of the pollen and the kinds of cells containing oils are similar to families in the Magnoliales-type families. Other characters are sufficient to classify the family in its own order.

The family has no known fossil record.

## ORDER ILLICIALES
### Family Illiciaceae (Star Anise)

Figure 9.10 Family Illiciaceae. *Illicium:* (a) leafy branch with a single flower; (b) stamen; (c) flower with perianth removed, showing individual stamens and pistils; (d) front view of flower. (*SOURCE:* Redrawn with permission from Cronquist, 1981.)

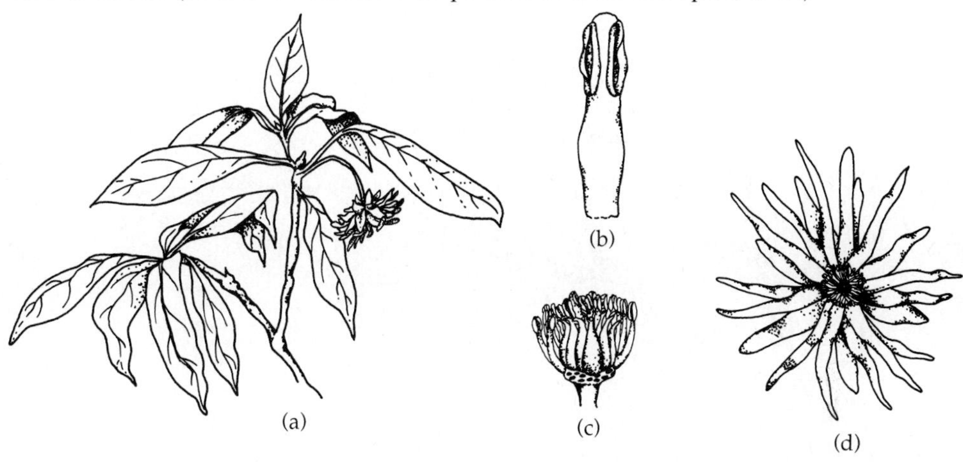


Shrubs or small trees with scattered ethereal and mucilage cells, very aromatic. Leaves evergreen, alternate, simple, entire, glabrous; stipules lacking. Flowers small, generally single or rarely 2–3 together, axillary, regular, bisexual and hypogynous. Perianth of 7–many undifferentiated tepals, arranged in several whorls. Stamens 4–many, spirally arranged; filaments short and thick, distinct; anthers basifixed, opening by longitudinal slits. Pistils simple of 5–20 individual carpels in a whorl; locule 1; ovule 1 and borne on a basal placenta; ovary superior. Fruit a follicle. Seeds shiny; embryo small, endosperm abundant (Fig. 9.10).

The family Illiciaceae consists of only the genus *Illicium* with about 40 species distributed in the southeast United States, West Indies, and Veracruz, Mexico, China, Japan, and southeast Asia.

Economically the family is important for the aromatic oils. *Illicium anisatum,* Japanese star anise, and *Illicium verum,* star anise, are sources of anethole, which is used in dentifrices, flavorings, and perfumes. Oil from the bark of *I. parviflorum,* yellow star anise, is used in flavorings.

The family has been included in the past with the Magnoliaceae and Winteraceae.

Fossil pollen similar to that of *Illicium* has been found in Upper Cretaceous deposits.

## ORDER NYMPHAEALES
### Family Nelumbonaceae (Lotus Lily)

Figure 9.11 Family Nelumbonaceae. *Nelumbo:* (a) seeds imbedded in the fleshy receptacle; (b) peltate leaves; (c) flower.

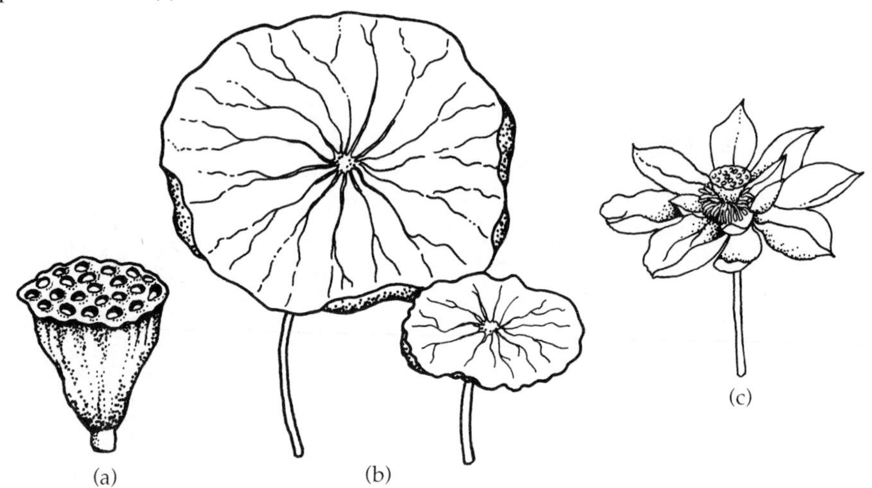


Herb, aquatic perennial from stout creeping rhizomes. Leaves usually with simple, elevated (floating) peltate blades on long petioles arising from the rhizome. Flowers large, showy, regular, hypogynous, perfect, solitary; completely rising above the water; insect pollinated. Sepals 2, distinct, green. Petals many, distinct in several whorls. Stamens many, distinct, spirally arranged. Pistil simple of many individual carpels embedded in a flat-topped receptacle; locule 1; ovule 1 and borne on an apical placenta, ovary superior; style lacking, stigma sessile. Fruit a nut that is loose in the cavity of the receptacle. Seed with a large embryo, very little endosperm present (Fig. 9.11).

The family Nelumbonaceae consists of only the genus *Nelumbo* and 2 species, *N. nucifera,* American lotus, in the eastern United States and *N. lutea,* Oriental lotus-lily of southeast Asia and Australia.

Economically the family is of minor importance. *Nelumbo lutea* is considered to have religious significance in the Orient. The seeds of *Nelumbo* are eaten by native peoples for food.

The seeds of *N. nucifera* are known to live over 3000 years in proper storage conditions.

Fossil pollen attributed to *Nelumbo* has been found in Eocene and more recent deposits.

## Family Nymphaeaceae (Water Lily)

Figure 9.12 Family Nymphaeaceae. *Brasenia:* (a) floating peltate leaves and flowers. *Nymphaea:* (b) floating cordate leaves with long petioles and flowers with long peduncles.

Herb, aquatic, perennial from a stout creeping rhizome or short caudex; vessels absent. Leaves alternate, simple, immersed or floating, cordate to orbicular with long petioles from the rhizome. Flowers large and showy, regular, perfect, hypogynous; solitary, rising to above the water, where insects are attracted by their fragrance. Sepals 4–many, distinct or connate, often petal-like. Petals 8–many, large, distinct or connate at the base, commonly passing into stamens. Stamens many, filaments flattened, usually distinct, spirally arranged, sometimes adnate to the perianth. Pistil compound of 3–many united carpels, pistil expanded at its apex into a disk with radiating stigmatic lines; locules 5–many; ovules 2–many and borne over the inner surface of the locule; ovary superior or inferior. Fruit spongy and berry-like. Seeds with a small embryo, endosperm present (Fig. 9.12).

The family Nymphaeaceae consists of 5 genera and 50 species distributed throughout the world.

Economically the family is used as cultivated "fish pond" plants, but is more important ecologically as a food for aquatic animals and birds, and in the early stages of aquatic lake succession. This vigorous early successional habit sometimes puts the plant in conflict with individuals wishing to have their ponds and lakes free of plants for fishing and recreational use. The South American *Victoria amazonica,* royal water lily, is known for its enormous floating leaves, which are known to support the weight of a small child. The seeds and rhizomes of *Nymphaea* and *Victoria* are sometimes eaten by native peoples.

The family is recognizable as a "dicot" but does have some characters similar to "monocots."

The family has no known fossil record.

## Family Ceratophyllaceae (Hornwort)

Figure 9.13 Family Ceratophyllaceae. *Ceratophyllum:* (a) stem section showing the dissected whorled leaves; (b) spiny fruit; (c) dichotomously branched leaves. (*SOURCE:* Redrawn with permission from Gleason, 1952.)

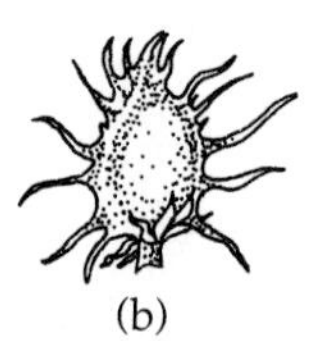


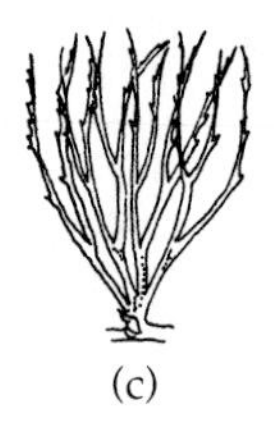


Herbs, rootless aquatic or submerged, weak stemmed. Leaves whorled, dichotomously dissected, usually serrulate; stipules lacking; typical stomata lacking. Flowers small, unisexual, hypogynous; the solitary flowers being a different sex at each node, with male flowers normally above the female. Sepals 8–15 in a whorl, connate at base. Petals absent. Stamens 10–many, filaments distinct and spirally arranged, anthers not easily distinguished from the short filaments, the connective between the anther sacs extended into short points. Pistil of a single carpel; locule 1; ovule 1 and borne on an apical placenta; ovary superior; style 1, stigma decurrent. Fruit an achene. Seed with a spine-like persistent style, lacking endosperm and perisperm (Fig. 9.13).

The family Ceratophyllaceae is comprised of only the genus *Ceratophyllum* and 6 species with a cosmopolitan distribution in fresh water.

The family is not economically important, but it can become a troublesome aquatic weed in freshwater lakes used for recreation.

The flowers of the Ceratophyllaceae are the most reduced and distinctive of all the flowers in the order Nymphaeales.

The family has no known fossil record.

## ORDER RANUNCULALES
### Family Ranunculaceae (Buttercup)

Figure 9.14 Family Ranunculaceae. *Ranunculus:* (a) plant flowering; (b) mature fruit (achene); (c) petal with nectary. *Aconitum:* (d) stem tip showing irregular flowers; (e) follicle fruits.

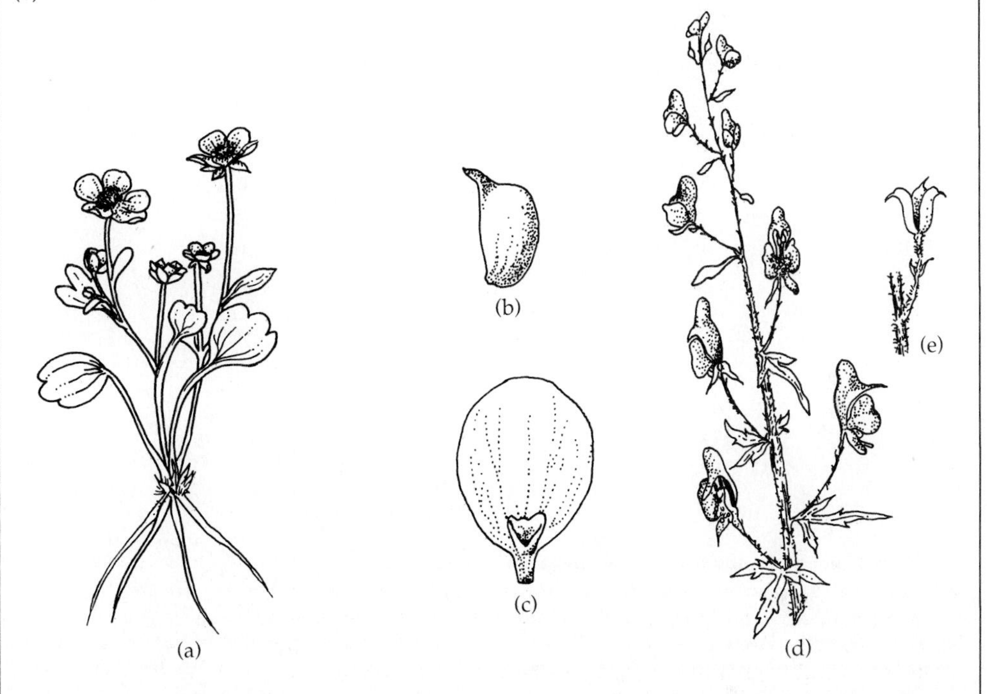


Herbs, soft woody shrubs or small trees (rarely vines). Leaves alternate (some opposite), simple or compound with netted venation; stipules lacking. Flowers typically regular or perfect (rarely irregular and unisexual), hypogynous; inflorescence of solitary flowers or flowers borne in cymes, panicles, or racemes. Perianth distinct, variable in number, attached to an elongate receptacle, petal-like and not differentiated into calyx or corolla, often with nectar glands present. Stamens many, filaments distinct, spirally arranged. Pistil simple of 3–many individual carpels, spirally arranged; locule 1; ovules 1–many and borne on parietal placentas, ovary superior; style 1, stigma 1- or 2-lobed. Fruit an achene, berry, follicle, or rarely a capsule. Seeds with small embryo, endosperm present (Fig. 9.14).

The family Ranunculaceae is relatively large and consists of about 50 genera and 2000 species distributed chiefly in cooler temperate regions of the Northern Hemisphere.

Economically the Ranunculaceae is of value as a source of many beautiful garden ornamentals, such as *Aquilegia,* columbine; *Anemone,* pasque flower, wind flower; *Aconitum,* monk's hood; *Caltha,* marsh marigold; *Clematis,* virgin's bower; *Delphinium,* larkspur; and *Ranunculus,* buttercup. Some members are important livestock poisoners, such as *Delphinium* and *Aconitum* in western North America.

The family is so large and diverse that at least 6 subfamilies have been recognized by some botanists. Others prefer to separate some genera into a number of smaller families.

The family has no known fossil record.

## Family Berberidaceae (Barberry)

Figure 9.15 Family Berberidaceae. *Podophyllum:* (a) leafy plant bearing flower. *Berberis:* (b) branch terminous with odd pinnately compound leaf and berry fruits. (*SOURCE:* Redrawn with permission from Hitchcock et al., 1963.)

(a)

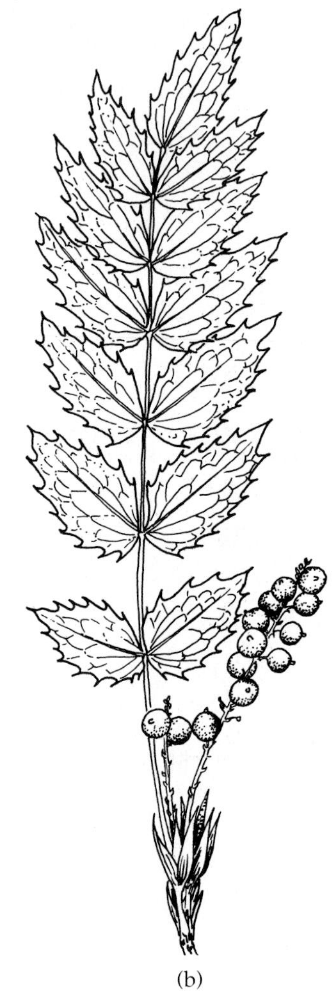

(b)

Herbs, shrubs, or small trees. Leaves alternate (rarely opposite or basal), deciduous or evergreen, simple or pinnately compound; generally lacking stipules. Flowers commonly small, regular, perfect, hypogynous; inflorescence of solitary flowers or the flowers borne in cymes, panicles, racemes, or spikes. Sepals 4–many, distinct. Petals 4–many (rarely lacking), sometimes in several whorls. Stamens 4–18 (mostly 6), filaments distinct; anthers opening by pores (rarely by slits). Pistil compound of 2–3 united carpels (appearing as 1 carpel); locule 1 (rarely 2); ovules 1–many and borne on a basal or parietal placenta; ovary superior; style 1, stigma sessile, sometimes 3-lobed. Fruit a berry. Seeds with small embryo and abundant endosperm (Fig. 9.15).

The family Berberidaceae consists of approximately 13 genera and 650 species widespread in temperate climates. The genera *Berberis,* barberry (about 500 species), and *Mahonia,* Oregon grape (100 species), make up most of the family's species. In eastern North American, *Achlys,* vanilla leaf, and *Caulophyllum,* blue cohosh, are common deciduous forest understory plants. *Berberis vulgaris,* common barberry, is an alternate host in the life cycle of wheat rust, *Puccinia graminis.*

Economically the family is of little value, except for cultivated ornamental shrubs.

*Podophyllum,* may apple of eastern North America, and the monotypic genus *Jeffersonia* (*J. diphylla*), twin leaf, are sometimes placed in their own family, Podophyllaceae, but the traditional classification is followed here.

The family has no known fossil record.

## Family Menispermaceae (Moonseed)

Figure 9.16 Family Menispermaceae. *Menispermum:* (a) twining vine with leaf and flowers; (b) male flower; (c) pistil with perianth removed showing three individual carpels; (d) female flower with staminodes. (*SOURCE:* Redrawn with permission from Cronquist, 1981.)

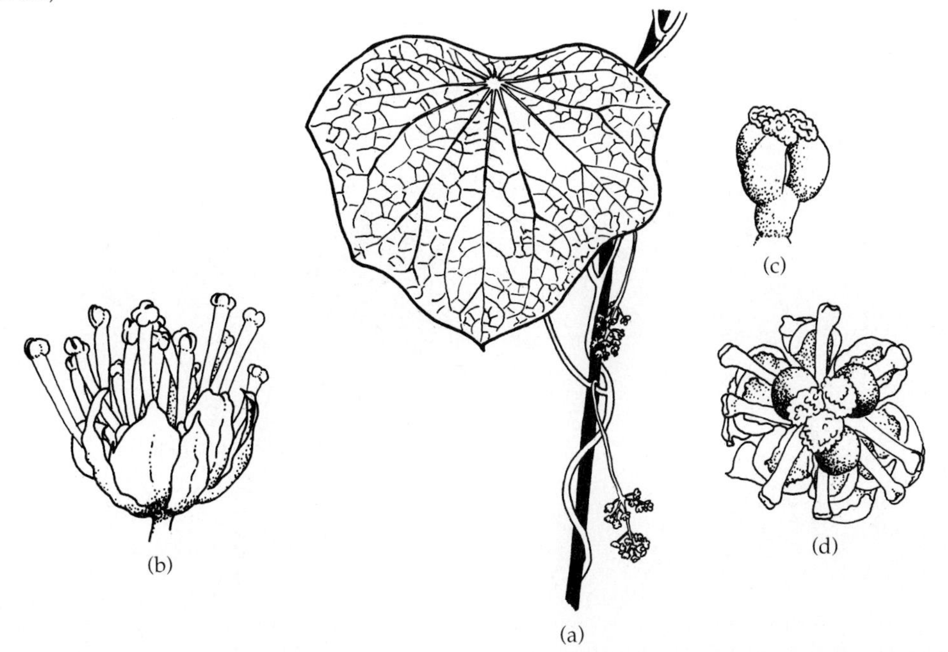


Herbs, shrubs, small trees, or vines, generally with poisonous bitter alkaloids. Leaves alternate, simple (rarely trifoliate) with palmate venation; stipules lacking. Flowers small, regular, unisexual (commonly dioecious), hypogynous; inflorescence of axillary cymes, panicles, or solitary clusters; nectaries lacking. Sepals usually 6 (less commonly 3 or more) in 2 whorls, distinct. Petals usually 6 (less commonly fewer or more or lacking), distinct. Staminate flowers generally with 6 stamens (as few as 3 or as many as 40), opposite the petals. Pistillate flowers usually with 3–6 separate carpels; locule 1; ovules 2, aborting to 1 and borne on parietal placenta; ovary superior; style 1 or lacking, stigma sessile. Fruit a drupe or achene with a hard, bony endocarp. Seeds and embryo usually curved or horseshoe-shaped, endosperm present or absent (Fig. 9.16).

The family Menispermaceae consists of about 70 genera and 400 species distributed mostly throughout the pantropical regions. A few genera are found in southern Europe and 4 species of 3 genera are found in temperate North America: *Calocarpum lyonii* reaches Illinois and Kansas, *Cocculus carolinus* extends to southern Indiana and Missouri, and *Menispermum canadense* goes as far north as Quebec and Manitoba in Canada. The largest genera are *Stephania* (40 species) and *Tinospora* (35 species).

Economically the family is of little importance. A few species are cultivated as ornamentals. *Chondrodendron tomentosum* is the source of curare (tubocurarine chloride), which is used as a muscle relaxant during neurosurgical procedures.

The classification of the Menispermaceae is presently debated by botanists and is unsettled. It is agreed, however, that the family belongs in the order Ranunculales.

Wood resembling *Cocculus* has been found in Upper Cretaceous deposits in California, while other fossils have been reported from Eocene and more recent deposits.

## ORDER PAPAVERALES
### Family Papaveraceae (Poppy)

Figure 9.17 Family Papaveraceae. *Papaver:* (a) flower showing the lobed stigmas of the pistil; (b) fruit (capsule). *Eschscholzia:* (c) mature plant bearing flower and fruit.

Herbs or soft-wooded shrubs, producing whitish or colored latex, often in laticifers; vascular bundles of stem in 1, 2, or more rings. Leaves generally alternate, entire to lobed to dissected; stipules lacking. Flowers generally large, regular, perfect, hypogynous (rarely pergynous); inflorescence of solitary flowers or flowers borne in cymes, panicles, or umbels. Sepals 2–4, distinct, enclosing the bud before flowering and deciduous when flowers open. Petals usually 4 or 6 (rarely 8, 12, 16, or lacking), distinct. Stamens numerous, distinct, often in multiples of 2 or 3, nectaries lacking; anthers opening by longitudinal slits. Pistil compound of 2 or more united carpels; usually unilocular; ovules many and borne on parietal placentas but sometimes with partitions that extend into the locule making the ovary axile; ovary superior; style 1–many or absent. Stigma 1 or sessile or lobed. Fruit a capsule, opening by pores or dehiscing longitudinally or splitting crosswise into individual nutlets. Seeds with small straight embryo, oily endosperm present (Fig. 9.17).

The family Papaveraceae consists of 25 genera and approximately 200 species distributed mostly in the subtropical and temperate Northern Hemisphere. The largest genus, *Papaver*, comprises about half the species, including *P. somniferum*, the opium poppy, which is a valuable economic plant legally for medicine and illegally for drugs. A number of species of the North American genus *Eschscholzia*, California poppy, are cultivated as ornamentals.

The family is sometimes classified in another order and associated with the families Brassicaceae and Capparidaceae.

The family Papaveraceae at present has no known fossil record.

## Family Fumariaceae (Fumitory)

Figure 9.18 Family Fumariaceae. *Corydalis:* (a) upper stem of plant bearing fruits; (b) flower. *Dicentra:* (c) mature plant bearing flowers; (d) fruits; (e) base of plant with small tubers. (*SOURCE:* Redrawn with permission from Hitchcock et al., 1963.)

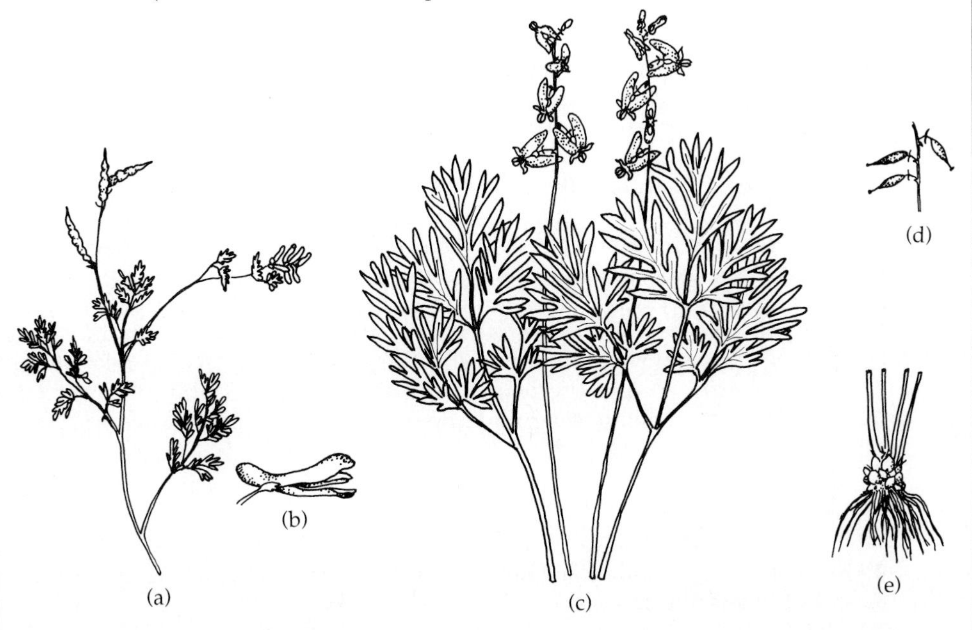


Herbs, annual or perennial, produce various alkaloids, lack latex or colored juice. Leaves alternate or basal, compound; stipules lacking. Flowers regular to strongly irregular, perfect, hypogynous; inflorescence a cyme or raceme. Sepals 2, small, bract-like. Petals 4 in 2 whorls, with 2 inside and smaller, and 2 outside, sometimes with a saclike pouch or spur. Stamens generally 6 (rarely 4) and diadelphous, filaments distinct or connate at the base; anthers dimorphic; nectaries present at base of the stamens. Pistil compound of 2 united carpels; locule 1; ovules 2–many and borne on parietal placentas; ovary superior; style 1, stigma lobed. Fruit a capsule. Seeds small; embryo small, some species with only 1 cotyledon, endosperm present (Fig. 9.18).

The family Fumariaceae consists of approximately 19 genera and 400 species distributed mostly throughout the North Temperate region, but with a few species in southern Africa. The two largest genera are *Corydalis* (300 species) and *Fumaria* (50 species).

Economically the family is not very important, except for the species *Dicentra spectabilis,* bleeding heart, used as a garden ornamental. Some species, such as *Corydalis sempervirens* and *Dicentra cucullaria,* Dutchman's breeches, are common members of the spring flora in the deciduous forest of eastern North America.

The family is included by some authors in the Papaveraceae, but it lacks the colored latex and has a distinctive, more complex flower.

The family has no known fossil record.

**Subclass II. Hamamelidae** The Hamamelidae consist of 11 orders, 24 families, and approximately 3400 species. The Urticales have over two thirds of the species, with about one fourth in the Fagales. (See Table 9.2.) Most plants are woody, though some herbaceous taxa are present in the Urticales. Leaves are usually simple, less commonly compound. The flowers are generally reduced in size and mostly unisexual and monoecious (less commonly dioecious), with a poorly developed perianth (or lacking), and arranged in catkins. The anthers are large and produce large quantities of light, smooth pollen; wind pollinated. The female flowers usually produce few seeds per ovary. The fruit is an achene, nut, or samara.

In some of the older classifications systems, the catkin-bearing groups were placed together and considered "simple" because of the superficial appearance of the catkin being similar to the "cones" of some pinophytes. The obvious conclusion was to assume the group to be the "most primitive" of the flowering plants. Today we know that the lack of flower parts is a reduction of parts only and that not all catkin-bearing groups are related.

Fossils attributed to the Hamamelidae have been found in Lower Cretaceous deposits, making the group one of the first groups of flowering plants observed in the fossil record.

**TABLE 9.2 A list of orders and family names of flowering plants found in the subclass Hamamelidae.**

Subclass II. Hamamelidae
- Order A. Trochodendrales
  - Family 1. Tetracentraceae
  - 2. Trochodendraceae
- Order B. Hamamelidales
  - Family 1. Cercidiphyllaceae
  - 2. Eupteliaceae
  - 3. Platanaceae*
  - 4. Hamamelidaceae*
  - 5. Myrothamnaceae
- Order C. Daphniphyllales
  - Family 1. Daphniphyllaceae
- Order D. Didymelales
  - Family 1. Didymelaceae
- Order E. Eucommiales
  - Family 1. Eucommiaceae
- Order F. Urticales
  - Family 1. Barbeyaceae
  - 2. Ulmaceae*
  - 3. Cannabaceae*
  - 4. Moraceae*
  - 5. Cecropiaceae
  - 6. Urticaceae*
- Order G. Leitneriales
  - Family 1. Leitneriaceae
- Order H. Juglandales
  - Family 1. Rhoipteleaceae
  - 2. Juglandaceae*
- Order I. Myricales
  - Family 1. Myricaceae*
- Order J. Fagales
  - Family 1. Balanopaceae
  - 2. Fagaceae*
  - 3. Betulaceae*
- Order K. Casuarinales
  - Family 1. Casuarinaceae*

*NOTE:* Family names followed by an asterisk are discussed and illustrated in the text.

## ORDER HAMAMELIDALES
### Family Platanaceae (Plane Tree)

Figure 9.19 Platanaceae. *Platanus:* (a) twig with winter buds; (b) male flower; (c) female flower; (d) leafy branch with round female inflorescences; (e) lateral bud encased in enlarged base of leaf petiole; (f) leaf and mature ball-like fruit. (*SOURCE:* Redrawn with permission from R. J. Preston, *North American Trees.* Ames, IA: Iowa State University Press, 1989.)

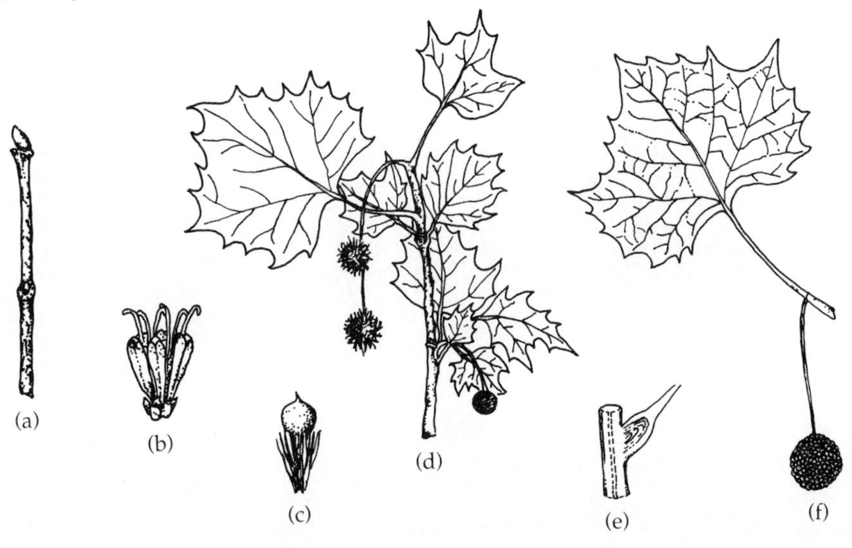


Trees, large, attractive with peeling, scaly bark leaving a patchy surface; all plant parts with branching hairs. Leaves alternate, simple, palmately lobed (except 1 pinnate species) and veined, deciduous; petiole long with an enlarged base enclosing the axillary bud; stipules large but soon falling. Flowers small, regular, unisexual and monoecious, hypogynous; wind pollinated, clustered in dense round heads. Perianth 3- to 8-merous, distinct or connate at the base, sepals may be lacking in female flowers. Stamens same number as perianth, filaments short or lacking, sometimes as staminodes, especially in pistillate flowers; anthers opening by longitudinal slits. Pistil simple of 3–9 separate carpels in 2–3 whorls, sometimes imperfectly sealed; locule 1; ovules 1–2 and borne on apical placentas; ovary superior; stigma decurrent down inside of style. Fruit a globose (round) head of hairy achenes (Fig. 9.19).

The family Platanaceae is comprised of the genus *Platanus* and 6–10 species found chiefly from Canada to Mexico and in the eastern Mediterranean region to the Himalaya Mountains of southern Asia.

Economically the family is of some importance. Plane trees are also called sycamores in North America. The fine, closed-grained wood is used in veneer, and the trees are widely grown as ornamentals in urban areas, especially *Platanus acerifolia*, the London plane tree.

Certain characters shared between the Platanaceae and Hamamelidaceae, such as anatomical features, etc., indicate that the families should be classified closely together.

Fossil leaves and pollen attributed to *Platanus* have been recorded from the Upper Cretaceous, with some "plane tree-like" plant material occurring in Lower Cretaceous deposits.

## Family Hamamelidaceae (Witch Hazel)

Figure 9.20 Family Hamamelidaceae. *Liquidambar:* (a) branch with leaves and inflorescence head; (b) Inflorescence head. *Hamamelis:* (c) apical buds of winter twig; (d) leaf attached to end of twig showing venation; (e) flowers; (f) fruits. (*SOURCE:* Redrawn with permission from Gleason, 1952.)

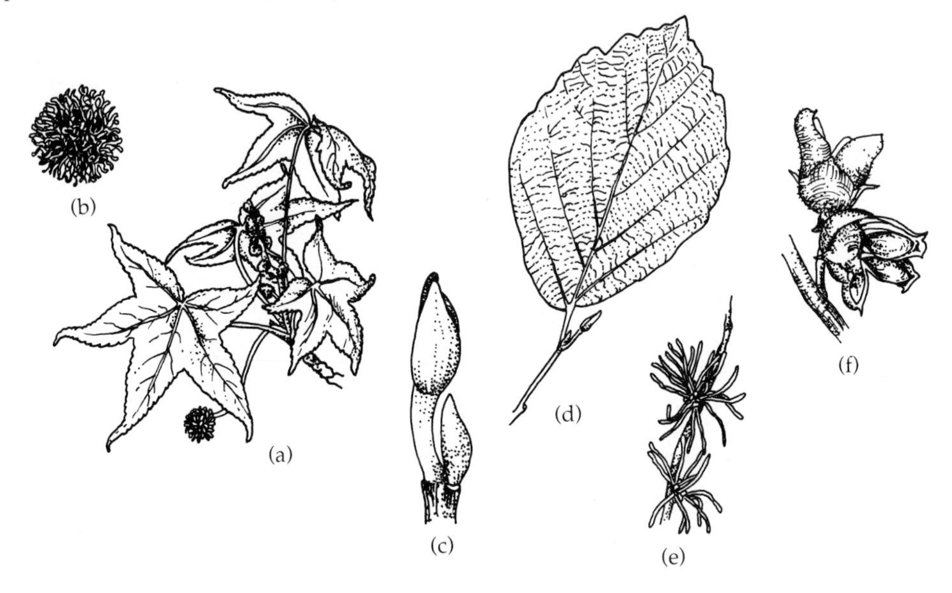


Shrubs or trees. Leaves alternate, simple, often palmately lobed with stellate hairs; stipules present. Flowers regular (rarely irregular), perfect or unisexual (monoecious or dioecious), epigynous; wind or insect pollinated; sometimes with colored bracts at the base; inflorescence a head or spike. Calyx 4–5, distinct or connate (sometimes lacking). Petals 4–5 (rarely lacking), distinct. Stamens 4–5 (rarely more or less), filaments distinct and alternate the petals. Pistil compound of 2 united carpels; locules 1; ovules 1–many on axile placentas; ovary superior or inferior. Fruit a woody capsule. Seeds with large straight embryo, endosperm present (Fig. 9.20).

The family Hamamelidaceae consists of approximately 26 genera and 100 species distributed widely in the Old and New Worlds, mostly in warm-temperature, subtropical regions. *Corylopsis* (30 species) is the largest genus and is found in eastern Asia. The North American species *Hamamelis virginiana,* witch hazel, is one of the last shrubs to bloom in the late autumn.

Economically the family is of some importance. *Liquidambar,* sweet gum, has a hard wood that is used for furniture, a fragrant resin used in making perfume, and is cultivated as an ornamental. The American sweet gum, L. *styraciflua,* and *Hamamelis* have a gum that is used to produce tonics, lotions, and astringents. Witch-hazel twigs are preferred by water diviners for their dowsers.

The family has a fair number of genera that fit in the group very loosely and are considered by some botanists to be placed in their own or other families.

Fossil leaves and pollen attributed to the Hamamelidaceae have been found in Upper Cretaceous deposits in Europe.

## ORDER URTICALES
### Family Ulmaceae (Elm)

Figure 9.21 Ulmaceae. *Ulmus:* (a) twig with leaves; (b) cluster of flowers; (c) male flower; (d) fruit. *Celtis:* (e) leafy branch bearing drupe fruits.

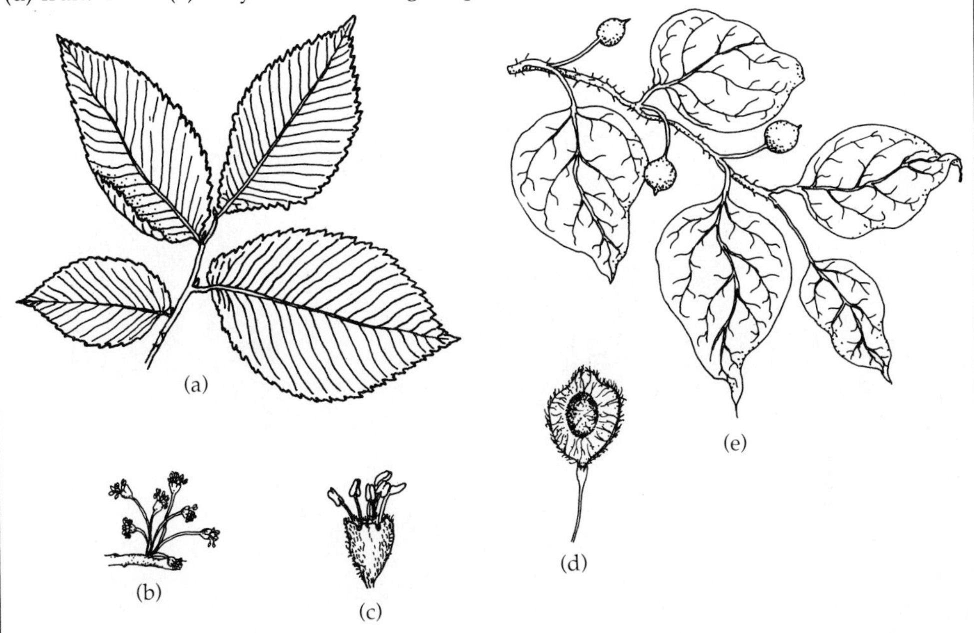


Shrubs or trees, without laticifers and latex. Leaves alternate (rarely opposite), simple, bases often oblique with prominent pinnate veins; stipules present but deciduous as the new leaf enlarges. Flowers inconspicuous and small, generally regular, perfect or unisexual and monoecious, hypogynous or perigynous; inflorescence of flowers in axillary clusters. Sepals 2–9, distinct or connate. Petals lacking. Stamens generally same number as sepals and opposite them, sometimes adnate to the sepals. Pistil compound of 2 united carpels; locules 1 or 2; ovules 1 per locule and borne on an apical placenta; ovary superior; separate decurrent stigmas and styles. Fruit a drupe, nut, or samara. Seed with a straight or curved embryo, endosperm present or lacking (Fig. 9.21).

The family Ulmaceae is comprised of about 18 genera and over 150 species distributed in the Tropics and North Temperate regions. The largest genera are *Celtis* (70 species), *Trema* (30 species), and *Ulmus* (20 species).

Economically the family is important for timber and as a shade ornamental. Most *Ulmus* produce a fine timber used in furniture and for posts and underwater pilings. *Ulmus americana*, American elm, was used extensively in North America as a shade tree until it was decimated in recent years by the introduced Dutch elm disease. The mucilaginous inner bark of *U. rubra*, slippery elm, provides a medicinal product. Some species of *Aphananthe*, *Celtis*, *Trema*, and *Zelkovia* produce good timber.

The family is sometimes divided into two subfamilies: Celtidoideae, with drupelike fruits and leaves with 3 main palmate veins, and Ulmoideae, with dry samara fruits and prominent pinnate veins.

*Ulmus*-like pollen has been found in Upper Cretaceous deposits and fossil wood has also been collected from Eocene strata of Yellowstone National Park, Wyoming in the United States.

## Family Cannabaceae (Hemp)

Figure 9.22 Cannabaceae. *Humulus:* (a) branch showing leaves and mature fruits. *Cannabis:* (b) female flower; (c) fruit; (d) male flower; (e) upper stem of staminate plant.

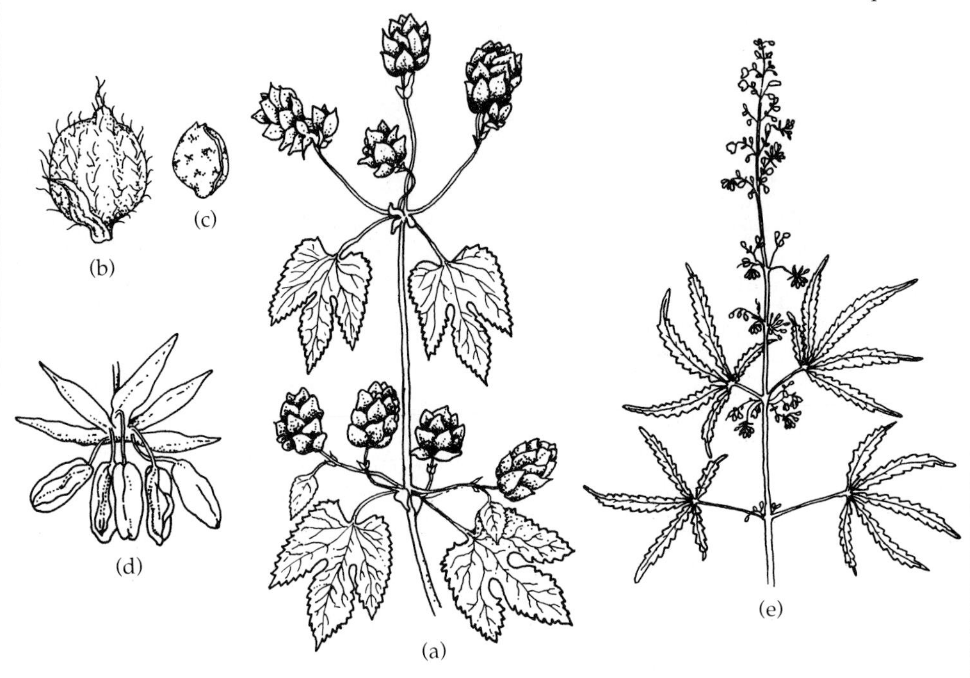


Herbs, twining or erect, producing strong stem fibers, lacking latex. Leaves opposite or alternate, palmately lobed or compound; stipules present; cystolith crystals commonly present; glandular hairs may be present, containing aromatic or psychotropic compounds. Flowers small, regular, unisexual (monoecious or dioecious), hypogynous, wind pollinated; inflorescences of axillary panicles or spikes among the upper leaves. Sepals 5, distinct. Petals lacking. Male flowers with 5 stamens, filaments distinct, erect, opposite the sepals. Pistil of female flowers compound of 2 united carpels; locule 1; ovule 1 and borne on an apical placenta; ovary superior and surrounded by a thin calyx tube enclosing the ovary. Fruit an achene. Seeds with a curved embryo, endosperm scanty (Fig. 9.22).

The family Cannabaceae contains only 2 genera and 3 species: *Cannabis sativa* and 2 species of *Humulus* native to North Temperate regions.

Economically the family is important for *Humulus lupulus,* hops, which imparts a "bitter" taste to beer, and for the psychotropic drugs (marijuana, hashish) and fiber (hemp) from *Cannabis sativa.*

The above taxa are sometimes placed in the fig family (Moraceae), but they differ in lacking latex and having a perianth of 5 instead of 4 points.

The family has no known fossil record.

## Family Moraceae (Fig)

Figure 9.23 Moraceae. *Morus:* (a) twig with leaves and multiple fruit. *Ficus:* (b) twig with leaves and synconium fruit, showing one fruit in longitudinal section.

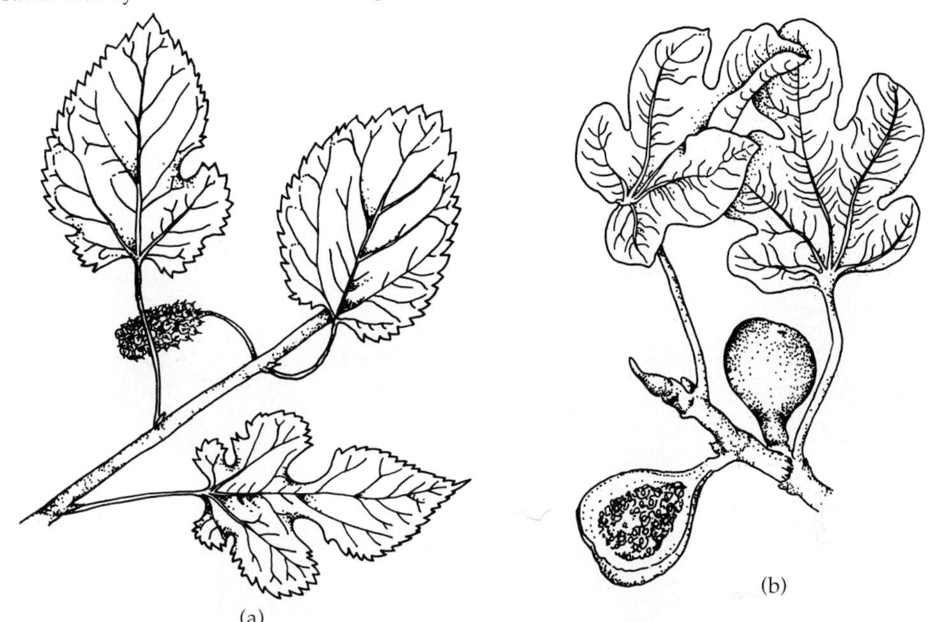


Shrubs, trees, and woody vines with laticifers containing milky juice, glandular hairs. Leaves alternate or opposite with entire or lobed margins; sometimes with cystolith crystals; stipules present. Flowers small, unisexual (monoecious or dioecious); inflorescences of various types; the axis of the flower cluster often enlarged; wind pollinated or insect pollinated. Sepals 4–6, distinct or connate. Petals lacking. Stamens of male flowers straight or reflexed in bud, same number as the sepals and opposite them. Pistil of female flowers compound of 2 united carpels; locule 1; ovule 1 and borne on an apical placenta; ovary superior or inferior; styles 2, branched. Fruit an achene, multiple, or synconium. Seeds with embryo curved or straight, endosperm present or lacking (Fig. 9.23).

The family Moraceae is comprised of 40 genera and approximately 1000 species distributed most commonly in subtropical to tropical climates, but with some taxa extending into temperate regions. The largest genus is *Ficus* (fig) with more than 500 species.

Economically the family is important for its fruits. Some of the more well known are *Ficus carica,* fig; *Morus,* mulberry; *Artocarpus faltilis,* breadfruit of Mutiny on the Bounty and Captain Bligh fame; and house plants such as *Ficus benjhamina,* weeping fig, and *Ficus elastica,* Indian rubber.

The number of genera and species is highly variable, depending on how the family is classified and what is included.

Fossil material attributed to the Moraceae and collected from the Upper Cretaceous and Paleocene are being questioned today as to their validity, but plant material from Eocene deposits are valid.

## Family Urticaceae (Nettle)

Figure 9.24 Urticaceae. *Urtica:* (a) habit of plant; (b) male flower with extended anthers; (c) female flower with stinging hairs; (d) nodes showing male catkins (below) and female (above).

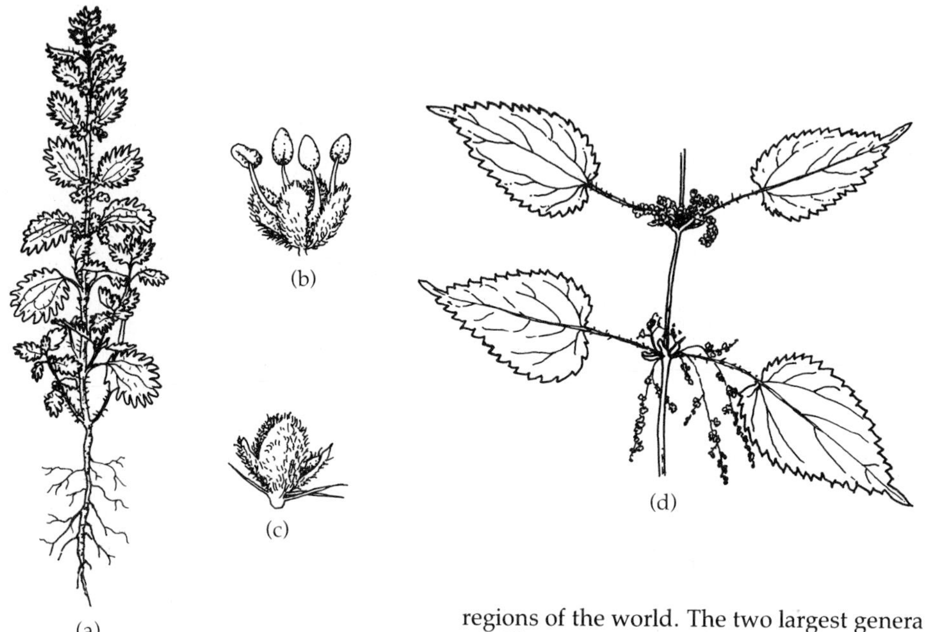


Herbs, shrubs, or trees with soft wood, often with specialized siliaceous stinging hairs. Leaves alternate or opposite, simple; epidermis commonly with calcium carbonate cystoliths; stipules generally present. Flowers small, greenish, generally unisexual, (monoecious or dioecious), hypogynous, apetalous; inflorescence in axillary cymes, panicles, or spikes. Male flowers with sepals 4–5, same number as stamens and opposite them. Stamens in bud inflexed and spring outward when mature in an exploding manner; anthers open by longitudinal slits; rudimentary pistil often present. Female flowers with 4–5 sepals or perianth lacking. Pistil simple of 1 individual carpel; locule 1; ovule 1 and borne on a basal placenta; ovary superior; style 1 or stigma sessile. Fruit an achene, drupe, or small nut. Seed with embryo straight, endosperm present (Fig. 9.24).

The family Urticaceae consists of approximately 50 genera and 700 species distributed widely in temperate and tropical regions of the world. The two largest genera are *Elatostema* (350 species) in the Old World Tropics and *Pilea* (200 species), a pantropical genus. In the temperate cooler regions *Urtica,* stinging nettle, is more common in waste areas and rich alluvial places. *Dendrocnide,* the stinging tree of Australia and southeast Asia, is notorious for its painful stings, which have caused death to animals and humans in some cases. The vegetative morphology of many taxa of the family is greatly influenced by environmental conditions.

Economically the family is of some importance. Fibers from *Boehmeria nivea,* ramie, from Southeast Asia are used in textiles and ropes; species of *Elatostema, Helxine,* and *Pilea* are grown as ornamentals; and *Urtica dioica* is a bothersome weed.

The family is distinctive from other families of the order because of the specialized stinging hairs, with their irritating solution, and the explosive nature of the anthers.

Leaf imprints of *Urtica* have been found in Eocene and more recent deposits. Fossils from older deposits are presently being questioned.

## ORDER JUGLANDALES
### Family Juglandaceae (Walnut)

Figure 9.25 Juglandaceae. *Juglans:* (a) winter twig showing chambered pith; (b) fruit; (c) pinnately compound leaf; (d) female flower. *Carya:* (e) leafy branch with male flowers in catkin.

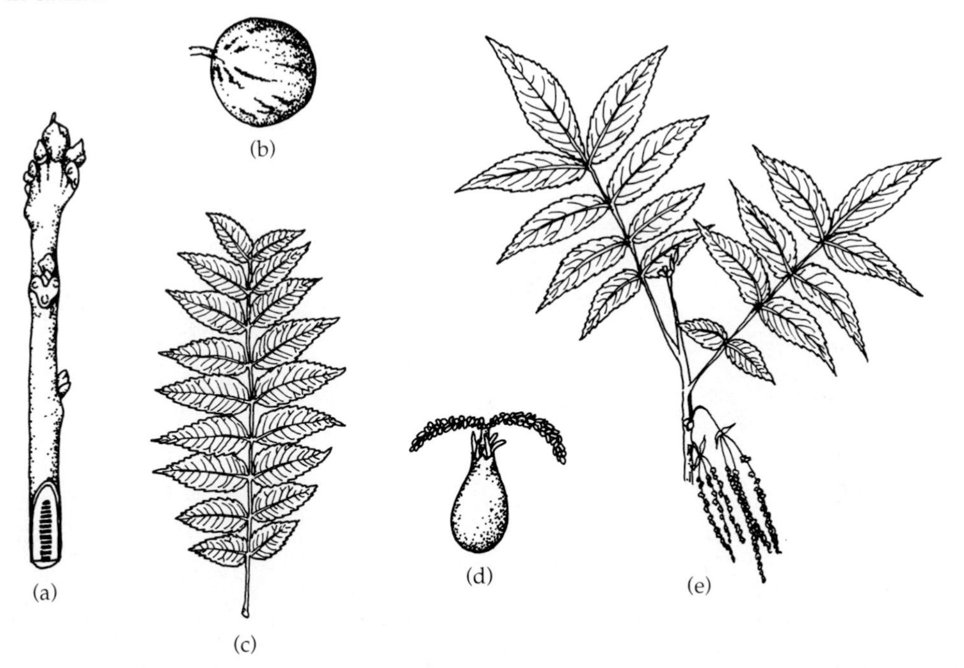


Trees, small to large, shoots bearing hairy prominent buds. Leaves alternate (rarely opposite), pinnately compound with resinous, aromatic glands; stipules lacking. Flowers greenish, small, regular, bracteate, unisexual and monoecious, apetalous; male flowers in pendulous catkins, female flowers usually erect on new twig growth. Perianth is usually 4 lobed (rarely 3–5), reduced or absent. Stamens in male flowers 3–many in several whorls; filaments distinct, short; anthers opening by longitudinal slits; rudimentary pistil present. Female flower with pistil compound of 2 (rarely 3) united carpels; locules 2 (rarely 3); ovule 1 and borne on a basal placenta; ovary inferior; style 1, short, stigmas 2. Fruit a drupe or nut. Seed with large, massive cotyledons, endosperm lacking (Fig. 9.25).

The family Juglandaceae is comprised of 7 genera and about 60 species distributed mostly in North Temperate or Subtropical regions, but with a few species extending into the South American Andes and southwest Pacific.

Economically the family is of some importance. The family is well known for its edible nuts, as in *Carya pecan,* the pecan; *Juglans regia* and *J. nigra,* the walnut; and *Carya ovata,* the hickory nut, and for the fine timber of its wood. *Juglans nigra* wood of North America is highly prized for furniture and a large tree cut into logs may bring $12,000 (U.S.) depending on the grain of the wood. The oil from the nuts has been used in soaps, paints, and cosmetics and is sold in health food stores for human consumption.

That the family is distinctive has not been of concern to botanists, but how it should be classified in relationship to other families is still debated.

Fossil pollen attributed to the Juglandaceae is known from the Paleocene with wood from Eocene deposits of Yellowstone National Park, Wyoming, U.S.A.

## ORDER MYRICALES
### Family Myricaceae (Bayberry)

Figure 9.26 Myricaceae. *Myrica:* (a) leafy branch with female inflorescence; (b) two male flowers and subtending bracts; (c) female flower; (d) leafless branches with fruits.

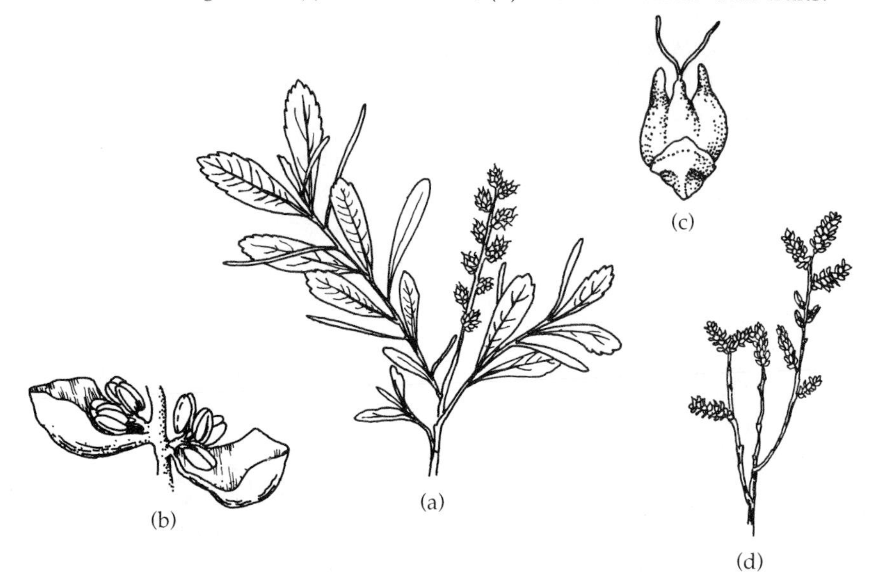


Shrubs or small trees having the ability to harbor nitrogen-fixing bacteria on their roots. Leaves, deciduous or evergreen, alternate, simple with pinnate veins and generally stipules lacking. Flowers small, perfect or unisexual and monoecious, subtended by small bracts; both staminate and pistillate inflorescences borne in catkin-like spikes; perfect flowers with 6 stamens on top of the ovary; male flowers usually with 4 stamens (sometimes up to 10), filaments distinct or connate, subtended by bracts and with a disk at the base of the stamen; female flowers with 2 or more bracteoles, which may look like a perianth. Pistil compound of 2 united carpels; locule 1; ovule 1 and borne on a basal placenta; ovary superior or inferior (in perfect flowers); style 1, stigma 2-branched. Fruit a drupe, sometimes with a thick, waxy covering. Seeds with embryo straight, endosperm lacking (Fig. 9.26).

The family Myricaceae is comprised of 2–3 genera and approximately 50 species distributed in almost all parts of the world, except Australia and some warm parts of the Old World. Most of the species belong to *Myrica,* which includes *M. cerifora* and *M. pensylvanica,* bayberry, from which the wax on the outside of the fruit is removed to make candles. Some species are planted as ornamental shrubs.

The family has many characters that are shared with both the Betulaceae and Juglandaceae families but is sufficiently distinctive in some features to be grouped in its own order.

Fossil wood attributed to the family has been found in Eocene deposits of Yellowstone National Park, Wyoming, U.S.A., with some microfossils being present in Oligocene formations.

## ORDER FAGALES
### Family Fagaceae (Beech)

Figure 9.27 Fagaceae. *Castanea:* (a) leafy branch with staminate inflorescences; (b) burr-like fruits. *Quercus:* (c) leafy branch with female flower; (d) leafy branch with staminate catkins; (e) fruits (acorns).

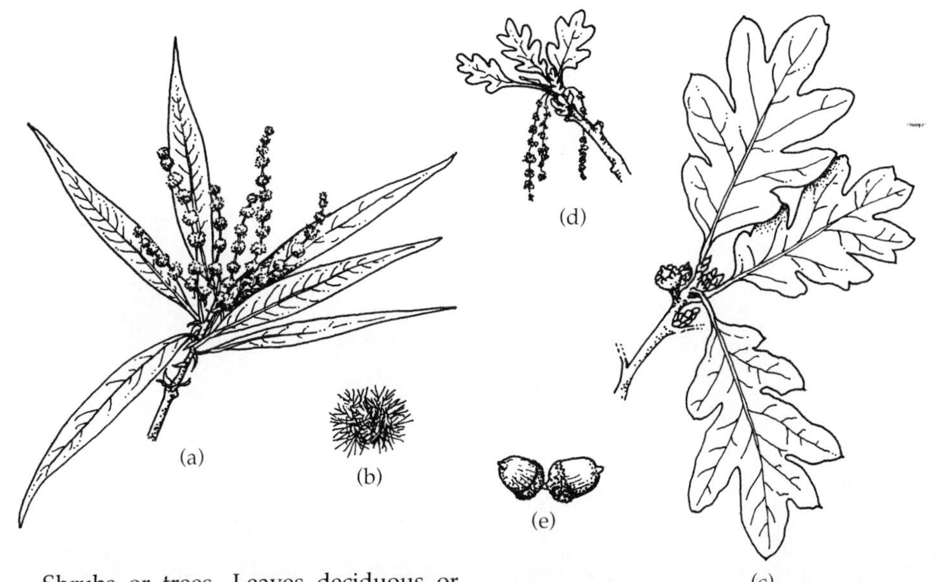


Shrubs or trees. Leaves deciduous or evergreen, alternate (rarely opposite or whorled), simple, pinnate veined, entire, toothed to lobed margins; stipules deciduous. Flowers small, unisexual and monoecious, apetalous; wind or insect pollinated; male flowers single or more commonly arranged in catkins, small heads, or spikes; female flowers usually 1–7, surrounded by a basal involucre positioned at the base of male catkins or from separate axils, may have rudimentary staminodes. Sepals 4–7, commonly 6 and bract-like. Stamens of male flowers 4–many, filaments distinct, slender; anthers opening by longitudinal slits. Pistil of female flowers compound of 3–7 united carpels; locules 3–7; ovules 2 per locule with all aborting but 1; ovary inferior; styles 3–7; involucre at maturity forming a "cap" or "cupule" around the base or whole of the ovary. Fruit a nut or acorn. Seed single, embryo straight, endosperm lacking (Fig. 9.27).

The Fagaceae is comprised of 6–8 genera and about 800–1000 species with a cosmopolitan distribution, except for parts of tropical South America and Africa, and southern Africa. *Quercus,* oaks; *Castanea,* chestnut; and *Fagus,* beech, are dominant components of the deciduous forests of Europe and North America. In south New Zealand, South America, and Tasmania, *Nothofagus,* the southern beech, are dominant members of the forests.

Economically the family is rather important. The wood of most members of this family is highly prized for building materials, furniture, whiskey barrels, and in years past for building sailing ships. The bark of *Quercus suber,* Mediterranean cork oak, is gathered for commercial cork. Various species of *Castanea,* chestnut, are grown for their edible nuts.

The nature of the cupule of "cap" in the family has been debated by botanists. One idea is that it indicates a fusing together of reduced leaves and modified branches. Another theory is that the cupule is part of the pedicle.

Pollen considered to be *Nothofagus* has been found in Upper Cretaceous and more recent deposits. Pollen of *Quercus* and *Castanea,* and beech-like wood, date from Eocene sediments.

## Family Betulaceae (Birch)

Figure 9.28 Betulaceae. *Alnus:* (a) leafy branch with cone-like fruits. *Betula:* (b) bract; (c) winged fruit (samara); (d) bark with horizontal lenticles; (e) twig with staminate catkins.

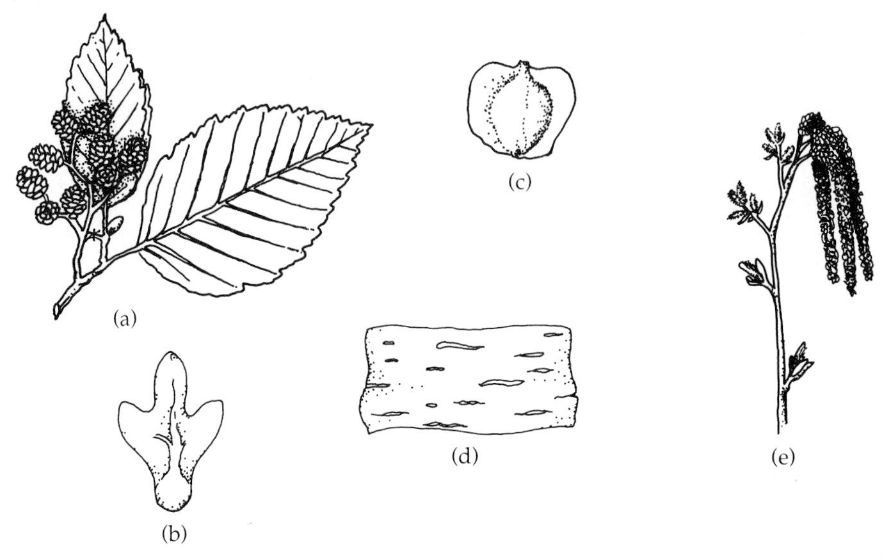


Shrubs or trees often having pealing, thin-layered bark with horizontal lenticles. Leaves deciduous, alternate, simple, toothed with pinnate veins. Flowers small, unisexual and monoecious; wind pollinated; inflorescence of catkins with male flowers in pendulous aments, while the female flowers are in cymes, subtended by bracts, and usually erect on a stiff axis. Calyx of 1–6 distinct, scale-like segments. Petals lacking. Stamens of male flowers 2–8, vestigial pistil lacking. Female flower compound of 2 united carpels; locule 1 above and 2 locules below; ovules 1–2 per locule and borne on axile placentas; ovary inferior or superior; styles 2. Fruit a nut or winged samara, subtended or enclosed by an enlarged bract. Seed 1, embryo large and straight, endosperm lacking (Fig. 9.28).

The family Betulaceae is comprised of 6 genera and 120–150 species distributed mostly in the cooler parts of the Northern Hemisphere, except for those species occurring in the Andes of South America. It is sometimes divided up into 3 separate families. The largest genus is *Betula,* birch (60 species).

Economically the family is of some importance. *Betula papyrifera* wood is used for making boxes, plywood, and wood pulp for newsprint paper. Its bark was used by North American Indians for making canoes. Two other species, *B. lutea* and *B. lenta,* provide quality wood for home products such as doors, flooring, and furniture. The wood of *Alnus rubra* in western North America looks like mahogony and is used for spools. *Corylus,* the hazelnut, produces an edible nut, while *Ostrya,* ironwood or hornbeam, wood is very hard and is used for posts and mallets. Many members of the family are cultivated as ornamentals.

The three tribes of the family are sometimes separated into three smaller families, but recent research supports keeping the group as one family.

Fossil pollen attributed to the family has been found in Upper Cretaceous deposits, while *Alnus* and *Carpinus* woods are known from Eocene sediments of Yellowstone National Park, Wyoming, U.S.A.

## ORDER CASUARINALES
### Family Casuarinaceae (She Oak)

Figure 9.29 Casuarinaceae. *Casuarina:* (a) end of branch with whorls of male flowers; (b) cone-like fruit; (c) branch with jointed secondary branches and cone-like fruit; (d) two nodes of jointed stem; (e) reduced whorled leaves.

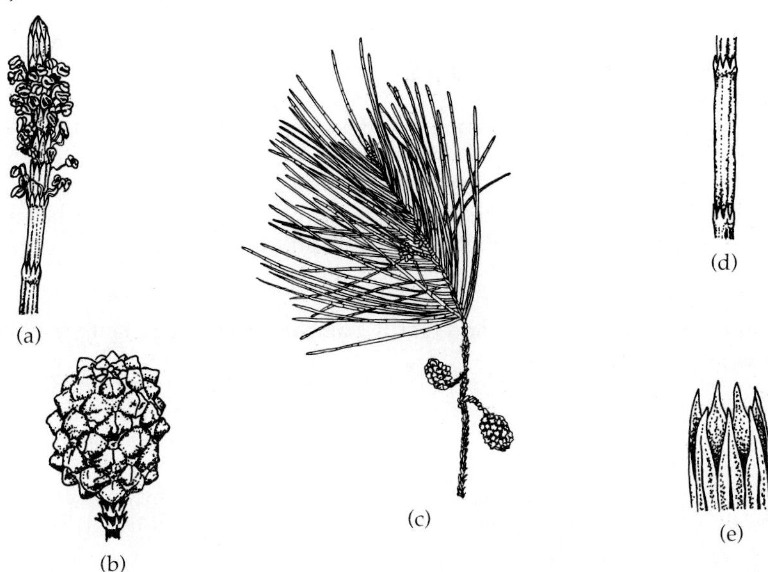


Trees (less commonly shrubs) with jointed, slender drooping branches that are round, grooved, and have short internodes; epidermis usually with calcium oxalate crystals; roots commonly have nitrogen-fixing bacteria associated with them. Leaves evergreen, reduced, and scale-like, whorled at the nodes. Flowers small, without a perianth, unisexual, with male and female flowers formed on different parts of the tree. Male flowers in whorls at nodes in aments at end of lateral branches, 1 stamen per flower, each subtended by a bract and 2 pairs of scales or bracteoles. Female flowers in dense heads, naked, subtended by a leaf-like bract, which becomes woody; pistil compound of 2 united carpels; locules 2 but one aborts; ovules 2 per locule; ovary superior; style 1, stigma decurrent on style branches. Fruit a cone-like multiple, each ovary producing a 1-seeded samara. Seed with large straight embryo, endosperm lacking (Fig. 9.29).

The family Casuarinaceae is comprised of only 1 genus, *Casuarina,* Australian oak or she oak, with about 65 species distributed natively in Australia and with some species in Fiji, Malaysia, Mascarene Islands, New Caledonia, and Southeast Asia. It has been commonly introduced into many tropical to subtropical dry regions, such as the Bahamas and California, where it is so common as to appear native.

Economically the family is of some value. The wood of various species is hard, nicely grained, and valued for making furniture. *Casuarina equisetifolia* is cultivated widely for its wood, as a windbreak, and as an ornamental. Two other commonly used Australian species are *C. stricta* and *C. cunninghamiana* (she oak).

Superficially the family looks like a gymnosperm, especially *Ephedra,* because of the jointed branches. Closer inspection, however, reveals that the group is much like many of the other apetalous flowering trees.

Fossil pollen attributed to *Casuarina* has been found in Paleocene deposits, with macrofossils known from Eocene and Miocene sediments of Australia and Argentina, respectively.

**Subclass III. Caryophyllidae** The subclass Caryophyllidae is not as large as some of the other subclasses. It consists of 3 orders, 14 families, and approximately 11,000 species. (See Table 9.3.) The order Caryophyllales makes up about 90% of the taxa. The plants are more commonly herbaceous, with the woody representatives lacking the typical secondary stem growth. Anatomically the sieve tubes and embryological development is distinctive. The stamens originate in a centrifugal fashion. The petals are distinct or less commonly lacking. Ovules are attached by basal or free-central placentas. The pollen grains are mostly trinucleate, less commonly with 2 nuclei. The seeds have perisperm in place of endosperm.

Fossils for the group have been found in Upper Cretaceous deposits, but in spite of this, the lineage of the families is highly speculative because of the morphological diversity of the families.

**TABLE 9.3 A list of orders and family names of flowering plants found in the subclass Caryophyllidae.**

| Subclass III. Caryophyllidae | |
|---|---|
| Order A. Caryophyllales | Order B. Polygonales |
| Family 1. Phytolaccaceae* | Family 1. Polygonaceae* |
| 2. Achatocarpaceae | Order C. Plumbaginales |
| 3. Nyctaginaceae* | Family 1. Plumbaginaceae* |
| 4. Aizoaceae* | |
| 5. Didiereaceae | |
| 6. Cactaceae* | |
| 7. Chenopodiaceae* | |
| 8. Amaranthaceae* | |
| 9. Portulacaceae* | |
| 10. Basellaceae | |
| 11. Molluginaceae | |
| 12. Caryophyllaceae* | |

*NOTE:* Family names followed by an asterisk are discussed and illustrated in the text.

## ORDER CARYOPHYLLALES
### Family Phytolaccaceae (Pokeweed)

Figure 9.30 Phytolaccaceae. *Phytolacca:* (a) upper stem with flowers and fruits; (b) face view of flower.

Herbs, shrubs (rarely trees) with glabrous somewhat succulent stems. Leaves alternate, simple, and stipules lacking. Flowers small, regular, perfect (rarely irregular or unisexual), hypogynous; inflorescence usually borne in axillary or terminal panicles, racemes, or spikes. Sepals mostly 4–5, distinct and persistent. Petals usually absent. Stamens distinct or connate at the base, mostly same number as calyx lobes and opposite them. Pistil comprised of 1 to many separate or united carpels; ovary superior, sometimes on a raised stalk with as many locules as carpels, and a single basal ovule for each carpel; styles slender and may be as many as the carpels. Fruit a berry, capsule, or nut. Seeds with a curved peripheral embryo and mealy perisperm (Fig. 9.30).

The family Phytolaccaceae is comprised of 18–25 genera and 125 species, with most taxa in the New World but widespread in other subtropical to tropical climates. In temperate North America *Phytolacca americana,* pokeweed, is commonly found.

Economically the family is of minor importance. *Trichostigma peruvianum* is grown in greenhouses along with *Rivina humilis,* blood berry plant. A red dye is obtained from the berries of *R. humilis.* In South America *Petiveria alliacea* is used medicinally and smells like onions. Various species of *Agdestis* and *Phytolacca* are used in treating syphilis. Still other species of *Phytolacca* are used for dyes and medicinal uses.

The family has been divided into 4 subfamilies. These subfamilies are diverse enough that some botanists have proposed making them separate families.

The family has no known fossil record.

## Family Nyctaginaceae (Four O'clock)

Figure 9.31 Nyctaginaceae. *Mirabilis:* (a) upper stem showing flowers with involucral bracts. *Abronia:* (b) creeping stem with flowers and fruits.

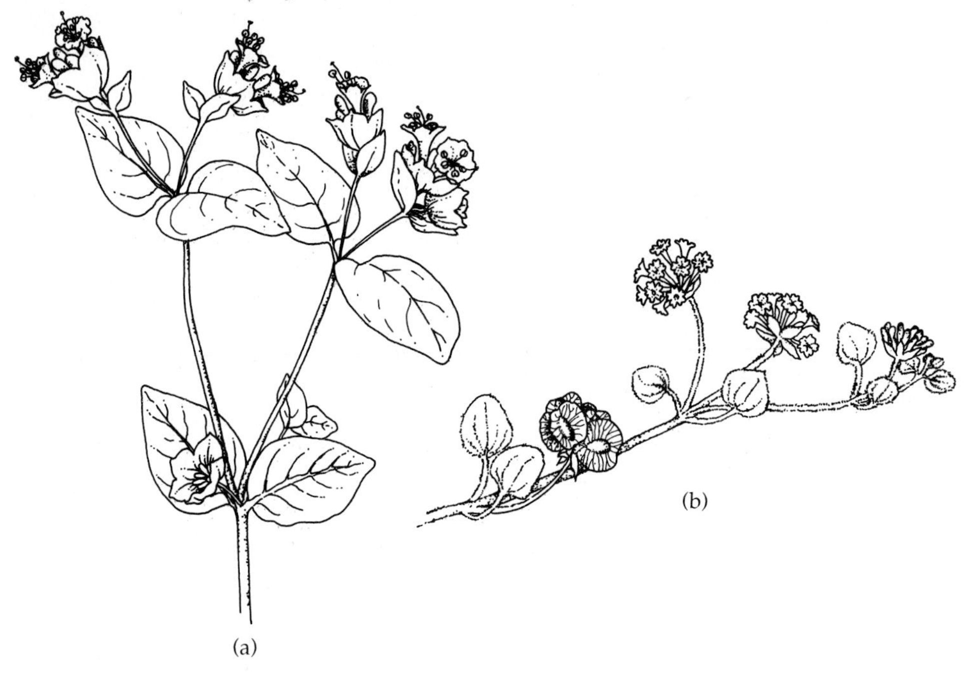


Herbs, shrubs, or trees. Leaves alternate or opposite, simple; lacking stipules. Flowers regular (rarely irregular), perfect or unisexual; hypogynous; inflorescence in cymes or head-like, often surrounded by a colored involucre of bracts. Sepals 5, tubular and petal-like, commonly surrounding the fruit. Petals absent. Stamens 5 (may be 1–many), filaments distinct or connate at the base, alternate with perianth lobes; anthers opening by longitudinal slits. Pistil simple of 1 carpel; locule 1; ovule 1 and borne on a basal placenta; ovary superior; style 1, long and slender. Fruit an achene or nut, often enclosed by the calyx. Seed with a large curved or straight embryo, endosperm present or absent (Fig 9.31).

The family Nyctaginaceae is comprised of about 30 genera and 200–300 species distributed mostly pantropically, but most common in the New World. Only a few species are temperate, such as *Mirabilis jalapa.* Various taxa of the South American *Bougainvillea* are grown as decorative hedges in tropical regions and in temperate greenhouses. Various cultivars of *Mirabilis jalapa,* common four o'clock, are commonly grown in flower gardens. They open their flowers in late afternoon, and thus have been named four o'clock.

Classifying the family in the order Caryophyllales has sometimes been questioned, but recent data from anatomy and embryology support its present classification.

The family has no known fossil record.

## Family Aizoaceae (Fig Marigold)

Figure 9.32 Aizoaceae. *Conophytum:* (a) fleshy plant bearing flowers; (b) flower from top; (c) longitudinal section showing stamens and perianth; (d) pistil.

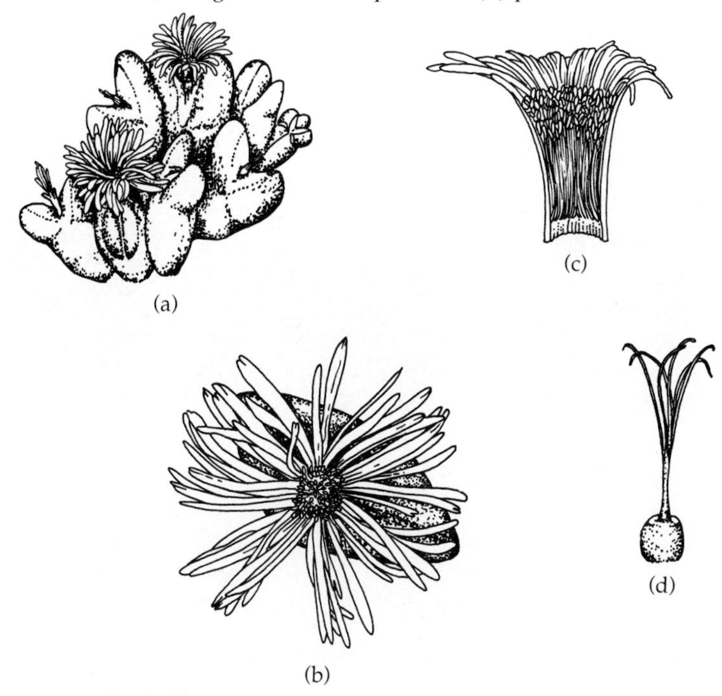


Herbs or small shrubs, succulent, annual or perennial. Leaves alternate or opposite, mostly simple; without spines and stipules. Flowers regular, perfect or unisexual (monoecious), perigynous; inflorescence of solitary flowers or flowers in cymes. Sepals 5 (rarely 3–8), succulent. Petals many (rarely lacking), of staminodial origin in 1–6 whorls, distinct or connate. Stamens 4–5 or 8–10 or many, filaments distinct or connate, adnate to the hypanthium. Pistil compound of 2–many united carpels; locules 2–many; ovules many and borne on apical, axile, basal, or parietal placentas; ovary superior or inferior; styles 2–many. Fruit commonly a loculicidal capsule or berry (rarely a nut). Seed with a large curved embryo, perisperm present, endosperm lacking (Fig. 9.32).

The family Aizoaceae is comprised of 12–143 genera and 2300–2500 species distributed pantropically, with its center of diversity in South Africa. The largest genera of the family are *Mesembryanthemum* (2000 species), sometimes put in its own family complex, and *Conophytum* (250 species). *Lithops*, the stone plants or living stones, are sometimes associated with a particular type of rock or rock formation to form a "mimicry" with the rocks. For example, *Titanopsis* has a white covering on the leaves and it blends in with the surrounding quartz rocks.

Economically the group is of minor value as a greenhouse ornamental and summer bedding plant in warmer regions. Some species help stabilize sand dunes and road banks in Southern California and Mediterranean areas.

The Azioaceae share many morphological and physiological features with other unrelated families of plants that have adaptations to dry habitats such as the Cactaceae and Crassulaceae. These features include fleshy leaves, some succulent roots, reduced leaf size, and the crassulacean type of metabolism.

The family has no known fossil record.

## Family Cactaceae (Cactus)

Figure 9.33 Cactaceae. *Opuntia:* (a) longitudinal section through inferior ovary and flower; (b) fleshy, jointed stem bearing flowers. *Mammillaria:* (c) habit of small plant; (d) two areoles with spines.

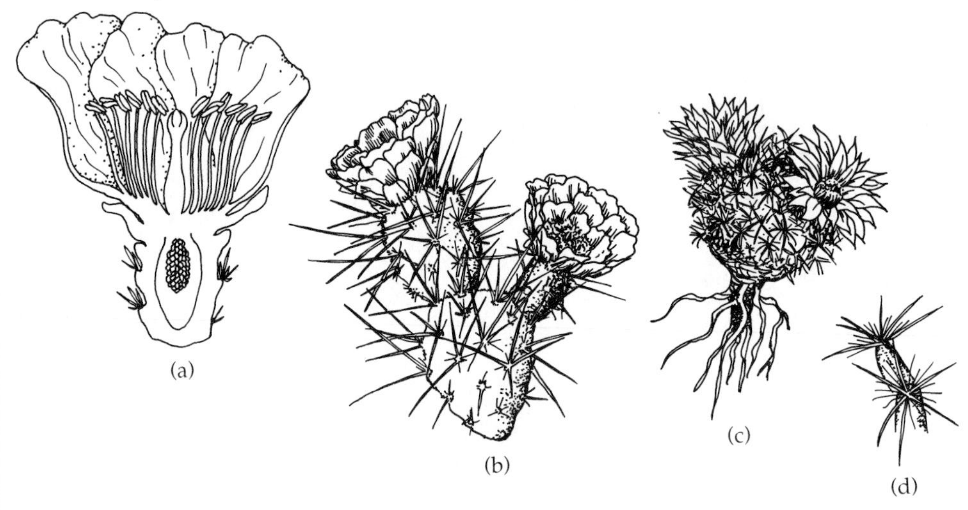


Perennial herbs or woody, xerophytic, succulent stemmed plants with spines; spines and flowers arising from little cushions or **areoles** (really reduced branches); root system widely spread but shallow; plants with crassulacean metabolism. Leaves absent, reduced, or coming from the areole. Flowers large, showy, usually regular, perfect (rarely unisexual), mostly solitary or at the ends of branches; all colors of the rainbow except blue; pollinated by bats, bees, hawkmoths, or hummingbirds. Sepals many. Petals many. Stamens many, filaments distinct. Pistil compound of 2–many united carpels; locule 1; ovules numerous and borne on parietal placentas; ovary inferior; style 1, stigmas 2–many, branched. Fruit a fleshy berry (rarely dry). Seeds many with a straight or curved embryo, perisperm and endosperm present or absent (Fig. 9.33).

The family Cactaceae is comprised of 30–200 genera and 1000–2000 species native to the semiarid and arid regions of the New World. Many taxa have become naturalized in warm areas of the world (Australia, Mediterranean, South Africa, etc.). *Opuntia,* prickly-pear (300 species) cacti, are most common in Greece and have become a pest in Australia. Biological control techniques have reduced the problem considerably in Australia as compared to a few years ago.

Economically many cacti are highly prized by amateur gardeners and by people of northern climates who travel south in the wintertime (called "snowbirds") to a warmer climate. As a result, demand for new, different and rare cacti has risen greatly. Therefore, "cactus rustling" is a multimillion dollar business in North America alone. This pressure and modifying of desert habitat has forced more cactus species to be placed on the North American endangered and threatened species list than any other group of plants. Except for their horticultural appeal, most cacti have little use. An exception to this is *Opuntia* (prickly pear), whose young stem joints are used for food and whose fruits ("apples") are harvested for juice to make jelly. Some central and South American people use cacti plants for hedging and fences.

The classification of the Cactaceae in the order Caryophyllales has been supported by recent anatomical, embryological, and pollen research.

The family has no known fossil record.

## Family Chenopodiaceae (Goosefoot)

Figure 9.34 Chenopodiaceae. *Chenopodium:* (a) habit of plant; (b) fruit enclosed by calyx; (c) pistil; (d) leaf showing venation. *Salicornia:* (e) close-up of inflorescence; (f) habit of plant.

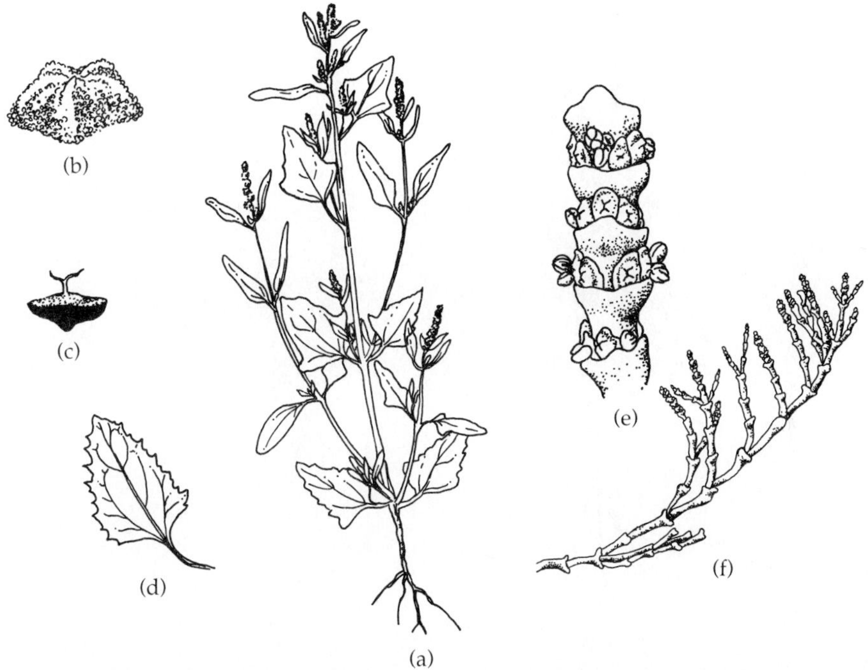


Herbs or shrubs (rarely trees), many deep rooted, annual or perennial or adapted to live with saline soils (called **halophytes**). Leaves alternate, simple, often mealy covered **(scurfy)**; stipules lacking. Flowers inconspicuous, greenish, regular, perfect (rarely unisexual), hypogynous; inflorescence of cymes, panicles, or spikes; wind pollinated. Sepals 2–5, distinct or connate, many times enclosing the fruit. Petals absent. Stamens same number as sepals and opposite them, filaments distinct or connate at the base and adnate to the calyx; anthers opening by longitudinal slits. Pistil compound of 2–3 united carpels, locules 1, ovule 1 and borne on a basal placenta, ovary superior (rarely inferior), styles 2–3. Fruit an achene or nut, often associated with a persistent calyx. Seed lens shaped, with a curved or spiral embryo around the periphery of seed, perisperm usually present, endosperm lacking (Fig. 9.34).

The family Chenopodiaceae is made up of about 100 genera and 1500 species distributed widely but most commonly in semiarid to arid temperate and subtropical saline habitats. The largest genera are *Chenopodium* (200 species), and *Atriplex* (150 species). Some species are adapted to growing in soils high in mineral salts, while others are common weeds of cultivated fields, gardens, and waste areas, such as *Chenopodium album,* lambs quarters, and *Salsola kali,* Russian thistle or tumbleweed. The latter was made famous in early western North American folksongs.

Economically the two most important species are *Beta vulgaris,* the beet, and *Spinacia,* spinach, which are common food plants. Various varieties of the table beet include the sugar beet, Swiss chard, and mangel-wurzels. Seeds and leaves of *Chenopodium quinoa* are eaten by Peruvians in the Andes.

The arrangement of the subfamilies of the Chenopodiaceae has been debated by some botanists, with no common consensus at present.

Fossil pollen attributed to the family has been found in Upper Cretaceous deposits.

## ɩaranthaceae (Amaranth)

ɹ.35 Amaranthaceae. *Amaranthus:* (a) habit of plant; (b) male flower; (c) longi-
ʌdınal section of flower with spiny bract.

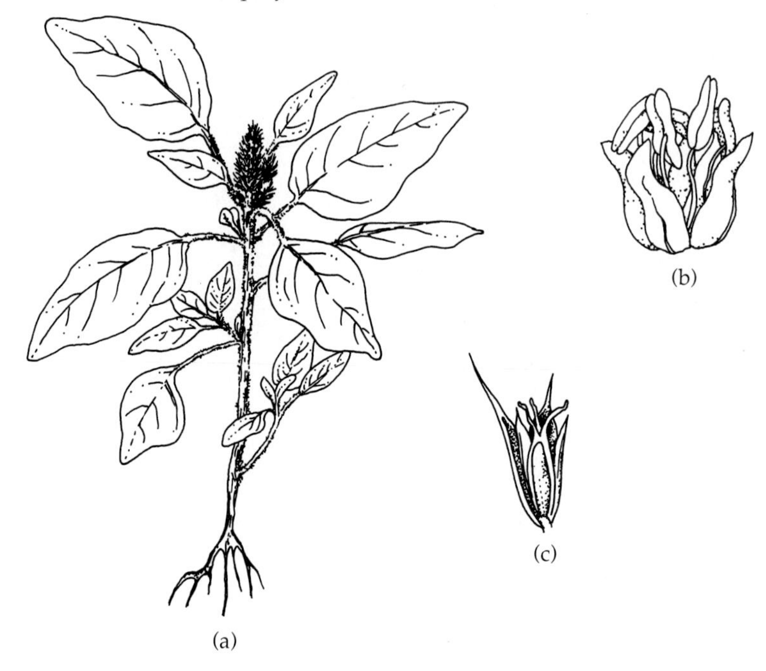


Herbs or shrubs (rarely vines). Leaves alternate or opposite, simple; without stipules. Flowers generally perfect, hypogynous and apetalous; inflorescence a solitary cluster, more commonly in axillary or terminal spikes or heads, bracteate (sometimes colored) inflorescences; wind or insect pollinated; sterile flowers may develop laterally as bristles or hooks and subtend normal flowers. Sepals 4–5, segments distinct or connate. Stamens usually the same number as sepals and opposite them, filaments free or connate at the base. Pistil compound of 2–3 united carpels; locule 1; ovules many and borne on basal placenta; ovary superior or inferior; styles 2–3. Fruit an achene, berry, circumscissile capsule, or utricle. Seeds usually shiny, embryo curved, perisperm present, endosperm lacking (Fig. 9.35).

The family Amaranthaceae is comprised of 65 genera and 900 species with a cosmopolitan distribution, but with more species in the warmer regions. The largest genera are *Alternanthera* (170 species) and *Gomphrena* (100 species).

Economically the family is of some importance. Many species of *Amaranthus,* pigweed, are weeds, such as *A. retroflexus,* but others have edible seeds that are high in protein. *Celosia cristata,* cockscomb, with its bright yellow to scarlet bracts, is used extensively as a bedding plant. South American *Iresine herbstii* and *I. linderii* are grown for their colorful leaves. Some taxa are said to have medicinal properties.

The family is similar in superficial appearance to the Chenopodiaceae but differs in having the stamens commonly fusing into a tube and in the presence of scarous, spiny bracts.

The family has no known fossil record.

## Family Portulacaceae (Purslane)

Figure 9.36 Portulacaceae. *Claytonia:* (a) plant bearing flowers and leaves growing from corm-like underground stem; (b) flower; (c) seed. *Lewisia:* (d) leafy flowering plant growing from a branched caudex.

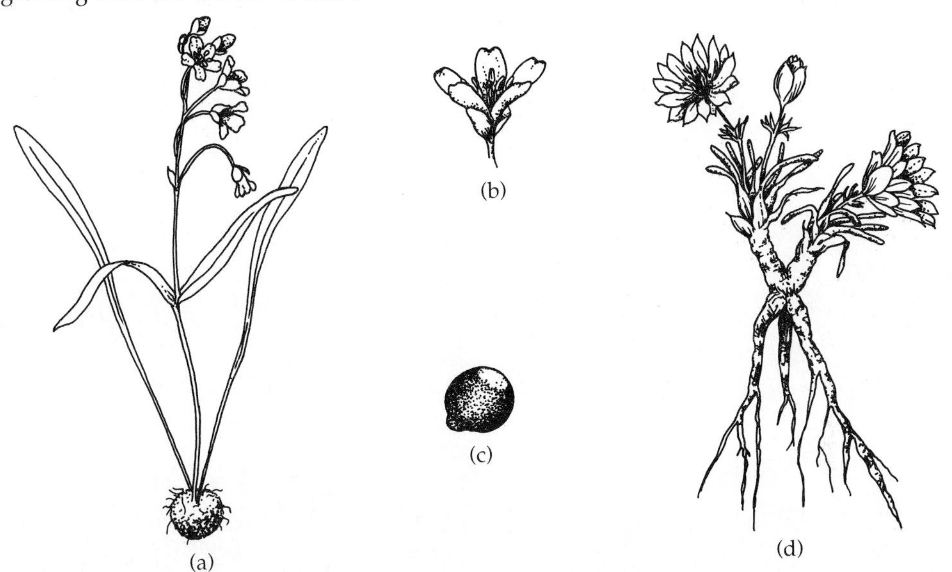


Herbs (rarely shrubs), annual or perennial, somewhat succulent. Leaves alternate or opposite, simple; stipules absent or as scarous membranes or hairs. Flowers regular, perfect, hypogynous; inflorescence of solitary flowers or flowers in cymes or racemes. Sepals 2, green. Petals 5 (rarely less or more in number), distinct and somewhat overlapping. Stamens commonly the same number as petals and opposite them, filaments distinct or connate at the base, often adnate to the petals; anthers opening by longitudinal slits. Pistil compound of 2–3 united carpels; locule 1; ovules 1–many and borne on basal or free-central placenta (placentation basal with 1 ovule or free central with 2–many ovules); ovary superior; styles usually distinct (rarely single). Fruit a circumscissile or loculicidal capsule. Seeds lens-shaped, smooth, and shiny; embryo large, coiled, endosperm lacking (Fig. 9.36).

The family Portulacaceae consists of approximately 20 genera and 500 species with a cosmopolitan distribution, but is well developed in the Andes, North America, and South Africa. Over half of the species belong to only 3 genera: *Calandrinia* (150 species), *Portulaca* (100 species), and *Talinum* (50 species).

Economically the family is of little value. *Portulaca oleracea,* common purslane, is a weed but is cultivated as a pot herb, while another species, *P. grandiflora,* rose moss, is commonly used as a bedding plant. Various species of *Montia,* miner's lettuce, were used as a salad green by early pioneers in western North America. Various species of *Lewisia* are spectacular alpine and rock garden plants. The root stalks of *L. rediviva,* bitterroot, was a staple of various North American Plains Indians. Several species of *Claytonia,* spring beauty, are common members of the eastern North American spring flora, with *C. virginica* having a wide range of chromosome numbers (12–almost 200) in the same plant, in different parts of the plant, and during different times of the growing season. This genetic phenomenon is called **chromosomal drift.**

Recent investigations seem to indicate that the 2 sepals are really bracteoles, the petals are really sepals, and the petals are lacking.

The family has no known fossil record.

## Family Caryophyllaceae (Pink)

Figure 9.37 Caryophyllaceae. *Stellaria:* (a) habit of plant; (b) seed; (c) sepal and notched petal; (d) capsular fruit and calyx; (e) longitudinal section of ovary showing free central placentation. *Silene:* (f) calyx; (g) habit of plant; (h) bilobed petal with ligules.

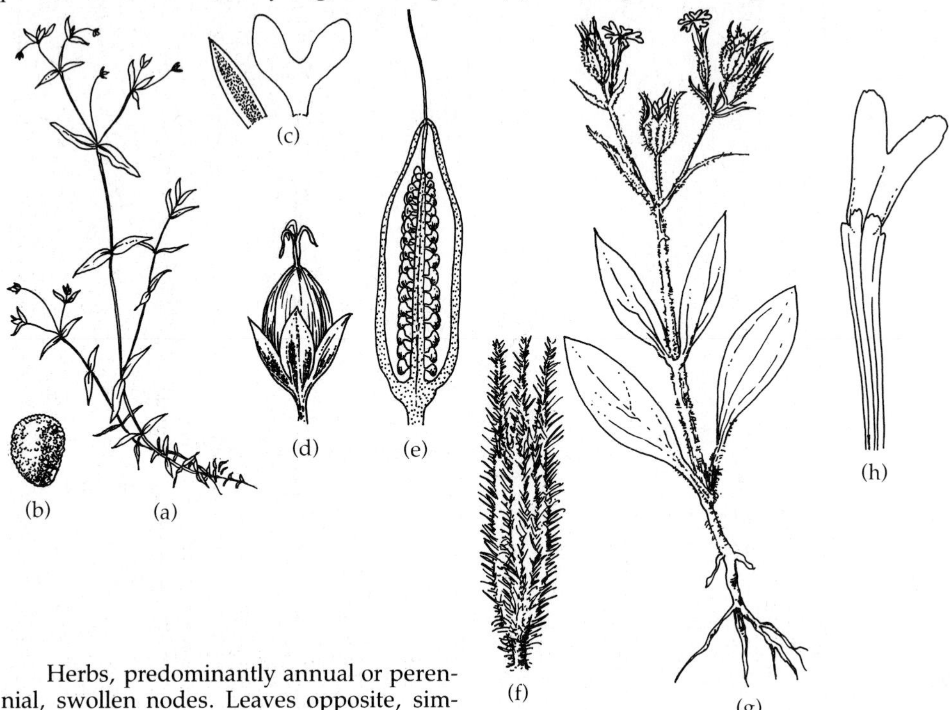


Herbs, predominantly annual or perennial, swollen nodes. Leaves opposite, simple, with the base of the paired leaves often connected; stipules generally lacking, but if present then scarous. Flowers regular (rarely irregular), perfect (rarely unisexual), hypogynous (rarely perigynous), perianth 5-merous (rarely 4-merous). Sepals distinct or connate, sometimes subtended by bract. Petals present (occasionally absent), distinct, sometimes developed into a blade and claw, blade may be bilobed and have ligule outgrowths where blade and claw join. Stamens 5 or more, filaments distinct, commonly twice the number of the petals; anthers opening by longitudinal slits. Pistil compound of 2–5 united carpels; locule 1; ovules many and borne on free-central or basal placentas; ovary superior; styles 2–5, distinct. Fruit usually a capsule, dehiscing by means of apical teeth. Seeds ornamented on the surface, embryo curved, endosperm lacking (Fig. 9.37).

The family Caryophyllaceae consists of 75–80 genera and about 2000 species distributed in all temperate parts of the world. The largest genera are *Silene,* campion or catch fly (400 species), and *Dianthus,* the pinks (300 species).

A large number of species of Caryophyllaceae are cultivated as garden ornamentals. The most important is *Dianthus caryophyllus,* carnation, used in bouquets. Other *Dianthus; Gypsophila,* baby's breath; *Agrostemma,* corn cockle; and *Lychnis,* catch fly, are also used. Various species of *Stellaria* and *Cerastium* are troublesome weeds.

The family Caryophyllaceae is very characteristic of the order because of its flower morphology, especially the basal and free-central placentation. The Caryophyllaceae and Molluginaceae differ from the other families of the order, however, in producing anthocyanin compounds instead of betalains.

Fossil pollen attributed to the family are known from Oligocene and more recent deposits.

## ORDER POLYGONALES
## Family Polygonaceae (Smartweed)

Figure 9.38 Polygonaceae. *Polygonum:* (a) habit of plant; (b) ocrea with fringed hairs; (c) flower; (d) achene fruit. *Eriogonum:* (e) mature plant in flower and lacking ocrea; (f) flower subtended by an involucre.

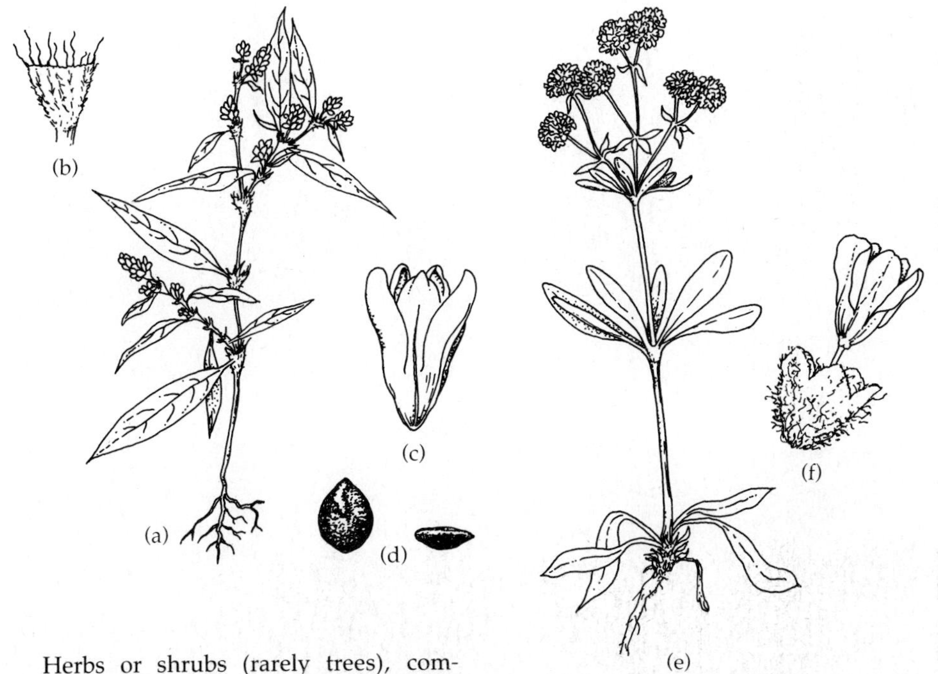


Herbs or shrubs (rarely trees), commonly with swollen nodes. Leaves alternate or whorled (rarely opposite); base of petiole a sheathing stipule around the stem, called an **ocrea** (maybe lacking in *Eriogonum*). Flowers small, white, yellow, greenish, or reddish; regular, perfect (rarely unisexual); flowers in various inflorescence types. Sepals 3–6, sometimes in 2 series and enlarged in fruit. Petals absent. Stamens usually 6–9 in 2 or 3 whorls of 3, filaments opposite the sepals, a nectar disk often surrounding the base of the ovary; anthers opening by longitudinal slits. Pistil compound of 2–4 united carpels; locule 1; ovule 1 and borne on a basal placenta; ovary superior; styles 2–3. Fruit a lens-shaped or triangular-shaped achene or nut. Seeds hard, embryo straight or curved, endosperm abundant (Fig. 9.38).

The family Polygonaceae consists of about 30 genera and 1000 species, with the most common genera being *Eriogonum,* false buckwheat (250 species); *Polygonum,* smartweed (200 species); and *Rumex,* dock (200 species). Many species of *Polygonum* and *Rumex* are common weeds. In particular, *R. acetosilla,* sheep sorrel, is now widespread into almost all climates.

Economically the family has some plants of importance. The young petioles of *Rheum rhaponticum,* rhubarb, are grown in gardens for food and *Fagopyrum esculentum,* buckwheat, is widely cultivated as a cover crop, green manure, and for its seeds. Some species are used as rock garden plants, such as various *Eriogonum* species, *Muehlenbeckia axillaris,* and *Atraphaxis frutescens.* The berries of *Coccoloba uvifera,* seaside grape, are eaten in the West Indies.

The family Polygonaceae is distinctive, with its single basal ovule and unilocular ovary, from the families in the order Caryophyllales.

Pollen attributed to the Polygonaceae has been found in Paleocene deposits.

## ORDER PLUMBAGINALES
### Family Plumbaginaceae (Leadwort)

Figure 9.39 Plumbaginaceae. *Armeria:* (a) habit of plant; (b) individual flower; (c) corolla. *Limonium:* (d) end of branch showing flowers; (e) individual flower; (f) leaf.

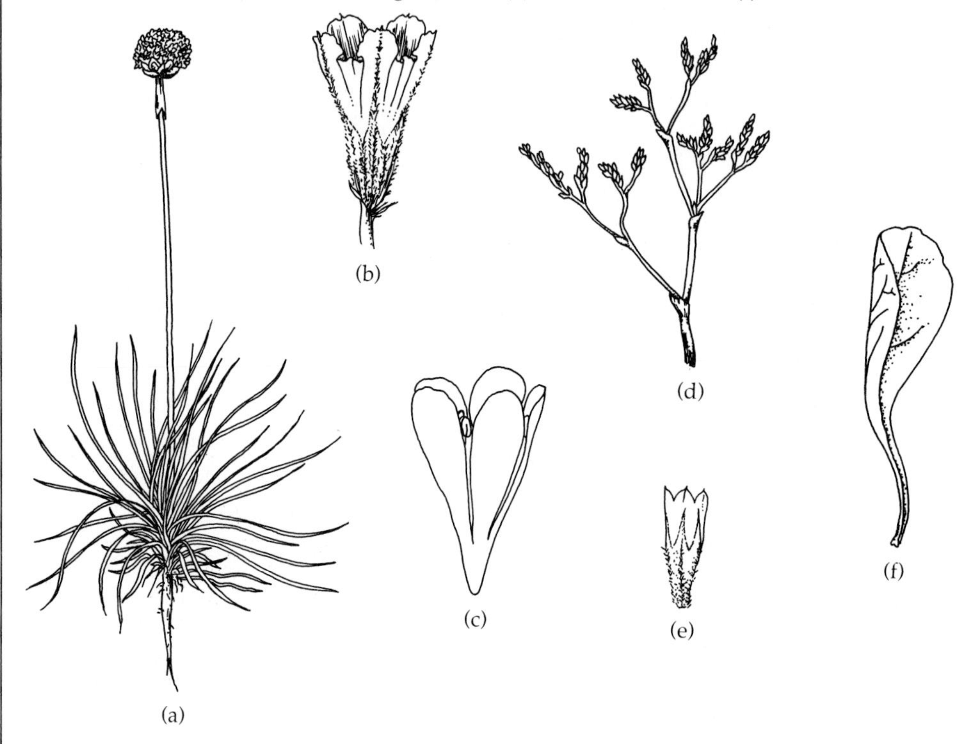


Herbs, shrubs, or vines. Leaves alternate or basal, simple; lacking stipules. Flowers regular, perfect, hypogynous, 5-merous, sometimes with bracts forming an involucre; inflorescence of cymes, heads, panicles, or racemes. Sepals connate into a calyx tube with 5 lobes. Corolla sympetalous (rarely distinct), petals clawed. Stamens 5, filaments distinct, opposite the corolla lobes, adnate to the corolla tube; anthers opening by longitudinal slits. Pistil compound of 5 united carpels; locule 1; ovule 1 and borne on a basal placenta; superior ovary; styles 1–5, distinct. Fruit an achene or various types of capsule, enclosed by the persistent calyx. Seed with a large straight embryo and no perisperm, endosperm present or absent (Fig. 9.39).

The family Plumbaginaceae is comprised of 10–12 genera and 400–560 species with a cosmopolitan distribution, but is most abundant in dry or saline habitats of Central Asia and the Mediterranean region. The main genera are *Acantholimon* (120 species) and *Armeria* (80 species). A number of species are important medically. Bronchial hemorrhages are treated with root extracts from *Limonium vulgare* in Europe. *Plumbago europaea* and *P. scandens* are used to treat dental ailments. Various *Armeria,* sea pink, such as *A. maritima,* are grown in rock gardens and species of *Limonium,* sea lavender, and *Plumbago* are cultivated as ornamentals.

Data about the Plumbaginaceae gives mixed results as to how the family should be classified. How the information is interpreted and used will reflect the bias of the individual.

Fossils of the family have been found in the fossil record of Middle Miocene deposits.

**Subclass IV. Dilleniidae** The subclass Dilleniidae is a large subclass with about 23,000–25,000 species, 78 families, and 13 orders. (See Table 9.4.) The group is a somewhat natural one with a cosmopolitan distribution, but is better represented in warmer climates than in temperate regions.

The family includes both herbaceous and woody species, as well as epiphytes and insectivorous plants. The leaves are mostly simple, rarely compound. The flowers are both polypetalous or sympetalous. The stamens may be many in number, and if so then the development is **centrifugal** (stamen development begins at the middle of the flower and proceeds away from the center axis). Commonly the stamens have various structural differences or modifications. The pistil is usually compound of fused carpels (**syncarpous**). The ovary is both superior and inferior, with axile, basal, free-central, and parietal placentation. Ovules are generally many in number, with only 1–2 ovules per locule being much less common. The fruit types vary. The seeds lack a perisperm, and the endosperm may be present or absent.

**TABLE 9.4 A list of orders and family names of flowering plants found in the subclass Dilleniidae.**

Subclass IV. Dilleniidae
- Order A. Dilleniales
  - Family 1. Dilleniaceae*
  - 2. Paeoniaceae*
- Order B. Theales
  - Family 1. Ochnaceae*
  - 2. Sphaerosepalaceae
  - 3. Sarcolaenaceae
  - 4. Dipterocarpaceae*
  - 5. Caryocaraceae
  - 6. Theaceae*
  - 7. Actinidiaceae*
  - 8. Scytopetalaceae
  - 9. Pentaphylacaceae
  - 10. Tetrameristaceae
  - 11. Pellicieraceae
  - 12. Oncothecaceae
  - 13. Marcgraviaceae
  - 14. Quiinaceae
  - 15. Elatinaceae
  - 16. Paracryphiaceae
  - 17. Medusagynaceae
  - 18. Clusiaceae (Guttiferae)*
- Order C. Malvales
  - Family 1. Elaeocarpaceae*
  - 2. Tiliaceae*
  - 3. Sterculiaceae*
  - 4. Bombacaceae*
  - 5. Malvaceae*
- Order D. Lecythidales
  - Family 1. Lecythidaceae*
- Order E. Nepenthales
  - Family 1. Sarraceniaceae*
  - 2. Nepenthaceae*
  - 3. Droseraceae*
- Order F. Violales
  - Family 1. Flacourtiaceae*
  - 2. Peridiscaceae
  - 3. Bixaceae
  - 4. Cistaceae*
  - 5. Huaceae
  - 6. Lacistemataceae
  - 7. Scyphostegiaceae
  - 8. Stachyuraceae
  - 9. Violaceae*
  - 10. Tamaricaceae*
  - 11. Frankeniaceae
  - 12. Dioncophyllaceae
  - 13. Ancistrocladaceae
  - 14. Turneraceae*
  - 15. Malesherbiaceae
  - 16. Passifloraceae*
  - 17. Achariaceae
  - 18. Caricaceae*
  - 19. Fouquieriaceae*
  - 20. Hoplestigmataceae
  - 21. Cucurbitaceae*

*(continued)*

*TABLE 9.4 (continued)*

Order F. Violales (cont.)
  Family 22. Datisaceae
    23. Begoniaceae*
    24. Loasaceae*
Order G. Salicales
  Family 1. Salicaceae*
Order H. Capparales
  Family 1. Tovariaceae
    2. Capparaceae*
    3. Brassicaceae (Cruciferae)*
    4. Moringaceae
    5. Resedaceae*
Order I. Batales
  Family 1. Gyrostemonaceae
    2. Bataceae
Order J. Ericales
  Family 1. Cyrillaceae
    2. Clethraceae*
    3. Grubbiaceae
    4. Empetraceae*
    5. Epacridaceae*
    6. Ericaceae*
    7. Pyrolaceae*
    8. Monotropaceae*
Order K. Diapensiales
  Family 1. Diapensiaceae*
Order L. Ebenales
    1. Sapotaceae*
    2. Ebenaceae*
    3. Styracaceae*
    4. Lissocarpaceae
    5. Symplocaceae*
Order M. Primulales
  Family 1. Theophrastaceae*
    2. Myrsinaceae*
    3. Primulaceae*

*NOTE:* Family names followed by an asterisk are discussed and illustrated in the text.

## ORDER DILLENIALES
### Family Dilleniaceae (Dillenia)

Figure 9.40 Dilleniaceae. *Hibbertia:* (a) branch of plant showing bud, flowers, and fruit; (b) longitudinal section of ovary of individual carpel; (c) front view of stamens and pistils. (*SOURCE:* Redrawn with permission from Cronquist, 1981.)

Herbs, shrubs, woody vines (lianas), or trees. Leaves alternate, simple, usually leathery and evergreen, having prominent veins; with or without stipules; often with crystal sand in the parenchymous tissue. Flowers white or yellow, regular, usually perfect (rarely unisexual), hypogynous; inflorescence of solitary flowers or flowers in cymes. Sepals and petals 5, imbricate (overlapping). Stamens usually many but may be reduced to a few in number, filaments distinct or connate, spirally arranged. Pistil simple of many numerous carpels, free or slightly fused at the base in 1 or 2 whorls, sometimes the carpels are not completely sealed; locule 1; ovules 1–many and borne on marginal placentas; ovary superior; style 1, stigma capitate. Fruit berry-like or a follicle. Seed with an aril and very small, embryo straight, endosperm present (Fig. 9.40).

The family Dilleniaceae is comprised of 10–18 genera and 350–530 species distributed largely pantropically and best represented in the Australasia region. About one third of the species are in the genus *Hibbertia.*

*Dillenia* and *Hibbertia* are cultivated as ornamentals in the southern United States. The flowers of *Dillenia* are like gigantic "buttercup" flowers; *D. obovata* is yellow and about 15 cm (6 in.) across and *D. indica* is 20 cm (8 in.) in diameter and white.

How the family Dilleniaceae should be classified is debated by botanists. The members are very similar to the Magnoliidae in having distinct carpels, some of which are not completely sealed, many stamens, and small embryos with much endosperm. They differ in various chemical features and in how the many stamens are formed.

The family has no known fossil record, but many Dilleniaceae have features that are considered "primitive" by botanists.

## Family Paeoniaceae (Peony)

Figure 9.41 Paeoniaceae. *Paeonia:* (a) habit of leafy flowering stem; (b) follicle fruits; (c) stamen; (d) longitudinal section of pistil.

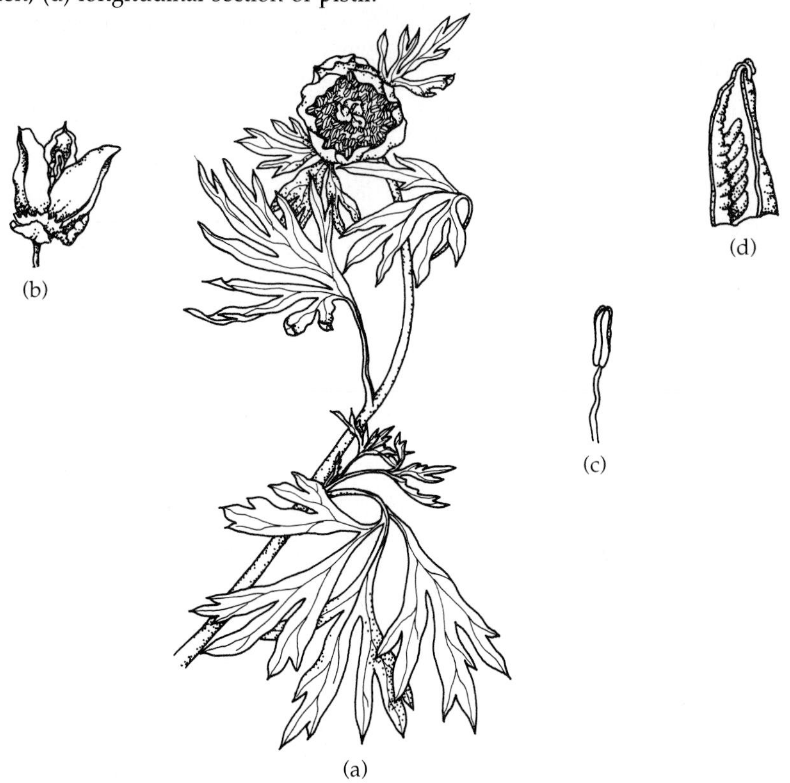


Herbs or soft shrubs, perennial, stems usually thick, arising from soft tubers or rhizomes. Leaves alternate, pinnately or ternately compound to highly dissected or lobed; stipules lacking. Flowers large, conspicuous, regular, bisexual, hypogynous, somewhat round in appearance. Sepals usually 5, green, may be irregular, persistent and leathery in fruit. Petals 5–10, large, distinct, colored white to purple but rarely yellow. Stamens many, filaments distinct, spiral, centrifugally attached to a fleshy disk, which surrounds the carpels; anthers opening by longitudinal slits. Pistil simple of 5 (sometimes 2–8) fleshy carpels; locules 1 per carpel; ovules 2–many and borne on marginal placentas; ovary superior. Fruit of leathery follicles. Seeds several, large, red turning black at maturity; embryo very small, endosperm abundant (Fig. 9.41).

The family Paeoniaceae consists of 1 monotypic genus, *Paeonia,* and 33 species. Except for several species in the western United States, most species are found in temperate regions of Asia and Southern Europe.

Economically the family is of value for the many ornamental peonies cultivated for their attractive flowers. *Paeonia officinalis* may have flowers over 15 cm (6 in.) in diameter.

*Paeonia* at one time was included in the Ranunculaceae, but recent evidence supports classifying the group as a distinct family.

The family has no known fossil record.

## ORDER THEALES
### Family Ochnaceae (Ochna)

Figure 9.42 Ochnaceae. *Ochna:* (a) branch with flower; (b) stamen; (c) pistil; (d) fruit. (*SOURCE:* Parts (b), (c), and (d) redrawn with permission from Cronquist, 1981.)

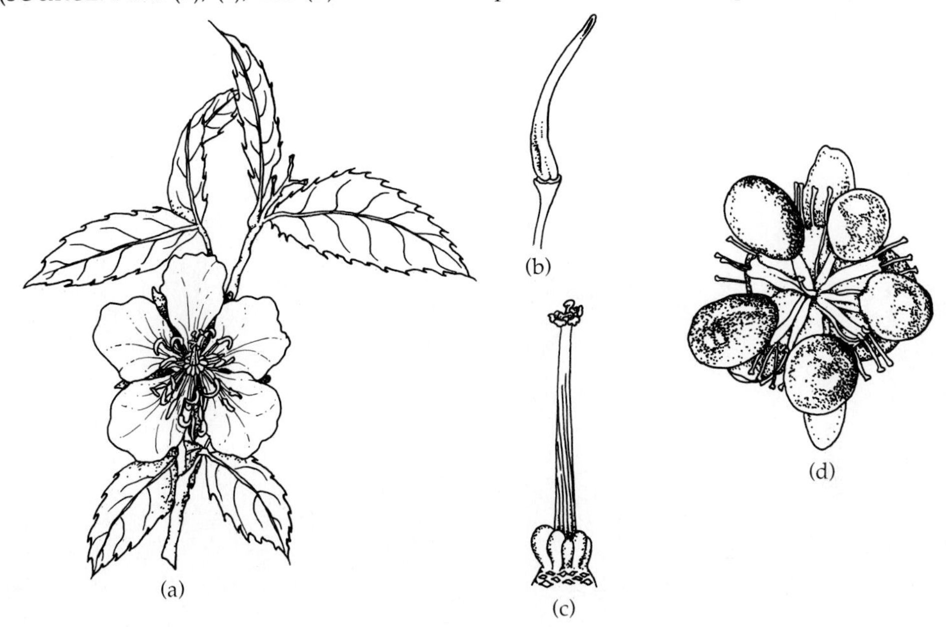


Shrubs or trees. Leaves alternate, simple (rarely pinnately compound), leathery; with stipules. Flowers regular, perfect, hypogynous; inflorescence arranged in panicles, racemes, or false umbels. Sepals generally 5, distinct or connate at the base. Petals usually 5, (rarely 4–10), sometimes contorted. Stamens 5–many, filaments distinct or slightly connate, anthers opening by pores or longitudinal slits. Pistil compound of 2–5 (rarely more) united carpels that are distinct at the base but connate with a common style; locules 2–5; ovules 1–many per locule and borne on axile or parietal placentas; ovary superior; style 1. Fruit a berry or aggregate of drupes (rarely a capsule), the receptacle of some genera enlarges and becomes red and fleshy. Seed with a straight embryo, endosperm present or lacking (Fig. 9.42).

The family Ochnaceae consists of about 40 genera and 600 species distributed pantropically, chiefly in South America and Brazil. The largest genus is *Ouratea* (300 species).

Economically the family is of minor importance. *Ochna multiflora* is cultivated in the southern United States, while *O. kibbiensis* and *O. flava* are cultivated in temperate greenhouses.

Some botanists feel this diverse family is most similar in character to the Dipterocarpaceae, while others have separated out a number of smaller families.

The family has no known fossil record.

## Family Dipterocarpaceae (Meranti)

Figure 9.43 Dipterocarpaceae. *Dipterocarpus:* (a) branch with flowers; (b) longitudinal section of flower; (c) stamen; (d) fruit with persistent winged calyx. (*SOURCE:* Redrawn with permission from H. Keng, *Orders and Families of Malayan Seed Plants*. Singapore: Singapore University Press, 1978.)

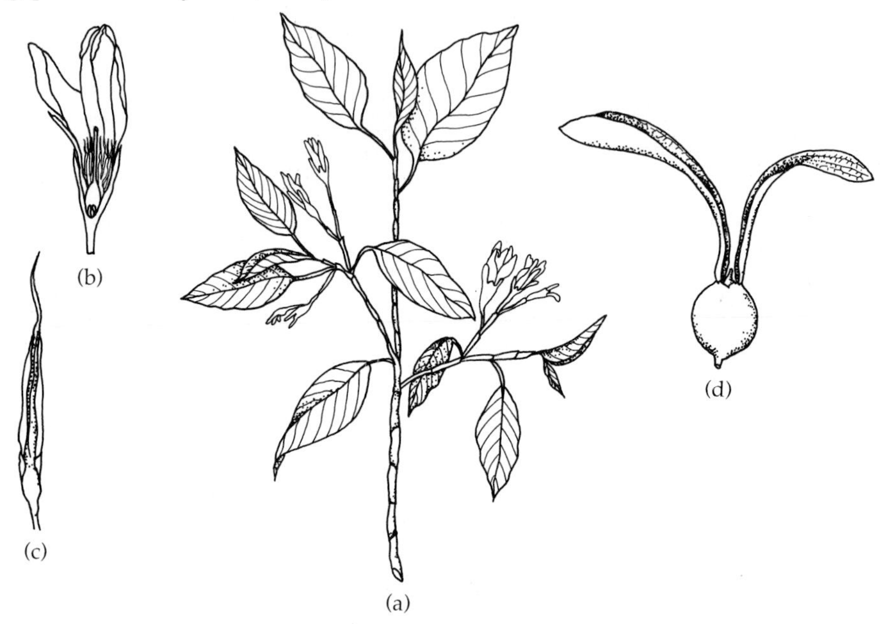


Trees, small to large with buttressed bases and smooth unbranched trunks up to the crown; all tissues have special resin canals. Leaves alternate, simple, generally evergreen, entire; stipules present frequently with **domatia** (special depressions, sometimes housing small insects or mites). Flowers fragrant, regular, perfect, perigynous or epigynous; inflorescence in axillary or terminal racemes or panicles. Sepals 5, distinct or connate, sometimes enlarged and winglike at fruiting time. Petals 5, distinct or connate at the base and spirally twisted, often leathery. Stamens 5–many, filaments free or connate at the base; anthers distinctive with sterile tips. Pistil compound of 3 united carpels, locules 3; ovules 2 per locule, but only 1 ovule develops and is borne on an axile placenta; ovary superior or inferior. Fruit a single-seeded nut but with persistent calyx in fruit forming wing-like appendages that help in wind dispersal. Seed with endosperm lacking (Fig. 9.43).

The family Dipterocarpaceae consists of about 15–16 genera and 600 species distributed heavily in tropical Asia and Indomalaysia, with a smaller representation in tropical Africa and South America. The largest genera are *Shorea* (150 species) and *Dipterocarpus* (75 species). Species of this family make up much of the monsoon forests in regions of Burma, India, and the evergreen forests of Malaysia.

Economically the family is of some importance. Wood from dipterocarp trees are the world's main source of tropical hardwood timber. The species *Shorea robusta* and *Dipterocarpus tuberculatus* form almost pure forest stands in Burma. With the rapid depletion of the world's tropical rain forests, many of these hardwoods are doomed by the end of this century unless conservation methods are followed.

The family has characters similar to the Ochnaceae and also to the families Elaeocarpaceae and Tiliaceae in the order Malvales.

Pollen and macrofossils attributed to the family have been found in Oligocene deposits, with abundant pollen found in Miocene sediments in Borneo.

## Family Theaceae (Tea, Camellia)

Figure 9.44 Theaceae. *Franklinia:* (a) branch with flowers; (b) Cross section of ovary; (c) stamen; (d) pistil. (*SOURCE:* Parts (b) and (d) redrawn with permission from Cronquist, 1981.)

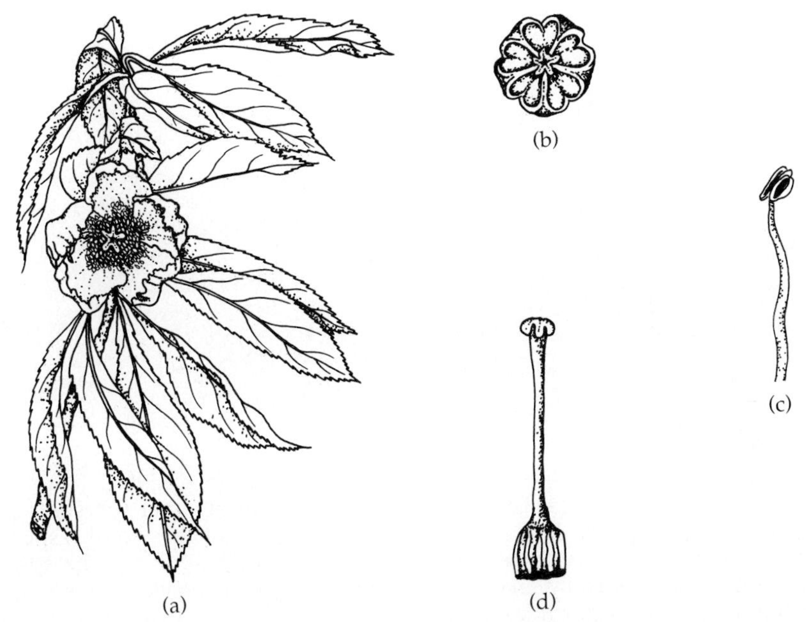


Shrubs or trees. Leaves alternate, simple, entire or toothed; usually evergreen and lacking stipules. Flowers commonly large and showy, regular, perfect or rarely unisexual (dioecious), hypogynous to epigynous; inflorescence of solitary flowers or flowers arranged axillary or in terminal racemes or panicles, subtended by a pair of bracts. Sepals 4–7, distinct or connate, persistent in fruit. Petals 4–7 (rarely many), distinct or connate at the base. Stamens many, filaments distinct or connate at the base into bundles or a ring alternating with the petals. Pistil compound of 3–5 (rarely 6–many) united carpels; locules as many as carpels; ovules usually many (rarely 1 or 2) and borne on axile placentas; ovary superior (rarely inferior); styles 3–5 (rarely 6–many), distinct or connate. Fruit a capsule or less commonly an achene or berry. Seed with a straight or curved embryo, endosperm present or absent (Fig. 9.44).

The family Theaceae is comprised of 16–40 genera and 500–1100 species distributed widely throughout the tropics but centered chiefly in Asia and the American tropical regions. The largest genera are *Ternstroemia* (130 species), *Camellia* (including *Thea,* 82 species), *Eurya* (70 species), and *Laplacea* (40 species).

Economically the family is of some importance. *Camellia japonica* is a well-known ornamental cultivated in the Far East for its beautiful, scented flowers. *Camellia sinensis,* the tea plant, a native of Southeast Asia, has long been cultivated as a stimulating beverage and medicinal plant. *Gordonia* and *Stewartia,* which are sometimes called "camellias," are found in the southern United States. *Franklinia alatamaha,* the Franklin tree, originally native and described from near Forth Barrington, Georgia, U.S.A., is now known only from cultivation.

The family is very similar to the Dilleniaceae but differs in various flower and seed characters.

Fossil pollen and some questionable macrofossils attributed to the Theaceae have been found in Eocene and more recent deposits.

## Family Actinidiaceae (Chinese Gooseberry)

Figure 9.45 Actinidiaceae. *Actinidia:* (a) leafy branch bearing pendulous flowers; (b) longitudinal section through ovary; (c) stamen; (d) front view of flower. (*SOURCE:* Redrawn with permission from Cronquist, 1981.)

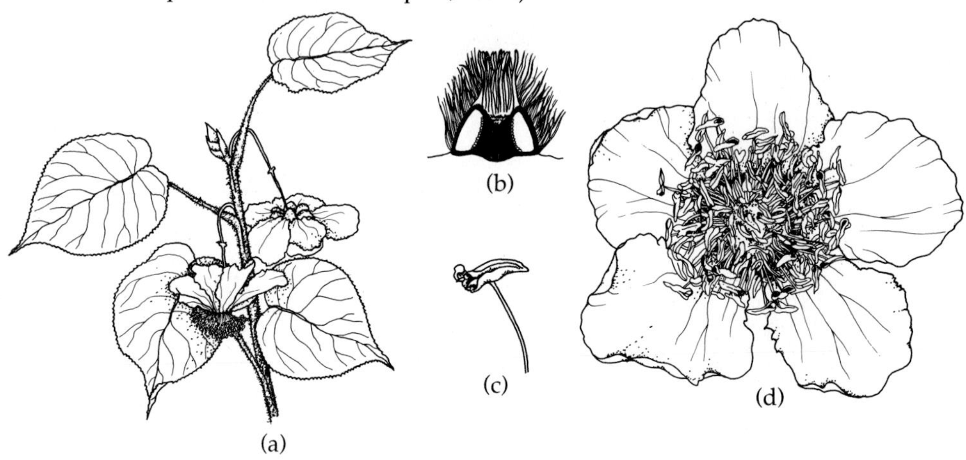


Shrubs, trees, or lianas. Leaves alternate, simple; lacking stipules. Flowers regular, perfect or unisexual (plants dioecious), hypogynous, often subtended by small bracteoles; inflorescence of axillary cymes (rarely heads). Sepals 5, distinct. Petals 5 (rarely fewer or more), distinct or connate at the base. Stamens 10–many, filaments distinct, commonly adnate to the petals; anthers opening by apical pores or short longitudinal slits. Pistil compound of 3–many united carpels; locules as many as the carpels; ovules 10 or more per locule and borne on axile placentas; ovary superior; styles 5–many, usually persistent. Fruit a berry or loculicidal capsule. Seeds small, embryo curved or straight, endosperm oily (Fig. 9.45).

The family Actinidiaceae consists of 3 or 4 genera and about 300–350 species distributed widely throughout the subtropical to tropical regions of the world, especially in the Asiatic tropics. Most of the species belong to the genus *Saurauria* (250 species), with about one third of them in the American tropics.

Economically the family is of minor importance. *Actinida* has about 35 species, one of which *A. chinesis,* Chinese gooseberry or kiwi fruit, is imported into North America and Europe from New Zealand. The species is native to China, but when grown in the northern New Zealand climate, the plants do very well as a commercial fruit crop.

The family has various features with the Dilleniaceae, Ericaceae, and Theaceae, and at one time or another has been placed in each of these groups.

The family has no known fossil record.

## Family Clusiaceae (Guttiferae) (Mangosteen)

Figure 9.46 Clusiaceae. *Hypericum:* (a) habit of plant; (b) cross section of ovary; (c) stamen; (d) clustered stamens and pistil; (e) face view of flower.

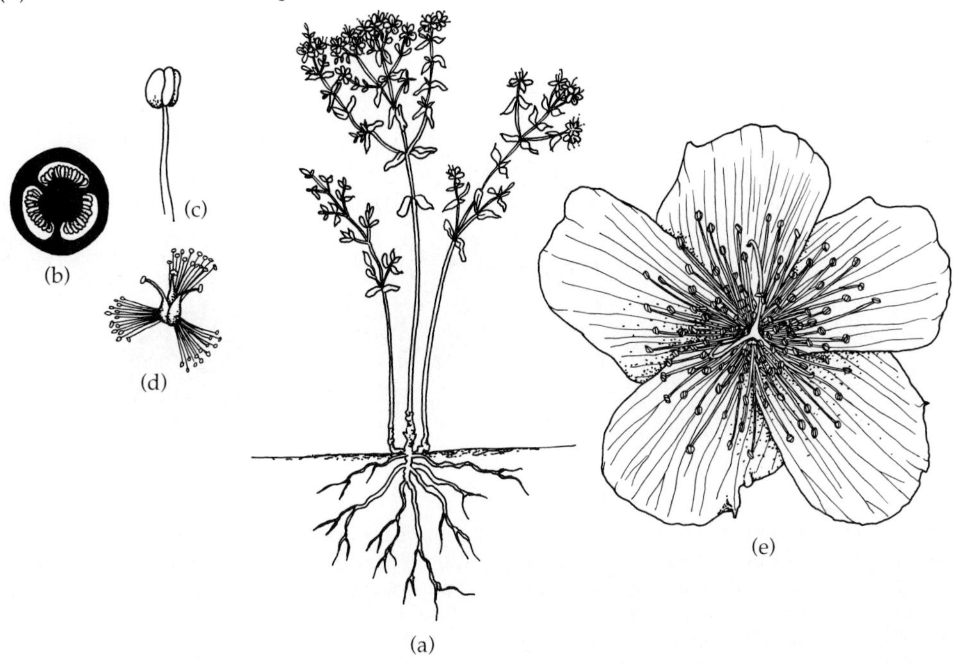


Herbs, shrubs, or trees (rarely lianas). Leaves usually opposite or whorled, simple, entire; lacking stipules; often dotted with resin cavities, secretory cavities, or canals, commonly scattered throughout the tissues and filled with white or yellow fluid. Flowers regular, perfect, or unisexual, hypogynous; inflorescence of solitary flowers or flowers in terminal cymes. Sepals 4–5, distinct, often with bracts just below. Petals 3–6 (rarely 4–14), distinct or connate at the base, often yellow. Stamens many, grouped in bundles, opposite the petals. Pistil compound and comprised of 3–5 (rarely more or less) united carpels; locules 1 or 3–5; ovules many and borne on parietal or axile placentas; ovary superior; styles the same number as the carpels. Fruit a capsule but sometimes a berry or drupe. Seeds often with wings or an aril, embryo straight, endosperm lacking (Fig. 9.46).

The family Clusiaceae is comprised of 40–50 genera and 1000–1200 species distributed widely but most common in the moist tropics. The largest genera are *Hypericum*, St. John's wort (350 species); *Garcinia*, mangosteen (220 species); and *Clusia*, the autograph tree (200 species).

Economically the family is of some importance. *Garcinia mangostana*, the mangosteen, is a very well-known tropical fruit. *Mammea americana*, the mammey apple, is commonly cultivated in the American tropics. Drugs and cosmetics are obtained from *Harungana madagascariensis*, *Hypericum* ssp., and *Mesua ferrea*. Wood for timber is harvested from species of *Calophyllum*, *Cratoxylum*, *Mesua*, and *Platonia*. Species of *Hypericum (H. perforatum, H. japonicum)* are weeds in more temperate and tropical climates.

The subdivision of the family into 2–5 subfamilies has led to confusion as to the correct name for the family. If the -aceae ending is followed, the correct name is Clusiaceae with the various subfamilies segregated from one another on sexual, fruit, and seed characters.

Clusiaceae-like pollen has been found in Eocene and more recent deposits.

## ORDER MALVALES
### Family Elaeocarpaceae (Elaeocarpus)

Figure 9.47 Elaeocarpaceae. *Muntingia:* (a) branch with flowers; (b) flower lacking perianth; (c) flower; (d) cross section of ovary. (*SOURCE:* Part (b) redrawn with permission from Cronquist, 1981.)

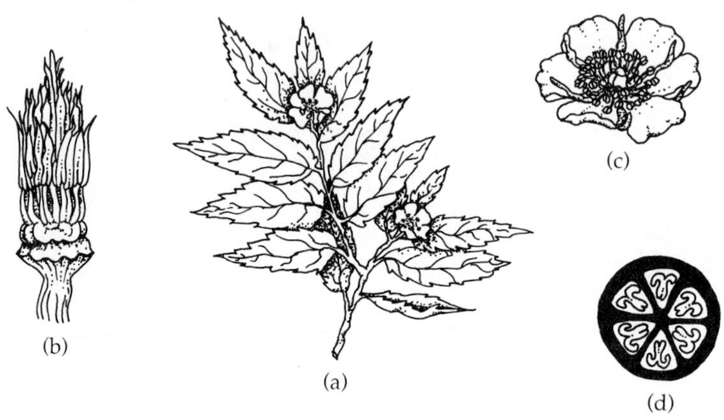


Shrubs or trees. Leaves alternate or opposite, simple; stipules present. Flowers regular, perfect; inflorescence of cymes, panicles, or racemes. Sepals and petals 4–5 or absent altogether, usually free, petals often fringed at the tip. Stamens numerous, free, adnate to a disk; 2 locules of the anthers opening by pores or short slits. Pistil compound of 2–many united carpels; locules 2–many; ovules 2 or more and borne on axile placenta; ovary superior; style simple to lobed. Fruit a capsule or drupe. Seeds with a straight embryo, endosperm abundant and oily (Fig. 9.47).

The family Elaeocarpaceae is comprised of about 10 genera and 400 species that are widespread in tropical and subtropical areas, especially eastern Asia to Australasia to Pacific regions, the West Indies, and South America; missing from Africa. Over one-half the species are found in the genus *Elaeocarpus* (250 species), with another 100 species in *Sloanea*.

The economic uses of species in this family are primarily as ornamental shrubs or as fruits used locally for food. *Elaeocarpus cyaneus* of Australia and *E. dentatus* from New Zealand are cultivated in Europe. In Chile, *Aristotelia chilensis* berries are made into wine and reported to have medicinal properties.

The family appears to be related to the Tiliaceae, and some botanists have even combined the two families together.

Fossil fruits attributed to the family have been found in Eocene deposits, and fossil wood has been collected from Paleocene sediments in Patagonia.

## Family Tiliaceae (Linden)

Figure 9.48 Tiliaceae. *Tilia:* (a) branch with flowers and subtending bracts; (b) winter twig and buds; (c) flower; (d) cross section of ovary; (e) fruits and bract.

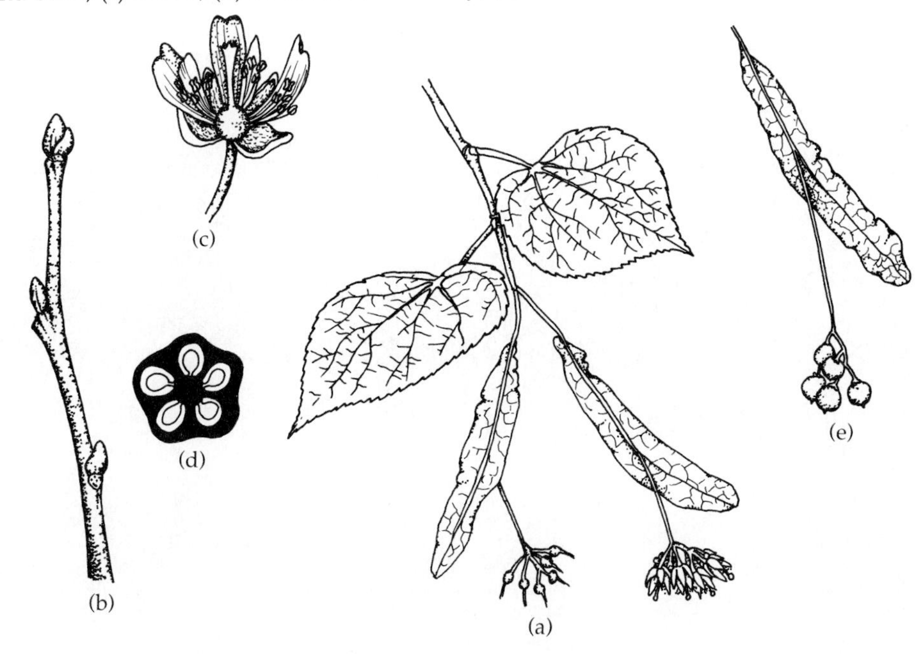


Shrubs or trees (rarely herbs). Leaves alternate (rarely opposite), simple, deciduous; asymmetrical bases with stipules and branched hairs. Flowers regular, generally perfect, hypogynous, sometimes with an epicalyx; inflorescence of cymes. Sepals and petals 3–5 (rarely lacking), distinct or connate at the base; nectaries often as glandular hairs at the base. Stamens 10–many, filaments distinct or connate into groups of 5 or 10, adnate to the base of the petals; anthers opening by longitudinal slits (rarely by apical pores). Pistil compound of 2–many united carpels; locules as many as carpels; ovules 1–many per locule and borne on axile placentas; ovary superior; style single, stigma lobed or capitate. Fruit fleshy or dry of various types. Seeds with straight embryos, endosperm present (Fig. 9.48).

The family Tiliaceae consists of 41–50 genera and 400–450 species widely distributed throughout the tropical world with a few *Tilia* species extending into the temperate regions of Europe and North America. The largest genera are *Grewia* (150 species) and *Triumfetta* (70 species).

Economically the family is of some importance. Some *Tilia* tree species are valuable for timber. *Tilia americana,* American basswood, and *T. cordata,* European lime or linden, are good for making furniture and musical instruments, and as ornamental trees. Bees make a superb honey from linden flowers. Jute is made from the fibers of the herbaceous genus *Corchorus.*

The family is similar in many respects to the Malvaceae but differs in various stamen characters.

The first substantial fossil record of leaves, flowers, and fruits of the Tiliaceae came from the Upper Oligocene or Lower Miocene.

## Family Sterculiaceae (Cacao)

Figure 9.49 Sterculiaceae. *Melhania:* (a) leafy branch showing buds. *Fremontia:* (b) longitudinal section of filament tube showing pistil; (c) front view of flower. *Hannafordia:* (d) fruit.

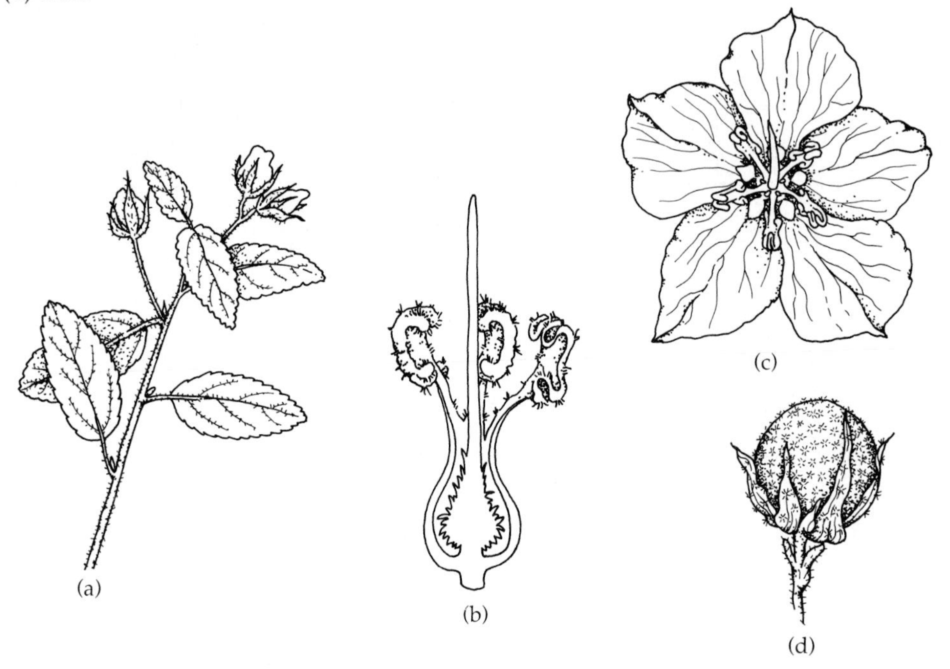


Shrubs or trees (few herbs and vines), wood soft. Leaves alternate, simple; stipules present, many species with stellate hairs. Flowers regular, perfect or unisexual on the same plant, hypogynous; inflorescence in complex cymes. Sepals 3–5, more or less connate, sometimes subtended by an epicalyx. Petals 5, distinct or connate, often absent or very small in size. Stamens 5 or 10 in two whorls, the outer ring often reduced to staminoids, the inner whorl bears the anthers; anthers open by small apical slits or pores. Pistil compound of 2–12 united carpels; locules as many as carpels; ovules 2 or more per locule and borne on axile placenta; ovary superior; style 1, 1–many branched. Fruit a berry, capsule, follicle, or schizocarp. Seeds with embryo straight or curved, little or no endosperm (Fig. 9.49).

The family Sterculiaceae consists of 60–65 genera and 700–1000 species distributed pantropically, extending into subtropical regions. The largest genera are *Dombeya* (350 species) in Africa, Madagascar, and the Mascarene Islands and *Sterculia* (300 species), named after the Roman god of privies, Sterculius, because many of the flowers and leaves smell like dung.

Economically the family is important for two taxa: the genus *Cola* with 125 African species includes *C. nitida*, cola, and *Theobroma* (30–50 species), native to America, of which *T. cacao* provides us with cocoa.

The family is very diverse morphologically, resulting in differences of opinion as to what should or should not be included in the family.

Fossils attributed to the family appear first in the upper Cretaceous, with leaves and pollen occurring in the Eocene and Paleocene deposits, respectively.

## Family Bombacaceae (Kapok Tree)

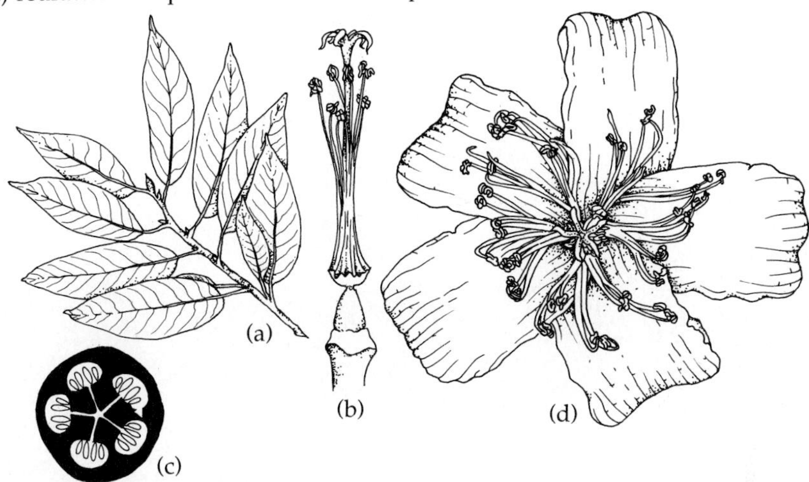


Figure 9.50 Bombacaceae. *Durio:* (a) leafy branch. *Bombax:* (b) pistil and stamens with perianth removed; (c) cross section of ovary; (d) front view of flower. (*SOURCE:* Parts (b) and (d) redrawn with permission from Cronquist, 1981.)

Trees, often large, commonly with soft, light wood, generally with a peculiar thickened, barrel-shaped or bottle-shaped trunk, an adaptation to arid environments for water storage forming at the end of the rainy season. Leaves alternate, entire, simple or palmately compound; deciduous stipules present. Flowers large and showy, regular, open during the leafless period, perfect, hypogynous or slightly perigynous; inflorescence of solitary flowers or flowers in axillary clusters; the entire flower is subtended by sepal-like appendages, the epicalyx. Sepals 5, distinct or connate. Petals 5, distinct or rarely lacking. Stamens 5–many, filaments connate into a tube and adnate to the base of the petals; anthers opening by longitudinal slits. Pistil compound of 2–5 united carpels; locules the same number as carpels; ovules 2–many per locule and borne on axile placentas; ovary superior; style 1, elongate, stigma lobed. Fruit a capsule. Seeds smooth, often with long hairs, embryo curved, endosperm scanty or lacking (Fig. 9.50).

The family Bombacaceae consists of 20–30 genera and about 180–200 species confined to warm climates, but most common in the American tropics. The largest genus is *Bombax* (about 60 species).

Economically the family is of some importance. The cottony hairs around the seeds are used in cushions and the soft wood carved into dug-out canoes. *Bombax ceiba*, the silk tree from tropical Asia, and *Ceiba pentandra* are the source of kapok, a filling material for commerce. Several species of *Adansonia*, the baobab of Africa, are ant pollinated and have ants living within their spines. Baobab is an important tree in African society. *Ochroma pyramidale* of tropical America is the source for balsa wood. The foul-smelling but delicious fruit of *Durio zibethinus*, the durian, is extensively eaten in Malaya and Southeast Asia.

The Bombacaceae is similar to and sometimes has been classified in the Malvaceae. Its distinctive smooth pollen is in contrast to the rough Malvaceae pollen.

Fossil pollen attributed to the family has been found in Upper Cretaceous and more recent deposits in various parts of the world.

## Family Malvaceae (Mallow)

Figure 9.51 Malvaceae. *Malva:* (a) leafy stem with flowers and fruits; (b) pistil; (c) longitudinal section through stamen tube; (d) fruit with subtending calyx. (*SOURCE:* Parts (b), (c), and (d) redrawn with permission from Cronquist, 1981.)

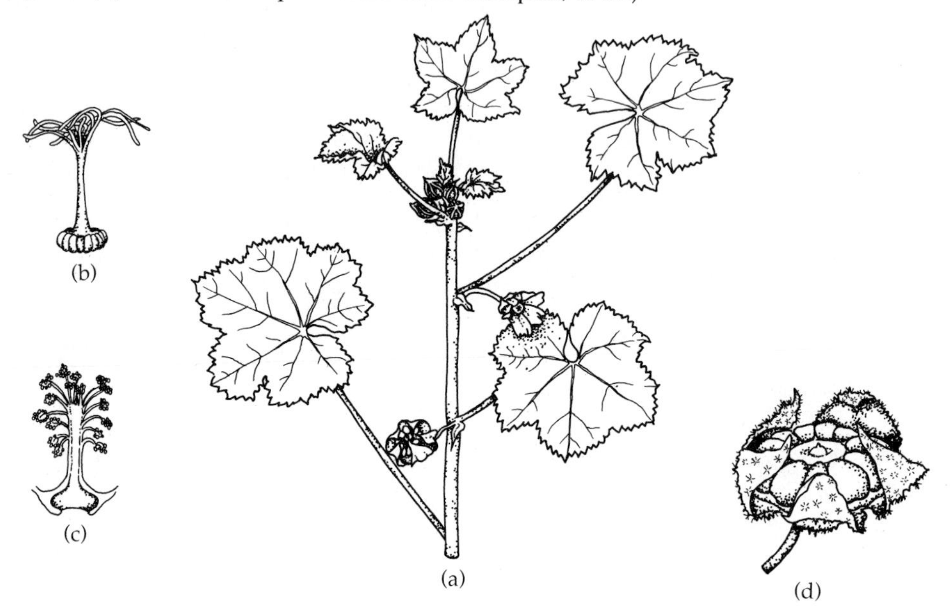


Herbs, shrubs, or rarely small trees. Leaves simple, alternate, generally palmately veined with stipules; epidermis having stellate hairs. Flowers regular, hypogynous, generally perfect; solitary or in axillary to cymose inflorescences. Sepals 5, sometimes joined at base and often subtended by an epicalyx thought to be fused stipules or bracteoles. Petals 5, distinct but often adnate to the filament tube. Stamens many, monadelphous, and united by their filaments into a tube; anthers with 1 locule, opening by longitudinal slits. Pistil compound of 2–many united carpels; locules as many as carpels; ovules 1–many per locule and borne on axile placentas; ovary superior; style branched. Fruit a loculicidal capsule or schizocarp (rarely a berry). Seeds 1–many, often covered with hairs, embryo curved or straight and with little or no endosperm (Fig. 9.51).

The family Malvaceae is comprised of 75–80 genera and 1000–1500 species cosmopolitan in distribution but is best developed in the South American tropics. The main genera are *Hibiscus* (200+ species), *Sida* (175+ species), *Pavonia* (150+ species), and *Abutilon* (100+ species).

The Malvaceae has some important economic food and ornamental plants. These include *Gossypium,* cotton; young fruits of *Hibiscus esculentus,* okra; *Abutilon avicennae,* China jute; *Alcea rosea,* hollyhock; and *Hibiscus syriacus,* rose of Sharon.

The classification of the family is open to question, and there are various opinions as to delimiting genera and how the family is related to the Bombacaceae and Tiliaceae.

Fossil pollen attributed to the Malvaceae was first found in upper Eocene deposits.

## ORDER LECYTHIDALES
### Family Lecythidaceae (Brazil Nut)

Figure 9.52 Lecythidaceae. *Couroupita:* (a) leafy branch; (b) androecium; (c) front view of flower; (d) inflorescence. (*SOURCE:* Redrawn with permission from Cronquist, 1981.)

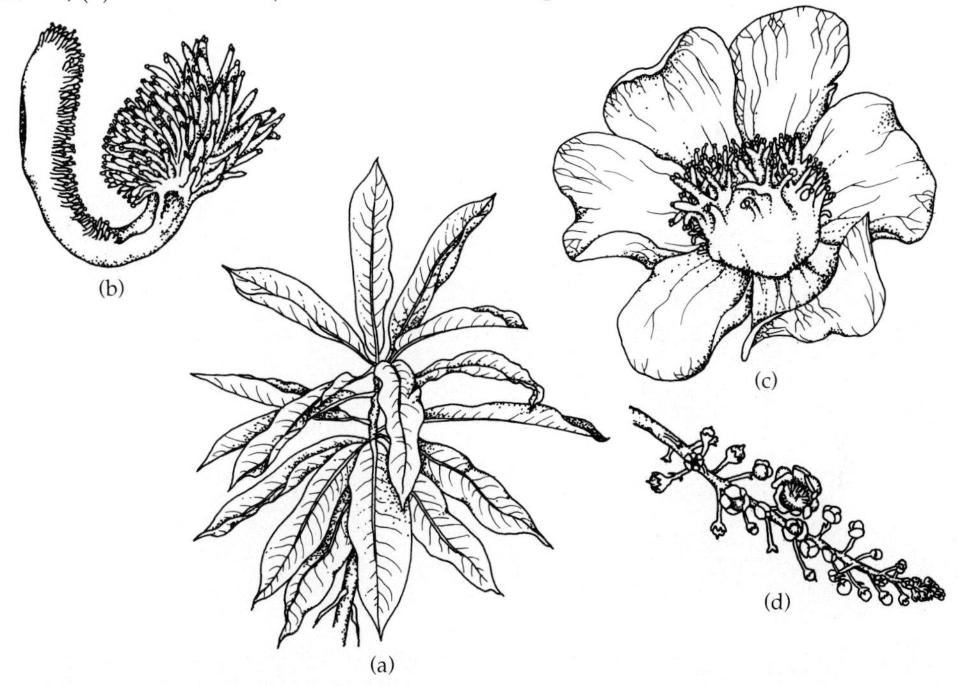


Shrubs or trees. Leaves alternate, simple, clustered at the tips of the twigs; generally lacking stipules. Flowers generally large and showy, commonly red, pink, yellow or white, sweet smelling; opening at night and pollinated by bats and night-flying insects; perfect, regular or irregular, epigynous; inflorescence of solitary flowers or flowers in spikes, racemes, or panicles and borne on the trunk or older stems. Sepals generally 4–6, distinct or connate. Petals 4–6, distinct or connate or absent. Stamens many, filaments distinct above and connate below, joined to the petals at their base in one or more rings. Pistil compound of 2–6 united carpels; locules same number as carpels; ovules 1–many and borne on axile placentas; ovary inferior; style long, simple, stigma lobed or capitate. Fruit a capsule (rarely a drupe or berry), usually large with hard inner layers that open by a lid, hence a "pot." Seeds large, woody, generally lacking endosperm (Fig. 9.52).

The family Lecythidaceae consists of about 20 genera and 400–450 species confined to the tropics, especially in South America. The largest genera are *Eschweilera* (100 species) and *Gustavia* (40 species).

The most economically important tree is *Bertholletia excelsa,* the Brazil nut. *Couroupita guianensis,* the cannonball tree, is native to South America and is cultivated as a striking ornamental. Its large (10 cm) sweet-smelling red and yellow, waxy flowers produce large round fruits ("cannon balls") 15–20 cm in diameter from the trunk. Useful timber is obtained from the species.

The family has been grouped in the past in the order Myrtales. However, when its many distinctive characters are considered together, it seems best to include the Lecythidaceae in its own order.

Fossil pollen attributed to the family has been found in Lower Eocene to more recent deposits.

## ORDER NEPENTHALES
### Family Sarraceniaceae (Pitcher Plant)

Figure 9.53 Sarraceniaceae. *Sarracenia:* (a) mature plant in flower showing pitcher-like leaves; (b) longitudinal section through flower. *Darlingtonia:* (c) pitcher-like leaves.

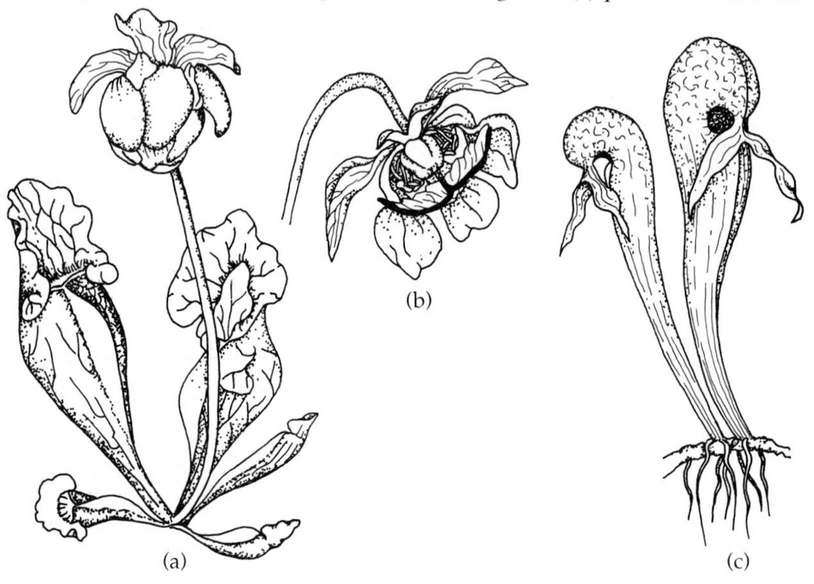


Insectivorous herbs, perennial. Leaves are highly modified organs, alternate, present in a rosette originating directly from a rhizome, most leaves with hollow elongate petioles filled with water, blades reduced to hood-like lids. Insects are attracted to the mouth of the leaf pitcher by strong odors, nectar glands, window-like perforations or colors. Inside the pitcher, insects encounter sharp downward pointed hairs that act as a slide. Drowned insects are digested by leaf-secreted enzymes. Flowers large, solitary on a scape or in racemes, regular and nodding, perfect, hypogynous, arising from the middle of the rosette. Sepals 3–6, distinct or overlapping, often colored. Petals 5 or lacking, distinct or overlapping, colored and deciduous. Stamens many, short. Pistil compound of 3–5 united carpels; locules same number as carpels; ovules many per locule and borne on axile placentas; ovary superior. Fruit a loculicidal capsule. Seeds small, numerous, embryo straight, endosperm present (Fig. 9.53).

The family Sarraceniaceae has only 3 genera and 17 species confined to the New World: *Darlingtonia californica* from near the Pacific Coast of Oregon and California, *Heliamphora* from the Guayana Highlands of South America, and *Sarracenia* found in bogs of eastern North America.

The family is economically important only as unique greenhouse cultivated ornamentals.

Some botanists believe the New World Sarraceniaceae correlates with the Old World Nepenthaceae in various anatomical and morphological features.

The family has no known fossil record.

## Family Nepenthaceae (East Indian Pitcher Plant)

Figure 9.54 Nepenthaceae. *Nepenthus:* (a) leaves with pitchers; (b) pitcher; (c) longitudinal section through pitcher; (d) male flower.

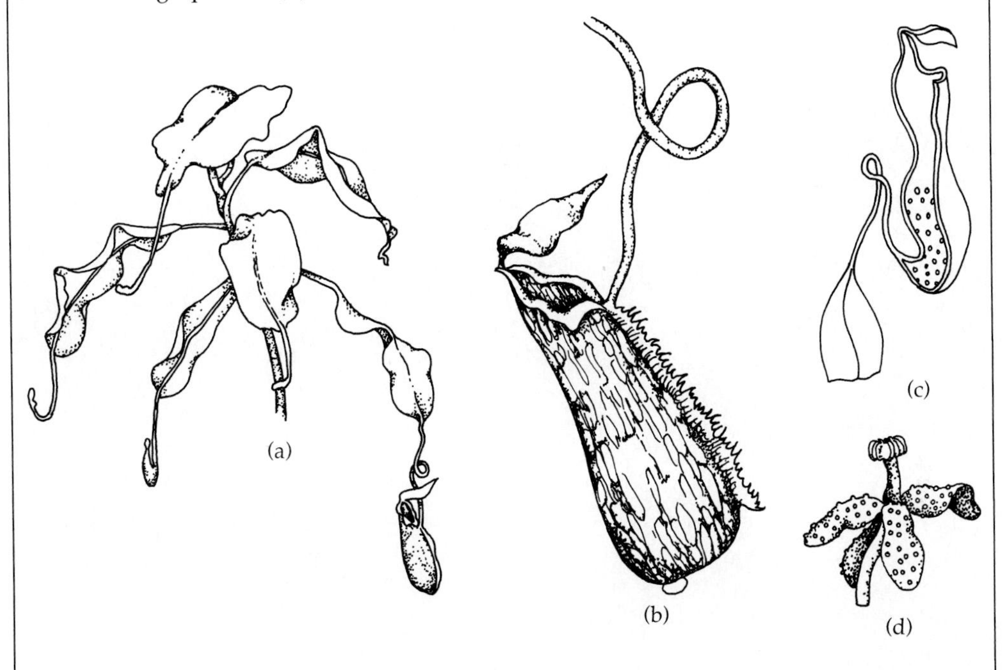


Insectivorous herbs, shrubs, and climbing vines; often epiphytic. Leaves alternate, without stipules, generally comprised of a petiole and blade. Tendril from the blade apex becomes swollen and develops into a pitcher, with a lid that projects over the mouth of the pitcher. Honey glands surround the mouth and the slippery interior surface, causing insects attracted to the honey glands to slip, fall into the water collected in the pitcher, and be absorbed as they decay. Flowers small, regular, unisexual and dioecious, hypogynous; inflorescence in spike-like to mixed panicles. Sepals 3–4, distinct to slightly connate at the base; nectaries and glands within. Petals lacking. Stamens in male flowers 4–many, filaments fused into a column with anthers crowded at the top. Pistil of female flowers compound of 3–4 united carpels; locules same number as carpels; ovules many per locule and borne on axile placentas; ovary superior; style short or lacking, stigma disk-like. Fruit a loculicidal capsule. Seeds with tufted hairs on the end, embryo straight, endosperm present (Fig. 9.54).

The family Nepenthaceae is comprised of the single genus *Nepenthus* and about 75 species distributed from Northern Australia and the East Indies to Madagascar. The family requires very moist, humid, rain forest habitats from sea level to over 2500 m (8000 ft) elevation.

Economically the family is important only as a greenhouse ornamental, and the stems of some species, like *N. reinwardtiana,* are used locally in Malaysia for basketmaking and for cord.

The family is similar to the Droseraceae and Sarraceniaceae in having the insectivorous habit. In the past, all three families were classified in three different orders.

The family has no known fossil record.

## Family Droseraceae (Sundew)

Figure 9.55 Droseraceae. *Drosera:* (a) plant showing inflorescence and leaves with sticky glandular hairs; (b) flower; (c) leaf with enlarged glandular hairs. *Dioneae:* (d) plant with trap leaves.

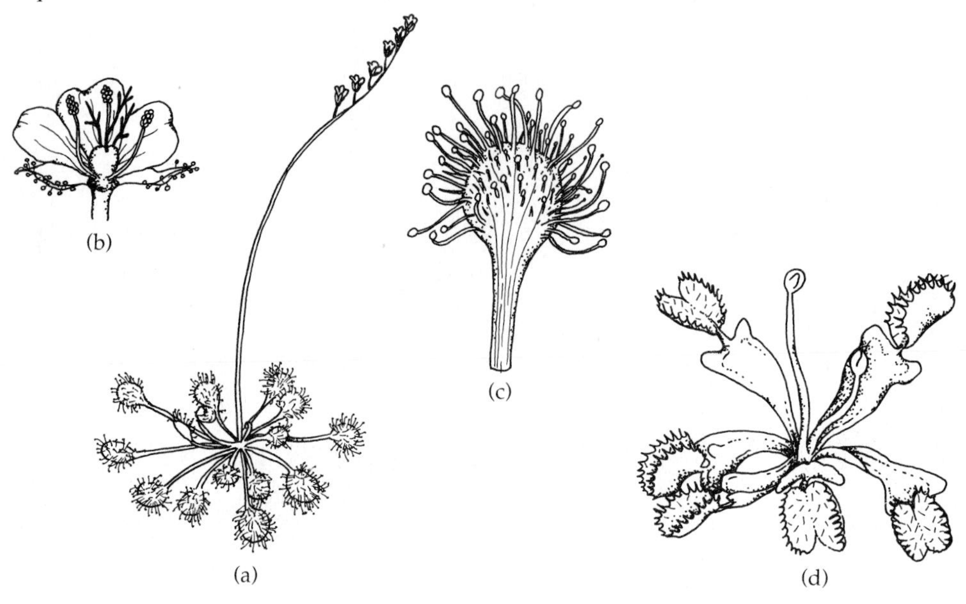


Insectivorous herbs or subshrubs, annual or perennial. Leaves alternate or in basal rosettes, with or without stipules, covered with sticky glands or the blade modified into a hinged trap. Flowers regular, perfect, hypogynous; inflorescence of solitary flowers or flowers in cymes. Sepals and petals 4–5, each distinct or connate at the base. Stamens 4–many, filaments distinct or connate at the base. Pistil compound of 2–5 united carpels; locule 1; ovules many and borne on parietal or basal placentas; ovary superior; styles 1–5. Fruit a loculicidal capsule. Seeds usually many, embryo straight, endosperm present (Fig. 9.55).

The family Droseraceae is comprised of 4 genera and 103 species with a cosmopolitan distribution. *Drosera,* the sundew, is the largest genus (100 species).

Economically the family is of interest for ornamental and greenhouse use only.

The family's greatest diversity is found in Australia and New Zealand. The leaves of *Drosera* and *Drosophyllum* are covered with long, red, gland-tipped hairs that hold and digest insects by secreting ribonucleases and enzymes. In *Dionaea muscipula,* the Venus flytrap in the southeastern United States, the leaves are divided into two halves that can come together very quickly like a trap when sensitive hairs are touched by an unsuspecting insect. The insect is digested by secretory glands on the leaf surface. Most species inhabit habitats that contain little or no available nitrogen, such as bogs and waterlogged or wet, sandy soils.

Fossils attributed to the family have been found in Eocene and more recent deposits.

## ORDER VIOLALES
### Family Flacourtiaceae (Flacourtia)

Figure 9.56 Flacourtiaceae. *Flacourtia:* (a) branch with flowers and fruit. *Azara:* (b) stamen; (c) cross section of ovary; (d) flower. (*SOURCE:* Parts (b), (c), and (d) redrawn with permission from Cronquist, 1981.)

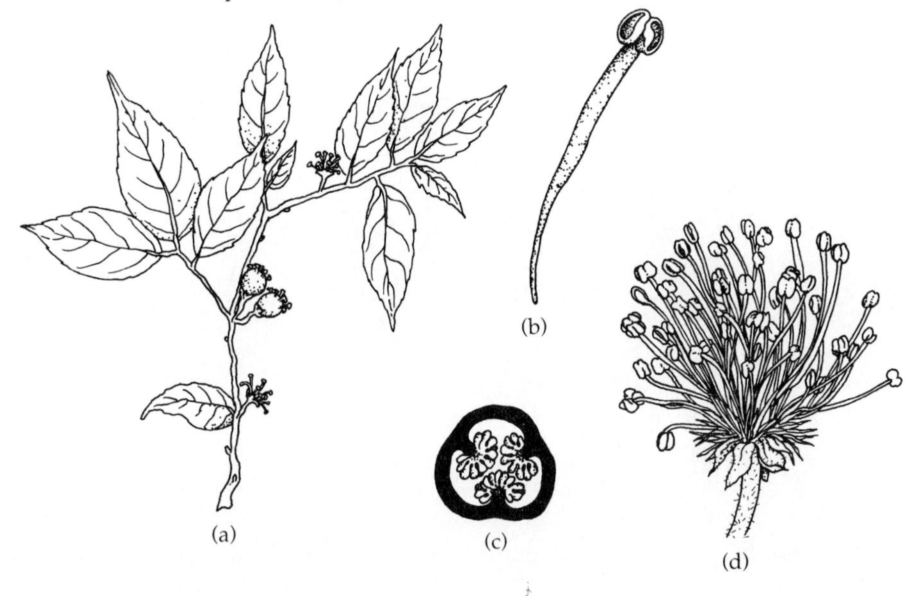


Shrubs or trees. Leaves deciduous or evergreen, alternate (rarely opposite or whorled), simple, commonly pinnately veined, entire or glandular toothed; stipules present. Flowers regular, perfect or unisexual and dioecious, hypogynous (rarely perigynous); inflorescence of solitary flowers or flowers in various types of inflorescences. Sepals 3–16, distinct or connate at the base. Petals 3–16, generally small, distinct or connate at the base, alternating with the sepals. Stamens as few as 4 but frequently many or same number as and opposite the petals; filaments distinct or grouped in clusters; anthers opening by longitudinal slits, less commonly by pores. Pistil simple or compound of 2–10 distinct or united fused carpels; locule 1; ovules 2–many and borne on each parietal placenta; ovary superior to inferior; styles 2–10, distinct or fused. Fruit a berry, loculicidal capsule, or drupe. Seed sometimes with arils or hairs, embryo straight, endosperm abundant (Fig. 9.56).

The family Flacourtiaceae consists of 85–89 genera and 800–1250 species distributed throughout the tropical and subtropical world. The largest genera are *Homalium* (200 species), *Casearia* (160 species), and *Xylosoma* (100 species).

The family has few economically important plants. Seeds of *Hydnocarpus wightiana* and *Taraktogenos kurzii* of Southeast Asia yield chaulmoogra oil used in treating leprosy. A few other species are used for ornamentals and timber.

The family is considered by many botanists to be the least advanced in characters in the order Violales. However, this does not mean that it is the ancestor to the other families.

Fossil pollen attributed to the Flacourtiaceae were first found in Upper Miocene deposits.

**Family Cistaceae (Rockrose)**

Figure 9.57 Cistaceae. *Helianthemum:* (a) branch with flowers; (b) front view of flower (c) cross section of ovary; (d) stamen. (*SOURCE:* Redrawn with permission from Cronquist, 1981.)

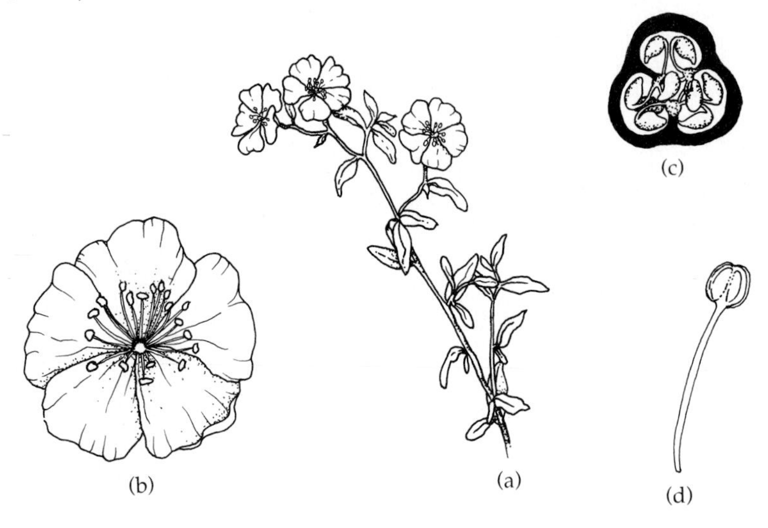


Herbs or shrubs, characteristic of sunny dry habitats. Leaves opposite (rarely alternate or whorled), simple; with or without stipules; often with oil glands or glanular hairs. Flowers regular, hypogynous, perfect; solitary or flowers in cymes or axillary inflorescences. Sepals 5, distinct and often unequal. Petals 5 (rarely 3 or missing), distinct, commonly overlapping. Stamens many, filaments distinct, borne on a nectary disk, and sensitive to touch in some species; anthers opening by longitudinal slits. Pistil compound of 3–10 united carpels; locule 1; ovules few to many and borne on parietal placentas; ovary superior; style single, stigma simple or lobed. Fruit a capsule. Seeds 3–many, small, embryo curved or straight, endosperm hard (Fig. 9.57).

The family Cistaceae consists of 8 genera and 165–200 species distributed in temperate regions, especially in eastern North America, the Mediterranean area and southern Europe. The genus *Helianthemum* (about 70 species) is the largest group in the family and is comprised of small shrubs.

Economically the family is of minor importance. The species *H. nummularium* and its cultivars make attractive garden plants. The leaves of *Cistus,* especially *C. ladanifer* and *C. ncanus* subsp. *creticus,* exude an aromatic resin, ladanum, used in making perfume and formerly used in medicine.

The family superficially resembles the Papaveraceae.

Pollen microfossils attributed to the Cistaceae have been found in Lower Miocene and more recent strata.

## Family Violaceae (Violet)

Figure 9.58 Violaceae. *Viola:* (a) plant with flowers; (b) cross section of ovary; (c) spurred stamen; (d) face view of flower; (e) fruit.

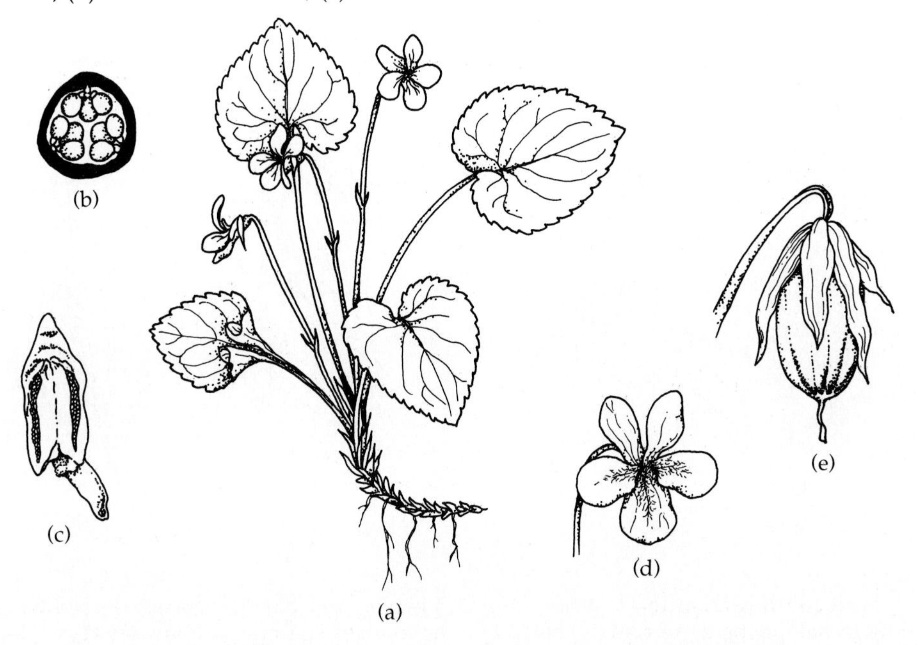


Herbs or shrubs (rarely lianas or small trees). Leaves alternate (rarely opposite), simple with stipules. Flowers regular or irregular, perfect (rarely unisexual), hypogynous, occasionally some are **cleistogamous** (having flowers that never open but set seed); inflorescence of solitary flowers or flowers in various inflorescences. Sepals and petals 5, distinct or overlapping, lowermost petal in irregular flowers prolonged into a sac or spur. Stamens 5, filaments short, distinct or connate at the base, often with a nectar gland or spur attached to 2 of them; anthers often clustered around style. Pistil compound of 3 united carpels; locule 1; ovules numerous and attached to parietal placentas; ovary superior; style 1. Fruit a berry or capsule (rarely a nut). Seeds with a straight embryo, endosperm present (Fig. 9.58).

The family Violaceae consists of 16–22 genera and about 800–900 species with a cosmopolitan distribution, but chiefly temperate and in the mountains of the tropics. *Viola* (about 400 species) is the largest genus of North Temperate herbs.

The family is of little commercial value, except for *Viola,* which contains V.x. *wittrockiana,* the garden pansies, a hybrid group. Many other species are grown as ornamentals. *Viola odorata* produces oils used in the perfume industry and is grown in southern France. The roots of *Anchietea salutaris, Corynostylis hybanthus,* and *Hybanthus ipecacuanha* are used medicinally as an emetic.

The genus *Rinorea* (300 species) is a pantropical shrub group that has many characters shared with the Flacourtiaceae, indicating a possible relationship.

The family has no known fossil record.

## Family Tamaricaceae (Tamarix)

Figure 9.59 Tamaricaceae. *Tamarix:* (a) end of flowing branch; (b) inflorescence; (c) flower; (d) vegetative branch.

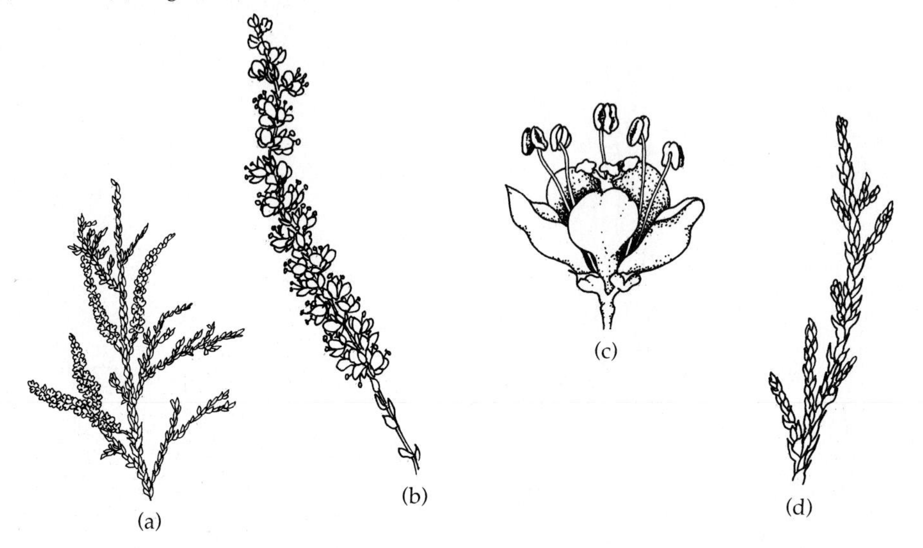


Small heath-like shrubs or trees, commonly in halophytic or xerophytic habitats. Leaves alternate, small, sessile and scale-like; without stipules. Flowers small to minute, regular, perfect, hypogynous; solitary or commonly in dense scaly spikes, racemes, or panicles. Sepals and petals 4–5, distinct with the petals alternate the sepals. Stamens 5–10 or many, distinct or connate at the base, attached to a fleshy nectar disk; anthers opening by longitudinal slits. Pistil compound of 2–5 united carpels; locule 1; ovules few to many and borne on parietal or basal placentas; ovary superior; style usually free but may be absent with the stigma sessile. Fruit a loculicidal capsule. Seeds with a coma or covered with hair, embryo straight, commonly lacking endosperm (Fig. 9.59).

The family Tamaricaceae consists of 4–5 genera and 100–120 species distributed in temperate to subtropical climates of the Mediterranean, North Africa, and Southeastern Europe to central Asia and Southwestern Africa; introduced widely in southwestern North America and Mexico. The largest genus is *Tamarix* (90 species).

The genus *Tamarix* has halophytic shrubs with various species of economic value. The twigs of *T. mannifera* produce a "manna," a sweet, white, gummy substance. The wood of *T. articulata* is used in home construction in North Africa. Dyes, medicinal extracts, and tannins are obtained from insect galls found on *T. articulata* and *T. gallica.*

The family is considered by many botanists as being similar to the Frankeniaceae.

The family has no known fossil record.

## Family Turneraceae (Turnera)

Figure 9.60 Turneraceae. *Turnera:* (a) branch with flower; (b) cross section of ovary; (c) pistil with brush-like stigmas; (d) flower with perianth partially removed.

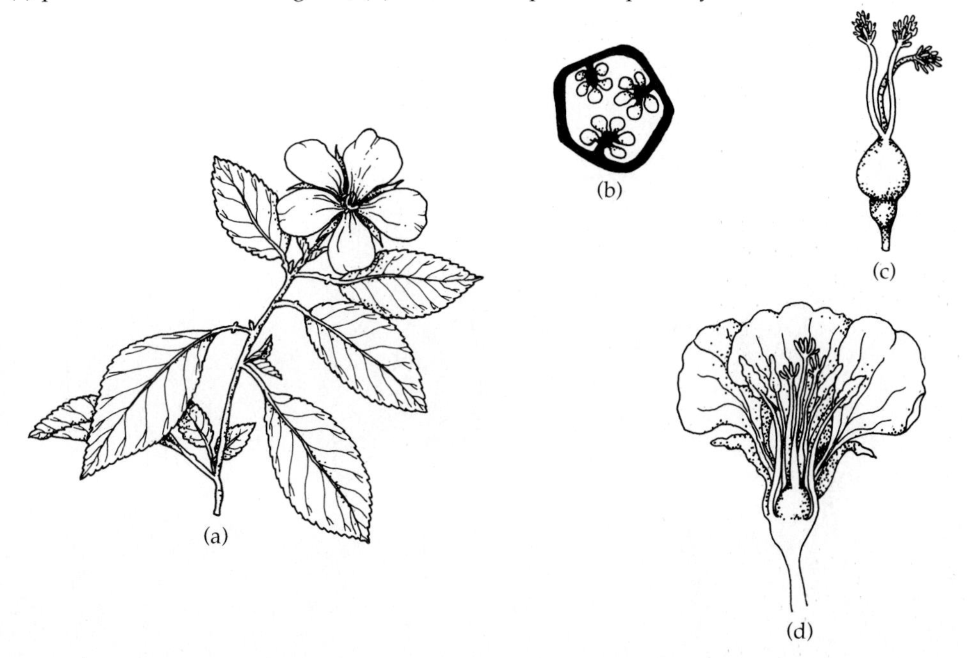


Herbs, shrubs, or small trees. Leaves alternate, simple, entire, glandular toothed or lobed, commonly with a pair of glands at the base of the blade; stipules small or lacking. Flowers regular, perfect, perigynous with the hypanthium often subtended with 2 bracteoles; inflorescence of solitary flowers or flowers in various inflorescences. Sepals 5, overlapping, usually with swellings on the inside surface. Petals 5, distinct, clawed at the base and attached to the lip of the hypanthium. Stamens 5, alternate with the petals and attached to the hypanthium wall; anthers opening by longitudinal slits. Pistil compound of 3 united carpels; locule 1; ovules 3–many and attached to parietal placentas; ovary superior or partially inferior; styles 3, distinct, often with brush-like stigmas. Fruit a loculicidal capsule. Seeds 3–many, each with an aril, embryo curved or straight, endosperm present (Fig. 9.60).

The family Turneraceae consists of 8 genera and 100–120 species native to the subtropical and tropical regions of the New World, Africa and Madagascar. The largest genus *Turnera* (60 species) has 1 species in Texas. *Turnera ulmifolia* native of the West Indies has been introduced into Florida.

The family is of little economic value, except for the leaves of some *Turnera* species, which are used locally in Mexico for medicine, as a substitute for tea, and to flavor wine.

The family has some characters similar to the families Malesherbiaceae and Passifloraceae and therefore is classified near these families in the order.

The family has no known fossil record.

## Family Passifloraceae (Passion Flower)

Figure 9.61 Passifloraceae. *Passiflora:* (a) section of stem with characteristic flower and tendrils; (b) cross section of ovary; (c) back side of anther; (d) pistil. (*SOURCE:* Parts (b), (c), and (d) redrawn with permission from Cronquist, 1981.)

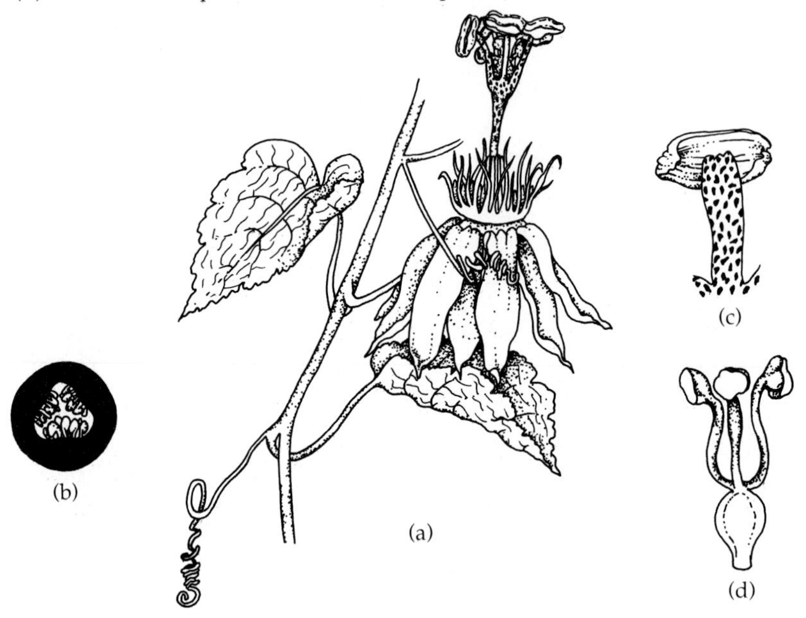


Herbs or woody vines (less commonly shrubs or trees), climbing by axillary tendrils that are modified inflorescences. Leaves alternate, simple, entire or palmately lobed (rarely compound); with small, generally deciduous stipules. Flowers regular, perfect (less commonly unisexual), perigynous with a saucer-shaped hypanthium; inflorescences of solitary flowers or flowers in a cyme or raceme. Sepals 5, distinct or connate at the base, often petal-like. Petals 5, alternate the sepals, distinct or connate at the base; corona almost always present, with numerous lobes developed between the petals and stamens. Stamens usually 5, filaments distinct or connate into a tube, a staminode-formed nectar disk often around the ovary; anthers opening by longitudinal slits. Pistil compound of 3–5 united carpels; locule 1; ovules many and attached to parietal placentas; ovary superior; style distinct or connate, the 3–5 stigmas are discoid or capitate. Fruit a capsule or berry. Seeds compressed with a fleshy aril, embryo large and straight, endosperm present (Fig. 9.61).

The family Passifloraceae consists of 16–20 genera and 600–650 species native to the tropics and subtropics, with two thirds of the species in the New World tropics. *Passiflora* is the largest genus with 400–500 species mostly in the American tropics. *Adenia* (80–100 species) is found in tropical Africa and Asia.

Economically the family is of some importance for food and as ornamentals. Over 50 species of *Passiflora* have edible fruits. The passion-fruit drink industry of Hawaii uses mostly *P. edulis* var. *flaviocarpa,* yellow passion fruit, while *P. edulis* and *P. maliformis* are used in Australia and southern Asia, and the West Indies, respectively. Other species are cultivated for their attractive flowers.

The Passifloraceae is very similar to some species in the family Flacourtiaceae. The genus *Paropsis* is included in either family, depending on the author. The family also shares characters with and is similar to the Cucurbitaceae and Loasaceae.

The family has no known fossil record.

## Family Caricaceae (Papaya)

Figure 9.62 Caricaceae. *Carica:* (a) habit of plant; (b) longitudinal section of male flower; (c) cross section of ovary; (d) female flower.

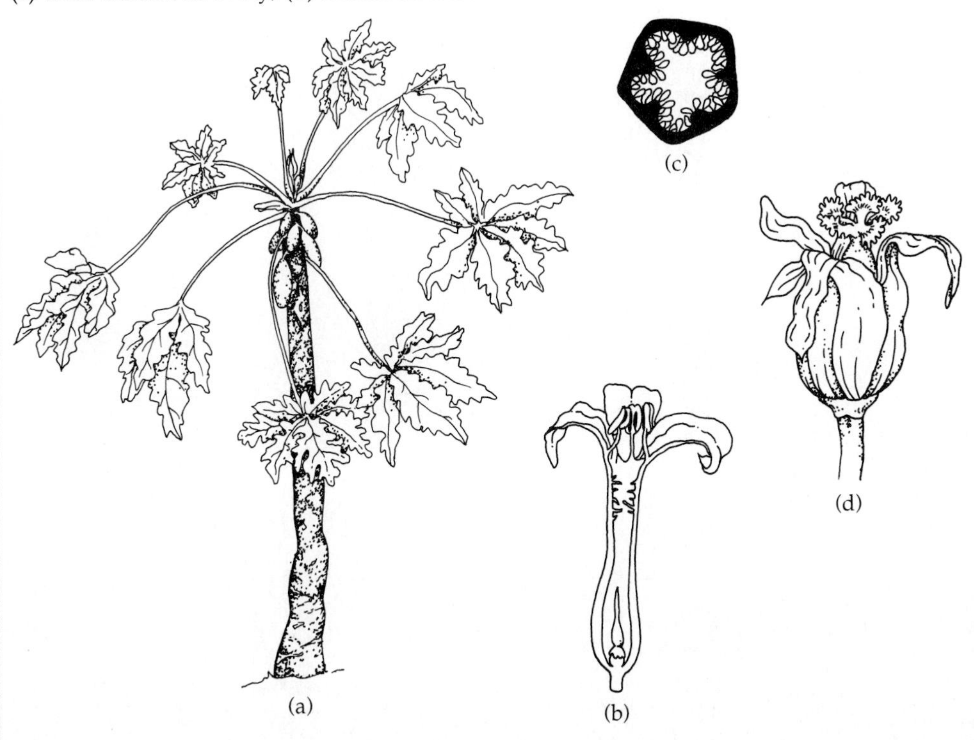


Small trees, soft stemmed, generally with an unbranched trunk containing a milky latex. Leaves alternate, mostly palmately veined, lobed, or compound, clustered at the stem apex; stipules usually lacking or spine-like. Flowers regular, perfect or unisexual, hypogynous; inflorescence of a solitary flower in leaf axils or in cymes. Sepals 5, connate into a lobed or toothed calyx. Petals 5, sympetalous with the tube short in female flowers and long in male flowers. Stamens 10, filaments distinct or connate at the base, adnate to the corolla tube; anthers opening by longitudinal slits. Pistil compound of 5 united carpels; locule 1; ovules many attached to intruded parietal or axile placentas; ovary superior; style distinct. Fruit a berry. Seeds many with a gelatinous covering, embryo straight, endosperm present (Fig. 9.62).

The family Caricaceae consists of 4 genera and 30 species native mainly in subtropical and tropical America, West Indies, and West Africa. The largest genus is *Carica* (about 22 species).

Economically the family is of some importance. The most well-known species in the family is *C. papaya,* the papaya. This species is nocturnal flowering and gives off a sweet scent to attract moths. The green fruit, besides being delicious to eat, produces a latex that is the source of the proteolytic enzyme, papain, used as a meat tenderizer. High-grade papain will digest 35 times its own weight of lean meat.

The family has sometimes been classified with the Cucurbitaceae or the Passifloraceae.

The family has no known fossil record.

## Family Fouquieriaceae (Ocotillo)

Figure 9.63 Fouquieriaceae. *Fouquieria:* (a) spiny branch with flowers; (b) cross section of ovary; (c) longitudinal section of flower; (d) flower; (e) habit of plant. (*SOURCE:* Parts (b), (c), and (d) redrawn with permission from Cronquist, 1981.)

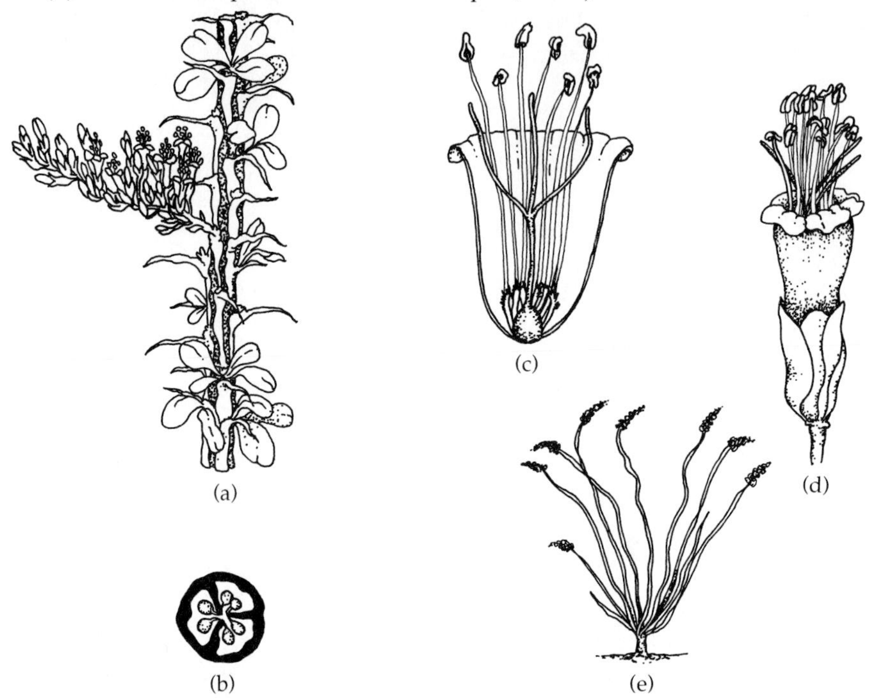


Spiny shrubs or trees that are xerophytic. Leaves small, simple, alternate, usually entire; quickly produced following moisture and deciduous as the soil becomes dry; a portion of the petiole persisting as a spine. Flowers regular, perfect, hypogynous, sympetalous; borne in axillary or terminal inflorescences. Sepals 5, distinct, the outer 2 smaller than the 3 inner ones. Petals 5, connate and forming a tubular corolla. Stamens 10–many in one or more series, filaments free or slightly connate, hairy at the base; anthers opening by longitudinal slits. Pistil compound of 3 united carpels; locule 1; ovules 2–many per locule and borne on axile placentas and partitions at the ovary base and on parietal placenta above; ovary superior; style single below then branched above the middle, stigmas terminal. Fruit a loculicidal capsule. Seeds flattened and winged, embryo straight, with or without endosperm (Fig. 9.63).

The family Fouquieriaceae consists of only the genus *Fouquieria* and 11 species native to the southwestern deserts of North America.

Economically the family is of little importance. *Fouquieria splendens,* the ocotillo, is common in the Sonoran and Mojave deserts. It is sometimes planted close together for a spiny hedge. Wax is also obtained from the stem of some species.

The classification affinity of this family is presently disputed by botanists; some place the group in the subclass Asteridae near the family Polemoniaceae, while others place it in the order Ebenales.

The family has no known fossil record.

## Family Cucurbitaceae (Cucumber)

Figure 9.64 Cucurbitaceae. *Echinocystis:* (a) segment of leafy stem bearing flowers; (b) face view of female flower; (c) longitudinal section of female flower. (*SOURCE:* Part (b) redrawn with permission from Cronquist, 1981.)

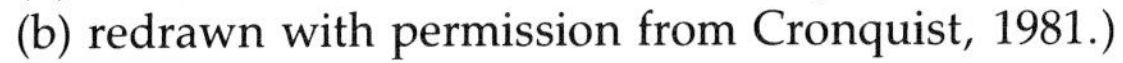

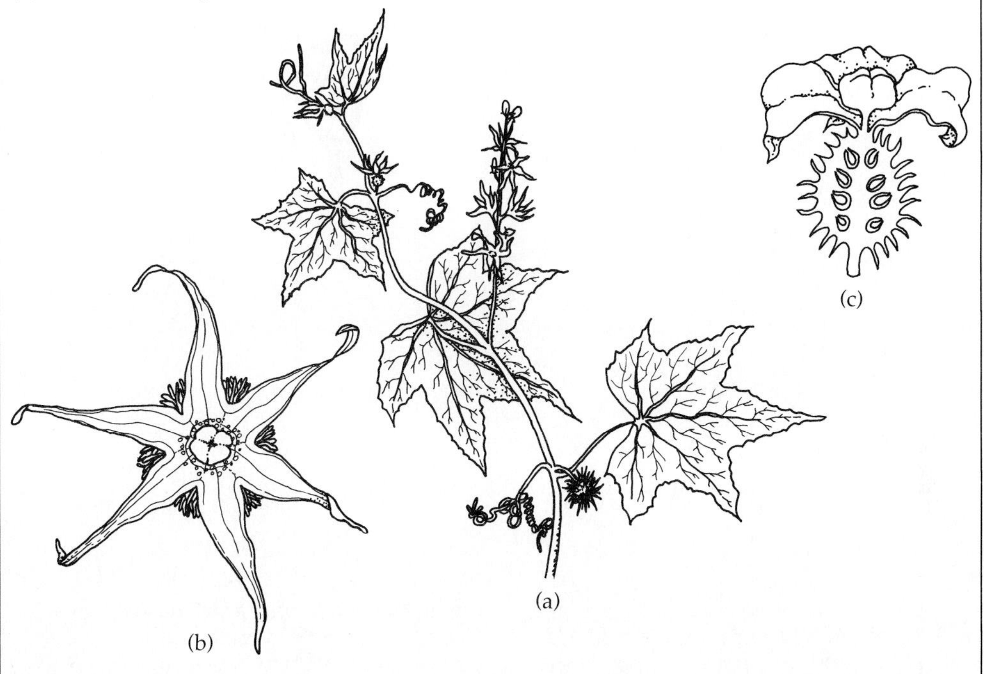


Herbs or soft woody climbing or trailing plants with spiral tendrils. Leaves alternate, mostly palmately veined, lobed, or compound; without stipules. Flowers regular, epigynous, unisexual (monoecious or dioecious), rarely perfect; inflorescence of solitary flowers or flowers in axillary cymes, racemes, or panicles. Sepals usually 5, distinct or connate, often reduced and borne at the top of the cup or hypanthium. Petals 5, commonly yellow or white, sympetalous, sometimes unlike in male and female flowers. Stamens 5, filaments distinct or connate, adnate toward the base of the hypanthium; anther pollen sacs may vary from 4 to 2 and open by longitudinal slits. Pistil compound of 3 united carpels; locules 1–3; ovules many and borne on enlarged parietal or axile placentas; ovary inferior. Fruit a berry, capsule, or pepo. Seeds many, large, flattened; embryo straight, endosperm lacking (Fig. 9.64).

The family Cucurbitaceae consists of 90–121 genera and 700–735 species, well represented in the subtropical and tropical regions of the world; some are found in semiarid regions, with species in Australasia and in temperate climates.

The family is very important economically for its food plants. These include *Cucurbita* (27 species), the gourds, pumpkins, and squash; *Cucumis* (30 species), cantaloupe, honeydew melon, and cucumber *(C. sativus)*; and *Citrullus lanatus*, watermelon. Some are cultivated as ornamentals such as *Cucurbita pepo*, ornamental gourds. The dried skeleton of the fruit of *Luffa cylindrica* is the source of loofa sponges.

The classification of the Cucurbitaceae in relation to other families is presently debated and poorly understood. The family is sometimes placed in its own order, Cucurbitales.

Fossil leaves attributed to the Cucurbitaceae were first found in Paleocene deposits with pollen collected from the Oligocene strata.

## Family Begoniaceae (Begonia)

Figure 9.65 Begoniaceae. *Begonia:* (a) leafy stem with flowers; (b) cross section of ovary; (c) stamen; (d) flower.

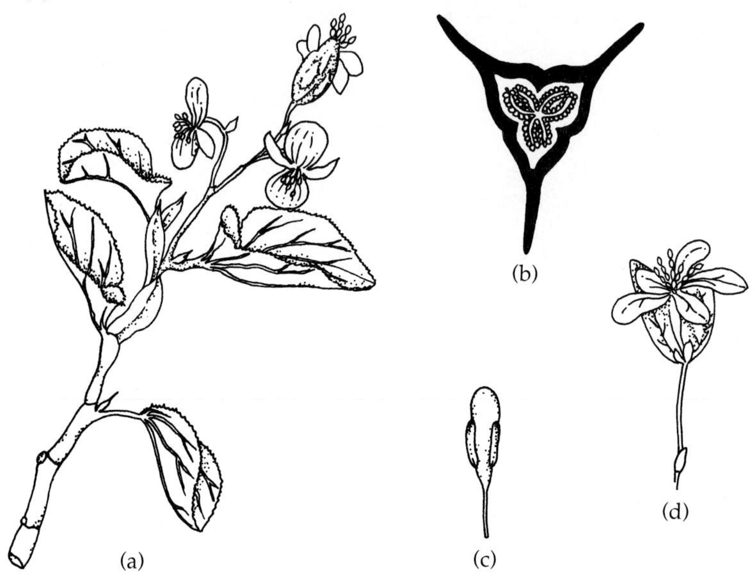


Fleshy herbs or shrubs, often with jointed stems, commonly with thick rhizomes or tubers. Leaves alternate, simple, usually palmately veined and commonly palmately lobed; asymmetrical with prominent stipules. Flowers regular or irregular, imperfect, unisexual (mostly monoecious); inflorescence of solitary flowers or flowers in cymes (rarely racemes). Sepals 2, petaloid, 2 in staminate flowers, 2–10 in the pistillate ones, distinct. Petals 2 or absent in the staminate flowers, lacking in the pistillate ones. Stamens 4–many, centripetal, distinct or connate at the base, regularly arranged or on one side of the flower; anthers opening by longitudinal slits (rare terminal pores). Pistil compound of 2–6 united carpels; locules 1 or 3; ovules many and borne on axile or parietal placentas, ovary inferior and often winged; style distinct or connate at the base, stigmas often twisted. Fruit a loculicidal capsule (rarely a berry). Seeds small, numerous; embryo straight, endosperm lacking (Fig. 9.65).

The family Begoniaceae consists of 5 genera and over 900–1000 species distributed pantropically, except for the tropics of Australia and Polynesia. Most of the species belong to the genus *Begonia,* with less than 20 species in the remaining genera.

*Begonia* species are economically important as popular ornamentals. Hybrids of *B. rex* are used as pot plants, while *B. semperflorens* are the common glabrous bedding plants. The leaves of *B. tuberosa* are eaten for food in the Moluccas Islands.

The classification position of this family is presently disputed by botanists and needs further investigation.

The family has no known fossil record.

## Family Loasaceae (Loasa)

Figure 9.66 Loasaceae. *Mentzelia:* (a) stem tip with flower; (b) fruit with persistent calyx; (c) flower; (d) longitudinal section through flower.

Herbs, shrubs, or rarely small trees; often with rough or stinging hairs. Leaves alternate or opposite, simple, entire or lobed; stipules lacking. Flowers regular, perfect, epigynous; inflorescence of cymes or heads. Sepals 4–7, distinct or connate at their base, persistent in fruit. Petals 4–7 (many), distinct or connate at their base. Stamens many (rarely 5), filaments distinct or connate at their base, sometimes modified into staminodes that have nectar; anthers opening by longitudinal slits. Pistil compound of 3–5 united carpels; locule 1 (rarely 2–3); ovules many and borne on intruding parietal (rarely axile) placentas; ovary inferior; style simple. Fruit an achene or capsule. Seeds many, embryo straight, endosperm present or absent (Fig. 9.66).

The family Loasaceae consists of 14–15 genera and 200–250 species widespread in temperate and tropical New World. Only two species of the unique genus *Fissenia* (*Kissenia*) are found in Africa and Arabia. The largest genera—*Loasa* (75 species), *Mentzelia* (50 species), the blazing star, and *Caiophora* (50 species)—have large showy flowers.

Economically the family is of little importance, except for a few species of *Loasa, Mentzelia,* and *Blumenbachia* grown as garden ornamentals.

The classification of the Loasaceae is still being debated by botanists and is uncertain, with some classifying the group near the Ericaceae and others placing the family in the subclass Asteridae.

The family has no known fossil record.

## ORDER SALICALES
### Family Salicaceae (Willow)

Figure 9.67 Salicaceae. *Populus:* (a) branch with leaves; (b) female catkin; (c) male catkin. *Salix:* (d) branch with leaves and stipules; (e) female catkin; (f) female flower; (g) male catkins.

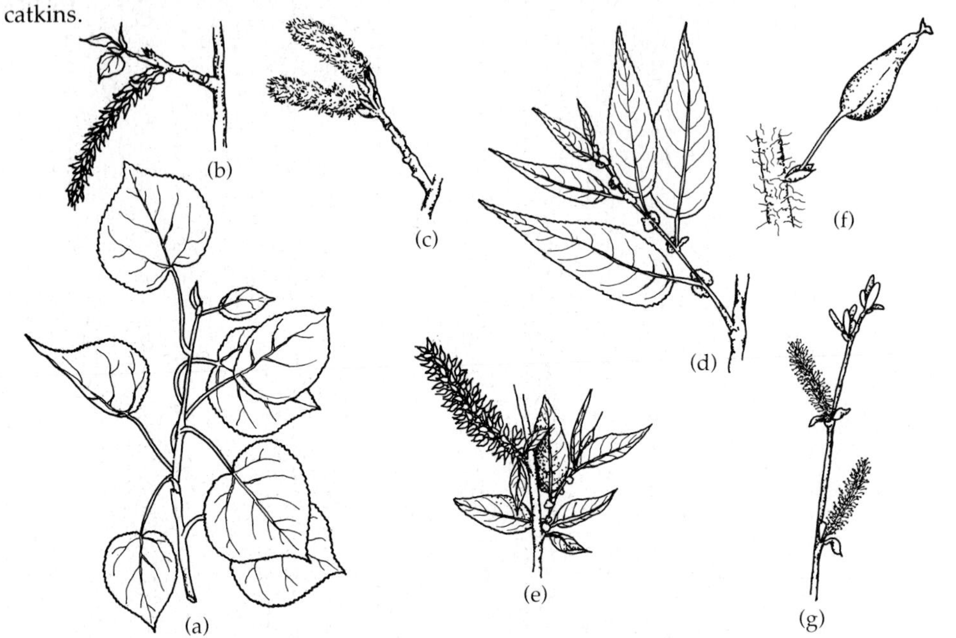


Shrubs or trees. Leaves alternate, deciduous, simple; stipules often fall off easily. Flowers imperfect, unisexual and dioecious, subtended by small bracts; inflorescence a catkin, usually appearing before or at the time of the leaves in the spring; wind or insect pollinated. Sepals lacking or fused and reduced to a saucer-shaped disk. Petals absent. Stamens 2–many, filaments distinct or connate at the base; anthers opening by longitudinal slits. Pistil compound of 2–4 united carpels; locule 1; ovules 2–many and borne on parietal placentas; ovary superior; style distinct, either long or short, stigma often lobed. Fruit a capsule. Seeds small, many, wind dispersed via attached long hairs, embryo straight, endosperm lacking (Fig. 9.67).

The family Salicaceae consists of 4 genera and about 350 species, distributed commonly in the North Temperate regions, with a few species extending into the tropics and Southern Hemisphere in Africa and South America. The largest genera are *Salix,* the willows (about 300 species), and *Populus,* the aspen, cottonwood, or poplar (40 species). *Salix* is common along streams and in mountainous areas. The genus extends into the alpine and arctic regions, sometimes as miniature shrubs that are only a few centimeters high.

Economically the family is of some importance. *Salix babylonica,* weeping willow, is a commonly planted ornamental tree. *Populus* prefers moist habitats and are fast-growing trees. Some like *Populus alba,* white poplar; *P. deltoidea,* cottonwood; and *P. nigra,* the black poplar are planted as ornamental shade trees. Hybrids are common in species of willows and poplars. The flexible twigs of *Salix* are used in making baskets, and some willow bark is used for medicine. Aspirin was originally obtained from the bark of *S. alba* and is still used in its natural form in many countries.

In earlier classifications the Salicaceae was considered a "primitive" family because of its reduced flowers. This idea is no longer held by most botanists.

Fossil leaves attributed to the family were first found in Eocene deposits.

## ORDER CAPPARALES
### Family Capparaceae (Caper)

Figure 9.68 Capparaceae. *Cleome:* (a) face view of flower; (b) cross section of ovary; (c) fruit; (d) stem with leaves and flowers.

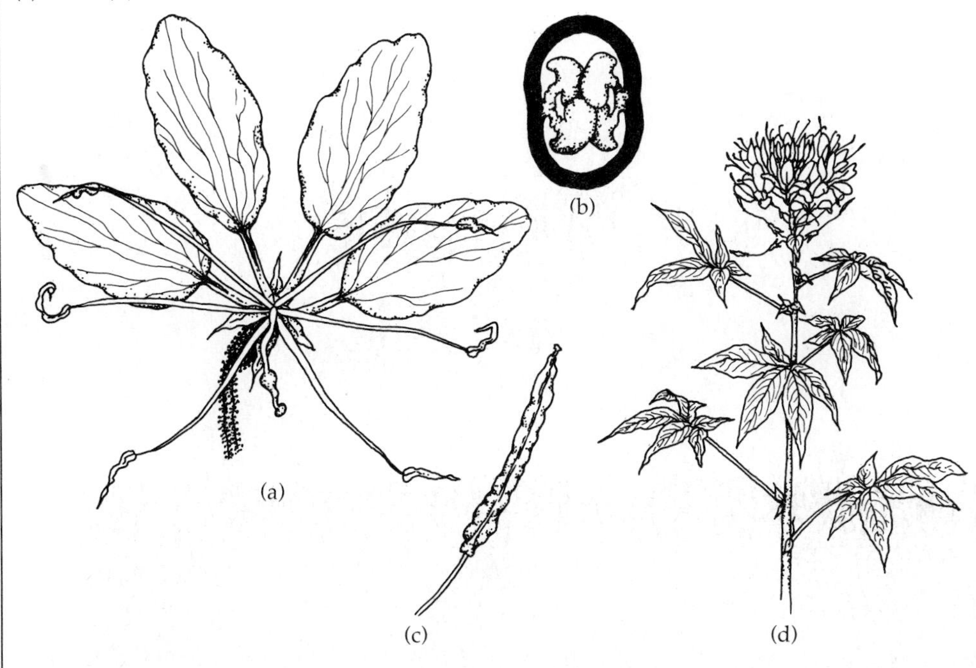


Herbs or shrubs (rarely trees). Leaves alternate (rarely opposite), simple or palmately compound; stipules usually present, small sometimes reduced to glands or spines; hairs diverse, simple, forked, stellate or peltate. Flowers irregular, perfect (rarely unisexual), hypogynous; inflorescence of solitary flowers or flowers in axillary clusters, racemes, or corymbs. Sepals 4 (rarely 2–8), distinct or connate at the base. Petals 2–many (rarely lacking), distinct often with a basal claw. Stamens 4–many, filaments distinct, centrifugal; anthers opening by longitudinal slits. Pistil compound of 2–12 united carpels; locule 1; ovules 1–many and borne sometimes on protruding parietal placentas; ovary superior often on a stipe; style 1, stigma capitate or bilobed. Fruit often stalked, of various types (berry, drupe, capsule, nut, samara, silicle, or silique). Seeds 1–many, commonly kidney-shaped; embryo curved, endosperm usually absent (Fig. 9.68).

The family Capparaceae consists of 40–50 genera and 700–800 species widespread in subtropical and tropical regions. Most of the species belong to the genera *Capparis* (350 species), the capers, and *Cleome* (200 species), the spider plant.

The economic value of the family is low, except for a few cultivated as garden plants. *Cleome spinosa*, the spider flower, is strongly scented, and pickled flower buds of *Capparis spinosa*, capers, are used for seasoning.

This family has many characteristics in common with the Brassicaceae but can be easily distinguished by characters of the stamens.

Fossils attributed to the Capparaceae were first observed in Eocene deposits.

## Family Brassicaceae (Cruciferae) (Mustard)

Figure 9.69 Brassicaceae (Cruciferae). *Brassica:* (a) stem tip with flowers and fruits; (b) longitudinal section of flower; (c) fruit (silique). *Capsella:* (d) fruit (silicle). *Dentaria:* (e) cross section of ovary; (f) face view of flower.

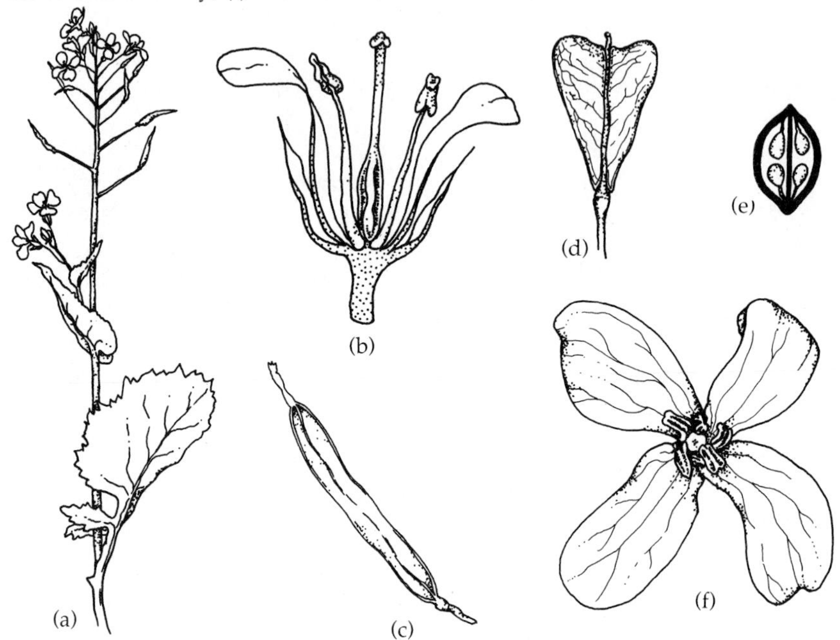


Herbs or shrubs (rarely vines). Leaves alternate (rarely opposite or whorled), simple to pinnately compound; stipules lacking; hairs diverse simple, forked, stellate or peltate. Flowers regular (rarely irregular), perfect, hypogynous, lacking bracts; inflorescence a corymb or raceme. Sepals 4, distinct. Petals 4 (rarely lacking), distinct, in the form of a cross. Stamens 6 (rarely 2, 4, or 16), usually **tetradynamous** (the inner with long filaments, 2 outer ones with short filaments); anthers opening by longitudinal slits. Pistil compound of 2 united carpels; locules usually 2 due to the formation of a **false partition** or septum formed by outgrowth of the placentas; ovules few to many and attached to parietal placentas; ovary superior; style 1, stigma capitate or lobed. Fruit a silicle or silique, occasionally indehiscent. Seeds few to many, attached to the margin of the false partition, which often persists after seeds are shed; embryo curved, endosperm usually absent (Fig. 9.69).

The family Brassicaceae consists of 350–380 genera and 3000 species with a cosmopolitan distribution, but found mostly in temperate climates of the Northern and Southern Hemispheres. The largest genera are *Draba* (300 species), *Cardamine* (150 species), and *Lepidium*, peppergrass (130 species).

Economically the family is of major importance for ornamentals and especially for human consumption, as well as livestock feed and forage crops. Familiar ones include *Brassica oleracea*, broccoli, brussels sprouts, cabbage, and cauliflower; *B. nigra*, black mustard; *B. campestris*, oilseed rape; and *Raphanus sativus*, radish. Many species have escaped and are common weeds in various regions of North America and Europe.

The fruit of the Brassicaceae makes the family unique because of its false partition or septum, and due to the two halves of the mature ovary **(valves)** falling away, leaving the seeds attached to the margin of the partition. The variation in fruit morphology is useful in the classification of various genera and species. The Brassicaceae family shares many characteristics with the Capparaceae.

The fossil record of the Brassicaceae was first observed in the Oligocene.

## Family Resedaceae (Mignonette)

Figure 9.70 Resedaceae. *Reseda:* (a) leafy branch with flowers and fruits; (b) flower; (c) cross section of ovary; (d) fruit.

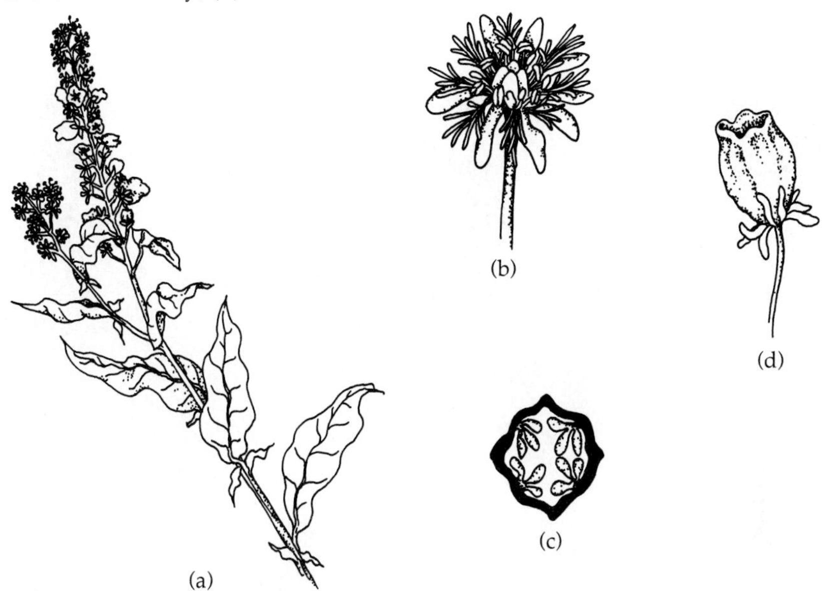


Herbs or shrubs. Leaves alternate, simple, entire to pinnatifid; stipules small often reduced to glands. Flowers irregular, usually perfect, hypogynous; inflorescence a raceme or spike. Sepals 2–8, distinct or connate at the base, sometimes unequal. Petals 2–8 (rarely absent), distinct, unequal with some having fringed appendages or cut blades. Stamens 3–many, filaments distinct, sometimes arranged on one side of the flower; anthers opening by longitudinal slits. Pistil compound of 2–7 united or partially united carpels, carpels often open near the apex; locule 1; ovule 1 (rarely 2) and borne on axile, basal or parietal placentas; ovary superior; styles 2–7 and same number as carpels, stigma dry. Fruit a berry, capsule, or a cluster of follicles. Seeds kidney-shaped, embryo curved, endosperm usually lacking (Fig. 9.70).

The family Resedaceae consists of 6 genera and 75 species distributed mostly in the Northern Hemisphere, concentrated in the Mediterranean area to India; also with species in southwestern North America and South Africa. The largest genus is *Reseda* (55 species).

Economically the family is of minor importance. *Reseda odorata,* mignonette, is grown as a garden ornamental for its fragrance and for a perfume oil, and *R. luteola* is grown as a source for a reddish-yellow dye.

The family has characters that make it similar to the Brassicaceae and the Capparaceae.

The family has no known fossil record.

## ORDER ERICALES
### Family Clethraceae (Clethra)

Figure 9.71 Clethraceae. *Clethra:* (a) branch with flowers; (b) cross section of ovary; (c) pistil; (d) stamen; (e) inflorescence; (f) flower.

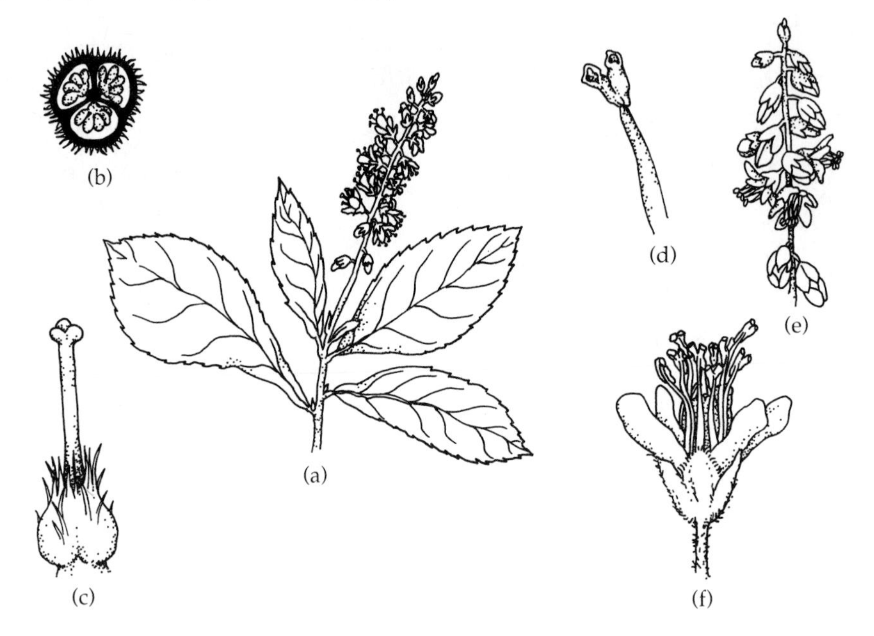


Shrubs or small trees. Leaves deciduous or evergreen, alternate, simple, entire or toothed; stipules lacking; commonly with stellate hairs. Flowers regular, perfect, hypogynous, white; inflorescence of terminal panicles or racemes. Sepals 5, connate at the base into a tube. Petals 5, distinct or connate at the base. Stamens 10, filaments distinct; anthers opening by apical pores. Pistil compound of 3 united carpels; locules 3; ovules many and borne on axile placentas; ovary superior; style 3-lobed. Fruit a loculicidal capsule. Seeds many, commonly winged, embryo cylindrical, endosperm present (Fig. 9.71).

The family Clethraceae consists of the genus *Clethra* and 65–120 species distributed in the subtropical and tropical New World north to the southeastern United States and also in Southeast Asia and the East Indies. The name *Clethra* comes from the Greek and means alder, because some species resemble alders *(Alnus)*.

Economically the family is known for a few ornamentals, in particular *C. arborea,* the lily-of-the-valley tree, which has drooping, bell-like, fragrant flowers. Other species cultivated are *C. acuminata,* white alder, and *C. alnifolia,* the sweet pepper bush.

In terms of its classification, the Clethraceae has many characters like the Ericaceae.

The fossil record of the family is most inconclusive.

## Family Empetraceae (Crowberry)

Figure 9.72 Empetraceae. *Empetrum:* (a) leafy branch; (b) front view of female flower; (c) fruit; (d) cross section of ovary; (e) male flower.

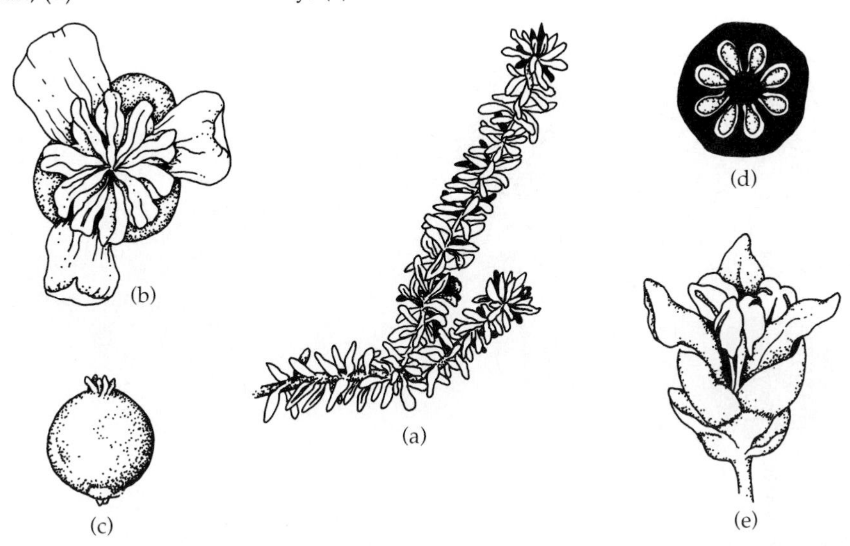


Small shrubs. Leaves small, linear, alternate, simple, evergreen; lacking stipules. Flowers 1–3, small, regular, unisexual and dioecious, hypogynous, subtended by 1–several bracts; inflorescence of axillary heads or racemes. Sepals 2–3, distinct, petal-like. Petals 3–4, overlapping and similar to the sepals. Stamens 2–4, filaments distinct, alternate with the petals, anthers small; opening by longitudinal slits and without apical appendages. Pistil compound of 2–9 united carpels; locules 2–9; ovule 1 per locule and borne on axile placentas; ovary superior; lacking a nectary disc, style short, stigma lobed with 2–9 branches. Fruit a dry or fleshy drupe. Seeds 1 per locule, embryo straight, endosperm present (Fig. 9.72).

The family Empetraceae consists of 3 genera and 4–6 species distributed throughout the cool North Temperate regions, especially eastern North America and southwestern Europe. It also occurs in southern South America.

*Empetrum nigrum*, crowberry, and *E. rubrum* do exist in large ground-cover populations in Eastern Canada. Its edible fruit is used locally for jams and pastries. The description in early Norse documents of the extensive populations and morphology of *Empetrum,* and the use of the fruit for pastry, has helped anthropologists to determine the early landing of Europeans on North America before Columbus. *Corema* and *Empetrum* species are cultivated as rock garden plants.

The Empetraceae resembles the Ericaceae in many features but differs in its stamen morphology.

The family has no known fossil record.

## Family Epacridaceae (Epacris)

Figure 9.73 Epacridaceae. *Styphelia:* (a) leafy branch bearing flowers; (b) anther; (c) cross section of ovary; (d) front view of flower. (*SOURCE:* Parts (c) and (d) redrawn with permission from Cronquist, 1981.)

Heath-like shrubs or small trees commonly found in acidic, low-nutrient soils. Leaves usually alternate, simple, narrow, stiff, xeromorphic; lacking stipules. Flowers small, regular, perfect, hypogynous (rarely epigynous), with bracts; inflorescence of solitary flowers or flowers borne in panicles, racemes, or spikes. Sepals 4–5, distinct and persistent. Petals sympetalous, tubular with 4–5 lobes. Stamens 4–5, filaments distinct and attached to the corolla or below the ovary, arranged alternate the corolla lobes and with glands or hairs between the filaments; anthers opening by longitudinal slits and mostly without appendages. Pistil compound of 2–5 united carpels; locules 5 (rarely 2–10); ovules 1–many and attached to intruding parietal placentas, which then appear as axile placentas; ovary superior; style simple, stigma capitate. Fruit a loculicidal capsule or drupe. Seeds 1–many, embryo cylindrical and straight, endosperm present (Fig. 9.73).

The family Epacridaceae consists of about 30 genera and 400 species distributed mostly in Australia and New Zealand, with some species found in the East Indies, Philippines, Hawaii, and Patagonia. It is a dominant family in the heathlands "down under," playing a similar ecological role to that of the Ericaceae family in heaths of the Northern Hemisphere. The largest genera are *Leucopogon* (140 species), *Styphelia* (40 species), *Epacris* (40 species), and *Dracophyllum* (35 species).

Economically the family is of minor importance. Some species of *Dracophyllum, Epacris,* and *Styphelia* are grown as ornamentals. The roots of *Styphelia malayana* are used for medicine, while the inner bark is used for caulking to make canoes waterproof.

The Epacridaceae is very similar to the Ericaceae, but differs in the anthers opening by longitudinal slits and lacking appendages.

The family has no known fossil record.

## Family Ericaceae (Heath)

Figure 9.74 Ericaceae. *Gaultheria:* (a) leafy branch with flowers; (b) flower; (c) longitudinal section through flower. *Vaccinium:* (d) fruits; (e) stamen. *Leucothoe:* (f) cross section of ovary.

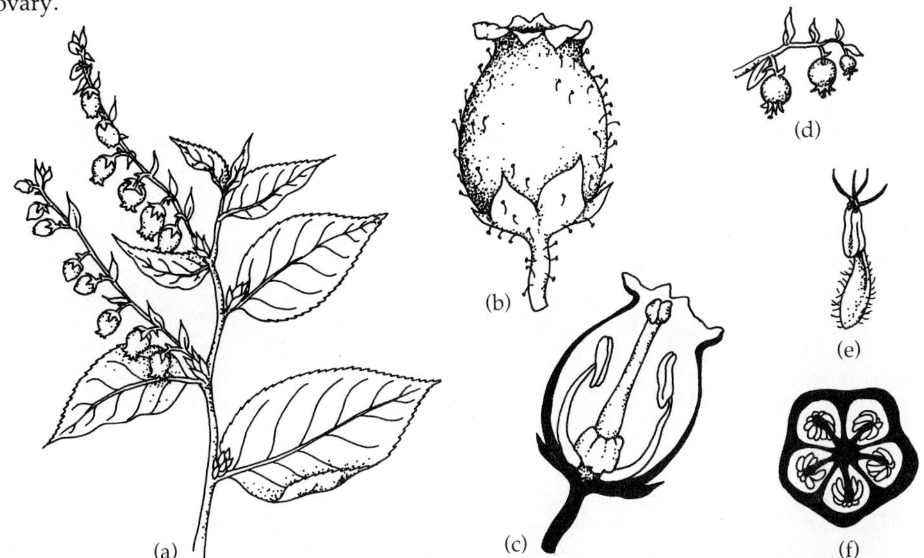


Shrubs, small trees, or lianas generally growing in acid soil with low nutrients, having considerable morphological variation; successful growth is dependent on a mycorrhiza fungal relationship. Leaves deciduous or evergreen, usually alternate, simple, sometimes needle or scale-like; pinnate venation; stipules lacking. Flowers regular or slightly irregular, usually perfect, hypogynous to epigynous; solitary or in various axillary or terminal inflorescences. Sepals 5 (rarely 4–7), distinct or connate at the base. Petals 5 (rarely 4–7), sympetalous (rarely distinct), commonly urn- or bell-like with short lobes. Stamens 10 (4–many), usually twice the number of the petals, filaments distinct, attached to the receptacle not the corolla; anthers opening by pores or short apical slits and often with appendages. Pistil compound of 2–10 (mostly 5) united carpels; locules 1–10; ovules many and borne on axile placentas; ovary superior or inferior; a nectary disc present; style single, stigma capitate. Fruit a berry, capsule, or drupe. Seeds small, many, embryo straight, endosperm present (Fig. 9.74).

The family Ericaceae consists of 100–125 genera and 3000–3500 species distributed widely throughout the cooler, temperate, or subtropical regions. It is mostly absent from Australia, where it is replaced in similar habitats by representatives of the Epacridaceae. The largest genus, *Rhododendron* (850–1200 species), the rhododendrons, is well developed in the Himalayan and Malaysian regions. *Erica* (500–600 species), the heaths, has many species in South Africa, a large number of which are confined to the Cape Province.

The genus *Vaccinium,* blueberry (450 species), is cultivated for its edible fruit. Many species of this family are cultivated as ornamentals. These include *Arbutus,* the madrone; *Calluna,* heather; *Erica,* the heath; and *Rhododendron. Gaultheria procumbens* produces wintergreen oil.

The subfamily classification of the Ericaceae is unsettled. Some authors wish to divide the family into smaller families (i.e., Vacciniaceae, etc.), while others prefer one large family with five subfamilies.

Paleocene deposits in England have revealed seeds that are Ericaceae-like, while pollen similar to Ericaceae or related families dates from the Upper Cretaceous.

## Family Pyrolaceae (Shinleaf)

Figure 9.75 Pyrolaceae. *Pyrola:* (a) individual plant bearing leaves and flowers; (b) side view of flower; (c) stamens; (d) cross section of ovary.

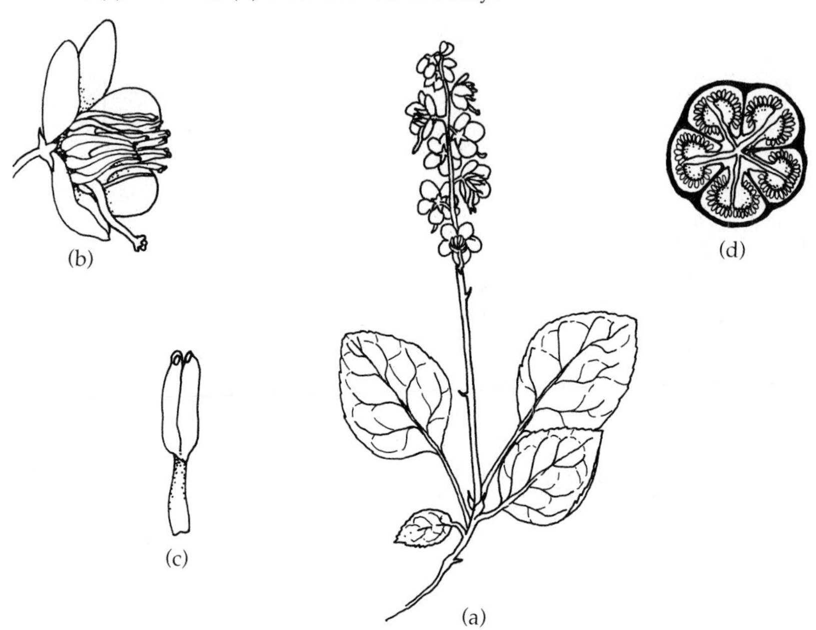


Herbs or small shrubs, from slender rhizomes; successful growth is dependent on a mycorrhizal fungal relationship. Leaves mostly evergreen, alternate (rarely opposite or whorled) or basal, simple; stipules lacking. Flowers regular, perfect, hypogynous; inflorescence of solitary flowers or flowers in corymbs or racemes. Sepals 5, distinct or connate at the base, persistent. Petals 5, distinct or connate at the base. Stamens twice the number of the petals, filaments distinct and attached to the receptacle; anthers opening by apical pores and with the pollen grains in tetrads. Pistil compound of 5 united carpels; locules 5; ovules many and borne on intruded parietal placentas (looking axile) that do not fuse in the middle; ovary superior; style single, straight or curved. Fruit a loculicidal capsule. Seeds many, small embryo not differentiated, endosperm abundant (Fig. 9.75).

The family Pyrolaceae consists of 4 genera and 30–45 species distributed in cooler more temperate regions of the Northern Hemisphere; most abundant in the boreal and Arctic areas. The largest genus is *Pyrola* (20–35 species), the shinleaf, followed by *Chimaphila* (4–8 species), the prince's pine. *Orthilia* and *Moneses*, one-flowered pyrola, are monotypic genera.

The family has very little economic value, except for a few species cultivated as ornamentals, and the leaves of some are used for healing wounds.

The Pyrolaceae is classified by some botanists as a subfamily of the Ericaceae or as including the Monotropaceae within it. The family differs from the Ericaceae in being mostly herbaceous and in its embryo lacking differentiated cotyledons.

The family has no known fossil record.

## Family Monotropaceae (Indian Pipe)

Figure 9.76 Monotropaceae. *Monotropa:* (a) cluster of stems bearing flowers; (b) anther of stamen; (c) cross section of ovary.

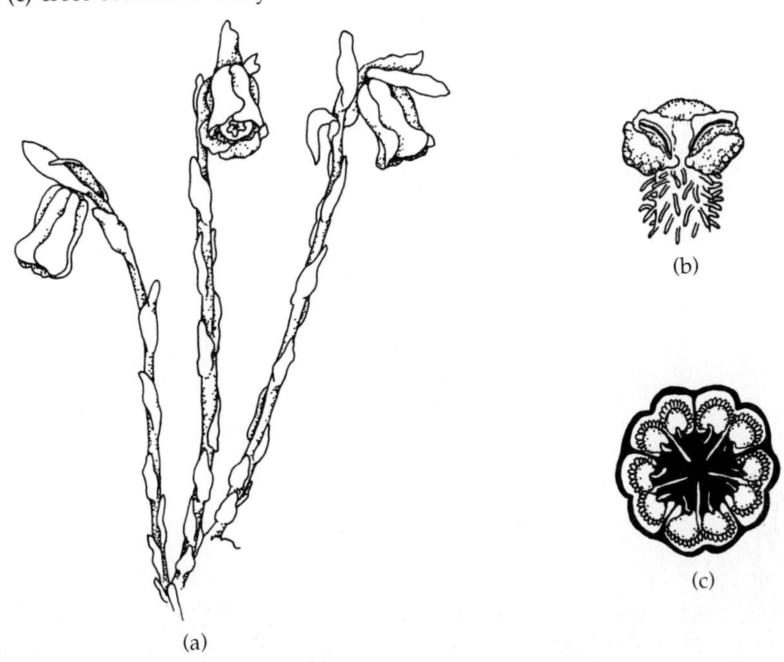


Perennial herbs, somewhat fleshy, lacking chlorophyll and nongreen, brown, purple, reddish, or yellow; dependent on mycorrhiza fungi for food. Leaves alternate, small and reduced, scale-like. Flowers regular, perfect, hypogynous; inflorescence of solitary flowers or flowers borne in racemes. Sepals usually 4–5, distinct. Petals usually 4–5 (rarely lacking), distinct or connate at the base. Stamens 6–12, filaments distinct or connate at the base and attached to the receptacle; anthers opening by longitudinal slits or apical pores, occasionally spurred at the base, pollen grains single. Pistil compound of 4–6 united carpels; locules 4–6; ovules many and borne on axile placentas or locule 1 with many ovules and borne on 1 parietal placenta; ovary superior; style single, stigma capitate or lobed. Fruit a berry or capsule. Seeds many, small, embryo without 2 cotyledons, endosperm present or lacking (Fig. 9.76).

The family Monotropaceae consists of 10 genera and 12 species found mostly in the cooler, more temperate regions of the Northern Hemisphere. It extends south in the mountains to Malaya and to Colombia. Most genera are monotypic and include *Allotropa virgata,* candy stick; *Pterospora andromedea;* pinedrops; and *Monotropa uniflora,* the Indian pipe, which turns black when dry.

The family has no real economic value.

The Monotropaceae has been classified within the Ericaceae but differs in being herbaceous, lacking chlorophyll, being completely reliant on a mycorrhizal fungus for food, its gamapetalous corolla, anthers opening by slits, and pollen grains being single. It has also been classified by some botanists within the Pyrolaceae.

The family has no known fossil record.

## ORDER DIAPENSIALES
### Family Diapensiaceae (Diapensia)

Figure 9.77 Diapensiaceae. *Diapensia:* (a) leafy plant with flowers; (b) flower; (c) longitudinal section through flower; (d) cross section through ovary.

Herbs or small shrubs; growth is dependent on a mycorrhizal fungal relationship. Leaves evergreen, alternate or in a rosette, simple; stipules lacking. Flowers white, regular, perfect, hypogynous; inflorescence of solitary flowers or flowers in racemes. Sepals 5, distinct or connate and forming a tube. Petals 5, distinct or sympetalous. Stamens 5, fertile ones opposite the sepals and adnate to the corolla; staminodes commonly 5 and opposite the corolla lobes; anthers opening by longitudinal slits, in a few species with appendages. Pistil compound of 3 (rarely 5) united carpels; locules 3 (rarely 5); ovules few to many and attached to axile placentas; ovary superior; style single, stigma 3-lobed. Fruit a loculicidal capsule. Seeds small, few to many, embryo curved or straight, endosperm present (Fig. 9.77).

The family Diapensiaceae consists of 6–7 genera and 18–20 species with a circumboreal and circumpolar distribution.

All genera are small but attractive, with some cultivated as rock garden plants, especially *Galax aphylla* and species of *Diapensia, Pyxidanthera, Schizocodon,* and *Shortia.*

The family resembles the Ericaceae and has been classified within the order Ericales by some botanists. However, the Diapensiaceae differs in various stamen and embryological characters.

The family has no known fossil record.

## ORDER EBENALES
### Family Sapotaceae (Sapodilla)

Figure 9.78 Sapotaceae. *Palaquium:* (a) leafy branch; (b) longitudinal section through flower; (c) flower; (d) cross section of ovary; (e) stamen.

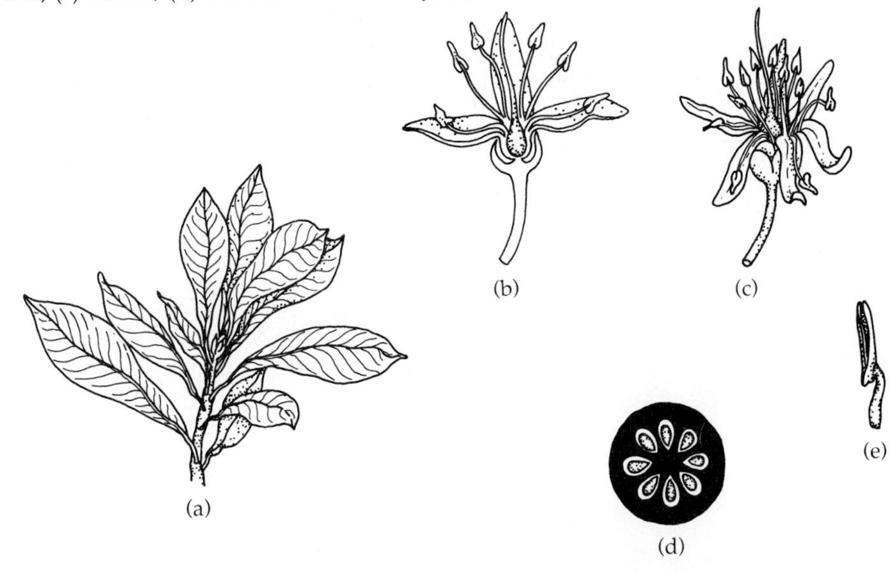


Shrubs or trees with milky juice and 2 armed hairs. Leaves alternate (rarely opposite), simple, entire; with or without stipules. Flowers rather small, regular or irregular, perfect, hypogynous; inflorescence of solitary flowers or flowers in axillary cymes; white or cream colored, commonly nocturnal and bat pollinated. Sepals 4–12, distinct or connate at the base in two or more whorls. Petals 4–12, sympetalous, usually same number as the sepals, sometimes with dorsolateral appendages. Stamens 3–many, in 1–3 whorls, filaments distinct, sometimes with staminoids alternate the corolla lobes, filaments attached to the corolla tube. Pistil compound of 2–many united carpels; locules 2–many; ovules 1 per locule and attached to axile or basal placenta; ovary superior and usually hairy; style single, stigma capitate or lobed. Fruit a berry. Seeds single in each locule, embryo large, endosperm present or absent (Fig. 9.78).

The family Sapotaceae consists of 35–75 genera and approximately 800 species, distributed pantropically with few species in temperate regions. The largest genera are *Pouteria* (150 species), *Palaquium* (115 species), and *Planchonella* (100 species).

The family is becoming very important for timber production in Malaya. *Manilkara zapota* of Central America provides chicle, the elastic substance in chewing gum. Species of *Palaquium* give gutta-percha, a latex substance used in early golf balls and submarine telephone cables as insulation and today used as dental stoppings. Popular edible fruits come from *Achras zapota,* sapodilla plum, and *Chrysophyllum cainito,* the star apple, native of tropical America.

Along with the Ebenaceae, the Sapotaceae make up most of the species of the order Ebenales. The generic limits are, at best, difficult to determine, and the opinions of botanists vary on how genera should be distinguished; therefore, there is a variation in the number of genera.

Fossil pollen thought to be from the Sapotaceae has been found in Upper Cretaceous and more recent deposits.

## Family Ebenaceae (Ebony)

Figure 9.79 Ebenaceae. *Diospyros:* (a) leafy branch; (b) cross section of ovary; (c) flower; (d) stamen; (e) longitudinal section of flower; (f) fruit.

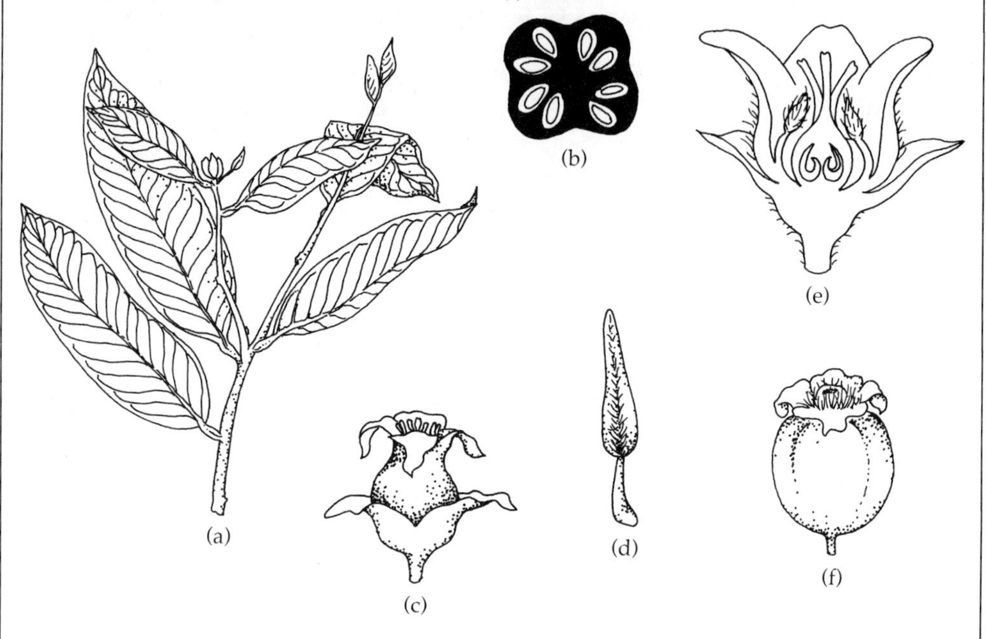


Shrubs or trees usually with black charcoal-like bark, lacking milky juice. Leaves alternate, simple, entire; lacking stipules. Flowers regular, usually unisexual, hypogynous, staminate flowers with a rudimentary pistil, pistillate flowers commonly with staminoids; inflorescence of axillary cymes or of solitary flowers. Sepals 3–7, fused and persistent. Petals 3–7, sympetalous, with the lobes contorted. Stamens 3–many in two whorls, usually twice the number as the corolla lobes, distinct or filaments adnate to the corolla tube; anthers opening by longitudinal slits (rarely by apical pores). Pistil compound of 3–8 united carpels; pluriocular, each locule with 2 ovules attached from the top but each ovule separated by a partition, therefore having apical-axile placentas; ovary superior; style fused at least at the base, stigmas the same number as the locules. Fruit a berry, often with enlarged calyx attached. Seed with curved or straight embryo, endosperm commonly irregularly grooved or ridged (Fig. 9.79).

The family Ebenaceae consists of 2–5 genera and 400–500 species widespread in the subtropics and tropics, with a few species in the temperate regions of North America and Australia. Most species belong to the genus *Diospyros* (over 400 species).

The family is economically important for hard, black wood: *Diospyros ebeneum* and *D. reticulata,* both species called ebony; and for cultivated fruits: *D. kaki,* the Chinese or Japanese date plum, *D. lotus,* date plum, and *D. virginiana,* the persimmon of North America.

Along with the Sapotaceae, the Ebenaceae make up most of the species of the order Ebenales.

Fossil leaves similar to *Diospyros* have been found in Upper Cretaceous and more recent deposits, with wood found in Oligocene to recent strata.

### Family Styracaceae (Storax)

Figure 9.80 Styracaceae. *Styrax:* (a) leafy branch bearing flowers; (b) cross section of ovary; (c) longitudinal section of flower; (d) fruit.

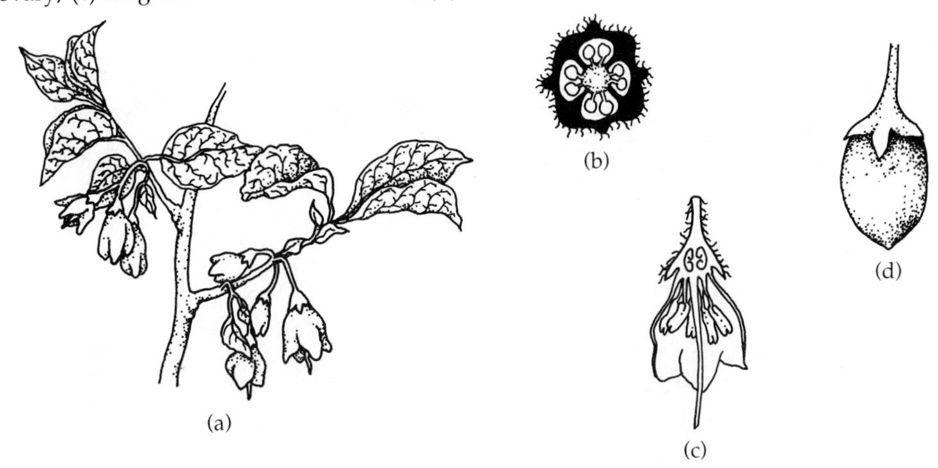


Shrubs or trees, lacking latex but often with the wood having many resin canals; pubescence of stellate or peltate hairs. Leaves alternate, simple, usually entire and lacking stipules. Flowers regular, perfect (rarely unisexual), hypogynous or epigynous; inflorescence of solitary flowers or flowers in cymes, racemes, or panicles. Sepals 4–8, connate. Petals 4–8, distinct or sympetalous. Stamens equal to or double the number of corolla lobes, in 1 whorl, filaments distinct or connate, adnate to the corolla tube; anthers opening by longitudinal slits. Pistil compound of 3–5 united carpels; locules 3–5; ovules 1–many per locule and attached by axile placentas; ovary superior or inferior; style single, stigma capitate or lobed. Fruit a capsule or drupe with a persistent calyx. Seeds 1–many, embryo curved or straight, endosperm present (Fig. 9.80).

The family Styracaceae consists of 10–12 genera and 150–180 species distributed in three regions: southeastern North America to South America, the Mediterranean, and southeastern Asia to Malaysia.

The largest genus *Styrax* (120–130 species) is also the most important economically. *Styrax benzoin* produces a resin that is traded as benzoin and used as gum benjamin, and *S. officinale* is the source of storax, a resin used in medicine as an antiseptic, expectorant, and inhalant, and in incense. Common ornamentals include *Halesia,* the silver bell tree, and some *Styrax,* the snowball trees.

The genus *Afrostyrax* of tropical West Africa is sometimes placed in the Styracaceae or in the family Huaceae.

Fossils attributed to the Styracaceae have been found in Tertiary and more recent deposits.

## Family Symplocaceae (Sweetleaf)

Figure 9.81 Symplocaceae. *Symplocos:* (a) leafy branch with fruits; (b) side view of flower; (c) longitudinal section of flower; (d) cross section of ovary; (e) connate stamens.

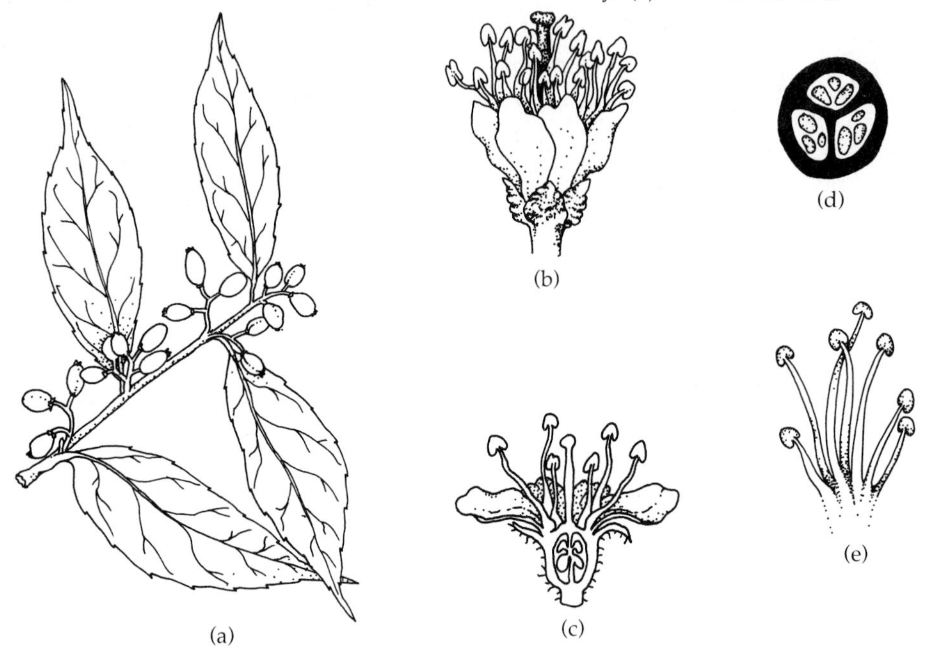


Shrubs or trees. Leaves evergreen (rarely deciduous), alternate, simple; lacking stipules; often with large bladder-like water cells that stick out from the epidermis; commonly sweet-tasting. Flowers regular, perfect (rarely unisexual and polygamous), epigynous; inflorescence of axillary racemes, panicles, or spikes. Sepals 5, connate at the base, persistent. Petals 5 or 10, connate to form a short corolla tube. Stamens 4–many, filaments distinct or connate, adnate to the corolla tube, usually in 2 or more whorls; anthers opening by longitudinal slits. Pistil compound of 2–5 united carpels; locules 2–5; ovules 2–4 in each locule and borne on axile placentas; ovary inferior; style single, often surrounded by a nectary disk, stigma capitate or lobed. Fruit a berry or drupe with persistent calyx. Seeds with curved or straight embryo, endosperm present (Fig. 9.81).

The family Symplocaceae consists of the genus *Symplocos* and 300–400 species common throughout the subtropical to tropical New World, eastern and southern Asia, East Indies, and Australia. It is absent from Africa, Europe, and western Asia.

Economically the family is not very important, however, from *Symplocos tinctoria,* found in the southeastern United States, a yellow dye is extracted from the bark and leaves.

The Symplocaceae family is included within the Theaceae by some botanists because of the whorls of stamens.

Pollen referred to the family has been found in Upper Cretaceous deposits.

## ORDER PRIMULALES
### Family Theophrastaceae (Theophrasta)

Figure 9.82 Theophrastaceae. *Jacquinia:* (a) leafy branch bearing fruits; (b) flower; (c) longitudinal section through flower; (d) longitudinal section through pistil; (e) fruit.

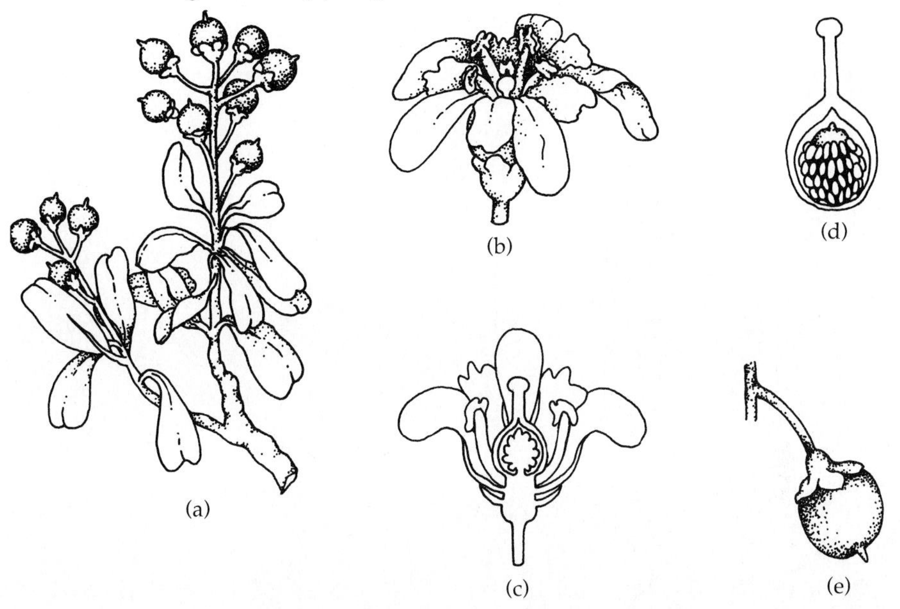


Shrubs or trees, commonly with palm-like habit. Leaves alternate, simple, the blade often with a submarginal fibrous strand; stipules lacking. Flowers large, regular, perfect (rarely unisexual), hypogynous; inflorescence of a solitary flower or flowers arranged in corymb, raceme, or panicle; the perianth marked with glandular dots. Sepals 5, distinct or connate at the base. Petals 5, fleshy, sympetalous, corolla tube short. Stamens 5, filaments distinct or connate near the base, adnate to the corolla tube, fertile stamens opposite the corolla lobes, petal-like staminoids if present alternate the lobes; anthers opening by longitudinal slits. Pistil compound of 5 united carpels; locule 1; ovules few to many and attached to free-central or basal placentas; ovary superior; style single, stigma capitate to lobed. Fruit large, 1–several seeded, yellow or reddish-orange berry or drupe. Seeds large, yellow to red-orange; embryo straight, endosperm present (Fig. 9.82).

The family Theophrastaceae consists of 4 genera and about 100 species distributed in the tropical Americas from southern Florida and Mexico to northern South America. The largest genera are *Clavija* and *Jacquinia* with almost 50 species each.

The family is not very important economically, even though a few species are cultivated as ornamentals. In the West Indies, *Jacquinia barbasco* produces a toxin used as a fish poison.

The genera of the Theophrastaceae are sometimes put in the family Myrsinaceae. The family differs, however, in the leaves lacking glands and resin ducts and having staminoids that alternate with the corolla lobes.

The family has no known fossil record.

## Family Myrsinaceae (Myrsine)

Figure 9.83 Myrsinaceae. *Ardisia:* (a) leafy branch bearing fruit; (b) flower showing staminodes; (c) longitudinal section through flower.

Shrubs or trees (rarely lianas). Leaves evergreen, alternate, simple, commonly dotted with glands or prominent resin ducts; stipules lacking. Flowers small, regular, perfect (rarely unisexual and dioecious), hypogynous to partially epigynous; arranged in various axillary or terminal inflorescences. Sepals 4–6, distinct or connate at the base. Petals 4–6, sympetalous (rarely distinct). Stamens as many as the corolla lobes and opposite them, filaments fused to the corolla tube (rarely distinct), lacking staminodes; anthers opening by longitudinal slits or apical pores. Pistil compound and comprised of 3–6 united carpels; locule 2–5; ovules few to many and attached to axile or free-central placentas; ovary superior or slightly inferior; style single, stigma capitate or lobed. Fruit a berry or drupe. Seeds relatively small, usually 1 per locule, embryo curved or straight, endosperm generally present (Fig. 9.83).

The family Myrsinaceae consists of 30–32 genera and about 1000 species distributed widely throughout the tropics and warmer parts of North America and the Old World. Almost two thirds of the species belong to the genera *Ardisia* (250 species), *Rapanea* (150 species), *Embelia* (130 species), and *Maesa* (100 species).

The family is of little economic value, except for a few species grown as ornamentals. *Ardisia crispa,* from Malaya, is grown for its red fruits and is eaten for stomach ailments. *Myrsine africana,* from Africa, is cultivated for its lovely bluish-purple fruits.

Sometimes the genera that comprise the Theophrastaceae are put in the Myrsinaceae, and the genus *Aegiceras* is placed in its own family.

The family has no known fossil record.

## Family Primulaceae (Primrose)

Figure 9.84 Primulaceae. *Dodecatheon:* (a) plant with flowers; (b) flower showing reflexed corolla. *Lysimachia:* (c) flower; (d) fruit.

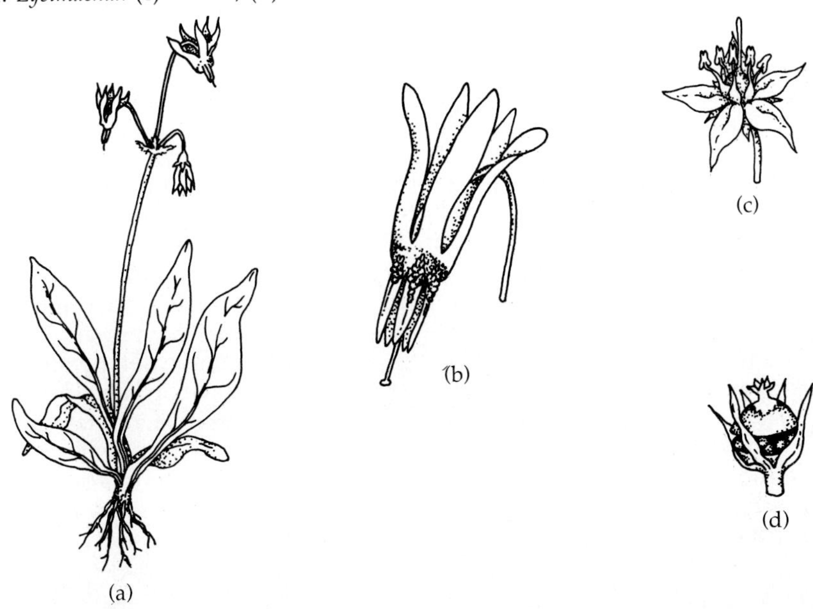

Herbs (rarely semishrubs). Leaves alternate, opposite, whorled, or all in a basal rosette, simple (rarely pinnately compound), often with glands or glandular hairs; lacking stipules. Flowers usually regular, perfect, hypogynous (rarely semiepigynous); inflorescence of solitary flowers or flowers in heads, panicles, racemes, or umbels; often heterostylic. Sepals 5 (rarely 4); connate, lobed, or toothed and persistent. Petals 5, (rarely 4), sympetalous, the lobes sometimes reflexed. Stamens as many as the corolla lobes and opposite them, the filaments attached to the corolla tube; anthers opening by longitudinal slits or less commonly by apical pores. Pistil compound, comprised mostly of 5 united carpels; locule 1; ovules few to many and attached to free-central placentas; ovary superior or semiinferior; style single with a head-like stigma. Fruit a capsule, sometimes circumscissle. Seeds few to many, with 2 cotyledons (sometimes monocotyledonous), embryo straight, endosperm present but not of starch (Fig. 9.84).

The family Primulaceae consists of 28–30 genera and nearly 1000 species that are cosmopolitan in distribution but most are common in the temperate and cooler regions of the Northern Hemisphere; also found in the mountains of the tropics. Over three fourths of the species are found in only 3 genera: *Primula,* the primrose (about 500 species); *Lysimachia,* loosestrife (150–200 species); and *Androsace,* androsace (100 species).

The family is important economically, mainly for its garden ornamentals. Many species of *Primula,* the primroses, are cultivated in rock gardens (i.e., *P. auricula*), as border plants (i.e., *P. denticulata*), and as house plants (i.e., *P. obconica*). *Cyclamen persicum* is a common winter pot plant in North America and Europe. *Anagallis arvensis* and *Cyclamen purpurascens* have medicinal uses as well as a poisonous glycoside. Some *Primula* cause dermatitis.

The family has many characters that are similar to the family Caryophyllaceae, especially features of the pistil. Other flower characters indicate its present classification in its own order.

The family has no known fossil record.

**Subclass V. Rosidae** The subclass Rosidae contains the largest number of families of any subclass of flowering plants. The subclass consists of over 58,000 species, 114 families, and 18 orders. It has about the same number of species as the subclass Asteridae. Five orders, Fabales (14,000 species), Myrtales (9000 species), Euphorbiales (7600 species), Rosales (6600 species), and Sapindales (5400 species), contain about 75% of the species in the subclass. (See Table 9.5.)

A group this large is very diverse in many different ways. Some groups within the subclass exhibit features found with some families of the subclass Magnoliidae, while others share characters with members of the Asteridae. Ecologically the family is found in all habitats and regions of the world.

The Rosidae contain herbs, shrubs, trees, lianas, and parasites. Compound leaves are as common as simple leaves. The flowers are predominantly perigynous or epigynous, with hypogynous flowers having an ovary associated with a nectary disk. The perianth is commonly polypetalous (rarely sympetalous). The stamens when many generally develop **centripetally** (from the outside toward the center of the axis). The number of stamens ranges from fewer than the petals to many more than the petals. Pistils range from simple pistils of few individual carpels to compound pistils of 3 or more united carpels. Placentation is usually axile or marginal, seldom basal, free-central or parietal, with the ovules only 1 or 2 per locule.

**TABLE 9.5 A list of orders and family names of flowering plants found in the subclass Rosidae.**

Subclass V. Rosidae
- Order A. Rosales
  - Family 1. Brunelliaceae
  - 2. Connaraceae*
  - 3. Eucryphiaceae
  - 4. Cunoniaceae*
  - 5. Davidsoniaceae
  - 6. Dialypetalanthaceae
  - 7. Pittosporaceae*
  - 8. Byblidaceae
  - 9. Hydrangeaceae*
  - 10. Columelliaceae
  - 11. Grossulariaceae*
  - 12. Greyiaceae
  - 13. Bruniaceae
  - 14. Anisophylleaceae
  - 15. Alseuosmiaceae
  - 16. Crassulaceae*
  - 17. Cephalotaceae
  - 18. Saxifragaceae*
  - 19. Rosaceae*
  - 20. Neuradaceae
  - 21. Crossosomataceae
  - 22. Chrysobalanaceae*
  - 23. Surianaceae
  - 24. Rhabdodendraceae
- Order B. Fabales
  - Family 1. Mimosaceae*
  - 2. Caesalpiniaceae*
  - 3. Fabaceae*
- Order C. Proteales
  - Family 1. Elaeagnaceae*
  - 2. Proteaceae*
- Order D. Podostemales
  - Family 1. Podostemaceae
- Order E. Haloragales
  - Family 1. Haloragaceae*
  - 2. Gunneraceae
- Order F. Myrtales
  - Family 1. Sonneratiaceae
  - 2. Lythraceae*
  - 3. Penaeaceae
  - 4. Crypteroniaceae
  - 5. Thymelaeaceae*
  - 6. Trapaceae

7. Myrtaceae*
8. Punicaceae
9. Onagraceae*
10. Oliniaceae*
11. Melastomataceae*
12. Combretaceae*

Order G. Rhizophorales
Family 1. Rhizophoraceae*

Order H. Cornales
Family 1. Alangiaceae
2. Nyssaceae*
3. Cornaceae*
4. Garryaceae*

Order I. Santalales
Family 1. Medusandraceae
2. Dipentodontaceae
3. Olacaceae*
4. Opiliaceae
5. Santalaceae*
6. Misodendraceae
7. Loranthaceae*
8. Viscaceae*
9. Eremolepidaceae
10. Balanophoraceae

Order J. Rafflesiales
Family 1. Hydnoraceae
2. Mitrastemonaceae
3. Rafflesiaceae*

Order K. Celastrales
Family 1. Geissolomataceae
2. Celastraceae*
3. Hippocrateaceae*
4. Stackhousiaceae
5. Salvadoraceae
6. Aquifoliaceae*
7. Icacinaceae*
8. Aextoxicaceae
9. Cardiopteridaceae
10. Corynocarpaceae
11. Dichapetalaceae

Order L. Euphorbiales
Family 1. Buxaceae*
2. Simmondsiaceae
3. Pandaceae
4. Euphorbiaceae*

Order M. Rhamnales
Family 1. Rhamnaceae*
2. Leeaceae
3. Vitaceae*

Order N. Linales
Family 1. Erythroxylaceae*
2. Humiriaceae
3. Ixonanthaceae
4. Hugoniaceae
5. Linaceae*

Order O. Polygalales
Family 1. Malpighiaceae*
2. Vochysiaceae*
3. Trigoniaceae
4. Tremandraceae
5. Polygalaceae*
6. Xanthophyllaceae
7. Krameriaceae*

Order P. Sapindales
Family 1. Staphyleaceae*
2. Melianthaceae
3. Bretschneideraceae
4. Akaniaceae
5. Sapindaceae*
6. Hippocastanaceae*
7. Aceraceae*
8. Burseraceae*
9. Anacardiaceae*
10. Julianiaceae
11. Simaroubaceae*
12. Cneoraceae
13. Meliaceae*
14. Rutaceae*
15. Zygophyllaceae*

Order Q. Geraniales
Family 1. Oxalidaceae*
2. Geraniaceae*
3. Limnanthaceae
4. Tropaeolaceae
5. Balsaminaceae*

Order R. Apiales
Family 1. Araliaceae*
2. Apiaceae (Umbelliferae)*

---

*NOTE:* Family names followed by an asterisk are discussed and illustrated in the text.

## ORDER ROSALES
### Family Connaraceae (Connarus)

Figure 9.85 Connaraceae. *Angelaea:* (a) leafy branch with flowers; (b) cross section of ovary; (c) flower; (d) pistil with carpels.

Shrubs, small trees, or lianas. Leaves alternate, pinnately compound or trifoliate; stipules lacking. Flowers small, regular or slightly irregular, perfect to unisexual (commonly dioecious), hypogynous or slightly perigynous; inflorescence of axillary or terminal racemes or panicles. Sepals 5, distinct or connate at the base, persistent on the bottom of the developing fruit. Petals 5, distinct or only slightly connate at the base. Stamens usually 5 or 10 (rarely 4 or 8), attached in 2 series, filaments distinct and attached to a nectary disk. Pistil simple of 1–5 individual carpels; ovules 2 to each carpel, commonly the carpel not fully closed; ovary superior; style terminal, stigma capitate. Fruit a 1-seeded follicle. Seed often with an aril, endosperm present or absent (Fig. 9.85).

The family Connaraceae consists of 16–24 genera and 300–400 species, distributed pantropically, especially in the Old World. The largest genera are *Connarus* (100 species) and *Rourea* (100 species).

The family is economically important for some timber, medicines, and foods. Zebra wood is obtained from *Connarus guianensis.* Medicinally, the leaves of *Agelaea villosa* are used in West Africa to treat dysentery, while those of *A. emetica* cause vomiting. Leaves of *Cnestis corniculata* and *C. ferruginea* are used as an astringent and a laxative, respectively, while *C. Platantha* of Malaysia has edible seeds. In Central America, the bark of *Rourea glabra* is used to impart a dark blue or purple color to animal skins.

The classification position of the family Connaraceae in the order Rosales is open to debate by botanists because of its pinnately compound leaves, free carpels, and seeds with arils.

The family has no known fossil record.

## Family Cunoniaceae (Cunonia)

Figure 9.86 Cunoniaceae. *Weinmannia:* (a) leafy branch bearing a terminal inflorescence; (b) flower. *Ceratopetalum:* (c) fruit; (d) longitudinal section through flower. *Callicoma:* (e) node showing stipules.

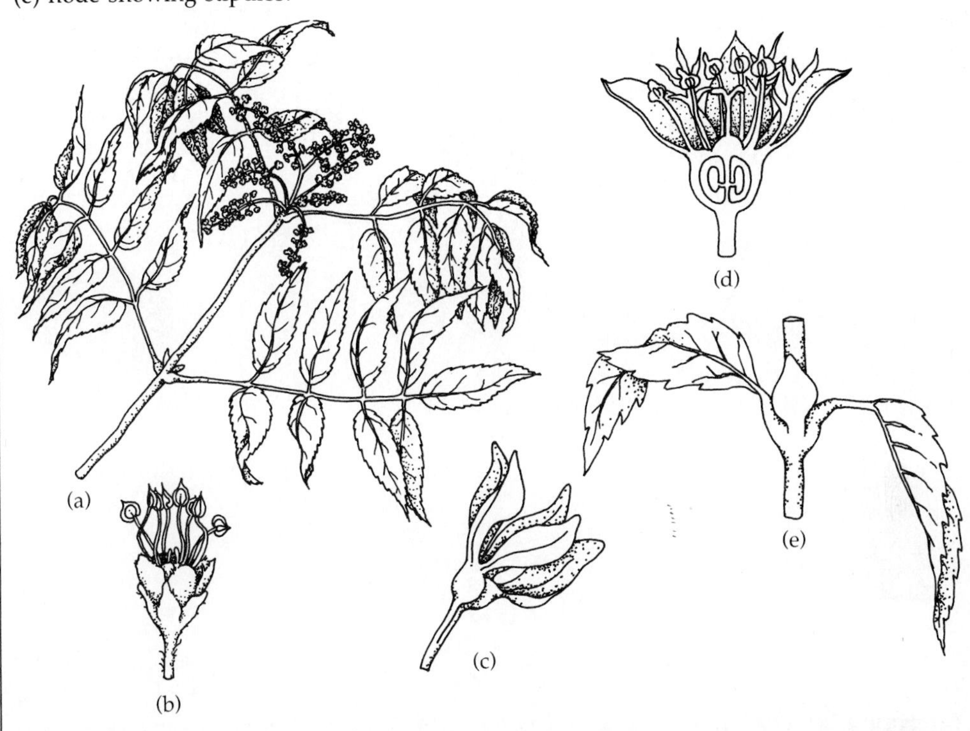


Shrubs or trees (rarely lianas). Leaves leathery, opposite or whorled, usually pinnately compound or trifoliate (rarely simple), often glandular and toothed; stipules present, may be large or fused in pairs. Flowers small, regular, perfect (rarely unisexual and dioecious), hypogynous to slightly epigynous; inflorescence of heads, panicles, or racemes. Sepals 3–6, distinct or connate at the base. Petals 3–6 (or lacking), alternate and shorter than the sepals. Stamens 4–many, distinct, attached to a nectary-disk, which surrounds the ovary; anthers opening by longitudinal slits. Pistil of 2–5 simple or fused carpels; locules 2–5; ovules 2–many in each locule and borne on axile placentas; ovary superior or partly inferior; styles terminal on each carpel, stigma capitate. Fruit a capsule or nut (rarely a drupe or follicle). Seeds small, sometimes winged, embryo straight, endosperm present (Fig. 9.86).

The family Cunoniaceae consists of 25–26 genera and 250–350 species distributed throughout the tropical regions of the Southern Hemisphere, especially in Australasia and the Pacific regions. The largest genus is *Weinmannia* (170 species), followed by *Pancheria* (25 species).

The family is of little economic value, except for the tall tree *Ceratopetalum apetalum*. Its wood is light in color and used in cabinetmaking, paneling, and plywood.

The Cunoniaceae has been classified by some botanists along with the Saxifragaceae.

The fossil record of the family is scattered among leaf and pollen fossils found in various Tertiary deposits.

## Family Pittosporaceae (Pittosporum)

Figure 9.87 Pittosporaceae. *Pittosporum:* (a) leafy branch bearing flowers; (b) fruits; (c) stamen; (d) cross section of ovary; (e) flower.

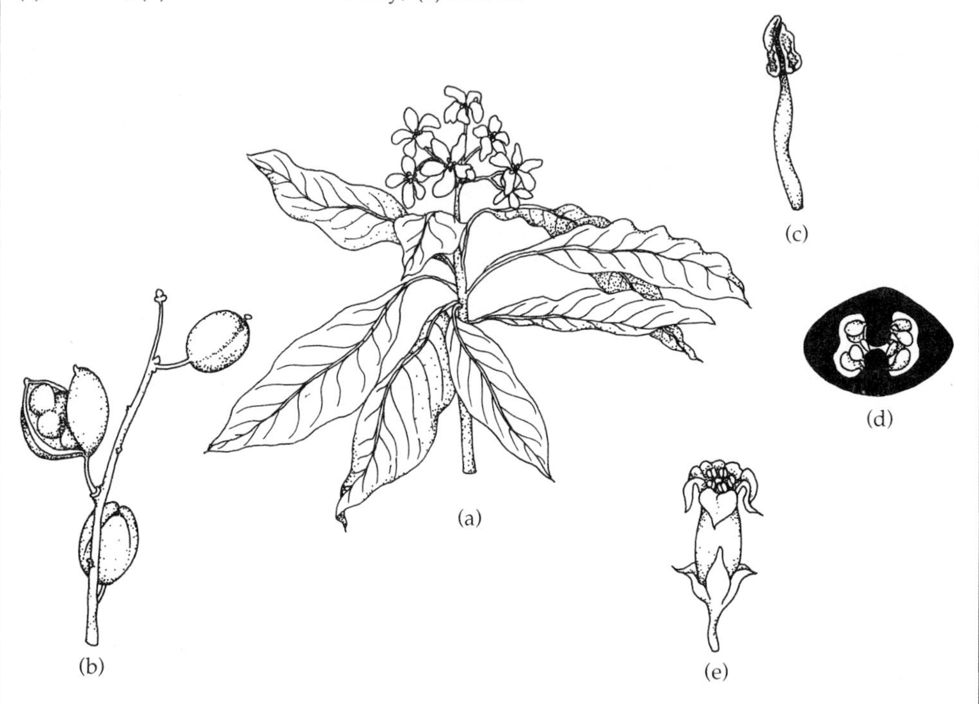


Shrubs or small trees with well-developed resin canals in the stem and bark. Leaves evergreen and leathery, alternate, simple, usually entire margined; stipules lacking. Flowers regular (rarely irregular), perfect (rarely unisexual and plants polygamous), hypogynous, subtended by 2 small bracts. Sepals 5, distinct or connate at the base. Petals 5, distinct or connate at the base. Stamens 5, filaments distinct or connate at the base and attached to the sepals, alternate the petals; anthers opening by longitudinal slits or apical pores. Pistil compound of 2 united carpels; locule usually 1; ovules few to many, attached to the intruding parietal placentas; ovary superior; style single, stigma capitate or lobed. Fruit a berry or loculicidal capsule. Seeds few to many, often covered with a resin-like pulp; embryo small, sometimes with 3–5 cotyledons, endosperm abundant (Fig. 9.87).

The family Pittosporaceae consists of 9 genera and 200–240 species distributed widely in the Old World tropics and the warm temperate areas of Australasia. The main genus is *Pittosporum* (140–200 species).

The economic value of the family is limited mostly to a few species of *Pittosporum* cultivated for their fragrant flowers: *P. crassifolium, P. tenuifolium,* and *P. eugenioides* from New Zealand; *P. tobira* from China and Japan; and *P. undulatum* from Australia.

The family Pittosporaceae has some characters that are similar to those found in the Saxifragaceae and has been classified in that family by some botanists.

Except for some flowers attributed to the Pittosporaceae found in some Miocene amber, the family lacks a fossil record.

## Family Hydrangeaceae (Hydrangea)

Figure 9.88 Hydrangeaceae. *Hydrangea:* (a) leafy branch with inflorescence; (b) flower; (c) cross section of ovary.

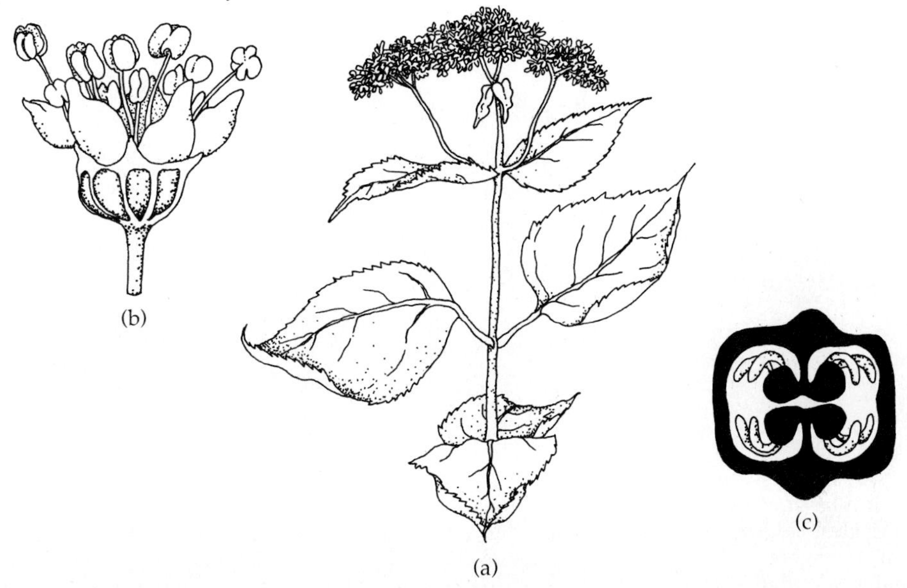


Herbs, shrubs, trees, or lianas. Leaves simple, opposite (rarely alternate); stipules lacking. Flowers regular or the marginal ones in the inflorescence irregular, perfect, usually perigynous, hypanthium present, may or may not be fused to the ovary; inflorescence of complex cymes, corymbs, or panicles. Sepals 4–5 (rarely 6–10), distinct or connate. Petals 4–5 (rarely 6–10) distinct. Stamens 4–many, filaments distinct or connate at the base. Pistil compound of 2–7 united carpels; locules 2–7; ovules many and attached to axile placentas, or locule 1 and the ovules on intruding parietal placentas; ovary inferior (rarely superior); style simple or connate at the base, stigma elongate, capitate, or lobed. Fruit a loculicidal or septicidal capsule (rarely a berry). Seeds 1–many, embryo straight, endosperm present (Fig. 9.88).

The family Hydrangeaceae consists of 17 genera and 170 species distributed widely in the Northern Hemisphere in both temperate and subtropical climates. The largest genera are *Deutcia* (50 species); *Philadelphus,* the mock orange (50 species); and *Hydrangea,* the hydrangeas (24 species).

Various species of the above genera are important as cultivated ornamentals. There are many cultivars and hybrids derived from *Hydrangea macrophylla* and from *Philadelphus coronarius.* Some hydrangea species are used as a source for hydrangin, a compound used in medicine; others as a poison.

The genera of the Hydrangeaceae are sometimes placed in a number of small families or in the Saxifragaceae. There is much morphological evidence to keep the Hydrangeaceae as an individual family.

The family has no known fossil record.

## Family Grossulariaceae (Currant)

Figure 9.89 Grossulariaceae. *Ribes:* (a) leafy branch with flowers and fruit; (b) saucer-shaped flower; (c) flower with long hypanthium.

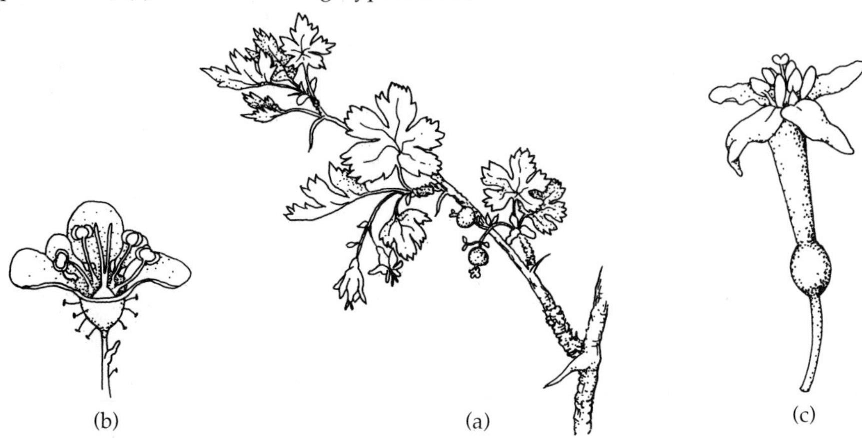


Shrubs or trees, with or without spines or prickles. Leaves alternate, simple, commonly palmately veined; usually lacking stipules or stipules very small and deciduous. Flowers regular, perfect (rarely unisexual), perigynous to epigynous, hypanthium sometimes very prominent, saucer-shaped to tubular-shaped; inflorescence of solitary flowers or flowers in axillary cymes, racemes, or panicles. Sepals 4–5, distinct or connate, forming a petal-like extension beyond the hypanthium. Petals 4–5 or lacking, distinct or connate at the base with the lobes alternate the calyx lobes. Stamens 4–5 (rarely 8–10), filaments distinct, sometimes 4–5 staminoids present, which alternate with the sepals; anthers opening by longitudinal slits. Pistil compound of 2 (less commonly 1–6) united carpels; locule 1 with ovules on parietal placentas or locules 2–6 and ovules many on axile placentas; ovary superior to inferior; style single or fused, stigma capitate to lobed. Fruit a berry, follicle, or capsule. Seeds many, often with an aril, embryo variable, endosperm present (Fig. 9.89).

The Grossulariaceae consists of 25 genera and 350 species and is cosmopolitan in distribution. The largest genus, *Ribes (Grossularia),* the currants (150 species), is common in the Northern Hemisphere, with a few species extending south into South America. *Polyosma* (60 species) is a tropical genus, while the genus *Escallonia* (50 species) is found in temperate climates of the Southern Hemisphere.

Economically, the genus *Ribes* is the most important group, which includes the edible fruits called currants or gooseberries. *Ribes* shrubs are the alternate host for the white pine blister rust, a fungus disease that does much damage to white pine trees.

The family Grossulariaceae is comprised of very diverse genera, so much so that some botanists have separated out as many as 8 families and put *Ribes* in the Saxifragaceae.

Fossil wood attributed to the Grossulariaceae has been found in Upper Cretaceous deposits, while leaves and fossil pollen have been found in various Tertiary deposits.

## Family Crassulaceae (Stonecrop)

Figure 9.90 Crassulaceae. *Sedum:* (a) plant with flowers and fruits; (b) fruit; (c) flower. *Kalanchöe:* (d) leafy stem with flowers; (e) flower; (f) longitudinal section through flower.

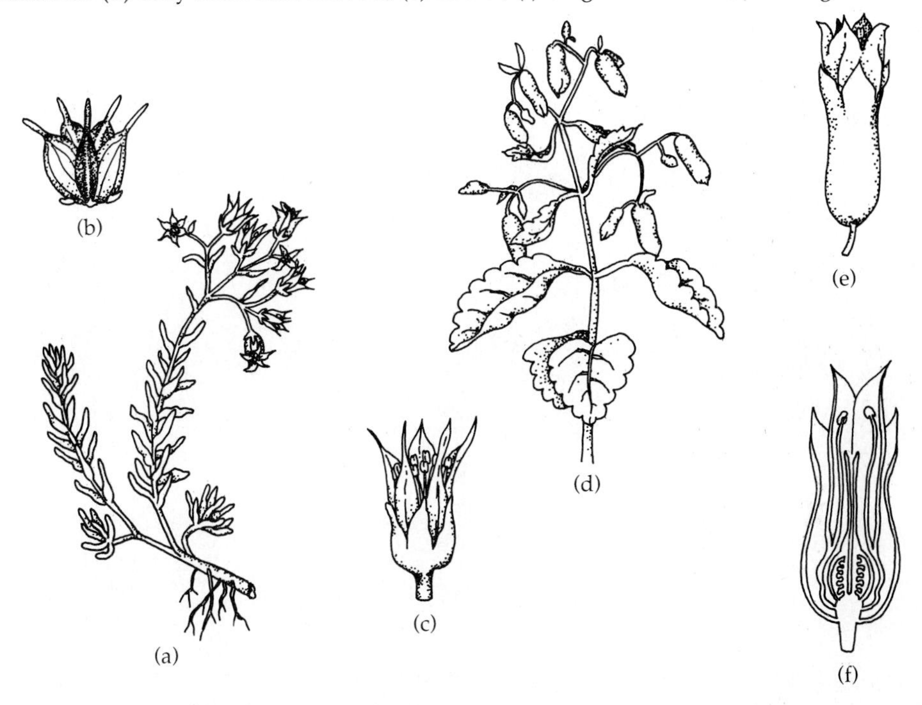


Herbs or small shrubs, usually succulent. Leaves fleshy, alternate, opposite (rarely whorled) or in rosettes, simple, entire; stipules lacking. Flowers small, regular, perfect (rarely unisexual), hypogynous or slightly perigynous; inflorescence of solitary flowers or flowers grouped into showy cymes, corymbs, or panicles. Sepals usually 4–6, distinct or connate. Petals 4–6, distinct or sympetalous. Stamens the same number as or twice the number of petals, filaments distinct (rarely connate) and adnate to the corolla tube; anthers opening by longitudinal slits. Pistil simple or compound with the same number of carpels as the petals, distinct or fused only at the base; locules 1–5; ovules few to many in each carpel and attached to parietal (rarely axile) placentas; ovary superior; style short or long, stigma small and moist. Fruit a capsule or cluster of follicles. Seeds small, endosperm present or absent (Fig. 9.90).

The family Crassulaceae consists of 25–35 genera and 900–1500 species with a cosmopolitan distribution, except for the Australia-Pacific region. The species are commonly in warm, dry habitats. The greatest diversity is found in South Africa, but it is common in the mountains of Asia and North America, including Mexico. The most common genera are *Sedum* (300 species) and *Crassula* (250 species), the stonecrops, and *Kalanchöe* (120 species).

The family is economically valuable as garden ornamentals, especially for rock gardens. The most used species are of *Sedum*, the stonecrop; *Simpervivum*, the house leeks; and in the greenhouse or home *Echevevia*, *Kalanchöe*, and *Rochea*.

The family Crassulaceae is similar in various ways to the Saxifragaceae and has been classified in this family in the past.

Pollen attributed to the Crassulaceae has been found in Miocene and more recent deposits.

## Family Saxifragaceae (Saxifrage)

Figure 9.91 Saxifragaceae. *Saxifraga:* (a) plant with flowers; (b) flower; (c) longitudinal section of flower; (d) stem with mature fruits.

Herbs, perennial (rarely annual). Leaves alternate, opposite, or basal, simple (rarely compound); stipules lacking. Flowers regular (rarely irregular), perfect (rarely unisexual), perigynous to epigynous, hypanthium well developed; inflorescences of various types. Sepals 5 (rarely 3–10), often as hypanthium lobes. Petals 5 (rarely 3–10 or lacking), small, alternate the sepals, sometimes clawed or cleft. Stamens usually twice the number of the petals and in 2 whorls, filaments distinct; anthers opening by longitudinal slits. Pistil compound of 2–4 united carpels (rarely distinct); locules 1–5 (commonly 1–2); ovules mostly many in each locule and attached to axile or parietal placentas; ovary superior or inferior; style single from each carpel, often curved and horn-like, stigma capitate. Fruit a septicidal capsule, splitting above the fused base. Seeds many, small, embryo small, straight, endosperm abundant to lacking (Fig. 9.91).

The family Saxifragaceae consists of approximately 40 genera and 700 species and is cosmopolitan in distribution, but with its greatest diversity in the arctic, boreal, and montane regions of the Northern Hemisphere. The largest genus is *Saxifraga,* the saxifrages (about 300 species), which is very common in arctic and alpine floras.

The family is important economically as garden ornamentals. The genus *Saxifraga* is a very diverse group, and many species are grown as rock garden plants. *Heuchera,* alum root or coral bells; *Mitella,* bishop's cap; and *Tiarella* are native woodland plants grown in gardens.

The classification of the Saxifragaceae has been interpreted various ways by different botanists. Some would include in the family the shrub genera of the Hydrangeaceae and Grossulariaceae, while others would split the family into as many as 15 families in 2 different orders.

Fossils attributed to the Saxifragaceae previously collected from Eocene deposits of Europe are now having the determinations being questioned by present-day paleobotanists.

## Family Rosaceae (Rose)

Figure 9.92 Rosaceae. *Prunus:* (a) leafy branch bearing fruits; (b) front view of flower showing hypanthium; (c) branch with flowers. *Rubus:* (d) leafy branch with fruits. *Rosa:* (e) leafy stem segment bearing flower, stipules, and prickles; (f) fruit (a hip).

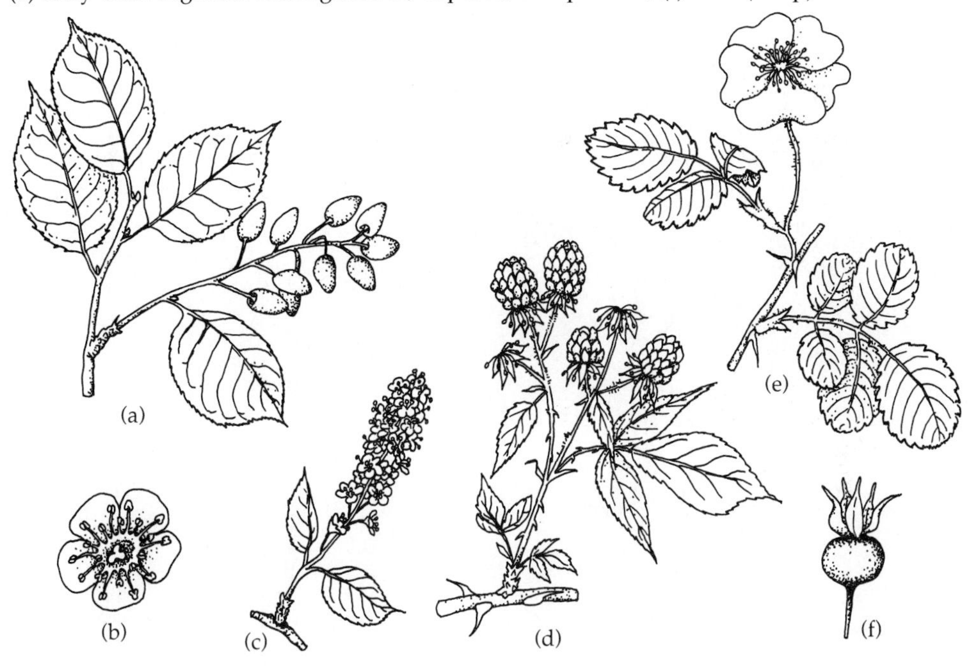


Herbs, shrubs, or trees. Leaves alternate (rarely opposite), simple or compound; usually with paired stipules; commonly with paired glands at the base of the blade. Flowers generally large, regular, perfect (rarely unisexual), hypogynous to perigynous to epigynous, hypanthium usually prominent; predominantly insect pollinated. Sepals 5, commonly subtended by an epicalyx of 5 bracts. Petals 5 (rarely lacking), distinct, often large. Stamens many (rarely less), filaments distinct, slender, attached to the hypanthium where the inside is commonly nectar covered; anthers opening by longitudinal slits (rarely by apical pores). Pistil simple of many single carpels or compound of 2–5 united carpels; in simple locules the ovules are on parietal placentas, while in pistils with 2–5 locules the ovules are attached to axile placentas; ovary superior or inferior; style 1 per carpel, distinct or connate, stigma capitate. Fruit of diverse types. Seeds with embryo straight or bent, endosperm scanty or lacking (Fig. 9.92).

The family Rosaceae is a large, diverse (sometimes treated as many families) family of 100–122 genera and 3000–3400 species distributed worldwide, but most common in the temperate regions of the Northern Hemisphere. Genera with many species include *Potentilla,* the cinquefoils (300 species); *Prunus,* cherries (200 species); *Rosa,* the rose (100 species); and *Spiraea,* spiraea (70–100 species).

The family is very important economically for fruits and ornamentals. Examples of these include *Crataegus,* the hawthorns; *Fragaria,* strawberry; *Pyrus (Malus),* apple; *Rubus,* blackberry and raspberry; *Sorbus,* mountain ash; and *Spiraea,* spiraea.

The diverse nature of the genera of the Rosaceae has led to equally diverse opinions among botanists about how to classify the family. The family is considered to be made up of 4–6 subfamilies, depending on interpretations.

The fossil record of the Rosaceae extends from various periods of the Tertiary.

## Family Chrysobalanaceae (Cocoa Plum)

Figure 9.93 Chrysobalanaceae. *Chrysobalanus:* (a) leafy branch with flowers and fruit; (b) flower; (c) longitudinal section through ovary. *Licania:* (d) longitudinal section through flower.

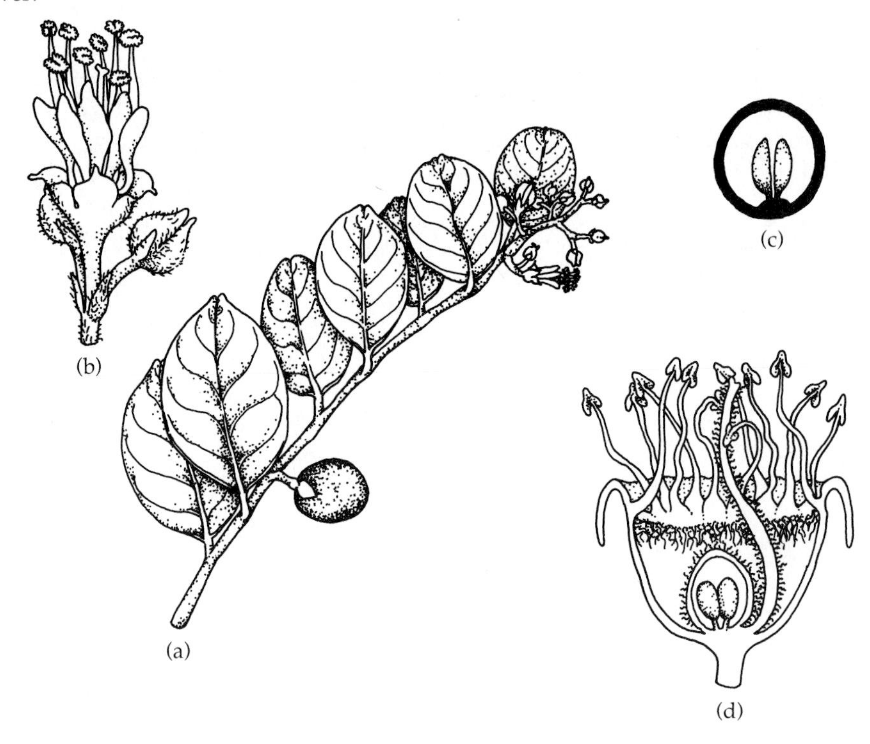


Shrubs or trees. Leaves alternate, simple, entire; stipules present. Flowers small, regular or irregular, usually perfect, strongly perigynous; arranged in various types of inflorescences. Sepals 5. Petals 5 or lacking. Stamens 2–many, filaments distinct or connate, attached around the hypanthium or in irregular flowers to one side, a circular nectary at the base of the stamens; anthers opening by longitudinal slits that split toward the center axis of the flower. Pistil compound of 3 carpels but only 1 carpel develops, the other 2 abort; ovules 2 attached basally marginally in the locule, ovary superior, style single and **gynobasic** (a style that comes from the base of the lobed ovary), stigma simple or lobed. Fruit a dry or fleshy drupe. Seeds large, embryo large, endosperm lacking (Fig. 9.93).

The family Chrysobalanaceae consists of 17 genera and 400–450 species distributed in pantropical lowlands, especially in the New World tropics; found also in the southern United States. The largest genera are *Licania* (160 species), *Hirtella* (85 species), *Couepia* (55 species), and *Parinari* (50 species).

The economic value of the family is mostly local. Species of *Maranthes* and *Parinari* are important genera of timber trees in the Solomon Islands. Some species are cultivated for their fruit, with *Chrysobalanus icaco*, the coco plum, most widely used. In Brazil, oil from the seeds of *Licania rigida*, oiticica, is obtained and used in the candle and soap industry.

The Chrysobalanaceae in the past was included within the Rosaceae as a subfamily. It differs in many ways morphologically and especially in its distinctive anatomy.

The family has no known fossil record.

## ORDER FABALES
### Family Mimosaceae (Mimosa)

Figure 9.94 Mimosaceae. *Mimosa:* (a) leafy branch with flowers in heads; (b) fruits; (c) flower.

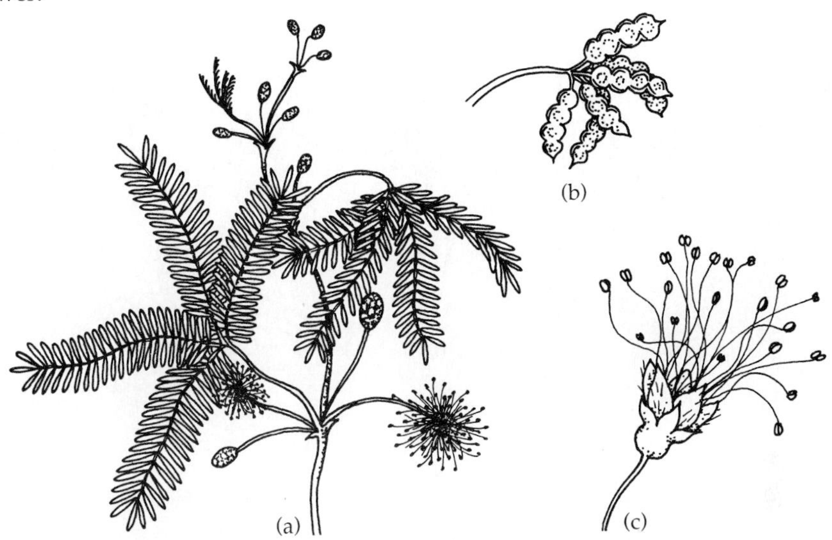


Shrubs or trees (rarely herbs), commonly with nitrogen-fixing bacteria associated in root nodules. Leaves alternate (rarely opposite), usually bipinnately (rarely once pinnate) compound; the compound blade is sometimes lacking, with an enlarged flattened petiole or phyllode functioning as a leaf; basal **pulvinus** (cluster of cells that govern orientation of the leaf and leaflets) located at base of leaflets and petiole. Flowers small, usually regular, perfect (rarely unisexual), hypogynous (rarely perigynous), with the cluster of flowers being showy; inflorescence of heads, racemes, or spikes. Sepals 5, connate, lobes somewhat reduced or absent. Petals 5, distinct or connate. Stamens 10–many, filaments distinct or connate, usually colored and exserted; anthers small, opening by longitudinal slits, often with a gland at the tip. Pistil simple of 1 carpel (rarely 2 or more), locule 1, ovules 2–many and borne on a marginal placenta; style terminal. Fruit a legume or loment. Seeds 2–many, with a hard seed coat, flattened, embryo large and straight, endosperm in a small amount or lacking (Fig. 9.94).

The family Mimosaceae consists of 50–56 genera and about 3000 species distributed mostly pantropically throughout the subtropical and tropical regions. Large genera include *Acacia,* the acacia or wattle (700–800 species); *Mimosa,* mimosa (450–500) species); *Albizia,* (150 species); and *Calliandra* (150 species).

The family has many species, with a wide range of economic uses. Various African species, such as *Acacia senegal* and *A. stenocarpa,* produce gum arabic. In Australia, *A. melanoxylan* and *A. visco* are important for timber. Many *Albizia* species are valuable timber species. *Albizia julibrissin* is a cultivated ornamental as far north as Michigan and the northwestern United States. The pods and seeds of *Prosopis juliflora,* mesquite native to southwestern North America, is ground up and used as animal feed.

The Mimosaceae separated into various smaller tribes is sometimes classified as a subfamily of the family Fabaceae (Leguminosae). There is evidence for both ways of classification.

Fossil pollen attributed to the Mimosaceae has been found in middle and upper Eocene deposits in the United States, with more recent finds from Africa.

## Family Caesalpiniaceae (Caesalpinia)

Figure 9.95 Caesalpiniaceae. *Caesalpinia:* (a) leafy branch with terminal inflorescence. *Bauhinia:* (b) leafy branch with terminal inflorescence; (c) fruit.

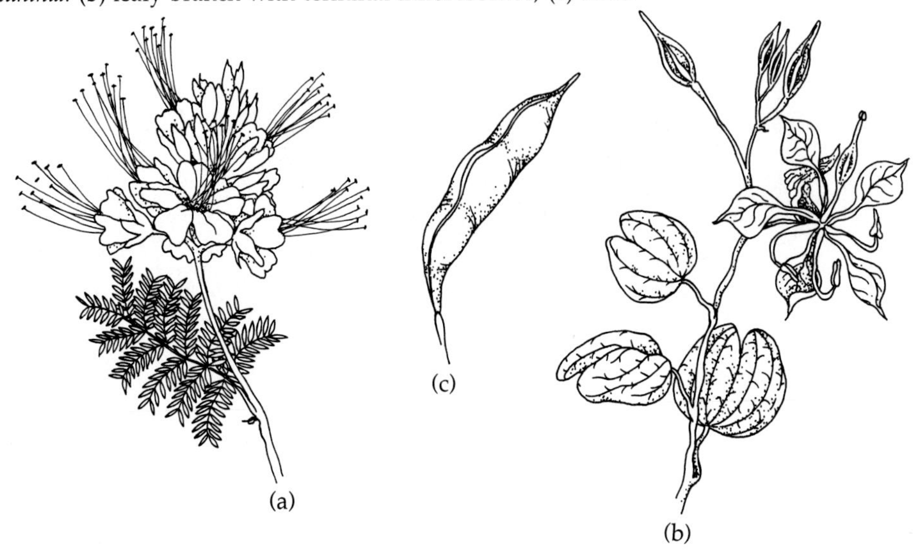


Herbs, shrubs, or trees (rarely lianas), with or without nitrogen-fixing bacteria associated in root nodules. Leaves alternate, usually bipinnate to pinnately compound (rarely simple) or modified into tendrils; stipules present. Flowers large or small, regular or irregular, perfect, slightly to strongly perigynous (rarely hypogynous); inflorescence of cymes, racemes, or spikes. Sepals 5, distinct or connate or upper 2 sepals connate. Petals 5, distinct, usually irregular with the uppermost petal **(banner or standard)** inside the two adjacent petals **(wings).** Stamens usually 10 or fewer, filaments distinct or connate in different ways; anthers opening by longitudinal slits or pores; nectar gland in a ring on the receptacle. Pistil simple of 1 carpel; locule 1; ovules 2–many and borne on a marginal placenta; style terminal. Fruit a dry or fleshy legume or loment. Seeds few to many, commonly with a hard coat, embryo straight, endosperm in small amount or lacking (Fig. 9.95).

The family Caesalpiniaceae consists of 150–180 genera and 2200–3000 species distributed throughout the subtropics and tropical regions, with only a few species found in temperate climates. The largest genera, with about 250 species each, are *Bauhinia, Chamaecrista,* and *Senna.*

This family contains some economically important species. Several *Caesalpinia* species are used for timber and dyes; *C. pulcherrima,* the pride of Barbados, is a common tropical ornamental as are various *Cassia, Senna,* and *Delonix (Poinciana) regia.* In temperate regions *Cercis canadensis,* the Judas tree or redbud; *Gleditsia tricanthos,* honey locust; and *Gymnocladus dioica,* Kentucky coffee tree, are common landscaping trees.

The Caesalpiniaceae, separated into various small tribes, is sometimes classified as a subfamily of one large family, Fabaceae (Leguminosae). There is evidence for both ways of classification and how it is done will probably depend on personal opinion.

Fossil pollen attributed to the Caesalpiniaceae appear first in Upper Cretaceous deposits from various parts of the world.

## Family Fabaceae (Leguminosae) (Bean)

Figure 9.96 Fabaceae (Leguminosae). *Lathyrus:* (a) leafy stem with tendrils and flowers; (b) fruit; (c) fruit partially open, showing seeds. *Ulex:* (d) longitudinal section through flower. *Trifolium:* (e) plant with head inflorescence; (f) flower.

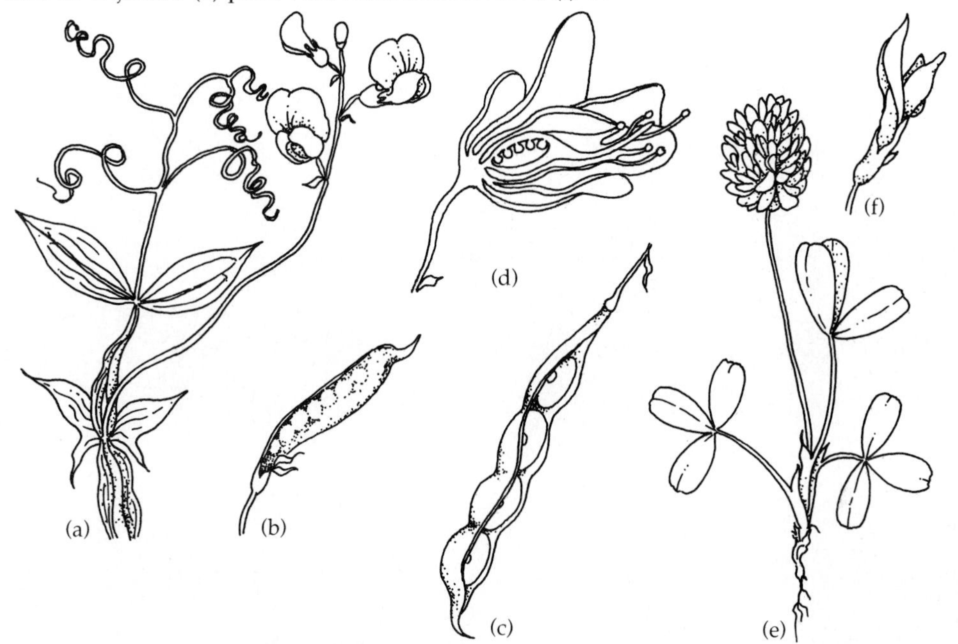


Herbs, shrubs, trees, or lianas, commonly with nitrogen-fixing bacteria associated in root nodules. Leaves alternate (rarely opposite), most commonly bipinnate to pinnately or palmately compound; leaflets sometimes modified into slender tendrils; stipules present. Flowers irregular, perfect, perigynous; arranged in heads, racemes, or spikes. Sepals 5, connate. Petals 5 (rarely fewer), distinct or the lower 2 petals with their bases distinct and blades fused; uppermost petal called the **banner** or **standard,** 2 lateral petals called **wings,** 2 lower petals connate and called the **keel,** and enclosing the stamens and pistil. Stamens 10, filaments distinct or diadelphous, nectar gland in a ring on the receptacle; anthers opening by longitudinal slits. Pistil simple, carpel 1; locule 1; ovules 2–many on a marginal placenta; ovary superior; style terminal. Fruit a dry or fleshy legume or loment. Seeds few to many, commonly with a hard seed coat, embryo usually curved, endosperm in small amount or lacking (Fig. 9.96).

The family Fabaceae consists of about 440 genera and 12,000 species cosmopolitan throughout the world. Some genera are very large: *Astragalus,* locoweed (2000 species); *Indigofera,* indigo (500 species); *Trifolium,* the clovers (300 species); *Phaseolus,* bean (200 species); and *Lupinus,* lupine (200 species).

Economically the Fabaceae ranks second in value next to the Poaceae (Graminaea), grasses. Food plants include *Arachis hypogaea,* the peanut; *Cicer arietinum,* chick pea; *Glycine max,* soybean; *Glycyrrhiza glabra,* licorice; *Lens culinaris,* lentil; and *Pisum sativum,* common pea. Forage plants include *Medicago sativa,* alfalfa; *Melilotus,* sweet clover; and *Trifolium,* the clovers. Ornamental garden plants include *Cytisus,* broom; *Erythrina; Laburnum; Lathyrus,* sweet pea; *Lupinus,* lupine; and *Wisteria,* wisteria.

The family Fabaceae in some classifications includes the Caesalpiniaceae and Mimosaceae as subfamilies.

Fossil pollen attributed to the family first appears in Upper Miocene deposits.

## ORDER PROTEALES
### Family Elaeagnaceae (Oleaster)

Figure 9.97 Elaeagnaceae. *Elaeagnus:* (a) leafy branch with flowers; (b) longitudinal section through flower; (c) surface feature of peltate scales and stellate hairs; (d) front view of flower.

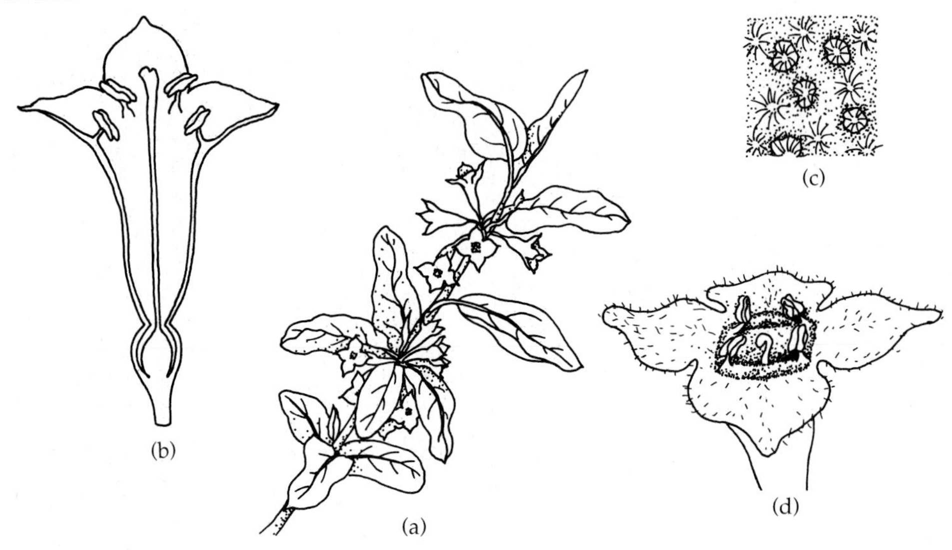


Shrubs or trees, covered with a dense pubescence of scales or stellate hairs, often thorny, commonly associated with nitrogen-fixing bacteria in nodules attached to the roots. Leaves alternate or opposite, simple, entire; stipules lacking. Flowers regular, perfect or unisexual, perigynous with prominent hypanthium; in perfect and female flowers the receptacle is usually tubular, while in male flowers the receptacle is usually saucer-shaped; inflorescence a solitary flower in leaf axils, or flowers in racemes or umbels. Sepals 2 or 4, petal-like, distinct or connate into a calyx tube. Petals lacking. Stamens 4 or 8, filaments very short, attached in the throat of the hypanthium, alternate with the calyx lobes; anthers opening by longitudinal slits. Pistil simple of 1 carpel; locule 1; ovule 1 and borne on a basal placenta; ovary superior; style terminal, slender and elongate, stigma capitate. Fruit a drupe-like achene with the persistent base of the hypanthium surrounding it. Seed with a hard seed coat, embryo straight, endosperm of small amount or lacking (Fig. 9.97).

The family Elaeagnaceae consists of 3 genera and 50 species distributed in the subtropical and temperate regions of the Northern Hemisphere and eastern Australia. The largest genus is *Elaeagnus,* the oleaster (45 species).

Economically the family is important mostly as ornamentals. These include *Elaeagnus angustifolia,* the Russian olive or oleaster, which is planted commonly in poor soils and dry regions of the world, especially western North America. Other planted species include *E. macrophylla, E. pungens,* and *E. umbellata.* The last species has become an introduced pest in abandoned fields and roadsides in parts of North America. *Hippophäe rhamnoides,* the sea buckthorn, is grown for its attractive orange fruits, from which jelly and a sauce are made. In Japan, *E. multiflora* is used to make an alcoholic drink.

Fossil pollen attributed to the Elaeagnaceae has been found in Paleocene deposits.

## Family Proteaceae (Protea)

Figure 9.98 Proteaceae. *Grevillea:* (a) leafy branch with terminal flowers. *Hakea:* (b) fruit; (c) flower; (d) cross section of ovary; (e) longitudinal section through unopened flower.

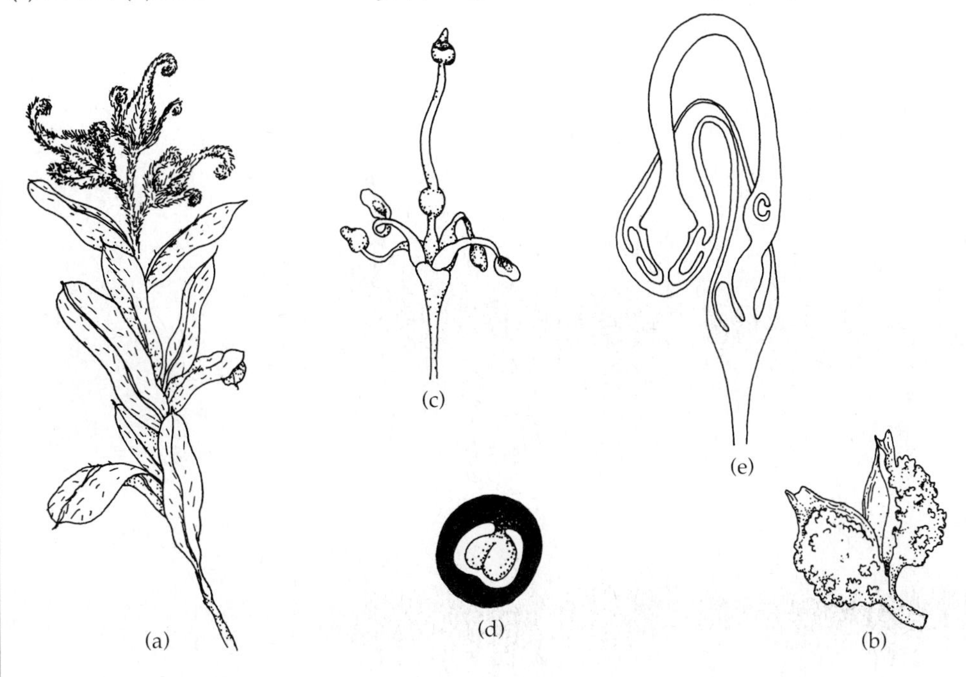


Shrubs or trees (rarely herbs), commonly hairy. Leaves usually leathery and evergreen, alternate (rarely opposite or whorled), simple to pinnately compound; stipules lacking. Flowers regular or irregular, perfect (rarely unisexual), hypogynous to perigynous; inflorescence of solitary flowers or flowers in heads, racemes, or spikes, sometimes very showy with up to 1000 flowers in an inflorescence; pollinated by birds, insects, small marsupials, or mice. Sepals 4, petal-like, distinct or connate. Petals lacking or as 2–4 small scales. Stamens 4, filaments broad and adnate to the calyx, opposite the sepals. Pistil simple, sometimes on a stipe; locule 1; ovules 1 or 2 and borne on an apical or parietal placentas; ovary superior; style elongate, commonly modified to aid pollination, stigma capitate or attached laterally. Fruit an achene, drupe, follicle, or nut. Seeds usually 1, commonly winged, embryo straight, cotyledons usually 2 (maybe 3–8), endosperm generally lacking (Fig. 9.98).

The family Proteaceae consists of 62–75 genera and over 1000 species distributed prominently in the warmer regions of the Southern Hemisphere, especially in Australasia and South Africa. The largest genera are *Grevillea* (250 species), *Hakea* (125 species), and *Protea* (110 species). The family is a dominant one in the coastal heathlands of Australia in low-nutrient, poor soils.

Economically the family is used primarily as ornamentals. Examples are various species of *Banksia, Grevillea, Protea,* and *Telopea. Grevillea robusta,* the silk oak, is now widely cultivated in California, while the large flowers of *Protea cynaroides,* the giant protea, and of *Hakea* species are commonly dried and sold for dried flower arrangements. *Macadamia integrifolia,* the macadamia nut or Queensland nut, has become an important crop in Hawaii.

Fossil pollen attributed to being Proteaceae has been found in Upper Cretaceous deposits, while leaves have been collected from Southern Hemisphere Paleocene deposits.

# ORDER HALORAGALES
## Family Haloragaceae (Milfoil)

Figure 9.99 Haloragaceae. *Myriophyllum:* (a) stem showing the dimorphism of submerged versus emergent whorled leaves; (b) male flower; (c) fruit; (d) longitudinal section through female flower; (e) female flower. *Haloragis:* (f) fruit.

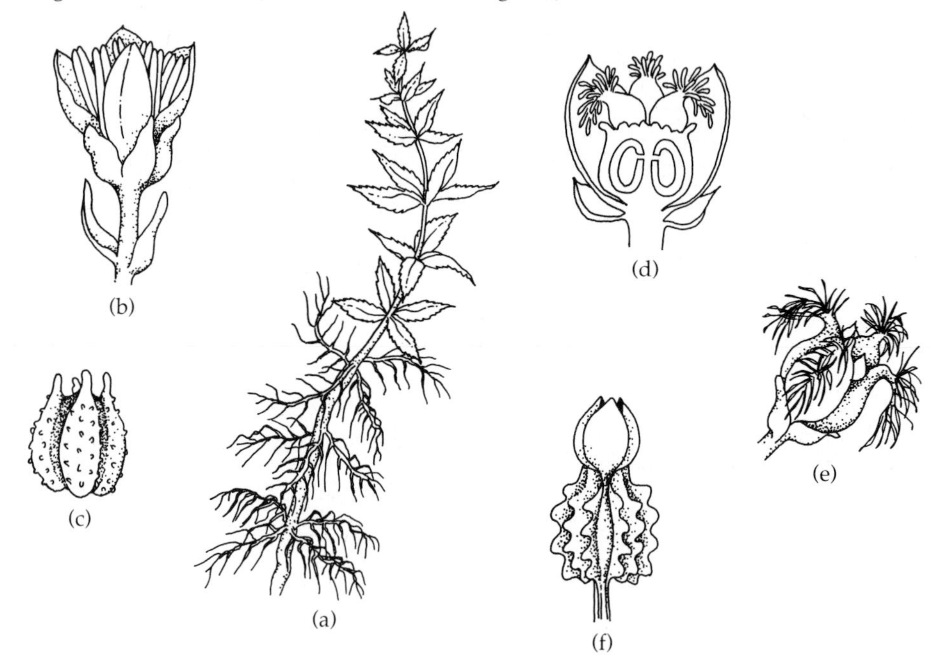


Herbs (rarely shrubs or trees), mostly aquatic. Leaves alternate, opposite or whorled, simple to pinnately dissected; stipules lacking. Flowers small, regular, perfect or unisexual, epigynous, subtended by a pair of bracts; inflorescence of solitary, axillary flowers or flowers in panicles, racemes, or spikes; wind pollinated. Sepals 3–4, persistent in fruit. Petals 3–4, or lacking, distinct, commonly deciduous. Stamens 2–8 in 2 whorls, filaments distinct, short; anthers opening by longitudinal slits. Pistil compound of 2–4 united carpels; locules 2–4 with 1 apical to axile ovule in each locule; ovary inferior, styles 2–4, distinct, feathery. Fruit small, a drupe or nut. Seeds 1 per locule, embryo straight, endosperm abundant (Fig. 9.99).

The family Haloragaceae consists of 8 genera and about 100 species that are cosmopolitan in distribution but are most developed in Australia and the Southern Hemisphere. The main genera are *Gonocarpus* (36 species), *Haloragis* (26 species), and *Myriophyllum* (about 20 species).

The family is of little economic value, except for *Myriophyllum aquaticum,* water milfoil, which is sold as an aquarium plant. On the negative side, *M. spicatum* was introduced into North America and has become a noxious aquatic weed in lakes and streams, inhibiting recreational sports.

The genus *Hippuris* of the family Hippuridaceae and the genus *Gunnera* of the family Gunneraceae have been included in the Haloragaceae by some botanists.

Fossil pollen attributed to the Haloragaceae has been recorded from Paleocene and more recent deposits with questionable pollen determinations from the Upper Cretaceous.

## ORDER MYRTALES
### Family Lythraceae (Loosestrife)

Figure 9.100 Lythraceae. *Lythrum:* (a) upper leafy stem showing flowers; (b) flower. *Didiplis:* (c) fruit; (d) leafy plant bearing fruits; (e) cross section of ovary.

Herbs, shrubs, or trees. Leaves opposite or whorled (rarely alternate), simple, entire; stipules small or lacking. Flowers regular or irregular, perfect, perigynous with a spurred hypanthium; inflorescence of solitary flowers or flowers in cymes, panicles, or racemes. Sepals 4–8, connate, sometimes with an epicalyx of bracts outside the sepals. Petals 4–8 (rarely lacking), alternate the sepals, distinct, usually attached to the rim of the hypanthium. Stamens mostly 8–16, distinct, usually twice the number of perianth parts, in 2 whorls, attached inside the hypanthium; nectary glands present; heterostyly common; anthers opening by longitudinal slits. Pistil compound of 2–6 united carpels; locules 2–6; ovules few to many in each locule and attached to axile placentas; ovary superior; style slender, stigma capitate. Fruit a capsule opening longitudinally or transversally. Seeds 1–many, embryo straight, endosperm usually lacking (Fig. 9.100).

The family Lythraceae consists of 22–24 genera and 450–500 species distributed widely in tropical regions, but with some herbaceous species in temperate regions. The largest genus is *Cuphea* (200 species).

The Lythraceae is known economically as a prime source for native dyes and as ornamentals. *Lawsonia inermis,* the migonette tree, is well known as the source for henna. *Woodfordia fruticosa* leaves give a red color; the bark of *Lafoensia pacari* gives a red dye. Some species are grown as ornamentals, such as *Cuphea,* a pot plant, and *Lythrum salicaria* in the perennial garden. Introduced from Eurasia into North America, *Lythrum salicaria* has become a noxious weed in moist waste areas and wetlands. *Lagerstroemia indica,* the crepe myrtle, is a widely cultivated ornamental in warm regions.

Some genera such as *Alzatea* have characters that make it difficult to determine in which family they should be classified.

Fossils attributed to the Lythraceae first appear in Eocene deposits in Europe and India, and continue to the present time.

## Family Thymelaeaceae (Mezereum)

Figure 9.101 Thymelaeaceae. *Pimelea:* (a) upper leafy stem with flowers; (b) female flower; (c) longitudinal section through hypanthium.

Shrubs or trees (rarely herbs or lianas) that are poisonous. Leaves deciduous or evergreen, alternate or opposite, simple, entire; stipules lacking. Flowers regular (rarely irregular), perfect (rarely unisexual), perigynous, hypanthium prominent and commonly colored; inflorescence axillary or terminal of heads, racemes, or umbels, often subtended by an involucre of bracts. Sepals 4–5, connate, petal-like. Petals 4–5 (8 or more or 0), small, distinct, often scale-like, attached inside the hypanthium. Stamens 2–10 (rarely many), filaments distinct, if same number as sepals then opposite them, attached in the hypanthium; nectary glands or disk usually present. Pistil compound of 1–5 united carpels; locules 2–5; ovules 1–2 per locule and borne on apical to axillary placentas; ovary superior; style usually long and slender, stigma usually capitate. Fruit variable as an achene, berry, capsule, drupe, or nut. Seeds with straight embryo, endosperm present in small amounts or absent (Fig. 9.101).

The family Thymelaeaceae consists of 45–50 genera and 500 species and is widespread, with many taxa in tropical Africa and Australia. The largest genera are *Gnidia* (100–140 species), *Pimelea* (80 species), *Wikstroemia* (70 species), and *Daphne* (50–70 species).

Economically the family is of minor importance. Some species of *Daphne,* daphne, are cultivated as fragrant ornamental shrubs. *Lagetto lintearia* in the West Indies produces the ornamental lace bark. The bark of some Mediterranean *Daphne* is used to stupefy fish. *Daphne genkwa* is used in China to induce abortion.

The family Thymelaeaceae is distinct as a unit but how it is classified in relation to other families is debatable.

Fossils attributed to the family have been found in Oligocene and Lower Eocene deposits.

## Family Myrtaceae (Myrtle)

Figure 9.102 Myrtaceae. *Callistemon:* (a) leafy branch with flowers and fruits. *Eucalyptus:* (b) fruit lacking calyptra; (c) umbel of fruits with calyptra; (d) fruit; (e) longitudinal section of flower; (f) cross section of ovary.

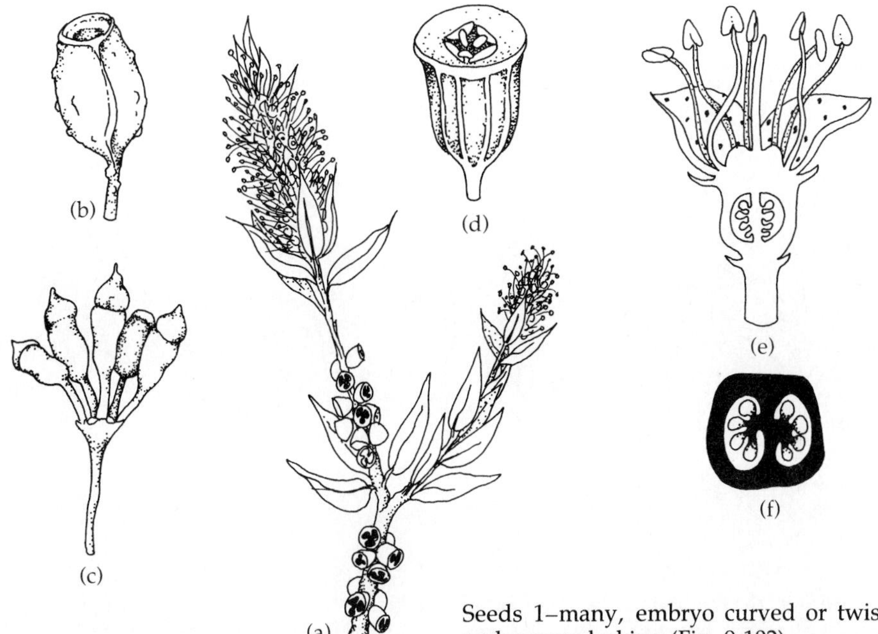


Shrubs or trees. Leaves alternate, opposite or whorled, simple, entire, strongly scented, bearing oil glands; stipules lacking. Flowers regular, perfect (rarely unisexual), epigynous or perigynous, hypanthium usually adnate to the ovary; inflorescence of solitary flowers or flowers arranged in cymes or racemes; nectary glands common; pollinators are various animals, including long-billed birds and bats. Sepals 4–5 (rarely 0), distinct or connate forming a cap **(calyptra).** Petals 4–5 (rarely 0), small, distinct, or connate. Stamens usually numerous, filaments distinct or connate in bundles opposite the petals, attached to the upper surface of the hypanthium or the top of the ovary; anthers opening by longitudinal slits or apical pores. Pistil compound of 2–5 united carpels; locules same number as carpels; ovules 2–many in each locule and attached on axile or parietal placentas; ovary inferior (rarely superior); style terminal, slender, stigma capitate. Fruit a berry, drupe, loculicidal capsule, or nut. Seeds 1–many, embryo curved or twisted, endosperm lacking (Fig. 9.102).

The family Myrtaceae consists of 100–150 genera and 3000–3500 species distributed mostly pantropically, especially in Australia. The largest genera are *Eugenia* (600 species), *Eucalyptus* (500 species), *Myrica* (300 species), and *Syzygium* (200 species). They are planted extensively in California, East Africa, Portugal, and similar climates.

Economically the family has many important species. Species of *Eucalyptus* (gum trees), especially *E. diversicolor,* make fine timber, while others provide eucalyptus oil. Spices come from *Syzygium aramaticum,* clove; *Melaleuca leucadendron,* cajeput oil; and *Pimenta dioica,* allspice or pimento. Ornamentals include *Callistemon,* bottlebrushes, and *Myrtus communis,* common myrtle. Edible fruits come from *Eugenia, Syzygium,* and *Psidium guajava,* the guava from the American tropics. Some species of *Eucalyptus* (i.e., *E. regnans*) reach over 100 m in height and are the world's tallest flowering plants.

Fossil examples attributed to the Myrtaceae have been found in the Upper Cretaceous and various Tertiary deposits.

## Family Onagraceae (Evening Primrose)

Figure 9.103 Onagraceae. *Epilobium:* (a) upper leafy stem with flowers; (b) cross section of ovary; (c) fruit; (d) face view of flower. *Clarkia:* (e) stem bearing flowers with 4-lobed stigma. *Fuchsia:* (f) longitudinal section through flower.

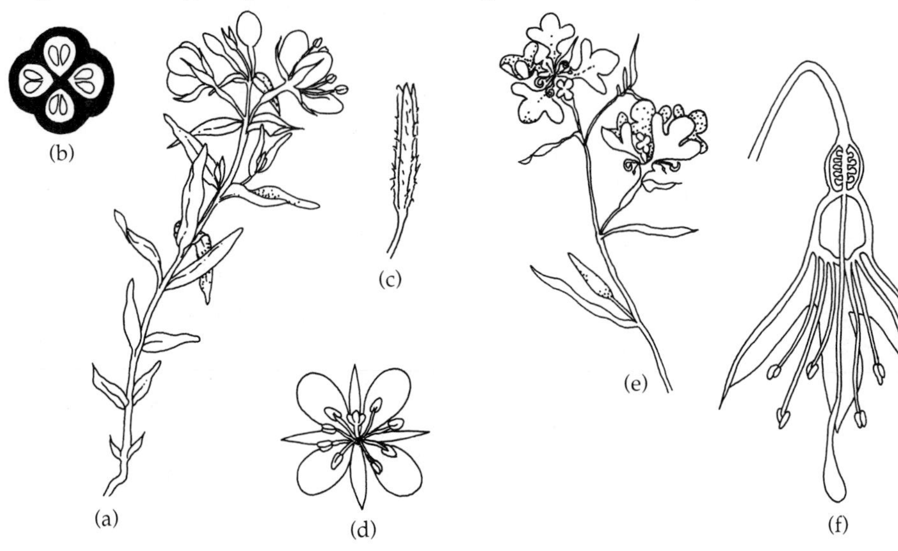


Herbs or shrubs (rarely trees). Leaves alternate, opposite, or whorled; usually simple (rarely pinnate); stipules present or absent. Flowers regular (less commonly irregular), perfect (rarely unisexual), epigynous with a prominent colored hypanthium, nectary glands inside toward the base; inflorescence of solitary flowers in axiles of leaves or in panicles, racemes, or spikes; pollinated by animals other than bats, beetles, or wind, commonly self-pollinated. Sepals 4–5 (rarely 2, 3, or 6), distinct or connate, occasionally petal-like and as lobes on the hypanthium. Petals 4–5 (rarely 2, 3, 6, or lacking), distinct sometimes clawed. Stamens 4–10 (rarely 2), filaments distinct, attached inside the hypanthium; anthers sometimes with cross-partitions and opening by longitudinal slits. Pistil compound of 4 (rarely 2 or 5) united carpels; locules same number as carpels; ovules few to many and borne on axile or parietal placentas; ovary inferior; style slender, stigma capitate to distinctively 4-lobed. Fruit a berry or loculicidal capsule or nut. Seeds usually many (rarely few or one), commonly each with a tuft of hair attached, which helps the seed to be carried in the air; embryo straight, endosperm lacking but diploid during development (Fig. 9.103).

The family Onagraceae consists of 17–18 genera and 640–675 species and is cosmopolitan in distribution but very diverse in western North America. The largest genera are *Epilobium,* the willowherb (200 species); *Oenothera,* evening primrose (125 species); and *Fuchsia,* fuchsia (90 species).

The family is important economically for its cultivated ornamentals. Annual ornamentals are found in the genera *Clarkia,* the clarkias, and *Oenothera,* while shrubs of *Fuchsia* are commonly grown in greenhouses.

The genera *Clarkia, Epilobium,* and *Oenothera* have been studied extensively by botanists for many years, and much cytogenetic information has been accumulated. The characters of the family make it distinctive from other families.

Fossils attributed to the Onagraceae are first found in Upper Cretaceous deposits and extend through various periods of the Tertiary to the present.

## Family Melastomataceae (Melastome)

Figure 9.104 Melastomataceae. *Rhexia:* (a) upper leafy stem with fruits; (b) stamen; (c) face view of flower; (d) longitudinal section through flower; (e) cross section of ovary.

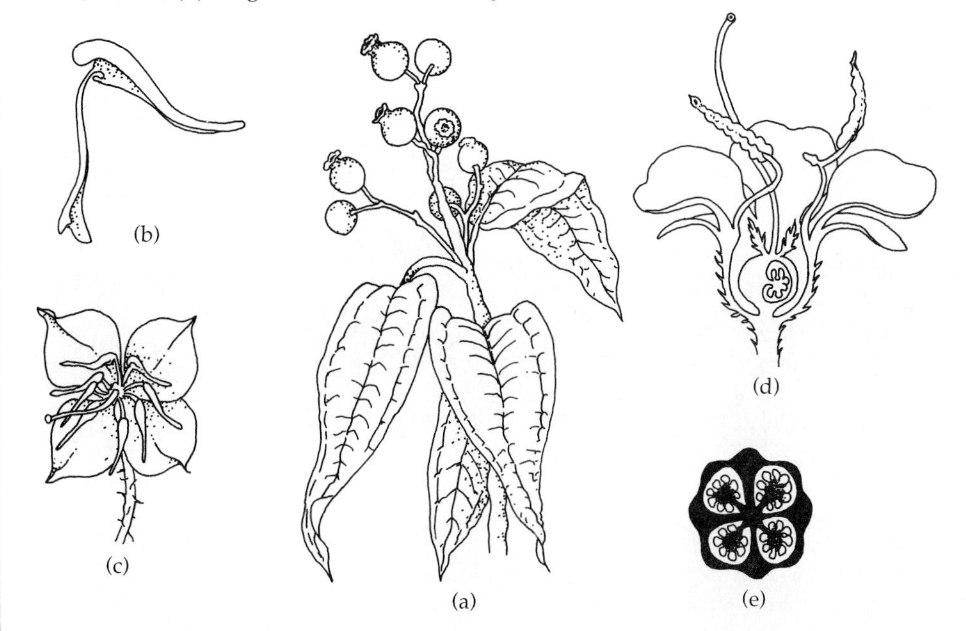


Herbs, shrubs, trees, or lianas, sometimes as epiphytes, commonly with a 4-sided stem. Leaves opposite (rarely alternate or whorled), simple, usually entire, main veins 3–9, usually palmate and parallel with many cross-connecting veins; stipules lacking. Flowers commonly large and showy, regular, perfect (rarely unisexual), epigynous or perigynous, hypanthium persistent; inflorescence of various types. Sepals 4–5, connate. Petals 4–5, distinct (rarely connate at the base). Stamens 4–10, commonly twice the number of the petals, filaments distinct but bent with sterile appendages attached, often twisted to place all stamens to one side; anthers usually opening by an apical pore (rarely longitudinal slits); nectary glands usually lacking, but pollinated by pollen gathering insects or mice. Pistil compound of 2–15 united carpels; locules 2–15 (rarely 1); ovules 1–many per locule and borne on axile, basal, or parietal placentas; ovary inferior (rarely superior); style single, terminal with a capitate to lobed stigma. Fruit a berry or loculicidal capsule. Seeds small, usually many, endosperm lacking (Fig. 9.104).

The family Melastomataceae consists of 200–240 genera and 3000–4000 species distributed throughout the subtropics and tropics, with only a few species temperate. They are most common in South America. The largest genera are *Miconia* (1000 species), *Medinilla* (300 species), and *Tibouchina* (250 species).

In spite of its size, the family has relatively few species of economic importance, except as ornamentals. These include *Dissotis grandiflora*, *Medinella curtisii*, *M. magnifica*, *Melastoma malabathricum*, and *Tibouchina urvilleana*, which is used locally for timber or for supplying dyes (yellow dye, *Memecylon edule*).

The size of the Melastomataceae allows it to be divided into 3 subfamilies and various tribes. The characters of the stamens set the group apart from other families.

Fossil examples attributed to the family have been found in Eocene and Oligocene deposits.

## Family Combretaceae (Indian Almond)

Figure 9.105 Combretaceae. *Combretum:* (a) leafy flowering branch; (b) flower; (c) longitudinal section through flower; (d) winged fruit.

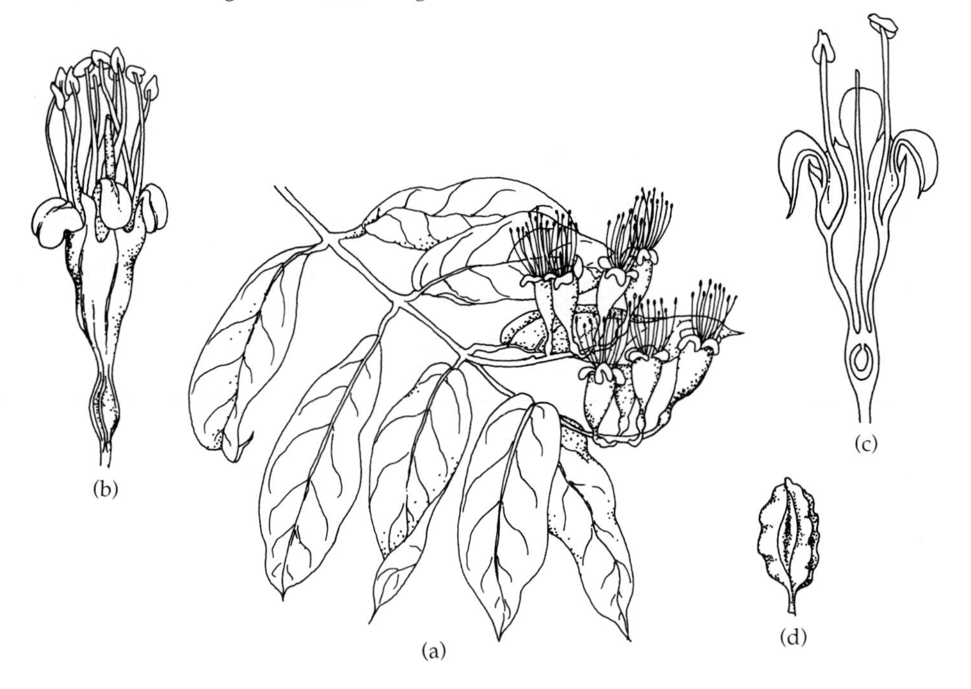


Shrubs, trees, or lianas, often with secretory cells in the stem tissue and having peculiar unicellular hairs. Leaves alternate or opposite (rarely whorled), simple, entire; lacking stipules. Flowers usually small, regular, perfect (rarely unisexual), epigynous, hypanthium present; inflorescence an axillary or terminal head, raceme, or spike. Sepals 4–5, distinct or connate. Petals 4–5 (rarely lacking), small, alternating with the sepals, distinct, attached at the top of the hypanthium. Stamens in 2 whorls of 4 or 5 (rarely many), sometimes outer stamens reduced to staminodes; filaments distinct, inflexed in bud, commonly exserted; anthers opening by longitudinal slits, much nectar produced. Pistil compound of 2–5 united carpels, locule 1, ovules usually 2 and attached by apical placentas, ovary inferior, style 1. Fruit a berry (rarely a capsule), 1 seeded, seed with oily embryo, endosperm lacking (Fig. 9.105).

The family Combretaceae consists of 20 genera and 400–475 species distributed pantropically, with many species in Africa. The largest genera are *Combretum* (150–200 species) and *Terminalia* (100–150 species).

Economically valuable tree species are found in the genus *Terminalia.* In West Africa, *T. ivorensis* and *T. superba* are important sources of timber. Cultivated in the tropics for its edible seed is *T. catappa,* the Indian almond. *Quisqualis indica,* the Rangoon creeper, is a cultivated ornamental in the tropics.

The family Combretaceae is distinctive from other families in the Myrtales because of the apical ovules, 1-locule ovary, and a special type of pubescence.

Fossils attributed to the Combretaceae have been found in Eocene deposits.

## ORDER RHIZOPHORALES
## Family Rhizophoraceae (Red Mangrove)

Figure 9.106 Rhizophoraceae. *Rhizophora:* (a) leafy branch, flowering and fruiting; (b) cross section through ovary; (c) longitudinal section through flower.

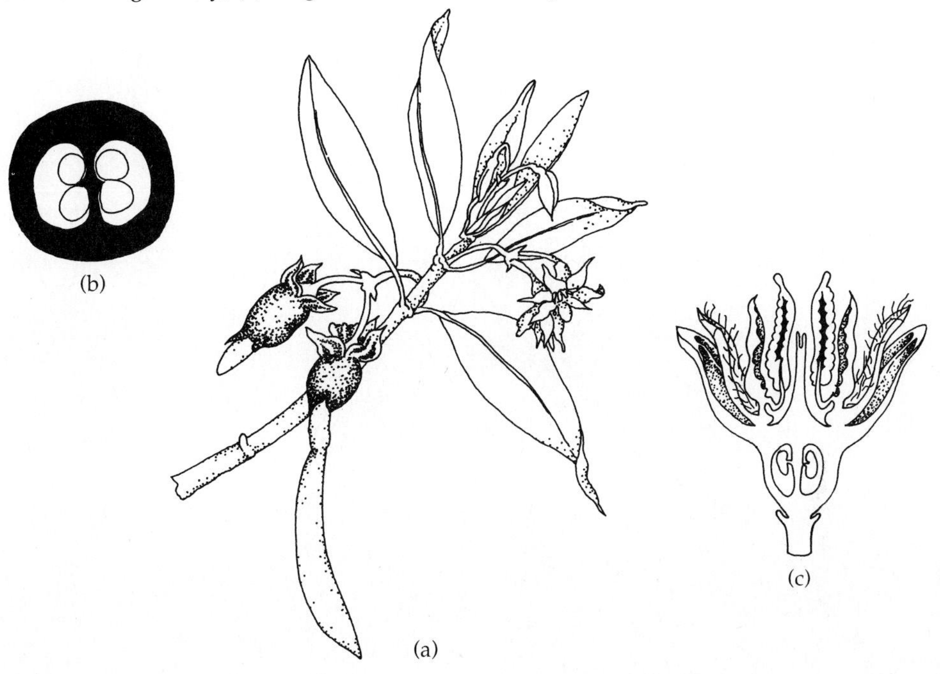


Shrubs, trees, or lianas. Leaves usually opposite, simple, entire; stipules large but easily falling from plant. Flowers regular, perfect (rarely unisexual), hypogynous to epigynous, hypanthium present or absent; inflorescence of a solitary flower or flowers borne in axillary cymes or racemes. Sepals 3–16, often fleshy or leathery, distinct. Petals 3–16, often fleshy, commonly shorter than sepals. Stamens 8–10 (rarely many), filaments distinct or connate, usually attached to a nectary disk; anthers opening by a longitudinal valve. Pistil compound of 2–12 united carpels, locules 2–12 (rarely 1), ovules usually 2 in each locule and borne on apical-axile placentas, ovary superior or inferior, style single or lobed, stigma capitate or lobed. Fruit a berry or drupe (rarely a capsule). Seeds 1 per locule sometimes with an aril, often germinating on plant, embryo straight, endosperm present or absent (Fig. 9.106).

The family Rhizophoraceae consists of 14–16 genera and 100–120 species pantropical in distribution. The largest genus is *Cassipourea* (about 70 species). The mangroves include the genera *Avicennia* (14 species), *Bruguiera* (6 species), *Ceriops* (2 species), and *Rhizophora* (6–9 species).

Economically the family has many uses. In timber production, the wood of *Ceriops tagel* is the most durable of all mangrove wood. The bark of the mangroves is high in tannin and is used in the tanning industry. *Rhizophora mucronata* is used especially for timber and pulp. Various species are used locally for medicine and food. Mangrove habitats are important environmentally as ecological filters.

The classification of the family Rhizophoraceae is unsettled and is debated by botanists.

The fossil record of the family is represented by pollen found in Upper Eocene deposits to the present.

## ORDER CORNALES
### Family Nyssaceae (Sour Gum)

Figure 9.107 Nyssaceae. *Nyssa:* (a) leafy flowering branch; (b) male flower; (c) fruits; (d) female flower.

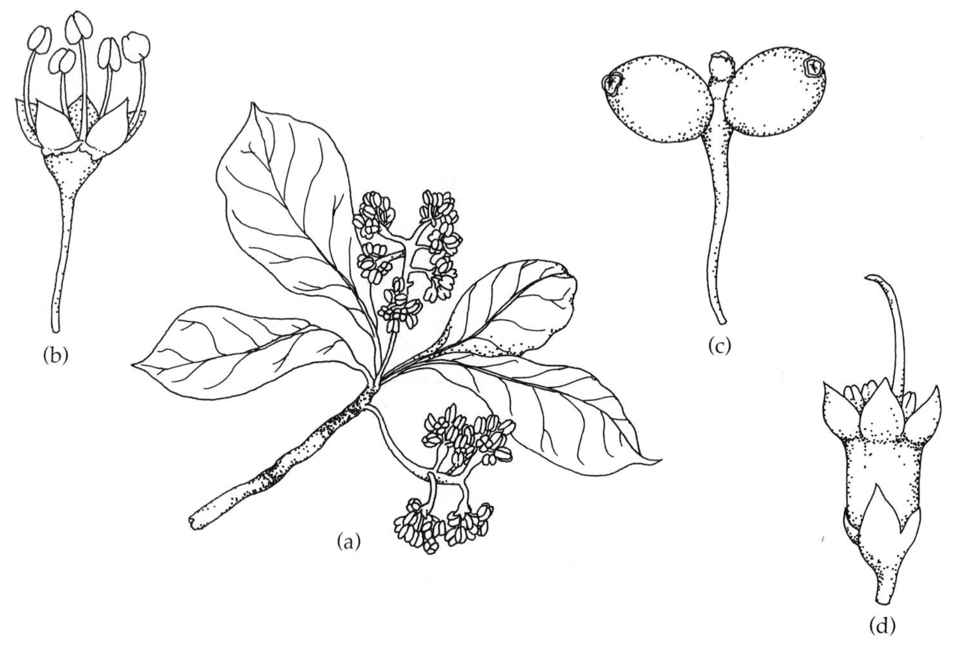


Shrubs or trees. Leaves deciduous, producing bright autumn colors, alternate, simple, entire or toothed; stipules lacking. Flowers small, regular, perfect or unisexual, borne in different inflorescences on the same plant or on different plants, epigynous; inflorescence a solitary flower or flowers in heads, racemes, or umbels. Sepals 5 or lacking, small. Petals 5 (rarely 4–8). Stamens 10 (less commonly 8–16), filaments of staminate flowers in 2 rows, filaments of perfect flowers as many as petals and alternate with them; anthers small, nectary disk present. Pistil comprised of 1 simple or 2 united carpels; locule 1 (rarely more); ovule 1 and attached on an apical placenta; ovary inferior; style simple or with 2 united styles, erect or twisted. Fruit a drupe or samara. Seed 1, embryo large and straight, endosperm present (Fig. 9.107).

The family Nyssaceae consists of 3 genera and 7–8 species distributed in eastern North America and Eastern Asia. The largest genus is *Nyssa,* the sour gum (5–6 species).

Economically the family is of minor importance. *Nyssa sylvatica,* the black gum or tupelo, produces wood used for timber and edible fruits; it also is a fine ornamental tree. *Davidia involucrata,* the handkerchief tree, and *Camptotheca acuminata* are also cultivated as ornamentals.

The classification of the family is fairly stable and without real controversy. The genus *Davidia* is felt by some botanists to belong in its own family.

Fossil fruits of the Nyssaceae are easy to recognize from lower Eocene deposits to the present. Earlier fossils attributed to the Nyssaceae from Cretaceous deposits are questionable.

## Family Cornaceae (Dogwood)

Figure 9.108 Cornaceae. *Cornus:* (a) leafy branch bearing fruit; (b) longitudinal section through flower; (c) leafy branch with flowers that have large bracts; (d) fruits; (e) flower without prominent bracts; (f) apical bud on winter twig.

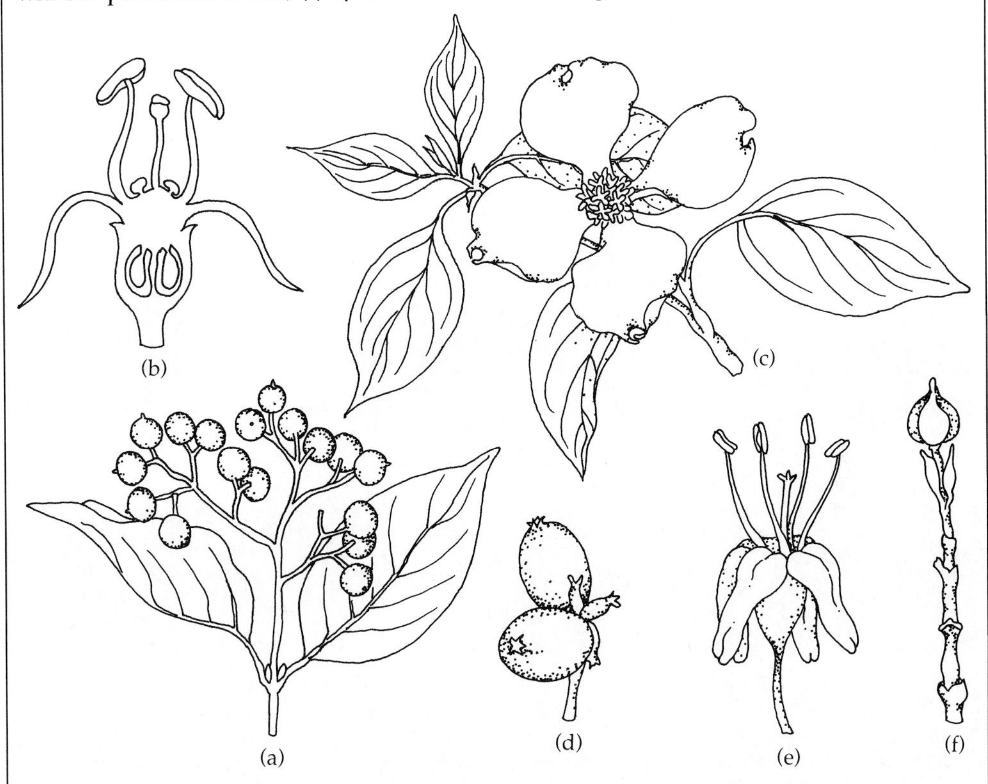


Shrubs or trees (rarely herbs). Leaves alternate, more commonly opposite, simple, pinnately (rarely palmately) veined; stipules lacking. Flowers regular, perfect (rarely unisexual), epigynous; inflorescence various of corymbs, cymes, panicles, or umbels, commonly surrounded by showy petal-like bracts. Sepals 4–5, distinct or connate. Petals 4–5, distinct, alternate the sepals. Stamens 4–5, alternate the petals, usually attached to the edge of the epigynous disk; anthers opening by longitudinal slits. Pistil compound of 2–4 united carpels; locules 2–4 (rarely 1); ovule 1 per locule and borne on an apical-axile placenta; ovary inferior; style single, stigma capitate or lobed. Fruit a berry, drupe, or multiple. Seeds 1–2, small; endosperm present (Fig. 9.108).

The family Cornaceae consists of 11–13 genera and 100 species and is distributed widely in the temperate Northern Hemisphere and in warmer regions of the Southern Hemisphere. The largest genus is *Cornus,* the dogwood (50 species).

Economically the family is mostly used as ornamentals, especially species of *Cornus* and *Griselina. Cornus florida* is a small tree planted in lawns in North America, and *C. nuttallii* is the emblem for the Canadian province of British Columbia. In France, an alcoholic drink, *vin de cornouille,* and preserves are made from the fruit of *Cornus.* Several species of *Cornus* produce wood used for furniture.

The genera of the Cornaceae are sometimes classified as individual families, but it seems best to treat the group as one family at the present time.

Fossil leaves attributed to the Cornaceae are found in Eocene and more recent deposits.

## Family Garryaceae (Silk Tassel)

Figure 9.109 Garryaceae. *Garrya:* (a) leafy branch bearing catkins; (b) female inflorescence; (c) anther; (d) longitudinal section through pistil.

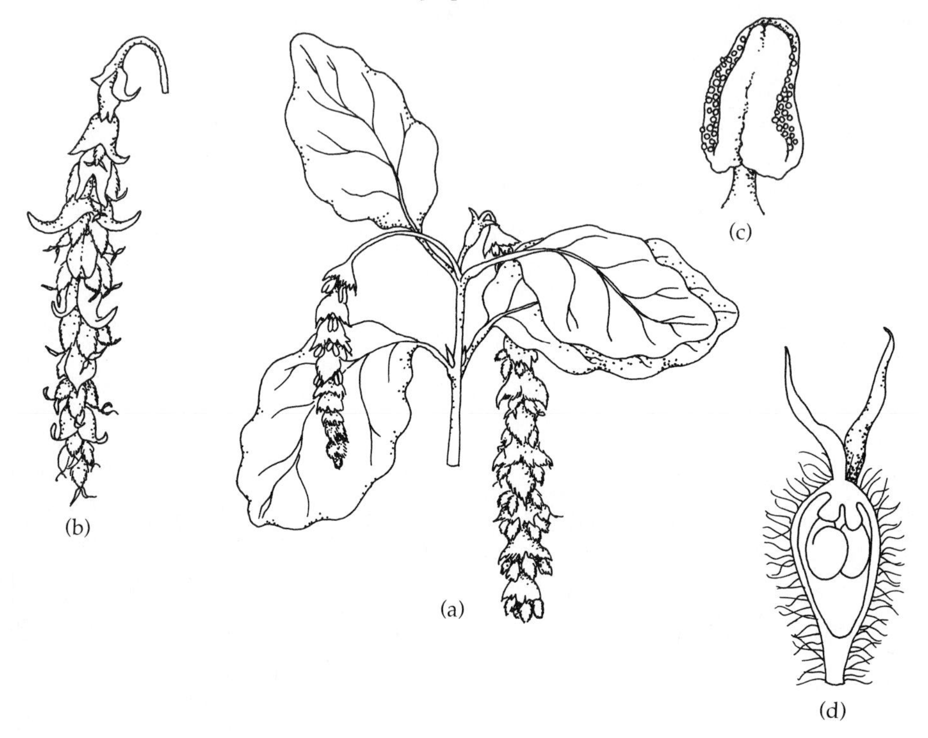


Shrubs or small trees. Leaves evergreen, leathery, opposite, simple, entire or wavy; stipules lacking. Flowers small, unisexual and dioecious; inflorescence an axillary or terminal pendulous catkin or raceme. Sepals 4, distinct in staminate flowers, pistillate flowers, sepals 2–4 or lacking, connate and adnate to the ovary. Petals absent. Stamens 4, filaments short, distinct, alternate with the sepals; anthers opening by longitudinal slits. Pistil compound of 2–3 united carpels, locule 1, ovules 2 and borne on apical-parietal placentas, ovary inferior; styles 2–3, slender, distinct, spreading and persistent. Fruit a 1- to 2-seeded berry. Seeds small, straight with abundant endosperm (Fig. 9.109).

The family Garryaceae consists of 1 genus, *Garrya,* with 13 species distributed in western North America from Washington state south to Guatemala and the West Indies.

The family is of little economic value, except for several species cultivated as ornamental shrubs.

In some nineteenth century classifications, the Garryaceae was classified with other families like the Salicaceae. This was due to the unisexual catkin-like inflorescences. Today botanists believe this view is incorrect and have placed the family in the order Cornales.

Fossil leaves and seeds attributed to the Garryaceae are known from Miocene and more recent deposits in western North America.

## ORDER SANTALALES
### Family Olacaceae (Olax)

Figure 9.110 Olacaceae. *Schoepfia:* (a) leafy branch with flowers and fruit; (b) cross section through ovary; (c) longitudinal section through flower. *Olax:* (d) fruit.

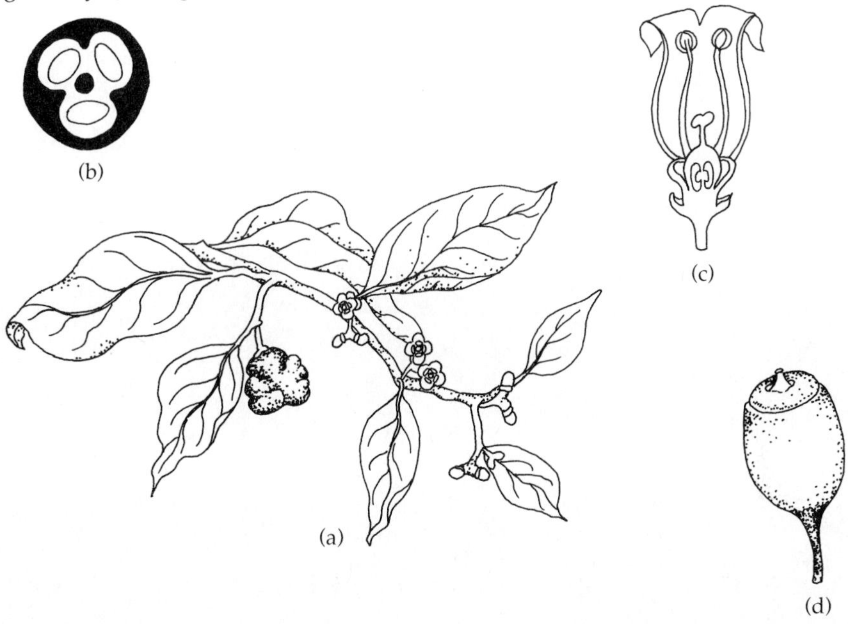


Shrubs, trees, or lianas; many are semiparasitic and attached to the roots of other plants. Leaves usually evergreen, alternate, entire, pinnately veined, rough to touch; stipules lacking. Flowers small, regular, perfect (rare unisexual and dioecious), hypogynous, epigynous or perigynous; inflorescence axillary in panicles or racemes, colored green or white. Sepals small, connate, 3–6 lobed (rarely lacking), often persistent after flowering. Petals 3–6, distinct or connate, alternate with the calyx lobes. Stamens up to twice as many as the petals and opposite them, filaments distinct or connate, in 1 whorl; nectary disk present; anthers opening by longitudinal slits or pores. Pistil compound of 2–5 united carpels; locules 2–5; ovule 1 in each locule and attached to axile or free-central placentas; ovary superior or inferior; style terminal, stigma 2–5 lobed. Fruit a drupe or nut, often with the calyx attached. Seed 1, small, embryo small and straight, endosperm present (Fig. 9.110).

The family Olacaceae consists of 25–30 genera and 250 species distributed pantropically, especially in the Old World tropics. The family is also well developed in the New World tropics, extending north into the southern United States. The largest genera are *Olax* (50 species) and *Schoepfia* (40 species).

The family is somewhat important economically in tropical regions. *Ximenia americana,* the hog plum or tallow wood of South America, is known for its timber, edible fruits, and oil-rich seeds. In West Africa, seeds of *Olax gambecola* are used as a condiment. *Coula edulis,* the African walnut, has edible fruits that look like walnuts and a strong wood that is valued in house building.

Because of considerable morphological variation, the family Olacaceae has been split into many smaller families. The final classification of the family is still to be determined.

The family has no known fossil record.

## Family Santalaceae (Sandalwood)

Figure 9.111 Santalaceae. *Comandra:* (a) leafy stem with terminal inflorescence; (b) fruit; (c) fruit with perianth scar. *Santalum:* (d) longitudinal section through flower.

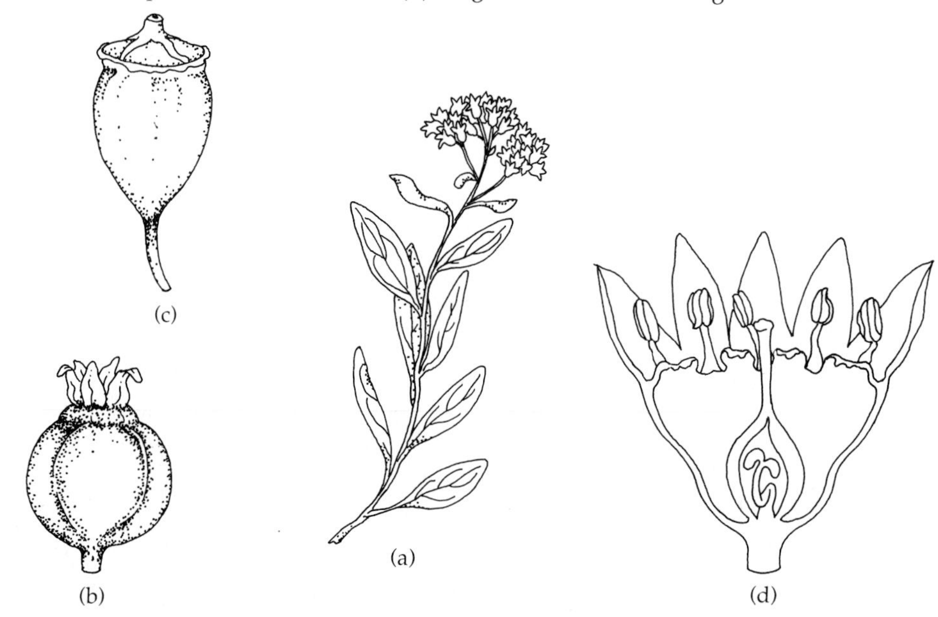


Herbs, shrubs, or trees, green but usually semiparasitic on the roots (rarely on branches) of host plants. Leaves alternate or opposite, simple, entire, well-developed blades or reduced and scale-like; stipules lacking. Flowers usually small, regular, perfect or unisexual, hypogynous, epigynous or perigynous; inflorescence a solitary flower or flowers in cymes, heads, racemes, or spikes. Sepals 3–6, distinct or connate, often petal-like. Petals lacking. Stamens 3–6, the same number as the perianth lobes and opposite them, filaments distinct but adnate to the sepals; anthers opening by longitudinal slits or an apical pore; nectary disk present. Pistil compound of 3–5 (rarely 2) united carpels; locules 1 or with septa at the base only; ovules 1–4 attached to free-central placentas; ovary inferior to less commonly superior; style 1, terminal, stigma capitate or lobed. Fruit a drupe or nut. Seed 1, viscid; embryo straight, abundant endosperm present (Fig. 9.111).

The family Santalaceae consists of 35 genera and 400 species with a cosmopolitan distribution, but well developed in the subtropical to tropical and arid climates. The largest genus is *Thesium* (325 species), found extensively in Africa and the Mediterranean region.

Economically the most important plant in the family is *Santalum album,* the sandalwood tree. This tree yields a valuable timber wood and sandal oil, used in making cosmetics, perfume, soap, and incense for religious worship in the Buddhist, Hindu, and Muslim religions, and for anointing the body. The fruits of *Acanthosyris falcata* and *Exocarpos cupressiformis* are used for food.

The family Santalaceae is sometimes divided into 3 tribes based on the diversity within the various genera.

The family has no known fossil record.

## Family Loranthaceae (Showy Mistletoe)

Figure 9.112 Loranthaceae. *Amyema:* (a) leafy branch with flowers in bud; (b) fruits; (c) flower.

Shrubby parasites (rarely terrestrial shrubs, trees, or lianas), mostly green, attached to the branches or roots of the host plant, often forming an enlargement of tissue where the parasite's adventitious roots enter the host's tissue. Leaves usually evergreen, opposite (rarely whorled), well developed or reduced to scales, simple, entire; stipules lacking. Flowers often large, regular or irregular, perfect (rarely unisexual and monoecious), epigynous; inflorescence is basically a cyme that appears as a head, raceme, spike, or umbel, often red or yellow colored; insect or bird pollinated. Sepals 4–6, short, connate or reduced to a cup around the top of the ovary. Petals 3–9, distinct or connate. Stamens the same number as the petals and opposite them, filaments attached to the corolla; nectary disk present or absent; pollen unusual, often in many cells of the anther; anthers opening by longitudinal slits. Pistil compound of 3–4 united carpels; locule usually 1 (rarely 4); ovules 4–12 and sunken into the free-central placentas; ovary inferior; style 1, short or long, stigma small, sometimes sessile. Fruit a berry or drupe. Seeds 1 (rarely 2–3), viscid, often with more than 1 large embryo, cotyledons initially 2 but later fusing, endosperm present (Fig. 9.112).

The family Loranthaceae consists of 60–70 genera and about 700–940 species distributed mostly in the subtropics and tropics and well represented in the Southern Hemisphere. Large genera are *Psittacanthus* and *Struthanthus*, each with 75 species.

Economically the family is of little importance, except as a parasitic pest on trees of economic value.

The family Loranthaceae has included the family Viscaceae in the past as a subfamily. Recent embryological studies support classifying the two subfamilies as separate families. The genus *Loranthus*, with over 600 species in the past, has been reduced to only *L. europaeus*, the European mistletoe, the other species being assigned to other genera.

The family has no known fossil record.

## Family Viscaceae (Christmas Mistletoe)

Figure 9.113 Viscaceae. *Viscum:* (a) branch with berry fruits. *Phoradendron:* (b) parasitic branch attached to host; (c) front view of male flower; (d) fruits; (e) cluster of male flowers.

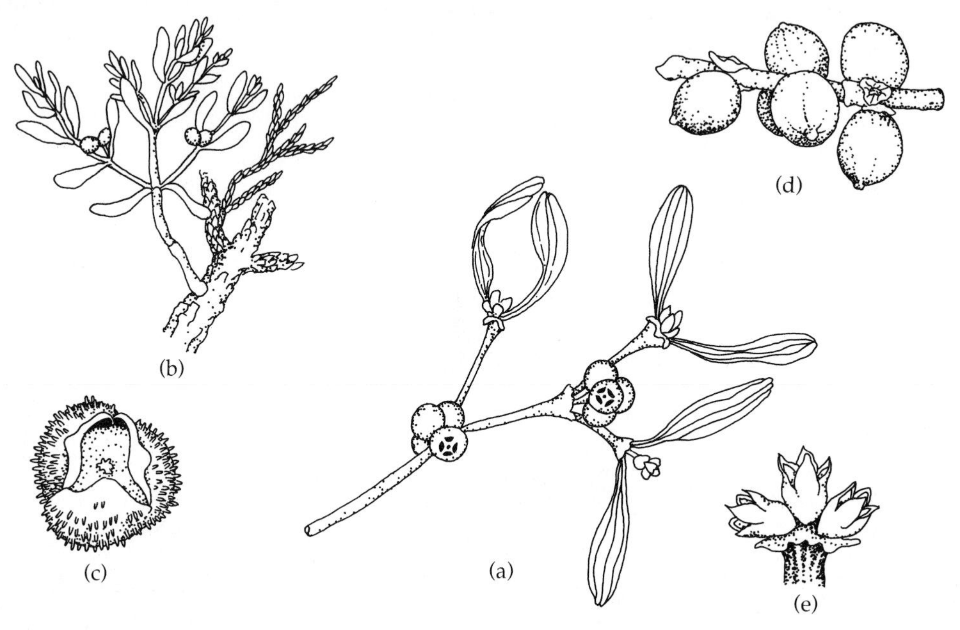


Herbs or shrubs, green photosynthetic parasites, attached to tree branches. Leaves well developed, scale-like or lacking, opposite (rarely alternate), simple, entire; stipules lacking. Flowers regular, unisexual (monoecious or dioecious), epigynous; inflorescence of small cymes or spikes, green or yellow colored; insect or wind pollinated. Sepals 2–4, connate, in staminate flowers filaments adnate to sepals, in pistillate flowers adnate to the ovary. Petals absent. Stamens 2–4, filaments distinct, often short, same number as the perianth lobes and opposite them; anthers commonly with many locules, opening by longitudinal slits or terminal pores. Pistil compound of 3–4 united carpels; locule 1 or lacking with the ovary being solid; ovules 2, appearing as embryo sacs embedded in the basal placenta; ovary inferior; style terminal, short, stigma small, sometimes sessile. Fruit a shiny berry. Seed 1, viscid, embryo large, endosperm present (Fig. 9.113).

The family Viscaceae consists of 7–8 genera and 350 species with a cosmopolitan distribution, but well developed in the tropics. The largest genus is *Phoradendron* (200 species).

The family is of little economic value directly, except for the use of *Viscum album,* European mistletoe, and *Phoradendron flavescens,* American mistletoe, during the Christmas season. Indirectly, the mistletoes do considerable damage to conifers in North America and hardwoods in India.

The family Viscaceae has been classified in the past as a subfamily of the Loranthaceae. Recent embryological studies support classifying the two groups as separate families.

The family has no known fossil record.

## ORDER RAFFLESIALES
### Family Rafflesiaceae (Rafflesia)

Figure 9.114 Rafflesiaceae. *Cytinus:* (a) longitudinal section through female flower. *Rafflesia:* (b) large male flower; (c) cross section through ovary; (d) longitudinal section through male flower, showing mycelia in host.

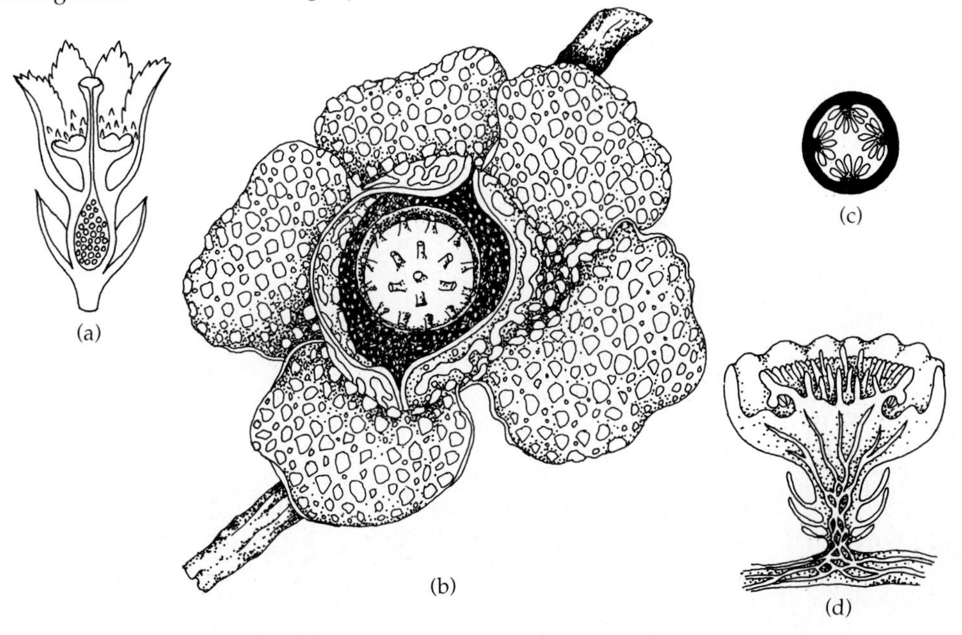


Internal, herbaceous, parasites on roots and stems, vegetative body fungus-like and filamentous, penetrating the host plant. Leaves alternate, opposite or whorled, scale-like or absent, without chlorophyll or stomates. Flowers sometimes small but usually enormous, fleshy, regular, unisexual, (rarely perfect), monoecious or dioecious, usually epigynous, mostly brightly colored; inflorescence of solitary flowers or flowers borne in racemes or spikes, originating inside the host and then pushing through to the outside. Sepals 3–10, distinct or connate, petal-like. Petals lacking. Stamens 5–many, filaments connate into a tube surrounding styler column or adnate to the stout columnar style; anthers opening by longitudinal or transverse slits or apical pores. Pistil compound of 4–many fused carpels; locule 1 or with scattered locules; ovules many and borne on parietal placentas; ovary inferior (rarely superior); style a stout column, expanded at the top into a flat stigmatic disk. Fruit a berry, capsule, or multiple. Seeds tiny, numerous, embryo of a few cells and undifferentiated, endosperm present (Fig. 9.114).

The family Rafflesiaceae consists of 7–9 genera and 50 species distributed pantropically. A few species are found in temperate climates, such as *Cytinus hypocistis* of the Mediterranean region, which is parasitic on members of the Cistaceae, and *Mitrastemon* from Japan, which parasitizes roots of *Quercus*, the oaks.

The species *Rafflesia arnoldii* from Sumatra is known for producing the largest flowers in the world, with some being up to 1 m (3 ft) across. With the rapid destruction of the tropical rain forest, a fair number of the species of this family are threatened with extinction or may already be extinct.

The family has no economic value but is of interest botanically because of its strange parasitic habit and unique beautiful flowers.

The family has no known fossil record.

## ORDER CELASTRALES
### Family Celastraceae (Bittersweet)

Figure 9.115 Celastraceae. *Euonymus:* (a) leafy branch with fruits; (b) longitudinal section through flower; (c) dehiscent fruit showing aril-covered seeds; (d) cross section through ovary. *Celastrus:* (e) fruit (a capsule); (f) flower.

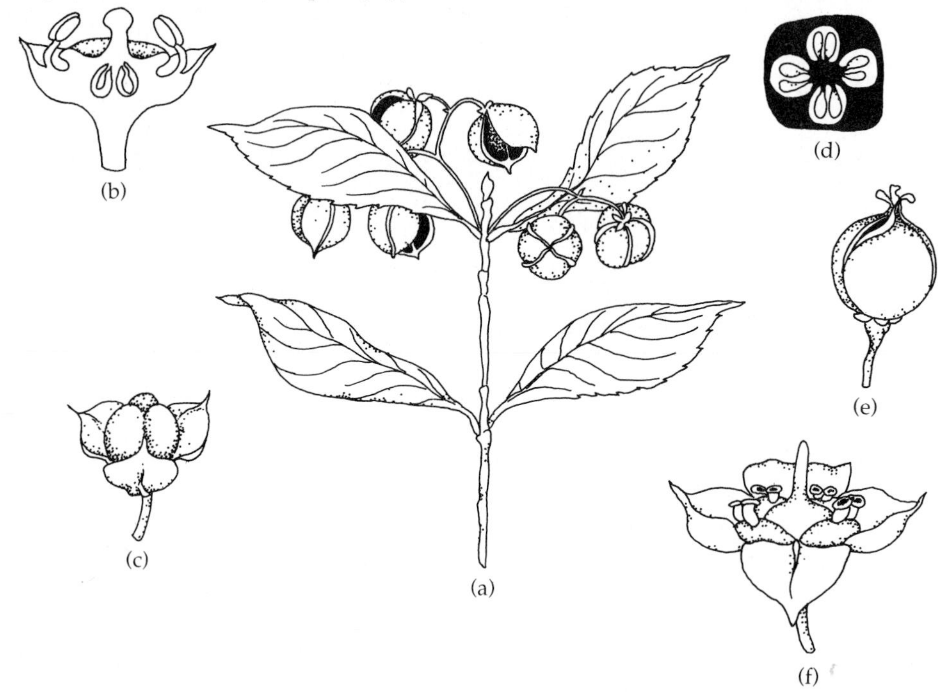


Shrubs, trees, or lianas. Leaves often leathery, alternate or opposite, simple; stipules small, dropping early or lacking. Flowers usually small, regular, perfect or unisexual; hypogynous, epigynous, or perigynous; hypanthium sometimes well developed; inflorescence a solitary flower or flowers in axillary or terminal cymes or racemes, commonly green or white in color. Sepals 3–5, small, distinct or connate. Petals 3–5 (rarely lacking), distinct. Stamens 3–5 (rarely 10), filaments distinct, alternate with the petals, attached on or outside the well-developed nectary disk; anthers open by longitudinal or transverse slits. Pistil compound of 2–5 united carpels, locules the same number as carpels, ovules 2–6 (rarely many) in each locule and attached to axile placentas; ovary superior or inferior; style terminal, short, stigma capitate or lobed. Fruit a berry, capsule, drupe, or samara. Seeds usually surrounded by a brightly colored aril, embryo large and straight, endosperm present (Fig. 9.115).

The family Celastraceae consists of 50–55 genera and 800–850 species distributed pantropically, but with a fair number of species in temperate regions. The largest genera are *Maytenus* (225 species) and *Euonymus* (200 species).

Economically the family is of minor importance; however, some species are cultivated as ornamentals. The vine *Celastrus scandens* is grown for its attractive colored fruits and seeds. Various shrub species of *Euonymus* are cultivated for their attractive foliage.

Fossils attributed to the Celastraceae are found in Cretaceous deposits, but their validity is presently being questioned. Pollen has been found in Oligocene and more recent deposits.

## Family Hippocrateaceae (Hippocratea)

Figure 9.116 Hippocrateaceae. *Hippocratea:* (a) leafy branch with flowers; (b) recurved stamens after shedding pollen, perianth lacking; (c) cross section through ovary; (d) flower; (e) fruit.

Shrubs, small trees, or lianas, commonly with well-developed latex canals in the stem and leaf tissues. Leaves opposite (rarely alternate), simple; stipules small or lacking. Flowers usually small, regular, perfect; hypogynous; inflorescence mostly a cyme. Sepals 5 (rarely 2–3), connate at the base. Petals 5 (rarely 2), distinct. Stamens usually 3 (rarely 2, 4, or 5), filaments connate and expanded at the base, aligned along the ovary sides, recurved outward at maturity; anthers opening by transverse slits; nectary disk is well developed and outside the stamens. Pistil compound of 3 united carpels; locules same number as carpels; ovules 2–10 (rarely many) in each locule and attached to axile placentas; ovary superior and often triangle-shaped; style terminal, slender, stigma as many lobed as carpels. Fruit a berry, capsule, or drupe; sometimes 3-lobed. Seed flattened, angular, or winged; embryo large, endosperm lacking (Fig. 9.116).

The Hippocrateaceae consists of 2 genera and 300 species distributed pantropically. The two genera are *Salacia* (200 species) and *Hippocratea* (100 species).

Few of the species are of any real economic value. However, liana species of *Hippocratea* are used in Africa to make "rope bridges." Some fruits of *Salacia* are eaten.

The classification of the family Hippocrateaceae is presently being debated. Some botanists would combine the family with the Celastraceae, while others consider it an unnatural group to be separated still further.

Fossil pollen attributed to the Hippocrateaceae has been found in Oligocene and more recent deposits.

## Family Aquifoliaceae (Holly)

Figure 9.117 Aquifoliaceae. *Ilex:* (a) leafy branch with fruits; (b) cross section through ovary; (c) front view of flower; (d) longitudinal section through flower.

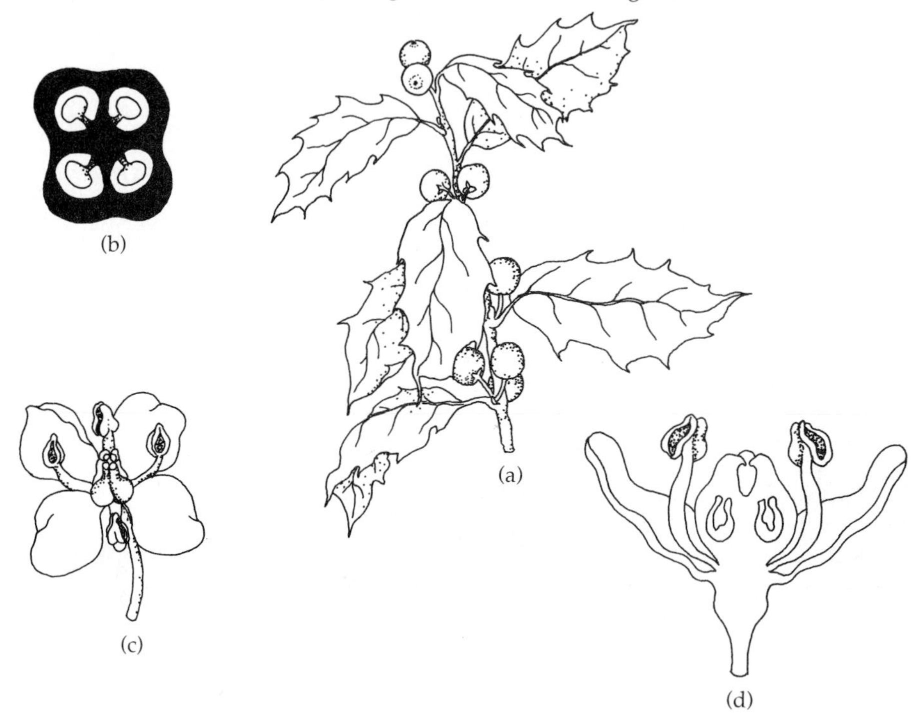


Shrubs or small trees. Leaves leathery and evergreen (less commonly deciduous), alternate (rarely opposite), sometimes appearing whorled, simple; stipules small or lacking. Flowers small, regular, unisexual and dioecious (rarely perfect), hypogynous; inflorescence a solitary flower or flowers in axillary or terminal cymes, panicles, racemes, or spikes. Sepals 4–6 (rarely 7–8), small, distinct or connate at the base. Petals 4–6 (rarely lacking), distinct or connate at the base. Stamens 4–12, usually as many as the petals and alternate with them, distinct or adnate to the corolla tube; anthers opening by longitudinal slits; nectary disk lacking. Pistil compound of 2–6 (rarely 8–many) united carpels; locules 2–many; ovules 1–2 per locule and attached to apical-axile placentas; ovary superior; style 1, terminal, short, stigma capitate or lobed. Fruit a berry or drupe. Seeds very tiny, hard, as many as carpels, embryo small, endosperm abundant (Fig. 9.117).

The family Aquifoliaceae consists of 3–4 genera and about 400 species with a cosmopolitan distribution, but are poorly represented in Africa and Australia. The largest genus is *Ilex,* the holly (almost 400 species).

Economically the family is important for a hard, white wood used in carving and for ornamentals, especially at Christmas time. The leaves of *Ilex paraguensis,* a native of South America, is used as a tea, called yerba maté.

The genus *Phelline* is sometimes classified in its own family, *Phellinaceae.* The holly family differs from other families in the order in lacking a nectary disk.

Fossils attributed to the Aquifoliaceae have been found in Upper Cretaceous deposits in Australia and in various Tertiary deposits elsewhere.

## Family Icacinaceae (Icacina)

Figure 9.118 Icacinaceae. *Villaresia:* (a) leafy branch with flowers; (b) front view of flower; (c) longitudinal section through flower; (d) longitudinal section through pistil.

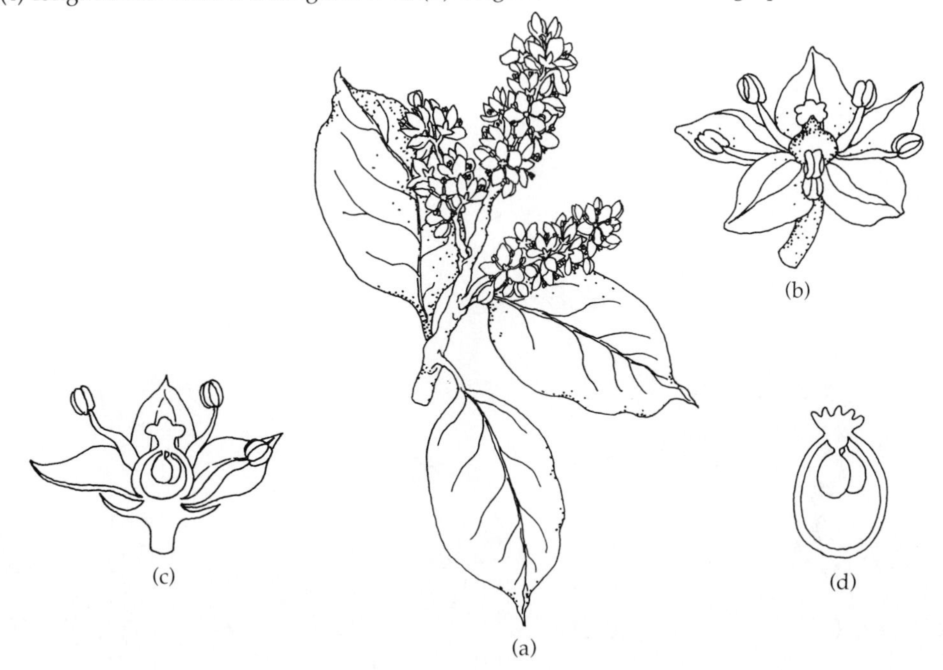


Shrubs, trees, or lianas. Leaves usually leathery, alternate (rarely opposite), simple; stipules lacking. Flowers regular, perfect or unisexual (mostly dioecious), hypogynous with the pedicel separating at the top; inflorescence of various types, usually a cyme. Sepals 4–5, small, connate into a tube. Petals 4–5 (rarely lacking), distinct or connate. Stamens 4–5, filaments distinct or attached to the corolla, alternate with the petals, often with hairs at the tip; anthers opening by longitudinal slits or apical pores; nectary disk usually lacking or as glands. Pistil compound of 3 (rarely 2 or 5) united carpels; locules functionally 1 (others aborted); ovules 2 and borne on apical-axile placentas; ovary superior; style terminal, simple and short; stigma 3–5 lobed. Fruit a drupe or samara. Seeds with embryo curved or straight, endosperm present or lacking (Fig. 9.118).

The family Icacinaceae consists of 50–56 genera and 400 species distributed pantropically, with only a few species reaching temperate climates. The largest genus is *Gamphanda* (about 50 species).

Economically the family is of moderate value. The timber of *Cantleya corniculata* is exported from Brunei as a substitute for sandalwood and is used in the marine industry. Seeds and tubers of *Icacina senegalensis* and *Humirianthera* provide a starchy flour. A yerba maté tea substitute is made from the leaves of *Citronella*. Bark and leaves of *Cassinopsis madagascariensis* produce an antidysentery product. Drinking water is obtained from the cut stems of *Miquelia* and *Phytocrene*.

The family Icacinaceae is distinctive in the order Celastrales, even though it has some features that are similar to those of other families within the group.

Fossils attributed to the family and found in Upper Cretaceous deposits are presently being questioned. Less doubtful fossils are found in Paleocene and Eocene deposits.

## ORDER EUPHORBIALES
### Family Buxaceae (Boxwood)

Figure 9.119 Buxaceae. *Buxus:* (a) leafy branch with flowers and fruits; (b) male flower; (c) cross section through ovary; (d) fruit. *Pachysandra:* (e) female flower.

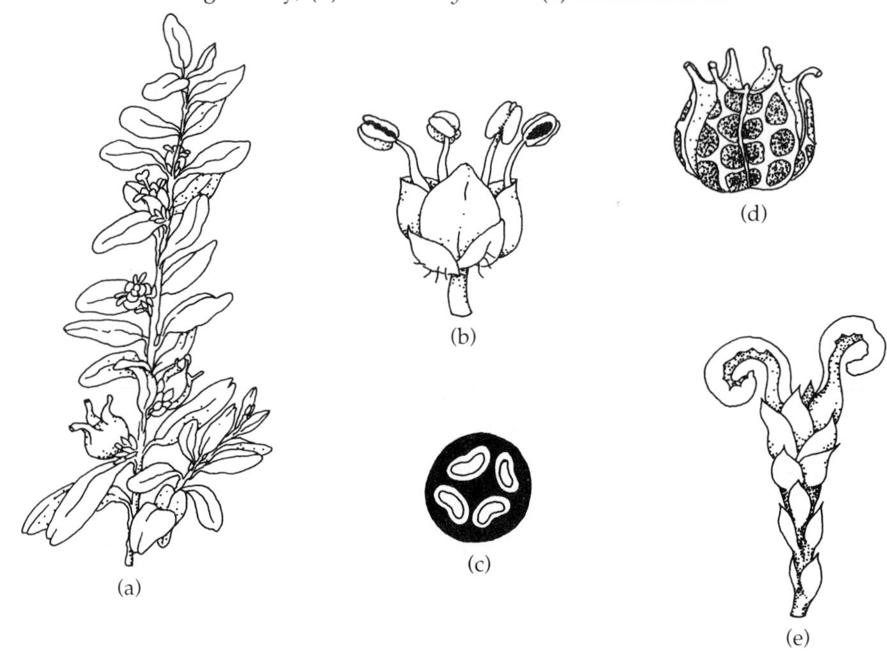


Shrubs (rarely herbs or trees). Leaves evergreen and leathery (rarely deciduous), alternate or opposite, simple; stipules lacking. Flowers small, regular, unisexual and monoecious (rarely dioecious or perfect), hypogynous; inflorescence of heads, racemes, or spikes. Sepals 4–6 (rarely more or lacking), distinct or connate. Petals lacking. Stamens 4–6 (rarely 8–many), filaments distinct, sometimes broad, if 4 then opposite the sepals; anthers large, opening by longitudinal slits; nectary disk lacking. Pistil compound of 3 (rarely 2, 4, or 6) united carpels; locules 3–6; ovules 1–2 in each locule and attached to axile placentas; ovary superior; styles 3 (rarely 2–6), terminal, distinct or connate at the base, stigma along inner surface of style. Fruit a loculicidal capsule or drupe. Seeds 2, black and shiny, embryo straight, endosperm abundant (Fig. 9.119).

The family Buxaceae consists of 5 genera and 60 species with a cosmopolitan distribution, but scattered. The largest genus is *Buxus,* the box (30 species).

Economically the family is important for ornamentals and some speciality woods. The most well-known species is *Buxus sempervirens,* the boxwood, which is used as an evergreen hedge, and as a special wood that is ideal for carving and for inlaying furniture and rulers. *Pachysandra procumbens* from eastern North America and *P. terminalis* from Japan are used extensively as groundcovers.

The family Buxaceae is distinctive within the order Euphorbiales because of its lack of stipules, its seed characters, and the lack of a milky juice.

Macrofossils attributed to the Buxaceae have been found in Miocene deposits.

## Family Euphorbiaceae (Spurge)

Figure 9.120 Euphorbiaceae. *Euphorbia:* (a) arborescent, fleshy stemmed plant; (b) cross section through ovary; (c) front view of cyathium; (d) upper stem with bracts and cyathia; (e) longitudinal section through cyathium.

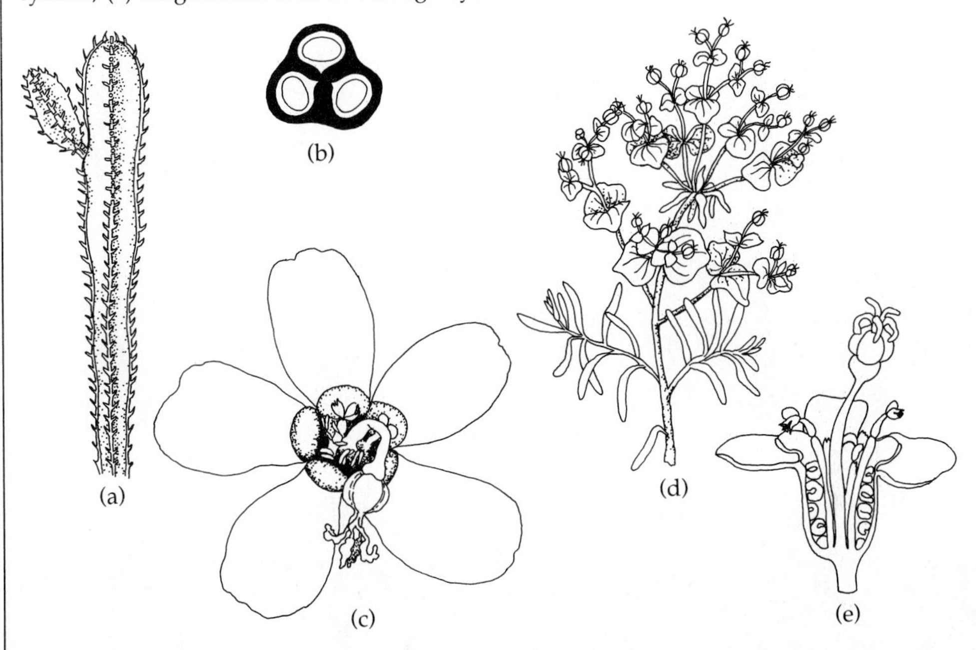


Herbs, shrubs, or trees with fleshy stems and milky or colored latex. Leaves alternate, opposite, or whorled, simple (rarely compound), venation pinnate or palmate; stipules present, large or small and gland-like (rarely lacking). Flowers regular, unisexual (monoecious or dioecious), hypogynous; inflorescence of various types, often compacted to form a special flower cluster called a **cyathium**. Perianth usually 5 (rarely 0–8), distinct or connate. Stamens 1–many, filaments distinct or connate; nectary disk sometimes present; anthers opening by longitudinal slits. Pistil compound of 3 (rarely 4–many) united carpels; locules usually the same number as the carpels; ovules 1–2 in each locule and attached to apical-axile placentas; ovary superior, commonly 3 lobed; styles 3 (rarely 1–4), distinct or connate into a single style. Fruit a capsule-like schizocarp (rarely a berry, drupe, or samara). Seeds often with a fleshy outgrowth, embryo curved or straight, endosperm abundant (rarely lacking) (Fig. 9.120).

The large family Euphorbiaceae consists of about 300 genera and 7500 species distributed extensively in the tropical and warmer regions of the world, but with some temperate species. The species-rich regions are Indomalaysia, the New World tropics, and Africa. The largest genera are *Euphorbia* (1500 species), *Croton* (750 species), *Phyllanthus* (400 species), and *Macaranga* (250 species).

The family is economically very important. A staple tropical food is *Manihot esculenta,* cassava or tapioca. *Hevea brasiliensis,* the para rubber tree native to Brazil, produces most of the world's natural rubber. Castor oil comes from *Ricinus communis.* The genus *Euphorbia* provides *E. pulcherrima,* the poinsettia, and the noxious weed *E. esula.* *Croton* has many species that are used as ornamentals. Many other uses include dyes, purgatives, and timber.

The Euphorbiaceae is most diverse morphologically and ecologically, with some species similar to cacti.

Fossils attributed to the Euphorbiaceae were first found in Paleocene and Eocene strata and then more recent deposits.

## ORDER RHAMNALES
### Family Rhamnaceae (Buckthorn)

Figure 9.121 Rhamnaceae. *Ceanothus:* (a) leafy branch with flowers; (b) fruit. *Rhamnus:* (c) longitudinal section through flower; (d) stamen; (e) cross section through ovary. *Adolphia:* (f) flower.

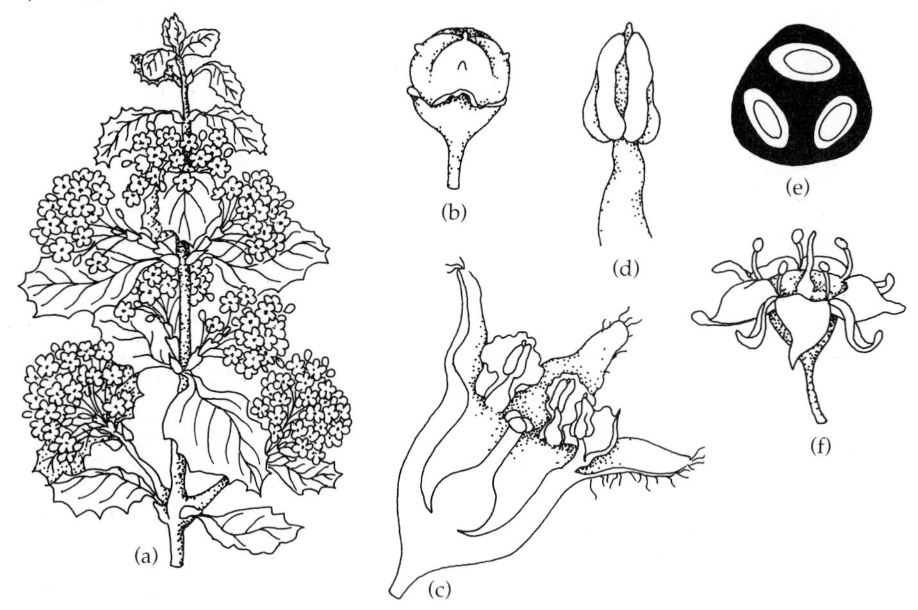


Shrubs, trees, or lianas (rarely herbs), sometimes thorny, occasionally associated with nitrogen-fixing organisms. Leaves alternate or opposite, simple; stipules present (rarely lacking), may be smaller and modified as spines. Flowers small, usually inconspicuous, regular, perfect (rarely unisexual), perigynous or epigynous, hypanthium present; inflorescence rarely a solitary flower, but in various axillary or terminal clusters. Sepals 4–5, connate. Petals 4–5 (rarely lacking), hooded over the anther. Stamens 4–5, filaments distinct, adnate to and opposite the petals; anthers opening by longitudinal slits; nectary disk between the filaments and adnate to the hypanthium. Pistil compound of 2–3 (rarely 5) united carpels; locules 2–3 (rarely 1); ovules 1 per locule and attached to a basal placenta; ovary superior to inferior; style terminal, lobed or cleft. Fruit a drupe or nut. Seed 1, embryo curved or straight, endosperm in small amounts or lacking (Fig. 9.121).

The family Rhamnaceae consists of 55–58 genera and 900 species distributed worldwide, but more common in subtropical and tropical regions. The largest genera are *Rhamnus,* the buckthorn (150 species); *Phylica* (150 species); and *Zizyphus* (100 species).

Some species of *Rhamnus* are used in producing dyes: sap green from *R. cathartica,* yellow from *R. infectoria* fruits, and Chinese green indigo from *R. chlorophora* bark. Many species are important medicinally. Fruits from *R. cathartica* and *R. purshiana* are strong laxatives; in Africa extracts of *Gouania* bark and leaves are used as wound dressings; and in Malaya *Ventilago oblongifolia* is used to treat cholera. Various species of *Ceanothus* are grown as ornamentals.

The family is associated with the Vitaceae because of many similar characters.

Fossil leaves attributed to the Rhamnaceae were first found in Eocene deposits, with both macro- and microfossils found in more recent deposits.

## Family Vitaceae (Grape)

Figure 9.122 Vitaceae. *Vitis:* (a) leafy branch with tendrils and flowers; (b) longitudinal section through pistil; (c) front view of perfect flower; (d) front view of male flower; (e) cross section through ovary.

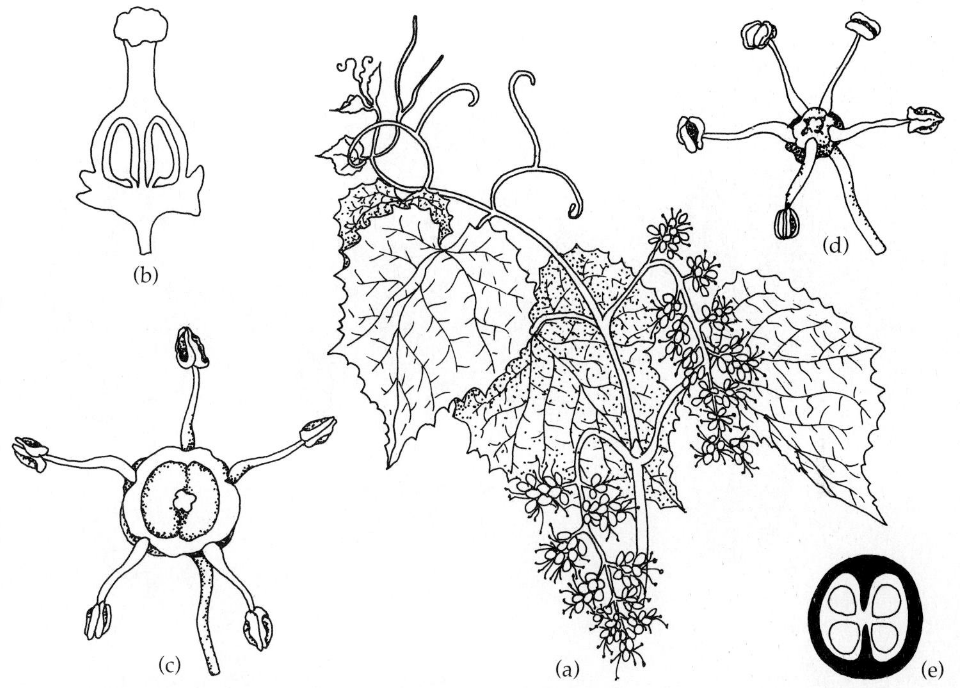


Commonly vines (rarely herbs or small trees), with tendrils opposite the leaves. Leaves alternate, simple, palmately veined or lobed or compound (rarely pinnately compound), commonly having specialized glands; stipules present but deciduous. Flowers small, regular, perfect or unisexual, hypogynous; inflorescence variable, terminal or opposite the leaves (rarely axillary). Sepals 4–5 (rarely 3, 6, or 7), small, connate, often lobed or toothed. Petals 4–5 (rarely 3, 6, or 7), distinct (rarely connate), often dropping as stamens develop. Stamens 4–5 (rarely 3, 6, or 7), filaments distinct and opposite the petals; anthers sometimes connate; nectary disk present. Pistil compound of 2 (rarely 3–6) united carpels; locules 2–6; ovules 2 per locule and attached to axile placentas; ovary superior; style 1, short, stigma capitate (rarely 4-lobed). Fruit a berry. Seeds 2, embryo small and straight, endosperm present (Fig. 9.122).

The family Vitaceae consists of 11–12 genera and 700 species distributed widely in warmer climates; there are also some temperate species. The largest genera are *Cissus* (300–350 species) and *Vitis* (50–65 species).

The family is important for the production of table grapes, raisins, and wine. Many cultivars of the wine grape, *Vitis vinifera,* which originated in southwest Asia, are planted throughout the temperate world. Other wine grape species are *V. aestivalis* and *V. labrusca* from North America. The fruits when dried are called raisins or, if from a seedless grape, sultanas. *Parthenocissus quinquefolia,* Virginia creeper, and *P. tricuspidata,* Boston ivy, are common ornamentals.

The genus *Leea* has been classified by some botanists in the Vitaceae but is put by others in its own family, Leeaceae.

Questionable fossils attributed to the Vitaceae were described from Upper Cretaceous deposits. Less questionable seed and pollen fossils are found in Eocene and more recent deposits.

## ORDER LINALES
## Family Erythroxylaceae (Coca)

Figure 9.123 Erythroxylaceae. *Erythroxylum:* (a) leafy branch with fruits; (b) front view of flower; (c) cross section through ovary; (d) petal with ligule appendage; (e) fruit.

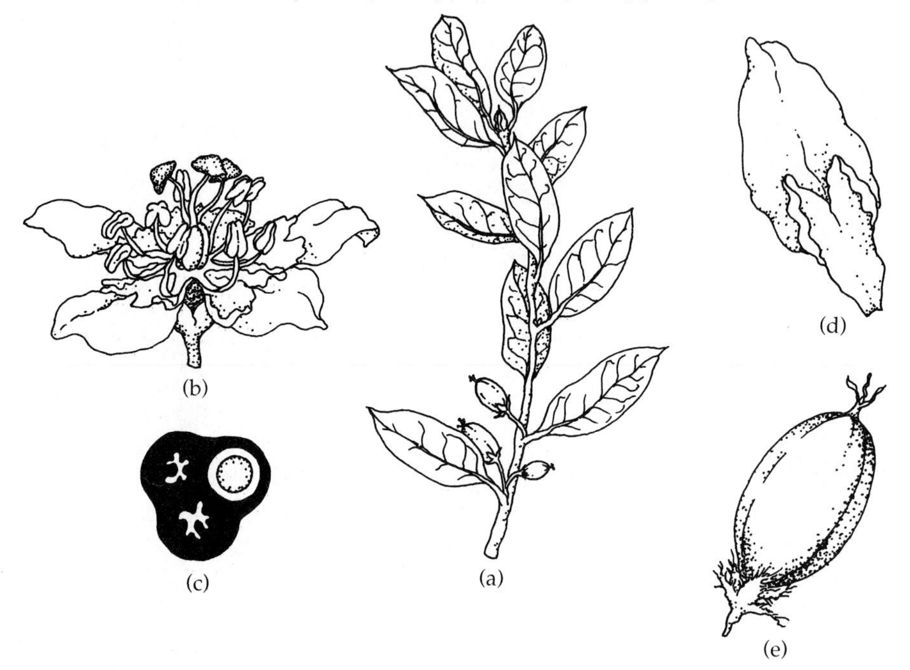


Shrubs or trees. Leaves alternate (rarely opposite), simple, entire, ovate; stipules within the petioles, commonly soon deciduous. Flowers small, regular, perfect (rarely unisexual and dioecious), hypogynous; inflorescence of a solitary flower or flowers borne in axillary clusters. Sepals 5, connate and bell-shaped. Petals 5, distinct but with a ligule appendage on the inside toward the base that falls off easily. Stamens 10, filaments connate forming a tube; anthers opening by longitudinal slits; nectary disk lacking. Pistil compound of 3–4 (rarely 2) united carpels; locules as many as the carpels; ovules 1–2 in only one fertile locule and borne on axile placentas; ovary superior; styles 3, distinct or connate. Fruit an oval-shaped drupe, calyx persistent. Seeds 1–2, embryo straight, endosperm present (Fig. 9.123).

The family Erythroxylaceae consists of 4 genera and 200–260 species distributed pantropically, but well developed in South America. The largest genus is *Erythroxylum* (190–250 species). The remaining three genera have only 10 species among them.

Economically the family is important mostly for cocaine. The leaves of *Erythroxylum coca* and *E. novagranatense* (both called coca) produce the alkaloid cocaine. This narcotic is widely used in medicine, but more recently it has become an important illegal drug in Western society. The centers for this narcotic trade are in northern South America, Java, and Sri Lanka. Other species of *Erythroxylum* provide dyes from the bark, wood for timber, oils, and medicinal products.

The family Erythroxylaceae, with its very distinctive characters, is sometimes included by some botanists in the family Linaceae.

Fossils attributed to the family have been found in Eocene deposits in South America.

## Family Linaceae (Flax)

Figure 9.124 Linaceae. *Linum:* (a) flowering plant; (b) cross section through ovary; (c) bud; (d) fruit; (e) stamens and pistil with perianth removed.

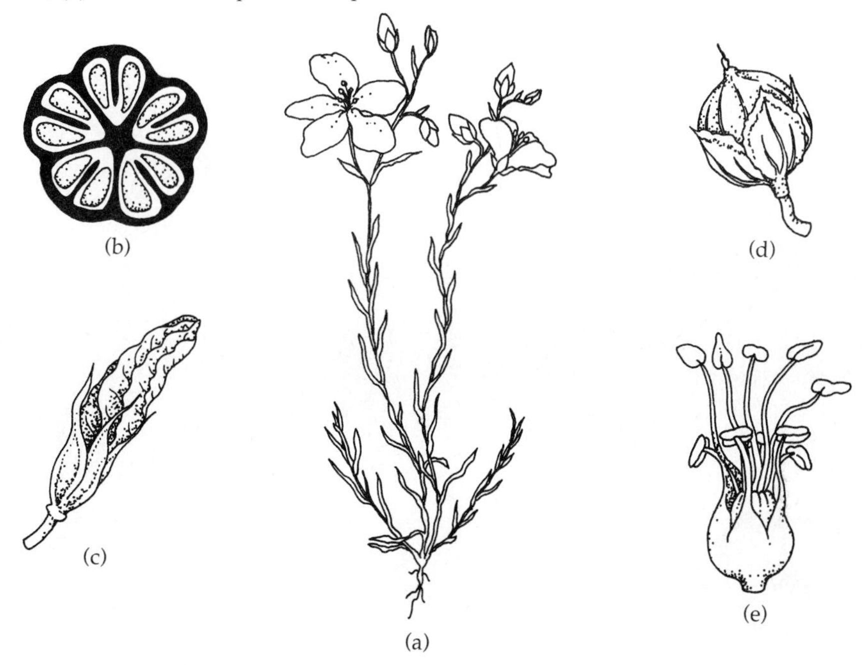


Herbs or shrubs. Leaves alternate or opposite, simple, entire; stipules small, modified into glands or lacking. Flowers regular, perfect, hypogynous; inflorescence a cyme, raceme, or spike. Sepals 5 (rarely 4), distinct or connate at the base. Petals 5 (rarely 4), distinct, sometimes clawed and falling easily. Stamens the same number as the petals and usually alternate with (rarely opposite) them; filaments distinct or basally connate, short, sometimes alternating with staminodes; anthers opening by longitudinal slits; nectary glands often outside the ring of stamens. Pistil compound of 3–5 united carpels; locules 3–10 due to development of incomplete septa; ovules 1–2 per locule and borne on axile placentas; ovary superior; styles 2–5, distinct or connate, stigma terminal and lobed. Fruit a septicidal capsule or drupe-like (rarely a schizocarp). Seeds usually 1, embryo straight, endosperm present or absent, the seed coat in some mucilaginous and swells when wet (Fig. 9.124).

The family Linaceae consists of 6–13 genera and 220–300 species distributed mostly in temperate climates, but with some tropical species. The largest genera are *Linum,* flax (200–230 species), and *Hugonia* (40 species).

The family is economically important for the genus *Linum. Linum usitatissimum,* flax, is a major source of fiber used in the manufacture of cigarette paper, fine writing paper, and linen. Extracted from the pressed seed is linseed oil, used extensively in paints, printing ink, and varnishes. The remaining material from the pressed seeds is used as animal feed.

The family Linaceae is well recognized for its distinctive flower characters, especially its deciduous petals.

The family has no known fossil record.

## ORDER POLYGALALES
### Family Malpighiaceae (Barbados Cherry)

Figure 9.125 Malpighiaceae. *Tristellateia:* (a) leafy branch with flowering and fruiting inflorescences. *Stigmaphyllon:* (b) front view of flower; (c) cross section through ovary. *Malpighia:* (d) stamens arranged as a stamen tube; (e) longitudinal section through flower.

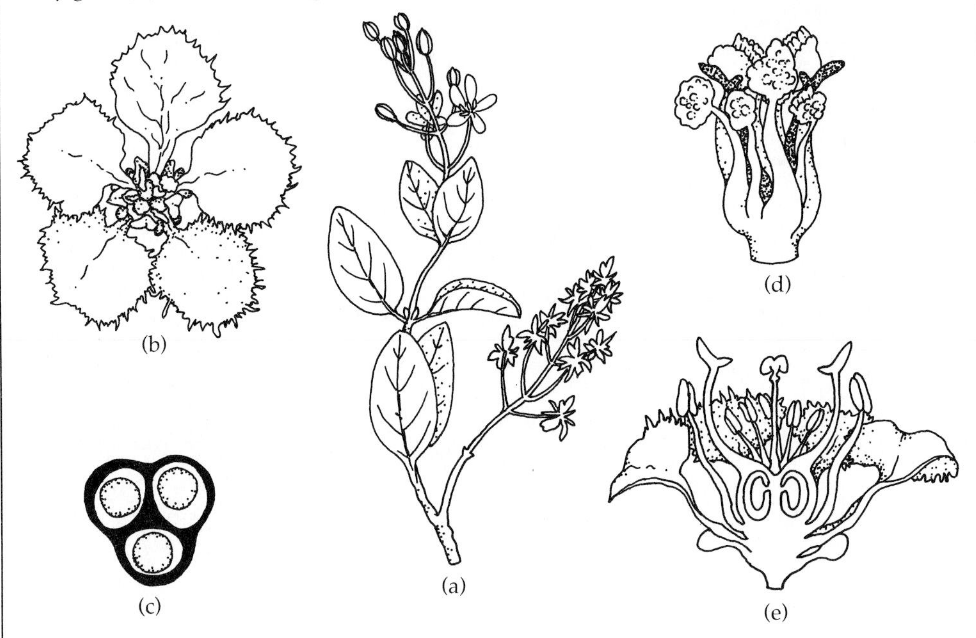


Shrubs, trees, or lianas. Leaves opposite (rarely alternate), simple and entire; often with large glands on the petiole at the base of the blade; commonly with unicellular 2-branched hairs (called **malpighian hairs**); stipules present or absent. Flowers often large, showy (rarely cleistogamous), regular to irregular, perfect (rarely unisexual), hypogynous; inflorescence of axillary or terminal cymes, panicles, or racemes. Sepals 5, distinct or connate, often with paired glands at the base of the sepals. Petals 5, distinct, usually clawed with fringed or toothed margins. Stamens 10 (rarely 5 or 15), filaments usually connate at the base into a tube and with some modified into staminodes; anthers opening by longitudinal slits (rarely pores); nectary disk lacking; pollinated by bees of the genus *Centris* that collect oil from the flowers. Pistil compound of 3 united carpels; locules 3; ovules 1 per locule and borne on an axile placenta; ovary superior; styles 1–3, usually distinct or connate toward the base, stigma terminal to subterminal. Fruit a drupe, nut, or schizocarp with winged mericarps. Seeds with embryo curved to straight, endosperm lacking (Fig. 9.125).

The family Malpighiaceae consists of 60 genera and 800–1200 species distributed pantropically, with its greatest development in South America. The largest genera are *Byrsonima* (150 species), *Heteropterys* (120 species), *Banisteriopsis* (100 species), and *Tetrapterys* (90 species).

Economically the family is of value in various ways. The leaves and shoots of several species of *Banisteriopsis* and of *Dipteropterys cabrerana* and *Banisteria caapi* produce hallucinogenic drugs. The fruits of various *Bunchosia* and *Malpighia* are cultivated in South America for their edible fruits, especially *M. glabra*, the Barbados cherry. A good number of various species are grown as ornamentals in the tropics.

The characteristics of the family are such as to allow it to be classified in the order Linales or the Polygalales.

The family has no known fossil record.

## Family Vochysiaceae (Vochysia)

Figure 9.126 Vochysiaceae. *Vochysia:* (a) leafy branch with flowers; (b) fruit (a samara); (c) cross section through ovary; (d) flower.

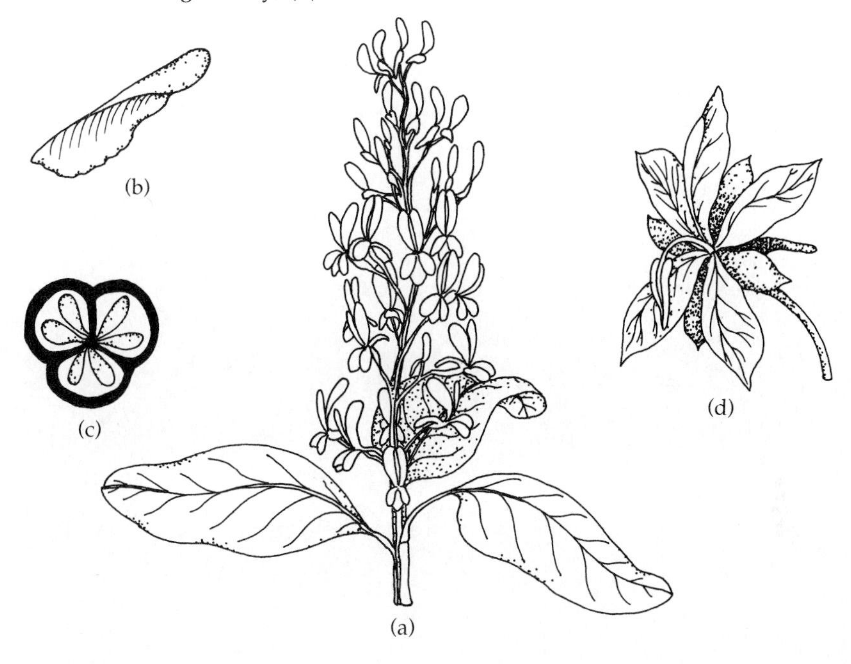


Shrubs, trees, or lianas with resin-like juice, unicellular 2-branched hairs (called malpighian hairs) usually lacking or of the stellate type. Leaves leathery, opposite or whorled (rarely alternate), simple, entire; stipules small, modified as glands or lacking. Flowers irregular, perfect, hypogynous, perigynous, or epigynous; inflorescence of racemes or panicles. Sepals 5, connate at the base, irregular, with one sepal larger and with a spur or gland at the base. Petals 1–5 (rarely lacking), irregular and unequal in size. Stamens 1–5, filaments distinct or connate, only 1 fertile with the others as staminoids, fertile one opposite the spurred sepal; anthers opening by longitudinal slits; nectary disk or glands lacking. Pistil compound of 3 united carpels, locules 1–3, ovules 1–2 attached to axile placentas, ovary superior or inferior and adnate to the calyx, style 1, stigma capitate or lateral. Fruit a loculicidal capsule or samara with the calyx attached. Seeds commonly winged and hairy, embryo straight, endosperm usually lacking (Fig. 9.126).

The family Vochysiaceae consists of 6–7 genera and 200 species distributed mostly in the New World tropics, except for 1 small genus in western Africa. The largest genera are *Vochysia* (100–105 species) and *Qualea* (60–65 species).

A few species have some economic value, especially *Vochysia tetraphylla* for timber and furniture and *V. hondurens* for fence posts and the construction of wooden boats. *Erisma calcaratum* seeds are a source of a tallow, called jaboty butter, used in making candles and soaps.

The family *Vochysiaceae* has some characters that are similar to the Polygalaceae, thereby classifying the groups near one another.

The family has no known fossil record.

## Family Polygalaceae (Milkwort)

Figure 9.127 Polygalaceae. *Polygala:* (a) plant in flower; (b) cross section through ovary; (c) flower; (d) split tube of stamens; (e) longitudinal section through ovary.

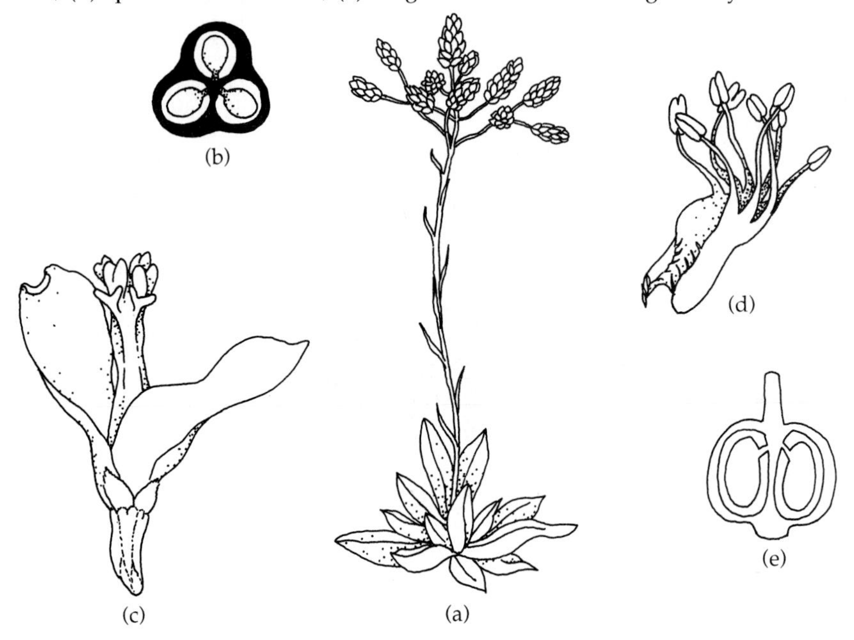


Herbs, shrubs, trees, or lianas, sometimes saprophytic. Leaves alternate (rarely opposite or whorled), simple, entire or reduced to scales; stipules present as glands or lacking. Flowers irregular, perfect, hypogynous or perigynous with a cup-like hypanthium, subtended by a bract and 2 bracteoles; inflorescence in axillary or terminal racemes, panicles, or spikes; sometimes cleistogamous flowers produced. Sepals 5, distinct or connate, usually lower 2 sepals connate or some petal-like. Petals 3 (rarely 5), distinct, the lower petal commonly fringed and saucer-shaped. Stamens usually 8 (rarely 3–10), in 2 whorls, filaments connate into a tube split on one side and adnate to the base of the corolla; anthers opening by a short slit or apical pore; pollen wall with distinctive sculpturing; a nectary disk occasionally between the stamens. Pistil compound of 2–5 united carpels; locules 2 (rarely 1–5); ovules 1 per locule and borne on apical-axile placentas; ovary superior; style single and curving, stigma capitate. Fruit a loculicidal capsule, drupe, nut, or samara. Seeds 1, usually hairy, embryo straight, endosperm present or absent (Fig. 9.127).

The family Polygalaceae consists of 12–17 genera and 750–1000 species with a cosmopolitan distribution, except for New Zealand, southern Pacific islands, and the extreme Northern Hemisphere. The largest genus is *Polygala* (500 species). The flowers superficially look like members of the bean family, Fabaceae.

The family is of minor importance economically. A few species are used locally for folk medicines. *Polygala senega* from eastern North America is used for treating snake bite. Some *Polygala* are used as dyes, while *P. butyracea* from Africa produces a strong fiber.

The characters of the various genera are rather diverse, allowing for botanists to place some genera in small families; *Xanthophyllum* in the Xanthophyllaceae, for example.

Fossil fruits attributed to the Polygalaceae were found in Eocene deposits, with micro- and macrofossils being found in Miocene and more recent deposits.

## Family Krameriaceae (Krameria)

Figure 9.128 Krameriaceae. *Krameria:* (a) plant with flowers; (b) pistil; (c) spiny fruit; (d) flower; (e) longitudinal section through flower.

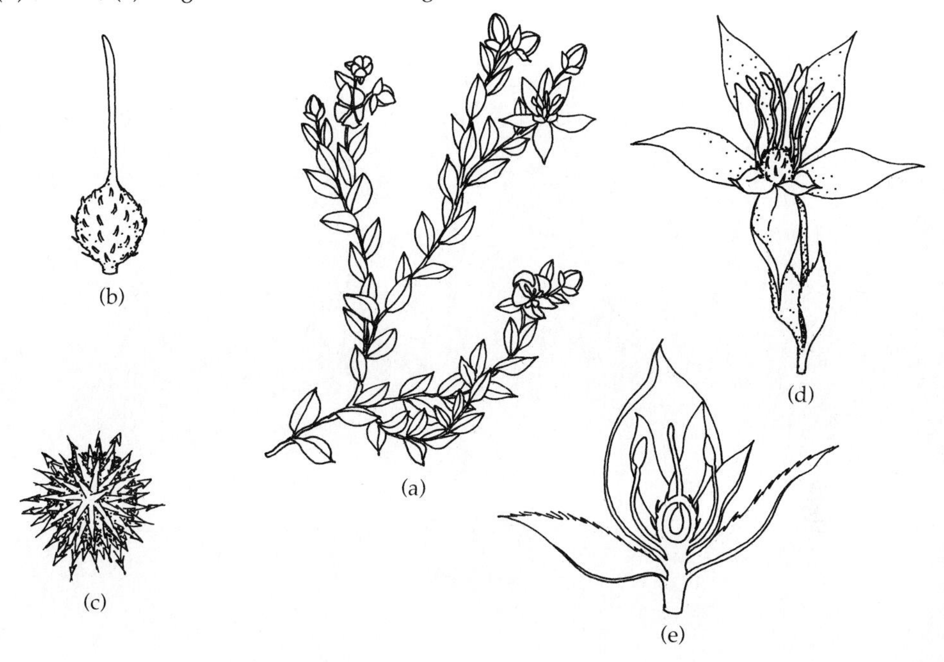


Herbs or shrubs (rarely small trees), semiparasitic on the roots of nearby plants; commonly covered with soft hairs. Leaves alternate, simple, entire (rarely trifoliate); stipules lacking. Flowers irregular, perfect, hypogynous; inflorescence an axillary, solitary flower or flowers in terminal racemes; pedicels with 2 leaf-like bracteoles. Sepals 4–5, distinct, petal-like and irregular. Petals 4–5, distinct, upper 3 are larger and usually connate by their claws, lower 2 are smaller, broader and thicker, and nectar producing. Stamens 3–4, distinct or connate, attached to the receptacle or to the claws of the petals, alternating with the upper petals; anthers opening by short slits or apical pores; nectary disk lacking. Pistil compound of 2 united carpels, but with only 1 fertile and 1 locule, the other empty and reduced; ovules 2 and borne on axile placentas; ovary superior; style single and terminal, stigma disk-shaped. Fruit a barbed or spiny pod. Seeds 1, embryo straight, endosperm lacking (Fig. 9.128).

The family Krameriaceae consists only of the genus *Krameria* and 12–25 species distributed in the arid regions of southwestern United States, Mexico to Central and South America, and the West Indies.

Economically the family is of minor importance. It is used locally for dyes, tanning, and medicinal properties. An extract obtained from *Krameria triandra* roots has astringent properties and is used in tooth preservation. A dye obtained from *K. parvifolia* is used to color fabrics, and the root of *K. tomentosa* is used in tanning leather in the American tropics.

In earlier classifications *Krameria* was classified in the Caesalpiniaceae or Polygalaceae. However, its distinctive characters are sufficient to justify putting it in its own family.

The family has no known fossil record.

## ORDER SAPINDALES
### Family Staphyleaceae (Bladdernut)

Figure 9.129 Staphyleaceae. *Staphylea:* (a) branch with flowers; (b) longitudinal section through flower; (c) cross section through ovary; (d) fruits.

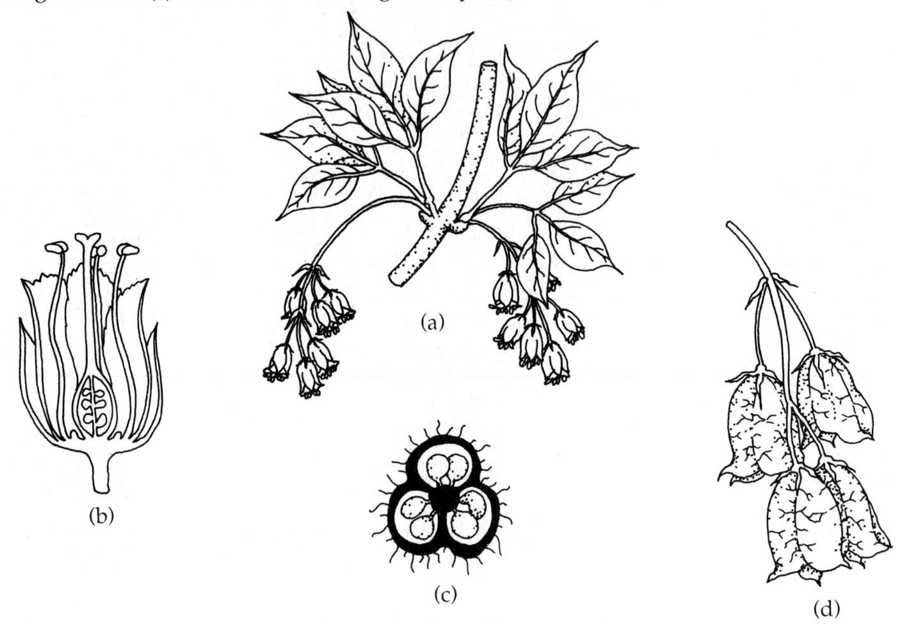


Shrubs or small trees. Leaves opposite (rarely alternate), pinnately compound or trifoliate (rarely simple); stipules present but falling off easily (or lacking). Flowers small, regular, perfect or unisexual (rarely dioecious); inflorescence an axillary or terminal panicle or raceme. Sepals 5, distinct or connate, sometimes petal-like. Petals 5, distinct or overlapping. Stamens 5, filaments distinct, sometimes flattened and alternate the petals; nectary disk present or absent; anthers opening by longitudinal slits. Pistil compound of 2–4 united (rarely distinct) carpels; locules 2–4; ovules 1–few per locule and attached to axile placentas; ovary superior; styles 2–4 distinct or connate, stigma lobed. Fruit a berry, drupe, or more commonly a capsule with inflated carpels that open at the top. Seeds 1–2 per carpel, embryo straight, endosperm present (Fig. 9.129).

The family Staphyleaceae consists of 5 genera and 50–60 species distributed in the north temperate regions, Southeast Asia and Central and northwest South America. The largest genera are *Turpineaz* (35 species) and *Staphylea* (10 species).

The family is of minor importance economically. *Staphylea colchica, S. holocarpa* var *rosea* from China, and *S. pinnata* from Europe have been grown as ornamental trees. In China and Japan, *Euscaphis japonica,* the hung-liang or gonzui zoku tree, is commonly grown as a medicinal plant.

The classification of the Staphyleaceae is not agreed upon by botanists, and the family is placed in different orders in various classification schemes.

Fossils attributed to the family were first found in Eocene deposits of the Rocky Mountains of North America.

## Family Sapindaceae (Soapberry)

Figure 9.130 Sapindaceae. *Koelreuteria:* (a) leafy branch with inflorescence of flowers; (b) cross section through ovary; (c) flower. *Litchi:* (d) pistil; (e) fruits.

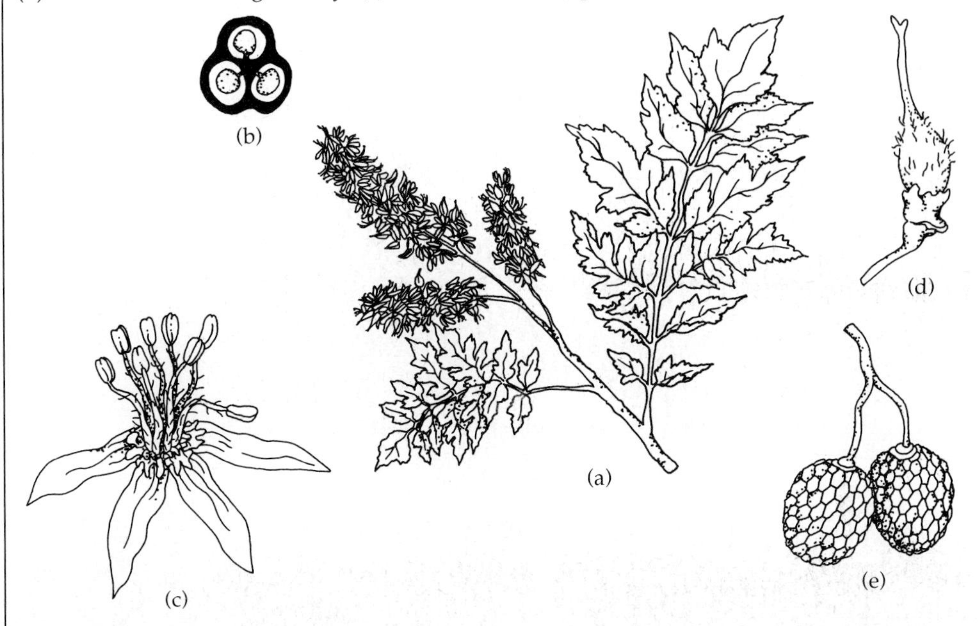


Shrubs, trees, and herbaceous or woody lianas (rarely herbs). Leaves alternate (rarely opposite), pinnately or bipinnately compound or trifoliate (rarely simple); stipules usually lacking. Flowers small, regular or irregular, perfect or unisexual (monoecious or dioecious); inflorescence rarely a solitary flower, usually an axillary or terminal cyme or cyme-like panicle. Sepals 4–5 (rarely lacking), distinct and commonly clawed. Stamens 4–10 (rarely many), filaments distinct and usually hairy; anthers opening by longitudinal slits; nectary disk present. Pistil compound of 3 (rarely 2 or 4–6) united carpels; locules 3 (rarely 2–6); ovules 1 (rarely 2–several) per locule and attached on axile placentas; ovary superior; style usually 1 (rarely 2–6), stigma lobed. Fruit various and often red, a berry, capsule, drupe, nut, samara, or schizocarp. Seeds 1, often with an aril; embryo curved, endosperm lacking (Fig. 9.130).

The family Sapindaceae consists of 140–150 genera and 1500–2000 species distributed pantropically, with a few species found in temperate climates. The largest genera are *Serjania* (220 species), *Allophylus* (190 species), and *Paullinia* (150–180 species). About 300 species in the family are lianas.

Economically the family is moderately important, especially for its edible fruits. *Litchi chenensis,* the litchi, or lychee, is native to southern China and is grown throughout the tropics for the sweet-tasting aril in the fruit. The fruit of *Nephelium lappaceum* is sought after in the Old World tropics. *Sapindus saponaria* berries from the New World tropics produce a lather with water and are used as a soap by native peoples. *Paullinia cupuna* is used in Brazil to make a caffeine-rich drink called *yoco.* A number of species are used as ornamentals.

The small family Ptaeroxylaceae is now included by some botanists in the family Sapindaceae.

Fossil pollen attributed to the family has been recorded from Upper Cretaceous deposits.

## Family Hippocastanaceae (Horse Chestnut)

Figure 9.131 Hippocastanaceae. *Aesculus:* (a) leafy branch with terminal inflorescence; (b) flower; (c) fruit; (d) seed; (e) cross section through ovary.

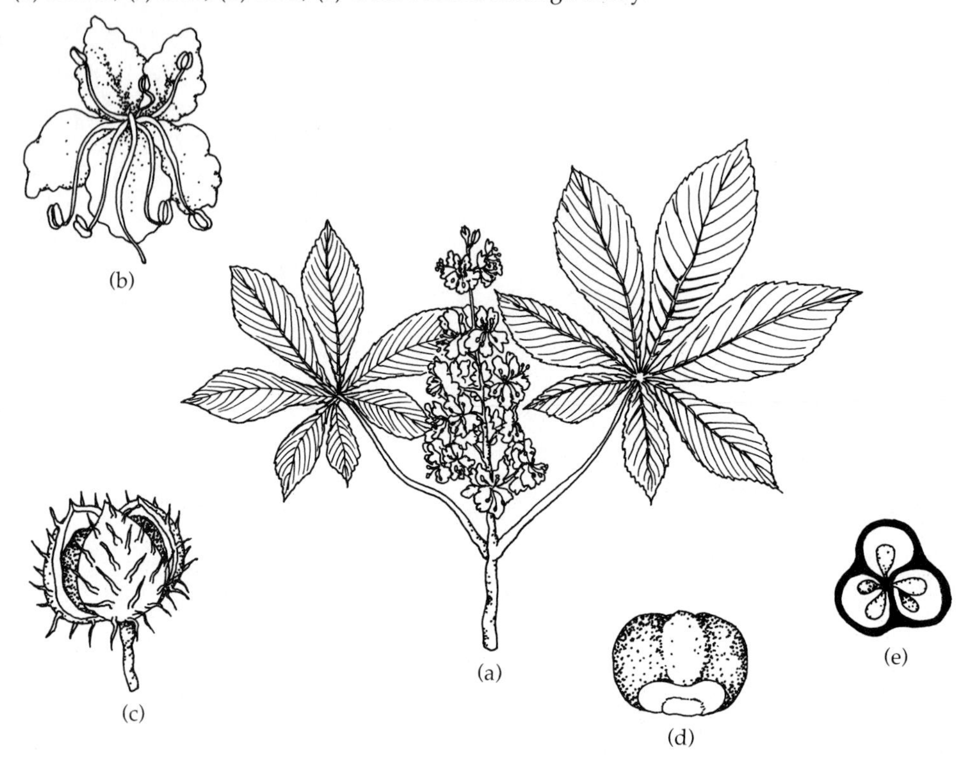


Shrubs or trees. Leaves deciduous or evergreen, opposite, deeply palmately lobed or compound; stipules lacking. Flowers somewhat large and showy, irregular, perfect or some staminate in function, hypogynous; inflorescence a terminal panicle or raceme. Sepals 4–5, distinct or connate at the base. Petals 4–5, white, yellow or red color, distinct, clawed, and unequal. Stamens 5–8, filaments distinct; anthers opening by longitudinal slits; nectary disk present between the petals and stamens. Pistil compound of 3 (rarely 2 or 4) united carpels; locules 3 (rarely 2 or 4); ovules 2 per locule and attached on axile placentas; ovary superior; style 1, terminal stigma small and slightly lobed. Fruit a leathery, loculicidal capsule. Seeds large, commonly 1 through abortion, embryo large and curved, endosperm lacking (Fig. 9.131).

The family Hippocastanaceae consists of only 2 genera: *Aesculus,* the buckeyes and horse chestnuts, has 13 species distributed mostly in the North Temperate region, while *Billia,* with 3 species, extends from southern Mexico south to northern South America.

The family is of some economic value. *Aesculus hippocastanum,* the horse chestnut, is grown ornamentally for its attractive inflorescenses, and its seeds, which are used as "conkers" by children in games and also in some folk medicines. Seeds from *A. californica,* the Californian buckeye, were eaten by native peoples of California; *Aesculus octandra,* the yellow buckeye of the southeastern United States, is a valuable timber tree.

The family Hippocastanaceae is distinctive in the order Sapindales because of its palmate leaves, leathery fruit, and large seeds.

The family has no known fossil record.

## Family Aceraceae (Maple)

Figure 9.132 Aceraceae. *Acer:* (a) leafy branch segment with winged fruits; (b) perfect flower; (c) male inflorescence; (d) female inflorescence; (e) winter twig showing buds.

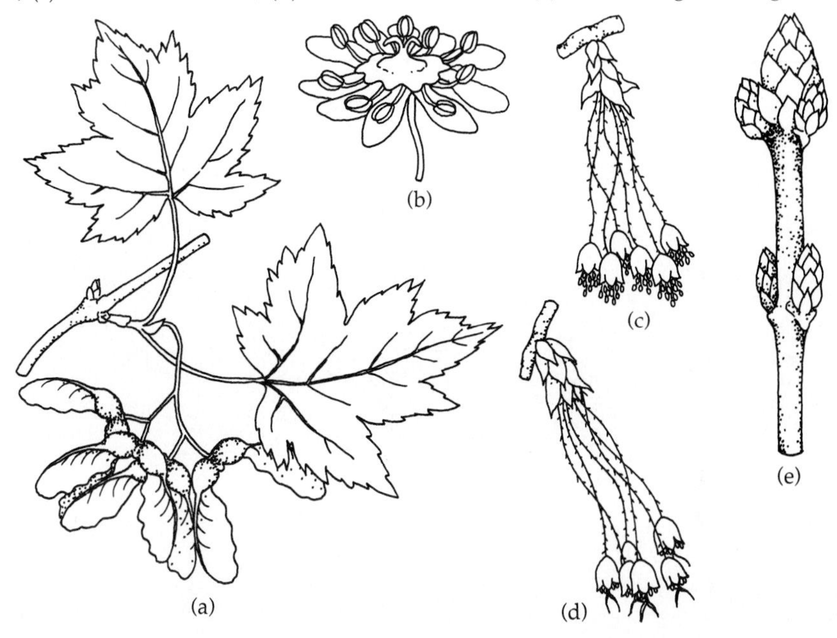


Shrubs or trees, commonly with sugar stored in the sap during the winter. Leaves deciduous or evergreen, opposite, usually simple (rarely palmately or pinnately compound) and palmately veined; stipules lacking. Flowers small, regular, perfect or unisexual (dioecious, monoecious or polygamous), hypogynous or perigynous; inflorescence various in corymbs, panicles, racemes, or umbels; insect or wind pollinated. Sepals 5 (rarely 4 or 6), distinct or connate at the base. Petals 5 (rarely 4, 6, or lacking), distinct. Stamens usually 8 (less commonly 4–12), filaments distinct; anthers opening by longitudinal slits; nectary disk usually present. Pistil compound of 2 united carpels; locules 2 (rarely 3 or more); ovules 2 per locule and attached to axile placentas; ovary superior; flattened at right angles to the partition, style 1 or 2 divided, stigma elongate along the inner side of styles. Fruit a winged schizocarp (also called a double samara). Seeds 1, embryo present, endosperm lacking (Fig. 9.132).

The family Aceraceae consists of 2 genera and 102–152 species distributed widely in the temperate Northern Hemisphere deciduous forests. The largest genus is *Acer*, the maples (100–150 species), while *Dipteronia* (2 species) is small and confined to central China.

Economically the family is important for various reasons. Many species of *Acer* are grown as ornamental trees (i.e., *A. palmatum*, Japanese maple; *A. platanoides*, Norway maple; *A. rubrum*, red maple; and *A. saccharum*, sugar maple), especially for the foliage and autumn coloration. Maple syrup and sugar products are obtained in the spring from the sap of *A. nigra* and *A. saccharum*. Some species produce timber for wood products: *A. pseudoplatanus* for violin backs and *A. saccharum* for fine furniture.

The family shares various characters with the Hippocastanaceae and Sapindaceae.

Fossils attributed to the Aceraceae are common in deposits dating from the Eocene.

## Family Burseraceae (Frankincense)

Figure 9.133 Burseraceae. *Boswellia:* (a) overall habit of tree; (b) flower; (c) cross section through ovary; (d) leaves at end of branch; (e) terminal inflorescence of fruits.

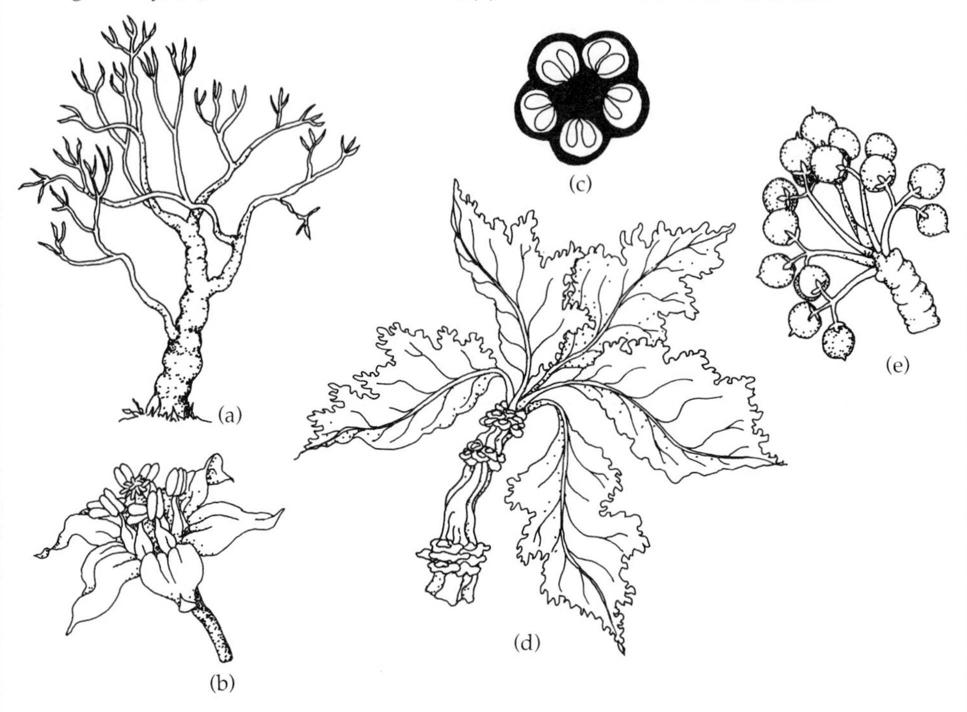


Shrubs or trees with resin, especially in the wood and bark. Leaves alternate (rarely opposite), pinnately compound or trifoliate, usually crowded near the tip of the twig; stipules present or lacking. Flowers small, cream or greenish colored, regular, perfect or unisexual and dioecious, hypogynous or perigynous; inflorescence of heads, panicles, or racemes. Sepals 3–5, connate. Petals 3–5 (rarely lacking), distinct. Stamens 3–5 or double the number of petals, filaments distinct (rarely connate); anthers opening by longitudinal slits; nectary disk present. Pistil compound of 2–5 united carpels; locules 2–5; ovules 2 (rarely 1) and borne on axile placentas; ovary superior; style 1, terminal stigma capitate or lobed. Fruit a capsule or drupe. Seeds usually 1 (less commonly 2–5) per locule, embryo curved or straight, endosperm lacking (Fig. 9.133).

The family Burseraceae consists of 16–20 genera and 500–600 species distributed pantropically, with many species in tropical America, northeastern Africa, and Malaysia. The largest genera are *Bursera* (100 species), *Commiphora* (100 species), and *Protium* (80 species).

Economically the family is of some importance. Two are well known by common name: frankincense from *Boswellia carteri* (Somaliland) and some other species, and myrrh from *Commiphora abyssinica* and *C. molmol* (Arabia and Ethiopia), the latter used in incense and perfume production. *Aucoumea kleineana* and *Canarium schweinfurthii* of Africa and *C. littorale* and *Santiria laevigata* of Malaysia produce valuable timber for construction.

The families Burseraceae and Anacardiaceae both have extensive resin and resin canals.

Fossils attributed to the family were first found in Eocene deposits.

## Family Anacardiaceae (Cashew)

Figure 9.134 Anacardiaceae. *Rhus:* (a) leafy branch with terminal inflorescence of fruits; (b) longitudinal section through male flower with aborted pistil in middle. *Toxicodendron:* (c) leafy branch with flowers and fruits; (d) female flower with perianth parts in longitudinal section.

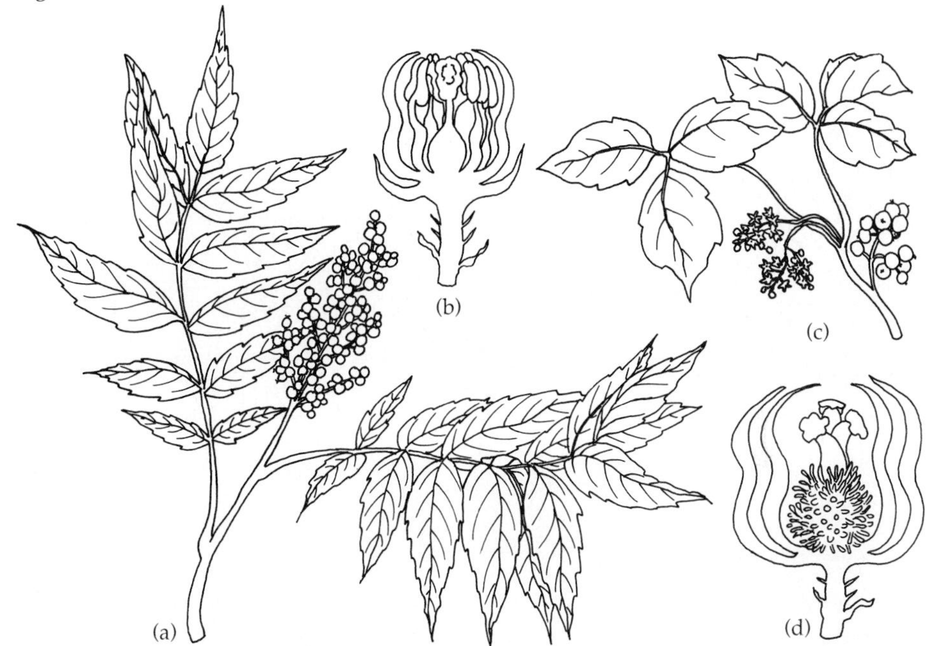


Shrubs, trees, or lianas, commonly with resin that is irritating or poisonous to touch for some people. Leaves alternate (rarely opposite), usually pinnately compound or trifoliate (rarely simple); stipules lacking. Flowers small, regular, perfect or unisexual, hypogynous, perigynous, or epigynous; inflorescence a complex panicle. Sepals 5 (rarely 3–7), connate toward the base. Petals 5 (rarely 3–7), distinct. Stamens 5–10 (rarely 1 or many), filaments distinct (rarely connate at the base); anthers opening by longitudinal slits; nectary disk present. Pistil compound of 1–5 (rarely 12) united carpels (rarely distinct); locules 1 (rarely 4–5); ovules usually 1 per locule and borne on an axile placenta; ovary superior or inferior; styles 1–3, distinct or connate, stigma capitate. Fruit a berry or drupe. Seeds 1, embryo curved or straight, endosperm lacking (Fig. 9.134).

The family Anacardiaceae consists of 60–80 genera and about 600 species distributed pantropically, but with a few genera in northern temperate regions. The largest genus is *Rhus,* the sumac (100 species).

Economically the family is important for various reasons. The fruits and nuts of some species are important: *Anacardium occidentale,* cashew nuts; *Mangifera indica,* the mango; *Pistacia vera,* pistachio nuts; and various species of *Spondias* (hogplum, Jamaica plum, and Otaheite apple). The mango and pistachio nut are harmless, but the cashew nut must be roasted before becoming safe to eat. The genus *Toxicodendron* causes dermatitis in many people and includes poison ivy, poison oak, and poison sumac. Various species of *Cotinus* and *Rhus* are used as ornamentals and as sources for tannins used in treating leather.

The family Anacardiaceae is similar to the Burseraceae and Julianiaceae in having resin and resin canals.

Fossils, both micro and macro, attributed to the family have been found in Paleocene and more recent deposits.

## Family Simaroubaceae (Quassia)

Figure 9.135 Simaroubaceae. *Ailanthus:* (a) leafy branch with large inflorescence; (b) fruits; (c) male flower; (d) functionally female flower; (e) winter twig.

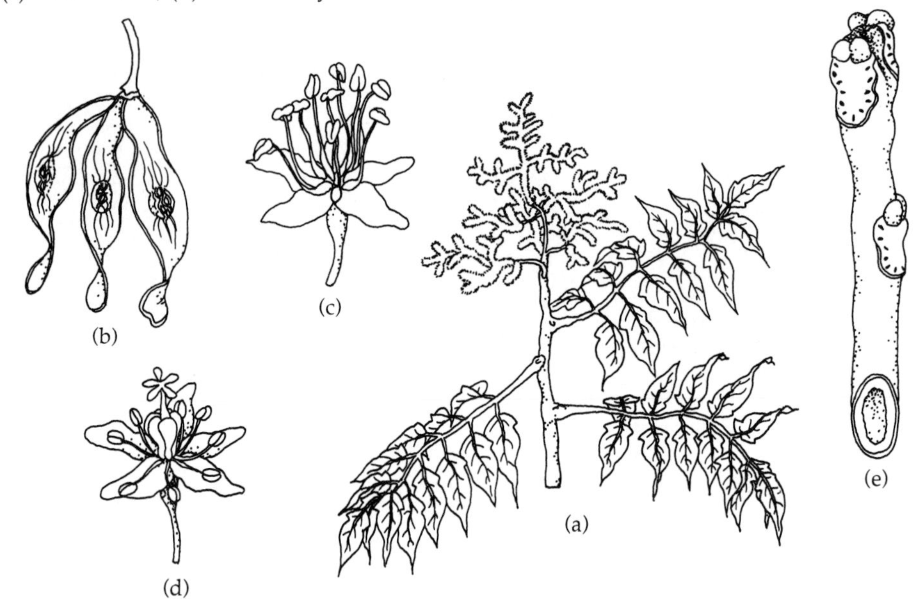


Shrubs or trees, commonly having very bitter bark, seeds, and wood. Leaves alternate (rarely opposite), pinnately compound (rarely simple), not dotted with glands; stipules lacking. Flowers small, regular, perfect or unisexual, hypogynous; inflorescence a cyme, panicle, or spike. Sepals 3–8, distinct or connate. Petals 3–8 (rarely lacking), distinct. Stamens 6–16, equal to and alternate the petals or double the number of petals, filaments distinct but commonly with a basal appendage; anthers opening by longitudinal slits; nectary disk present. Pistil compound of 2–5 (rarely more) united carpels, locules 1–10, ovules 1–2 per locule and borne on axile placentas, ovary superior; style 1–8, distinct or connate; stigma sessile. Fruit a capsule, samara, or schizocarp (rarely a berry or drupe). Seeds with embryo curved or straight, endosperm present or absent (Fig. 9.135).

The family Simaraoubaceae consists of 20–25 genera and 120–150 species distributed pantropically, but with some species being more temperate to subtropical in their distribution. The largest genera are *Picramnia* (50 species) and *Quassia* (40 species).

Economically the family is of moderate importance. *Ailanthus altissima,* the tree of heaven from Siberia, is a fast-growing ornamental tree that can become an introduced pest because of extensive root sucker development. In the New World various medicinal products, such as antimalarial medicines, are made from the bitter bark of *Quassia amara* and some species of *Picramnia* and *Simarouba.* The bitter bark and leaves of *Picramnia antidesma* have the flavor of licorice.

The family Simaroubaceae has many characters that are similar to the Rutaceae but differs in lacking the resin glands in the leaves. Some genera, such as *Irvingia, Suriana,* and *Kirkia,* are sometimes classified in their own families.

The family has no known fossil record.

## Family Meliaceae (Mahogany)

Figure 9.136 Meliaceae. *Melia:* (a) leafy branch with flowers and fruit; (b) flower; (c) cross section through ovary; (d) longitudinal section through flower, showing stamen tube.

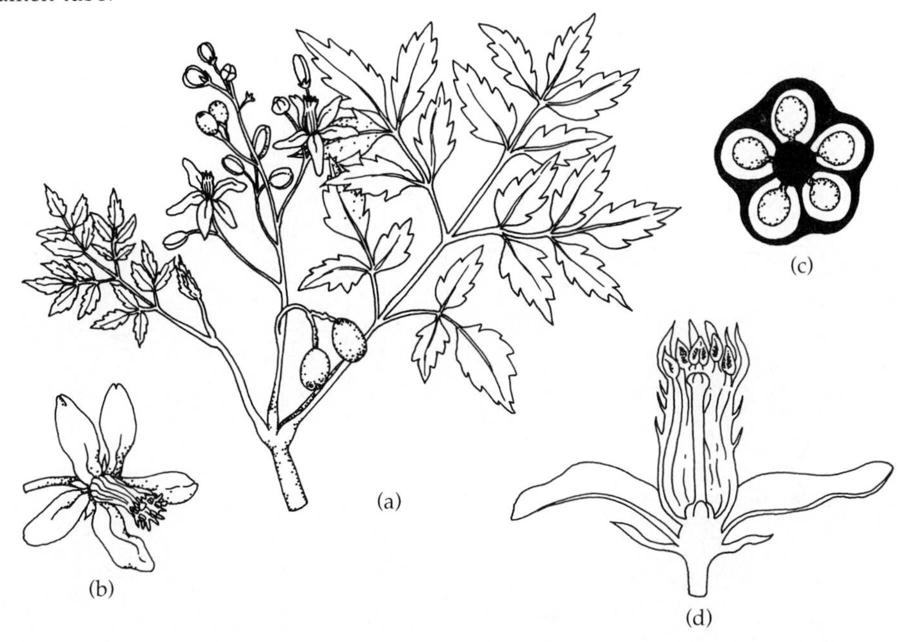


Shrubs or trees (rarely herbs). Leaves alternate (rarely opposite), usually pinnately compound (rarely simple), resin cells within the leaf tissues; stipules lacking. Flowers usually small, regular, perfect or unisexual, hypogynous; inflorescence of various axillary or terminal types or borne along the trunk or branches. Sepals 3–5 (rarely 2 or 7), distinct or connate at the base. Petals 3–5 (rarely more), distinct or connate at the base, alternate the sepals. Stamens 3–10 (rarely more), filaments distinct or connate into a filament tube, sometimes adnate to the petals, often with small appendages or teeth; anthers opening by longitudinal slits; nectary disk present. Pistil compound of 2–5 (rarely 1 or up to 20) united carpels; locules usually 2–5 (rarely 1 or up to 20); ovules 1–2 per locule and attached to axile placentas; ovary superior; style 1, stigma capitate. Fruit a berry, capsule, drupe, or nut. Seeds winged or with an aril, embryo curved or straight, endosperm present or absent (Fig. 9.136).

The family Meliaceae consists of 50–51 genera and 550 species distributed pantropically, with most species ecologically as rain forest understory trees. The largest genera are *Aglaia* (100 species), *Trichilia* (65 species), and *Dysozylum* (60 species).

Economically the family is highly prized for its fine true mahogany woods. These include *Swietenia,* in particular *S. mahogani* of the West Indies; *Entandrophragma, Khaya,* and *Lovoa* of Africa; *Cedrela,* in particular *C. odorata,* and *Toona* of Australasia. Some species are important for oils in making soap, for food, and as ornamentals. Species of *Azadirachta* and *Melia* are used to produce insecticides.

The family stands out from the other families of the order Sapindales because of the distinctive stamens.

Microfossils attributed to the family have been found in Oligocene and more recent deposits.

### Family Rutaceae (Rue)

Figure 9.137 Rutaceae. *Citrus:* (a) leafy branch with fruit; (b) longitudinal section through flower; (c) flower. *Ruta:* (d) flower; (e) cross section through ovary; (f) leafy branched stem with flowers.

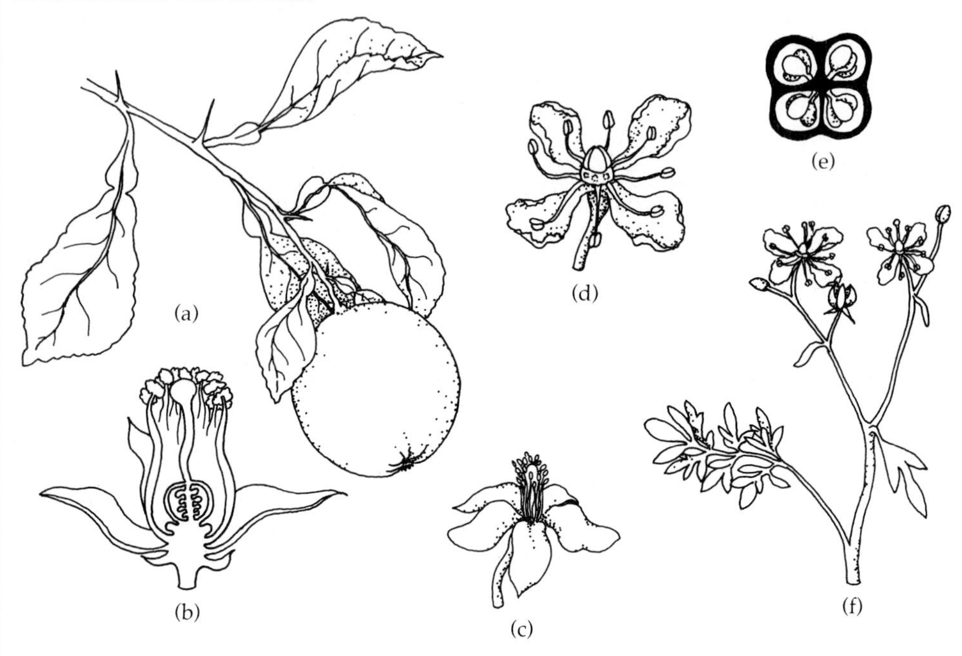


Shrubs or trees (rarely herbs). Leaves alternate (rarely opposite), simple or pinnately compound; usually with resin or oil glands or dots on the leaves, commonly giving off a strong aroma; stipules lacking. Flowers greenish yellow in color, regular (rarely irregular), perfect (rarely unisexual), hypogynous or perigynous; inflorescence of solitary flowers or flowers borne in cymes or racemes. Sepals 4–5, distinct or connate. Petals 4–5 (rarely lacking), alternate the sepals, distinct or connate at the base. Stamens 4–10 (rarely many), filaments distinct or connate toward the base; anthers opening by longitudinal slits and gland-tipped; nectary disk present. Pistil compound of 2–5 (rarely 1 or 6–many) united carpels; locules 2–5 (rarely 1 or 6–many); ovules 1–several per locule and attached to axile or parietal placentas; ovary superior and lobed; style 1, slender, stigma small. Fruit a berry, drupe, hesperidium, or schizocarp. Seeds with embryo curved or straight, endosperm present or absent (Fig. 9.137).

The family Rutaceae consists of 150 genera and 900–1500 species distributed in warm temperate and tropical regions, with the greatest species diversity in Australia and South Africa. The largest genus is *Zanthoxylum* (over 200 species).

Economically the family is very important, with the genus *Citrus* (60 species) the most important, for its fruits. Species cultivated include *C. aurantium,* sweet orange; *C. aurantifolia,* lime; *C. limon,* lemon; *C. media,* sour or Seville orange; *C. paradisi,* grapefruit; *C. reticulata,* mandarins and tangerines; and *C. sinensis,* sweet orange. Other species are used as ornamentals, such as *Ruta graveolens,* the ruta.

The classification of the Rutaceae in the order Sapindales is agreed upon by most botanists. The one character that distinguishes the Rutaceae from the other families of the order is the aromatic oil glands in the leaves.

The family has no known fossil record.

## Family Zygophyllaceae (Creosote Bush)

Figure 9.138 Zygophyllaceae. *Tribulus:* (a) leafy stem segment with flowers and spiny fruit; (b) spiny fruit; (c) longitudinal section through flower; (d) cross section through ovary. *Larrea:* (e) branch with flowers and fruits.

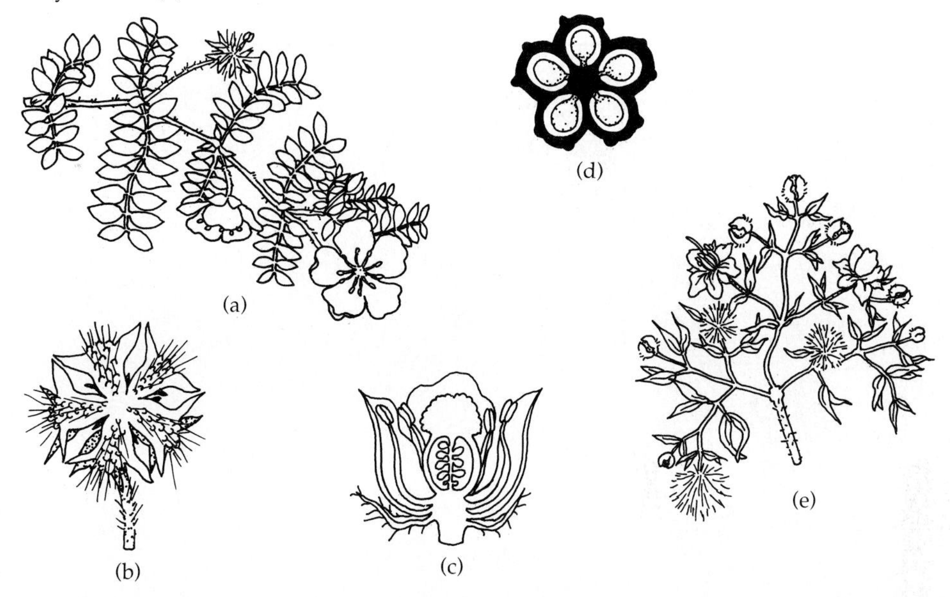


Herbs, shrubs, or trees, sometimes branches with prominent node joints. Leaves fleshy or leathery, opposite (rarely alternate), simple to even pinnately compound; stipules present, commonly spiny. Flowers regular (rarely irregular), perfect (rarely unisexual), hypogynous; inflorescence of solitary flowers or flowers borne in cymes or racemes. Sepals 4–5, distinct or connate. Petals 4–5 (rarely lacking), usually distinct. Stamens 10 (rarely a different number) in 2 or 3 (rarely only 1) whorls of 5, being opposite the petals, filaments distinct, each with a gland or appendage at the base; anthers opening by longitudinal slits; nectary disk present. Pistil compound of 5 (rarely 2–6) united carpels; locules usually 5 (rarely 2–6); ovules 1–many per locule and borne on axile placentas; ovary superior; styles 1, short, stigma capitate or lobed. Fruit a berry, drupe, capsule, or schizocarp. Seeds 1–many, embryo curved or straight, endosperm present or absent (Fig. 9.138).

The family Zygophyllaceae consists of 25–30 genera and 240–250 species distributed mostly in subtropical and tropical arid regions, often in saline habitats. The largest genus is *Zygophyllum,* with 80–90 species occurring in North Africa to central Asia.

Economically the family itself is of minor importance. A well-known and troublesome introduced European weed in dry waste areas in the southern United States is *Tribulus terrestria,* the puncture vine or goat head; its spiny fruit plays havoc with bicycle tires. The wood of *Guaiacum officinale* and *G. sanctum* of tropical Central America and the West Indies is a highly prized timber. Useful timber and oils for perfume come from *Bulnesia arborea,* the Maracaibo lignum vitae, and *B. sarmienti,* the Paraguay lignum vitae. Seeds of *Peganum harmala* are the source of the dye turkey red.

The family includes some genera, such as *Balanites* and *Nitraria,* that have been classified by botanists into their own families. How the group should be classified remains to be determined.

The family has a questionable fossil record.

## ORDER GERANIALES
### Family Oxalidaceae (Wood Sorrel)

Figure 9.139 Oxalidaceae. *Oxalis:* (a) leafy stem with flower and fruits; (b) fruit; (c) cross section through ovary; (d) seed; (e) flower.

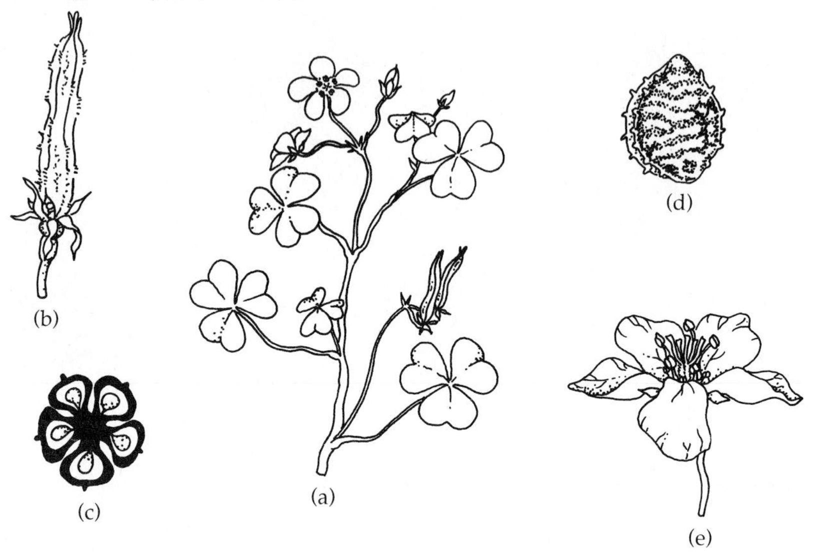


Herbs (rarely shrubs or trees). Leaves alternate or all basal, pinnately or palmately compound or trifoliate, often the leaflets drooping or folding in cool weather or at night; stipules very small or lacking. Flowers regular, perfect (rarely unisexual), hypogynous; inflorescence a solitary flower or flowers borne in cymes or umbel on the end of a long peduncle, with occasionally cleistogamous apetalous flowers. Sepals 5, distinct. Petals 5, distinct or connate at the base. Stamens usually 20 (rarely 15) and arranged in 2 whorls, the outer whorl shorter than the inner and opposite the petals; filaments distinct or connate at the base; anthers opening by longitudinal slits; nectar glands present. Pistil compound of 5 (rarely 3) united carpels, locules 5 (rarely 3), ovules 1–many in each locule and attached on axile placentas; styles 3–5, sometimes showing heterostyly; stigmas capitate or lobed. Fruit a loculicidal capsule (rarely a berry). Seeds 1–many, often with an aril around the base, which helps in dispersing the seed explosively from the fruit, embryo large and straight, endosperm present (Fig. 9.139).

The family Oxalidaceae consists of 7–8 genera and 900 species distributed widely in the subtropical and tropical regions, with some taxa temperate. The largest genera are *Oxalis*, the wood sorrel (about 800 species), and *Biophytum* (70 species).

Economically the family is of minor importance. Some species of *Oxalis* are cultivated as house plants or grown as rock garden species; others have become bothersome weeds in temperate climates. The leaves of *O. acetosella* are eaten as a salad green. The tubers of *O. deppei*, native of Mexico, are cultivated for food in parts of Europe; similarly the leaves and tubers of *O. crenata* are eaten in the Andes, especially in Peru. *Averrhoa carambola*, the carambola or star fruit, is cultivated widely in tropical countries for its edible fruit.

The genera *Hypseocharis*, and *Lipidobotrys* are sometimes separated into their own families, especially the woody genera *Averrhoa* and *Sarcotheca*.

The family has no known fossil record.

## Family Geraniaceae (Geranium)

Figure 9.140 Geraniaceae. *Geranium:* (a) plant showing flowers and fruits; (b) cross section through ovary; (c) fruit; (d) seed; (e) pistil; (f) longitudinal section through flower.

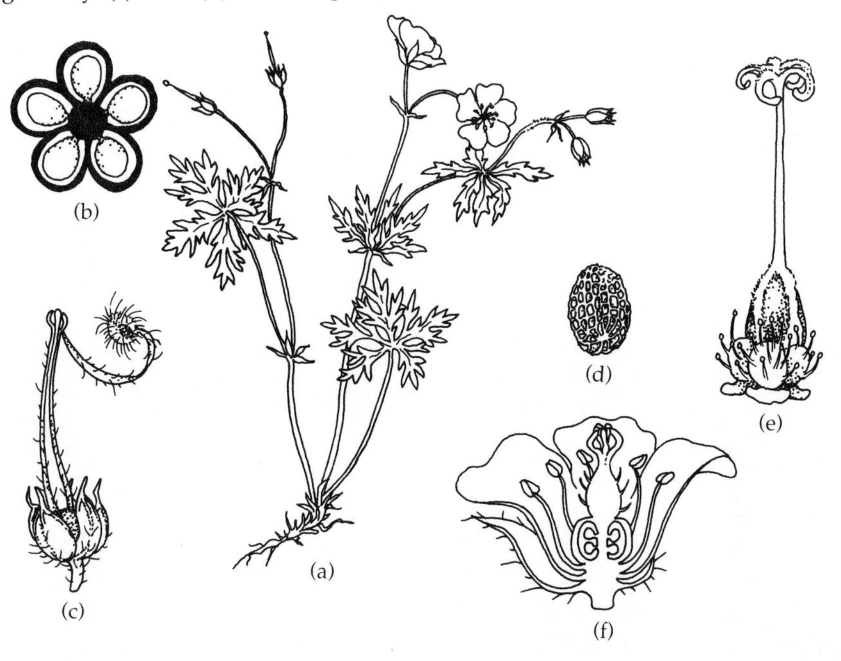


Herbs or shrubs, commonly with aromatic glandular hairs and stems with swollen jointed nodes. Leaves alternate or opposite, usually simple and either pinnate or palmately lobed, or pinnately to palmately dissected or compound; stipules usually present. Flowers regular or slightly irregular, perfect (rarely unisexual), hypogynous; inflorescence of cyme-like umbels (rarely solitary). Sepals 5 (rarely 4), distinct or connate at the base. Petals 5 (rarely 4 or 8 or lacking), distinct, usually brightly colored. Stamens mostly 10 (rarely 5 or 15) in successive whorls of 5 stamens, filaments connate at the base; anthers opening by longitudinal slits; nectary glands often at the base of the stamens. Pistil compound of usually 5 (rarely 3 or 8) united carpels; locules 5 (rarely 3 or 8); ovules 1–2 per locule and attached on axile placentas; ovary superior and 5 lobed; style 5 (rarely 3–8), connate, commonly elongating at maturity, stigmas lobed or separate. Fruit a schizocarp (rarely a capsule) splitting elastically from the base of the ovary toward the style apex. Seeds 1 per mericarp, embryo curved or straight, endosperm present or absent (Fig. 9.140).

The family Geraniaceae consists of 11 genera and 700–750 species distributed widely in subtropical and temperate areas in both hemispheres, even in the arctic and antarctic. The largest genera are *Geranium*, the wild geranium or cranesbill (300 species); *Pelargonium*, cultivated geraniums (250 species); and *Erodium*, the storksbill (75 species).

Economically the family is important mostly for ornamentals. The South African genus *Pelargonium* contains many horticultural hybrids. Other species, such as *P. capitatum*, *P. graveolens*, *P. odoratissimum*, and *P. radiata*, are grown for various geranium oils used in cosmetics and perfume.

Some of the smaller genera of the family, such as *Biebersteinia*, *Dirachma*, or *Viviania*, have been grouped in their own families by botanists.

The family has no known fossil record.

## Family Balsaminaceae (Touch-me-not)

Figure 9.141 Balsaminaceae. *Impatiens:* (a) flowering stem segment; (b) fruit; (c) cross section through ovary; (d) flower with curved spur.

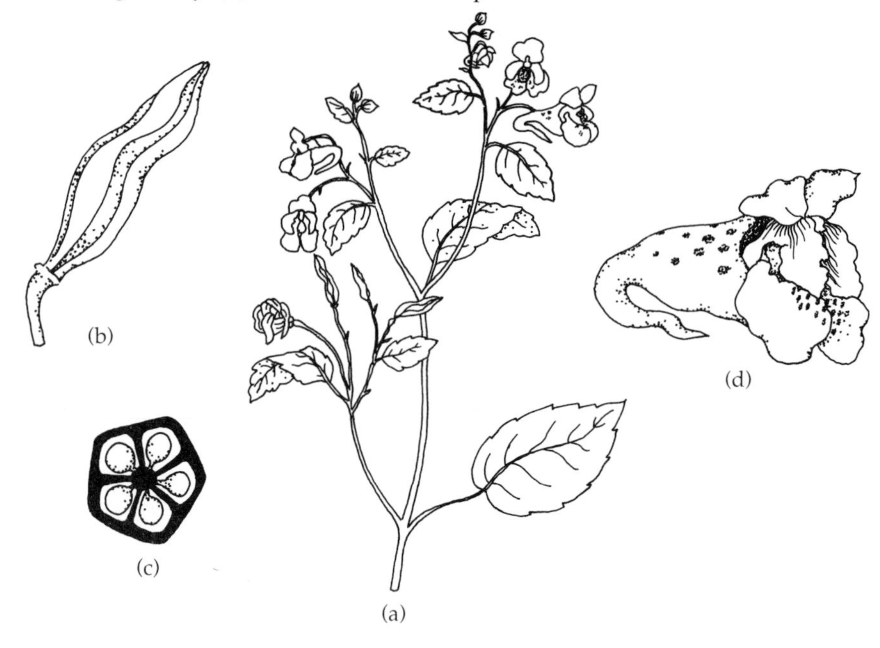


Herbs (rarely shrubs) with fleshy, translucent stems. Leaves alternate, opposite, or whorled, simple with pinnate venation; stipules as paired glands or lacking. Flowers irregular, perfect, hypogynous; inflorescence a solitary flower or flowers borne in axillary cymes, panicles, or racemes, cleistogamous nonopening, self-pollinating flowers sometimes produced. Sepals 3–5, distinct, the lower are commonly petal-like with a nectar spur or pouch. Petals 5, distinct or the lateral ones connate as 2 unequal pairs. Stamens 5, filaments short, flattened, connate toward the top, anthers somewhat fused forming a lid (calyptera) over the ovary; anthers opening by longitudinal slits; nectary disk lacking. Pistil compound of 5 united carpels; locules 5; ovules 1–many and attached on axile placentas; ovary superior, style 1, short stigmas 1–5. Fruit an explosive capsule (rarely a berry-like drupe). Seeds 1–many, embryo straight, endosperm lacking (Fig. 9.141).

The family Balsaminaceae consists of 2–4 genera and 450–600 species distributed mainly in subtropical to tropical Africa and Asia, and in temperate regions of the New and Old World. The largest genus, *Impatiens,* the touch-me-nots or balsams (450–600 species), has its greatest diversity in south and southeast Asia into Malaysia.

Economically the family is important only as ornamentals grown in the garden or greenhouse or as pot herbs. These include *Impatiens balsamina, I. parviflora,* and hybrids between *I. holstii* and *I. sultanii.*

The presence of a spur or nectar pouch has led some botanists to associate the family with other families having similar structures, but the spur originates from the calyx, while in other families the spur originates from receptacle tissue.

The family has no known fossil record.

## ORDER APIALES
### Family Araliaceae (Ginseng)

Figure 9.142 Araliaceae. *Aralia:* (a) plant in flower; (b) side view of flower; (c) front view of flower; (d) pistil; (e) cross section through ovary.

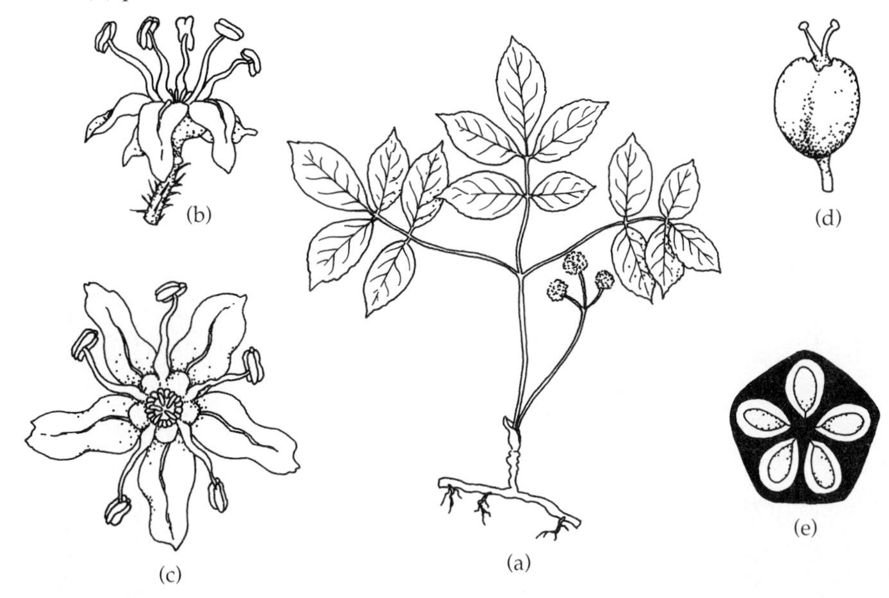


Herbs, shrubs, trees, or lianas commonly with prickly or stellate hairs. Leaves often large, alternate (rarely opposite or whorled), simple, and lobed or more commonly pinnately or palmately compound; petiole base often sheathing; stipules small. Flowers usually small, regular, perfect or unisexual and dioecious, epigynous (rarely hypogynous); inflorescence in umbels, less commonly in heads, racemes, or spikes. Sepals 5 (rarely lacking), distinct but reduced to small teeth. Petals 5 (rarely 3–12), distinct (rarely connate). Stamens 5 (rarely 3–12), usually same number as the petals and alternate them, filaments distinct, attached to a nectary disk, which lies on top of the ovary; anthers opening by longitudinal slits. Pistil compound of 2–15 (rarely 1) united carpels; locules 2–15 (rarely 1); ovules 1 per locule and borne on apical-axile placentas; ovary inferior (rarely superior); styles 2–5, distinct or connate and surrounded by a stylopodium. Fruit a berry or drupe (rarely a schizocarp with a carpophore). Seeds with a small embryo, endosperm present (Fig. 9.142).

The family Araliaceae consists of 55–70 genera and 700 species with a cosmopolitan distribution, with more species in subtropical and tropical regions. The largest genera are *Schefflera* (150 species) and *Oreopanex* (100 species).

Economically the family is important for ornamental and medicinal uses. The root of *Panax quinquefolia*, ginseng, is highly valued by the Chinese and others as a stimulant, tonic, and supposed aphrodisiac. Medicinal products are produced also from *Aralia cordata* and *A. racemosa*. Many cultivars of *Hedera helix*, ivy, are grown as house plants or groundcovers. Various *Schefflera* are cultivated as tropical ornamentals or temperate house plants.

The family Araliaceae shares many characters with some Apiaceae and in the past included the Apiaceae. Most botanists today, however, consider the two families separate.

Fossils attributed to the Araliaceae were found in Upper Cretaceous deposits and throughout other Tertiary deposits.

## Family Apiaceae (Umbelliferae) (Carrot)

Figure 9.143 Apiaceae (Umbelliferae). *Lomatium:* (a) leafy plant with compound umbel inflorescence; (b) winged fruit; (c) spiny fruit; (d) flower; (e) split schizocarp of two mericarps on carpophore; (f) pistil. *Eryngium:* (g) stem apex with head inflorescence and involucral bracts.

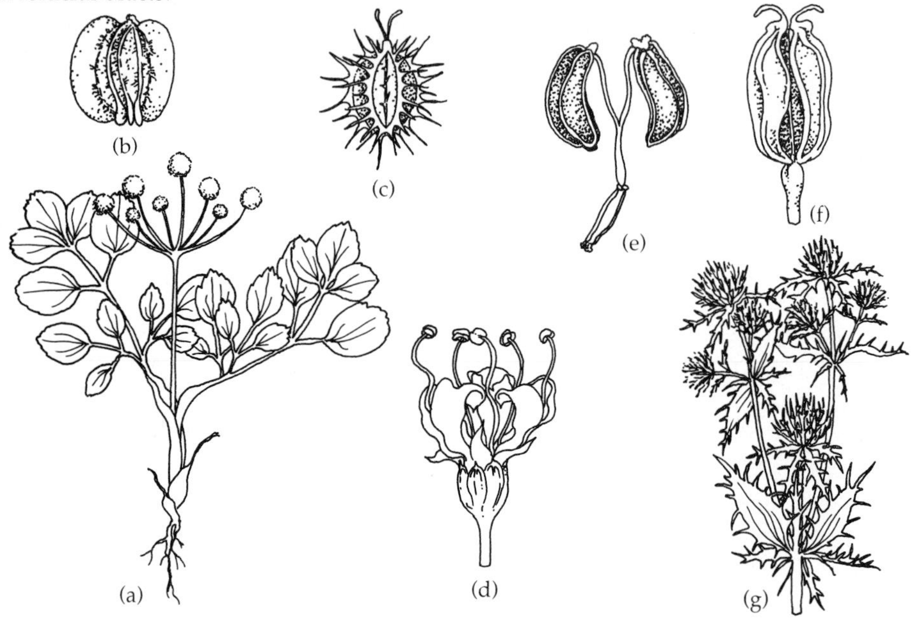


Herbs (rarely woody), with hollow internodes, commonly aromatic and poisonous. Leaves alternate (rarely opposite) or basal, simple or more commonly pinnately or palmately lobed, compound or dissected, petioles with a sheathing base; stipules lacking. Flowers small, regular (rarely irregular), perfect (rarely unisexual), epigynous; inflorescence usually a compound umbel, occasionally in heads or simple umbels, often subtended by an involucre of bracts. Sepals 5, distinct, small, or absent. Petals 5 (rarely lacking), distinct. Stamens 5, filaments distinct, attached to the epigynous nectary disk; anthers opening by longitudinal slits. Pistil compound of 2 united carpels; locules 2; ovules 1 per locule and borne on apical-axile placentas; ovary inferior; styles 2, often subtended by an enlarged stylopodium. Fruit a schizocarp comprised of 2 mericarps, attached to one another by a common stalk **(carpophore);** mostly ribbed, winged, or covered with bumps or prickles. Seeds with a small embryo, endosperm present (Fig. 9.143).

The family Apiaceae consists of 300 genera and 2500–3000 species with a cosmopolitan distribution, mostly in the north temperate regions and in tropical highlands.

Many species are grown for food and spices. *Daucus carota,* the carrot, and *Pastinaca sativa,* parsnip, are root crops. *Anthriscus cerefolium,* chervil; *Anethum graveolens,* dill; *Apium graveolens,* celery; *Carum carvi,* caraway; *Petroselinum crispum,* parsley; and *Pimpinella anisum,* anise; are used as flavorings, spices, or vegetables. Some poisonous species are *Aethusa, Cicuta* (*C. maculata,* said to be most poisonous of all north temperate plants), *Conium (C. maculatum,* poison hemlock, said to have killed Socrates), and *Oenanthe.*

The family has been classified in the past as part of the family Araliaceae. However, the family is considered by most botanists as a unit with 3 subfamilies; Apioideae, Hydrocotyloideae, and Saniculoideae.

Microfossils attributed to the Apiaceae were found first in Eocene deposits.

**Subclass VI. Asteridae** The subclass Asteridae is very large, with approximately 60,000 species scattered among 49 families and 11 orders. (See Table 9.6.) The family Asteraceae (Compositae) comprises about one third of the species, making it the largest family of the Magnoliopsida (dicotyledons).

**TABLE 9.6 A list of orders and family names of flowering plants found in the subclass Asteridae.**

|  |  |
|---|---|
| Subclass VI. Asteridae | |
| Order A. Gentianales | Family 1. Loganiaceae* |
| | 2. Retziaceae |
| | 3. Gentianaceae* |
| | 4. Saccifoliaceae |
| | 5. Apocynaceae* |
| | 6. Asclepiadaceae* |
| Order B. Solanales | Family 1. Duckeodendraceae |
| | 2. Nolanaceae |
| | 3. Solanaceae* |
| | 4. Convolvulaceae* |
| | 5. Cuscutaceae* |
| | 6. Menyanthaceae |
| | 7. Polemoniaceae* |
| | 8. Hydrophyllaceae* |
| Order C. Lamiales | Family 1. Lennoaceae |
| | 2. Boraginaceae* |
| | 3. Verbenaceae* |
| | 4. Lamiaceae (Labiatae)* |
| Order D. Callitrichales | Family 1. Hippuridaceae |
| | 2. Callitrichaceae* |
| | 3. Hydrostachyaceae |
| Order E. Plantaginales | Family 1. Plantaginaceae* |
| Order F. Scrophulariales | Family 1. Buddlejaceae* |
| | 2. Oleaceae* |
| | 3. Scrophulariaceae* |
| | 4. Globulariaceae* |
| | 5. Myoporaceae |
| | 6. Orobanchaceae* |
| | 7. Gesneriaceae* |
| | 8. Acanthaceae* |
| | 9. Pedaliaceae |
| | 10. Bignoniaceae* |
| | 11. Mendonciaceae |
| | 12. Lentibulariaceae* |
| Order G. Campanulales | Family 1. Pentaphragmataceae |
| | 2. Sphenocleaceae |
| | 3. Campanulaceae* |
| | 4. Stylidiaceae* |
| | 5. Donatiaceae |
| | 6. Brunoniaceae |
| | 7. Goodeniaceae* |
| Order H. Rubiales | Family 1. Rubiaceae* |
| | 2. Theligonaceae |
| Order I. Dipsacales | Family 1. Caprifoliaceae* |
| | 2. Adoxaceae |
| | 3. Valerianaceae* |
| | 4. Dipsacaceae* |
| Order J. Calycerales | Family 1. Calyceraceae |
| Order K. Asterales | Family 1. Asteraceae (Compositae)* |

*NOTE:* Family names followed by an asterisk are discussed and illustrated in the text.

The Asteridae include more herbs than woody plants, even though all habit types are found. Flowers usually are 4- or 5-merous with sympetalous corollas. The inflorescences are of various types, especially the head type in the Asteraceae (Compositae), in which many small flowers are grouped close together and appear and function as a unit. The stamens are usually the same number as the corolla lobes (or fewer), alternate with them, and having the filaments individually adnate to the corolla. More than any other subclass, the Asteridae has the most varied kinds of pollinators, and some of the most unique types of pollination. Pistil compound and most commonly comprised of 2 (sometimes 3–5) united carpels. The position of the ovary is both superior and inferior, and lacking a separate hypanthium. The fruits commonly are achenes, berries, or capsules. The group as a unit often produces chemicals called iridoid compounds but lack the betalains, benzyl-isoquinoline alkaloids, ellagic acid, and tannins found in other subclasses.

Fossils attributed to the Asteridae were first found in Tertiary deposits, with fossil pollen from the Paleocene placed in the Apocynaceae. Most fossils attributed to the Asteridae, however, are not observed until Oligocene strata.

## ORDER GENTIANALES
### Family Loganiaceae

Figure 9.144 Loganiaceae. *Fagraea:* (a) leafy branch bearing fruit; (b) cross section through ovary; (c) flower. *Gelsemium:* (d) stamens adnate to opened corolla; (e) pistil.

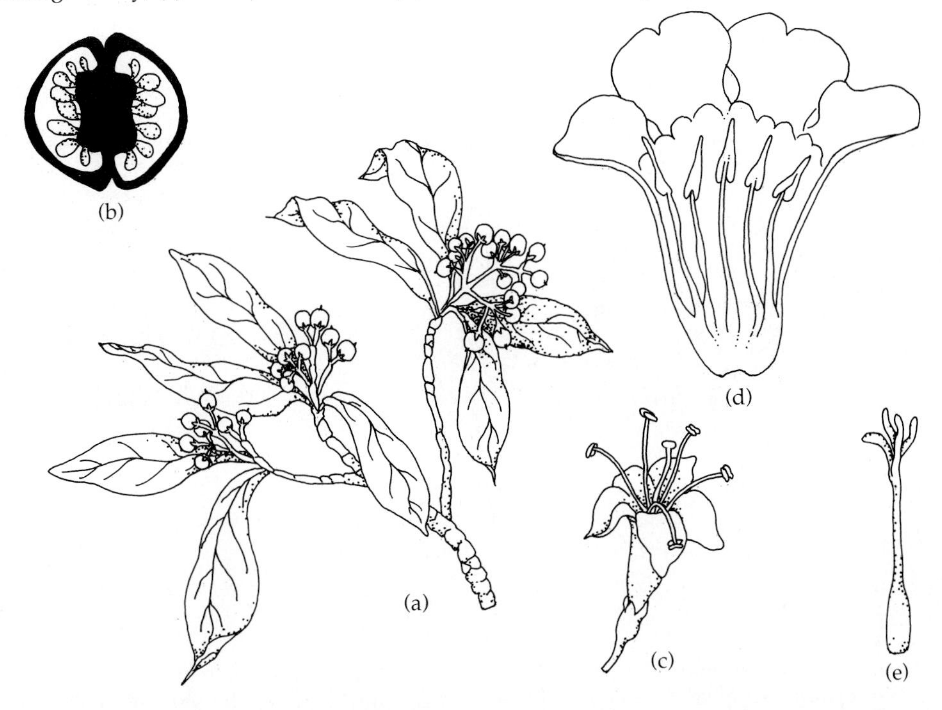


Herbs, shrubs, trees, or lianas, often with bitter-tasting compounds. Leaves opposite (rarely alternate or whorled), simple, entire (rarely toothed); stipules present. Flowers generally showy, regular, perfect, hypogynous (rarely epigynous); inflorescence a solitary flower or a terminal cyme. Sepals 4–5, connate (rarely only 2 lobed) and overlapping. Petals 4–5 (rarely 8–16), sympetalous. Stamens 4–5 (rarely 1–16), usually as many as and alternate with the corolla lobes; filaments distinct, adnate to the corolla tube; anthers opening by longitudinal slits; nectary disk usually lacking. Pistil compound of 2–3 (rarely 5) united carpels; locules 2–3 (rarely 5); ovules usually many (rarely few) and borne on axile placentas; ovary superior (rarely half inferior); style 1, sometimes lobed. Fruit a capsule (rarely a berry or drupe). Seeds commonly winged, embryo straight, endosperm present (Fig. 9.144).

The family Loganiaceae consists of 20–30 genera and 500–600 species distributed pantropically, with a few species in temperate regions. The family is widespread but normally not in great abundance in any area. The largest genus is *Strychnos* (150–200 species).

Economically the family is of moderate importance, especially for poisons, ornamentals, and timber. The well-known poison strychnine is extracted from *Strychnos nux-vomica,* while curare alkaloids are found in *S. toxifera.* Species of *Fagraea* are harvested for timber in Asia (i.e., *F. crenulata, F. elliptica,* and *F. fragrans*) or cultivated as ornamentals (i.e., *F. auriculata* and *F. fragrans*).

The family is a loose-knit group, with some genera being placed in their own families. In the past, *Buddleja* was placed in the family Loganiaceae, but today it is classified in its own family Buddlejaceae in the order Scrophulariales.

The family has no known fossil record.

## Family Gentianaceae (Gentian)

Figure 9.145 Gentianaceae. *Gentiana:* (a) stem tip with flowers; (b) cross section through ovary; (c) anther; (d) longitudinal section through corolla.

Herbs (rarely shrubs); sometimes having mycorrhizal fungi associated with the rhizome and bitter-tasting compounds in the tissues. Leaves opposite or whorled (rarely alternate), simple, entire or scale-like; stipules lacking. Flowers regular, perfect (rarely unisexual), hypogynous; inflorescence a solitary flower or cyme. Sepals 4–5 (rarely more), connate. Petals 4–5 (rarely more), sympetalous into a long corolla tube, sometimes having appendages inside the tube. Stamens 4–5, filaments distinct, adnate to the corolla tube and alternate the petals, staminodes rarely present; anthers opening by longitudinal slits or apical pores; a nectary disk or nectar glands present. Pistil compound of 2 united carpels, locule 1 (rarely 2), ovules many and borne on parietal (rarely axile) placentas, ovary superior, style 1, stigma 2-lobed. Fruit a septicidal capsule (rarely a berry). Seeds small, embryo straight, endosperm present (Fig. 9.145).

The family Gentianaceae consists of 75–80 genera and 900–1000 species with a cosmopolitan distribution, with many species in arctic and mountainous regions. The largest genera are *Gentiana*, the gentians (400 species), *Gentianella* (125 species), *Sebaea* (100 species), and *Swertia* (100 species).

Economically the family is of some importance as ornamentals or for medicinal purposes. Many species of *Gentiana* and *Sabatia*, rose pinks, are cultivated as garden ornamentals and rock garden plants. In South Africa, a remedy for gout, called portland powder, is made from a mixture of *Gentiana* rhizome, *Centaurium*, centaury, and three other plants. A popular before-dinner alcoholic drink, Suze, is made in France from *Gentiana*. In South Africa a mixture of fried *Chironia baccifera* and butter is used as a cathartic, applied to sores, and said to be good for improving the complexion. Seeds of *Blackstonia perforiata* yield a yellow dye.

The limits of the genera in the family are debated by botanists and are subject to interpretation. The family Gentianaceae has many characters in common with the Loganiaceae.

The family has no known fossil record.

## Family Apocynaceae (Dogbone)

Figure 9.146 Apocynaceae. *Apocynum:* (a) leafy flowering branch; (b) fruit. *Nerium:* (c) apex of style with specialized structure below stigmas; (d) longitudinal section through flower; (e) cross section through ovary.

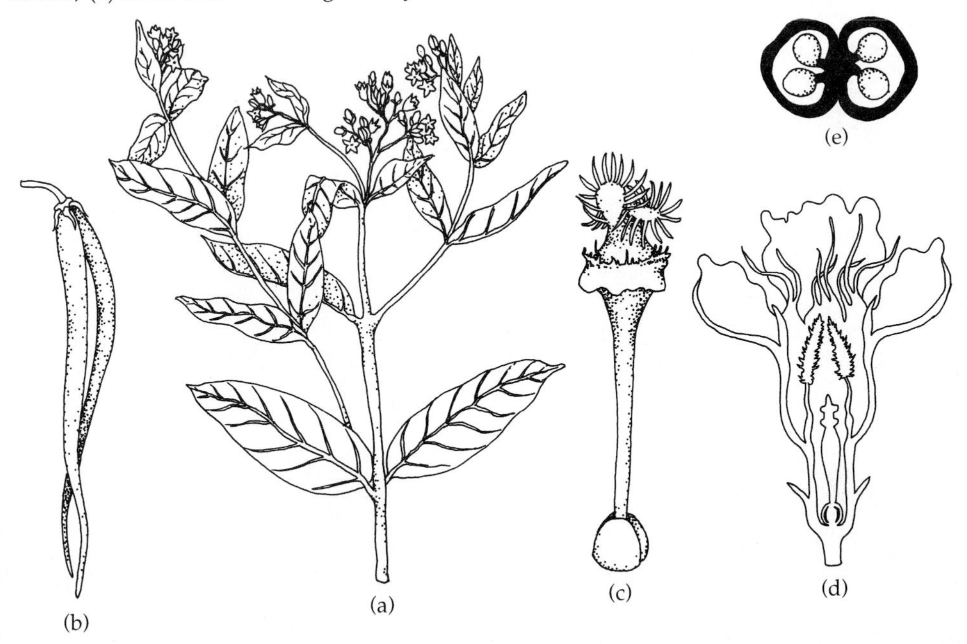


Herbs, shrubs, trees, or lianas, commonly with milky juice. Leaves evergreen (less commonly deciduous), opposite or whorled, simple; stipules lacking (rarely present). Flowers often large and showy, regular, perfect, hypogynous or epigynous; usually fragrant. Sepals 5, connate, sometimes with appendages inside. Petals 5, sympetalous, often with scales inside the corolla tube. Stamens 5, filaments distinct, as many as and alternate with the corolla lobes, adnate to the corolla tube; anthers clustered around the style, opening by longitudinal slits and with appendages; nectar glands 5 or a nectary disk present. Pistil compound of 2 united carpels; locules 1–2; ovules 2–many and borne on axile or parietal placentas; ovary superior or part-inferior; styles 1–2, distinct or connate at the top, stigmas various. Fruit a berry, drupe, 2 follicles, or schizocarp. Seeds 2–many, often with hairs at one end **(comose),** embryo straight, endosperm present or lacking (Fig. 9.146).

The family Apocynaceae consists of 180–200 genera and 1500–2000 species distributed pantropically, but with a few temperate herbs. The largest genera are *Tabermaemontana* (140 species), *Mandevilla* (115 species), *Rauwolfia* (100 species), and *Parsonia* (100 species).

Economically the family is important for drugs, medicines, and ornamentals. Heart stimulants are obtained from species of *Apocynum, Cerbera, Strophanthus,* etc. The medical alkaloid reserpine, used as a tranquilizer and to lower blood pressure, is obtained from *Rauwolfia.* The milky juice (latex) can be used as a source of commercial rubber. *Allamanda; Nerium,* oleander; *Plumeria,* frangipani; and *Vinca,* the periwinkles are common ornamentals.

The Apocynaceae is divided into two subfamilies, Apocynoideae and Plumerioideae, which some botanists prefer to treat as families.

The family has no known fossil record.

## Family Asclepiadaceae (Milkweed)

Figure 9.147 Asclepiadaceae. *Asclepias:* (a) leafy stem having bud and flower inflorescences; (b) single corona and horn; (c) cross section through ovary; (d) longitudinal section through flower; (e) flower; (f) two pollinia attached to gland by translator arms; (g) seed; (h) fruit (follicle).

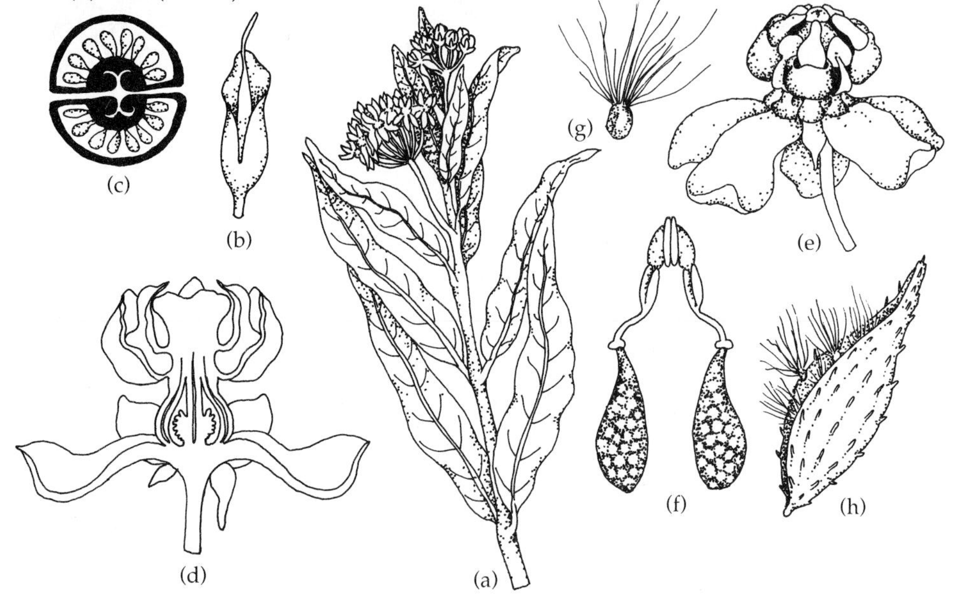


Herbs, shrubs, trees, or lianas, sometimes succulent, usually with milky juice. Leaves opposite or whorled (rarely alternate), simple and entire (rarely lobed or toothed); stipules lacking or very small. Flowers somewhat unusual, regular, perfect, hypogynous or epigynous; inflorescence a solitary flower or flowers borne in cymes, racemes, or umbels. Sepals 5, connate toward the base, sometimes reflexed. Petals 5, sympetalous, lobes spreading or reflexed and commonly with appendages. Stamens 5, filaments connate (rarely distinct), adnate to the stigma, forming a structure called a **gynostegium**, and with hood-like **corona** and/or **horn** appendages; the corona alternates with the petals and is adnate to the corolla; anthers have 2 locules, each filled with a waxy mass of pollen (a **pollinium**); each pollinium is attached to the adjacent anther pollinium by a **translator arm**, and each arm is joined to the other pollinium translator by a **gland**, thus allowing a visiting insect to remove the entire pollinia by the translator; nectary disk lacking. Pistil almost distinct of 2 carpels, free at the base but connate at the stigma; locule 1, ovules many and borne on parietal placentas, ovary superior or partly inferior, style 2 fused as 1 at the stigma. Fruit of 2 follicles (usually 1 matures). Seeds flat with a coma of hairs at one end, embryo straight, endosperm usually present (Fig. 9.147).

The family Asclepiadaceae consists of 250 genera and 1800–2000 species distributed mostly in warm regions, with a few species in temperate habitats. The largest genera are *Asclepias, Ceropegia,* and *Hoya* (150 species each).

Economically the family is used mostly as ornamentals, especially *Asclepias,* milkweeds; *Hoya,* wax plants; and many succulent groups, such as *Ceropegia* and *Stapelia.* Commercial rubber may be obtained from some *Cryptostegia* species. Some species are poisonous and bitter tasting.

The stamen arrangement makes the family distinctive in the order, but the milky juice is similar to the Apocynaceae.

The family has no known fossil record.

## ORDER SOLANALES
## Family Solanaceae (Nightshade, Potato)

Figure 9.148 Solanaceae. *Datura:* (a) leafy stem having flowers and fruit. *Lycopersicon:* (b) flower; (c) longitudinal section through flower; (d) fruits; (e) cross section through ovary.

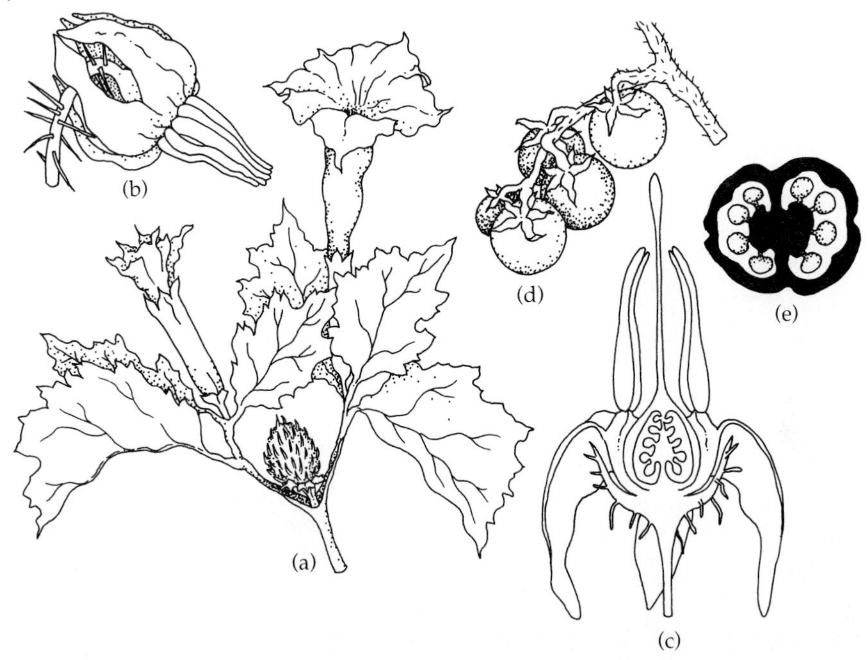


Herbs, shrubs, small trees, or lianas; many poisonous. Leaves alternate, entire to cleft, simple to pinnately compound; stipules lacking. Flowers regular, perfect, hypogynous; inflorescence a single flower or flowers borne in a cyme. Sepals 5 (rarely 3–10), connate, often persistent in fruit. Petals 5 (rarely to 10), sympetalous. Stamens 5 (rarely 2 or 4–7), filaments distinct, adnate to the corolla tube and alternate the lobes; anthers opening by longitudinal slits or apical pores; nectary disk usually present. Pistil compound of 2 united carpels; locules 2 (sometimes more); ovules 1–many per locule and borne on axile placentas; ovary superior (rarely inferior); style 1, stigma 2-lobed. Fruit a berry or capsule (rarely a drupe). Seeds many, embryo straight, endosperm present (Fig. 9.148).

The family Solanaceae consists of 85–90 genera and 2800–3000 species with a cosmopolitan distribution; many species in Australia and Central and South America. The largest genera are *Solanum* (1400–1700 species), *Centrum* (150 species), and *Physalis* (100 species).

Many species are grown for food: cultivars of *Capsicum annuum* var *annuum*, peppers; *Lycopersicon esculentum*, tomato; *Physalis peruviana*, Cape gooseberry; and *Solanum tuberosum*, potato. *Nicotiana tubacum*, which contains the toxic alkaloid nicotine, is grown for chewing, smoking, or snuff. It can be used as a strong insecticide. Some species of *Browallia*, *Nicotiana*, *Petunia*, and *Solanum*, night shade, are cultivated for their attractive flowers. Many species are very poisonous and/or used medicinally: *Atropa belladonna*, deadly nightshade (atropine); *Datura stramonium*, jimsonweed (stramonium); *Mandragora officinarum*, mandrake (hyoscyamine); and *Hyoscyamus niger*, black henbane (a hypnotic drug).

The family has many characters similar to the Scrophulariaceae but differs in its anatomy and in usually having regular flowers.

The family has no known fossil record.

## Family Convolvulaceae (Morning Glory)

Figure 9.149 Convolvulaceae. *Ipomoea:* (a) segment of leafy stem bearing flower; (b) cross section through ovary; (c) pistil and nectary disk; (d) stamens; (e) fruit subtended by calyx.

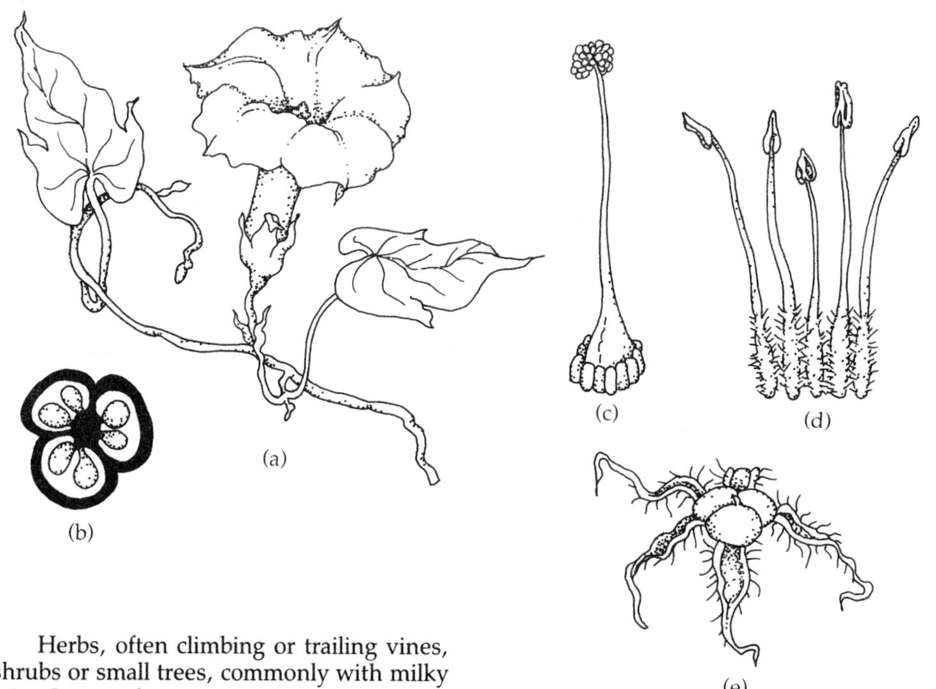


Herbs, often climbing or trailing vines, shrubs or small trees, commonly with milky juice. Leaves alternate, simple, entire to lobed or compound; stipules lacking. Flowers usually large, showy, regular (rarely slightly irregular), perfect (rarely unisexual), hypogynous; inflorescence a solitary flower in the axils or flowers borne in cymes, racemes, or panicles, commonly with a pair of involucre bracts. Sepals 5, distinct or connate at the base. Petals 5, sympetalous, mostly funnelform. Stamens 5, filaments distinct and unequal, adnate to the base of the corolla tube and alternate the corolla lobes; anthers opening by longitudinal slits; nectary disk present. Pistil compound of 2 (rarely 3–5) united carpels; locules 2 (rarely 3–5); ovules usually 2 (rarely 1 or 4) per locule and borne on basal-axile placentas; ovary superior; style 1–2, simple or 2-lobed, stigma capitate or lobed. Fruit a capsule (rarely a berry or a nut). Seeds often hairy, embryo large, curved or straight, endosperm present (Fig. 9.149).

The family Convolvulaceae consists of 50 genera and 1500–1800 species with a cosmopolitan distribution, but with most species in subtropical and tropical climates. The largest genera are *Ipomoea*, morning glory (400 species), and *Convolvulus*, bindweed (250 species).

Economically the family is used for food and as ornamentals. The most important plant is *Ipomoea batatas*, the sweet potato. Its many cultivars are cultivated widely in the tropics and especially in Japan. A drug used as a laxative is obtained from the roots of *Convolvulus scammonia*, scammony, and *Ipomoea purga*, jalap. *Ipomoea purpurea*, morning glory, is a commonly grown ornamental. Various *Convolvulus* species, bindweeds, are noxious weeds in grain fields and waste areas in Canada and the United States.

The genera *Dichondra*, *Falkia*, and *Humbertia* are sometimes placed by botanists in the families Dichondraceae and Humbertiaceae.

The family has no known fossil record.

## Family Cuscutaceae (Dodder)

Figure 9.150 Cuscutaceae. *Cuscuta:* (a) segment of twining flowering stems; (b) front view of flower, showing stamens and staminodes; (c) cross section through ovary; (d) stamen; (e) opened corolla, revealing stamens and staminodes.

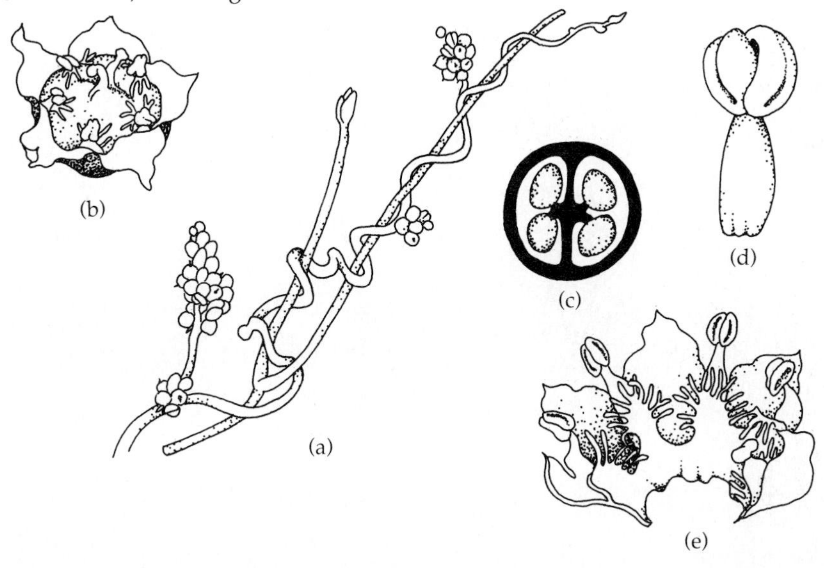


Herbs, twining, parasitic and lacking chlorophyll, usually orange or yellow colored. Leaves small scale-like or lacking. Flowers small, regular, perfect; inflorescence a cyme, head, or spike. Sepals 5 (rarely 3 or 4), distinct or connate at the base, sometimes persistant. Petals 5 (rarely 4), sympetalous; the corolla tube fringed with scale-like appendages, pink or white colored. Stamens 4–5, filaments distinct, adnate to the corolla tube and alternate the corolla lobes; anthers opening by longitudinal slits. Pistil compound of 2 (rarely 3) united carpels; locules 2 (rarely 3) sometimes with unfused septa; ovules 2 per locule and borne on axile placentas; ovary superior; style 1 or 2, single or 2-lobed, stigma various. Fruit a berry, or circumscissile or irregular dehiscent capsule. Seeds 2 per locule, embryo curved or spiral-shaped, sometimes lacking cotyledons, endosperm present (Fig. 9.150).

The family Cuscutaceae consists only of the genus *Cuscuta* and 150 species with a cosmopolitan distribution, with many species in the warmer regions of the New World.

The family is indirectly of some importance. The group is entirely parasitic and includes some species that are moderately host-specific, that is, parasitic on Asteraceae (Compositae) or Chenopodiaceae. This ability can reduce the yield of certain agricultural crops. Costs are incurred to eradicate these plants from the crop. On the positive side, plant pathogists have used the stems and haustoria fastened to the host to transmit and study virus diseases of plants.

The genus *Cuscuta* is occasionally included in the family Convolvulaceae by some botanists.

The family has no known fossil record.

## Family Polemoniaceae (Phlox)

Figure 9.151 Polemoniaceae. *Phlox:* (a) flowering stems; (b) cross section through ovary; (c) longitudinal section through perianth. *Polemonium:* (d) longitudinal section through flower.

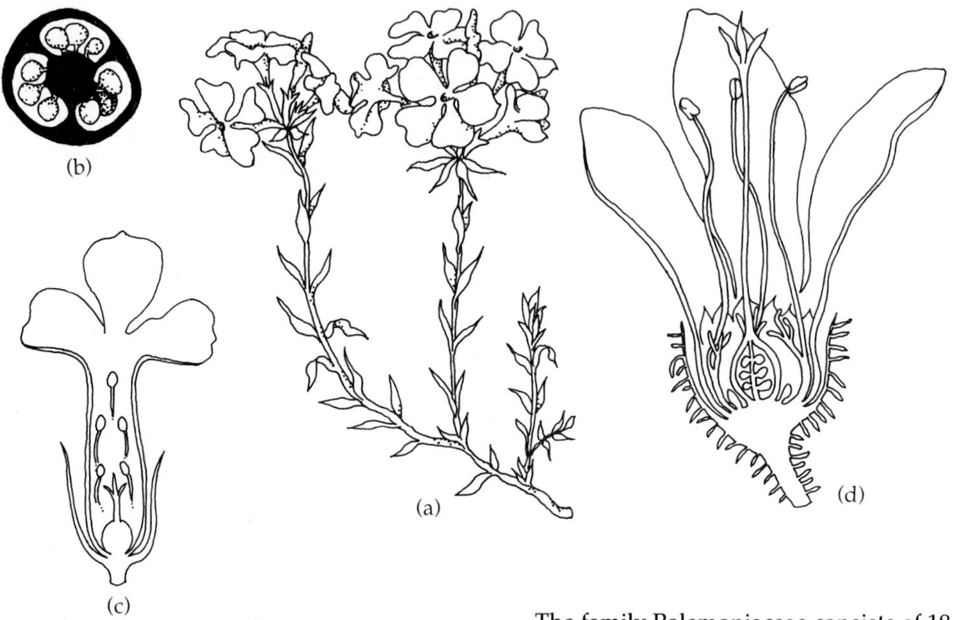


Herbs or shrubs (rarely trees or lianas); sometimes with a skunk smell. Leaves alternate, opposite (rarely whorled), simple to compound; stipules lacking. Flowers often showy, regular (rarely irregular), perfect, hypogynous; inflorescence a solitary flower or flowers in cymes or heads. Sepals 5 (rarely 4 or 6), distinct or connate. Petals 5 (rarely 4 or 6), sympetalous, commonly with a long corolla tube, variously colored. Stamens 5, filaments distinct, adnate to the corolla tube, often at different attachment levels and alternate with the corolla lobes; anthers opening by longitudinal slits; nectary disk present; interesting pollination biology includes various different insects, hummingbirds, and bats. Pistil compound of 3 (rarely 2 or 4) united carpels, locules 3 (rarely 2 or 4), ovules 1–many per locule and borne on axile placentas, ovary superior, style 1 and often long, stigma 3 (rarely 2 or 4) lobed. Fruit a capsule. Seeds 1–many, often sticky when wet, embryo curved or straight, endosperm present (rarely lacking) (Fig. 9.151).

The family Polemoniaceae consists of 18 genera and 300 species distributed extensively in North America, south to South America in the mountains, and in Europe to Asia. Many species are in the arid southwestern United States with over half of the taxa in California. The largest genera are *Gilia* (50 species) and *Phlox* (50 species).

Economically the family is of minor importance, except as ornamentals. Species of *Gilia, Phlox,* and *Polemonium,* Jacob's ladder, are cultivated as garden ornamentals, especially cultivars of *Phlox paniculata,* sweet William, and as rock garden plants.

The tropical genera are mostly woody, having relatively large flowers, and the seeds have wings, an embryo with large cotyledons, and lack endosperm. The temperate genera have flowers that are moderate to small in size, seeds that lack wings, and small cotyledons with endosperm present. The families Polemoniaceae and Hydrophyllaceae have various characters in common.

Microfossils attributed to the Polemoniaceae have been found in Miocene and more recent deposits, while megafossils are recorded from Pleistocene strata.

## Family Hydrophyllaceae (Waterleaf)

Figure 9.152 Hydrophyllaceae. *Hydrophyllum:* (a) leafy plant bearing flowers; (b) pistil; (c) cross section through ovary; (d) flower; (e) seed.

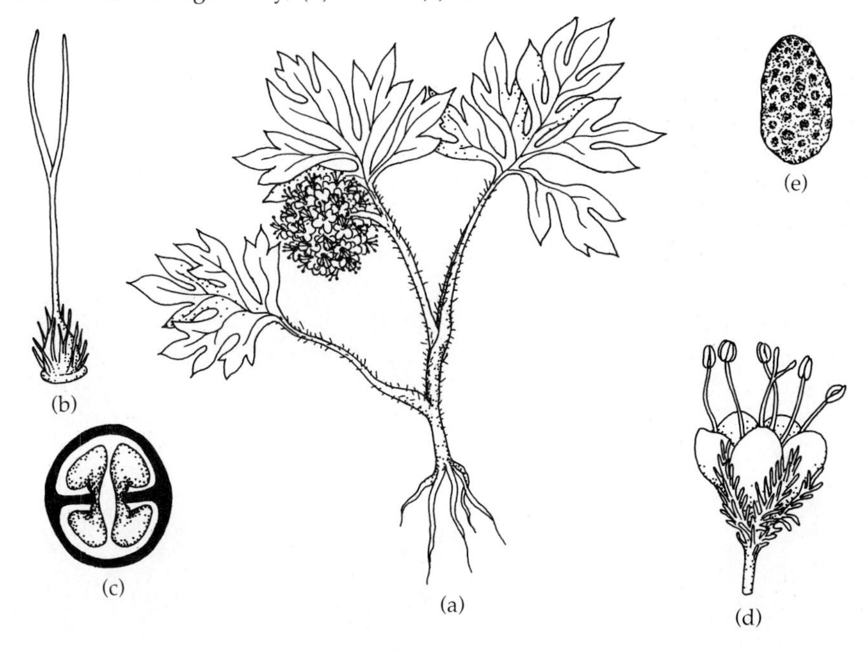


Herbs and small shrubs, commonly with stiff hairs, sometimes with a skunk smell. Leaves alternate (rarely opposite), simple to pinnate (rarely palmately) compound; stipules lacking. Flowers regular, perfect, hypogynous (rarely partly epigynous); inflorescence of a single flower or flowers borne in scorpioid cymes. Sepals 5 (rarely 10–12), distinct or connate, sometimes with auricles alternate the lobes. Petals 5 (rarely 10–12), blue- to purple-colored, sympetalous, bell or funnel-shaped, often having appendages inside. Stamens 5 (rarely 10–12), filaments distinct, adnate to the base of the corolla tube and alternate the corolla lobes, commonly with appendages attached to the filaments; anthers opening by longitudinal slits; nectary disk present. Pistil compound of 2 united carpels; locules 1 or 2; ovules 2–many and borne on axile or parietal placentas; ovary superior or partially inferior; styles 1 or 2, usually bilobed; stigmas capitate. Fruit a capsule. Seeds 2–many, embryo straight, endosperm present (Fig. 9.152).

The family Hydrophyllaceae consists of 18–20 genera and 250 species with a cosmopolitan distribution, with many species in the western United States. The largest genus is *Phacelia,* pussytoes (150 species).

Economically the family is of minor importance. *Phacelia* species are cultivated as garden plants in the United States, and *P. tanacetifolia* is cultivated in Europe as bee fodder. *Hydrophyllum virginianum,* the waterleaf, was eaten by Native Americans as a green and today is grown as an ornamental. Other cultivated ornamentals include *Nemophila menziesii,* baby blue-eyes; *Pholistoma auritum,* fiesta flower; and woody *Wigandia* species in warm climate regions.

The family Hydrophyllaceae is a relatively homogeneous group, with various characters in common with the Polemoniaceae.

The family has no known fossil record.

## ORDER LAMIALES
### Family Boraginaceae (Borage)

Figure 9.153 Boraginaceae. *Hackelia:* (a) plant with flowers; (b) partial longitudinal section through corolla; (c) fruit (nutlet). *Mertensia:* (d) leafy upper stem bearing flowers. *Cynoglossum:* (e) side view of 4-lobed ovary; (f) cross section through ovary; (g) four spiny fruits (nutlets).

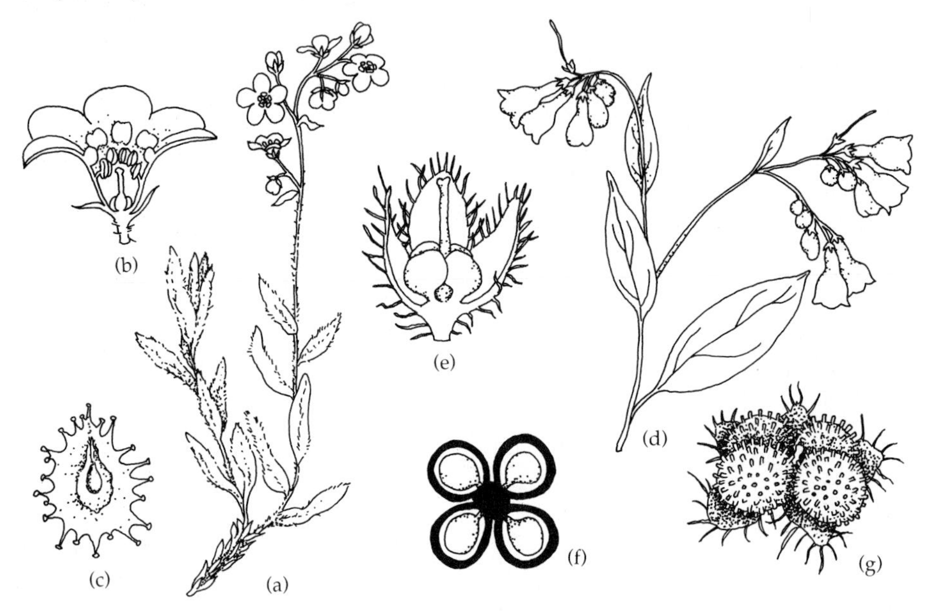


Herbs, less commonly shrubs or trees, often with stiff hairs. Leaves usually alternate (rarely opposite), simple, often entire; stipules lacking. Flowers regular (rarely irregular), perfect (rarely unisexual and pistillate flowers on separate plants), hypogynous; inflorescence usually scorpioid cymes (rarely solitary flowers). Sepals 5, distinct or connate. Petals 5, sympetalous, funnel-shaped to saucer-shaped, commonly with hairy scales at the mouth of the corolla tube. Stamens 5, filaments distinct, adnate to the corolla tube and alternate the corolla lobes; anthers opening by longitudinal slits. Pistil compound of 2 united carpels (rarely 4); locules twice the carpels due to a false partition splitting the primary locule; ovules 1 per locule and borne on an axile placenta, ovary superior, 4 lobed, style gynobasic; 2-lobed stigma capitate or lobed. Fruit a drupe or 2–4 nutlets, commonly covered with bumps, stiff hairs, or wrinkles. Seeds 1–4, embryo curved or straight, endosperm present or absent (Fig. 9.153).

The family consists of 100 genera and 2000 species with a cosmopolitan distribution. The largest genera are *Cordia* (200 species), *Heliotropium* (200 species), *Tournefortia* (200 species), and *Cryptantha* (150 species).

*Borago officinalis,* borage, and *Symphytum officinale,* comfrey, have been used medicinally, for flavorings, and as ornamentals. Other plants grown as ornamentals are *Heliotropium,* heliotrope; *Mertensia,* blue bells; and *Myosotis sylvatica,* forget-me-not. In western North America native peoples used *Lithospermum ruderale* as a contraceptive.

The family has some characteristics similar to the family Hydrophyllaceae and others similar to the Verbenaceae. The woody species are sometimes classified into the family Ehretiaceae.

Fossil pollen attributed to the Boraginaceae have been found in Oligocene and more recent deposits.

## Family Verbenaceae (Verbena)

Figure 9.154 Verbenaceae. *Lantana:* (a) leafy stem bearing axillary inflorescences; (b) longitudinal section through ovary; (c) cross section through ovary; (d) longitudinal section through corolla. *Verbena:* (e) stem with terminal inflorescence; (f) stamen; (g) flower and bract.

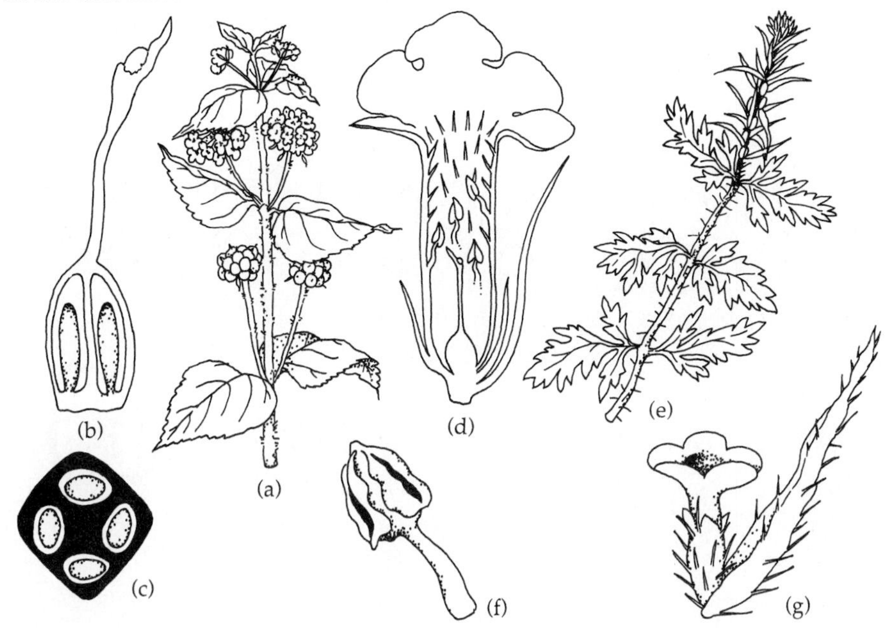


Herbs, shrubs, trees, or lianas. Leaves opposite (rarely alternate or whorled), simple to variously compound; stipules lacking. Flowers regular or irregular, perfect (rarely unisexual), hypogynous; inflorescence a cyme, head, raceme, or spike. Sepals 4–5 (rarely to 8), lobed or toothed, connate. Petals 4–5 (rarely to 8), sympetalous, commonly with a tubular corolla and with spreading or bilabiate lobes. Stamens usually 4 (rarely 2 or 5), filaments distinct, adnate to the corolla tube, alternate the corolla lobes; anthers opening by longitudinal slits. Pistil compound of 2 united carpels; locules beginning as 2 but becoming 4 due to false septa; ovules 1 per locule and borne on an axile placenta; ovary superior and commonly 4-lobed; style 1, terminal (rarely gynobasal), stigma sometimes bilobed. Fruit a drupe breaking into nutlets (rarely a capsule). Seeds 1–4, embryo straight, endosperm present or lacking (Fig. 9.154).

The family consists of 75–100 genera and 2600–3000 species distributed mostly pantropically, with a few species in temperate climates. The largest genera are *Clerodendron* (400 species), *Verbena* (250 species), *Vitex* (200 species), and *Premna* (200 species).

Economically the most valuable species is *Tectona grandis,* the teak of southeast Asia, a highly prized water-resistant timber. Others used for timber are *Petitea, Premna,* and *Vitex celebica. Citharexylum,* zither wood of Mexico and South America, is crafted into musical instruments. Oils are extracted from *Lippia citriodora* and *Vitex agnus-castus.* Some species are cultivated as ornamentals; some *Lantana* species have attractive flowers but have become noxious weeds in the tropics.

The family has some characteristics that are similar to the Lamiaceae. Some of the diverse genera, such as *Avicennia,* a mangrove, are classified into their own families.

The family has no known fossil record.

## Family Lamiaceae (Labiatae) (Mint)

Figure 9.155 Lamiaceae (Labiatae). *Monarda:* (a) leafy stem bearing terminal inflorescence; (b) flower; (c) anther. *Lamium:* (d) longitudinal section through flower; (e) lobed ovary and style; (f) cross section through ovary.

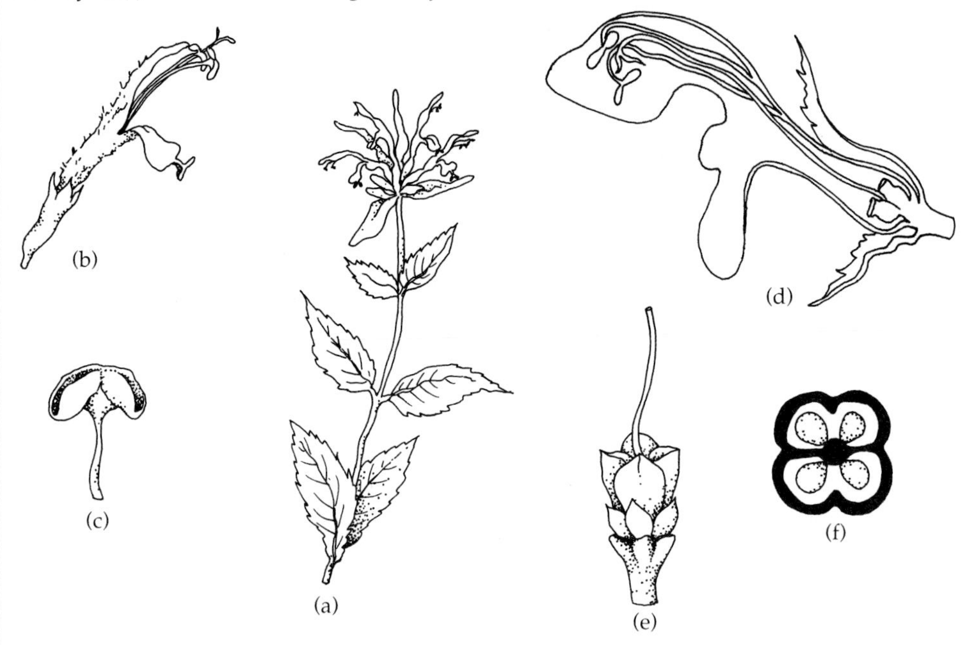


Herbs or shrubs (rarely trees), often covered with hairs and aromatic glands, usually with young stems 4 sided. Leaves opposite (rarely alternate or whorled), simple to pinnately compound; stipules lacking. Flowers irregular, often 2 lipped **(bilabiate),** perfect (rarely unisexual), hypogynous, usually having bracts; inflorescence of a solitary flower or flowers borne in axillary cymes or verticils, heads, panicles, or racemes. Sepals 5, connate. Petals 5, sympetalous, with the corolla bilabiate. Stamens 2 or 4, filaments distinct (rarely connate), sometimes didynamous and adnate to the corolla tube; anthers opening by longitudinal slits; nectary disk present. Pistil compound of 2 united carpels; locules 4; ovules 1 per locule and borne on basal-axile placentas; ovary superior; style 1, gynobasal (rarely terminal); stigma 2 lobed. Fruit of 4 (rarely 1–3), 1-seeded nutlets. Seeds with straight embryo, endosperm in small amounts or lacking. (Fig. 9.155).

The family Lamiaceae consists of 200 genera and 3000–3200 species distributed worldwide in all climates, with many taxa in the Mediterranean region. The largest genera are *Salvia* (500 species), *Hyptis* (350 species), *Scutellaria* (200 species), *Coleus* (200 species), *Plectranthus* (200 species), and *Stachys* (200 species).

Economically the family is important for ornamentals and cooking herbs. Some of the most widely used ones are *Ajuga,* bugle; *Coleus; Lavandula,* lavender; *Mentha,* mint; *Monarda,* bee balm or Oswego tea; *Nepeta,* catnip; *Ocimum,* basil and sweet basil; *Origanum,* oregano; *Salvia,* sage; and *Thymus,* thyme. *Pogostemon* species are grown for the perfume industry of southeast Asia.

The family Lamiaceae has many characters that are similar with the Verbenceae but differs in its deeply 4-lobed ovary, gynobasal, and aromatic oils.

The family has no known fossil record.

## ORDER CALLITRICHALES
## Family Callitrichaceae (Water Starwort)

Figure 9.156 Callitrichaceae. *Callitriche:* (a) leafy stem bearing male and female flowers; (b) female flower; (c) male flower; (d) cross section through ovary; (e) longitudinal section through fruit.

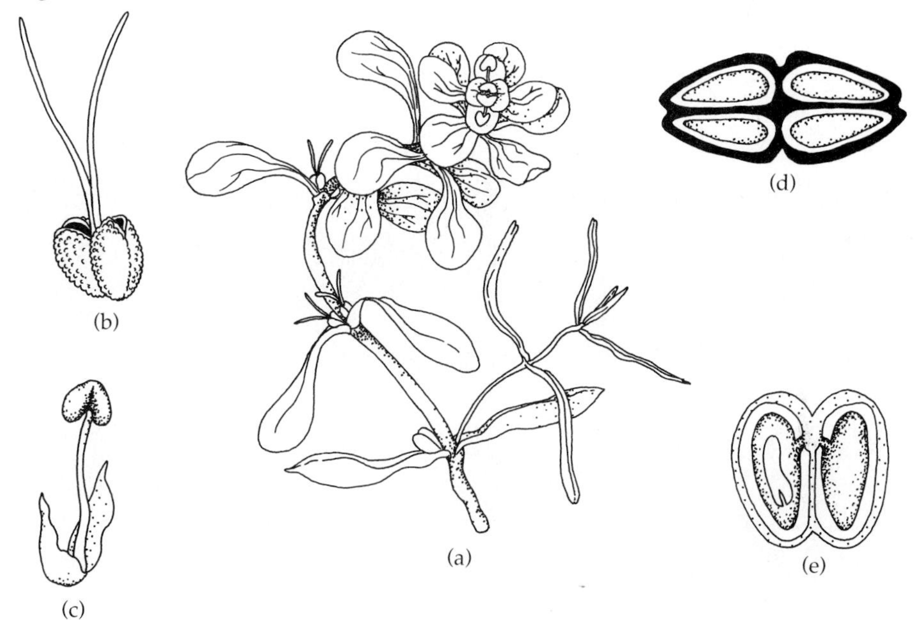


Herbs, aquatic (rarely terrestrial). Leaves small, opposite (rarely whorled), entire; stipules lacking. Flowers unisexual and monoecious, lacking a perianth but with some tiny bracts; inflorescence a solitary axillary flower. Stamens of male flower 1 (rarely 2–3), filaments distinct; anthers opening by longitudinal slits that join at the top of the anther. Pistil of the female flower compound of 2 united carpels; locules initially 2, becoming 4 by the development of false septa; ovules 1 per locule and borne on an axile placenta; ovary superior; style 1 or 2. Fruit of 4 nutlets each with 1 seed, lobed or with wings. Seeds with embryo curved or straight, endosperm present (Fig. 9.156).

The family Callitrichaceae consists of the genus *Callitriche* and 17–35 species distributed worldwide, but with most species in the temperate climates.

The family has no real economic uses. Several species are helpful, however, to ecologists as indicator species that reflect the level of water pollution.

The classification of the Callitrichaceae is debated by botanists, with some classifying the family in the order Lamiales. The vegetative morphology of the species is rather variable, leading to difficulty in species identification.

The family has no known fossil record.

## ORDER PLANTAGINALES
## Family Plantaginaceae (Plantain)

Figure 9.157 Plantaginaceae. *Plantago:* (a) plant with spiky inflorescences; (b) open corolla showing stamen attachment; (c) cross section through ovary; (d) pistil; (e) flower.

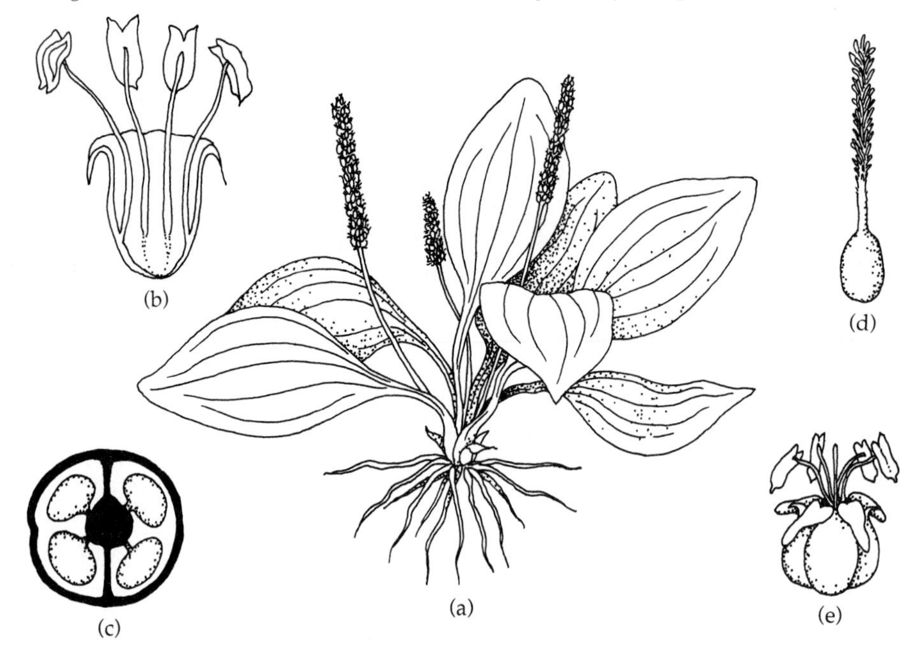


Herbs (rarely shrubs). Leaves usually all basal (rarely alternate or opposite), simple, parallel veined with sheathing petioles; stipules lacking. Flowers small and inconspicuous, regular, perfect (rarely unisexual), hypogynous; inflorescence a spike (rarely in 3-flowered clusters), wind pollinated but sometimes cleistogamous flowers produced. Sepals 4, distinct. Petals 4, sympetalous, scarious. Stamens 4 (rarely 1–2), filaments distinct, adnate to the corolla tube and alternate the corolla lobes; anthers exserted and opening by longitudinal slits. Pistil compound of 2 united carpels, locules 2, sometimes becoming 4 by the development of false septa, ovules 1–many and borne on axile placentas; ovary superior; style 1, stigma usually 2-lobed. Fruit a circumscissile capsule (rarely an achene or nut). Seeds 1–many, sometimes becoming mucilaginous when wet, embryo straight (rarely curved), endosperm present (Fig. 9.157).

The family Plantaginaceae consists of 3 genera and over 250 species with a cosmopolitan distribution. The largest genus is *Plantago*, the plantain (250 species).

Economically the family is not important, except in a negative way. Several species (i.e., *P. lanceolata* and *P. major*) can become troublesome weeds in lawn and waste areas, and therefore much money is spent on herbicides and eradication. Medicinally *P. afra* is used as a laxative, and the seeds of *P. ovata* are used to treat dysentery.

Where the family Plantaginaceae should be classified is debated today by botanists.

Microfossils attributed to the family have been found in Miocene and more recent deposits.

# ORDER SCROPHULARIALES
## Family Buddlejaceae (Butterfly Bush)

Figure 9.158 Buddlejaceae. *Buddleja:* (a) leafy stem with terminal inflorescence; (b) side view of flower; (c) cross section through ovary; (d) seed; (e) opened corolla tube; (f) front view of flower.

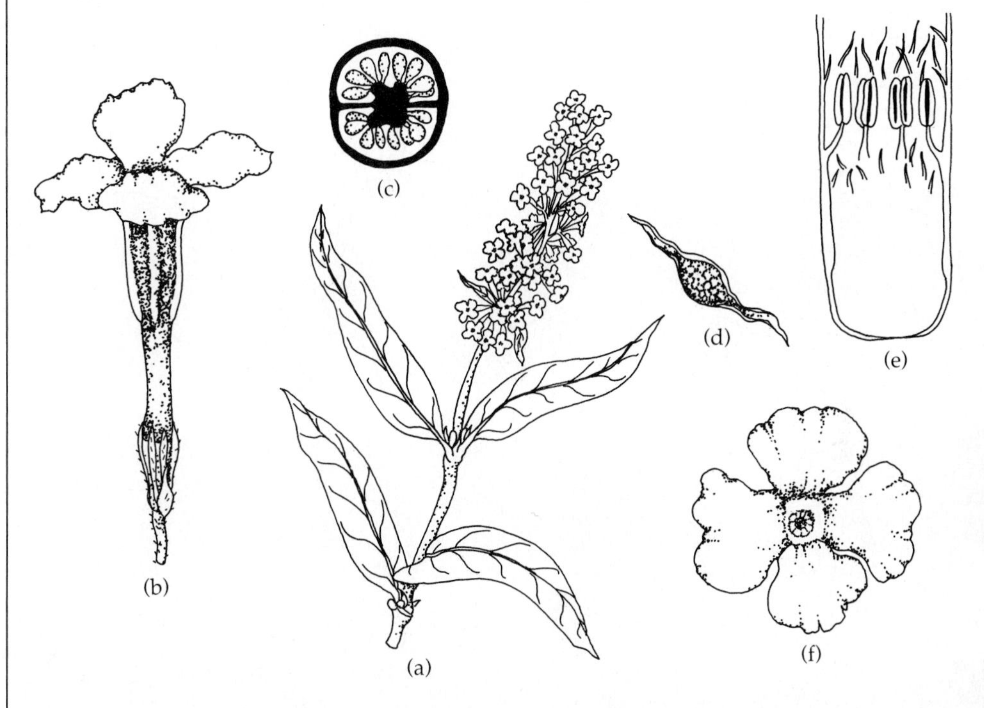


Shrubs or trees (rarely herbs), commonly with branched, stellate or glandular hairs. Leaves opposite or whorled (rarely alternate), simple; stipules greatly reduced or lacking. Flowers regular, perfect or functionally dioecious, hypogynous or epigynous; inflorescence of various types. Sepals 4, connate. Petals 4 (rarely 5), sympetalous. Stamens 4, filaments distinct, adnate to the corolla tube and alternate the corolla lobes; anthers opening by longitudinal slits. Pistil compound of 2 united carpels; locules 2; ovules many and borne on axile placentas; ovary superior or partially inferior; style 1, stigma capitate or 2-lobed. Fruit usually a septicidal capsule. Seeds many, commonly winged, embryo straight, endosperm present (Fig. 9.158).

The family Buddlejaceae consists of 10 genera and 150 species distributed in the subtropics and tropics, with a few taxa extending into temperate regions. The largest genera are *Buddleja,* the butterfly bush (100 species), and *Nuxia* (30 species).

Economically the family is of minor importance. *Buddleja davidii,* a shrub of China, is cultivated in Europe and in milder climatic regions of the United States. It has become rather weedy in England and invades waste areas. Butterflies are attracted to it in abundance. *Nuxia floribunda,* the kite tree, is cultivated as an ornamental in tropical regions. In Indonesia *B. asiatica* is used as a fish poison, while in Africa *B. salviifolia* provides timber.

The family Buddlejaceae has been classified by some botanists in the family Loganiaceae. It does share some characters with this family, but recent anatomical and chemical data indicates more shared characters with the family Scrophulariaceae.

The family has no known fossil record.

## Family Oleaceae (Olive)

Figure 9.159 Oleaceae. *Chionanthus:* (a) leafy branch bearing flowers; (b) cross section through ovary. *Osmanthus:* (c) longitudinal section through female flower; (d) leafy flowering branch; (e) female flower; (f) male flower. *Fraxinus:* (g) fruit.

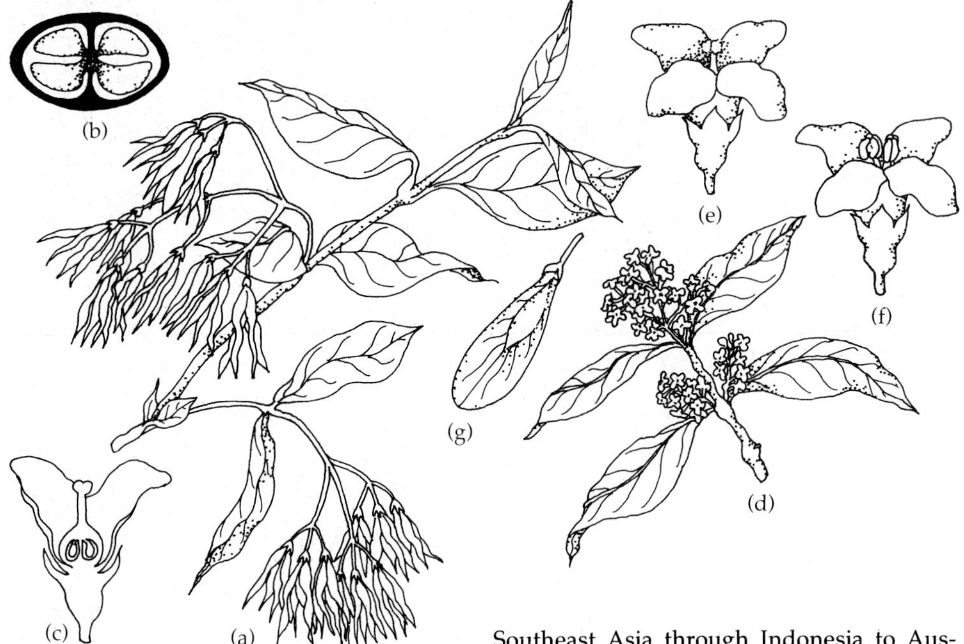


Shrubs, trees, or lianas, commonly with simple to specialized scale-like hairs. Leaves opposite (rarely alternate), simple to pinnately compound; stipules lacking. Flowers sometimes small, regular, perfect or unisexual and dioecious, hypogynous; inflorescence of a solitary flower or comprised of cymes, panicles, or racemes. Sepals 4 (rarely 5–15 or lacking), connate. Petals 4 (rarely to 12), distinct or sympetalous. Stamens 2 (rarely 4), filaments distinct, adnate to the corolla tube, usually alternate the corolla lobes; anthers opening by longitudinal slits; nectary disk present or absent. Pistil compound of 2 united carpels; locules 2; ovules 2 (rarely more) per locule and borne on axile placentas; ovary superior; style 1, stigma 1–2 lobed. Fruit a berry, capsule, drupe, or samara. Seeds with a straight embryo, endosperm present or lacking (Fig. 9.159).

The family Oleaceae consists of 29–30 genera and 600 species distributed mostly cosmopolitan with many diverse species in Southeast Asia through Indonesia to Australia. The most common genera are *Jasminum,* jasmine (200 species); *Chionanthus* (125 species); and *Fraxinus,* the ash (60 species).

Economically the family is of some importance. The most important species is *Olea europaea,* the olive. Valuable timber is obtained from *Fraxinus americana,* the American or white ash, and *F. excelsior,* the European ash. Ornamentals commonly planted include various *Forsythia* and *Jasminum,* the jasmines; *Ligustrum vulgare,* the common privet, used as hedging; and *Syringa vulgaris,* the common lilac. The fragrant flowers of *Jasminum sambac,* Arabian jasmine, and *Osmanthus fragrans* are grown for making perfume and to add fragrance to tea, respectively.

The classification status of the Oleaceae is much debated by botanists, and how it is placed in relation to other families is open to various opinions.

Microfossils attributed to the family have been found in upper Miocene and more recent deposits.

## Family Scrophulariaceae (Figwort)

Figure 9.160 Scrophulariaceae. *Penstemon:* (a) leafy stem bearing flowers; (b) front view of flower; (c) sterile stamen; (d) cross section through ovary; (e) anther; (f) pistil. *Castilleja:* (g) flowering stem; (h) bract; (i) galea (corolla); (j) calyx.

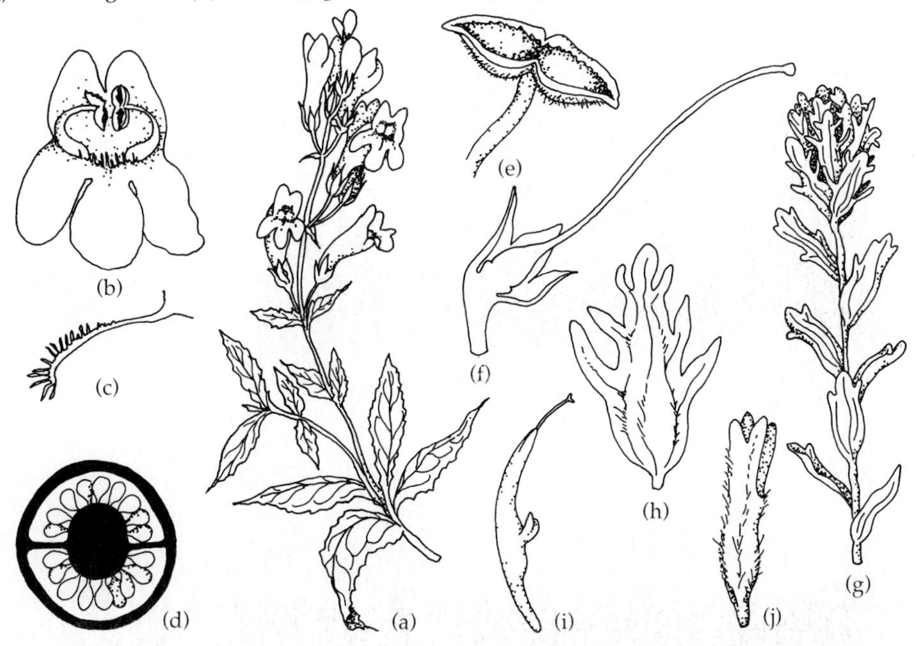


Herbs or shrubs (rarely trees), some are semiparasitic, taking some necessary nutrients from a host. Leaves deciduous and evergreen, alternate or opposite (rarely whorled), simple to pinnately compound; stipules lacking. Flowers usually irregular (rarely almost regular) perfect, hypogynous; inflorescence of various types, cymes, panicles, racemes, or spikes. Sepals 4–5, distinct or connate. Petals 4–5, sympetalous, commonly with a nectar sac or spur. Stamens usually 4 or 4 functional (sometimes 2 or 5), filaments distinct, adnate to the corolla tube and alternate the corolla lobes; anthers opening by longitudinal slits or by one slit extending across the pollen sacs of the anther; nectary disk present. Pistil compound of 2 united carpels; locules 2; ovules 2–many per locule and borne on axile placentas; ovary superior; style 1, long and slender; stigma simple or 2 lobed. Fruit a capsule (rarely a berry). Seeds smooth or with sculptured surface or winged, embryo curved or straight, endosperm present (Fig. 9.160).

The family Scrophulariaceae consists of 190–220 genera and 3000–4000 species with a cosmopolitan distribution, but mainly in temperate and mountain climates. The largest genera are *Pedicularis* (500 species), *Calceolaria* (300 species), *Verbascum* (300 species), *Veronica* (300 species), *Penstemon* (280 species), and *Castilleja*, Indian paintbrush (150 species).

Economically the family is of moderate importance. The well-known heart medicine digitoxin is extracted from *Digitalis purpurea*, the foxglove. Some genera used as ornamentals are *Antirrhinum*, snapdragons; *Calceolaria*, slipper flower; *Digitalis; Hebe; Mimulus*, the monkey flower; *Penstemon*, the beard tongue; and *Veronica*, the speedwell.

The family is a large distinctive family, with many different kinds of pollinators and pollination mechanisms.

The family has no known fossil record.

## Family Globulariaceae (Globularia)

Figure 9.161 Globulariaceae. *Globularia:* (a) plant with a head of flowers; (b) open perianth; (c) longitudinal section through ovary showing apical ovule; (d) flower. *Poskea:* (e) open corolla showing stamen attachment and corolla lobes.

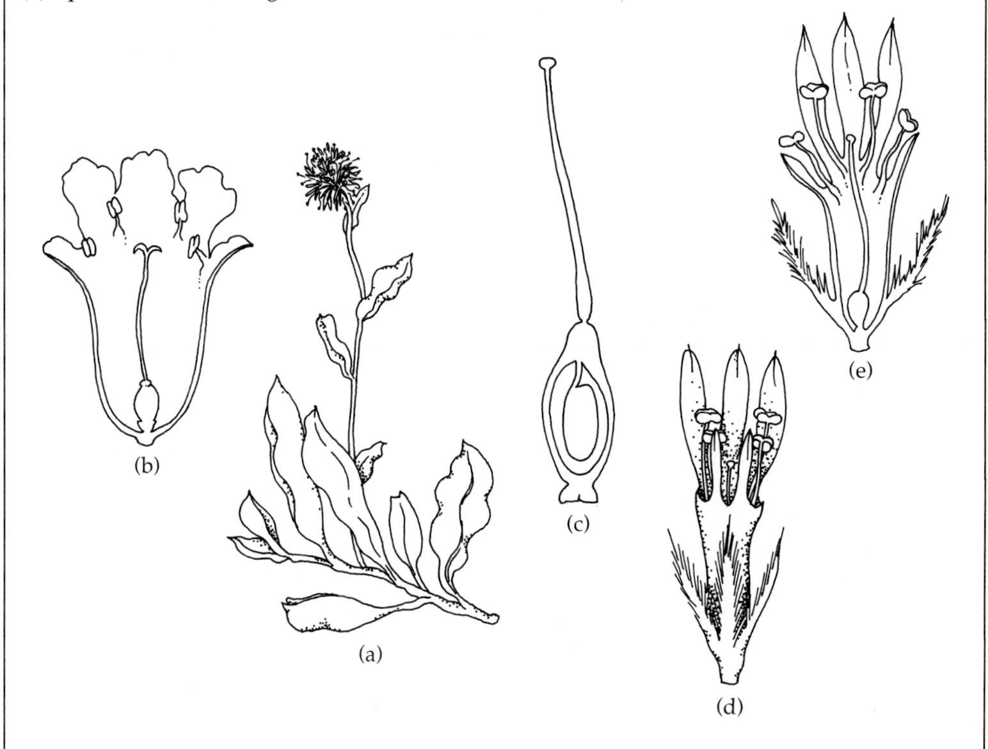


Herbs or shrubs having simple or glandular hairs. Leaves small, alternate (rarely opposite), simple; stipules lacking. Flowers individually small, irregular, perfect, hypogynous; inflorescence of corymbs, heads, or spikes subtended by an involucre of bracts. Sepals 5, connate, persistent on the fruit. Petals 5 (rarely 4), sympetalous, often 2 lipped (bilabiate). Stamens 4, filaments distinct, adnate to the corolla tube and alternate with the corolla lobes; anthers opening by 1 slit at the apex of the anther; nectary disk or nectar glands present. Pistil compound of 2 united carpels; locules 2; ovules 1 per locule and borne on an apical placenta; ovary superior; style 1, terminal, stigma capitate or 2 lobed. Fruit an achene or small nut. Seeds with a straight embryo, endosperm present (Fig. 9.161).

The family Globulariaceae consists of 10 genera and 300 species distributed in Europe, Western Asia, Africa, and Madagascar. The largest genera are Selago (180 species), *Habenstretia* (40 species), *Walafrida* (40 species), and *Globularia* (25 species).

Economically the family is of little value. Some species of *Globularia* are grown as rock garden plants.

Previous classifications of the family Globulariaceae included only 2–3 genera and about 30 species. Recent anatomical and pollen research indicated many similarities with the genera in the subfamily Selaginoideae of the family Scrophulariaceae. Therefore, the present view is to combine these two groups into one family, the Globulariaceae.

The family has no known fossil record.

## Family Orobanchaceae (Broomrape)

Figure 9.162 Orobanchaceae. *Orobanche:* (a) plant with flowers; (b) side view of flower; (c) cross section through ovary; (d) front view of flower; (e) anther; (f) glandular hairs.

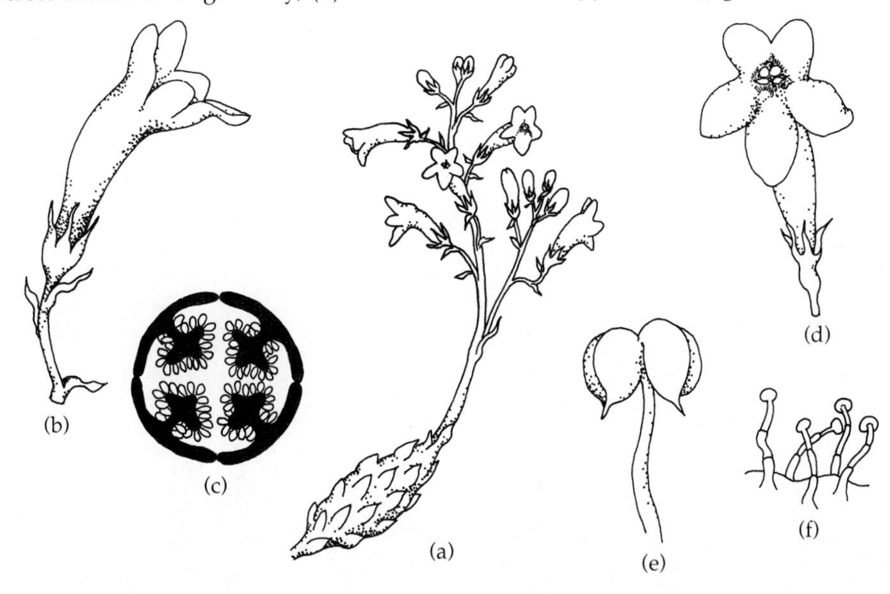


Herbs, fleshy, parasitic on roots, stems lacking chlorophyll, stalk of hairs 2–several celled. Leaves alternate, reduced, scale-like; stipules lacking. Flowers irregular, perfect, hypogynous; inflorescence a solitary flower or a terminal corymb, raceme, or spike. Sepals 2–5, distinct or connate, regular or irregular. Petals 5, sympetalous, 2-lipped corolla tube commonly curved. Stamens 4, (a fifth stamen is sterile and a staminode or lacking), filaments distinct, adnate to the corolla tube and alternate the corolla lobes; anthers opening by longitudinal slits. Pistil compound of 2 (rarely 3) united carpels; locule 1; ovules many and borne on 2 or 4 (rarely 6) parietal placentas; ovary superior; style 1, stigma capitate or 2–4 lobed. Fruit a loculicidal capsule. Seeds many, small, embryo minute, and poorly developed, endosperm present (Fig. 9.162).

The family Orobanchaceae consists of 14–17 genera and 150–180 species, distributed widely in the Northern Hemisphere, in the Old World subtropics and tropics, and in Australia. The largest genus is *Orobanche,* the broomrape (100 species).

Economically the family is of no importance, except that in the Mediterranean region *O. crenata* infects legume crops, such as beans and peas, causing some plant damage.

The family Orobanchaceae has many characters in common with the family Scrophulariaceae, which also has semiparasitic plants. Some botanists have recently suggested that the two groups should be combined.

The family has no known fossil record.

## Family Gesneriaceae (Gesneriad)

Figure 9.163 Gesneriaceae. *Aeschynanthus:* (a) leafy stem bearing flowers; (b) side view of flower; (c) touching anthers; (d) cross section through ovary; (e) glandular hairs.

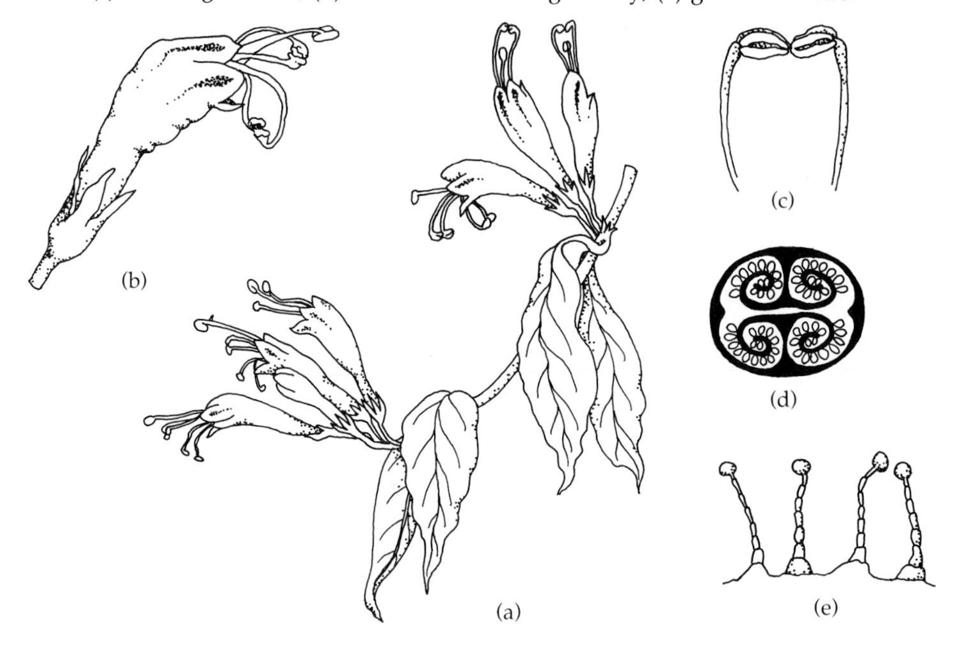


Herbs, shrubs, trees, lianas, or sometimes epiphytes. Leaves opposite (rarely alternate or whorled) or basal, simple (rarely pinnately compound); stipules lacking. Flowers mostly large, irregular, perfect, hypogynous to epigynous; inflorescence of solitary flowers or flowers borne in cymes or racemes. Sepals 5, distinct or connate. Petals 5, sympetalous and usually 2-lipped, sometimes with a nectar sac or spur at the base. Stamens 2 or 4, filaments distinct, adnate to the corolla tube and alternate the corolla lobes; anthers all pressed close together or connivent in pairs, and opening by longitudinal slits; nectary disk usually present (rarely lacking). Pistil compound of 2 united carpels; locules 1, less commonly 2–4 due to fusion of the placentas; ovules many and borne on 2 parietal placentas; ovary superior to inferior; styles 1, stigma usually 2-lobed. Fruit a capsule (rarely a berry). Seeds small, many, embryo straight, endosperm present or lacking. (Fig. 9.163).

The family Gesneriaceae consists of 120–125 genera and 2000–2300 species distributed pantropically, with a few species in temperate climates. The largest genera are *Cyrtandra,* (estimated 200–600 species), *Columnea* (150–250 species), and *Streptocarpus* (130 species).

Economically the family is important mostly as ornamentals. Many house and garden plants belong to the following genera: *Achimenes; Gesneria; Kohleria; Saintpaulia,* African violet; *Sinningia,* gloxinias; and *Streptocarpus,* Cape primrose. A few taxa are said to have some medicinal properties.

The family Gesneriaceae is sometimes said to be the tropical counterpart to the Scrophulariaceae, which is predominantly found in temperate regions. Many characters are shared between the two groups, but the gesneriads usually have 1 locule, parietal placentas, and sometimes an inferior ovary.

The family has no known fossil record.

## Family Acanthaceae (Acanthus)

Figure 9.164 Acanthaceae. *Justicia:* (a) leafy stem with flowers. *Ruellia:* (b) fruit with persistent calyx; (c) cross section through ovary; (d) split capsule; (e) pistil; (f) leaf epidermis showing cystolith crystals.

Herbs or shrubs (occasionally trees or lianas), commonly with different types of glandular and nonglandular hairs. Leaves opposite (rarely alternate), simple; stipules lacking, commonly with specialized crystals called **cystoliths.** Flowers irregular, perfect, hypogynous, usually subtended by prominant bracts, sometimes petal-like; inflorescence of solitary flowers or flowers borne in a cyme or raceme. Sepals 4–5, distinct or connate. Petals 5, sympetalous often 2-lipped. Stamens 2–4 (rarely 5), filaments distinct or connate in pairs, adnate to the corolla tube and alternate the corolla lobes; anthers opening by longitudinal slits and sometimes with one anther smaller than the other; nectary disk usually present. Pistil compound of 2 united carpels, locules 2 (1 when false septa have formed), ovules usually 2 (rarely 1–many) per locule and attached to parietal placentas, ovary superior, style 1, stigmas 2. Fruit a capsule that sometimes opens explosively or a drupe. Seeds usually become slimy when wet, embryo large, curved or straight, endosperm lacking (rarely present) (Fig. 9.164).

The family Acanthaceae consists of 250 genera and 2500 species with a cosmopolitan distribution, but with most species in the tropical regions. The largest genera are *Beloperone* (300 species), *Ruellia* (250 species), *Barleria* (250 species), *Strobilanthes* (200 species), and *Thunbergia* (200 species).

Economically the family is of some importance as cultivated ornamentals and for medicinal uses. Species of *Acanthus, Aphelandra* (i.e., *A. squarrosa*), *Barleria, Crossandra* (i.e., *C. nilotica*), *Justicia,* and *Thunbergia* (i.e., *T. alata*) are commonly cultivated. A cough medicine is made from boiled leaves of *Acanthus ebracteatus* in Malaya, and *A. mollis* is used to treat diarrhea in Europe.

Various characters of the Acanthaceae are similar to those of the family Scrophulariaceae, leading some botanists to suggest the possible joining of all or parts of these families together as one family.

Fossil pollen attributed to the family Acanthaceae has been found in Miocene and more recent deposits.

## Family Bignoniaceae (Bignonia)

Figure 9.165 Bignoniaceae. *Campsis:* (a) leafy stem segment bearing terminal flowers; (b) opened corolla with adnate stamens; (c) seed; (d) fruit; (e) cross section through ovary; (f) pistil.

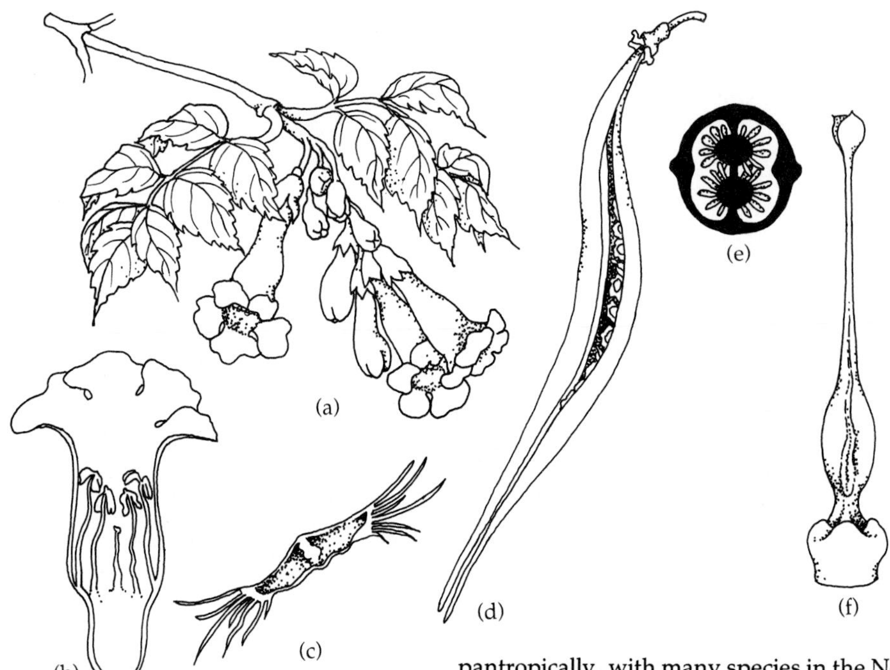


Shrubs, trees, or lianas (rarely herbs). Leaves opposite or whorled (rarely alternate), simple or compound, sometimes with the leaflets modified as a tendril; stipules lacking but commonly with glands at the base of the petiole. Flowers usually large, irregular to almost regular, perfect, hypogynous; inflorescence of a solitary flower or flowers in a cyme, raceme, or panicle. Sepals 5, connate. Petals 5, sympetalous and sometimes 2-lipped. Stamens 4 (rarely 2 or 5), occasionally reduced to staminodes, filaments distinct, adnate to the corolla tube; anthers opening by longitudinal slits; nectary disk usually present. Pistil compound of 2 united carpels; locules 2; ovules many and borne on axile or parietal placentas; ovary superior; styles 2, stigma 2-lobed. Fruit a capsule (rarely a berry). Seeds commonly flat and winged, embryo straight, endosperm lacking (Fig. 9.165).

The family Bignoniaceae consists of 100–120 genera and 650–800 species distributed pantropically, with many species in the New World tropics; only a few taxa are in temperate climates. The largest genus is *Tabebuia* (100 species) of the American tropics.

Economically the family is important for some spectacular ornamentals and for timber. In temperate climates *Paulownia,* the empress tree, and *Catalpa bignonioides* are commonly planted shade trees, which are also valued for fence posts and timber. Some attractive tropical ornamentals include *Bignonia; Campsis,* trumpet vine; *Crescentia cujete,* calabash tree; *Jacaranda; Kigelia,* sausage tree; *Spathodia,* African tulip tree; *Tabebuia,* poui; and *Tecamaria,* Cape honeysuckle. The wood of *Tabebuia* is prized for all kinds of wood uses.

The family has some characters in common with the Scrophulariaceae, with some genera having intermediate characters between the two families.

Fossil pollen, flowers, and fruit attributed to the Bignoniaceae have been discovered in Eocene deposits, with pollen discovered from more recent strata.

## Family Lentibulariaceae (Bladderwort)

Figure 9.166 Lentibulariaceae. *Utricularia:* (a) flowering stem bearing dissected "leaves" and insect-catching bladders; (b) side view of flower; (c) insect-catching bladder; (d) fruit; (e) stamen; (f) cross section through ovary; (g) longitudinal section through flower.

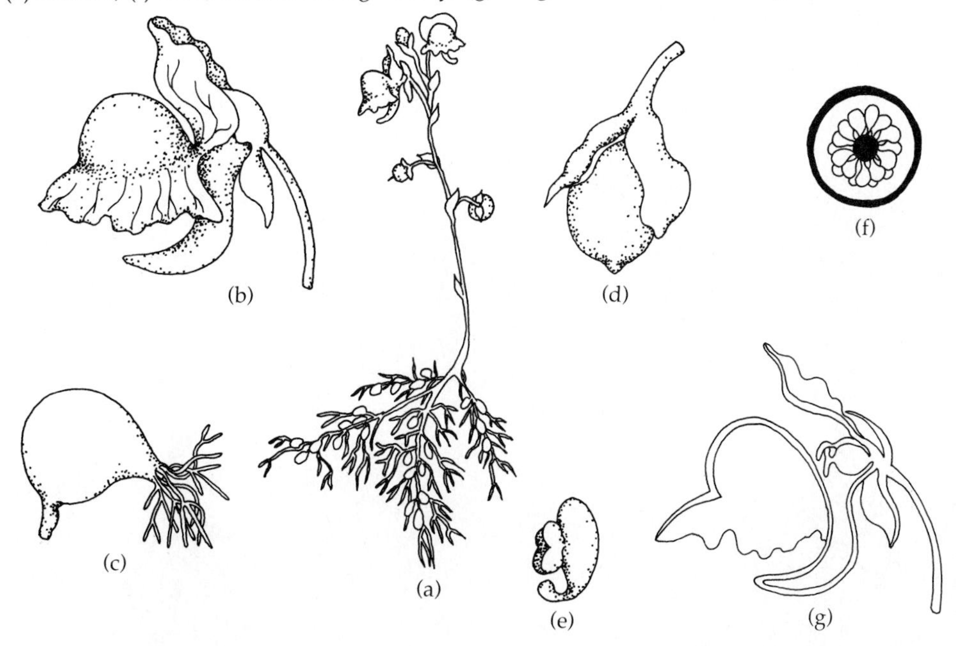


Herbs, usually insectivorous plants with specialized insect-trapping glands or structures; mostly of aquatic or wet habitats. Leaves alternate or basal (rarely lacking), simple or sometimes dissected (considered as leaves but really stem tissue), with insect-trapping bladders or pitchers attached. Flowers irregular, perfect, hypogynous; inflorescence a naked scape with a solitary flower or a raceme. Sepals 2–5, distinct or connate at the base. Petals 5, sympetalous, showy, 2 lipped with a spur or sac on the lower lip. Stamens 2, filaments distinct, adnate to the corolla tube; nectary disk lacking. Pistil compound of 2 united carpels; locule 1; ovules 2–many and borne on free-central placentas; ovary superior; style very short or lacking, stigma usually sessile with 2 unequal lobes. Fruit a capsule. Seeds small, embryo poorly developed, endosperm lacking (Fig. 9.166).

The family Lentibulariaceae consists of 4–5 genera and 180–200 species with a cosmopolitan distribution. The largest genera are *Utricularia,* bladderwort (150 species), and *Pinguicula,* butterwort (35 species).

Economically the family is of little importance. Some species of *Pinguicula* are cultivated by plant hobbiests who are interested in insectivorous plants. In rice-growing regions of the world, *Utricularia* sometimes becomes a noxious aquatic weed in rice fields.

Some botanists prefer to place the genera *Pinguicula* and *Utricularia* in their own families. The family is similar to the family Scrophulariaceae in various characters, but is markedly different in its insectivorous habit and in having parietal placentation.

The family has no known fossil record.

## ORDER CAMPANULALES
## Family Campanulaceae (Bellflower)

Figure 9.167 Campanulaceae. *Campanula:* (a) leafy plant bearing flowers; (b) flower; (c) longitudinal section through flower; (d) fruit with attached calyx; (e) cleistogamous flower. *Lobelia:* (f) flower; (g) cross section through ovary; (h) style sticking out of anther tube.

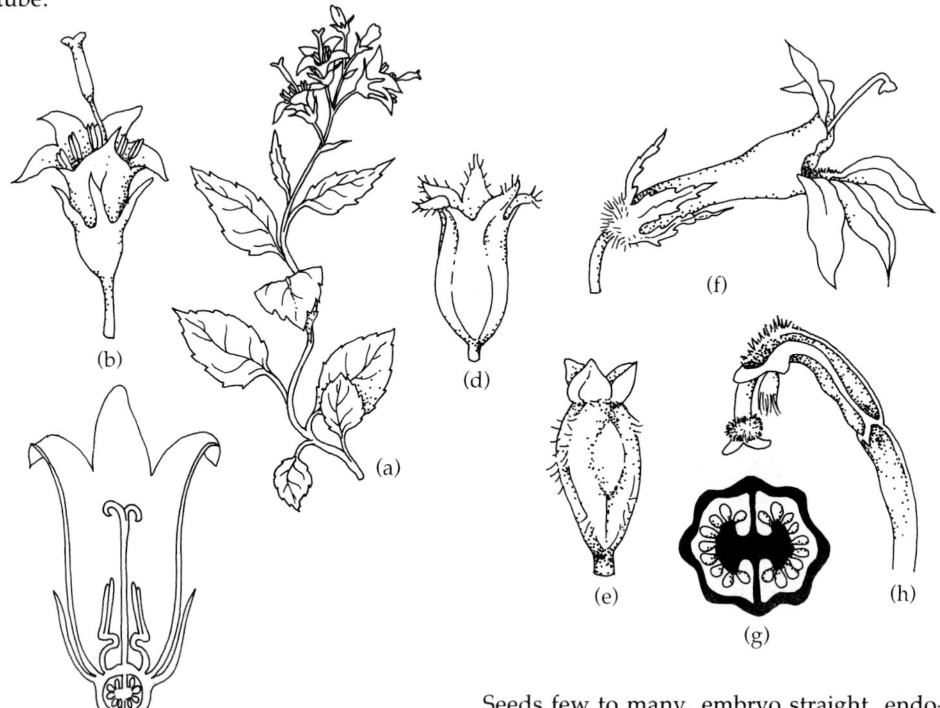


Herbs (rarely shrubs), commonly with latex in tissue and with a carbohydrate called inulin. Leaves alternate (rarely opposite or whorled), simple (rarely pinnately compound); stipules lacking. Flowers often large, blue colored, regular, perfect, epigynous (rarely hypogynous); inflorescence a single flower or flowers borne in a cyme or raceme. Sepals 5, distinct or connate, commonly with flaps of tissue between the calyx lobes. Petals 5, distinct or sympetalous, adnate where the calyx is attached. Stamens 5, filaments distinct or connate, adnate to the base of the corolla; anthers opening by longitudinal slits; nectary disk present. Pistil compound of 2–5 united carpels, locules 2–5 (rarely 10), ovules few to many and borne on axile placentas, ovary inferior (rarely superior), style 1, hairy, stigmas 2–5. Fruit a capsule (rarely a berry). Seeds few to many, embryo straight, endosperm present (Fig. 9.167).

The family Campanulaceae consists of 70 genera and approximately 2000 species distributed most extensively throughout the Northern Hemisphere. The largest genera are *Campanula,* bell flower (300 species); *Lobelia* (300 species); *Centropogon* (200 species); and *Siphocampylus* (200 species).

Economically the family is of moderate importance, mostly as garden ornamentals. Many species of *Campanula; Edraianthus; Lobelia; Platycodon,* balloon flower; and *Symphyandra* are cultivated as ornamentals.

The family sometimes includes the genus *Lobelia,* which is placed by some botanists in its own family, Lobeliaceae. How the various genera are grouped ultimately depends on the personal bias of the botanist. The family is a fairly distinct homogeneous group.

The family has no known fossil record.

## Family Stylidiaceae (Trigger Plant)

Figure 9.168 Stylidiaceae. *Stylidium:* (a) plant with many small flowers; (b) flower with exserted style, or column; (c) end of style, or column, showing two stamens and bilobed stigma; (d) longitudinal section through flower.

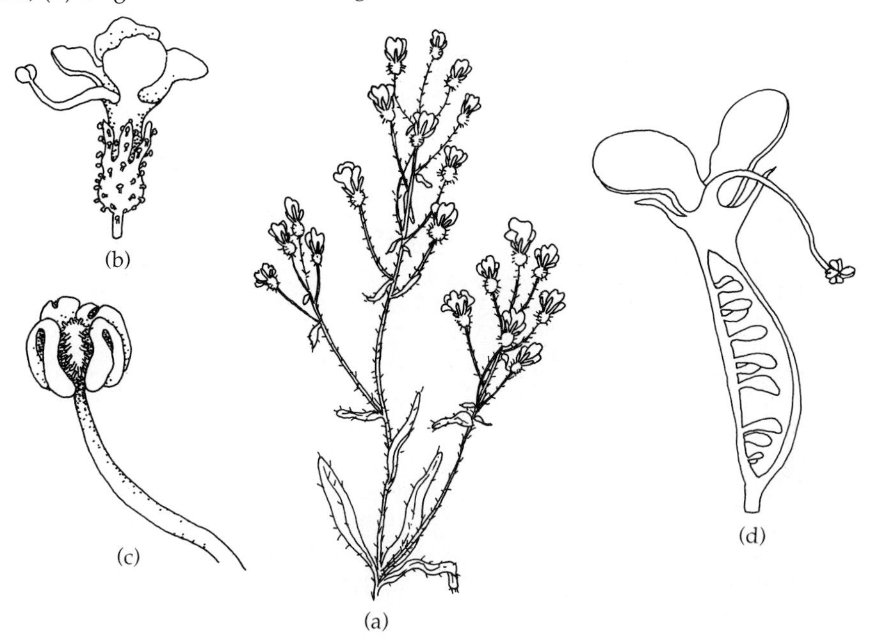


Herbs or shrubs, commonly xerophytic and having glandular hairs, with a head comprised of 2 or more cells. Leaves alternate or in a basal rosette, simple; stipules lacking. Flowers irregular (rarely regular), perfect or unisexual, epigynous; inflorescence of a solitary flower or flowers borne in terminal cymes or racemes. Sepals 5 (rarely 2–7), connate and persistant. Petals 5, sympetalous with 4 upright lobes and 1 lobe curved downward. Stamens 2 (rarely 3), filaments distinct and adnate to the style, which together comprises a movable column that moves abruptly up or down, transferring pollen to the visiting insect; anthers open by longitudinal slits that open outward; nectary disk or nectar glands present. Pistil compound of 2 united carpels; locules 1 or 2; ovules few to many and borne on axile to free-central placentas; ovary inferior; style 1, stigma entire or 2 lobed. Fruit a capsule. Seeds small, few to many; embryo small with 1 or 2 cotyledons, endosperm present (Fig. 9.168).

The family Stylidiaceae consists of 5–6 genera and 150–155 species distributed mostly in Australia but with some species in New Zealand, southern Asia, and the southern tip of South America. The largest genus is *Stylidium,* the trigger plant (about 140 species).

Economically the family is of minor importance. A few species are cultivated as ornamentals, especially some evergreen shrubs from western Australia, known in the horticulture trade as Candollea, and some *Forstera.*

How the family should be classified is open to debate. Some botanists have placed the family with the Saxifragaceae, while pollen features resemble other families in the Campanulales. The genus *Donatia* is sometimes placed in its own family, the Donatiaceae.

The family has no known fossil record.

## Family Goodeniaceae (Goodenia)

Figure 9.169 Goodeniaceae. *Scaevola:* (a) leafy stem segment bearing flowers and fruits; (b) side view of flower; (c) longitudinal section through flower; (d) cross section through ovary; (e) style surrounded by anthers.

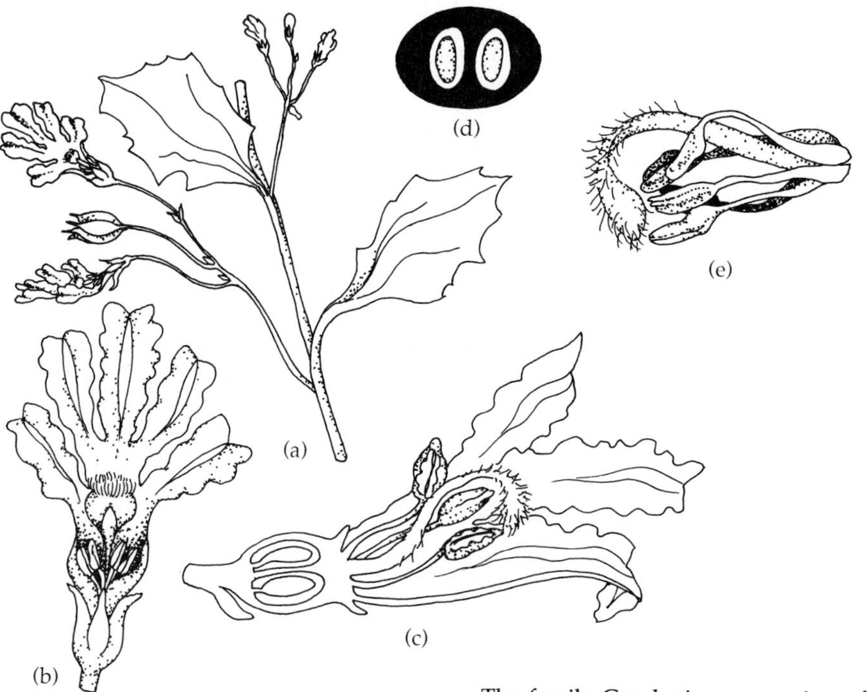


Herbs or shrubs. Leaves alternate (rarely opposite), simple to pinnately divided; stipules lacking. Flowers irregular, perfect, epigynous; inflorescence of solitary flowers or flowers borne in cymes, heads, or racemes. Sepals 5 (rarely 3), connate. Petals 5, sympetalous, slit on the upper side almost to the base giving a prominent 2-lipped appearance. Stamens 5, filaments distinct, sometimes adnate to the corolla tube and alternate the corolla lobes; anthers opening by slits and either free or connate into an anther tube around the style. Pistil compound of 2 united carpels, locules 2 (rarely 1 or 4), ovules 1–many and borne on axile or basal-axile placentas; ovary inferior; style has a cup-like outgrowth or indusium just below the usually 2-lobed stigma, that is fringed with hairs. Fruit a capsule (rarely a drupe or nut). Seeds 1–many, embryo straight, endosperm present (Fig. 9.169).

The family Goodeniaceae consists of 14 genera and 300 species distributed mostly in Australia, with a few species extending to New Zealand, eastern Asia, Africa, South America, and some Pacific Islands. The largest genera are *Goodenia* (100 species), *Scaevola* (80 species), and *Dampiera* (60 species).

Some species of *Goodenia, Dampiera, Leschenaultia* (especially *L. biloba* with large blue flowers), and *Scaevola* are grown as greenhouse plants or garden ornamentals in warm climates. The family is a major component of the spectacular spring flora of western Australia.

The reproductive mechanism and pollination biology of the Goodeniaceae flowers make the family distinctive. This did not go unnoticed by Charles Darwin, who wrote letters to Sir Joseph Hooker, Director of the Royal Botanic Gardens, Kew, about the strange reproductive system in the family.

Pollen attributed to the Goodeniaceae has been recorded from Oligocene and more recent deposits.

## ORDER RUBIALES
### Family Rubiaceae (Madder)

Figure 9.170 Rubiaceae. *Galium:* (a) flowering stem with whorled leaves; (b) fruits; (c) front view of flower. *Cephalanthus:* (d) leafy branch bearing heads of small flowers; (e) cross section through ovary; (f) longitudinal section through flower.

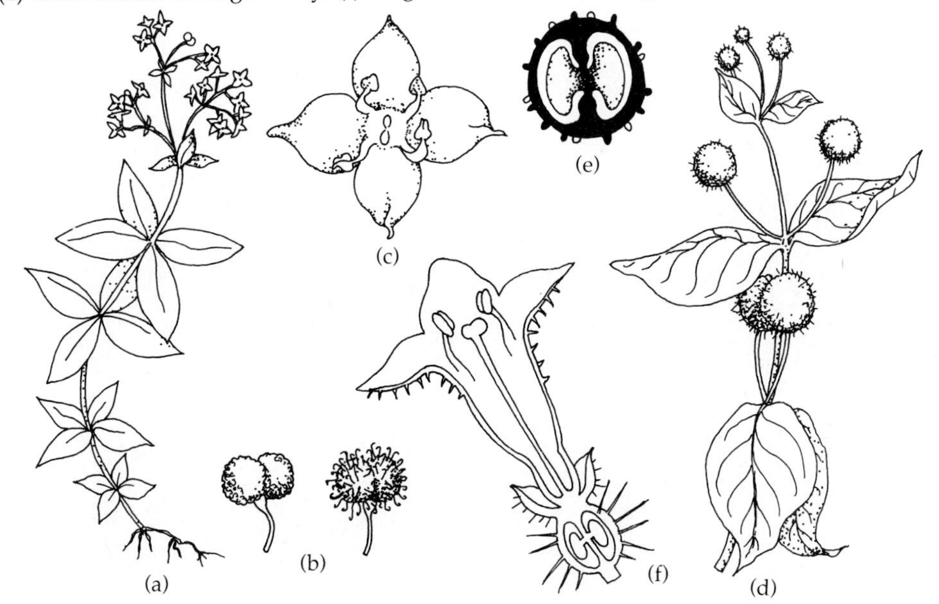


Herbs, shrubs, trees, or lianas. Leaves opposite or whorled, simple, entire (rarely lobed); stipules present, often fused at the node, leaf-like and appearing like whorled leaves. Flowers regular (rarely irregular), perfect (rarely unisexual), epigynous, commonly showing heterostyly; inflorescence usually a cyme, head, or panicle (rarely a solitary flower). Sepals 4–5, distinct or connate, lobes often greatly reduced in size or lacking. Petals 4–5 (rarely 8–10), sympetalous. Stamens 4–5, filaments distinct, adnate to the corolla tube and alternate the lobes; anthers opening by longitudinal slits; nectary disk usually present. Pistil compound of 2 (rarely 1–many) united carpels; locules 2 (rarely 1–many); ovules 1–many and borne on axile or parietal placentas; ovary inferior (rarely superior); style 1, slender, stigma capitate or lobed. Fruit a berry, capsule, drupe, or schizocarp. Seeds 1–many, embryo curved or straight, endosperm present or lacking (Fig. 9.170).

The family Rubiaceae consists of 450 genera and 6500–7000 species distributed worldwide, with the majority of the taxa being woody and found in subtropical and tropical regions. In temperate climates most species are herbaceous. Two of the largest genera are *Psychotria* (700 species) and *Galium* (300 species).

Economically the family has some very important species. The most important species is coffee, produced from *Coffea arabica* and *C. canephora*. Various species of *Cinchona* are the source of quinine. The drug ipecacuanha comes from *Cephaelis,* while the original source for the red dye alizarin is *Rubia tinctorum.* Commonly cultivated ornamentals include some *Galium,* bedstraw; *Gardenia;* and *Houstonia,* bluets.

The family as a whole is a fairly recognizable group. Debate among botanists arises as to what order the family should be placed in and the intrafamily arrangement of the various genera.

Pollen and leaf fossils attributed to the Rubiaceae have been collected from Eocene and more recent deposits.

## ORDER DIPSACALES
### Family Caprifoliaceae (Honeysuckle)

Figure 9.171 Caprifoliaceae. *Kolkwitzia:* (a) flowering branch with opposite leaves. *Lonicera:* (b) Leafy branch bearing flowers; (c) front view of flower; (d) side view of flower; (e) longitudinal section through corolla; (f) cross section through ovary.

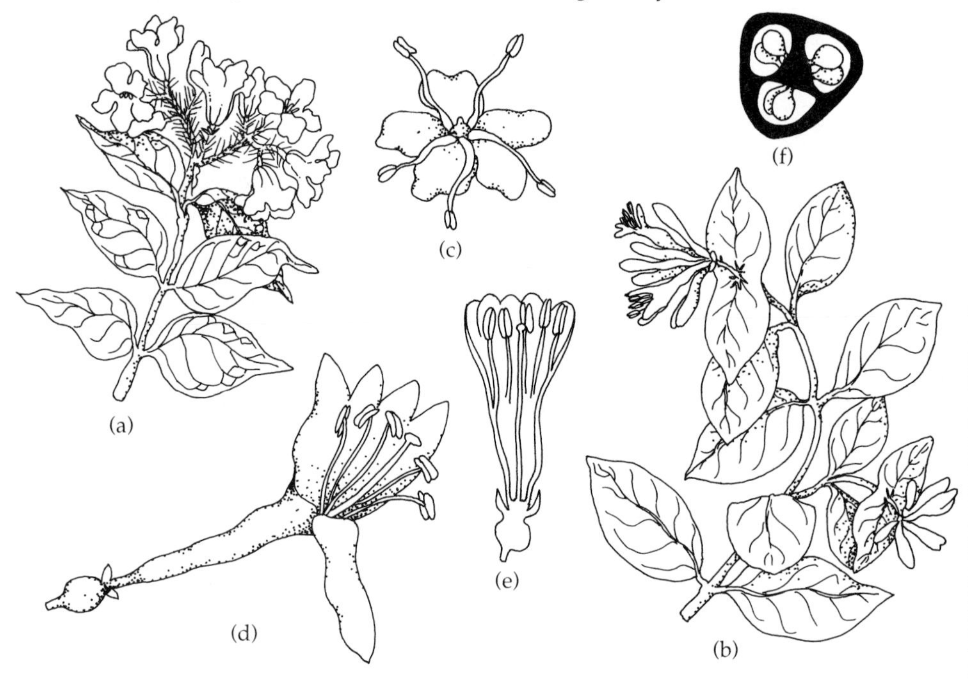


Herbs, shrubs, small trees, or lianas. Leaves opposite, simple or pinnately compound; stipules rarely present, usually lacking or as nonfloral nectar glands. Flowers regular or irregular, perfect, epigynous; inflorescence usually a cyme, corymb, panicle, or a pair of flowers, commonly bearing bracts. Sepals 5 (rarely 4), small, distinct or connate. Petals 5, sympetalous, sometimes 2 lipped; corolla tube commonly spurred or with a nectar gland at the base. Stamens 5 (rarely 4), filaments distinct, adnate to the corolla tube and alternate the corolla lobes; anthers opening by longitudinal slits. Pistil compound of 2–5 united carpels; locules 2–5; ovules 1–many and attached on axile placentas; ovary inferior; style 1, stigma capitate or lobed, sometimes sessile. Fruit an achene, berry, capsule, or drupe. Seeds 1–many, embryo small and straight, endosperm present (Fig. 9.171).

The family Caprifoliaceae consists of 15–18 genera and 400–500 species distributed most often in North Temperate climates and mountains of the tropics. The largest genera are *Lonicera,* the honeysuckles (150 species), and *Viburnum* (150 species).

Economically the family is important for its ornamentals. Most commonly grown shrubs are species of *Abelia; Lonicera; Sambucus,* elderberry; *Symphoricarpos,* snowberry; *Viburnum;* and *Weigela.* Wine and jelly are made from the fruits of *Sambucus.*

The genera *Sambucus* and *Viburnum* have various characters that make them not fit in the Caprifoliaceae very well. For example, *Sambucus* is distinctive in having pinnately compound leaves. How these well-known genera should be classified is presently open for discussion among botanists.

The family has no known fossil record.

## Family Valerianaceae (Valerian)

Figure 9.172 Valerianaceae. *Valeriana:* (a) leafy plant having terminal flowers; (b) winged fruit; (c) fruit having pappus-like calyx; (d) flower; (e) flower with open corolla; (f) cross section through fruit.

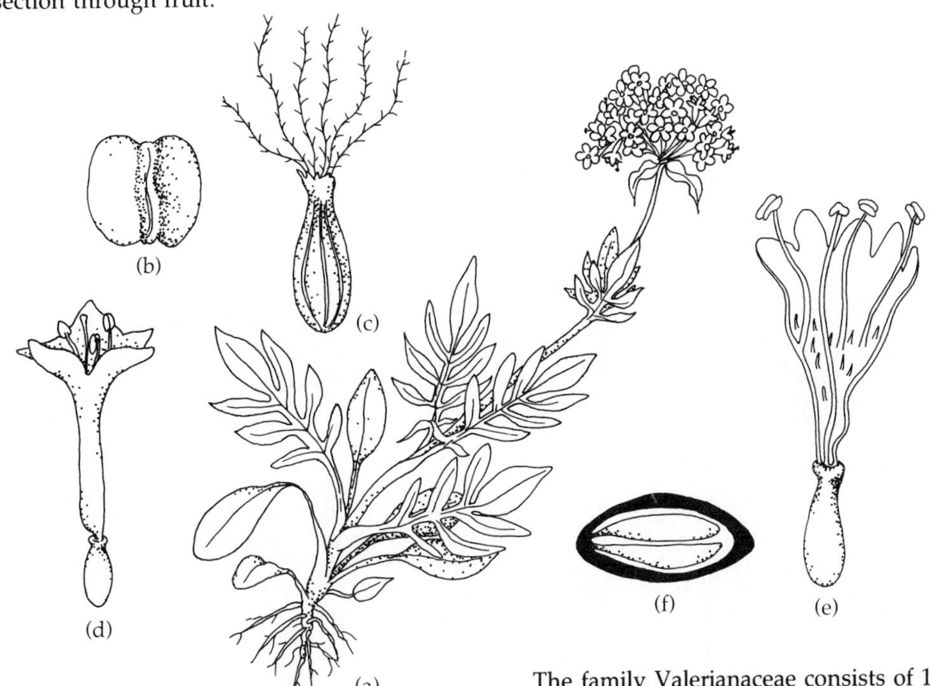


Herbs or shrubs, commonly with an offensive odor. Leaves opposite (rarely basal), simple to pinnately compound, often with clasping petioles; stipules lacking. Flowers regular or irregular, usually perfect (rarely unisexual), epigynous; inflorescence of various types of cymes. Sepals 5 (sometimes lacking), adnate to the top of the ovary, usually reduced at flowering and becoming pappus-like at fruiting. Petals 4–5, sympetalous, commonly with a long corolla tube; nectar spur often present; spurred flowers moth pollinated. Stamens 2–4 (rarely 1), filaments distinct, adnate to the upper part of the corolla tube and alternate with the lobes; anthers opening by longitudinal slits. Pistil compound of 3 (2 become reduced and abort) united carpels; locules 1; ovules 1 and borne on an apical placenta; ovary inferior; style 1, slender, stigma capitate or lobed. Fruit an achene, often with plumose hairs or winged. Seed 1, embryo large and straight, endosperm lacking (Fig. 9.172).

The family Valerianaceae consists of 13 genera and 300–400 species with a cosmopolitan distribution, with the greatest diversity in the north temperate regions, the Mediterranean, and the mountains of South America. The largest genera are *Valeriana,* valerian (250 species), and *Valerianella* (50 species).

Economically the family is of minor importance. The most commonly recognized ornamental is *Centranthus ruber,* red valerian. Some species are grown for dyes or perfume: *Nardostachys jatamansi,* spikenard, and species of *Valeriana.* Leaves and root extracts from *Valeriana officinalis* are used to treat nervous disorders. *Valerianella locusta,* corn salad or lamb's lettuce, is eaten in Europe as a pot herb.

The characters of the family Valerianaceae, when taken together, keep the family together as a unit in between the Caprifoliaceae and Dipsaceae. The genus *Triplostegia* is sometimes placed in its own family, the Triplostegiaceae.

The family has no known fossil record.

## Family Dipsacaceae (Teasel)

Figure 9.173 Dipsacaceae. *Dipsacus:* (a) stem bearing flowers in heads; (b) fruit; (c) side view of flower. *Scabiosa:* (d) front view of flower; (e) side view of flower; (f) longitudinal section through flower.

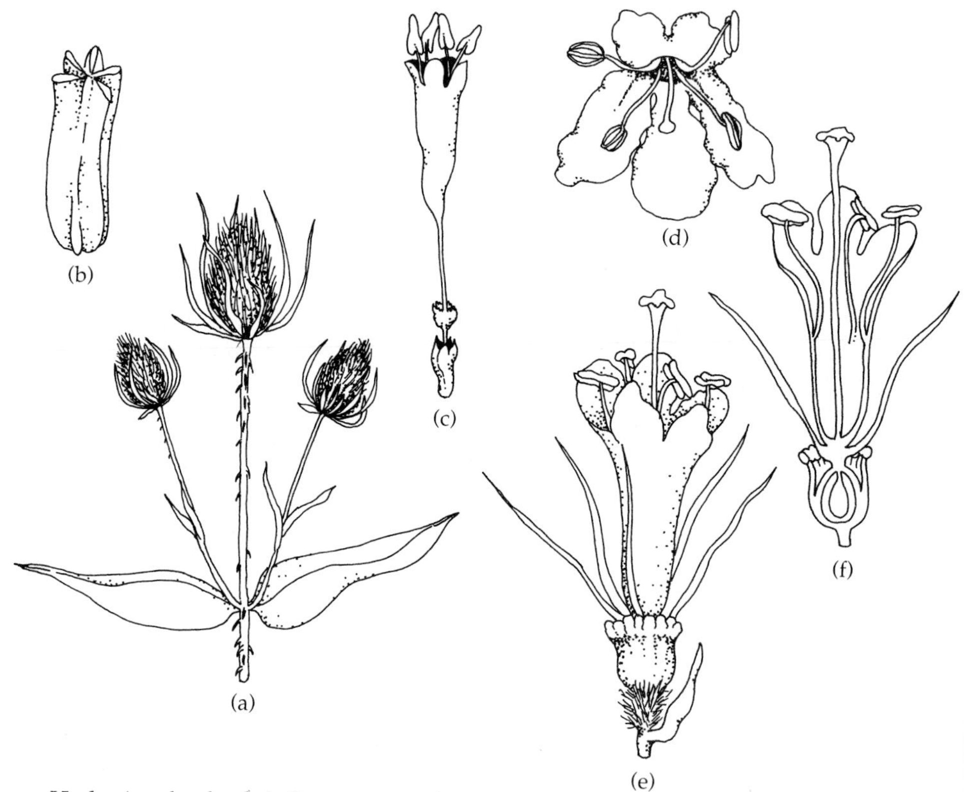


Herbs (rarely shrubs). Leaves opposite or whorled, simple to pinnately compound; stipules lacking. Flowers irregular to almost regular, perfect, epigynous; inflorescence a head or spike that is subtended by an involucre of bracts; each flower of the head subtended by an epicalyx. Sepals 5, small, connate, cup-shaped or the lobes separated into 5 or more bristles or teeth. Petals 4 or 5, sympetalous. Stamens 4 (rarely 2 or 3), filaments distinct, adnate to the upper part of the corolla tube; anthers opening by longitudinal slits; nectary disk present. Pistil compound of 2 united carpels but 1 carpel aborts; locules 1; ovules 1 and borne on an apical placenta; ovary inferior; style 1, stigma entire or 2 lobed. Fruit an achene, often surrounded by an epicalyx and with the calyx persistent. Seeds 1, embryo large and straight, endosperm present (Fig. 9.173).

The family Dipsacaceae consists of 10–11 genera and 270–350 species natively distributed in Europe to eastern Asia, central and southern Africa. The largest genera are *Scabiosa*, scabious (80 species); *Cephalaria* (65 species); and *Knautia* (60 species).

Economically, the family is of little importance. Some species of *Cephalaria, Pterocephalus,* and *Scabiosa* are cultivated as ornamentals. *Morina persica* seeds are eaten like rice in Iran. In the United States *Dipsacus sylvestris,* teasel, has become a tall, introduced noxious weed.

Superficially the family looks like the Asteraceae because of the head-type inflorescence and subtending bracts, but differs in its calyx, stamen, and fruit characters.

The family has no known fossil record.

## ORDER ASTERALES
### Family Asteraceae (Compositae) (Aster)

Figure 9.174 Asteraceae. *Aster:* (a) leafy plant bearing flowers in heads (ray and disk flowers). *Echinops:* (b) disk flower with awn pappus and bract at base. *Lactuca:* (c) mature fruit (achene) with elongate beak and pappus of capillary bristles. Stylized sketches: (d) side view of disk flower with barbed pappus; (e) bract from base of flower; (f) longitudinal section through disk flower; (g) longitudinal section through head of ray and disk flowers; (h) ray (strap) flower with bristle pappus; (i) head of ray and disk flowers surrounded by involucre of bracts. *(continued)*

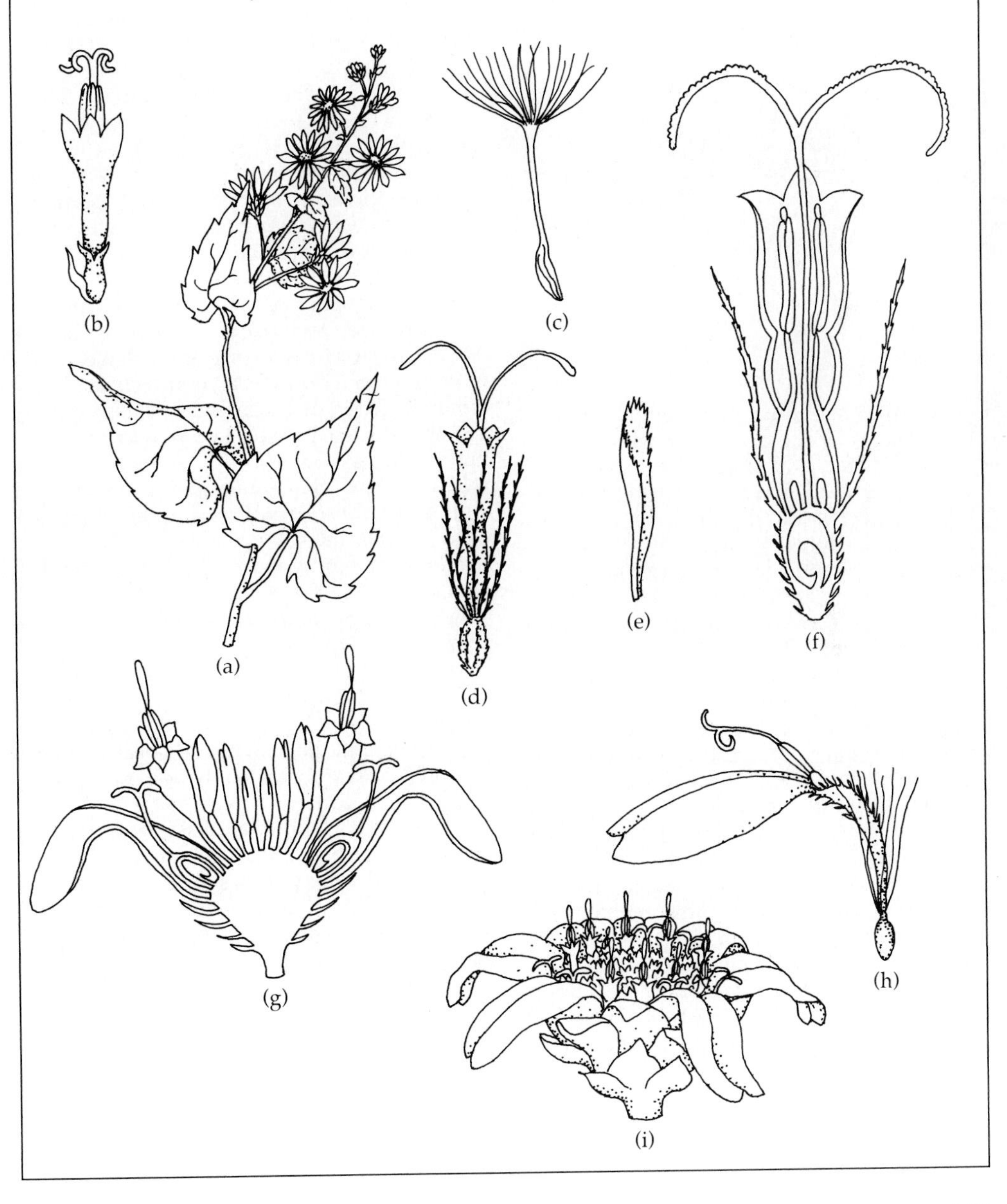

Figure 9.174 *(cont.)*
*Eupatorium:* (j) leafy plant bearing heads of disk flowers only; (k) opened head of disk flowers surrounded by bracts.

Herbs or shrubs (less commonly succulent trees or lianas), often with various types of glands and pubescence, sometimes with milky juice. Leaves alternate or opposite (rarely whorled), simple to pinnately compound; stipules lacking. Flowers regular or irregular, perfect or imperfect, epigynous; inflorescence a head, subtended by an involucre of *phyllaries,* or secondarily the heads arranged in corymbs, cymes, panicles, or racemes; flowers organized on the receptacle as **discoid** heads with only **disk flowers, radiate** heads with **ray (ligulate** or **strap) flowers** around the margin of the head, and disk flowers in the center and ray flowers toward the margin of the receptacle. Sepals lacking or as modified nongreen **pappus,** which may be as awns, bristles, feather-like, or scales, attached to the top of the ovary. Petals 3 or 5 (rarely 4), sympetalous and sometimes bilabiate (2-lipped) or with 1 lip. Stamens 5 (rarely 4), filaments distinct or connate, adnate to the corolla tube and alternate the corolla lobes; anthers opening by longitudinal slits and connate around the style, usually with some apical appendage; pollen released inside the anther tube and pushed out by the growing style. Pistil compound of 2 united carpels; locule 1; ovules 1 and borne on a basal placenta; ovary inferior; style 1 and 2 lobed, stigma various. Fruit an achene, often with pappus attached, seeds 1, embryo straight, endosperm usually lacking (Fig. 9.174).

The family Asteraceae is one of the largest plant families, with 1100 genera and about 20,000 species distributed worldwide, with the exception being mainland Antarctica. It is well represented in warm climates of the Mediterranean and subtropical regions and poorly represented in the tropical rain forest. The largest genera are *Senecio* (1500 species), *Vernonia* (900 species), *Hieracium* (over 800 species), and *Eupatorium* (600 species). The family includes many commonly encountered wild and cultivated plants.

Important food plants include *Cichorium endivia,* endive; *C. intybus,* chicory; *Lactuca sativa,* lettuce; *Helianthus tuberosus,* Jerusalem artichoke; and *Tragopogon porrifolius,* salsify. Some species are cultivated for their food- and oil-producing seeds: *Carthamus tinctorius,* safflower and *Helianthus annuus,* sunflower. Pyrethrum is an insecticide that comes naturally from *Tanacetum cinerariifolium.* The family is rich in many chemical compounds, and therefore many species are used in natural medicines. Some examples are *Anthemis nobilis* produces chamomile; and *Artemisia cina* and *A. maritima* produce santonin, that is used to treat intestinal parasites. Many ornamentals are grown in gardens: *Aster,* aster; *Dahlia,* dahlia; *Erigeron,* daisy; *Helianthus,* sunflower; *Leucanthemum,* ox-eye daisy; *Solidago,* goldenrod; and *Zinnia,* zinnia. *Ambrosia artemisiifolia* and *A. trifida,* the ragweed, have become noxious "hayfever" plants in North America. Other bothersome "weeds" are *Carduus* and *Cirsium,* thistle; *Sonchus,* sow thistle; *Taraxacum,* dandelion; and *Xanthium,* cocklebur.

In spite of the immense size of the family, the genera within it are easily recognized. The diversity allows the Asteraceae to be divided into two subfamilies, Cichorioideae and Asteroidea, each with a number of tribes. Debate is still ongoing among botanists as to the classification of the tribes, but the number of tribes ranges from 12 to 17.

Fossils attributed to the Asteraceae have been found relatively recently in the fossil record. Both fossil pollen and what appears to be a fossilized head have been found in Oligocene deposits.

## *SELECTED REFERENCES*


Barnes, B. V., and W. H. Wagner, Jr. 1981. *Michigan Trees.* Univ. Michigan Press, Ann Arbor, MI, 343 pp.

Benson, L. 1979. *Plant Classification.* D. C. Heath, Lexington, MA, 901 pp.

Cronquist, A. 1981. *An Integrated System of Classification of Flowering Plants.* Columbia Univ. Press, New York, 1262 pp.

Cronquist, A. 1988. *The Evolution and Classification of Flowering Plants,* 2nd ed. The New York Botanical Garden, Bronx, NY, 555 pp.

Degener, O. 1940. *Flora Hawaiiensis.* Vol. 4. Published by author.

Gleason, H. A. (ed.). 1952. *New Britton and Brown Illustrated Flora of the Northeastern United States and Adjacent Canada.* New York Botanical Garden, Bronx, NY.

Gleason, H. A. 1963. *New Britton and Brown Illustrated Flora of the Northeastern United States and Adjacent Canada.* 3 vols. Hafner, New York.

Heywood, V. H. (ed.). 1985. *Flowering Plants of the World.* Prentice Hall, Englewood Cliffs, NJ, 335 pp.

Hickey, M. and C. King. 1981. *100 Families of Flowering Plants.* Cambridge University Press, Cambridge, 567 pp.

Hitchcock, C. L., et al. 1955–1969. *Vascular Plants of the Pacific Northwest.* 5 parts. Univ. Washington Press, Seattle, WA.

Hutchinson, J. 1959. *The Families of Flowering Plants,* 2 vols. Oxford University Press, Oxford, 792 pp.

Hutchinson, J. 1964. *The Genera of Flowering Plants. Dicotyledons.* I. Oxford University Press, London, 516 pp.

Hutchinson, J. 1967. *The Genera of Flowering Plants. Dicotyledons.* II. Oxford University Press, London, 659 pp.

Keng, H. 1978. *Orders and Families of Malayan Seed Plants.* Singapore University Press, Singapore, 437 pp.

Kubitzski, K. (ed.). 1977. *Flowering Plants Evolution and Classification of Higher Categories.* Plant Systematics and Evolution, Suppl. I. Springer-Verlag, Vienna.

Lawrence, G. H. M. 1951. *Taxonomy of Vascular Plants.* Macmillan, New York, 823 pp.

Mabberley, D. J. 1987. *The Plant Book.* Cambridge University Press, Cambridge, 706 pp.

Metcalf, C. R. and L. Chalk. 1950. *Anatomy of the Dicotyledons.* 2 vols. Oxford University Press, Oxford, 1500 pp.

Preston, R. J. 1989. *North American Trees.* 4th ed. Iowa State Univ. Press, Ames, IA, 407 pp.

Shukla, P. and S. P. Misra. 1979. *An Introduction to Taxonomy of Angiosperms.* Vikas Publ. House, New Delhi, India, 546 pp.

Smith, J. P. 1977. *Vascular Plant Families.* Mad River Press, Eureka, CA, 320 pp.

Thorne, R. F. 1983. Proposed New Realignments in the Angiosperms. *Nord. J. Bot.* 3:85–117.

Walters, D. R. and D. J. Keil. 1988. *Vascular Plant Taxonomy,* 3rd ed. Kendall/Hunt, Dubuque, IA, 488 pp.

Willis, J. C. 1973. *A Dictionary of Flowering Plants and Ferns,* 8th ed. (Rev. by H. K. A. Shaw.) Cambridge University Press, London, 1245 pp.

Wood, C. E. 1974. *A Student's Atlas of Flowering Plants: Some Dicotyledons of Eastern North America.* Harper and Row, New York, 120 pp.

Zomlefer, W. B. 1983. *Common Florida Angiosperm Families.* Part I. Biological Illustrations, Gainesville, FL, 108 pp.

Zomlefer, W. B. 1986. *Common Florida Angiosperm Families.* Part II. Biological Illustrations, Gainesville, FL, 106 pp.

# 10

# *Families of Flowering Plants II. Liliopsida (Monocots)*

## DIVISION MAGNOLIOPHYTA (ANGIOSPERMS)

### Class Liliopsida (Monocotyledons)

The Liliopsida (monocots) and the Magnoliopsida (dicots) are different in a combination of various characters. Monocots are herbaceous, less commonly woody, and lack secondary tissues and the development of a vascular cambium and complete vascular bundles; the vascular bundles are scattered or in more than 1 ring, and vessels are usually lacking from the stem but are present in roots. The leaves are mostly with parallel or pinnate-parallel venation (rarely pinnate or palmate); the base of the leaf commonly forms a sheath around the stem. The perianth of the flowers is normally 3-merous (rarely 2 or 4), or is reduced in size or lacking. The stamens are 3 or 6, with pollen grains being usually uniaperturate or derived from uniaperturate (rarely triaperturate) pollen. The pistil is mostly compound of 3 united carpels, less commonly of few to many distinct carpels. Cotyledons 1.

The Liliopsida consists of 5 subclasses, 19 orders, 65 families, and approximately 50,000 species.

**Subclass I. Alismatidae** The Alismatidae consists of approximately 500 species and 16 families grouped in 4 orders. (See Table 10.1.) It is a group of semiaquatic or aquatic herbs, sometimes lacking chlorophyll and associated with mycorrhizal fungi. The leaves are usually alternate, simple, and parallel veined, with narrow to broad leaf blades. The flowers are small to relatively large, regular or irregular, perfect or unisexual, hypogynous or epigynous; inflorescence of various types and commonly subtended by a spathe. Sepals and petals 3 (rarely more or lacking). Stamens 1–many. Pistil simple of 1–many usually distinct carpels (less commonly 2–3 united carpels); locules 1; ovules 1–many; ovary superior or inferior. Fruit mostly a follicle, less commonly an achene or drupe. Seeds usually lacking endosperm.

**TABLE 10.1 A list of orders and family names of flowering plants found in the subclass Alismatidae.**

Division Magnoliophyta (Flowering Plants)
Class Liliopsida (Monocots)

Subclass I. Alismatidae

- Order A. Alismatales
  - Family 1. Butomaceae
  - 2. Limnocharitaceae
  - 3. Alismataceae*
- Order B. Hydrocharitales
  - Family 1. Hydrocharitaceae*
- Order C. Najadales
  - Family 1. Aponogetonaceae
  - 2. Scheuchzeriaceae
  - 3. Juncaginaceae*
  - 4. Potamogetonaceae*
  - 5. Ruppiaceae
  - 6. Najadaceae*
  - 7. Zannichelliaceae
  - 8. Posidoniaceae
  - 9. Cymodoceaceae
  - 10. Zosteraceae*
- Order D. Triuridales
  - Family 1. Petrosaviaceae
  - 2. Triuridaceae

*NOTE:* Family names followed by an asterisk are discussed and illustrated in the text.

Fossils attributed unquestionably to the Alismatidea are found first in Paleocene deposits.

## ORDER ALISMATALES
## Family Alismataceae (Water Plantain)

Figure 10.1 Alismataceae. *Sagittaria:* (a) plant bearing flowers; (b) pistil; (c) longitudinal section through pistil; (d) anther; (e) flower in fruit; (f) longitudinal section through flower; (g) face view of female flower; (h) face view of male flower.

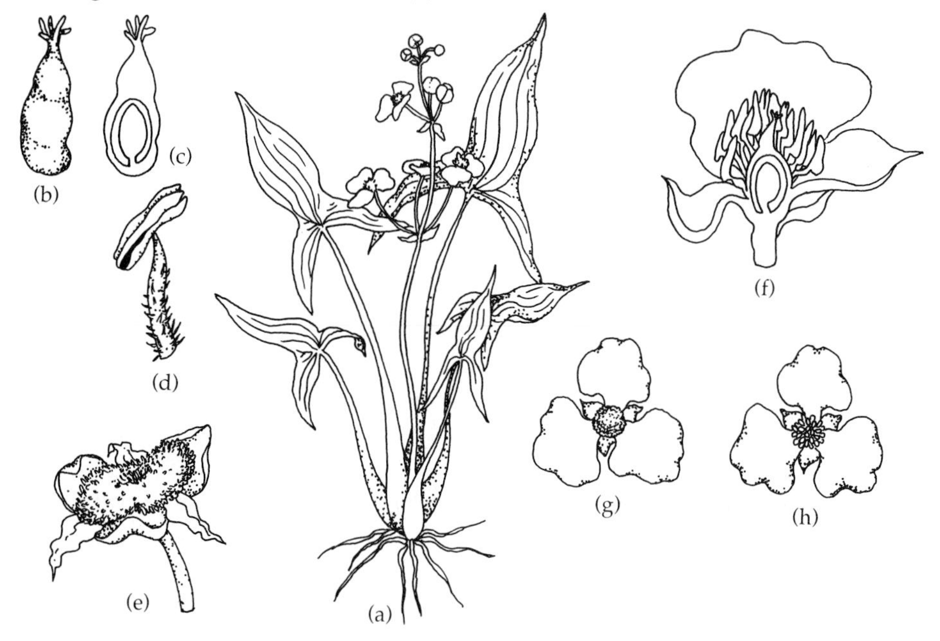


Herbs, semiaquatic or aquatic, glabrous. Leaves alternate and basal, simple, linear to ovate to sagittate or hastate, petiole distinct with a flattened sheathing base; stipules lacking. Stem a scape with pedicels usually in whorls of 3, modified stem cormlike or creeping. Flowers regular, perfect or unisexual and monoecious (rarely dioecious), hypogynous; inflorescence of a solitary flower or borne in whorled panicles, racemes, or umbels. Sepals 3, distinct, green and commonly persistent. Petals 3 (rarely lacking), distinct, lasting only for a day. Stamens 3, 6, 9–many, filaments distinct; anthers opening by longitudinal slits; nectaries present. Pistil simple of 3–many individual carpels, usually closed; locule 1 per carpel; ovules 1 (rarely 2) and borne on a basal placenta; ovary superior; style 1, terminal, stigma decurrent. Fruit an achene (rarely as follicles). Seeds with a curved or folded embryo, endosperm lacking (Fig. 10.1).

The family Alismataceae consists of 11–12 genera and 75–100 species with a cosmopolitan distribution, with the majority of species in the Northern Hemisphere. The largest genera are *Echinodorus,* burhead (25 species); *Sagittaria,* arrowleaf (20 species); and *Alisma,* water plantain (5–9 species).

Economically the family is of minor importance. The corms of *Sagittaria latifolia* were eaten by the North American Indians, while today *S. sagittifolia* is grown and eaten in China and Japan. A few species of *Alisma, Echinodorus,* and *Sagittaria* are cultivated as ornamental wetland and pond plants. Most species are important ecologically as a valuable food source for wildlife, especially migratory water birds.

Superficially the family looks like the Ranunculaceae, a Magnoliidae (dicot) family, but there are many characters in which the two groups differ.

Fossils attributed to the Alismataceae have been collected from Paleocene and Oligocene deposits.

## ORDER HYDROCHARITALES
### Family Hydrocharitaceae (Tape Grass)

Figure 10.2 Hydrocharitaceae. *Elodea:* (a) terminal stem segment bearing flowers. *Limnobium:* (b) plants arising from a creeping runner; (c) longitudinal section through female flower; (d) stigma; (e) cross section through ovary; (f) fruit; (g) male flower.

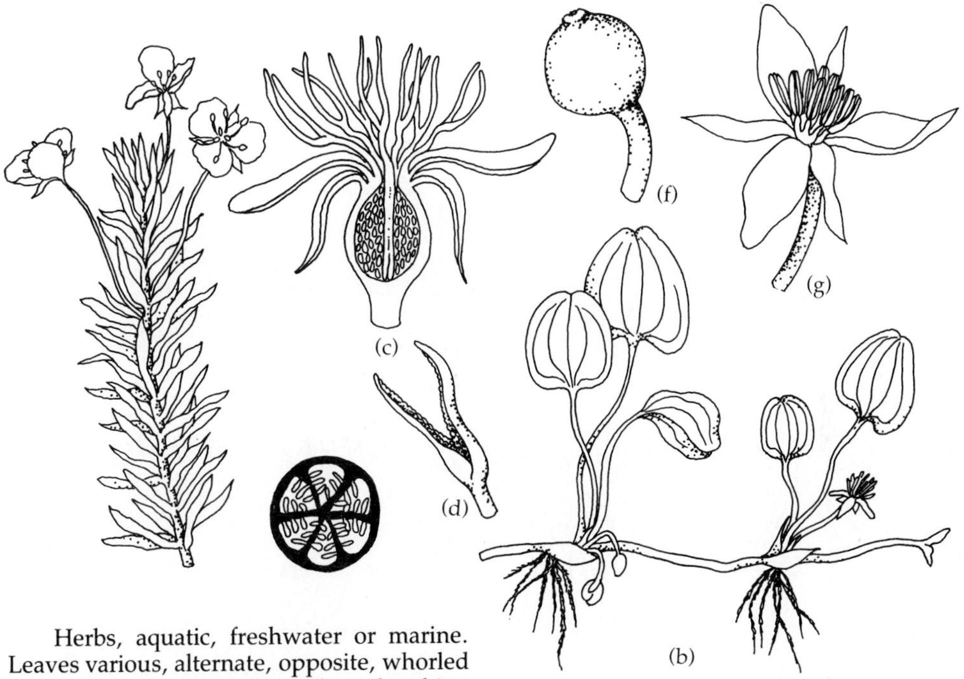


Herbs, aquatic, freshwater or marine. Leaves various, alternate, opposite, whorled or basal; simple, usually with a sheathing base. Flowers large to minute, regular to slightly irregular, perfect or unisexual, epigynous; inflorescence of a single flower (usually pistillate) or flowers borne in cymes or umbels, occasionally a slender hypanthium, subtended by 1 or 2 spathe-like bracts, sometimes male flowers become disconnected in bud and float on the water surface. Sepals 3 (rarely lacking), distinct and green. Petals 3 (rarely lacking), distinct, attached to the top of the ovary. Stamens 2–many, filaments distinct or connate in 1 to several whorls; anthers opening by longitudinal slits; pollen sometimes in chains; water pollinated. Pistil weakly compound of 2–20 partially united carpels; locules 1; ovules scattered over the inside wall of the ovary and borne on the intruding parietal partitions and laminar placenta; ovary inferior; styles 2–20, commonly bi- or trilobed. Fruit a berry or capsule. Seeds few to many, embryo straight, endosperm lacking (Fig. 10.2).

The family Hydrocharitaceae consists of 15 genera and 100 species distributed worldwide. The largest genera are *Elodea* (15 species) and *Vallisneria* (8 species).

The family is of some importance commercially as aquarium plants. Species of *Elodea* and *Vallisneria* are commonly sold for this use. Negatively, some taxa have become noxious aquatic weeds, requiring costly eradication. These include *Elodea canadensis*, Canadian waterweed, in Europe; *Hydrilla verticillata* in the United States; and *Lagarosiphon* in New Zealand.

Pollination occurs in, at, or above the water surface by chains of pollen grains. This is similar to the method used by the marine plant family Zosteraceae.

Most fossils attributed to the family have been found from lower Tertiary and more recent deposits. However, a fossil rhizome reported to be Hydrocharitaceae has been collected from Upper Cretaceous strata.

## ORDER NAJADALES
### Family Juncaginaceae (Arrow Grass)

Figure 10.3 Juncaginaceae. *Triglochin:* (a) flowering plant; (b) cross section through ovary; (c) base of leaf showing ligule; (d) fruits; (e) flower.

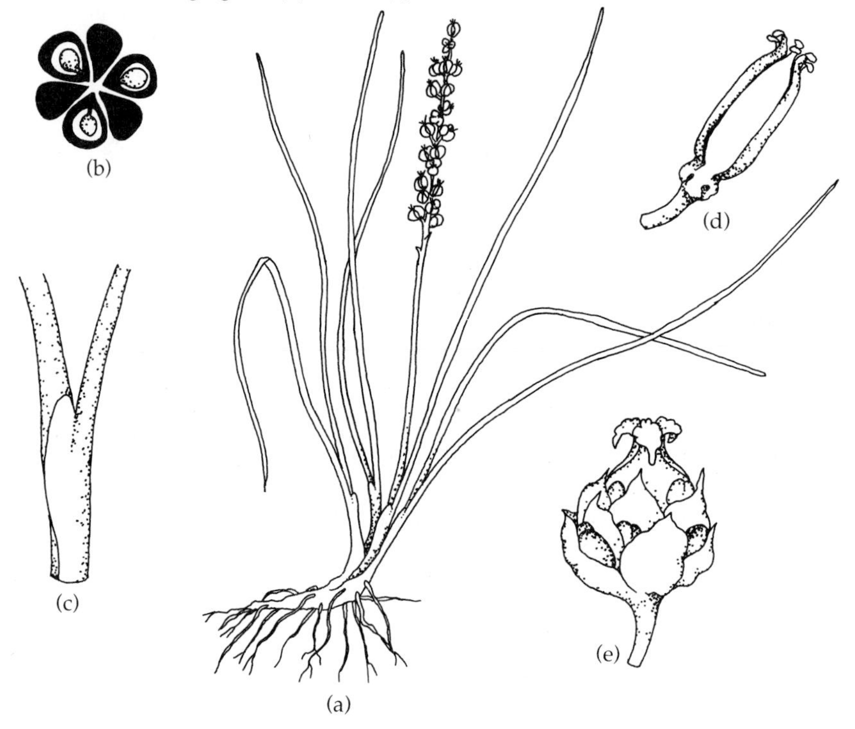


Herbs, mostly of bogs and similar habitats (rarely aquatic and submerged). Leaves alternate, basal, simple, with a prominent ligule where the base of the narrow leaf expands and sheathes the stem. Flowers small, regular, perfect or unisexual, hypogynous; inflorescence a raceme or spike. Perianth usually 6 (rarely 1, 3, or 4) tepals, distinct in 2 series, green or red colored. Stamens usually 6 (rarely 1, 3, 4, or 8), filaments distinct and short, adnate to the base of the sepal; anthers opening by longitudinal slits; wind pollinated. Pistil simple or compound or 4 or 6 (rarely 1), distinct or partially united carpels; locules 4 or 6 (rarely 1); ovules 1 per locule and borne on a basal placenta; ovary superior; style 1 per carpel, short or lacking, stigma sessile or feather-like. Fruit of achenes or follicles. Seeds 1 per locule, embryo straight, endosperm lacking (Fig. 10.3).

The family Juncaginaceae consists of 3–5 genera and 14–20 species distributed widely in marshy and bog habitats in the cool temperate regions of the Northern and Southern Hemisphere. The largest genus is *Triglochin* (12–15 species).

Economically the family is of little importance. The rhizome of *Triglochin procerum* is used for food by the aborigines of Australia. Livestock are sometimes poisoned by *Triglochin* leaves, which allegedly contain hydrogen cyanide. The fruit is occasionally sold as bird seed in France.

The characters of the family are similar in various ways to other Liliopsida (monocot) families that inhabit aquatic habitats, such as the Najadaceae. The monotypic family Scheuchzeriaceae is sometimes included by botanists with the Juncaginaceae. The genus *Lilaea* is sometimes placed in its own family Lilaeaceae.

The family has no known fossil record.

## Family Potamogetonaceae (Pondweed)

Figure 10.4 Potamogetonaceae. *Potamogeton:* (a) flowering plant with both broad floating leaves and narrow submerged leaves; (b) inflorescence; (c) two flowers on fleshy stem; (d) base of leaf and prominent sheathing ligule; (e) flower; (f) fruit.

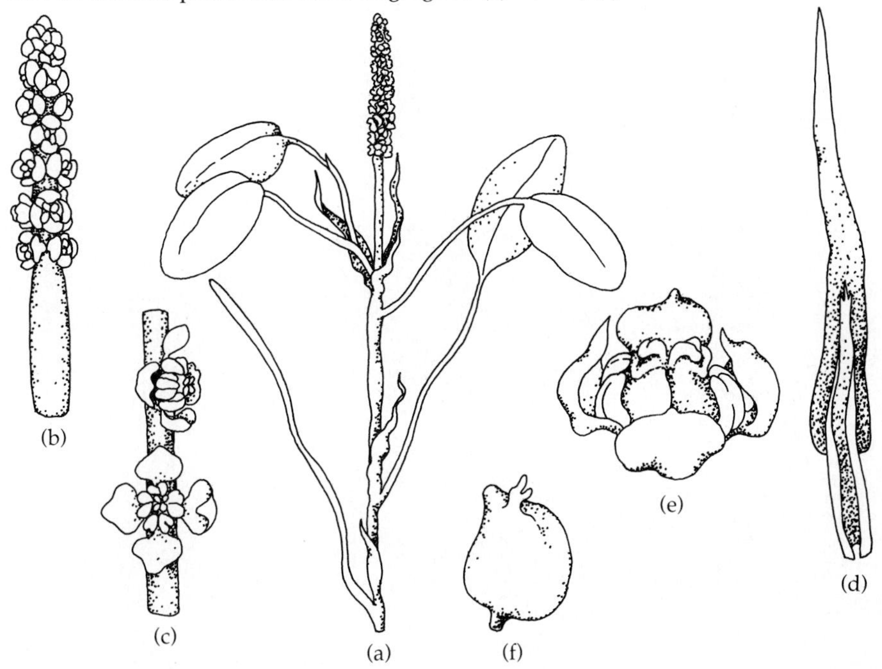


Herbs, perennial (rarely annual), aquatic, brackish or fresh water. Leaves alternate (rarely opposite or whorled), submerged or floating, simple, entire; sheathing stipule-like bases. Flowers small, regular, perfect, hypogynous; inflorescence an axillary or terminal fleshy spike; wind or water pollinated. Perianth of 4 tepals, distinct, fleshy, commonly clawed. Stamens 4, distinct, adnate to the perianth and opposite each segment; anthers sessile and opening by longitudinal slits. Pistil simple of 4 distinct carpels, alternate the stamens; locules 1 per carpel; ovule 1 per locule and borne on a basal-marginal placenta; ovary superior; style 1 per carpel, short or lacking, stigma sessile or on a short style. Fruit an achene or drupe. Seeds 1, embryo with prominent hypocotyl, endosperm lacking (Fig. 10.4).

The family Potamogetonaceae consists of 1–2 genera and about 100 species with a cosmopolitan distribution. The family consists of the genera *Potamogeton,* pondweed, and the monotypic *Groenlandia densa* (sometimes placed in *Potamogeton*).

Economically the family is of little importance but is of biological importance as food for aquatic birds and mammals, and as an oxygenator of water. Some species of *Potamogeton* have become noxious aquatic weeds in canals and irrigation ditches. The fleshy, starchy rootstocks can be eaten for food.

The genus *Groenlandia* differs from *Potamogeton* in having leaves opposite or whorled, lacking the sheathing leaf bases, and having an achene fruit. The genus *Ruppia* is sometimes included in the Potamogetonaceae or in its own family, Ruppiaceae.

Fossils attributed to the Potamogetonaceae have been found in Paleocene and more recent deposits.

## Family Najadaceae (Water Nymph)

Figure 10.5 Najadaceae. *Najas:* (a) leafy plant; (b) male flower; (c) seed; (d) female flower at node.

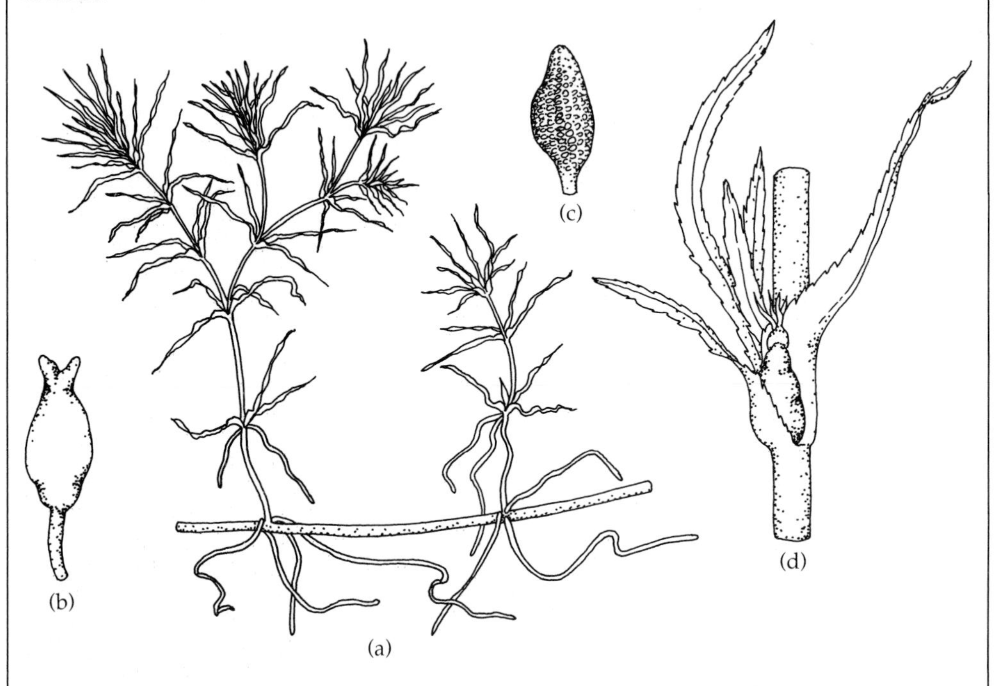


Herbs, submerged aquatic of brackish or freshwater, stems slender. Leaves opposite or whorled, simple, narrowly linear, toothed with sheathing base. Flowers small, irregular or naked, unisexual (monoecious or dioecious), hypogynous to partially epigynous; inflorescence of a solitary flower or flowers borne in small axillary clusters; pollinated under water. Perianth of male flowers flask-shaped, a membrane-like sheath (a spathe), 2 lipped, subtended by distinct or connate involucre scales or bracts. Perianth of female flowers lacking or connate, a sheath or spathe, sometimes adnate to the ovary. Stamen 1, filament lacking with a sessile anther but filament elongating with maturity; anther opening irregularly; pollen sacs 2 or 4, pollen lacking pores. Pistil simple of 1 carpel; locule 1; ovule 1 and borne on a basal placenta; ovary superior or inferior; styles 1, short, stigmas 2–4, linear. Fruit an achene. Seed 1, embryo straight, endosperm lacking (Fig. 10.5).

The family Najadaceae consists of 1 genus, *Najas,* and 35–50 species with a cosmopolitan distribution.

Economically the family is of little importance. In subtropical to tropical climates it is used as a green mulch. *Najas* can become a noxious aquatic weed in irrigation ditches, ponds, and rice paddies. Ecologically the group is a good oxygenator of water and is used by fish for food.

As with many aquatic plants, the morphology of the family is highly variable, making identification of taxa difficult.

Fossil seeds attributed to *Najas* have been found in Oligocene and more recent deposits.

## Family Zosteraceae (Eel Grass)

Figure 10.6 Zosteraceae. *Phyllospadix:* (a) plant bearing inflorescences; (b) male spadix; (c) fruit.

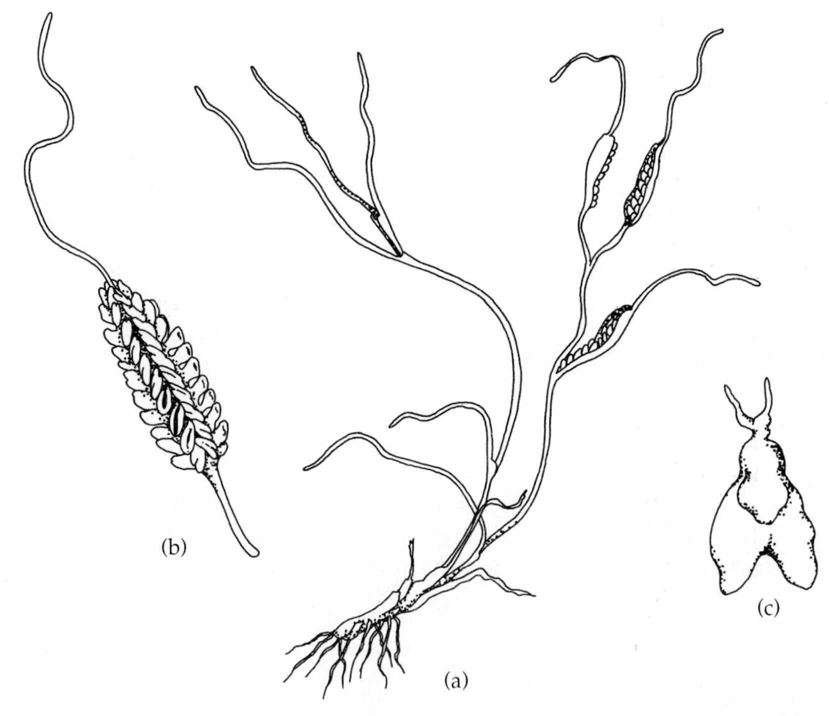


Herbs, grass-like from a creeping rhizome, floating or submerged and marine, but may be exposed at low tide. Leaves alternate, simple, linear and grass-like, with the bases sheathing the stem and stipule-like; sheath open or closed. Flowers small, unisexual (monoecious or dioecious), mostly naked, hypogynous; inflorescence a flattened spadix that is enclosed by the spathe-like uppermost leaf sheath. Perianth segment 1, bract-like or lacking. Stamen of male flower 1; pollen thread-like, water pollinated. Pistil of female flower simple or compound of 1 distinct or 2 united carpels; locule 1; ovule 1 and borne on an apical placenta; ovary superior; style 1, connate, stigmas 2, long. Fruit a small achene or drupe. Seed 1, endosperm lacking (Fig. 10.6).

The family Zosteraceae consists of 3 genera and 18 species distributed worldwide in the cooler, temperate marine waters of the Northern and Southern Hemispheres. The largest genus is *Zostera,* eel grass (12 species).

Economically the family is of minor importance. *Zostera maritima,* common eel grass, is widespread in the intertidal zone and provides shelter and is a food source for many small marine organisms and water birds. Fine-quality glass and instruments have been packed in the past in dry *Zostera* stems and leaves to prevent breakage.

As a result of certain morphological similarities, some botanists have placed the Zosteraceae in the family Potamogetonaceae.

The validity of older reports of fossils attributed to the Zosteraceae being collected in Upper Cretaceous and Eocene deposits are now being questioned by paleobotanists.

**Subclass II. Arecidae** The subclass Arecidae is a moderate-sized group with approximately 5600 species, 5 families in 4 orders. (See Table 10.2.) Over half of the species are found in the family Arecaceae, the palms. It is a very diverse group that includes the smallest flowering plant in the world, *Wolffia* (1–2 mm in diameter), and also the largest monocots in the world, the palms.

The Arecidae include herbs, shrubs, trees, and lianas. The body type ranges from the reduced, floating thallus family Lemnaceae, to the large, woody (a different secondary growth than seen in the dicots) trees of the Arecaceae (Palmae), the palms. The leaves are usually alternate (absent in the Lemnaceae) or basal with parallel veins, but some orders are palmately or pinnately veined, lobed, or compound. Flowers of many taxa are relatively small and crowded into a spadix and subtended by a leafy bract, the spathe (the family Araceae). Some flowers are 3-merous (3 sepals, 3 petals), but with others the perianth is reduced or lacking. The stamens range in number from 1 to many. Pistils are usually compound of 3 united carpels, but occasionally the carpels are distinct. The position of the ovary is superior, sometimes confused by the ovary being sunken into a fleshy receptacle. The fruits are berries, drupes, or a multiple. The endosperm of the seed is usually oily or of protein, rarely starchy.

There are more fossils attributed to the palms than any other group within the subclass, and these have been found in Upper Cretaceous deposits. In spite of this, few taxonomic conclusions can be made.

**TABLE 10.2 A list of orders and family names of flowering plants found in the subclass Arecidae.**

| Subclass II. Arecidae |
|---|
| Order A. Arecales |
| Family 1. Arecaceae (Palmae)* |
| Order B. Cyclanthales |
| Family 1. Cyclanthaceae* |
| Order C. Pandanales |
| Family 1. Pandanaceae* |
| Order D. Arales |
| Family 1. Araceae* |
| 2. Lemnaceae* |

*NOTE:* Family names followed by an asterisk are discussed and illustrated in the text.

## ORDER ARECALES
### Family Arecaceae (Palmae) (Palm)

Figure 10.7 Arecaceae (Palmae). *Sabal:* (a) tree habit; (b) side view of perfect flower; (c) longitudinal section through flower. *Pseudophoenix:* (d) tree habit; (e) fruits. *Roystonea:* (f) flower cluster of two male and one female flowers; (g) front view of male flower.

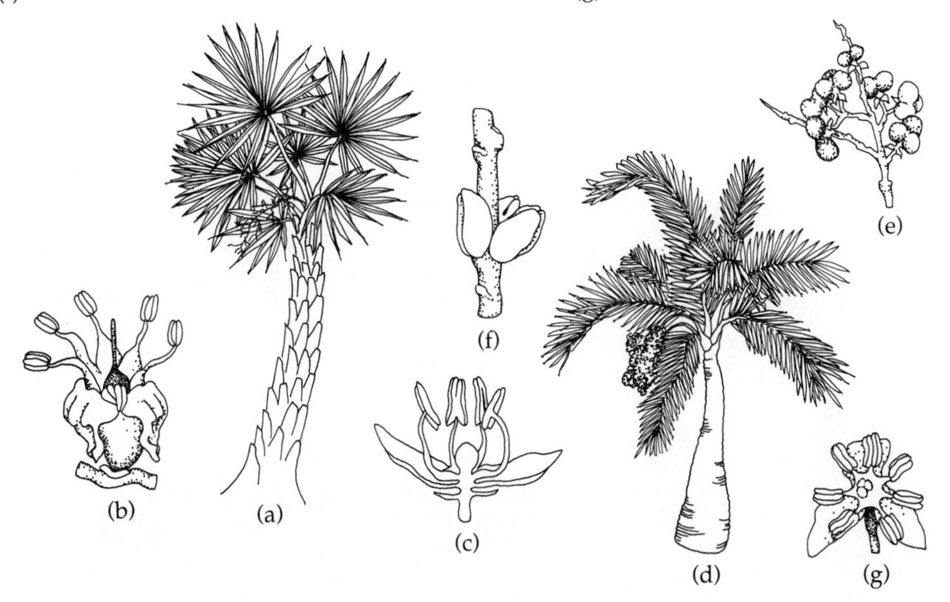


Shrubs, trees, or lianas with unbranched (rarely branched) trunk or stem, short or tall (to 60 m), often from creeping, underground rhizome. Leaves alternate or crowded at trunk apex, evergreen, large (to 20 m long), simple to palmately or pinnately lobed or compound; petioles long with sheathing base; stipules lacking. Flowers many, regular to slightly irregular, perfect or unisexual, hypogynous, sometimes with hypanthium; inflorescence of various types, subtended by 1–several spathes. Sepals 3 (rarely 2), distinct or connate. Stamens 6 (occasionally 3–many) in 2 whorls of 3, filaments distinct or connate or adnate to the petals; anthers opening by longitudinal slits; pollinated by insects or wind; staminodes commonly present; nectaries present or absent. Pistil simple or compound of 3 (rarely up to 7) distinct or united carpels; locules 1–3; ovules 1 per locule and borne on an axile, basal, or parietal placenta; ovary superior; style 3 (rarely 1–7), may be connate at base, stigmas sometimes sessile. Fruit a berry or drupe. Seeds 1, sometimes large, with small plumule and 1 radicle, endosperm present (Fig. 10.7).

The family consists of over 200 genera and about 3000 species distributed pantropically; a few species extend into warm temperate climates. The largest genera are *Calamus* (370 species) and *Bactris* (230 species).

*Cocos nucifera,* coconut palm; *Phoenix dactylifera,* date palm; *Elaeis guineensis,* oil palm; and species of *Metroxylon,* sago palm, are food sources. Strong fibers come from various parts of the plant. *Ceroxylon* and *Copernicia* produce valuable waxes. *Areca catechu* is the betel nut palm of Africa and Southeast Asia. Many species are fine ornamentals. The largest seeds in the world are produced by *Lodoicea maldivica,* the Seychelles palm or double coconut.

The "woody," arborescent large-leaved nature of the palms makes the family stand out among the Liliopsida. The group's complexity is reflected in the lack of agreement among botanists as to how the genera, tribes, and subfamilies should be classified.

Fossils reliably attributed to the family have been found in Upper Cretaceous deposits. Pre-Cretaceous reports are presently being questioned.

## ORDER CYCLANTHALES
## Family Cyclanthaceae (Panama Hat Plant)

Figure 10.8 Cyclanthaceae. *Carludovica:* (a) leafy plant bearing inflorescence; (b) male flower; (c) segment of spadix showing female and male flowers; (d) female flower with thread-like staminode.

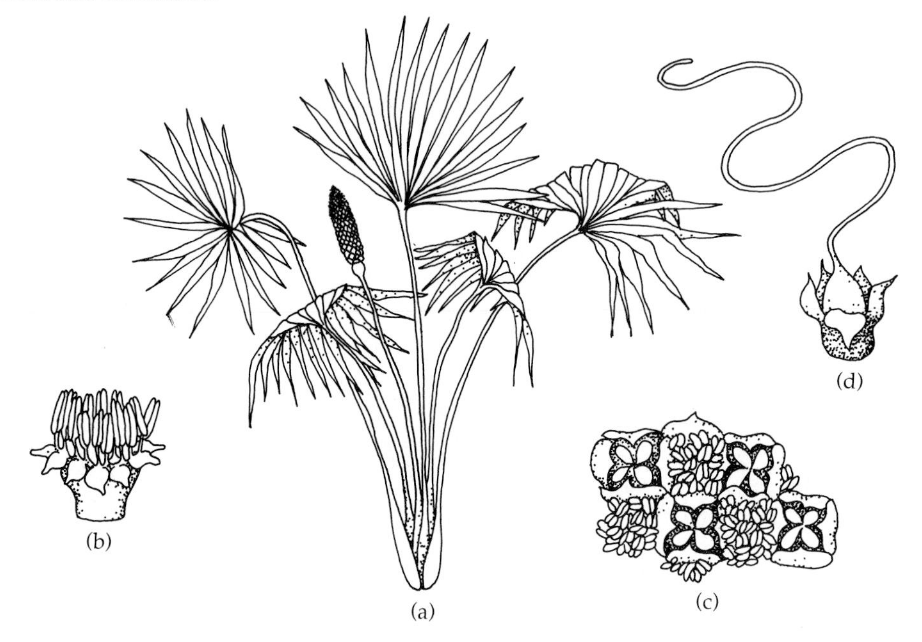


Herbs, lianas, or epiphytes, perennial from a rhizome, milky or watery juice present. Leaves alternate or spirally arranged, blade palm-like, deeply lobed or compound, petiole with sheathing base. Flowers small and numerous, unisexual and monoecious, sessile; inflorescence a spadix (that includes both flower types), usually enclosed by 2 or more bracts. Perianth of male flowers 4–many (rarely lacking) in 1 or 2 whorls, distinct or connate, cup-shaped; perianth of female flower with 4 tepals, partly adnate to 4 staminodes and opposite them, often flowers become fused together collectively. Stamens of male flowers 6–many, filaments connate at the base; anthers opening by longitudinal slits, nectar glands lacking; pollinated by beetles. Pistil of female flowers compound of 4 united carpels; locule 1; ovules few to many and borne on 1 or 4 apical or 4 parietal placentas; ovary superior but sunken into the axis; style 1, short, stigmas 1–4, spreading. Fruit a single berry or multiple of fused berries. Seeds few to many with a fleshy covering, embryo small, endosperm present (Fig. 10.8).

The family Cyclanthaceae consists of 11 genera and 180 species distributed in the New World tropical rain forests. The largest genus is *Asplundia* (89 species).

Economically the family is of some importance, especially *Carludovica palmata,* the Panama hat plant, from which young leaves are woven into Panama hats; the older leaves are made into baskets and mats. The leaves of *C. angustifolia* and *C. sarmentosa* are used for making thatched roofs and brooms, respectively.

The family has many characters similar to the Arecaceae, the palms, and some botanists have suggested the two families be combined as one.

Fossils attributed to the Cyclanthaceae have been identified from Eocene deposits.

## ORDER PANDANALES
### Family Pandanaceae (Screw Pine)

Figure 10.9 Pandanaceae. *Pandanus:* (a) leafy plant showing prop roots; (b) male flower; (c) longitudinal section through pistil; (d) multiple fruit; (e) female inflorescences.

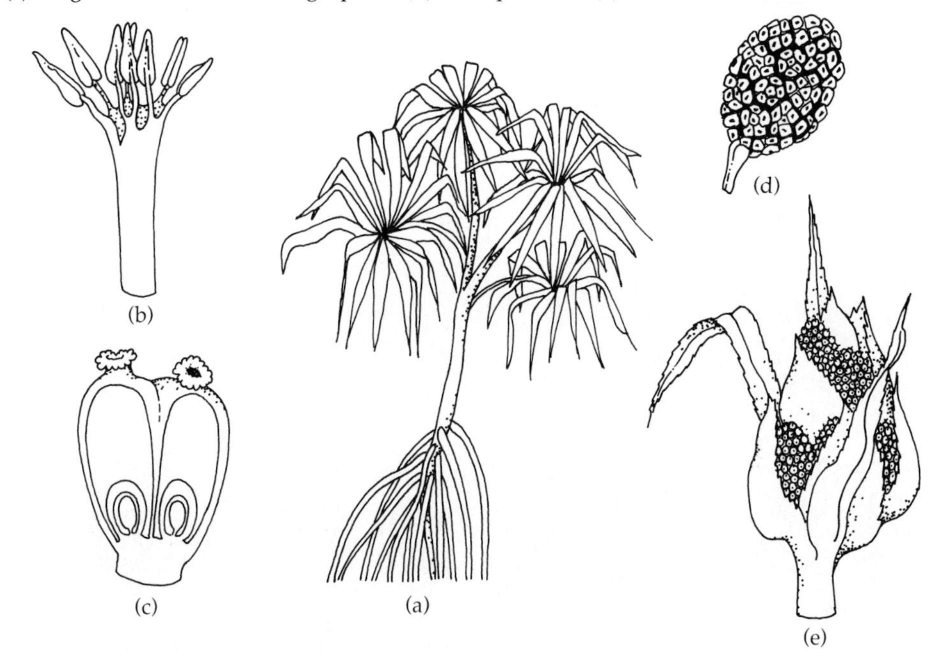


Shrubs, trees, or lianas with tall stems and often with stout, adventitous prop roots. Leaves alternate in 3–4 ranks, stem commonly twisted with leaves appearing in spirals, simple, long and stiff with spiny margins; base sheathing. Flowers very small, unisexual and dioecious, incomplete, hypogynous; inflorescence a spadix subtended by 1–several spathes, which in turn may be arranged in racemes or umbels; usually wind pollinated but less commonly insect, bat, or bird pollinated. Perianth of male and female flowers lacking. Stamens of male flower 1–many, filaments distinct or connate much of their length, sometimes staminodes in female flowers. Pistil of female flower simple or compound of 1–many distinct or fused carpels; locules 1–many depending on the degree of carpel fusion; ovules 1–many per locule and borne on basal or parietal placentas; ovary superior; style usually lacking, stigmas sessile. Fruit a berry or drupe. Seeds and embryo small, endosperm present (Fig. 10.9).

The family Pandanaceae consists of 3 genera and 680–780 species distributed in the Old World subtropics and tropics. The largest genus is *Pandanus,* screw pine (500–600 species).

Economically the family is of some importance, usually locally. Some species of *Pandanus* are eaten for food: *P. andamanensiu, P. leram,* and *P. utilis.* The leaves of *P. odoratissimus* are used for thatching and weaving. A number of species are cultivated as ornamentals, especially *Freycinetia banksii* from New Zealand.

The family shares various characters with the families Arecaceae and Cyclanthaceae.

Microfossils attributed to the family have been found in Upper Cretaceous formations, while fossil fruits have been observed from Eocene deposits.

## ORDER ARALES
### Family Araceae (Arum)

Figure 10.10 Araceae. *Lysichitum:* (a) plant bearing two inflorescences; (b) male flower; (c) spadix and spathe; (d) perfect flower; (e) female flowers; (f) schematic longitudinal section through spadix and spathe.

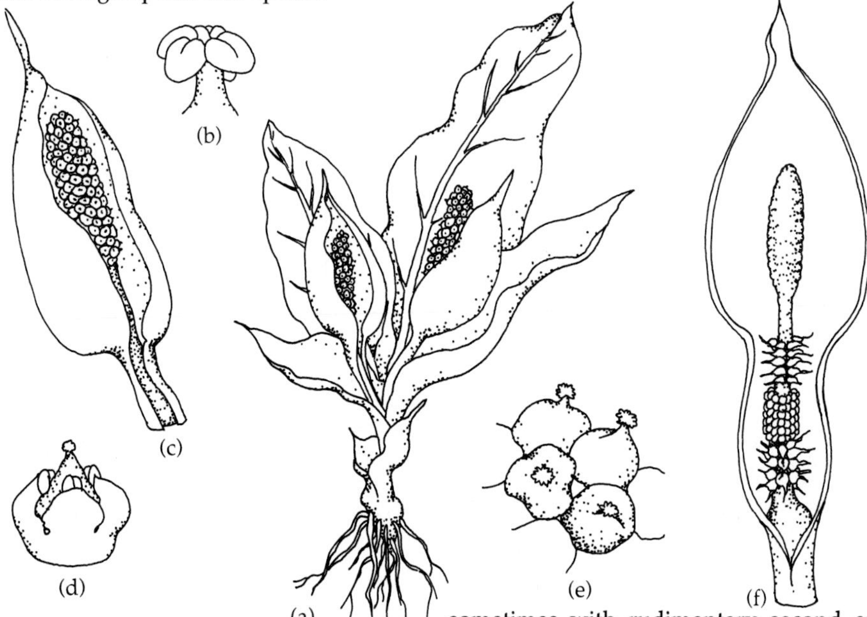


Herbs, shrubs, lianas, or epiphytes sometimes containing milky juice; some poisonous and with adventitous roots. Leaves alternate or basal, simple or compound, venation parallel, pinnate or palmate; petiole usually with a sheathing base often with a prominent ligule. Flowers small and many, regular or irregular, perfect or unisexual (usually monoecious and male flowers above female, rarely dioecious), hypogynous or epigynous; inflorescence a spadix surrounded or subtended by a spathe, often brightly colored. Perianth 4, 6, or lacking, tepals distinct or connate, usually present only in perfect flowers. Stamens 4 or 6 (rarely 1 or 8), filaments distinct or connate, short and broad, opposite tepals if present; anthers opening by apical pores or longitudinal slits. Pistil compound of 2–many united carpels; locules 1–many per carpel; ovules 1–many, borne on apical, axile, basal, or parietal placentas; ovary superior, sometimes embedded in spadix tissue; style 1, short or lacking, stigma sessile. Fruit a berry. Seeds 1–many, embryo large, curved or straight, sometimes with rudimentary second cotyledon, endosperm present or lacking (Fig. 10.10).

The family Araceae consists of 110 genera and 1800–2000 species with a cosmopolitan distribution, but with most taxa found in the subtropical and tropical regions. The largest genera are *Anthurium* (500 species) and *Philodendron* (250 species).

The tubers or roots of *Alocasia; Amorphophallus; Colocasia esculenta,* taro; *Cyrtosperma;* and *Xanthosoma,* tanier, are eaten. Some genera, such as *Dieffenbachia,* dumb cane, *Monstera,* and *Philodendron* are popular foliage house plants but are deadly poisonous if ingested. *Anthurium andraeanum* var. *lindenii* is cultivated for its brightly colored spathe.

The family is easily recognized by its distinctive spadix and spathe, the flowers of which are usually insect-pollinated; some are sweet-smelling, while others attract carrion beetles by smelling like rotting meat.

Fossils attributed to the Araceae have been found in Eocene and more recent deposits.

## Family Lemnaceae (Duckweed)

Figure 10.11 Lemnaceae. *Lemna:* (a) thallus-like stem bearing small roots; (b) fruit on thallus; (c) large root caps on end of root; (d) male flower; (e) one female flower and two male flowers. *Wolffia:* (f) round thallus floating under surface of water.

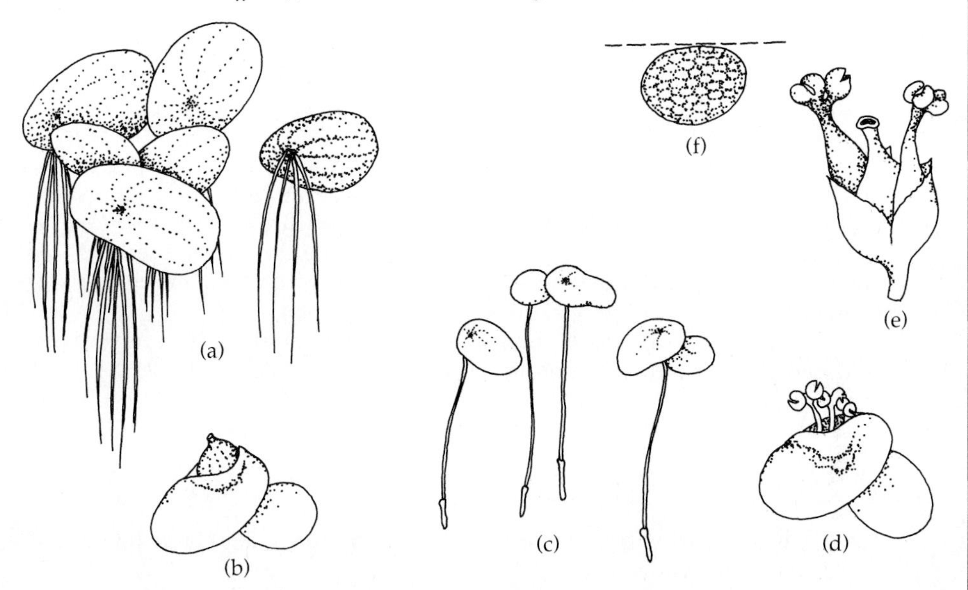


Herbs, floating or submerged aquatic, thallus-like, with or without 1–few simple roots, reproducing mostly by budding. Leaves lacking. Flowers unisexual and monoecious (rarely dioecious), borne in a pouch; inflorescence made up of 1 female flower and/or 2 male flowers that are naked and surrounded by a spathe. Male flower lacking a perianth, 1 or 2 stamens; anthers opening by longitudinal or transverse slits. Female flower lacking a perianth, pistil simple, of 1 carpel; locule 1; ovules 1–7 and borne on a basal placenta; ovary superior; style 1, short, stigma usually sessile. Fruit a utricle. Seeds 1–4, embryo relatively large, straight, endosperm present or lacking (Fig. 10.11).

The family Lemnaceae consists of 6 genera and 30 species with a cosmopolitan distribution. The largest genus is *Lemna*, duckweed (9 species). *Wolffia* (7 species) (especially *W. arrhiza*) is the world's smallest flowering plant (ranging in size from 1 to 2 mm).

Economically the family is of minor importance. Ecologically it is an important fish and waterbird food source, but sometimes becomes a noxious aquatic weed in still, nutrient-enriched water. Because of its prolific asexual budding, it is being studied by NASA as a possible food source when grown artificially during long space journeys.

The family is similar to the Araceae in having the flowers subtended by a sheath or spathe. The thallus-like plants rarely flower, and when they do produce flowers, it appears to be environmentally induced.

The family has no known fossil record.

**Subclass III. Commelinidae** The subclass Commelinidae consists of about 15,000–22,000 species, found in 16 families and 7 orders. (See Table 10.3.) The two families Cyperaceae and Poaceae (Gramineae) make up perhaps 75–80% of the total species.

The group is comprised of herbs (rarely woody plants). The leaves are alternate or basal, simple usually with a narrow blade, mostly with a sheath around the stem, the sheath being open or closed; a **ligule** may be present where the blade and sheath join. Flowers perfect or unisexual, hypogynous, usually lacking nectaries or nectar; pollinated by pollen-gathering insects or wind-pollinated, sometimes self-pollinated **(apomictic).** Perianth usually in 3s, reduced to scales or lacking. Stamens 3 or 6 (rarely 1–many). Pistil compound of 2 or 3 (rarely 4) carpels, various placentation types, ovules 1–many, ovary superior. Fruits usually dry. Seeds mostly have starch as the storage material.

Within the subclass is found considerable variation in flower structure and reduction, and in types of pollination, especially adaptation to wind pollination.

Fossil pollen attributed to the group has been found in Upper Cretaceous deposits.

**TABLE 10.3 A list of orders and family names of flowering plants found in the subclass Commelinidae.**

Subclass III. Commelinidae

- Order A. Commelinales
  - Family 1. Rapateaceae
  - 2. Xyridaceae*
  - 3. Mayacaceae
  - 4. Commelinaceae*
- Order B. Eriocaulales
  - Family 1. Eriocaulaceae*
- Order C. Restionales
  - Family 1. Flagellariaceae
  - 2. Joinvilleaceae
  - 3. Restionaceae*
  - 4. Centrolepidaceae
- Order D. Juncales
  - Family 1. Juncaceae*
  - 2. Thurniaceae
- Order E. Cyperales
  - Family 1. Cyperaceae*
  - 2. Poaceae (Gramineae)*
- Order F. Hydatellales
  - Family 1. Hydatellaceae
- Order G. Typhales
  - Family 1. Sparganiaceae*
  - 2. Typhaceae*

*NOTE:* Family names followed by an asterisk are discussed and illustrated in the text.

## ORDER COMMELINALES
## Family Xyridaceae (Yellow-eyed Grass)

Figure 10.12 Xyridaceae. *Xyris:* (a) plant bearing terminal inflorescences; (b) cross section through ovary; (c) inflorescence with flower; (d) stamens adnate to opened corolla; (e) pistil.

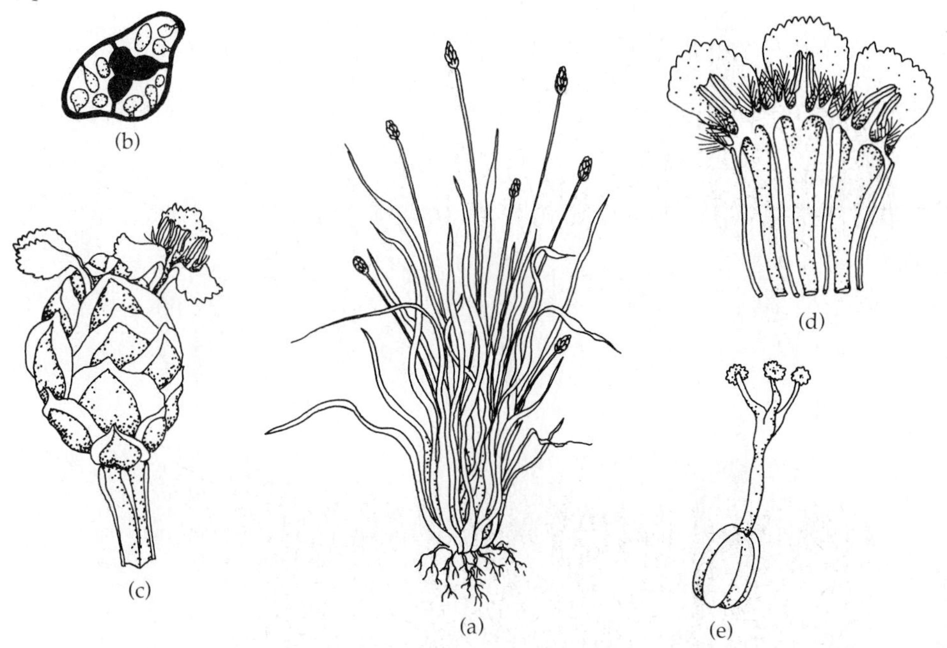


Herbs, mostly perennial of wet, marshy habitats. Leaves alternate or basal, simple, entire, blade flat, cylindric, or equitant, sheathing at the base. Flowers slightly irregular, perfect, sessile (rarely with a pedicle) in the axil of stiff bracts, hypogynous; inflorescence a round head or spike on the end of a scape; nectar or nectaries if present odorless; pollinated by pollen-gathering bees. Sepals 3, distinct, one forms a hood and may enclose the flower parts, remaining sepals boat-shaped and keeled. Petals 3, distinct or connate, clawed, usually yellow. Stamens 3 (rarely 6), filaments distinct, short, adnate to the corolla and opposite the petals, sometimes with 3 alternating staminodes; anthers opening by longitudinal slits. Pistil compound of 3 united carpels; locule 1 (rarely 3); ovules 1–many and borne on axile, free-central, or parietal placentas; ovary superior, style 1 or 3 lobed. Fruit a capsule sometimes with a persistent corolla tube. Seeds many, small, embryo small, may not be well differentiated, endosperm present (Fig. 10.12).

The family Xyridaceae consists of 4 genera and 200–240 species distributed widely in the subtropics and tropics, with few taxa in temperate regions. The largest genus is *Xyris* (180–220 species).

Economically the family is of little importance. Two North American species, *Xyris ambigua* and *X. caroliniana*, have been used to treat colds and skin ailments, respectively. Some species of *Xyris* are sold as aquarium plants.

The family has some characters that are similar to the families Commelinaceae and Eriocaulaceae.

The family has no known fossil record.

## Family Commelinaceae (Spiderwort)

Figure 10.13 Commelinaceae. *Commelina:* (a) leafy plant bearing flowers; (b) inflorescence; (c) cross section through ovary; (d) seed; (e) anther having filament hairs.

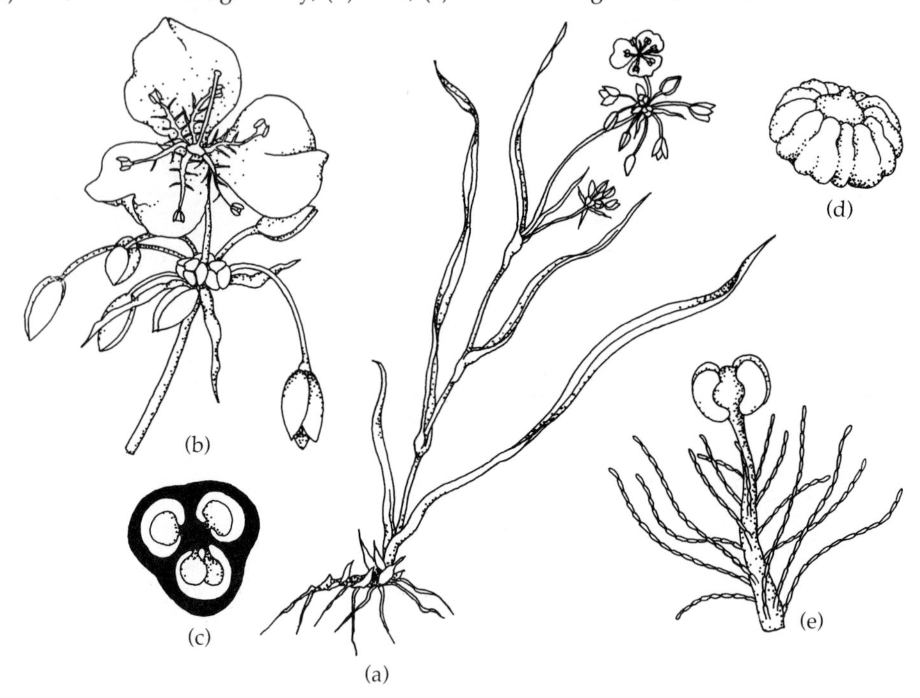


Herbs, commonly succulent with swollen nodes, often with 3-celled glandular hairs. Leaves alternate, simple, entire, sheathing at the base. Flowers regular or irregular, perfect (rarely unisexual), hypogynous; inflorescence a cyme borne at the end of the stem or in leaf axils subtended by 1 or more spathes; nectaries or nectar usually lacking but flowers still insect pollinated. Sepals 3, distinct or connate, green. Petals 3, distinct (rarely connate) usually blue, violet, or white. Stamens 6 (rarely 1) in 2 whorls of 3, filaments distinct, sometimes with brightly colored long hairs; anthers opening by pores or longitudinal slits. Pistil compound of 3 united carpels; locules 3 (rarely 2); ovules 1–few per locule and borne on axile placentas; ovary superior; style 1, stigma capitate or 3 lobed. Fruit a capsule (rarely a berry). Seeds with a roughened or ridged surface and sometimes covered with an aril; embryo location marked by a disk-like structure or swelling on the seed coat, and called the **embryostega;** endosperm present, mealy (Fig. 10.13).

The family Commelinaceae consists of 38–50 genera and 600–700 species distributed widely throughout the subtropical and tropical regions. A few species are found in the United States, China, Japan, and Australia. The largest genus is *Commelina* (150–200 species).

Economically the family is of minor importance, with its greatest use as ornamentals or house plants, including *Cyanotis; Dichorisandra; Commelina; Tradescantia,* spiderwort; and *Zebrina,* wandering jew. Not only is *T. virginiana* grown as an ornamental, but young leaves and shoots are eaten in salads. In Africa *Aneilema beninense* is used as a laxative.

How the genera of the family should be classified is open to opinion and interpretation. Similar morphological features such as sheathed leaves, embryo development, etc. are shared with other families.

The family has no known fossil record.

## ORDER ERIOCAULALES
### Family Eriocaulaceae (Pipewort)

Figure 10.14 Eriocaulaceae. *Eriocaulon:* (a) leafy plant bearing inflorescence at end of scape; (b) male flower; (c) seed; (d) female flower.

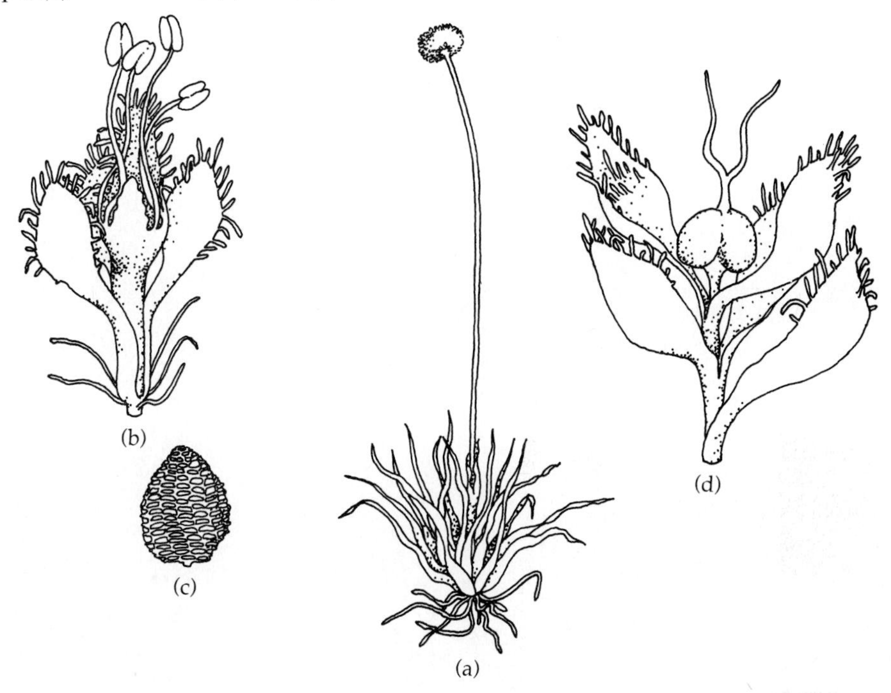


Herbs, perennial (rarely annual), small and growing in wet habitats or in shallow water. Leaves alternate and basal, crowded, simple, linear and narrow, grass-like, sheathing base lacking. Flowers tiny, regular or irregular, unisexual, hypogynous; inflorescence a head at the end of a scape, subtended by an involucre of bracts; nectar and nectaries usually lacking; wind or insect pollinated. Sepals 2 or 3, distinct or connate. Petals 2 or 3 (rarely lacking), distinct or connate. Stamens of male flowers 2–6, filaments distinct or connate, adnate to the petals and opposite them; anthers opening by longitudinal slits. Pistil of female flowers compound of 2 or 3 united carpels, stalked; locules 2–3; ovules 1 per locule and borne on a basal placenta; ovary superior; style 1, stigma 2 or 3 branched. Fruit a loculicidal capsule. Seeds small, embryo small, endosperm present, and mealy (Fig. 10.14).

The family Eriocaulaceae consists of 13 genera and 1200 species distributed widely in the subtropic and tropical regions, with only a few in temperate climates. The largest genera are *Paepalanthus* (500 species); *Eriocaulon*, pipewort (400 species); and *Syngonanthus* (200 species).

Economically the family is of almost no importance, except for the drying and selling of inflorescences as "everlasting." A few species are minor weeds in rice fields.

The family is similar to the family Xyridaceae in its head-type inflorescence at the end of a scape, basal leaves, and overall habit. It differs in various aspects of the flower, thereby emphasizing the group's distinctiveness.

The family has no unquestioned fossil record.

## ORDER RESTIONALES
### Family Restionaceae (Restio)

Figure 10.15 Restionaceae. *Restio:* (a) flowering stems from a creeping rhizome; (b) inflorescence; (c) longitudinal section through ovary; (d) female flower; (e) male flower.

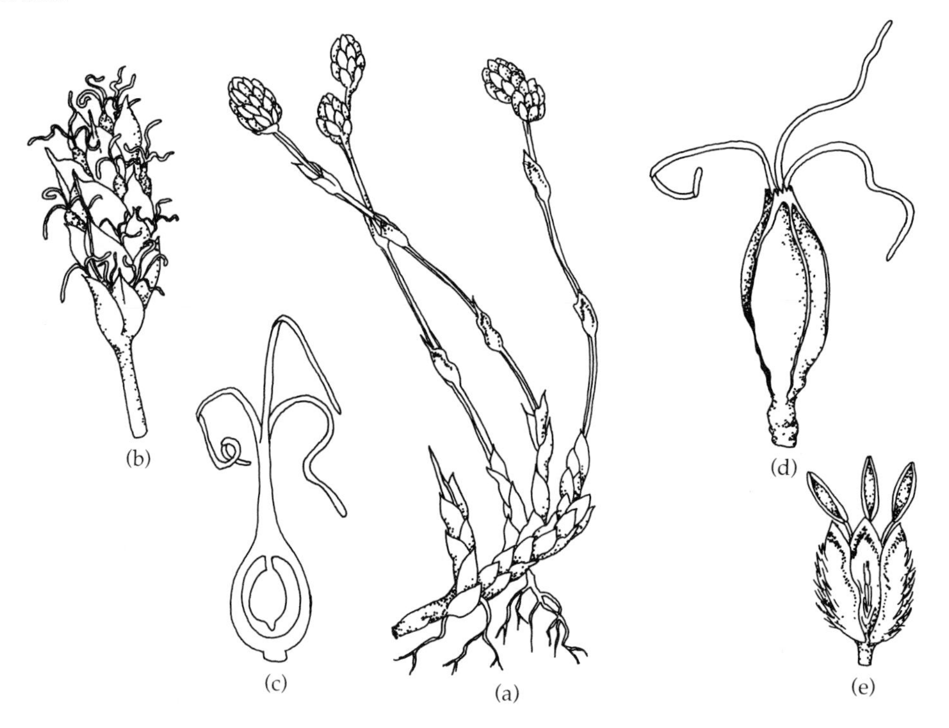


Herbs, perennial from rhizomes, rush-like in appearance with tough, elongate shoots that function photosynthetically. Leaves alternate, blades poorly developed with the sheathing leaf bases functioning as leaves, ligules lacking. Flowers small, regular, unisexual and dioecious (rarely monoecious), hypogynous, subtended by chaffy bracts; inflorescence of spikelets arranged in panicles, racemes, or spikes with a spathe at the base; wind pollinated. Perianth 3–6, dry, thin and chaffy, usually in 2 whorls of 3. Stamens of male flowers 3, filaments distinct (rarely connate); anthers opening by longitudinal slits. Pistil of female flowers compound of 3 united carpels; locules 3 (sometimes 2, rarely 1); ovules 1 per locule, and borne on an apical or apical-axil placenta; ovary superior; style 1–3, free or connate. Fruit a capsule or nut-like. Seeds with small embryo, endosperm present (Fig. 10.15).

The family Restionaceae consists of 30 genera and 320–400 species distributed widely in the Southern Hemisphere, subtropics, and tropics. The largest genus is *Restio* (over 100 species).

Economically the family is of little importance. Some native people do use the stems for brooms, matting, or thatching for their homes.

The family has various characters that are similar to the families Centrolepidaceae, Juncaceae, and Thurniaceae. Two small relatively distinct genera, *Anarthria* and *Ecdeiocolea,* are sometimes put in their own families by botanists.

Fossil pollen attributed to the Restionaceae have been found in Upper Cretaceous and more recent deposits.

## ORDER JUNCALES
### Family Juncaceae (Rush)

Figure 10.16 Juncaceae. *Juncus:* (a) flowering plant; (b) cross section through ovary; (c) anther; (d) flower. *Luzula:* (e) flowering plant from creeping rhizome; (f) seed.

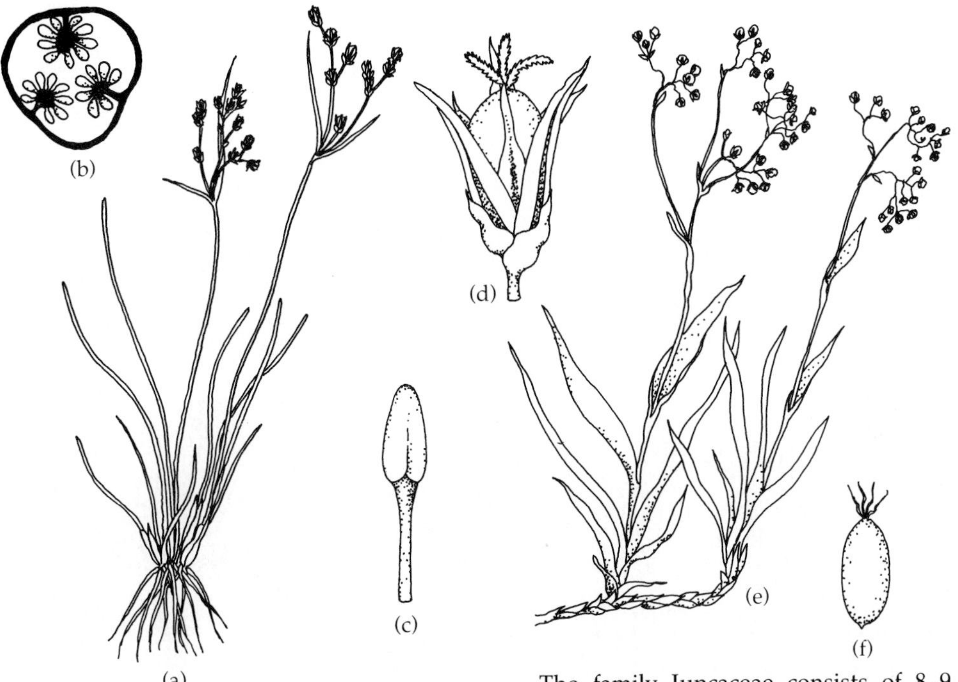


Herbs (rarely shrubs), annual or perennial, stems usually erect, cylindrical and clumped, commonly of cool, damp montane habitats. Leaves alternate and/or basal, simple, blades entire, usually 3 ranked, cylindric or flat, sheathing at the base or reduced to only a sheath, sheath open or closed. Flowers small and inconspicuous, regular, perfect (rarely unisexual), hypogynous; inflorescence of cymes, clustered heads, or panicles, rarely of solitary flowers; wind pollinated. Sepals 3, distinct, scale-like, mostly green, brown or black. Petals 3, distinct, much like the sepals. Stamens 3 or 6, filaments distinct or connate at the base, opposite the perianth parts; anthers opening by longitudinal slits. Pistil compound of 3 united carpels; locules 1 or 3; ovules 3–many and borne on axile, basal, or parietal placentas; ovary superior; styles 1–3, stigma 3. Fruit a loculicidal capsule. Seeds few to many, small, embryo straight, endosperm present (Fig. 10.16).

The family Juncaceae consists of 8–9 genera and 300–400 species distributed worldwide, but chiefly in cool, temperate climates and in montane tropical regions. The largest genera are *Juncus,* the rushes (225 species), and *Luzula,* the wood rushes (80 species).

Economically the family is of little importance. A few species, such as *Juncus effusus* and *J. squarrosus,* are used in making baskets and wicker bottoms for chairs. The group is rarely cultivated and the stiff stems are not very palatable to domestic livestock. Ecologically the rushes do provide waterfowl habitat and are good soil binders to prevent erosion.

The family has various features in common with the Cyperaceae, Liliaceae, Poaceae, and the Restionaceae. How the Juncaceae should be correctly classified is open to discussion by botanists.

Fossils correctly attributed to the family have been found in Miocene and more recent deposits.

## ORDER CYPERALES
### Family Cyperaceae (Sedge)

Figure 10.17 Cyperaceae. *Carex:* (a) upper stem with male and female spikes; (b) pistil; (c) perigynium surrounding pistil; (d) longitudinal section through perigynium showing pistil. *Scirpus:* (e) inflorescence of spikes; (f) pistil and bristle-like perianth. *Eriophorum:* (g) inflorescence; (h) perianth of many bristles; (i) fused sheath of leaf.

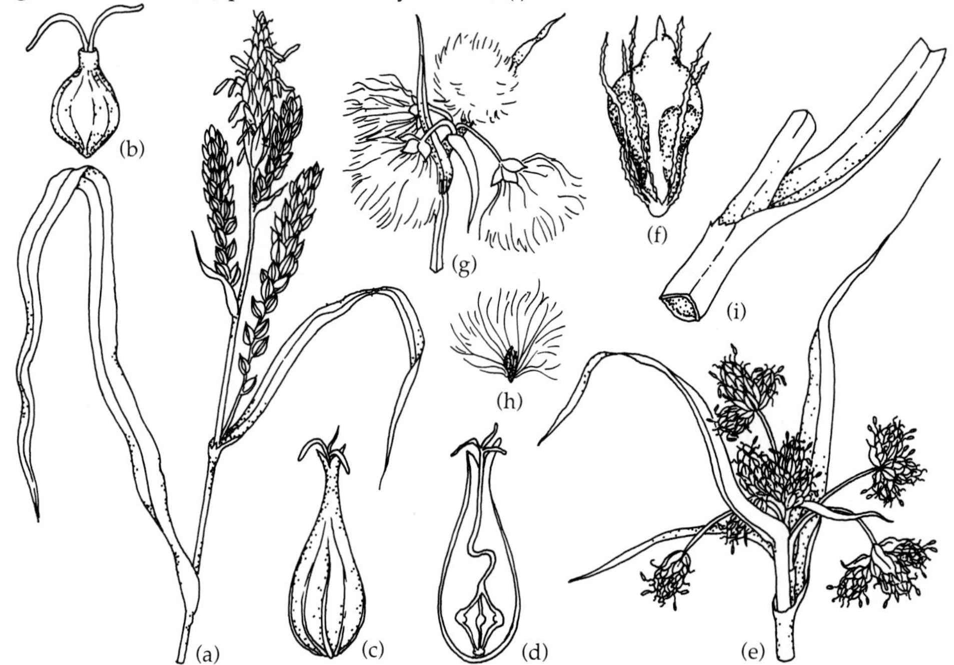


Herbs (rarely shrubs), mostly perennial, grass-like but with 3-sided, solid stems; commonly of wet, marshy habitats. Leaves alternate or basal, simple, 3 ranked, linear; blades round, triangle-shaped, flat, or lacking; sheathing at base, sheath usually closed; ligules present or absent; guard cells of stomata dumbbell-shaped. Flowers small, regular or naked, perfect or unisexual (usually monoecious); each flower subtended by bracts, male flower with 1 bract, female with 2, the second sac-like and surrounding the pistil (called a **perigynium**); arranged in spikes or spikelets, which are variously arranged. Perianth 1–many, reduced to bristles, scales or absent. Stamens usually 3 (rarely 1–6), filaments distinct; anthers opening by longitudinal slits. Pistil compound of 2 or 3 united carpels; locule 1; ovule 1, borne on a basal placenta; ovary superior; style 1 with 2 or 3 stigmatic branches. Fruit an achene, sometimes with a persistent style. Seeds lens- or triangle-shaped. Seed with small embryo, endosperm mealy (Fig. 10.17).

The family Cyperaceae consists of 70–90 genera and 4000 species distributed worldwide, in moist temperate habitats. The largest genera are *Carex,* sedge (1100 species); *Cyperus* (600 species); *Scirpus* (250 species); and *Rhynchospora* (250 species).

*Cyperus papyrus,* the paper or papyrus rush, provided the Egyptians with parchment and reed-constructed boats. Stems of other *Cyperus* are made into mats and material for mudbrick. The tubers of *C. esculentus,* nut sedge, and *Eleocharis tuberosa,* Chinese water chestnut, are eaten for food. *Cyperus esculentus,* introduced into North America, has become a noxious weed in some places.

Superficially the family looks grass-like, but the many distinctive characters (perigynia, etc.) make it stand alone.

Fossils attributed to the family have been found in Eocene deposits.

## Family Poaceae (Gramineae) (Grass)

Figure 10.18 Poaceae (Gramineae). *Poa:* (a) plant in flower with many spikelets; (b) spikelet made up of two glumes and three florets; (c) how spikelet is put together; (d) floret; (e) base of blade and open sheath. *Bromus:* (f) base of blade and closed sheath. *Calamogrostis:* (g) spikelet with bent awn from back of lemma. *Agropyron:* (h) spikelet with awns from end of lemma; (i) lodicules. *Poa:* (j) coma of hair at base of floret.

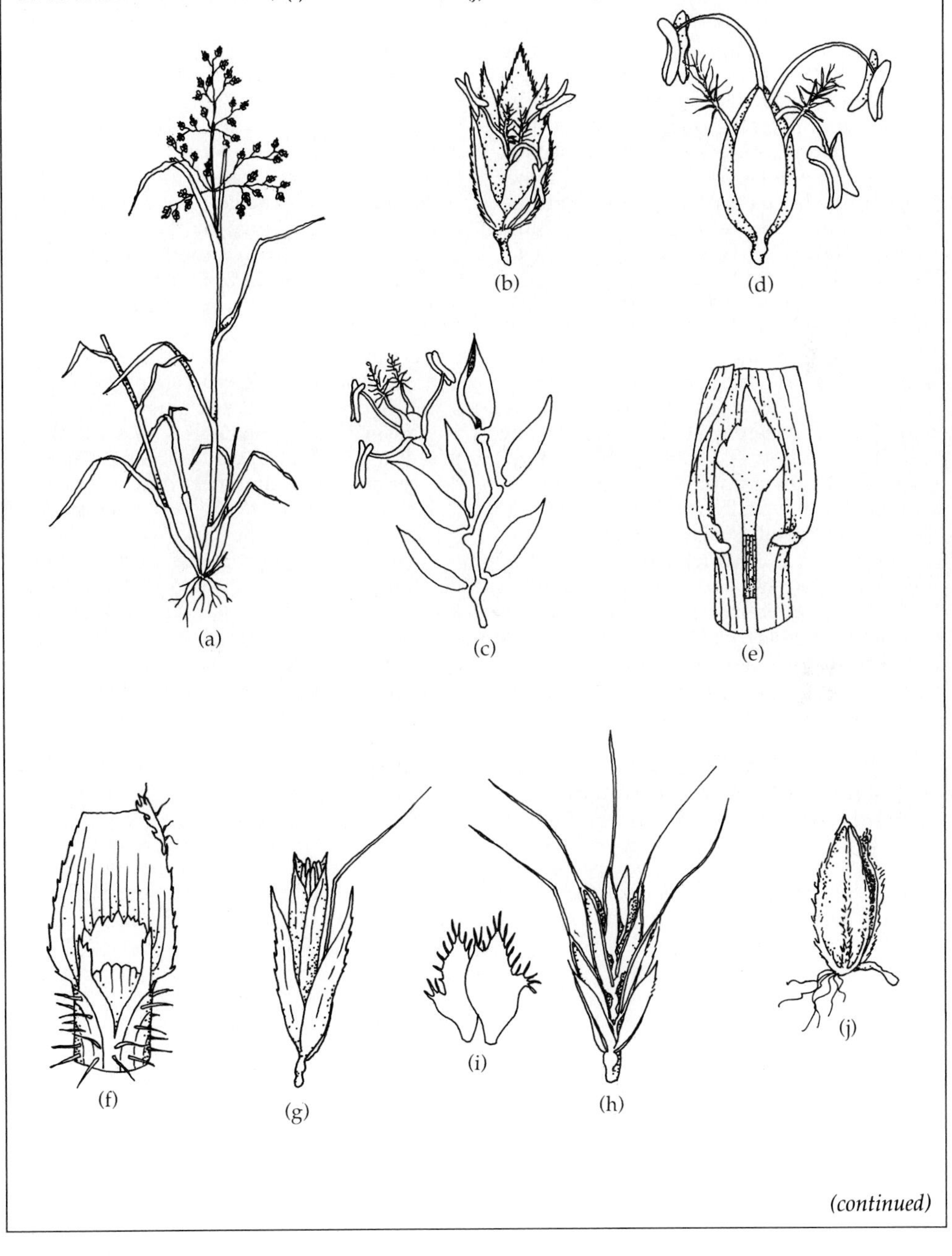


*(continued)*

## Family Poaceae (Gramineae) (Grass) *(continued)*

Herbs, or less commonly woody shrubs or trees, stems round or flat on one side, usually with hollow internodes and swollen nodes. Leaves alternate or basal (rarely lacking) in 2 rows (2 ranked), simple; blade narrow, linear; sheathing around the stem at the base, usually with open (rarely fused) overlapping margins; paired auricles and/or a ligule usually present where the blade, stem, and sheath meet. Flowers small, irregular, perfect or imperfect (monoecious or dioecious); inflorescence of **spikelets,** secondarily arranged in panicles, racemes, or spikes; each spikelet usually with 2 (0, 1–7) **glumes** (bractlets or scales) outside, and 1 or more **florets** (the flower), each enclosed by 2 bractlets, the **lemma** and **palea,** each spikelet attached to a central axis (the **rachilla**). A **fertile floret** contains a pistil or grain, and a **sterile floret** lacks a pistil or has only stamens; the glumes, lemma, and paleas often produce stiff bristles called **awns.** Perianth if present modified as 2 (rarely 3) **lodicules,** distinct and fleshy. Stamens 3 (rarely 1–6 or more), filaments distinct; anthers opening by longitudinal slits; usually wind pollinated with pollen short lived. Pistil compound of 2 or 3 fused carpels, locules 1, ovules 1 and borne on a basal placenta and adnate to the ovary wall, ovary superior, styles 2 (rarely 1 or 3), stigmas feather-like. Fruit a caryopsis or grain (rarely a drupe, nut, or utricle), often enclosed by a persistent lemma and palea. Seed 1, fused to the pericarp, embryo straight, endosperm present (Fig. 10.18).

The family Poaceae consists of 600–650 genera and 7500–10,000 species distributed worldwide in all climates and regions. Grasses are a major component of the various grasslands, savannas, and steppe regions of the world. It has been estimated that 20% of the world's vegetational cover is comprised of grasses. The largest genera are *Panicum* (400 species); *Poa,* the bluegrasses (300 species); *Eragrostis,* the love grasses (300 species); and *Stipa* (200 species).

Economically the Poaceae is the most important plant family to humans. It provides all of the cereal crops, including "the big three": *Triticum spp.,* the wheats; *Oryza sativa,* rice; and *Zea mays,* maize. Other food grasses include *Avena sativa,* oats; *Hordeum vulgare,* barley; *Pennisetum glaucum,* pearl millet; *Saccharum officinarum,* sugar cane; *Seceale cereale,* rye; and *Sorghum bicolor,* sorghum. Grasses make up the major component of grazing, lawn, and turf plants. The woody bamboos are a superb building material and are used in building construction in Asia for matting, thatch, and scaffolding. Various grasses have become noxious weeds in agricultural fields and waste areas. The light windblown pollens create problems for some allergy sufferers, and grasses are major "hayfever" plants. Some grasses are known to develop poisons like hydrocyanic acid under certain environmental conditions or to cause St. Anthony's fire, an abnormality caused by the ergot fungus *Claviceps.*

The classification of grasses into subfamilies and tribes is in a state of disarray at present. In previous classifications grasses were arranged according to characters of the spikelet. Today botanists are finding various micro- and macroanatomical features helpful in grass taxonomy. Therefore, the number of subfamilies, tribes, and genera varies with interpretations of botanists. The number of subfamilies ranges from 3–6, with over 50 tribes.

Fossil "grass-like" pollen has been found in Upper Cretaceous deposits, with fossil grass pollen and grains occurring in Paleocene, Eocene, and more recent deposits.

## ORDER TYPHALES
### Family Sparganiaceae (Bur Reed)

Figure 10.19 Sparganiaceae. *Sparganium:* (a) upper stem bearing axillary inflorescences—male above, female below; (b) stamen; (c) female flower; (d) female inflorescence.

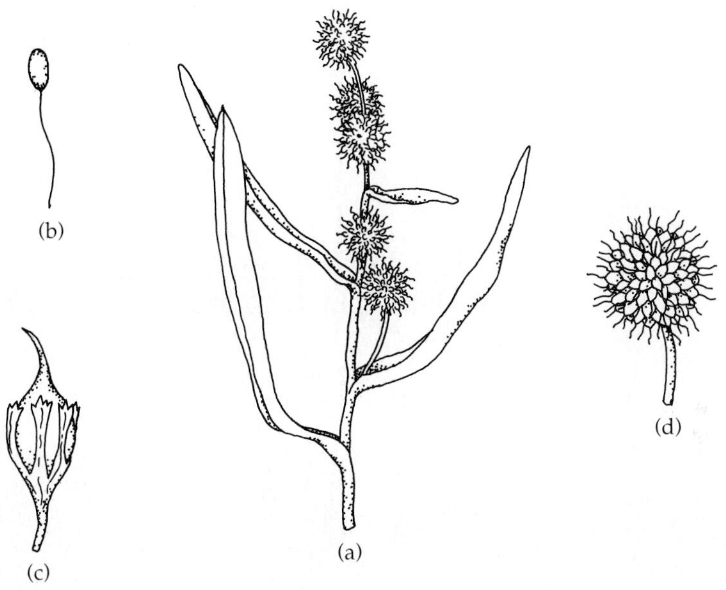


Herbs, perennial from a rhizome, aquatic or semiaquatic. Leaves alternate, 2 ranked, blades linear, entire, thick and spongy, sheathing at the base. Flowers regular, unisexual and monoecious, sessile and grouped into dense heads, attached to the main axis with the female (pistillate) heads below the male (staminate) heads; wind pollinated. Perianth tepals 1–6, distinct and scale-like. Stamens of male flower 1–8, filaments distinct, opposite the tepals when the same number. Pistil of female flower simple or compound of 1–2 (rarely 3) distinct or connate carpels; locules 1–3; ovule 1 per locule and borne on an apical or basal placenta; ovary superior; style 1 or 2, short, persisting as a beak on the fruit, stigma, elongate at the style tip. Fruit an achene subtended by persistent tepals and with a beaked style. Seed with straight embryo, endosperm present and mealy, distributed by water and by water birds internally (Fig. 10.19).

The family Sparganiaceae consists of the genus *Sparganium* and 13–15 species distributed widely throughout the north temperate to arctic zones, and in Australia and New Zealand in aquatic and marshy habitats.

Economically the family is of little importance. Ecologically the plants' stems and leaves are used for nesting material by water birds and their fruits as food for aquatic birds and mammals.

The family has various characters that are similar to the Typhaceae but differs in many other features. Interestingly, a parasitic fungus on the Sparganiaceae is also found on Araceae plants.

Fossils attributed to the Sparganiaceae have been found in Paleocene strata, while debatable fossils date from Upper Cretaceous deposits.

## Family Typhaceae (Cattail)

Figure 10.20 Typhaceae. *Typha:* (a) clone of plants; (b) stem tip bearing female (below) and male (above) inflorescences; (c) female flower; (d) male flower.

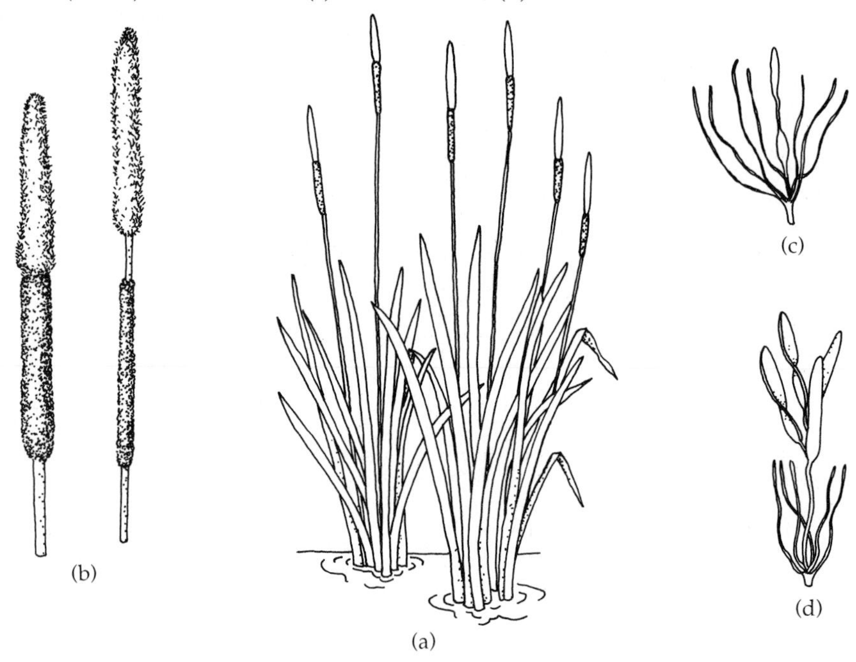


Herb, tall perennial from a stout rhizome, forming dense clones in wet, marshy soils. Leaves alternate and basal, simple, entire, long linear or strap-shaped, spongy textured, sheathing at the base. Flowers numerous, unisexual and monoecious; inflorescence on a long peduncle composed of a double, dense spadix, the lower spadix of female flowers, the terminal spadix of male flowers, each spadix subtended by a deciduous spathe; wind pollinated. Perianth of numerous bristles or scales. Male flowers with 2–5 stamens, filaments distinct or connate. Female flowers with a simple pistil of 1 carpel; locule 1; ovule 1 and borne on an apical placenta; ovary superior; elevated at maturity on a stipe, style 1, elongate, persistent on fruit, stigma elongate. Fruit an achene or follicle, small and wind dispersed. Seed small, embryo straight, endosperm present (Fig. 10.20).

The family Typhaceae consists of the genus *Typha* and 10–15 species distributed almost worldwide in semiaquatic and aquatic habitats. Absent from Australia, New Zealand, and most of South America.

Economically the family is of little value. The leaves of *Typha latifolia,* common cattail, are used for weaving baskets, chair bottoms, and mats. Ecologically the clonal nature of the species makes them good "soil binders," retarding erosion, and as good waterfowl habitat. The starchy rhizome can be used for food.

The Typhaceae, along with the Sparganiaceae, are two fairly distinct families, different enough to be placed together in a separate order.

Fossil pollen and fruits attributed to the family have been found in Oligocene deposits.

**Subclass IV. Zingiberidae** The subclass Zingiberidae is a small subclass consisting of approximately 3800 species, 9 families, and 2 orders. (See Table 10.4.) The 2 orders are about equal in size, with the Zingiberales comprised of 8 families and the Bromeliales made up of only the family Bromeliaceae.

The Zingiberidae consist of herbs (rarely small trees) that lack secondary growth. The leaves are alternate or basal, with the blades linear and strap-shaped or broad and with sheathing bases. The venation is parallel or pinnate-parallel. The flowers are regular or irregular, perfect or functionally unisexual, hypogynous or epigynous. The inflorescence often has very showy colored bracts subtending a single flower or flower cluster. Pollination is usually by bat, birds, or insects, rarely wind pollinated. The perianth is 3-merous, with the sepals distinct from the petals. The stamens range from 1–6 fertile ones; the others are lacking or petal-like staminodes. The pistil is compound of 3 united carpels with 3 locules and a superior or inferior ovary. The fruit is a berry or capsule with a starchy endosperm.

Because the orders of the Zingiberidae share characters with both the Commelinidae and Liliidae, the subclass has been classified in the past with either of the two groups. The feeling today among botanists is that the group is distinct in enough ways to warrant a subclass designation.

**TABLE 10.4 A list of orders and family names of flowering plants found in the subclass Zingiberidae.**

| Subclass IV. Zingiberidae | | |
|---|---|---|
| Order A. Bromeliales | | |
| | Family 1. | Bromeliaceae* |
| Order B. Zingiberales | | |
| | Family 1. | Strelitziaceae |
| | 2. | Heliconiaceae |
| | 3. | Musaceae* |
| | 4. | Lowiaceae |
| | 5. | Zingiberaceae* |
| | 6. | Costaceae* |
| | 7. | Cannaceae* |
| | 8. | Marantaceae* |

*NOTE:* Family names followed by an asterisk are discussed and illustrated in the text.

## ORDER BROMELIALES
### Family Bromeliaceae (Bromeliad)

Figure 10.21 Bromeliaceae. *Catopsis:* (a) flowering plant; (b) pistil. *Billbergia:* (c) flower. *Tillandsia:* (d) longitudinal section through flower; (e) cross section through ovary.

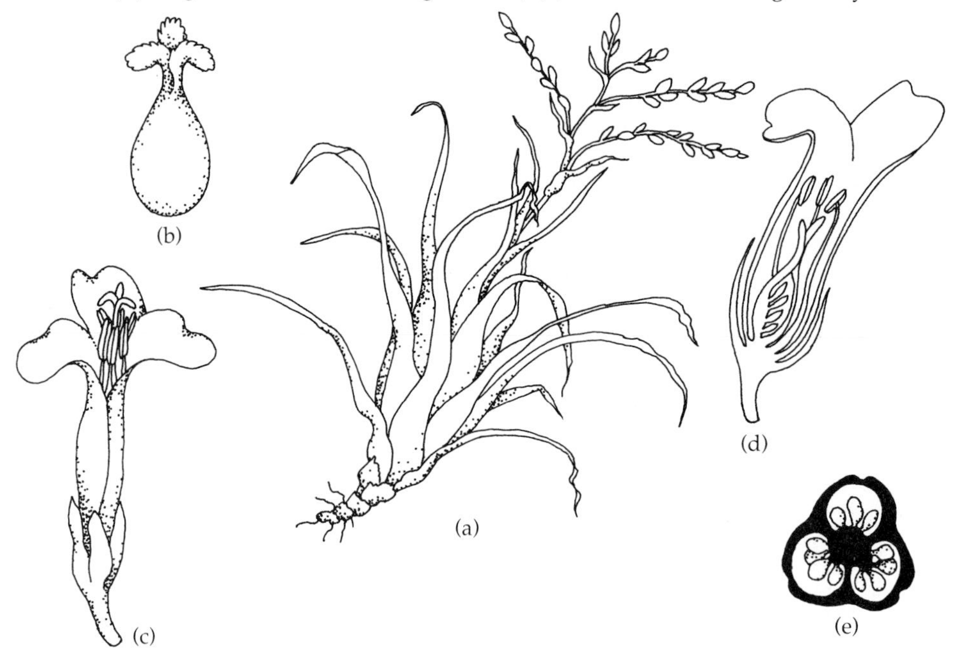


Herbs (rarely woody), mostly epiphytes with short stems, many xerophytic in habit. Leaves alternate, commonly developing, dense, stiff, basal rosettes, entire or serrate-spiny, with sheathing bases that often trap water. Flowers regular or irregular, perfect (rarely unisexual), hypogynous to epigynous; inflorescence a panicle, raceme, or spike, often with brightly colored bracts; pollinated by bats, birds, or insects (rarely wind-pollinated). Sepals 3, distinct or connate, green, sometimes petal-like. Petals 3, distinct or connate, usually brightly colored and commonly with nectaries as scale-like appendages. Stamens 6 in 2 whorls of 3, filaments distinct or connate at the base, sometimes adnate to the perianth; anthers opening by longitudinal slits. Pistil compound of 3 united carpels; locules 3; ovules few to many and borne on axile placentas; ovary superior or inferior; style 1, stigma 3. Fruit is a berry, capsule, or multiple. Seeds few to many, often plumed or winged. Embryo small to large, endosperm present (Fig. 10.21).

The family Bromeliaceae consists of 45–50 genera and about 2000 species distributed in the subtropics and tropics of the New World; 1 species is found in west tropical Africa. The largest genera are *Tillandsia* (400 species), *Pitcairnia* (250 species), and *Vriesia* (200 species).

Economically the family is of some importance. The most important species is *Ananas comosus*, pineapple, which is nutritionally high in vitamins A and B. Many species are cultivated inside as greenhouse ornamentals in temperate climates. The water in sheathing, water-catching leaf bases of some bromeliad species is a breeding area for malaria-carrying mosquitoes. Therefore, the presence of bromeliads has impeded the control of malaria in some New World tropical regions.

The family is so distinctive in its morphology in most cases that the family is recognized by the most casual observer.

The family has no known fossil record.

## ORDER ZINGIBERALES
### Family Musaceae (Banana)

Figure 10.22 Musaceae. *Musa:* (a) mature plant bearing fruiting inflorescence; (b) stamen; (c) cross section through ovary; (d) front view of female flower; (e) male flower; (f) young inflorescence.

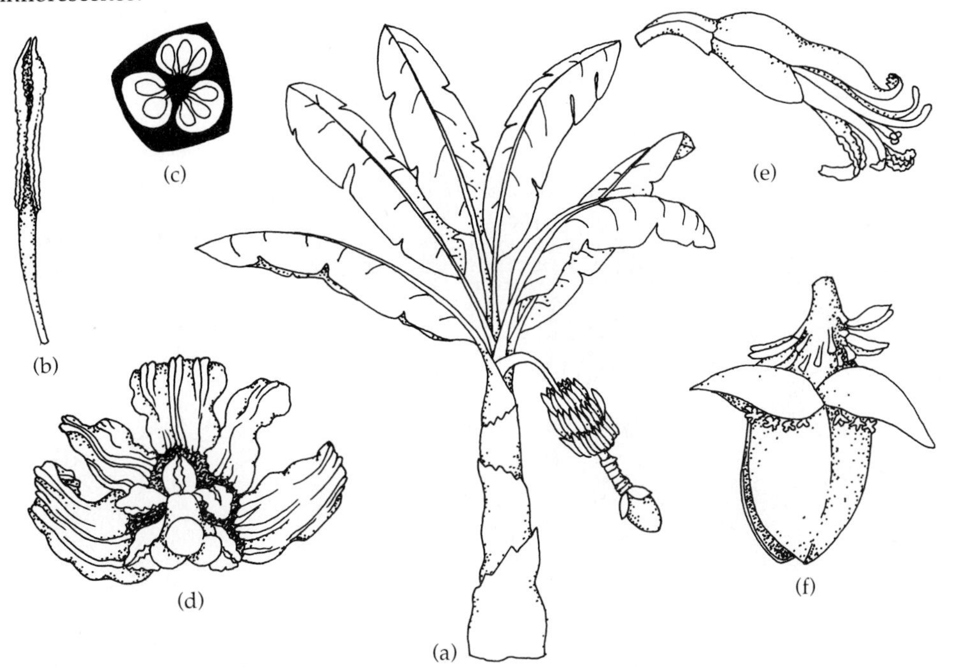


Herbs, usually large and tree-like, perennial from a massive corm, pseudostems large, comprised of large overlapping sheathing petioles, with milky juice. Leaves alternate, large, simple, entire; venation pinnate-parallel from a stout midrib; margin often torn to appear as pinnately compound, petioles long, sheathing at the base. Flowers irregular, unisexual and monoecious, epigynous, sometimes with a hypanthium; inflorescence a panicle-like cyme with 1–many spathes, the axis of the inflorescence arises from the growing points of the corm by growing up through the pseudostem. Female flowers clustered below the terminal male flowers; nectar glands give a strong odor; pollinated by bats and birds. Sepals 3, petal-like, different from the petals but adnate with 2 petals. Petals 3, irregular, the 1 nonadnate petal distinct. Stamens 5 or 6, filaments distinct, adnate to the corolla, usually only 5 functional with stamen 6 a staminode; anthers opening by longitudinal slits. Pistil compound of 3 united carpels; locules 3; ovules many and borne on axile placentas; ovary inferior; style 1, stigma 3-lobed. Fruit a long modified berry with a thick exocarp. Seeds few to many, embryo curved or straight, endosperm present (Fig. 10.22).

The family Musaceae consists of 2 genera and about 40 species distributed natively in the subtropical to tropical climates of the Old World; introduced into other tropical regions. The largest genus is *Musa* (35 species).

Economically the family is important mostly for *Musa* × *paradisiaca*, the cultivated banana, a sterile triploid, which is a cross of *M. acuminata* and *M. balbisiana* in Southeast Asia. Manila hemp comes from the fibers of *M. textilis*.

In some older classifications the family Strelitziaceae was included in this family.

The family has no known fossil record.

## Family Zingiberaceae (Ginger)

Figure 10.23 Zingiberaceae. *Hedychium:* (a) leafy flowering stem; (b) cross section through ovary; (c) flower.

Herbs, perennial with aromatic oils, stems short from fleshy rhizomes or tuberous roots. Leaves alternate, obviously 2 ranked (also called **distichous**), simple, entire with blades rolled up in bud; venation pinnate-parallel with a prominent midrib; leaf bases forming open sheaths, ligule present, petiole usually long (rarely lacking). Flowers sometimes showy, irregular, perfect, epigynous, somewhat complicated structurally; inflorescence of solitary flowers or flowers borne in terminal cymes, heads, racemes, or spikes; each flower subtended by a sheathing bract. Sepals 3, connate, usually green. Petals 3, connate. Stamens most distinctive feature of flower, 3 or 5, 2 or 4 sterile, as staminodes fused into a 3-lobed petal-like lip (called a **labellum**), remaining sixth stamen fertile, opposite the staminodes, adnate to the corolla; filaments often grooved, and surrounding the style. Pistil compound of 3 united carpels; locules 3 (rarely 1 or 2); ovules many and borne on axile or basal placentas; ovary inferior; nectaries 2, style 1, stigma extends beyond the anthers. Fruit usually brightly colored, a berry or capsule. Seeds large, round, often with a red aril, embryo straight, endosperm present (Fig. 10.23).

The family Zingiberaceae consists of 45–50 genera and 1000–1300 species distributed in the tropics of Africa, through Asia to the Pacific. The largest genera are *Alpina* (200–250 species) and *Zingiber* (80–90 species).

Economically the family is of some importance. Many species have aromatic, volatile oils and are used for dyes, medicinal purposes, and seasonings. The most commonly used is *Zingiber officinale,* ginger. Others of importance include *Curcuma domestica,* which provides turmeric, a yellow dye and spice, which is an ingredient of curry powder. The seeds of *Elettaria cardomomum* give the spice cardomom.

Along with the Cannaceae, Costaceae, Marantaceae, Musaceae, and Strelitziaceae, the Zingiberaceae make a fairly distinctive botanical group.

The family has no known fossil record.

## Family Costaceae (Costus)

Figure 10.24 Costaceae. *Costus:* (a) leafy flowering branch; (b) cut-away view of perianth showing the anthers, stigma, and style; (c) cross section through ovary; (d) side view of flower.

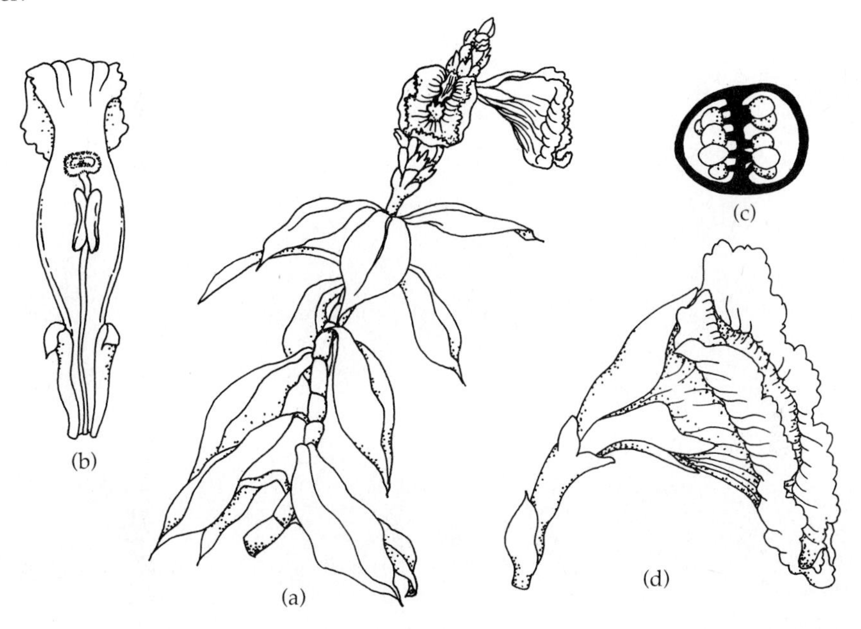


Herbs, perennial, nonaromatic from fleshy rhizomes or tuberous roots, stems commonly spirally twisted. Leaves alternate and spiral arranged, simple, entire with blade rolled up in bud; venation pinnate-parallel with a prominent midrib, leaf bases forming usually closed sheaths, ligule present, petiole short. Flowers irregular, perfect, epigynous; inflorescence a terminal head or spike (rarely a solitary flower); each flower subtended by a bract; 2 nectaries present; pollinated by bats or insects. Sepals 3, connate, green. Petals 3, connate toward the base with the median corolla lobe larger and hood-shaped. Stamens 6, very distinctive with 5 of the stamens fused into a large staminode of 3–5 lobes, petal-like lip (a labellum), usually only 1 fertile stamen adnate to the corolla; anthers opening by longitudinal slits. Pistil compound of 3 united carpels, locules 3 (rarely 1), ovules many and borne on axile or parietal placentas, ovary inferior, style 1, long, stigma extends beyond the anthers. Fruit a berry or capsule. Seeds many, often with a red aril, embryo straight, endosperm present (Fig. 10.24).

The family Costaceae consists of 4 genera and 150–200 species distributed pantropically, with many taxa in the New World. The largest genus is *Costus* (about 130 species).

Economically the family is of minor importance. Some species of *Costus* are used locally as ornamentals and for some medicinal uses.

The family Costaceae has sometimes been included in the Zingiberaceae but differs in its lack of aromatic oils, closed sheaths, and the labellum comprised of 5 staminodes instead of 3.

The family has no known fossil record.

## Family Cannaceae (Canna)

Figure 10.25 Cannaceae. *Canna:* (a) tip of plant bearing inflorescence; (b) anther clustered around style; (c) flower; (d) cross section through ovary; (e) fruit bearing persistent calyx; (f) fruits.

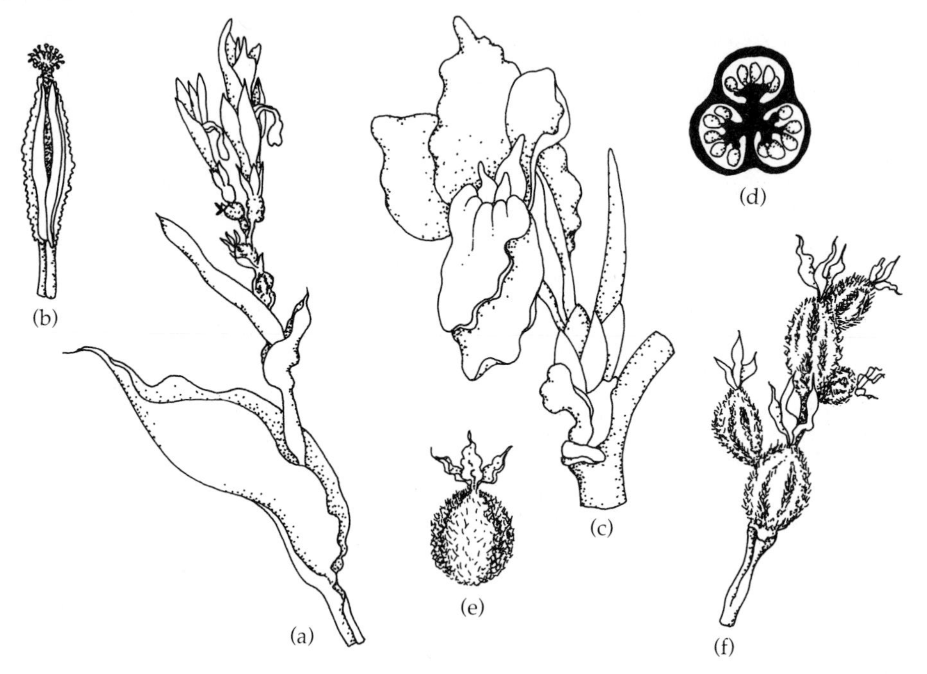


Herbs, perennial from a swollen tuberous rhizome, stem often with mucilage canals. Leaves generally large, alternate, spirally arranged, simple, entire with pinnate-parallel venation from a prominent midrib, petiole sheathing the stems, ligule lacking. Flowers large, irregular, perfect, epigynous, obliquely attached to the peduncle axis; inflorescence a panicle or raceme, each flower subtended by a bract. Sepals 3, green or purple, distinct, persistent in fruit. Petals 3, connate, one smaller than the other 2. Fertile stamens 1, all others sterile and petal-like, filaments often with a groove, connate, adnate to the corolla; anther with 1 pollen sac, surrounding the style. Pistil compound of 3 united carpels; locules 3; ovules many and attached to axile placentas; ovary inferior; nectaries present, style 1, petal-like. Fruit a warty capsule. Seeds many, without an aril, embryo straight, endosperm present (Fig. 10.25).

The family Cannaceae consists of the genus *Canna* and 30–55 species, distributed natively in the subtropical-tropical regions of the southeastern United States, Central America, and the West Indies.

Economically the family is of some importance. The starch, called purple or Queensland arrowroot, of the rhizome of *C. edulis* is easily digested; therefore, it is suitable for infants and speciality diets. It is grown commercially in Asia, Australia, and regions of the Pacific. Extracts from the rhizome of *C. gigantea* and *C. speciosa* are used in medicines. *Canna indica* and its hybrids are grown as ornamental plants.

The varied number of recognized species indicates the different views of botanists toward the plastic morphology of the species. The family differs from the Zingiberaceae in lacking ligules.

Fossils attributed to the Cannaceae have been found in Eocene deposits.

## Family Marantaceae (Prayer Plant)

Figure 10.26 Marantaceae. *Maranta:* (a) leafy flowering branch; (b) cross section through ovary; (c) staminode; (d) pistil; (e) stamen; (f) pair of flowers.

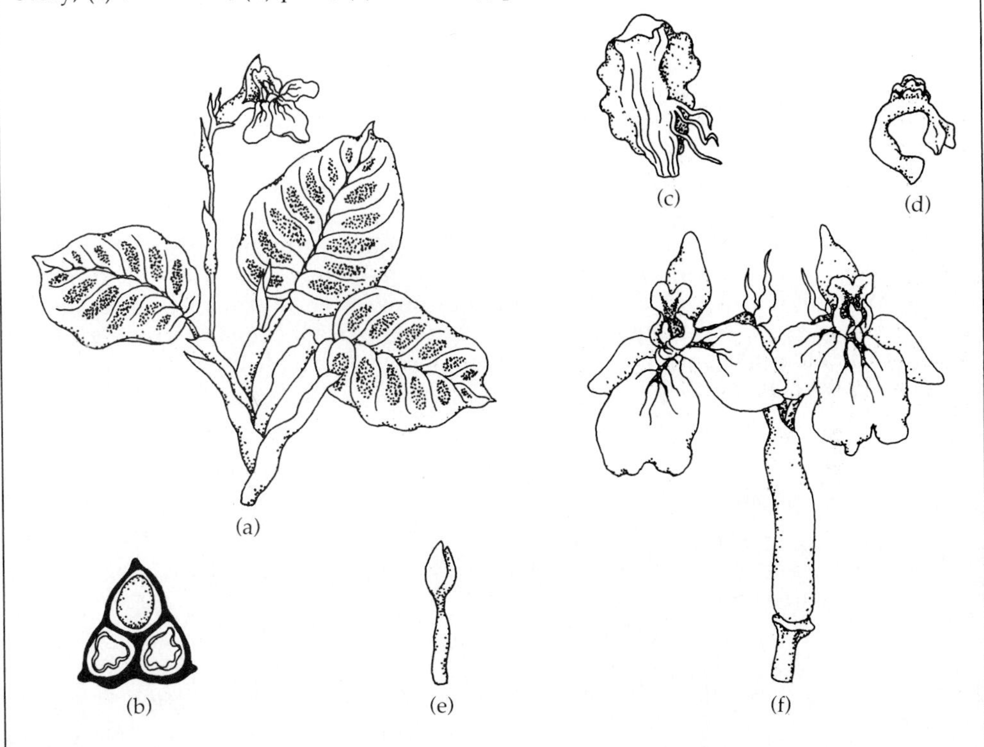


Herbs, perennial from a tuberous rhizome, stem without mucilage canals. Leaves alternate and usually 2-ranked, simple, entire; venation pinnate-parallel with a prominent midrib; petiole sheathing the stem and sometimes with wings, where petiole joins the blade is a swollen tissue, called the **pulvinus,** that controls leaf orientation. Flowers small, irregular, perfect, epigynous; inflorescence of 2-flowered cymes borne in panicles or spikes. Sepals 3, distinct, green. Petals 3, distinct or connate toward the base. Stamens all sterile and modified as staminodes, except 1 functional petal-like stamen with anther having 1 fertile pollen sac, filaments connate, adnate to the corolla. Pistil compound of 3 united carpels, locules 1 or 3 (with 2 aborted), ovules 1 per locule and borne on a basal placenta, ovary inferior, nectaries present, style 1, stigma 3-lobed; the corolla, stamens, and style forming a unique mechanism for insect pollination. Fruit a berry or capsule. Seed with an aril, embryo curved, endosperm present (Fig. 10.26).

The family Marantaceae consists of 30 genera and 350–400 species distributed pantropically, most abundantly in the New World tropics, and found also in southeastern United States. The largest genus is *Calathea* (150 species).

Economically the family is of some importance. The most important plant is *Maranta arundiacea,* the West Indian arrowroot, whose rhizome provides an easily digestible starch for food. The leaves of various *Calathea* are used for basket and thatch material. Species of *Calathea* and *Maranta* are prized for their foliage as greenhouse and house plants.

Many of the family characters are similar to other families of the order, but the distinctive stamen and pistil morphology make the family stand out.

Fossils attributed to the family have been found in Eocene deposits.

**Subclass V. Liliidae** The subclass Liliidae is the largest of the monocotyledon subclasses with 25,000–35,000 species, 19–25 families, and 2 orders. (See Table 10.5.) The subclass includes some of the most well-known ornamentals, including the very large and beautiful orchid family, the Orchidaceae.

The family includes herbs, shrubs, trees, vines, and epiphytes; some even lack chlorophyll. The leaves are alternate (rarely opposite or whorled), simple, entire; venation parallel or netted; usually with a basal sheath. The flowers are usually showy, regular or irregular, perfect (less commonly unisexual), hypogynous or epigynous; nectar glands usually present to attract bats, birds, or insects. There are 3 sepals usually petal-like, less frequently green. The petals are also 3, attractively colored, distinguished from petal-like sepals by being the inner tepals. The stamens are 6 or less. The pistil is compound of 3 united carpels, superior or inferior, and with axile or parietal placentation. The ovules are few to many. The fruit is most commonly a berry or capsule. The embryo of the seed is usually very small and not differentiated into specific parts.

**TABLE 10.5 A list of orders and family names of flowering plants found in the subclass Liliidae.**

| Order A. Liliales | Order B. Orchidales |
|---|---|
| Subclass V. Liliidae | |
| Family 1. Philydraceae | Family 1. Geosiridaceae |
| 2. Pontederiaceae* | 2. Burmanniaceae |
| 3. Haemodoraceae* | 3. Corsiaceae |
| 4. Cyanastraceae | 4. Orchidaceae* |
| 5. Liliaceae* | |
| 6. Iridaceae* | |
| 7. Velloziaceae | |
| 8. Aloeaceae* | |
| 9. Agavaceae* | |
| 10. Xanthorrhoeaceae* | |
| 11. Hanguanaceae | |
| 12. Taccaceae | |
| 13. Stemonaceae | |
| 14. Smilacaceae* | |
| 15. Dioscoreaceae* | |

*NOTE:* Family names followed by an asterisk are discussed and illustrated in the text.

## ORDER LILIALES
## Family Pontederiaceae (Water Hyacinth)

Figure 10.27 Pontederiaceae. *Eichhornia:* (a) plant having flowers and many feathery roots; (b) fruit with perianth attached at the top. *Pontederia:* (c) front view of flower.

Herbs, annual or perennial, semiaquatic or aquatic, submerged, floating, or emergent; commonly reproduced asexually by vegetative fragmentation of the stolons or stems. Leaves alternate, basal, opposite, or whorled, many with an open sheathing petiole, petiole commonly enlarged with air cells, often with stipule-like ligule. Flowers usually showy, regular or irregular, perfect, hypogynous; inflorescence of solitary terminal flower or flowers borne in panicles, racemes, or spikes; subtended by a spathe-like sheath. Perianth of 6 tepals in 2 whorls, connate at the base to form a perianth tube. Stamens 3 or 6 (rarely 1), filaments distinct, adnate to the perianth tube (hypanthium); anthers opening by longitudinal slits or terminal pores; heterostyly present and unusual of tristylic type. Pistil compound of 3 united carpels; locules 1 or 3; ovules many and borne on axile or parietal placentas; ovary superior; nectaries present. Style 1, stigma 1 or 3 lobed. Fruit an achene or loculicidal capsule. Seeds small and ribbed lengthwise, embryo straight, endosperm present (Fig. 10.27).

The family Pontederiaceae consists of 9 genera and 30–34 species distributed pantropically in freshwater locations. The largest genus is *Heteranthera* (8 species).

Economically the family is of little value directly. However, some species are bothersome weeds. *Eichhornia crassipes,* water hyacinth, is undoubtedly the most widely distributed and noxious aquatic weed in warm climates of the world. Other weedy species found in canals and rice fields are *Eichhornia natans* in Africa, *Heteranthera limosa* and *H. reniformis* in the United States, *Monochoria vaginalis* in Asia, and *Reussia rotundifolia* in South America. A redeeming aspect of *E. crassipes* is its ability to remove large amounts of dissolved inorganic pollutants in water.

The tristylic condition found in the genera *Eichhornia, Reussia,* and *Pontederia* is distinctive among flowering plants.

The family has no known fossil record.

## Family Haemodoraceae (Kangaroo Paw)

Figure 10.28 Haemodoraceae. *Anigozanthos:* (a) leafy plant bearing flowers; (b) flower; (c) cross section through ovary. *Lophiola:* (d) flower.

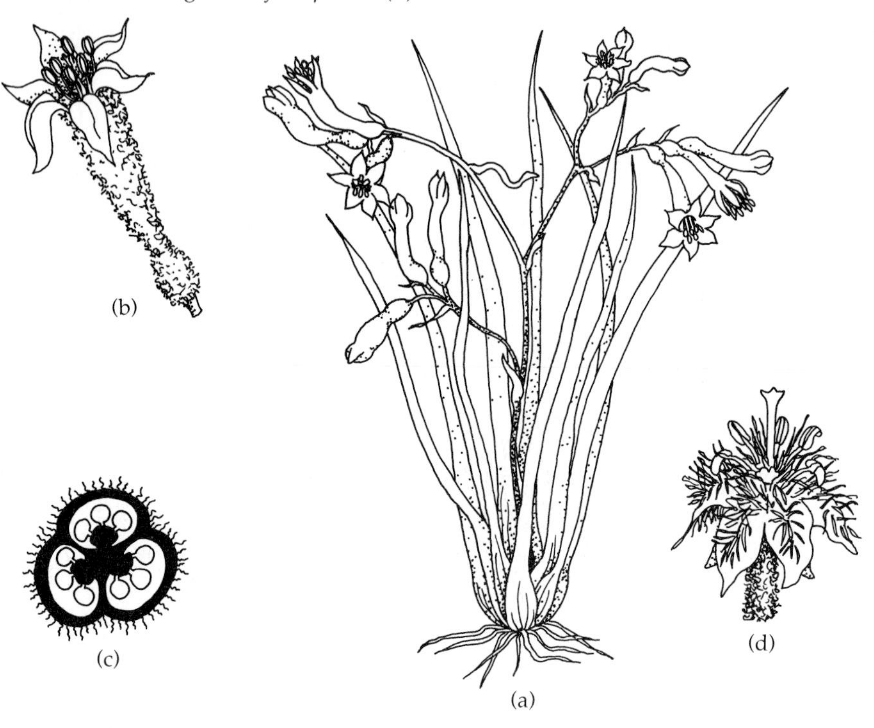


Herbs, strong fibrous roots from rhizomes, stolons, or tubers. Leaves basal and equitant, developing from the ground, long, linear with a sheathing base; venation parallel; cauline leaves small or lacking. Flowers regular (rarely irregular), perfect, hypogynous to epigynous; inflorescence of cymes, panicles, or racemes often covered with hairs. Perianth of 6 tepals in 1 whorl or in 2 whorls of 3, distinct or connate at the base, forming a curved or straight tube. Stamens 3 or 6, filaments distinct or adnate to the perianth tube, opposite the petals when 3; anthers opening by longitudinal slits. Pistil compound of 3 united carpels; locules 3; ovules 1–many in each locule and borne on axile placentas; ovary superior or inferior; style 1, slender, stigma capitate. Fruit a loculicidal capsule. Seeds with a small embryo, endosperm abundant (Fig. 10.28).

The family Haemodoraceae consists of 16–17 genera and about 100 species distributed in the Southern Hemisphere (excluding New Zealand), Central America, and North America in the eastern United States. The largest genera are *Conostylis* (25 species) and *Haemodorum* (20 species).

Economically the family is of little value, except as attractive ornamentals. Some species of *Anigozanthos*, kangaroo paw, are cultivated, with *A. manglesii*, the Western Australia state emblem, being very attractive.

The family has sometimes been classified by some botanists in the Liliaceae.

The family has no known fossil record.

## Family Liliaceae (Lily)

Figure 10.29 Liliaceae. *Lilium:* (a) flowering stem tip; (b) splitting fruit; (c) cross section through ovary; (d) longitudinal section through flower. *Allium:* (e) bulb epidermis cell pattern; (f) flowering plant from bulb.

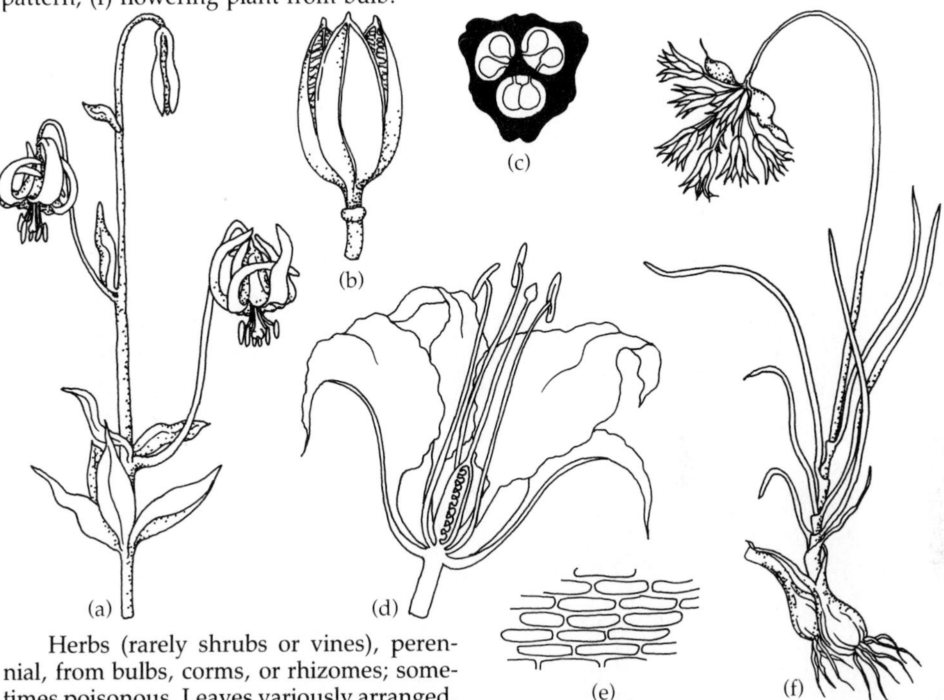


Herbs (rarely shrubs or vines), perennial, from bulbs, corms, or rhizomes; sometimes poisonous. Leaves variously arranged, simple, often sessile and with sheathing base; venation usually parallel (rarely netted). Flowers often showy, regular (rarely irregular), perfect, hypogynous or epigynous; hypanthium may be present; inflorescence of solitary flower or flowers borne in cymes, racemes, spikes, or umbels; nectaries present; insect pollinated. Sepals 3, (rarely 2–5), distinct or adnate to petals, often petal-like. Petals 3 (rarely 2–5) distinct or adnate to sepals, forming a perianth tube. Stamens 6 (rarely 3–12), filaments distinct or connate, free or adnate to hypanthium and opposite perianth lobes; anthers opening by longitudinal slits (rarely by apical pores.) Pistil compound of 3 (rarely 2 or 4) united or nearly distinct carpels; locules 3 (rarely 1–5); ovules 1–many per locule, borne on axile placentas (rarely 1 locule and intruded parietal placentas); ovary superior or inferior; style 1 (rarely 3), stigma capitate to 3-lobed. Fruit a berry, capsule, or samara. Seeds sometimes flat or with an aril, embryo straight, endosperm present (Fig. 10.29).

The family consists of about 280 genera and 4000 species distributed worldwide. The family is an important component of the early spring flora in the deciduous forests of eastern North America.

The young shoots of *Asparagus officinalis,* asparagus, are eaten, as are the bulbs and leaves of *Allium cepa,* onion; *A. porrum,* leek; and *A. sativum,* garlic. Among many well-known garden and house ornamentals are *Amaryllis,* amaryllis; *Hemerocallis,* day lily; *Hosta; Lilium,* lily; *Narcissus,* daffodil; *Trillium;* and *Tulipa,* tulip. *Veratrum,* hellebore, and *Zigadenus,* death camus, are poisonous and kill many domestic animals each year in the western United States.

The Liliaceae is a very diverse family that some botanists separate into numerous (recently divided into 27) smaller families, such as Amaryllidaceae, Alliaceae, Asparagaceae, Ruscaceae, and Trilliaceae.

Fossil pollen unquestionably attributed to the family has been found in Eocene and more recent deposits.

## Family Iridaceae (Iris)

Figure 10.30 Iridaceae. *Iris:* (a) leafy plant bearing flower; (b) cross section through ovary; (c) petal-like stigmas and accompanying stamens. *Sisyrinchium:* (d) flower; (e) flowering plant.

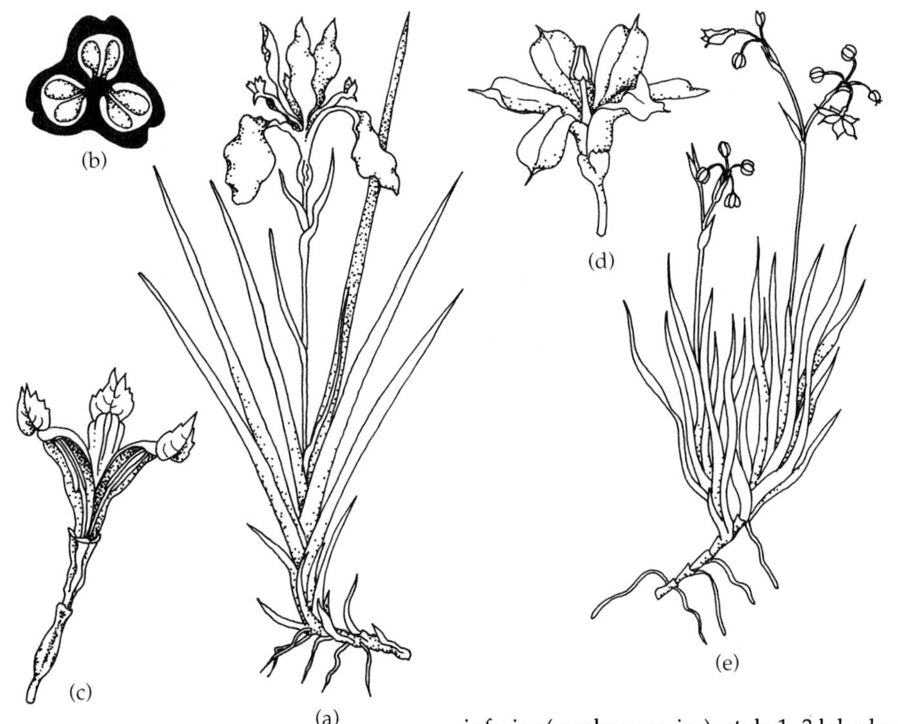


Herbs or shrubs, perennial from bulbs, corms, or rhizomes. Leaves alternate or basal, 2 ranked **(distichous),** often equitant with an open sheathing base; petiole usually lacking. Flowers mostly large and showy, regular, or irregular, perfect, epigynous; inflorescence of a solitary flower or flowers borne as cymes, panicles, spikes, or umbels; commonly subtended or enclosed by 1 or more sheaths (spathes); a hypanthium is usually well formed; nectaries present; bird, insect, or wind pollinated. Sepals 3, petal-like, distinct or connate. Petals 3, distinct or connate and adnate with the sepals forming a perianth tube. Stamens 3, filaments distinct or connate, sometimes adnate to the hypanthium; anthers opening by longitudinal slits (rarely apical pores). Pistil compound of 3 (rarely 2 or 4) united carpels; locules 3; ovules many per locule and borne on axile placentas; ovary inferior (rarely superior); style 1, 3 lobed and flattened; stigma capitate to petal-like, often on outside of style. Fruit a loculicidal capsule. Seeds with embryo straight, endosperm present (Fig. 10.30).

The family Iridaceae consists of 70–80 genera and 1500–1800 species distributed worldwide, but with much diversity in Africa, especially South Africa. The largest genera are *Iris* (200 species), *Gladiolus* (150 species), and *Moraea* (100 species).

Economically the family is of value as ornamentals. Many cultivars are grown of *Crocus,* crocus; *Freesia; Gladiolus,* gladiolus; *Iris,* iris; *Iria;* and *Tigridia.* The anthers of *Crocus sativus* provide dye and the world's most expensive spice, saffron. The root of *Iris florentina,* orris root, is used in making cosmetics and perfume.

The group is similar to the family Liliaceae in some respects but differs in having 3 stamens and petal-like stigmas.

The family has no known fossil record.

## Family Aloeaceae (Aloe)

Figure 10.31 Aloeaceae. *Aloe:* (a) plant bearing large inflorescence; (b) cross section through ovary; (c) pistil; (d) side view of flower; (e) anther.

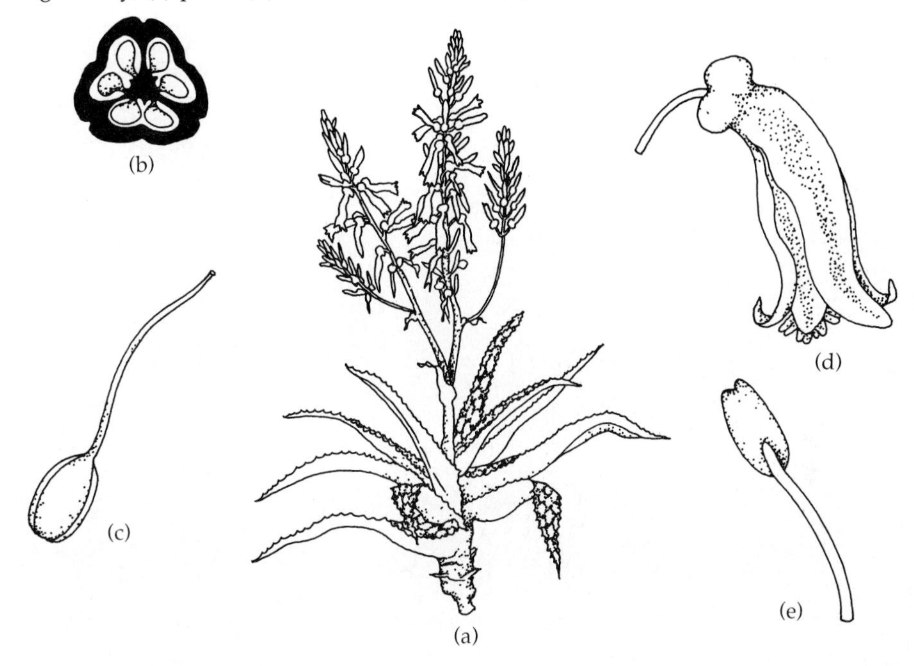


Herbs, shrubs, or trees, arising from fleshy caudex or rhizome. Leaves succulent, alternate at the base of stem in rosettes or on ends of branches, simple, sessile, commonly with prickly margins. Flowers often showy, regular or irregular, perfect, hypogynous; inflorescence of terminal panicles, racemes, or spikes; pollinated by birds and insects. Sepals 3, petal-like, distinct, commonly fleshy. Petals 3, distinct or connate. Stamens 6, filaments distinct; anthers opening by longitudinal slits; nectaries present. Pistil compound of 3 united carpels; locules 3; ovules many per locule and attached to axile placentas; ovary superior; style 1, stigma single to 3 lobed. Fruit a berry or loculicidal capsule. Seeds many, commonly flattened or winged, embryo straight, endosperm present (Fig. 10.31).

The family Aloeaceae consists of 5 genera and almost 700 species native to Africa, in general, Arabia, and Madagascar, with greatest diversity in South Africa. The largest genera are *Aloe* (350 species) and *Haworthia* (200 species).

Economically the family is of minor importance. Some species such as *Kniphofia martagon,* the red-hot poker, are grown as ornamentals. Various species of *Aloe* have medicinal uses, i.e., *A. vera.* Aloin, a purgative extractive from the juice of *Aloe* species, is called bitter aloe.

The family Aloeaceae is very similar in many ways to the Liliaceae. Some botanists therefore classify the Aloeaceae within the Liliaceae. However, to do this would create questions about how other families, such as the Agavaceae or Xanthorrhoeaceae, should be grouped. It seems best at this time to treat the Aloeaceae as a separate family.

The family has no known fossil record.

## Family Agavaceae (Century Plant)

Figure 10.32 Agavaceae. *Yucca:* (a) plant with flowers; (b) cross section through ovary; (c) front view of flower; (d) splitting fruits.

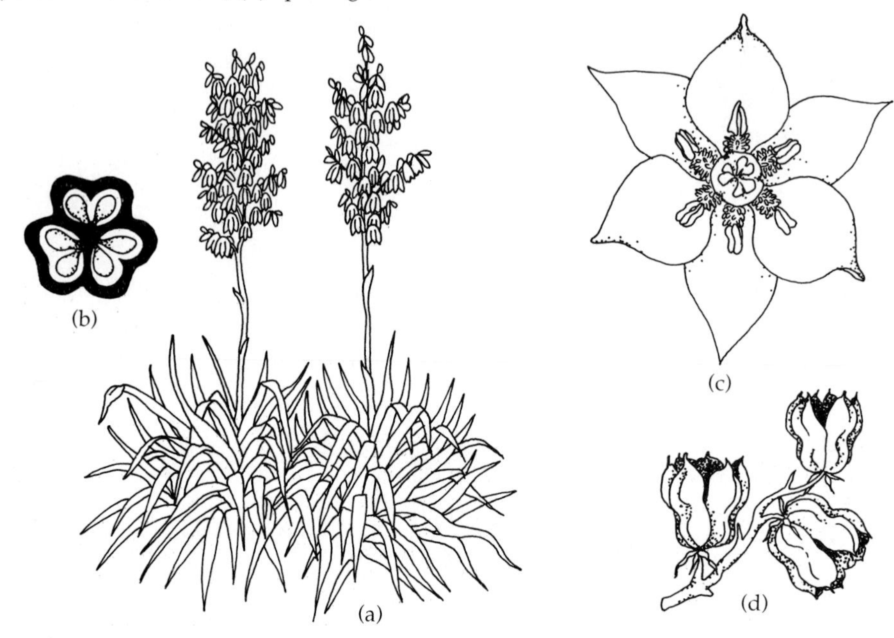


Herbs, shrubs, or trees. Leaves alternate or basal, sessile, forming dense rosettes at the plant base or at the end of the stem or branches, somewhat thick or fleshy (succulent), stiff, leathery, with strong anatomical fibers; linear, entire or with sharp prickles along the margin, apex sharp pointed. Flowers regular or slightly irregular, perfect or unisexual, a hypanthium sometimes present, hypogynous or epigynous; inflorescence can be massive and borne in heads, panicles, or racemes; nectaries present; pollinated by bats, birds, or insects. Sepals 3, petal-like, distinct or connate. Petals 3, distinct or connate sometimes into a short perianth tube. Stamens 6, filaments distinct, free or adnate to the hypanthium; anthers opening by longitudinal slits. Pistil compound of 3 united carpels; locules 3; ovules 1–many per locule and borne on axile placentas (rarely with locule 1 and parietal placenta); ovary superior or inferior; style 1, stigmas 3. Fruit a berry or loculicidal capsule. Seeds flattened, embryo straight, endosperm present (Fig. 10.32).

The family consists of 18–20 genera and 600–700 species distributed widely in warm, arid subtropical regions of the New and Old World and in Australia and New Zealand. The largest genera are *Agave*, century plant (300 species), and *Dracaena* (80 species).

Economically the family is of some importance. The leaves of *Agave* and *Yucca* produce fiber made into cord and rope. Various fermented drinks are produced: tequila, a type of mezcal, is made from *A. tequilana* in the city of Tequila, Mexico; mezcal and pulque are made mostly from *A. atrovirens, A. americana, A. ferox,* and *A. lurida.* The flowers of *Agave* are sometimes eaten by Mexicans for food, and the root of *Yucca* was used for soap by North American Plains Indians.

Some genera (such as *Yucca*) were previously classified in the family Liliaceae, while others *(Agave)* were placed in the Amaryllidaceae. The taxonomic status of the genera is open to debate by botanists.

Fossil pollen attributed to the family has been found in Eocene and more recent deposits.

## Family Xanthorrhoeaceae (Grass Tree)

Figure 10.33 Xanthorrhoeaceae. *Xanthorrhoea:* (a) large leafy plant bearing inflorescence. *Kingia:* (b) club-shaped inflorescence; (c) minute flower from inflorescence. *Calectasia:* (d) flower at end of short branch.

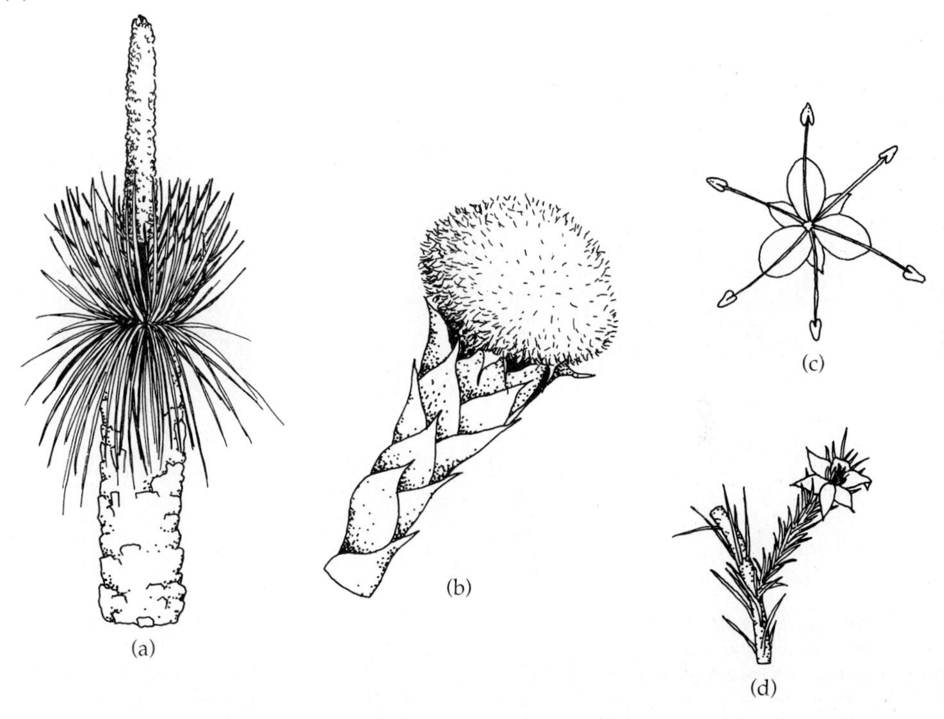


Herbs or shrubs, perennial, arising from a thick caudex or rhizome. Leaves alternate, simple, linear with sheathing base, leathery to prickly, old leaf bases often persistent. Flowers small, regular, perfect or unisexual (dioecious), commonly dry and papery, hypogynous; inflorescence usually a head, long panicle, or spike. Perianth in 2 whorls or 3, distinct or the inner whorl connate and the outer whorl distinct. Stamens 6, filaments distinct, commonly adnate to the base of the tepals; anthers opening by longitudinal slits. Pistil compound of 3 united carpels; locules 3 (rarely 1); ovules 1–many per locule and borne on basal or axile-basal placentas; ovary superior; styles 1 or 3, stigmas 3. Fruit a loculicidal capsule or nut. Seeds with a straight (rarely curved) embryo, endosperm present (Fig. 10.33).

The family Xanthorrhoeaceae consists of 8–9 genera and 55–66 species distributed in Australia, New Caledonia, New Guinea, and New Zealand, often as xerophytes; where frequented by fire, the stems are black. The largest genera are *Lomandra* (30 species) and *Xanthorrhoea*, grass tree (15 species).

Economically the family is of little use. The resin or gum from various *Xanthorrhoea* species is used in making varnish or lacquer. The stems of *X. preissii* are made into bowls and small containers.

The family has some features like the family Agavaceae but differs in having small, dry, and papery flowers.

The family has no known fossil record.

## Family Smilacaceae (Catbriar)

Figure 10.34 Smilacaceae. *Smilax:* (a) leafy branch segment bearing fruits; (b) female flower; (c) male flower; (d) cross section through ovary.

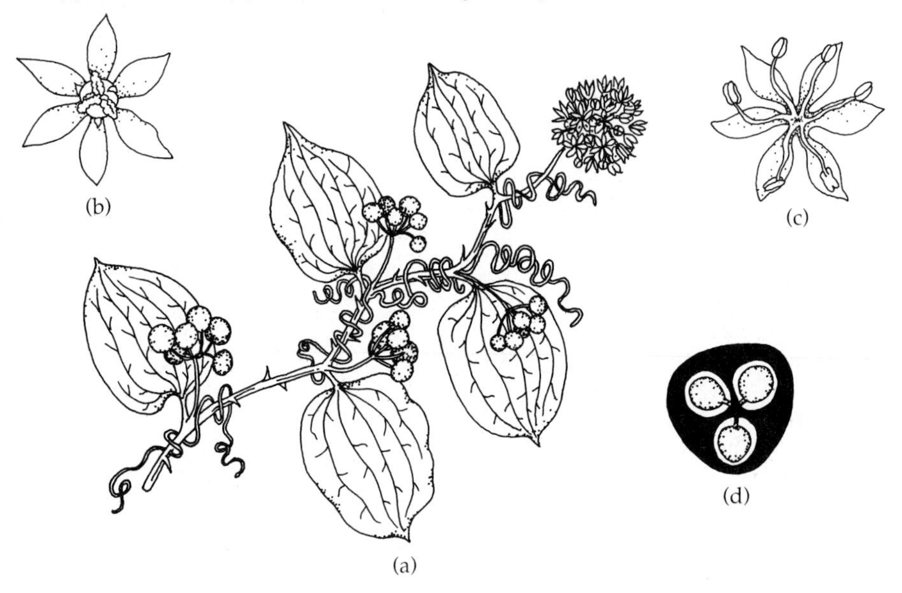


Herbs, shrubs, or woody vines, commonly prickly or twining with or without tendrils from a tuber-like, starchy rhizome. Leaves alternate (rarely opposite), simple, entire, venation 3–7 nerved, palmate-reticulate, rather leathery; petiole sheathing at the base, with the leaf bases developing into stipular tendrils. Flowers regular, perfect or unisexual (dioecious), hypogynous (rarely epigynous); inflorescence of solitary axillary flowers or flowers in racemes, spikes, or umbels. Perianth in 2 whorls of 3, distinct or connate at the base into a tube, petal-like. Stamens 6 (rarely 3 or more than 6), filaments distinct or occasionally adnate to the perianth tube; anthers opening by longitudinal slits; female flowers with staminodes; nectar glands present. Pistil compound of 3 united carpels; locules 1 or 3; ovules 1–many per locule and borne on parietal or axile placentas; ovary superior (rarely inferior); style 1 or 3, stigma capitate or 3 lobed. Fruit a berry. Seeds 1–3, embryo small, endosperm hard, no starch (Fig. 10.34).

The family Smilacaceae consists of 10–12 genera and 300–375 species distributed widely in the subtropical and tropical regions, especially in the Southern Hemisphere, but extending into northern temperate climates. The largest genus is *Smilax,* catbriar (300–350 species).

Economically the family is of minor importance. The starch-filled tubers of *Smilax* have been used by native peoples. The flavoring sarsaparilla is obtained from several species of *Smilax* and *Rhipogonum scandens* and is used to treat rheumatism and similar ailments. Extracts of *S. china,* China root, are used as stimulants.

In older classifications the Smilacaceae has been included within the family Liliaceae. It differs from most Liliaceae in its woody, viny habit, netted venation of the leaves, and some anatomical features.

Fossil leaves and pollen attributed to the family have been found in Eocene deposits.

## Family Dioscoreaceae (Yam)

Figure 10.35 Dioscoreaceae. *Dioscorea:* (a) leafy stem segment bearing female flowers; (b) front view of female flower; (c) leafy stem segment bearing male flower; (d) anther; (e) seed.

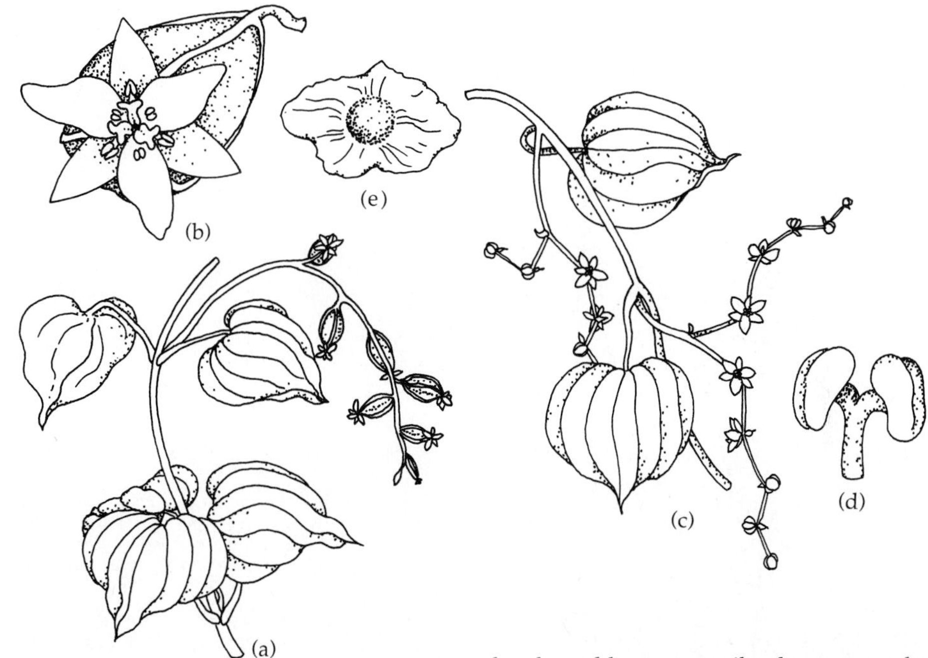


Herbs, shrubs, or herbaceous vines, sometimes prickly, arising from a large tuber or thick starch-filled rhizome; tendrils lacking. Leaves alternate (rarely opposite), simple, commonly cordate, entire to palmately cleft; venation parallel with 3–many curved-convergent veins, each connected by secondary veins, often with pits that harbor nitrogen-fixing bacteria, petiole usually twisted, nonsheathing and lacking stipules. Flowers small, regular, perfect or unisexual (dioecious), epigynous, short hypanthium usually present; inflorescence of panicles, racemes, or spikes. Sepals 3, distinct, green or petal-like. Petals 3, distinct, usually adnate to the sepals forming a short tube or hypanthium. Stamens 6 in 2 whorls of 3, filaments distinct or connate at the base and sometimes adnate to the hypanthium; anthers opening by longitudinal slits and separated by a broad connective. Pistil compound of 3 united carpels; locules 3; ovules 2 (rarely many) per locule and borne on axile placentas; styles 1 or 3, stigmas 1 or 3. Fruit a berry, capsule, or samara; commonly triangle-shaped and winged. Seeds usually winged, embryo small, sometimes with a rudimentary second cotyledon, endosperm present (Fig. 10.35).

The family consists of 6 genera and 630 species distributed pantropically, with few species in the temperate United States, Europe, and Southeast Asia. The largest genus is *Dioscorea* (about 600 species).

Some species of *Dioscorea*, yams, are cultivated for their edible tubers. Other species, such as *D. floribunda*, are cultivated for diosgenin, a precursor of cortisone and progesterone that is used in oral contraceptives.

The family is similar to the Smilacaceae in its twiny habit, prickly stem, and leaves, and certain anatomical features, but differs in a sufficient number of characters to be maintained as a family.

Fossil leaves attributed to the Dioscoreaceae have been found in Eocene and more recent deposits.

## ORDER ORCHIDALES
### Family Orchidaceae (Orchid)

Figure 10.36 Orchidaceae. *Calypso:* (a) single-leaved plant with flower. *Habinaria:* (b) leafy plant bearing an inflorescence of small flowers; (c) front view of flower showing spurs. *Oncidium:* (d) side view of flower; (e) pollinia; (f) cross section of ovary.

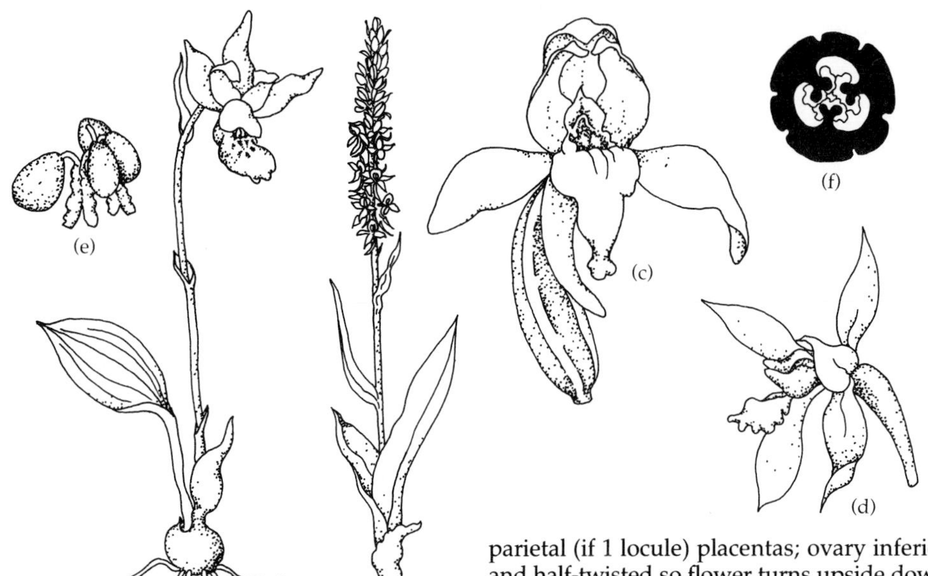


Herbs, terrestrial or epiphytic, mostly symbiotic with a fungus, perennial, sometimes saprophytic and lacking chlorophyll. Leaves alternate or basal (rarely opposite or whorled), sometimes reduced to scales, simple, entire, frequently fleshy and attached to an enlarged stem and sheathing at base. Flowers strongly irregular, perfect (rarely unisexual), epigynous; inflorescence of solitary flower or flowers borne in panicles, racemes, or spikes. Sepals 3, distinct or connate, green or petal-like but recognizable. Petals 3, distinct or connate, 2 similar and median one very different (called labellum, or lip) commonly with a nectar appendage, sac, or spur, often attractively colored. Stamens 1 or 2, adnate completely to style, forming a massive structure called a **column;** anthers opening by longitudinal slits; pollen united into 1–8 waxy masses called ***pollinia,*** often hidden behind a cap-like sterile lobe of the stigma, called a **rostellum.** Pistil compound of 3 united carpels, locules 1 or 3, ovules numerous, borne on axile (if 3 locules) or parietal (if 1 locule) placentas; ovary inferior and half-twisted so flower turns upside down during development; style 1, stigma various. Fruit a capsule. Seeds tiny, numerous; embryo minute; endosperm lacking (Fig. 10.36).

Orchidaceae is one of the largest plant families with 750–1000 genera and 18,000–20,000 species (some estimate 30,000 species) distributed worldwide, but most abundant and diverse in tropical rain forests. The largest genera are *Dendrobium* (1500 species), *Bulbophyllum* (1500 species), and *Pleurothallis* (1000 species).

The family is prized for its diverse and beautiful flowers. Many species are grown as ornamentals. These include species hybrids and cultivars of *Cattleya, Cymbidium, Dendrobium, Epidendrum,* and *Vanda.* Many have specialized or bizarre pollination mechanisms. *Vanilla* is the source of the flavoring, the only food product produced by the family.

The classification of the Orchidaceae is somewhat technical and centers mostly around the morphology of the column and pollinia. The family is usually considered to have 3 subfamilies, 6 tribes, and about 750–1000 genera. Species are easily recognized as members of the family.

The family has no known fossil record.

## *SELECTED REFERENCES*


Benson, L. 1979. *Plant Classification.* D. C. Heath, Lexington, MA, 901 pp.

Cronquist, A. 1981. *An Integrated System of Classification of Flowering Plants.* Columbia Univ. Press, New York, 1262 pp.

Cronquist, A. 1988. *The Evolution and Classification of Flowering Plants,* 2nd ed. The New York Botanical Garden, Bronx, New York, 555 pp.

Dahlgren, R. M. T. and H. T. Clifford. 1982. *The Monocotyledons: A Comparative Study.* Academic Press, New York, 378 pp.

Dahlgren, R. M. T., H. T. Clifford, and P. F. Yeo. 1985. *The Families of the Monocotyledons. Structure, Evolution and Taxonomy.* Springer-Verlag, New York, 520 pp.

Dressler, R. L. 1981. *The Orchids, Natural History and Classification.* Harvard Univ. Press, Cambridge, MA.

Heywood, V. H. (ed.). 1985. *Flowering Plants of the World.* Prentice Hall, Englewood Cliffs, NJ, 335 pp.

Hickey, M. and C. King. 1981. *100 Families of Flowering Plants.* Cambridge Univ. Press, Cambridge, 567 pp.

Hutchinson, J. 1959. *The Families of Flowering Plants.* 2 Vols. Oxford Univ. Press, Oxford, 792 pp.

Keng, H. 1978. *Orders and Families of Malayan Seed Plants.* Singapore University Press, Singapore, 437 pp.

Kubitzski, K. (ed.). 1977. *Flowering Plants, Evolution and Classification of Higher Categories.* Plant Systematics and Evolution, Suppl. I. Springer-Verlag, Vienna.

Lawrence, G. H. M. 1951. *Taxonomy of Vascular Plants.* Macmillan, New York, 823 pp.

Mabberley, D. J. 1987. *The Plant-Book.* Cambridge University Press, Cambridge, 706 pp.

Shukla, P. and S. P. Misra. 1979. *An Introduction to Taxonomy of Angiosperms.* Vikas Publ. House, New Delhi, India, 546 pp.

Smith, J. P. 1977. *Vascular Plant Families.* Mad River Press, Eureka, CA, 320 pp.

Thorne, R. F. 1983. Proposed New Realignments in the Angiosperms. *Nord. J. Bot.* 3:85–117.

Walters, D. R. and D. J. Keil. 1988. *Vascular Plant Taxonomy,* 3rd ed. Kendall/Hunt, Dubuque, IA, 488 pp.

Willis, J. C. 1973. *A Dictionary of Flowering Plants and Ferns,* 8th ed. (Rev. by H. K. A. Shaw.) Cambridge University Press, London, 1245 pp.

Zomlefer, W. B. 1983. *Common Florida Angiosperm Families* Part I. Biological Illustrations, Gainesville, FL, 108 pp.

Zomlefer, W. B. 1986. *Common Florida Angiosperm Families* Part II. Biological Illustrations, Gainesville, FL, 106 pp.

# 11

# *History and Development of Classification*

The arrangement of plants into an organizational scheme is called **plant classification.** Humans by nature are inquisitive and have always asked questions about the plants they encounter. Historical information supports the contention that some civilizations classified the plants they encountered. Early societies were dependent on some plant or plant part for food, shelter, weapons, or tools. No doubt some type of selection process occurred, with some plants being superior for a particular use, and these plants were given names to aid societies in communication.

The earliest records for naming and classification date back many millennia. At approximately 4000 B.P. (2000 B.C.) the *Atharva Veda* in ancient India provided medicinal uses of plants. In this work a "botanist" named Parashara wrote what was termed *Virkshayurveda.* This treatise dealt with the property of soils, described various forest types in India, described the morphology of leaves and cells and delimited a considerable number of present-day plant families. Other treatises on medicinal plants used by Aztecs, Assyrians, Chinese, and Egyptians have also been discovered.

Even in the early Judeo-Christian record, naming and classification was recorded. We read in Genesis 2:19, 20 "Now . . . the Lord God . . . brought them to the man (Adam) to see what he would name them; and whatever the man called each living creature, that was its name. So the man gave names to all the livestock, the birds of the air and all the beasts of the field" (New International Version).

In the western world the beginnings of botany as a science can be traced to the ancient Greek philosophers who lived between 370 and 285 B.C. An inspection of the historical development of plant classification can be grouped into the systems: the form system of plants, the sexual system of plants; early "natural" systems; post-Darwin "natural" systems (1860–1950); and contemporary classification systems.

## *Form System of Plants*

The early philosophers and medical personnel of Greece and Rome wrote some of the early botanical works published in Europe.

An outstanding student of Aristotle and Plato was **Theophrastus** (370–287 B.C.) He classified approximately 500 plants in his *Historia plantarum* on the basis of habit or form, notably herbs, shrubs, trees, and undershrub. He distinguished between annual, biennial, or perennial lifespan. He also provided the basis for floral morphology by distinguishing between superior and inferior ovaries, polypetalous and gamopetalous corollas, and determinate and indeterminate inflorescences. He probably gained much of his botanical knowledge from the cultivated plants grown in early Athenian gardens.

**Pliny the Elder** (23–79 A.D.), a Roman scholar and naturalist, wrote an encyclopedic work of books termed *Historia naturalis.* In this series of books he mentioned plants in terms of such diverse topics as horticultural practices, medicinal uses, plant anatomy, and trees used for timber. He used a similar classification as the Greeks and also incorporated many of his predecessors' errors. His work is considered one of the most important of the early botanical works. His life was terminated by an eruption of Mt. Vesuvius.

**Dioscorides,** a contemporary of Pliny, was a physician in Emperor Nero's army. He traveled extensively in the Mediterranean region studying plants. His *de Materia medicia* (first century A.D.) was illustrated and included medicinal information and descriptions of about 600 mostly Mediterranean plants. This work was widely used by Europeans for the next 1500 years.

In about 512 A.D., the *Anicia Julianna Codex* was prepared from material originally written by Dioscorides for the daughter of a Byzantine emperor. This work was illustrated in color and was copied and recopied many times without major revision during the Middle Ages.

**Albertus Magnus** (1193–1280) wrote *de Vegetabilis,* in which he described various garden vegetables. He accepted Theophrastus' classification but introduced for the first time the differences between monocotyledon and dicotyledon seeds.

Little information of real botanical significance was made following the fall of the Roman civilization. It was not until the fifteenth and sixteenth centuries that there was any real awakening. During this period, sometimes called the **Age of the Herbals,** many new plants were described and illustrated. This was enhanced by the new art of printing illustrations by woodcuts and moveable type. The authors of these books were concerned with providing information on medicinal plants and therefore these works became known as **herbals** and the authors as **herbalists.**

In Eastern Asia herbals were compiled before and during this beginning of botany in Europe. In China *Cheng Lei Pen Tsào* was written by Tàng Shen in 1108 and went through 12 editions until 1600. Hsu Yung wrote *Pen Tsà O Fa Hui* in 1450 and Li Shi Chen wrote *Pen Tsào Kangmu* in 1590.

During this herbalist period the information was not of much value from a classification standpoint. Some herbals were well known for their fine illustrations of living plants. These included works by **Otto Brunfels** (1489–1534), **Jerome Bock** (1498–1554), **Leonhart Fuchs** (1501–1566), **Charles Clusius**

(1526–1609), and **Mathias de l'Obel** (1515–1568), known as the Father of British Botany.

One significant taxonomic contribution is worth mentioning. A pupil of Fuchs, **Gaspard (Kaspar) Bauhin,** attempted to use a binomial system of nomenclature in several publications, such as *Prodromus theatri botanici* (1620) and *Pinax theatri botanici* (1623).

**Andraea Caesalpino** (1519–1603), an Italian physician, organized an herbarium of 768 dried and mounted plants in 1563, some of which are still preserved. His work, *De plantis* (1583), described 1520 species of plants, arranged into herbaceous and woody. He further realized the value of using flower and fruit characters over characters of form or habit. It is said that he had a good grasp of the concept of what we today call genera and greatly influenced later botanists.

The first real advance in taxonomy and classification in many years was made by an English blacksmith, **John Ray** (1628–1705). After graduating from Trinity College, he traveled extensively in Europe observing plants. In his two main works, *Methodus plantarum nova* (1682) and *Historia plantarum* (1686–1704), he expanded the principle that all parts of the plant should be considered for classification (Fig. 11.1).

Ray's classification system divided plants first in the old way of Theophrastus into herbs (Herbae), shrubs, and trees (Arborae). He further divided

**FIGURE 11.1 John Ray (1628–1705), an English blacksmith who wrote an early classification of plants.**

*SOURCE:* Courtesy of the Royal Botanic Gardens, Kew.

them into 25 classes of dicotyledons and four classes of monocotyledons. Some of his groupings represent some present-day families such as Cruciferae (Tetrapetalae), Labitate (Verticillatae), Leguminosae, and grasses (Staminae). His writings dealt with approximately 18,000 species.

A French contemporary of John Ray was **Pierre Magnol** (1638–1715). Finding Ray's classification too cumbersome, he grouped plants into 76 families in his *Prodromus historiae generalis, in qua familiae per tabulas disponutur* (1689). Magnol is therefore the first to use the family concept emphasized today. His name is honored by the beautiful woody genus *Magnolia.*

**Joseph Pitton de Tournefort** (1656–1708) was a student of Magnol and became a professor of botany at the Jardin de Roi. Tournefort traveled extensively in Europe and Asia Minor collecting plants; he even climbed Mt. Ararat. He was the first to give descriptions of genera. He recognized flowers with petals and without petals (apetalous), corollas with separate and united petals, and regular and irregular corollas. It is noteworthy that he still grouped the plants into herbs and trees, but he did not recognize plant sexuality. In his *Eléments de botanique* (1694), enlarged in 1700 to *Institutiones rei herbariae,* he described 698 genera and 10,146 species. Some of his genus names, such as *Abutilon, Acer, Betula, Quercus,* and *Ulmus,* were used by Linnaeus.

**Rudolf Jacob Camerarius** (1665–1721) was the director of the botanical garden at Tübingen, Germany. He is not known for his books. Instead he wrote letters to botanists reporting his crossing experiments between different plants. In a letter dated 25 August 1694 to a professor at Giessen, titled *De sesu plantarum epistola,* he described how pistillate flowers would not set seed without staminate flowers being present. He referred to the stamens as the male sex organs, and the style and ovary as the female sex organs, and pollen as necessary for seed development. For the first time, sexual reproduction was established for the flowering plants.

## *Sexual System of Plants*

By the turn of the eighteenth century, there had been a gradual accumulation of plant material. Much of this material did not fit the known classification schemes of the day. It was during this historical-botanical period that a young Swedish botanist, Carl Linnaeus (1707–1778) came to the classification scene (Fig. 11.2, p. 376). Linnaeus was born into the family of a poor clergyman, at Råshult, Sweden. During his early childhood years he showed an interest in the flowers of the garden and constantly asked for the names of the plants. He enrolled at the University of Uppsala, where he came under the guidance of an elderly professor, Olaf Rudbeck, who treated him like a son. Under Rudbeck's guidance he published his first paper on the sexuality of plants in 1729.

In 1732 he obtained a small travel grant of approximately $125 (U.S.) from the Academy of Sciences of Uppsala for a botanical exploration of Lapland. In 5 months Linneaus traveled over 7600 km and returned with over 537 specimens. The results of the excursion were published in *Flora lapponica* (1737).

**FIGURE 11.2 Carl Linnaeus (1707–1778), the Father of Biological Classification, as a young man holding the plant named *Linnaea borealis* (the genus was named after Linnaeus by Gronovius).**

*SOURCE:* Courtesy of the Royal Botanic Gardens, Kew.

Young Linnaeus was encouraged to study medicine in mainland Europe, and in 1735 he traveled to the Netherlands, where he obtained his degree at Harderwijk. While in the Netherlands he became closely associated with two Dutch botanists, Hermann Boerhaave and J. Gronovius. Their influence was most invigorating to Linnaeus, for in the span of 2 years he wrote *Systema naturae* (1735), which was the basis of Linnaeus' sexual classification of plants, animals, and minerals; *Critica botanica* (1737); *Flora lapponica* (1737); *Hortis cliffortianus* (1737); and *Genera plantarum* (1737). This latter work is important because of the 935 genera described; five revisions and two supplements later would bring the total to 1336 genera described.

Linnaeus took short visits to England and France. When his former teacher Rudbeck died in 1742, he returned to Uppsala as Professor of Medicine and later also became Professor of Botany. As Professor of Botany, in 1753 he published *Species plantarum.* This work described approximately 1000 genera and 7300 species, all given binomial Latin names and arranged according to the sexual system of *Systema naturae,* which was published earlier (Table 11.1). Linnaeus was not the first to use binomial nomenclature (as discussed earlier), but he was the first to do so consistently. This is why *Species plantarum* and its published date of 1 May 1753 is chosen by contemporary botanists as the beginning point for plant nomenclature. Carl Linnaeus can truly be called the Father of Biological Classification.

**TABLE 11.1 Outline of the classes of the classification used by Linnaeus in *Species plantarum.***

| *Class Number* | *Name* | *Number of Stamens* | *Present-Day Example* |
|---|---|---|---|
| 1. | Moandria | 1 | *Canna* |
| 2. | Diandria | 2 | *Veronica* |
| 3. | Triandria | 3 | *Poa* |
| 4. | Tetrandria | 4 | *Protea* |
| 5. | Pentandria | 5 | *Campanula* |
| 6. | Hexandria | 6 | *Lilium* |
| 7. | Heptandria | 7 | *Aesculus* |
| 8. | Octandria | 8 | *Vaccinium* |
| 9. | Enneandria | 9 | *Laurus* |
| 10. | Decandria | 10 | *Rhododendron* |
| 11. | Dodecandria | 11–19 | *Euphorbia* |
| 12. | Icosandria | ≤ 20 on calyx | *Opuntia* |
| 13. | Polyandria | ≤ 20 on receptacles | *Ranunculus* |
| 14. | Didynamia | Stamens didynamous | *Lamium* |
| 15. | Tetradynamia | Stamens tetradynamous | *Brassica* |
| 16. | Monadelphia | Stamens monodelphous | *Malva* |
| 17. | Didelphia | Stamen didelphous | *Faba* |
| 18. | Polyadelphia | Stamen polyadelphous | *Hypericum* |
| 19. | Syngenesia | Stamen syngenesious | *Aster* |
| 20. | Gynandria | Stamen united | *Cypripedium* |
| 21. | Monoecia | Plants monoecious | *Carex* |
| 22. | Dioecia | Plants dioecious | *Salix* |
| 23. | Polygamia | Plants polygamous | *Acer* |
| 24. | Cryptogamia | Nonflowering plants | *Pteris* |

Linnaeus' greatest contribution to botany was his new system of naming plants. He gave the general (genus) name and a trivial (specific epithet) name to each plant, references of publications, herbarium specimens seen, and geographical areas where the plant was found.

Many students studied with Linnaeus and spread the "Linnean gospel" throughout the world. In 1761 he was nobilized and was known from this point as Carl von Linné.

The last years of his life were spent in ill health and he died in 1778. After his death his widow sold his personal library and herbarium to an Englishman, Dr. James E. Smith, in 1784 for the sum of £1000. These collections later became the property of The Linnean Society of London, which was founded by Smith in 1788. Today they are carefully preserved at Burlington House, on Piccadilly, at the headquarters of the Society (Fig 11.3).

Linnaeus was not as rigid with his sexual system of classification as many have been led to believe. Many times he included species in a genus that did not follow his system definition. He would use the technical characters of a plant species to confirm their general arrangement. In this way many genera fit a more natural classification.

Early in his career Linnaeus believed that species were unique, genetically true-breeding, and monotypic. The variation he observed he attributed to climate or soil differences. In later years he began to maintain that many species and even genera had developed through hybridization, a concept not normally thought of as applying to Linnaeus. His attempt at bringing together species that were more alike is indicated by Linnaeus' sixth edition of *Genera plantarum* (1764), where genera in 58 "natural orders" were presented. His sexual system was retained for identification purposes only.

*Species plantarum* did not end with Linnaeus. It underwent various revisions until the beginning of the 1800s. It was published in six volumes

**FIGURE 11.3 The climate-controlled vault at The Linnean Society of London, where Linnaeus' original specimens are housed.**

*SOURCE:* Photo courtesy of the librarian, The Linnean Society of London.

between the years 1797 and 1830 by the German botanist Karl Ludwig Willdenow (1765–1812), who included species from all over the world. Linnaeus' home has been restored and is preserved in Sweden (Fig. 11.4).

## *Early "Natural" Systems*

During the latter part of the eighteenth century many explorers and botanists, armed with the tool of the sexual system, went forth to the four corners of the world. It soon became apparent that the wealth of plant specimens could not be classified satisfactorily with Linnaeus' system and a better one was desperately needed. Botanists began to realize that some natural affinity occurred among plants. Various botanists then began to incorporate a natural arrangement of the plants in their works.

**Michel Adanson** (1727–1806), a Frenchman exploring the African flora, rejected Linnaeus' sexual system in favor of a natural one for his groupings, which today correspond to orders and families. This was recorded in his work *Families des plantes* (1763).

A French contemporary of Linnaeus, **Bernard de Jussieu** (1699–1777), was not satisfied with Linnaeus' system either and attempted to improve

**FIGURE 11.4 The restored home of Carl Linnaeus in Sweden.**

*SOURCE:* Photo courtesy of W. F. Grant.

upon it, modifying it into a more natural arrangement. His system was never published, but his nephew **Antoine Laurent de Jussieu** (1748–1836) (Fig. 11.5) did publish it, along with his own concepts in *Genera plantarum secundum ordines naturalis disposita* (1789). The plants were classified into 15 classes and 100 orders, which are recognized today as families (Table 11.2). This first comprehensive attempt at natural classification is the beginning point for conserved family names of flowering plants, as stated according to the International Code of Botanical Nomenclature.

**Augustin Pyramus de Candolle** (1778–1841) (Fig. 11.6), the senior member of the famous botanical Swiss-French family, developed further the classification system of A.L. de Jussieu in his *Théorie élémentaire de la botanique* (1813), in which 135 orders (today's families) were described. Along with his son **Alphonse de Candolle** and grandson **Casimir,** he published the enormous *Prodromus systematis naturalis regni vegetabilis* (1824–1873), which included descriptions of almost 59,000 species of gymnosperms and dicotyledons. The ferns were placed with the monocots and the gymnosperms with the dicots, while the algae, mosses, liverworts, fungi, and lichens were placed in the Cellulares (plants without vascular tissue). Even though other natural systems were proposed during this time period, the de Candolle system was dominant until about 1860. The de Candolle library and specimens are housed at the Conservatoire et Jardin Botanique, Genève, Switzerland.

A final interesting note about the popularity of the de Candolle classi-

**FIGURE 11.5 Antoine Laurent de Jussieu (1748–1836), a French contemporary of Linnaeus.**

*SOURCE:* Courtesy of the Royal Botanic Gardens, Kew.

**TABLE 11.2 Outline of the de Jussieu classification of plants.**

| *Number Group* | *Modern Examples* |
|---|---|
| 1. Acotyledones | Algae, fungi, mosses |
| Monocotyledones | |
| 2. Stamens hypogynous | Cyperaceae, Gramineae (Poaceae) |
| 3. Stamens perigynous | Iridaceae |
| 4. Stamens epigynous | Orchidaceae |
| Dicotyledones, Apetalae | |
| 5. Stamens epigynous | Aristolochiaceae |
| 6. Stamens perigynous | Polygonaceae, Proteaceae |
| 7. Stamens hypogynous | Amaranthaceae, Nyctaginaceae |
| Dicotyledones, Monopetalae | |
| 8. Corolla hypogynous | Boraginaceae, Labiatae (Lamiaceae) |
| 9. Corolla perigynous | Ericaceae |
| 10. Corolla epigynous, anthers united | Cichoraceae |
| 11. Corolla epigynous, anthers distinct | Rubiaceae |
| 12. Stamens epigynous | Araliaceae, Umbelliferae (Apiaceae) |
| 13. Stamens hypogynous | Cruiciferae (Brassicaceae), Ranunculaceae |
| 14. Stamens perigynous | Leguminosae (Fabaceae), Rosaceae |
| 15. *Declines irregularis* | Amentiferous plants, conifers |

*SOURCE:* Modified from Naik, 1984; Porter, 1967.

**FIGURE 11.6 Augustin Pyramus de Candolle (1778–1841), the senior member of the famous de Candolle family of botanists.**

*SOURCE:* Courtesy of the Royal Botanic Gardens, Kew.

fication was its sequence, starting with the polypetalous, hypogynous Ranalian groups, similar to present-day systems (Davis and Heywood, 1973).

A Scottish botanist, **Robert Brown** (1773–1858), was known not for any classification system of his own, but for his observations on floral and seed morphology (Fig. 11.7). He was the first to show that the gymnosperms (Pinophyta) were a separate group of plants with naked ovules, in contrast to the angiosperms, having the ovules enclosed in ovaries. Brown was the first to establish the families Asclepiadaceae and Santalaceae, to write about the morphology of the grass flower, and to understand the nature of the cyathium of the Euphorbiaceae. Davis and Heywood (1973) have written "it is difficult to understand how Brown could have written as he did without some intimation of evolution."

During the 20 years between 1825 and 1845, there were at least 24 classification systems proposed (Lawrence, 1951). They all were nothing more than a minor revision of the concepts of de Candolle and Brown. Three individuals worth mentioning briefly were the French botanist **Adolphe T. Brongniart** (1801–1876), who considered the apetalous plants to be reduced from the polypetalous; **John Lindley** (1799–1865), an Englishman who felt plants had developed along lines of simple to more complex morphology; and the Viennese botanist **S.L. Endlicher** (1805–1849), who divided the plant kingdom into thallophytes (algae, fungi, and lichens) and cormophytes (mosses, ferns, and seed plants). Endlicher's system was widely used in Europe but was neglected in England and North America.

**George Bentham** (1800–1884) (Fig. 11.8) and **Sir Joseph Hooker** (1817–1911) (Fig. 11.9) were British botanists who worked out of the Royal Botanic Gardens at Kew, just outside of London. They published their classification

**FIGURE 11.7 The Scottish botanist Robert Brown (1773–1858), the first to recognize gymnosperms as a separate group of plants.**

*SOURCE:* Courtesy of the Royal Botanic Gardens, Kew.

**FIGURE 11.8 The famous British botanist George Bentham (1800–1884), who worked at the Royal Botanic Gardens with Sir Joseph Hooker.**

*SOURCE:* Photo courtesy of the Royal Botanic Gardens, Kew.

**FIGURE 11.9 The famous British botanist Sir Joseph Hooker (1817–1911), who was a close friend of Charles Darwin.**

*SOURCE:* Photo courtesy of the Royal Botanic Gardens, Kew.

system in a three-volume work called *Genera plantarum* (1862–1883). This work described 202 orders (present-day families) and included all known seed plants at that time (over 97,000) according to the Bentham and Hooker system. This system followed that of de Candolle but also differed somewhat. All genera were described in Latin from observed living material at Kew or from herbarium specimens Bentham and Hooker had seen. Their work gave a synopsis of each family at the beginning and gave the geographical range of each genus. This system was a real landmark (Davis and Heywood, 1973) in botany for its scholarship and quality.

It could be noted that the most famous North American botanist at this time, **Asa Gray** (1810–1888), combined the Bentham and Hooker system with that of de Candolle for his own use.

## *Post-Darwinian "Natural" Classification Systems (1860–1950)*

With the publication of Charles Darwin's *The Origin of Species* in 1859, the complete direction of classification and biological thought was altered (Fig. 11.10). The theories of Darwin seemed to bring together all of the dissatisfaction that botanists had held toward Linnaeus' sexual system and then later toward the de Candollean system.

The systems that developed are based on the theories of descent and

**FIGURE 11.10 Charles Darwin, whose ideas on the origin of species revolutionized botanists' view of the classification of plants (Darwin autographed this photo, "I like this photograph very much better than any other which has been taken of me. Ch. Darwin").**

*SOURCE:* Photo courtesy of the Royal Botanic Gardens, Kew.

evolution and that the present-day life forms are a product of natural evolutionary processes. The authors of the various classification systems attempted to organize the various plant groups from the most simple to the most complex and to demonstrate ancestral relationships. As more and more data became available about plants and more botanists began to study plants, it was inevitable that different views would be forthcoming and different classification systems would be produced.

**August Wilhelm Eichler** (1839–1887), a German, modified earlier systems to reflect a better relationship between plants. The plant kingdom was divided into nonseed plants (Cryptogamae) and seed plants (Phanaerogamae). The former group was divided into Thallophyta, Bryophyta, and Pteridophyta. The seed plants (Phanaerogamae) were divided into the gymnosperms and angiosperms, and the angiosperms were broken down into monocots and dicots. This breakdown by Eichler was based on the assumption that plants with a complex flower organization are more "advanced" in their evolutionary development.

Eichler's system was the foundation for Adolf Engler's system (see below) and was never widely accepted. The system of Bentham and Hooker remained the dominant system in North America and England.

Another German, **Adolf Engler** (1844–1930) (Fig. 11.11), and his associate, **Karl Prantl** (1849–1893), adopted the major features of Eichler's clas-

**FIGURE 11.11 Adolf Engler (1844–1930), who, with his associate Karl Prantl, became famous for completing the only detailed classification of plants from algae to flowering plants.**

*SOURCE:* Photo courtesy of the Royal Botanic Gardens, Kew.

sification and published a 20-volume work, *Die natürlichen Pflanzenfamilien* (1887–1899), which provided a way to identify all of the known genera of plants from algae to flowering plants. The work was illustrated, and included keys; these features, along with information on anatomy, embryology, morphology, and geography, descriptions, and good publicity, helped botanists to adopt the system within a few years.

Many of Engler's changes to Eichler's system show the influence of earlier systems, but he still followed the basic phylogenetic breakdown. A main feature of the system is that the monocotyledons are placed before the dicotyledons and the catkin-bearing (ament-type) plants are placed before other families. This indicated that these plants were considered to be more primitive than the rest.

A slightly revised edition of the system was last published in 1964 by Engler and **L. Diels** as *Syllabus der Pflanzenfamilien.*

Up to the present time, Engler's system is still used by most herbaria in their arrangement of specimens and is followed by writers of many manuals and floras; it is still the only system that treats all plants (algae to flowering plants) in such depth.

**Richard von Wettstein** (1862–1931), a botanist from Austria, published a system that is similar in many respects to Engler's (Fig. 11.12). In his *Handbuch der systematischen Botanik* (1901) and fourth revision (1930–1935), he rearranged many of the dicot families. He felt that the monocots were derived from the order Ranales in the dicots; that unisexual flowers lacking perianth were the simplest, with the perfect flowers derived from them; and he considered more contemporary literature. His system was more phylogenetic than Engler's system.

The system of Wettstein was not widely accepted, especially in North America. However, his theory that monocots were derived from dicots differed from Engler's ideas. Some of the present-day contemporary systems

**FIGURE 11.12 The famous Austrian botanist Richard von Wettstein (1862–1931), who published a classification system similar to Engler's.**

*SOURCE:* Photo courtesy of the Royal Botanic Gardens, Kew.

have adopted many of his more specific conclusions to phylogenetic relationships.

The first North American to make a contribution to the general system of classification was a student of Asa Gray, **Charles E. Bessey** (1845–1915) (Fig. 11.13). Bessey began his career at Iowa State University but after a short time went to the University of Nebraska, where he spent most of his career. In creating his system, entitled "The Phylogenetic Taxonomy of Flowering Plants," he was the first to develop a scheme that was "truly phylogenetic" (Lawrence, 1951). He rejected many of the Eichler–Engler ideas.

Bessey's system was founded on various guiding rules or "dicta," which he used to determine the level of being, simple or advanced, of a group of plants. Some of his 28 dicta are: "(1) Evolution is not always upward, but often it involves degradation and degeneration. (2) In general, homogeneous structures (with many and similar parts) are lower, and heterogeneous structures (with fewer and dissimilar parts) are higher. (9) Woody stems (as of trees) are more primitive than herbaceous stems, and herbs are held to have been derived from trees. (12) Historically simple leaves preceded branched ("compound") leaves. (16) Petaly is the normal perianth structure, and apetaly is the result of perianth reduction (aphanisis). (19) Hypogny is the more primitive structure, and from it epigyny was derived later. (24) In earlier (primitive) flowers there are many stamens (polystemonous) while in later flowers there are fewer stamens (oligostemonous). (27) Flowers with both stamens and carpels (monoclinous) precede those in which these occur on separate flowers (diclinous)." From these rules he constructed a phylogenetic diagram that has been called "Bessey's cactus" or "*Opuntia besseyi*" by many American botanists. His ideas differed from the Engler and Prantl system in

**FIGURE 11.13 Charles E. Bessey (1845–1915), whose ideas have provided the basis for some of the current systems of classification of plants.**

*SOURCE:* Iowa State University Library/ University Archives.

that 1) he thought angiosperms were derived from a cycad-type ancestor, similar to that of the fossil Bennettitales, whereas in the Engler and Prantl system, flowering plants originated from some unknown gymnosperm that was similar to a conifer; 2) he felt that angiosperms developed from the Ranales group from a plant with flora features that were similar to this group, while Engler and Prantl felt that the dicots and monocots were derived from some catkin-bearing (amentiferous), floral-reduced ancestor and then obtained a perianth; and 3) he made his main separation of subclasses based on the floral cup characters of hypogynous, perigynous, and epigynous, while Engler and Prantl felt the major separation should be based on apetalous, choripetalous, and sympetalous.

The ideas of Bessey have greatly influenced the thinking of authors of late twentieth century systems.

A Swedish botanist, **Carl Skottsberg** (1880–1963), developed a modified Engler system to the plant kingdom. He borrowed some concepts from Wettstein in his publication *Vaxternas Lir* in 1932–1940. He maintained the separation of the pteridophytes and gymnosperms but thought the class Gnetinae should include the Ephedrales and Welwitschiales. He also felt that within the flowering plants some undiscovered dicot gave rise to the monocots. He deviated from the ideas of Engler and Wettstein in proposing that the apetalous families originated many times independently and were reorganized throughout the dicots.

**Hans Hallier** (1868–1932), a German botanist, proposed a phylogenetic system that is similar to Bessey's. He believed that the dicots developed from Magnolia-like ancestors, with the monocots being more advanced and developing from an unknown ancestor. Both Bessey and Hallier developed their systems independently of one another.

The British botanist **John Hutchinson** (1884–1972) worked at the Royal Botanic Gardens, Kew (Fig. 11.14). The system he proposed in his *Families of Flowering Plants* (1926, 1934; 3rd ed. 1973) was somewhat like Bessey's but had some major differences. He considered the angiosperms to be monophyletic in origin from a hypothetical seed plant. He considered the herbaceous or woody habit of the plant to be very important. The woody dicot groups developed from the Magnolias and the herbaceous groups were supposedly derived from herbaceous members in the Ranales. The monocotyledons were thought to have evolved from ancestral individuals in the Ranales. Hutchinson also proposed more orders than usual with fewer families in each order. The orders were derived from ancestral precursors to the present-day orders, not from each order directly.

Hutchinson's system has not been accepted by many botanists, even though the arrangement of many families, especially within the monocotyledons, has been very helpful to botanists. The greatest criticism is leveled at his initial separation of herbaceous from woody groups. This has split some families that are similar in flower morphology and appear as natural units, except for the differences in habit: herbaceous versus woody.

A completely different approach to classification systems was used by the German **Karl Mez** (1866–1944). Mez believed that relationships between various large groups of flowering plants could be ascertained by using serological antigen-antibody reactions. By comparing the protein reactions of

**FIGURE 11.14 The British botanist John Hutchinson (1884–1972), who proposed a family classification for the flowering plants.**

*SOURCE:* Photo courtesy of the Royal Botanic Gardens, Kew.

different groups (i.e., genera, families, etc.), close or distant affinities could be determined. Mez produced a "family tree" for the plant kingdom but it was severely criticized. Few botanists today follow Mez's ideas.

**Oswald Tippo** (b. 1911) at the University of Illinois attempted to do a broad classification to the higher groups of the plant kingdom in 1942. He used detailed information from other authors, studying various groups (both living and fossil), and divided the plant kingdom into various subkingdoms and phyla. He theorized that the pteridophytes were not a homogenous group and that seed plants and pteridophytes lack a demarcation between them. The Magnoliales were thought by him to be the most primitive flowering plants.

## *Contemporary Classification Systems (1950 to Present)*

During the 33 years from 1950 to 1983 there was intense interest generated in developing a more acceptable classification for the plant kingdom. A few individuals, such as **Harold C. Bold** (1909–1987) and **Robert Whittaker** (1921–1980), confined their efforts to the larger groups of plants. Others concentrated on the angiosperms and, using Bessey's original ideas, attempted to construct a more natural system.

In 1950, an American, **A. Gundersen** (1877–1958), proposed a new classification system for the dicotyledons based on anatomy, cytology, and morphological data. The dicots were divided into 10 groups, which were further separated into 42 orders. He ignored previously used floral characters, such as polypetaly and sympetaly.

Another American botanist, **Lyman Benson** (b. 1909), proposed a modified system of rearranged orders and families, taking ideas from both Bessey and Engler. He published this in the form of a two-dimensional chart in 1957, which was revised in 1979. The chart covered angiosperm groups only. Benson did not feel that any contemporary plant order was derived from any other living order, but believed that they developed from precursor groups that are now extinct. He felt that how the flowering plants originated was uncertain but that it was in some woody Ranales. None of the orders were connected. The monocots, however, developed from the same general Ranales or Alismatales area. His separation of orders into Corolliflore and Ovariflorae groups was purely arbitrary and not natural.

Benson's system is easy to use when teaching students, but has not been followed by many botanists.

**G. Ledyard Stebbins** (b. 1906) of the University of California, Davis, has made many outstanding contributions to botany (Fig. 11.15). In 1974, he discussed the basis for classification of flowering plants. He applied information from genetics (mutation, population, recombination, etc.), geology, microevolution, natural selection, reproduction, paleobotany, and ecology to the relationships within angiosperms. Stebbins proposed no new classification to the higher groups (families, orders) but modified the system of Cronquist. His broad-disciplined approach allows the reader the freedom to view the complexity of data as applied to a natural classification system of flowering plants. This contemporary approach, allowing freedom for differing views and ideas toward the same data, or as new data become available, is refreshing as compared with the dictatoral "dicta" of Bessey.

**FIGURE 11.15 The outstanding American geneticist G. Ledyard Stebbins.**

*SOURCE:* Photo courtesy of G. L. Stebbins.

**Armen L. Takhtajan** (b. 1910), working in Leningrad and publishing in Russian, has developed over many years a phylogenetic system of the angiosperms. His system, published in English in 1961, 1964, and 1969 and later versions (published earlier in Russian), reflected the influence of the Hallier system. He felt that Hallier's system gave a better insight into flowering plant phylogeny than did Bessey's. Takhtajan subdivided the Magnoliophyta (angiosperms) into two classes: Magnoliopsida (dicots) and Liliopsida (monocots) (Magnoliatae and Liliatae in earlier versions). He divided these into various subclasses with endings of–anae, rejecting the –florae ending used previously. In the Magnoliopsida, he recognized 7 subclasses, 20 superorders, 71 orders, and 333 families, while in the Liliopsida were found 3 subclasses, 8 superorders, 21 orders, and 77 families. He considered the flowering plants to be monophyletic, not polyphyletic, in origin. He considered the order Magnoliales to be the most primitive, and from this branch the various angiosperm groups developed, and the Liliopsida (monocots) were derived from a precursor in the Nymphaeales (water lily) group. This latter idea has received the most criticism from botanists, who feel the two groups are similar due to convergent evolution instead of divergent change. In spite of these criticisms, his ideas have been relatively well accepted by the botanical community. Takhtajan's relationships among groups can be seen in "Takhtajan's flower garden" (Fig. 11.16, p. 392).

**Arthur Cronquist** (b. 1919) (Fig. 11.17, p. 393) at the New York Botanical Garden first presented a system for dicotyledons in 1957, and in succeeding years (1966, 1968) he refined it to include all flowering plants. His system is based on a vast literature search (including the Russian literature), personal communication with other botanists, and individual study in the field and with herbarium specimens. His system takes its initial basis from Bessey, followed by a modification and refinement of Takhtajan's phylogenetic system. His "evolutionary tree" (more like two small shrubs) derives the two classes Magnoliopsida (dicotyledons) and Liliopsida (monocotyledons) from the primitive Ranales group. In the class Magnoliopsida, there are 6 subclasses, 55 orders, and 352 families; the Liliopsida has 5 subclasses, 18 orders, and 61 families (Fig. 11.18, p. 394). Cronquist's system is based on the idea that the orders within the subclasses were derived from one another. However, it should be pointed out that Cronquist emphasizes continually that the correctness of some of his arrangements are somewhat arbitrary or open to reinterpretation and new information.

Most recently, Cronquist discussed the system in great depth in *An Integrated System of Classification of Flowering Plants* (1981). Here the student will find a discussion; keys to all classes, subclasses, orders, and families of flowering plants; a commentary, and vast reference sources. Due to its thoroughness, many botanists in the world today are widely following Cronquist's system or a slight modification of it. The author of this book has followed the Cronquist system in the arrangement of families in Chapters 9 and 10. This was done not because the author feels the system is the most phylogenetically correct of the present systems, but because 1) it gives a contemporary system based on all types of research evidence that reflects our best knowledge about flowering plant relationships, 2) it provides more in-depth literature on the various plant taxa than any of the other systems currently

**FIGURE 11.16 "Takhtajan's flower garden," which shows the putative relationships between the orders and subclasses of the flowering plants.**

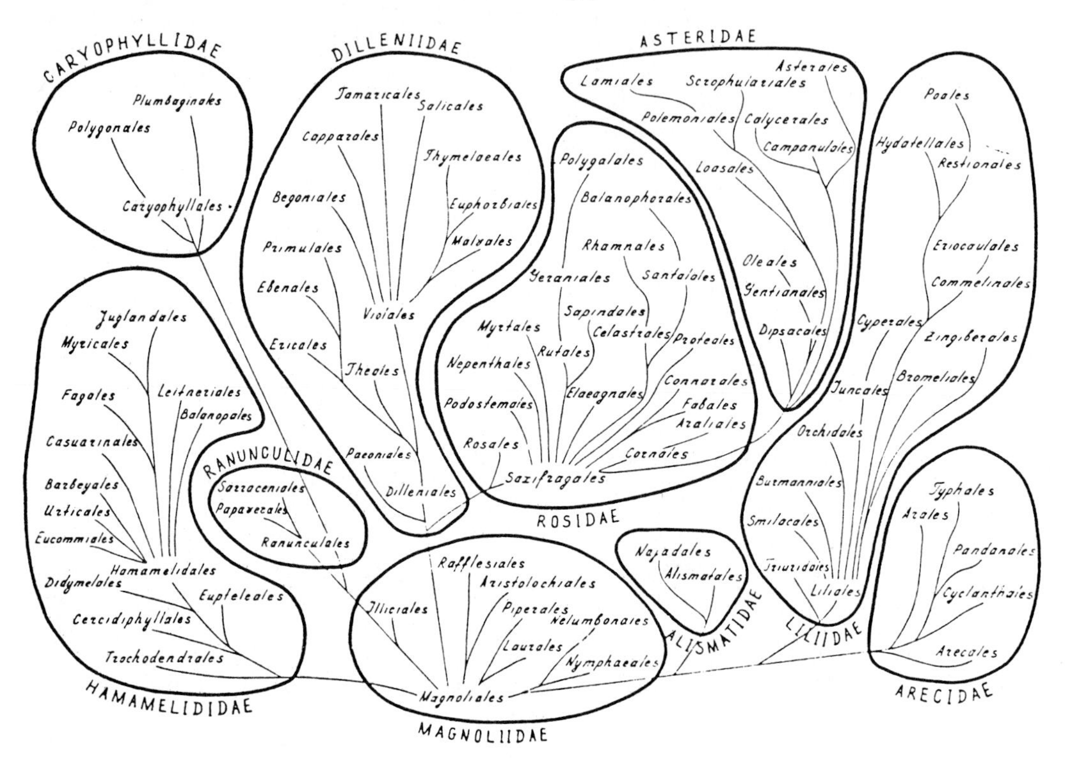


*SOURCE:* From Takhtajan, 1980.

being discussed and used by botanists today for the individual who wishes to search further, and 3) it provides a logical, orderly (if not natural) way for the student to learn about flowering plants.

A similar system to Takhtajan's and Cronquist's systems for classifying flowering plants has been proposed by a former curator of the Rancho Santa Ana Botanical Garden in California, **Robert F. Thorne** (b. 1920) (Fig. 11.19, p. 394). Thorne's system is centered around the angiosperms, which are divided into two monophyletic classes: Magnoliopsida (dicots) and Liliopsida (monocots). He next divides the classes into superorders. In Thorne's evolutionary shrub, he placed the superorders near the extinct precursor protoangiospermae in the center, where the "most primitive" characters are thought to have been retained. He next radiated out the various taxa from the "point of origin," with the orders that were most unlike the ancestral group placed furthest from the center (i.e., Asterales) (Fig. 11.20, p. 395).

Thorne states that his classification system, based on all obtainable information, is different from other systems because he stresses relationships and similarities of taxa rather than a classification based on the "importance" of presumed phylogenetically significant characters. In spite of Thorne's claims

**FIGURE 11.17 Two authors of well-known contemporary systems of flowering plant classification: left, Armen Takhtajan; right, Arthur Cronquist.**

*SOURCE:* Photo courtesy of A. Cronquist.

of a different system, many botanists consider the system to be similar to Cronquist's and Takhtajan's systems, but in actuality it is similar only in principle.

Another contemporary system needs to be considered. **Rolf M. T. Dahlgren** (1932–1987) (Fig. 11.21, p. 396) of Copenhagen presented a phylogenetic diagram. Recently his wife, Gertrud Dahlgren, has combined their ideas in a two-dimensional diagram (Fig. 11.22, p. 397). The class Magnoliopsida (angiosperms) is divided somewhat arbitrarily into two subclasses: Magnoliidae (dicots) and Liliidae (monocots). The Magnoliidae is broken down into 25 superorders, with the Liliidae having 10 superorders. Dahlgren considered flowering plants to be monophyletic in origin and believed that various specialized characters (i.e., presence of phloem companion cells, 8-nucleate embryosac, etc.) would only have developed once in the angiosperm precursor. The system as portrayed by Dahlgren shows irregular-shaped branches to be most similar when close to one another and the size relates to the number of species in each. He used the ending -florae, like Thorne, for superorders.

In spite of splitting up some orders into more families in the superorder Liliiflorae, the system is very similar to that of Takhtajan. To help in under-

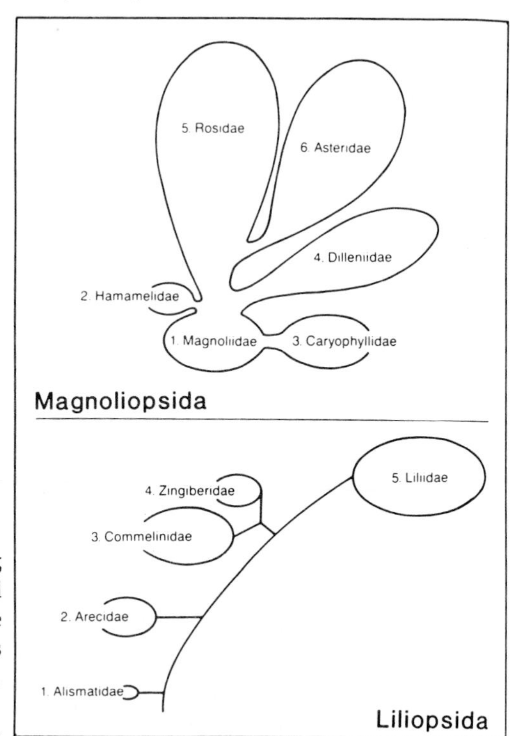


**FIGURE 11.18 Putative evolution among the subclasses of Magnoliopsida and Liliopsida according to Cronquist. The number of species in each group is proportional to the size of the balloons.**

*SOURCE:* From Cronquist, 1988.

**FIGURE 11.19 The American botanist Robert F. Thorne.**

*SOURCE:* Photo courtesy of R. F. Thorne.

FIGURE 11.20 Thorne's grouping of flowering plant groups in relation to a hypothetical extinct protoangiosperm.

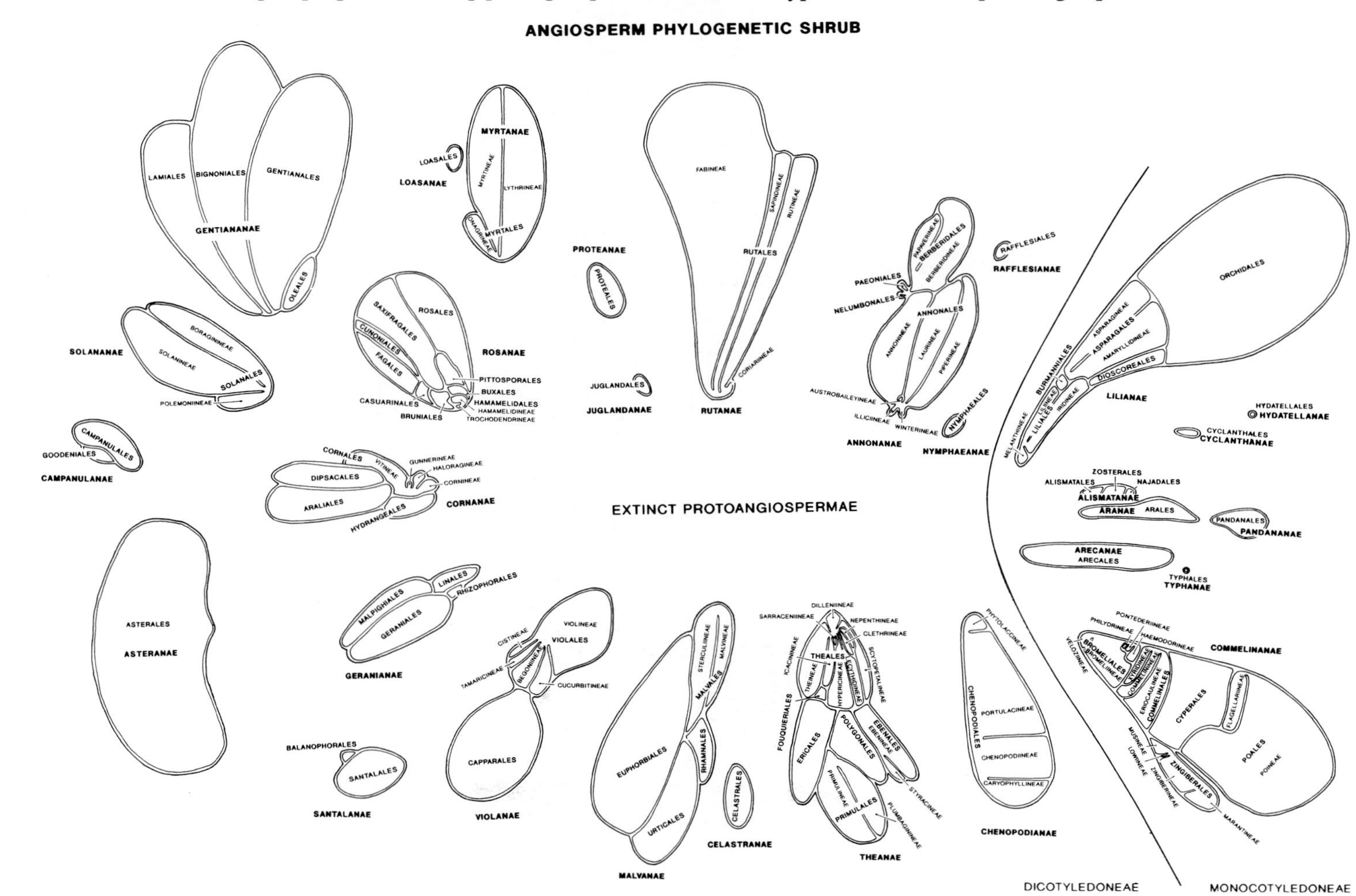


*SOURCE:* Unpublished diagram courtesy of R. F. Thorne.

**FIGURE 11.21 The Danish botanist Rolf Dahlgren (1932–1987).**

*SOURCE:* Photo courtesy of the Botanical Museum and Herbarium, Copenhagen.

standing how more recent contemporary classification systems have been influenced by previous workers, a flow chart showing the history of classification is given in Fig. 11.23 (p. 398). From this chart it can be seen that each author of a system had some influence on a succeeding system; a few individuals, such as Hallier and Bessey, profoundly influenced current contemporary systems; the current systems of Cronquist, Dahlgren, Stebbins, Takhtajan, and Thorne for the flowering plants have more similarities than differences; and the systems of today may be much different from the classification systems of tomorrow, as new data are forthcoming.

The student of systematics must remember that a classification system is one person's ideas as to possible relationships between plant groups. How data are used, perceived, interpreted, and applied depends on the investigator's background, environment, and personal bias. The student must not forget that the "natural picture" that is portrayed by the fossil record "data bank" at hand is most incomplete and sketchy. Therefore, botanists have to rely on extant micro- and macromorphological data to interpret the past. This can give incorrect results and conclusions. The point to be made is that just because a taxon is morphologically "similar" to another taxon does not mean that a relationship is assured. A case in point can be seen in Fig. 11.24 (p. 399). The late C. V. Morton, of the Smithsonian Institution in Washington, D.C., found an old leather bag of ancient paper clips left by his predecessor, Mr. Maxon. He arranged these on a fiberboard on his office wall, gave them descriptive names, and arranged them along morphological evolutionary lines

FIGURE 11.22 **Diagram, published after Dahlgren's death, showing his and his wife's views of angiosperm classification of superorders and orders.**

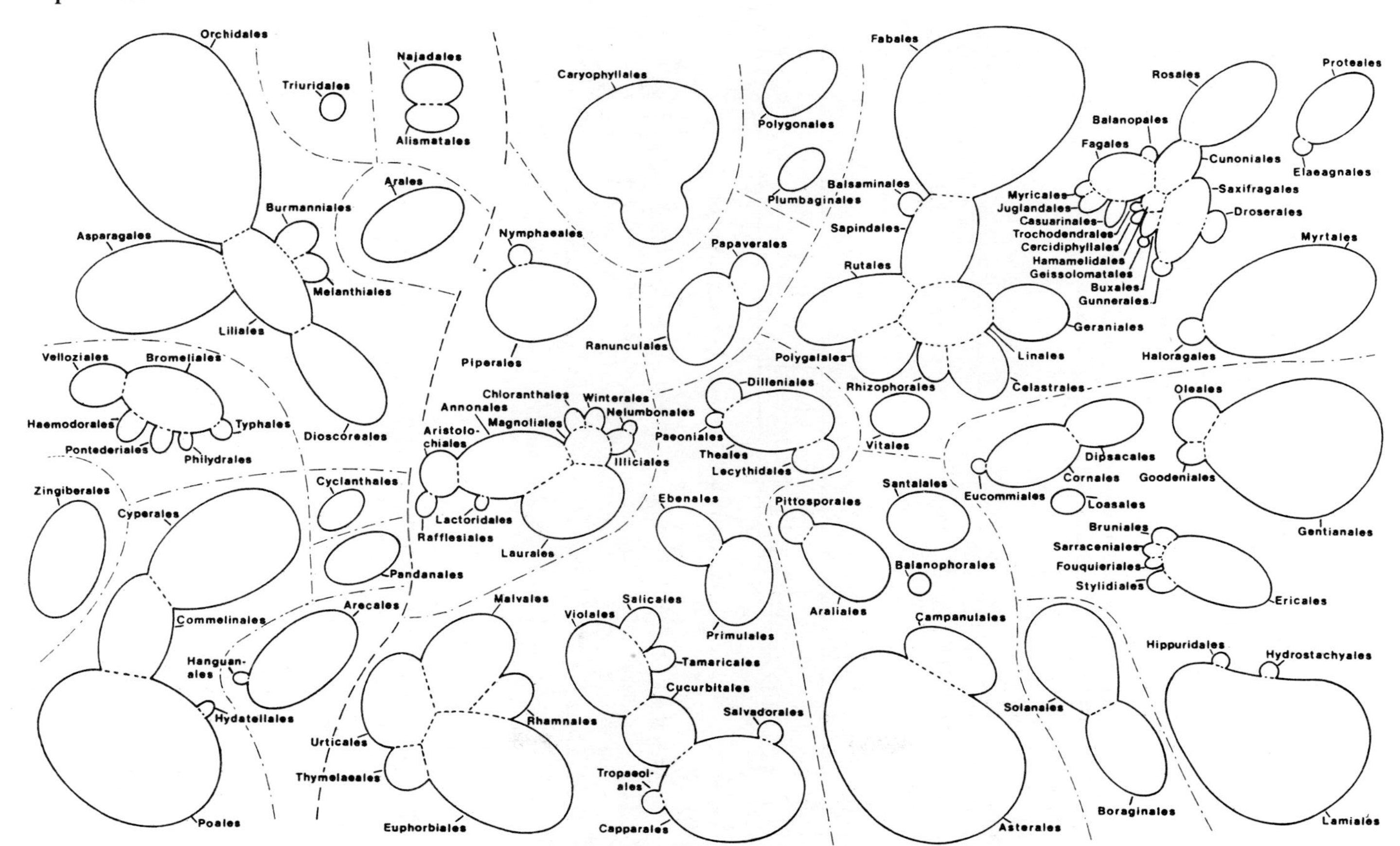


*SOURCE:* From G. Dahlgren, 1989. Used by permission of Academic Press Inc. (London) Ltd.

**FIGURE 11.23 Outline of various authors' attempts to develop a "natural" system of classification for the flowering plants. Arrows indicate the interrelationships of the systems.**

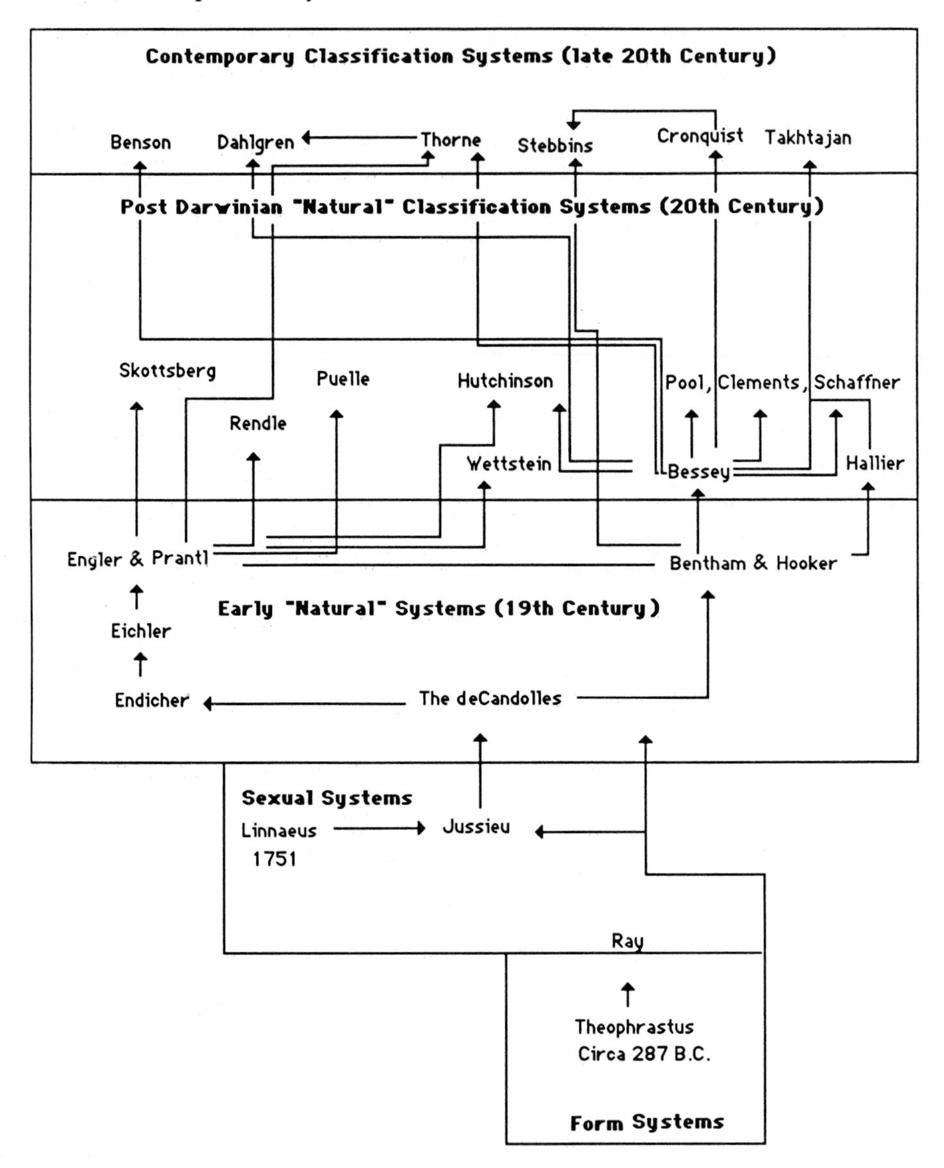


in a phylogenetic bush of the class Papyroclippopsida. Those paper clips manufactured before 1900 are below the recent fossil line; those manufactured after are above. Some evolutionary lines became extinct; others have no fossil record, such as the newer plastic ones. A predator enters the picture in the staple remover. A phylogenetic joke? Yes, but the above point is emphasized!

FIGURE 11.24 "Phylogenetic bush" of the class Papyroclippopsida by C. V. Morton.

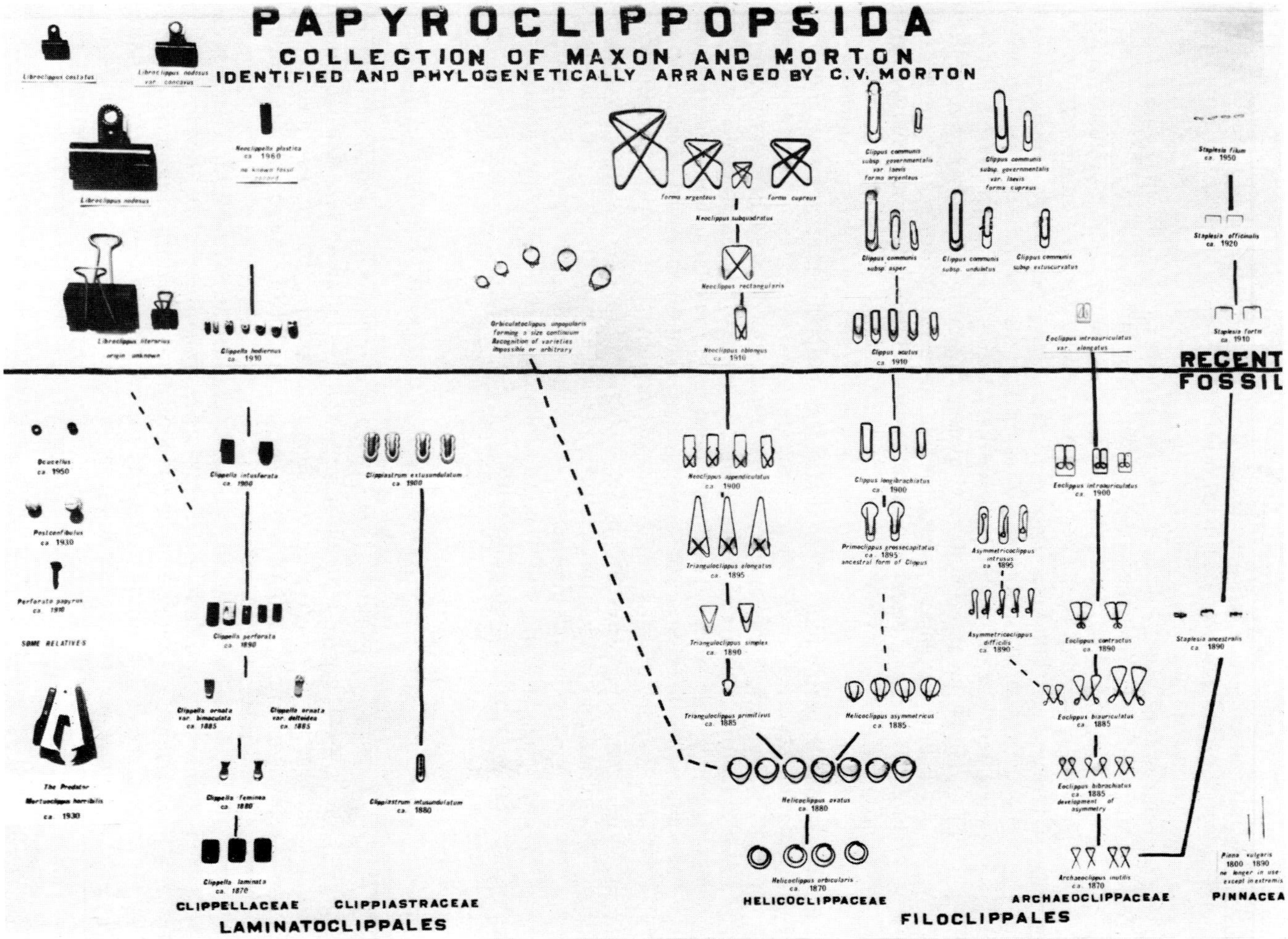
PAPYROCLIPPOPSIDA
COLLECTION OF MAXON AND MORTON
IDENTIFIED AND PHYLOGENETICALLY ARRANGED BY C.V. MORTON
RECENT
FOSSIL
SOME RELATIVES
CLIPPELLACEAE
CLIPPIASTRACEAE
LAMINATOCLIPPALES
HELICOCLIPPACEAE
ARCHAEOCLIPPACEAE
FILOCLIPPALES
PINNACEA

A reflection on the various classification systems reveals that for almost 300 years botanists have attempted to show a natural view of the botanical world. As time progressed, along with knowledge and evidence, various ideas and concepts were proposed, accepted and built upon, or rejected. Botanists working in different countries and disciplines added to the pool of information. Interestingly, different individuals working independently reached similar conclusions. This was to be expected, because each botanist should ultimately be using the same data base for information.

In spite of what appears to be many workers developing classification schemes, there are really very few individuals involved on a worldwide basis. Most younger botanists prefer to address less speculative issues. As a result, the "champions" of botanical classification may soon be leaving the botanical scene without many younger workers taking their place. Why? Further reflection may be in order. Today, more than at other times in botanical history, intelligent, inquisitive young people have many other areas of botany (or biology for that matter) in which to work. This was not the case in years past. Research funding from various granting agencies (whether private or public) also reflects this trend, with large sums of money going to support contemporary experimental research and less support being given for more speculative issues.

In spite of this contemporary trend, there are some encouraging signs: (1) During the past 35 years there has been an explosion of various types of botanical knowledge, which has been used in constructing classification systems. (2) The origin of the systems has taken on an international flair: American, Swedish, English, and Soviet Armenian. (3) A cooperative influence can be seen in each system, in how each author of a system has influenced another, and in their phylogenetic ideas and conclusions. The openness and undogmatic interchange of ideas and theories today can only enhance and encourage thinking young minds to begin to explore unanswered questions of how, when, where, and why.

## *Contemporary Phenetic Methods*

The classification systems discussed thus far in this chapter were based on observations of a small or large number of characters with the intuitive ordering of the individuals or taxa into groups that convey a relationship. However, most modern systematists prefer to follow a set methodological approach to classification.

One approach is the **phenetic classification method,** which seeks to express natural relationships based on similarities between characteristics of organisms. This is done without consideration of origin or evolutionary significance. The resulting similarity may appear to be practical and "natural," but may not reflect evolutionary or genetic history. The methods do not produce new data or new classification principles, but are methods of reorganization and presentation of information.

The beginning of this view to classification developed in the 1950s and 1960s around numerical methods of analyses and the development of computers, and became known as **numerical taxonomy** or **taximetrics.** The principles were championed by P. H. A. Sneath and R. R. Sokol in their work *Numerical Taxonomy* published in 1973.

Proponents of numerical taxonomy emphasize the following seven basic principles: (1) Systematics is no longer viewed as a deductive or interpretative science, but an empirical one. (2) Each character is considered to be of equal weight (*a priori*) with every other character when creating natural taxa. This is done to reduce subjectivity of the data. Some botanists claim that weighting of characters should be allowed because certain characteristics are obviously more important for plant survival than others (i.e., reproductive features are more important than the number of teeth on the margin of a leaf). (3) The greater the number of characters upon which the analysis is based, the better the resulting classification. Usually 100 or more characters are considered to be desirable, but studies using as few as 60 have been considered a minimum. (4) A function of the similarity of two entities will be reflected in their overall similarity. Care must be exercised that homologous characters are being compared. (5) Distinct taxa will be recognized because correlations of characters will differ in the taxa being studied. (6) From the character correlations of groups based on particular evolutionary assumptions, phylogenetic relationships can be made. (7) The resulting classifications will be based on phenetic similarity.

The basic units of numerical taxonomy are the **Operational Taxonomic Units** or **OTUs.** This is the term given to the lowest ranking taxon in any particular study. The investigator may choose individuals, species, genera, families, or other taxonomic units, which are as representative as possible of the unit chosen.

Characters of the OTUs should be chosen from as wide a range of variation as possible: preferably 100 or more are needed to develop a repeatable and reliable classification.

The equal-weighted characters selected are coded generally in a simple all-or-none way (+ or −; or 0 and 1), in which each character is said to exist in only two states. This codification is called a **two-state** or **binary system.** For example, leaves may have stipules present or absent, or the corolla with hairs in the throat versus corolla throat glabrous. Other characters are **multistate,** meaning having more than two continuous characters (i.e., corolla blue, orange, red, yellow, or white). A subjective decision must then be made to convert multistate characters into binary characters (i.e., corolla colored versus corolla white). This subjectivity detracts from the objectivity of the taximetric methodology.

The information is then arranged in a similarity or matrix diagram. Here the measurement of similarity (S) is calculated by a computer program that compares each OTU with the attributes of every other OTU. The computer **clusters** or sorts out the OTUs according to their most similar attributes. This is obtained by a table of similarity or dissimilarity coefficients. This process is called **cluster analysis.** The resulting visual cluster analysis is portrayed as a **dendrogram** or **phenogram** of phenetic relationships (Fig. 11.25, p. 402). The dendrogram allows for the visualization of the phenetic relationships, in which OTUs that are most similar are linked at higher coefficient levels (i.e., 0.9 or 90%), and less similar OTUs are linked at successively lower levels. Arbitrarily fixed coefficient levels of similarity applied to particular taxonomic

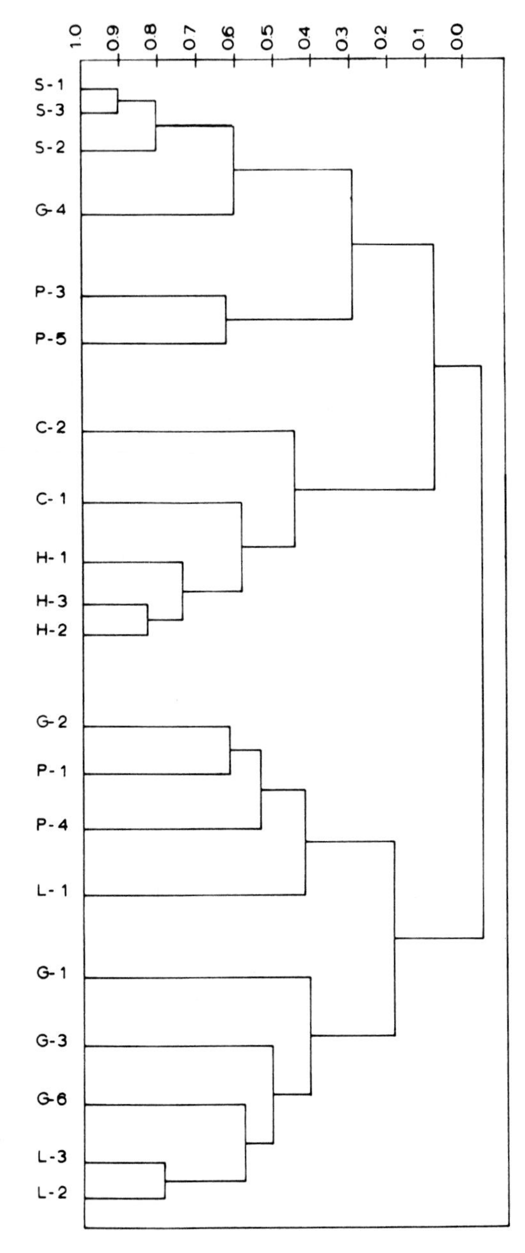


**FIGURE 11.25 Two-dimensional dendrogram showing clustering of biotypes of *Urtica* grown under natural field conditions. Abbreviations: C = *U. californica,* G = *U. gracilis,* H = *U. holosericea,* L = *U. lyallii,* P = *U. procera,* and S = *U. serra.* Numbers with the letters represent populations. The numbers at the top indicate the coefficient levels at which the populations are linked. The vertical lines connecting populations show the closest coefficient of association.**

ranks can be used to objectively classify taxa. This dendrogram shows that the populations of taxa S and H cluster together and are most similar, while the populations of the other taxa are more variable and less similar.

The above discussion has been greatly oversimplified in an attempt to introduce the basic concepts of phenetic classification. The interested student should refer to Sneath and Sokol (1973) for an in-depth discussion of the subject.

The methods of numerical taxonomy have been helpful in providing a refinement of existing classification systems and reclassification above the family level. Taximetrics has been criticized by many systematists, who are reluctant to allow a machine to make calculated taxonomic and biological judgments in place of the experience of a botanist. On the other hand, the methods of numerical taxonomy are becoming more and more helpful in comparing large sets of data in a precise manner and in computing phylogenetic hypotheses, rates, and trends within similar or between different taxonomic levels.

## *Contemporary Phylogenetic Methods*

At about the time Sneath and Sokol were developing their phenetic methods of numerical taxonomy, another numerical approach was hypothesized.

During the 1950s a German zoologist, Willi Hennig, proposed a new classification method. His new phylogenetic methods differed markedly from other concepts. The methodology, now known as **cladistics,** was a methodology that attempted to objectively analyze phylogenetic data in strict, repeatable methods. Independent of Hennig's work in Europe, the American botanist Warren H. Wagner was developing a convergent approach to construct phylogenetic trees, called **groundplan-divergence.** Wagner's effort was an attempt to understand the amount and direction of evolutionary change and the diverse branching portrayed. Both Hennig and Wagner's ideas appeared to be unrelated at first, but after the connection was made in the late 1960s, Wagner's concepts became the basis for the methodology.

From these two philosophies developed two basic approaches: (1) the **principle of parsimony,** whereby the most likely evolutionary route is the shortest hypothetical pathway of changes that explains the pattern under observation; and (2) **character compatibility** studies, in which each character is examined to determine the proper sequence of character state changes that take place as evolution progresses. Contemporary cladists determine ancestral and derived states of characters and define evolutionary lineages (or **clades**) by shared derived characters. The analyses can be undertaken within the group studied (called **in-group**) or conducted with outside relatives (**out-group**).

In parsimony and character compatibility methods, **cladists** (biologists who practice the methods of cladistics) consider the similarity as used by the pheneticists as not necessarily implying phylogenetic relationship. The classifications emphasize **monophyletic** groups (groups that have arisen through diversification of a simple ancestor) involving **homologous** shared and derived characters. A **polyphyletic** origin considers groups originating from two or more ancestral stacks and is not recognized by the cladists.

The cladistic relationships are presented in **cladograms,** a type of branching evolutionary diagram or phylogenetic tree. In Fig. 11.26 theoretical taxa A–G are all related by having a common ancestor at the stem point X. A

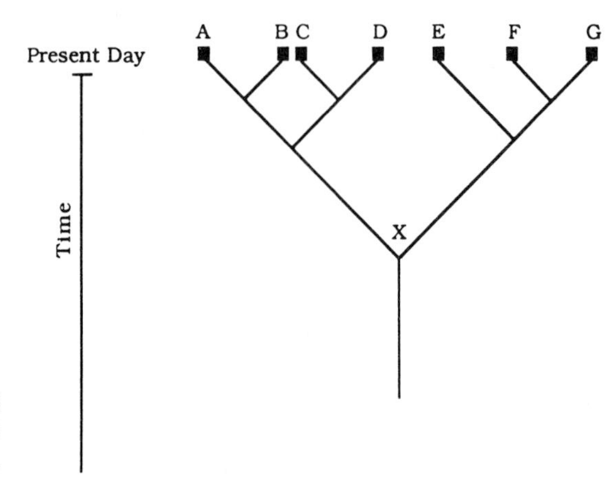


**FIGURE 11.26 Cladogram showing phylogenetic relationships between seven hypothetical taxa.**

different degree of cladistic relationship exists between each taxon. The paired taxa A and B, C and D, and F and G are separated by only one divergence before common ancestries are retraced. Proper interpretation of the cladogram takes into account the number of divergences connecting different branches and the time of origin of one taxon to another. Other methods may also be considered.

An application of cladistic methodology was recently applied to the relationships of various subgenera within the genus *Rhododendron,* tribe *Rhodoreae,* in the family Ericaceae (Kron and Judd, 1990). The resulting cladograms shown in Fig. 11.27 support (A) a monophyly of the taxa studied and (D) the polyphyly hypothesis. The numbers on the cladogram are the coded characters used in the analyses.

In the groundplan-divergence methods, Dr. Warren H. Wagner Jr., the originator of the method, hypothesizes an ancestor for the taxonomic group. Evolutionary modifications from this ancestor are then measured using (1) the number of changes from the ancestral condition, (2) the change of characters from one lineage to another, and (3) the sequences of cladistic branching.

A cladogram in the shape of a bull's eye with concentric rings has phylogenetic relationships plotted on it. The concentric rings represent stages in advancement from the hypothesized ancestor. Dots on the rings represent taxa, with more advanced taxa toward the outer rings. Lines between dots indicate lineages. The number of "advanced" character states (**apomorphy**) evolving with the taxa being studied are reflected in the distance of the taxon from the hypothesized ancestor. Taxa that advance independently show divergence. So-called primitive character states (**plesiomorphy**) may also be recognized.

In Fig. 11.28 (p. 406), taxa B, C, D, H, K, and L are the most equally advanced from the hypothetical ancestor, but each has its distinctive characters. Taxon F is less advanced than E, J is less than I, and I is less than H and K. The features of J and I are shared with taxa H and K. Taxa B and C and taxon L have no ancestor with recognizable features at the level indicated.

**FIGURE 11.27 Cladograms resulting from analysis of the tribe *Rhodoreae* of the genus *Rhododendron* supporting both monophyly (A) and polyphyly (D).**

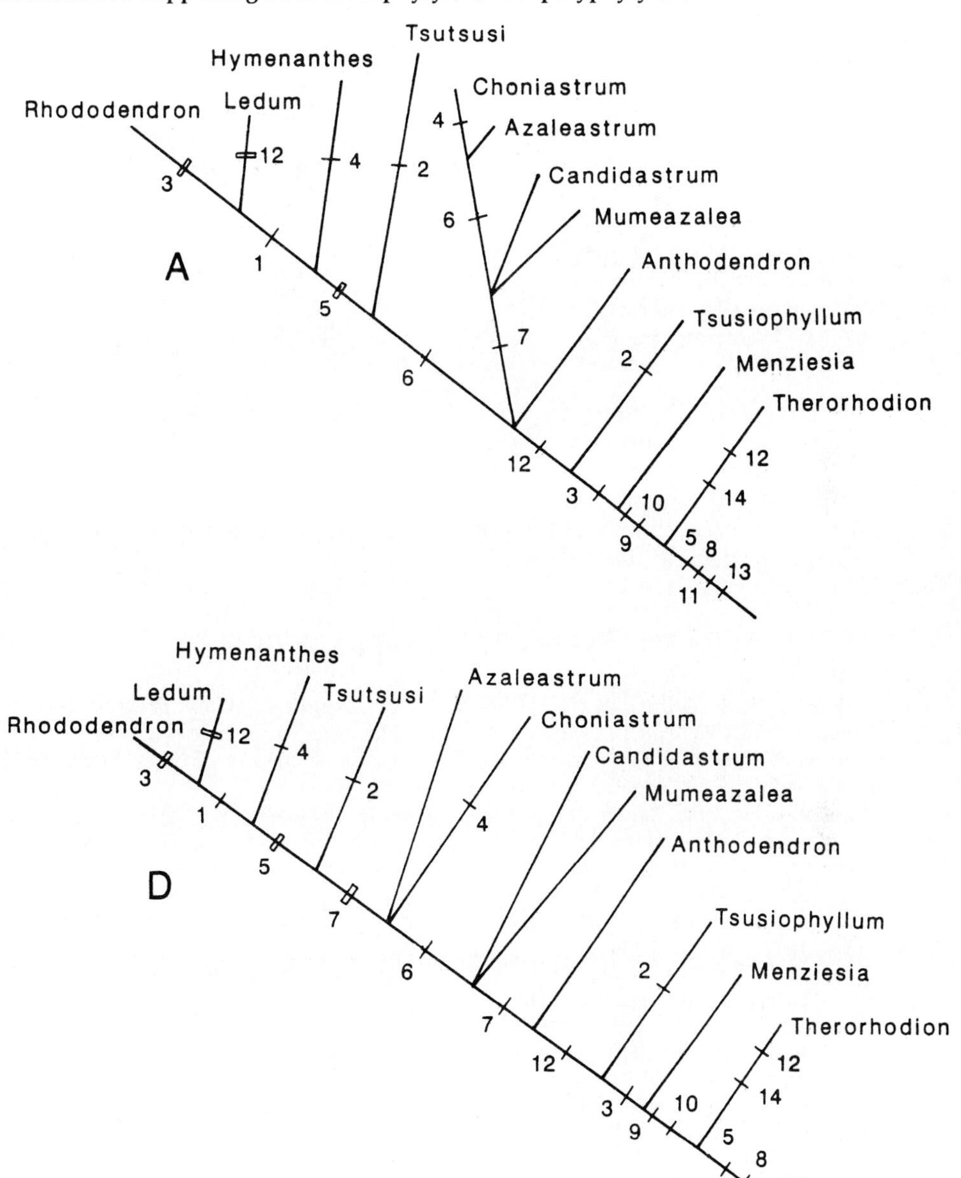


*SOURCE:* From Kron and Judd, 1990.

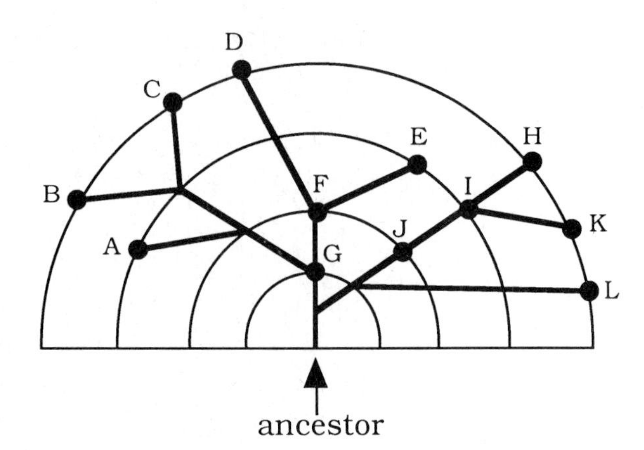


**FIGURE 11.28 Cladogram of a Wagner groundplan-divergence ("Wagner tree") showing the phylogenetic relationship between twelve taxa and the hypothetical ancestor taxon.**

The above methods are not without criticism on various argumentative points. However, any new method of classification that gives insight into a better understanding of evolutionary pathways is worthy of exploration. The student must remember that " . . . phylogenetic trees are hypotheses, not facts. Our ideas about the relationships among organisms changes with increasing understanding" (Wiley, 1981).

## *SELECTED REFERENCES*

Arber, A. 1938. *Herbals: Their Origin and Evolution.* Cambridge Univ. Press, Cambridge.

Benson, L. 1979. *Plant Classification,* 2nd ed. D.C. Heath, Lexington, MA, 901 pp.

Bentham, G. and J. D. Hooker. 1862–1863. *Genera Plantarum.* 3 vols. Reeve & Co., Williams & Norgate, London.

Bessey, C. E. 1915. The Phylogenetic Taxonomy of Flowering Plants. *Ann. Mo. Bot. Gard.* 2:108–164.

Bold, H. C. C. J. Alexopoulos, and T. Delavoryas. 1987. *Morphology of Plants and Fungi,* 5th ed. Harper & Row, New York, 912 pp.

Cain, A. J. 1958. Logic and Memory in Linnaeus' System of Taxonomy. *Proc. Linn. Soc. London* 169:144–163.

Cain, A. J. 1959. Deductive and Inductive Methods in post-Linnaean Taxonomy. *Proc. Linnaean Soc. London* 170:185–217.

Cronquist, A. 1957. Outline of a New System of Families and Orders of Dichotyledons. *Bull. Jard. Bot. Etat.* 27:13–40.

Cronquist, A. 1981. *An Integrated System of Classification of Flowering Plants.* Columbia Univ. Press, New York, 1261 pp.

Cronquist, A. 1988. *The Evolution and Classification of Flowering Plants,* 2nd ed. The New York Botanical Garden, Bronx, NY, 555 pp.

Cronquist, A., A. Takhtajan, and W. Zimmerman. 1966. On the Higher Taxa of Embryobionta. *Taxon* 15:129–134.

Crowson, R. A. 1970. *Classification and Biology.* Heinemann Educational Books, London, 350 pp.

Dahlgren, G. 1989. An Updated Angiosperm Classification. *J. Linn. Soc., Bot.* 100:197–203.

Dahlgren, R. M. T. 1977. A Commentary on a Diagrammatic Presentation of the Angiosperms in Relation to the Distribution of Character States In: K. Kubitzkiet (ed). *Plant Systematics & Evolution,* Suppl. 1. Springer-Verlag, New York, pp. 253–283.

Dahlgren, R. M. T. 1980. A Revised System of Classification of the Angiosperms. *J. Linn. Soc., Bot.* 80:91–124.

Dahlgren, R. M. T. 1981. Angiosperm Classification and Phylogeny—A Rectifying Comment. *J. Linn. Soc., Bot.* 82:89–92.

Dahlgren, R. M. T., H. T. Clifford, and P. F. Yeo. 1985. *The Families of the Monocotyledons. Structure, Evolution and Taxonomy.* Springer-Verlag, New York, 520 pp.

Danser, B. H. 1950. *A Theory of Systematics. Bibliotheca Biotheoretica.* Ser. D., Vol. IV. Pars 3, E.J. Brill, Leiden, p. 113–180.

Davis, P. H. and V. H. Heywood. 1973. *Principles of Angiosperm Taxonomy.* Robert E. Krieger, Huntington, NY, 558 pp.

Eid, S. E. 1970. *Evolutionary Taxonomy of Angiosperms.* Public Organization for Books and Scientific Appliances, Cairo Univ. Press, Cairo, 18 pp. plus chart.

Eldridge, N. and J. Cracraft. 1980. *Phylogenetic Patterns and the Evolutionary Process: Method and Theory in Comparative Biology.* Columbia Univ. Press, New York, 349 pp.

Funk, V. A. and T. F. Stuessy. 1978. Cladistics for the Practicing Taxonomist. *Syst. Bot.* 3:159–178.

Greene, E. L. 1983. In: *Landmarks of Botanical History* Parts. I, II. F. N. Egerton (ed). Stanford Univ. Press, Stanford, CA, 1139 pp.

Hallier, H. 1905. Provisional Scheme of the Natural (Phylogenetic) System of Flowering Plants. *New Phytol.* 4:151–162.

Hallier, H. 1908. On the Origin of Angiosperms. *Bot. Gaz.* 45:196–198.

Heywood, V. H. (ed). 1978. *Flowering Plants of the World.* Oxford Univ. Press, Oxford, 335 pp.

Heywood, V. H. and J. McNeill. 1964. *Phenetic and Phylogenetic Classification.* The Systematics Assoc. Publ. No. 6., London, 164 pp.

Hill, A. W. 1915. The History and Functions of Botanic Gardens. *Ann Missouri Bot. Gard.* 2:185–240.

Holmes, S. 1983. *Outline of Plant Classification.* Longman Group, London, 181 pp.

Hutchinson, J. 1926, 1934, 1973. *The Families of Flowering Plants, Vol. I Dicotyledons* (1926). *Vol. II. Monocotyledons* (1954). Macmillan, London 3rd ed. 1973, Clarendon Press, Oxford.

Hutchinson, J. 1967. *The Genera of Flowering Plants. Vol I. Dicotyledones.* 516 pp. Vol. II. 659 pp, Clarendon Press, Oxford.

Kron, K. A. and W. S. Judd. 1990. Phylogenetic Relationships within the Rhodorea (Ericaceae) with Specific Comments on the Placement of *Ledum. Syst. Bot.* 15:57–68.

Kubitzki, K. ed. 1977. *Flowering Plants—Evolution and Classification of Higher Categories.* Pl. Syst. Evol. Suppl. 1. Springer-Verlag, New York, 416 pp.

Lawrence, G. H. M. 1951. *Taxonomy of Vascular Plants.* Macmillan, New York, 823 pp.

Lawrence G. H. M. 1955. *An Introduction to Plant Taxonomy.* Macmillan, New York, 179 pp.

Majumdar, G. P. 1927. *Vanaspati—Plant and Plant Life as in Indian Treatise and Traditions.* Univ. Calcutta, India, 254 pp.

Majumdar, G. P. 1946. Genesis and Development of Plant Science in Ancient India. 13th All India Oriental Conf. Technical Science, Calcutta, pp. 47–120.

Margulis, L. and K. V. Schwartz. 1988. *Five Kingdoms. An Illustrated Guide to the Phyla of Life on Earth.* W. H. Freeman, New York, 376 pp.

Mez, C. 1936. Morphologie and Serodiagnostic. *Bot. Arch.* 38:86–104.

Naik, V. N. 1984. *Taxonomy of Angiosperms.* Tata McGraw-Hill, New Delhi, 304 pp.

Porter, C. L. 1967. *Taxonomy of Flowering Plants,* 2nd ed. W. H. Freeman, San Francisco, 472 pp.

Rensch, B. 1959. *Evolution Above the Species Level.* Columbia Univ. Press, New York, 419 pp.
Shukla, P. and S. P. Misra. 1979. *An Introduction to Taxonomy of Angiosperms.* Vikas Publishing House, New Dehli, 546 pp.
Sivarajan, V. V. 1984. *Introduction to Principles of Plant Taxonomy.* Oxford & IBH, New Delhi.
Skottsberg, C. 1932–1940. *Vaxternas Liv.* 5 vols., 2nd ed. 1953–1956. 11 vols., Stockholm.
Slobodchikoff, C. N. 1976. *Concepts of Species.* Dowden Hutchinson and Ross, (distrib. by Halsted Press, New York).
Soo, C. R. de. 1975. A Review of the New Classification Systems of Flowering Plants (Angiospermophyta, Magnoliophytina). *Taxon* 24:585–592.
Sporne, K. R. 1956. The Phylogenetic Classification of Angiosperms. *Biol. Rev.* 31:1–29.
Stearn, W. T. 1959. The Background of Linnaeus' Contribution to the Nomenclature and Methods of Systematic Biology. *Syst. Zool.* 8:4–22.
Stebbins, G. L. 1974. *Flowering Plants: Evolution Above the Species Level.* Belknap Press, Harvard Univ. Press, Cambridge, MA, 399 pp.
Takhtajan, A. 1958. *Origins of Angiospermous Plants.* (transl. from Russian by O.H. Gankin) Amer. Instit. Biol. Sci., Washington, D.C., 68 pp.
Takhtajan, A. 1964. The Taxa of Higher Plants Above the Rank of Order. *Taxon* 13:160–164.
Takhtajan, A. 1969. *Flowering Plants. Origin and Dispersal* (transl. from Russian by C. Jeffrey). Oliver and Boyd, Edinburgh, 310 pp.
Takhtajan, A. 1980. Outline of the Classification of Flowering Plants (Magnoliophyta). *Bot. Rev.* 46:225–359.
Takhtajan, A. 1986. *Floristic Regions of the World.* Univ. Calif. Press, Berkeley, CA, 522 pp.
Thomas, H. H. 1936. Paleobotany and the Origin of Angiosperms. *Bot. Rev.* 2:397–418.
Thorne, R. F. 1976. A Phylogenetic Classification of the Angiospermae In: M.K. Heckt, W.C. Steere, & B. Wallace (eds.). *Evolutionary Biology.* Vol. 9. Plenum Press, New York, pp. 35–106.
Thorne, R. F. 1983. Proposed New Realignments in the Angiosperms. *Nord. J. Bot.* 3:85–117.
Tippo, O. 1942. A Modern Classification of the Plant Kingdom. *Chron. Bot.* 7:203–206.
Wagner, W. H. Jr. 1980. Origin and Philosophy of the Groundplan/Divergence Method of Cladistics. *Syst. Bot.* 5:173–193.
Wettstein, R. von. 1935. *Handbuch der Systematischen Botanik,* 4th ed. Leipzig and Vienna.
Whittaker, R. H. 1959. On the Broad Classification of Organisms. *Quart. Rev. Biol.* 34:210–226.
Wiley, E. O. 1981. *Phylogenetics.* John Wiley, New York, 439 pp.

# 12

# *Contemporary Views of the Origin of Vascular Plants*

Throughout history humans have been interested in how organisms on this planet originated. The debate concerning the issue of origins appears in the earliest written and verbal records of humankind through to the present. Most ethnic groups attempt to explain the origin of living things by legends or stories.

## *Theories on Origins*

Obviously the method of the origin of the first living matter was not observed by humans. Therefore, it is no longer subject to nonquestionable observation and has resulted in various speculations, legends, and hypotheses as to the origin and development of life on earth. These hypotheses can be generally divided into two broad categories: origin by a supernatural force, and origin by spontaneous generation and organic evolution. While some individuals view these two categories as diametrically opposed to one another, others do not and blend various aspects of the categories.

The supernatural force hypotheses speculate that matter endowed with life originated by some unexplainable, supernatural, non-Earth force. This energy was exercised by a supernatural being or god with the ability to "call into existence" (create) the diverse physical and biological world as we know it. This diversity has therefore continued through time to the present. Extremely conservative views of this hypothesis hold that the present diversity has always been and that no new groups have arisen since the original creation. They also deny the existence of any natural relationships between the taxa. More progressive views hold that the supernatural force created the major diverse types of organisms and that from these basic types, present-day species have developed by speciation along natural lines, by means of so-called microevolution. Proponents of this theory usually support the idea that diversification has occurred over much shorter periods of time, according to the limits of demonstrated biological laws of change. Hypotheses of this nature are not testable by the scientific method and, therefore, are considered by most scientists to be nonscientific ideas and are not considered as a viable origin alternative.

An origin by the organic evolution hypothesis suggests that living matter formed from nonliving, simple chemical substances found in the solar system (such as ammonia, hydrogen, methane, and water) over long periods of time. Populations of living species at any given moment at any given location on the Earth represent to the organic evolutionist modified offspring of species present in ages past. It is also believed that the processes that brought about change in the past can be identified and continue to operate today, bringing about new species. These species, in turn, are variable and ever changing according to the limits of demonstrated biological laws of change.

As we view life on Earth, the evidence for change (evolution) is overwhelming and well substantiated. The diversity of taxa observed today in nature is explained by scientists to have occurred by mutations, genetic recombination, and subsequent populational segregation. Natural selection by the environment and adaptative radiation repeated many times has selected those taxa that are best adapted to particular environmental situations. Such changes may be induced experimentally and duplicated in living organisms in the laboratory by using chemical and/or physical means. These experimental changes are almost always in the lower taxonomic categories (i.e., species, subspecies, variety, etc.). Conversely, it is a very different matter to trace with unquestionable reliability the diversity and relationships between the higher taxonomic groups (divisions, classes, orders, families, genera) that by extrapolation are thought to have developed by the same biological processes of change demonstrated at lower taxonomic levels. The data extrapolated in this way are not as conclusive as a scientist would like them to be. Since no one was present when the event of change occurred, there are few satisfactory unquestionable alternatives open to explain the diversity in nature.

The resulting diagrams and evolutionary "trees" or "bushes" explaining the relationships between the higher taxonomic levels are therefore largely speculative and are subject to variation with the biases of the individual biologist who postulates them. The connections between branches can seldom be shown to be real. Therefore, these ideas will be subject to modification with the advent of new evidence.

The evidence upon which evolutionary relationships are hypothesized was first based upon the remains found in the fossil record and then the comparative data from both living and extinct organisms. The fossil record in almost all cases was very incomplete and inconclusive. Paleontologists hoped that a logical progression could be found throughout the fossil history of a group, but this was rarely the case. One next turned to the development and ontogenetic observations of comparative morphology. Here the botanist gathered clues that could be compared with biogeographic information to help construct a possible relationship hypothesis. Other clues may be forthcoming from the fields of comparative physiology, chemistry, genetics, ecology, and statistics, as applied to the extant or extinct groups (discussed in Chapter 13). The interpretation of the available evidence will vary with different individuals, and therefore different classification systems are usually proposed.

The various opinions encountered on subjects of origins may be disconcerting to the beginning student. Students should not forget that the progress of science revolves not around facts, but upon the essential parameters of theories, hypotheses, and speculations, which may be proven or disproven. The following discourse will introduce some current views held by botanists concerning the evolution of vascular plants.

## *Early Fossil Pteridophytes*

Knowledge about ancient vascular plants dates back to the middle of the nineteenth century, when Sir William Dawson (the principal of McGill University, Montreal) described a dichotomously branched vascular plant found from the Gaspé Peninsula of Quebec, Canada. These fossils had paired pendulous clusters of sporangia. They were named *Psilophyton princeps* and were collected from Devonian strata (see Fig. 12.1, pp. 412 and 413). Recent inspection of *Psilophyton princeps* fossils indicates that the remains were of several different taxa mixed together. Despite this mixing, the *Psilophyton* plants were the first Lower Devonian plants to be discovered.

These fossils were not seriously considered to be of much importance by paleobotanists until Kidston and Lang discovered (in 1917–1921) similar dichotomously branched fossils in the Lower Devonian chert deposits of Rhynie, Scotland. These "Rhynie chert plants" had naked stems with sporangia upright at the end of the stems (Fig. 12.2, p. 414).

Recent reexamination of the Rhynie chert fossils indicates that the original aboveground stems were more widely dichotomously branched [i.e., *Aglaophyton (Rhynia) major*] than previously thought. Each branch terminated in an upright sporangium.

At the present time, the smallest and oldest known vascular plant, *Cooksonia,* comes from mid-Silurian deposits in Wales, United Kingdom. *Cooksonia* had very slender, leafless dichotomously branched stems with small, round to reniform sporangia at the end (Fig. 12.3, p. 414), containing spores having Y-shaped ridges on the spore wall.

Other taxa have been named from Devonian deposits as well. It is not necessary to discuss all of them to emphasize that at the present time the first vascular plants to appear in the fossil record *(Cooksonia)* do so in the mid-Silurian, with many more taxa found in Lower and Upper Devonian strata. Most paleobotanists would agree that land plants were present at this time, but how long before is open to conjecture and speculation.

## *Early Fossil Pinophyta (Gymnosperms)*

As discussed thus far, the first type of vascular plants observed in the fossil record were producers of spores in sporangia. Within the Pinophyta is a characteristic not previously encountered, the ability to reproduce by seeds.

**FIGURE 12.1 Distribution of major plant groups as observed in the geologic column.**

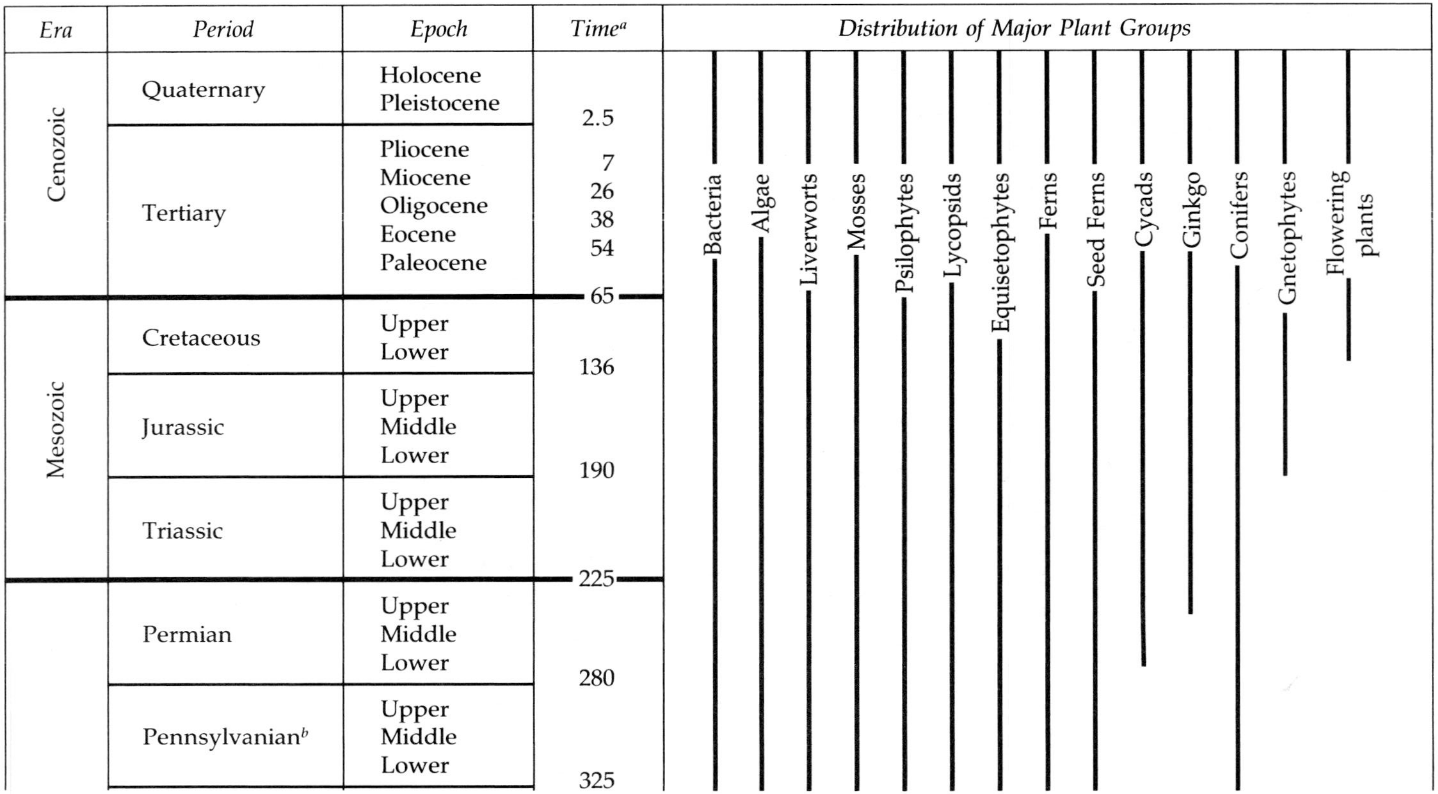

| *Era* | *Period* | *Epoch* | *Time*[a] |
|---|---|---|---|
| Cenozoic | Quaternary | Holocene<br>Pleistocene | 2.5 |
| | Tertiary | Pliocene<br>Miocene<br>Oligocene<br>Eocene<br>Paleocene | 7<br>26<br>38<br>54<br>65 |
| Mesozoic | Cretaceous | Upper<br>Lower | 136 |
| | Jurassic | Upper<br>Middle<br>Lower | 190 |
| | Triassic | Upper<br>Middle<br>Lower | 225 |
| | Permian | Upper<br>Middle<br>Lower | 280 |
| | Pennsylvanian[b] | Upper<br>Middle<br>Lower | 325 |

*Distribution of Major Plant Groups*

Bacteria, Algae, Liverworts, Mosses, Psilophytes, Lycopsids, Equisetophytes, Ferns, Seed Ferns, Cycads, Ginkgo, Conifers, Gnetophytes, Flowering plants

| Era | Period | Epoch | Time[a] | |
|---|---|---|---|---|
| Paleozoic | Mississipian[b] | Upper<br>Middle<br>Lower | 345 | |
| | Devonian | Upper<br>Middle<br>Lower | 395 | |
| | Silurian | Upper<br>Middle<br>Lower | 430 | |
| | Ordovician | Upper<br>Middle<br>Lower | 500 | |
| | Cambrian | Upper<br>Middle<br>Lower | 570 | |
| Precambrian | No general agreement on terminology | | 3100 | ? |

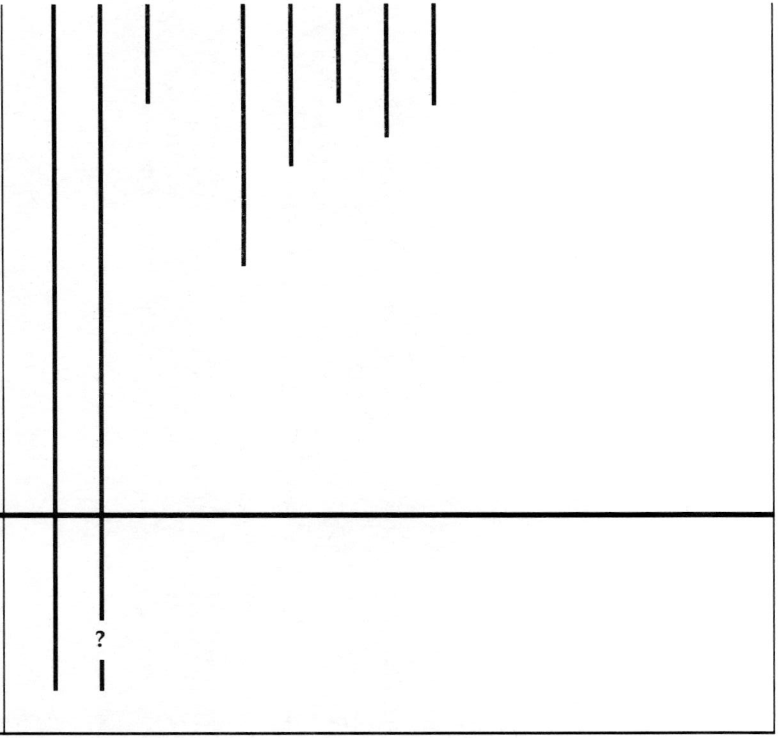

*SOURCE:* Figure modified from MORPHOLOGY OF PLANTS by Harold C. Bold. Copyright © 1957 by Harold C. Bold. Copyright renewed 1985 by Harold C. Bold. Reprinted by permission of HarperCollins Publishers.

[a]Estimated time in millions of radiometric years.

[b]When combined, designated as Carboniferous.

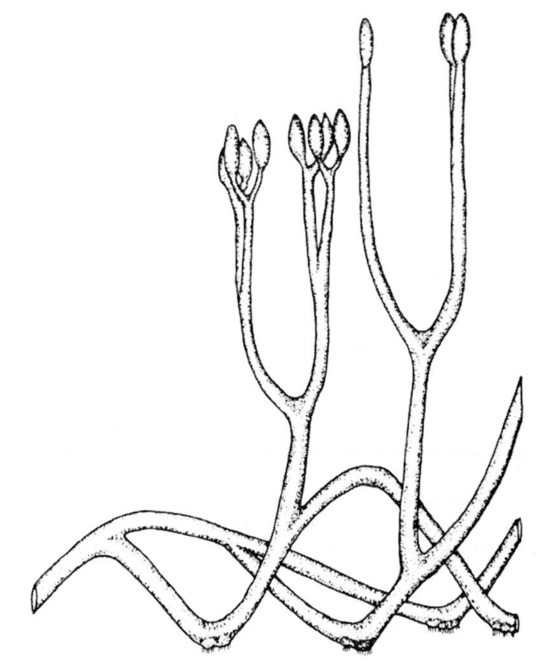

**FIGURE 12.2 A reconstruction of *Aglaophyton (Rhynia) major,* a plant from the Lower Devonian chert deposits of Rhynie, Scotland.**

*SOURCE:* From Edwards, 1986. Used by permission of Academic Press Inc. (London) Ltd.

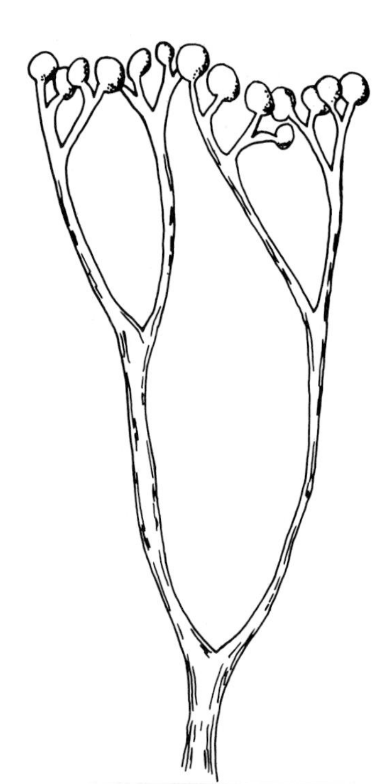

**FIGURE 12.3 A reconstruction of *Cooksonia caledonica,* the earliest known vascular plant from mid-Silurian deposits in Wales.**

*SOURCE:* Redrawn from Edwards, 1970.

Also now observed is the phenomenon of heterospory, seen in a few living pteridophytes such as *Isoetes* or *Selaginella,* which is consistent throughout all seed plants. Because there is little or no size difference between the megaspores and microspores, this is called **functional heterospory.** Furthermore, the megaspore is not shed from the megasporangium; instead, the female gametophyte develops within the confines of the megaspore wall. The immature male gametophyte is restricted within the microspore wall and is transported to the megasporangium area by external forces (wind, etc.). This is in contrast to the pteridophytes, in which the male gamete is released and usually swims through water to the megasporangium.

The Pinophyta (gymnosperms) include fossil representatives dating from the Upper Devonian. Understanding their origins is very difficult for two reasons. First, the fossil record is comprised of isolated fragments or portions of leaves, roots, spores, stems, and strobili, each given separate "organ genera" names, and second, living representatives found among the present-day flora are conspicuously absent. It therefore becomes a task of trying to match the correct plant organs to one another. One of the first reconstructions was the interconnection of *Archaeopteris* (presumed to be a fern frond) and *Callixylon* (the stem of a gymnosperm). This group of vascular plants, called Progymnospermopsida, combined the anatomy of typical gymnosperms with fern-like leaves that reproduced by sporangia and spores (Fig. 12.4, p. 416). The reconstruction of this plant proposed today is a tree of over 18 m in height and a trunk of 1–1.5 m in diameter.

A second example is the discovery of a seed-like structure named *Archaeosperma arnoldii* from Upper Devonian strata. This fossil consisted of a cluster of four cup-like structures. Each structure had a single thick-walled megaspore surrounded by a membrane and surrounding appendages (Fig. 12.5, p. 417). It should be pointed out that *Archaeosperma* (and other similar Paleozoic plants such as *Genomosperma)* was not a seed in the true sense. A conifer seed possesses a protective coat or integument, not just appendages, and also includes a female gametophyte retained within the integuments that develops ultimately into a miniature plant and not just a large megaspore, as in *Archaeosperma.* Nothing is known about the plant that bore this "seed," for this fossil was found detached from any stem.

The pteridosperms or "seed ferns" were plants with fern-like leaves and seeds attached to the leaves in place of sori. Seed fern fossils are found in Upper Devonian to Cretaceous strata. The fern-like fronds found so commonly in Carboniferous strata (sometimes designated the "Age of Ferns") may actually be seed fern leaves, raising the evolutionary question of where the seed method of reproduction came from and when. If seed ferns were contemporaneous with ferns, speculation has centered around various progymnosperms as being the ancestral group. The evidence is fragmentary, leaving the question still unanswered.

Cycads appear to have been much more widely distributed in the geological past. They were most abundant during the Mesozoic and then declined in number to the 10 genera found in the world today. Fossil evidence of true cycads first appeared in the fossil record as megasporophylls of the genus *Archaeocycas* in Lower Permian formations. *Archaeocycas* was a leaf-like megasporophyll with ovules on either side toward the base (Fig. 12.6, p. 418).

**FIGURE 12.4 A reconstruction of a vegetative lateral branch of *Archaeopteris* attached to a main branch.**

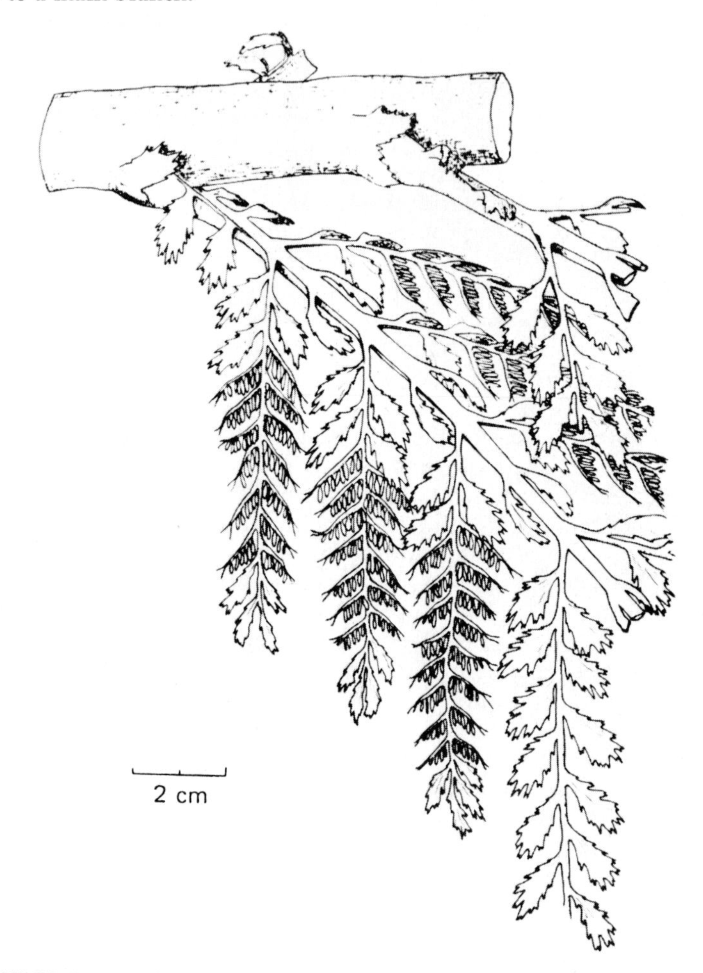


*SOURCE:* From Beck, 1971.

Most other cycad fossils do not appear until the Upper Triassic *(Leptocycas)* or Middle Jurassic *(Beania)*. An example of a common type form was *Leptocycas*, which had a slender trunk of 1.5 m topped with a crown of pinnate leaves, not unlike some cycads today.

Without doubt, cycads were present in the Lower Permian of the Paleozoic in recognizable cycad form, leaving the question unanswered as to the evolutionary lineage of the group. Some botanists have suggested development from some pteridosperm precursor. This is at best speculative. In reality, the origin of cycad diversity remains a mystery.

Living *Ginkgo* came originally from a small restricted area in China. The fossil record indicates, however, a much earlier worldwide distribution. Fossil

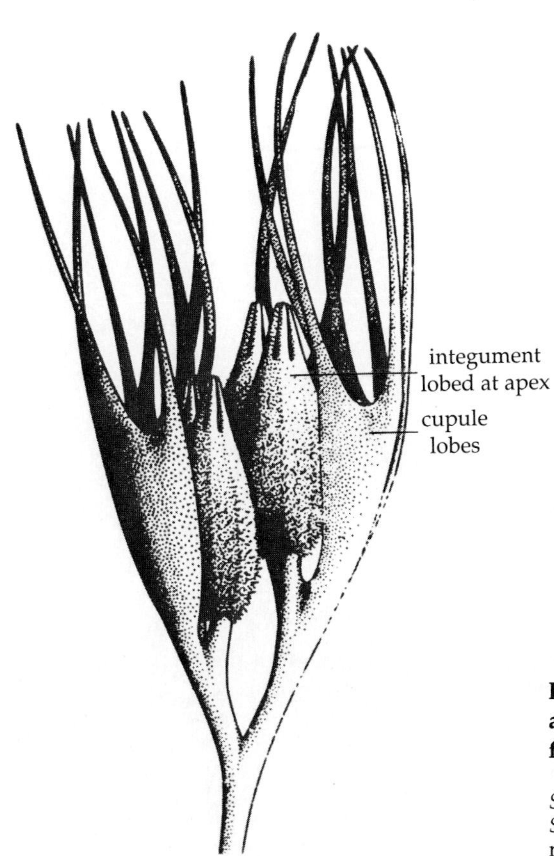


**FIGURE 12.5 Sketch of the cupule-like appendages of *Archaeosperma arnoldii* from Upper Devonian strata.**

*SOURCE:* From J. M. Pettit and C. B. Beck, *Science,* vol. 156, pp. 1727–1729, 1967. Copyright 1967 by the AAAS.

leaves dating from the Permian are more finely parted, with the more common Mesozoic specimens appearing more like variations of living *Ginkgo biloba.* Unquestionable information about reproduction in the fossils is lacking.

Today conifers are the most diverse and abundant existing gymnosperms. They are first noted as fossils from the Carboniferous period of the Paleozoic era, with various seeds and seed-bearing stalks common from the Mesozoic era. One of the best-known fossils from this geological era is called *Cordaites,* a tall tree reaching 30 m that is much branched with simple strap-shaped leaves (Fig. 12.7, p. 419). Its venation appeared at first to be parallel but was really dichotomously branching from the leaf base. The stem anatomy resembled that of conifers living today. The fertile structures consisted of bud-like short shoots in the axils of bracts, which in turn bore spirally arranged scales. The ovules were attached at the end of some of the scales.

Some present-day families can be traced back to the Triassic (Araucariaceae, Taxaceae) and Jurassic periods (Cupressaceae, Podocarpaceae, Pinaceae, and Taxodiaceae).

To summarize the fossil history of the Pinophyta (gymnosperms), we can say that some of the formerly abundant pteridosperms and cycads are

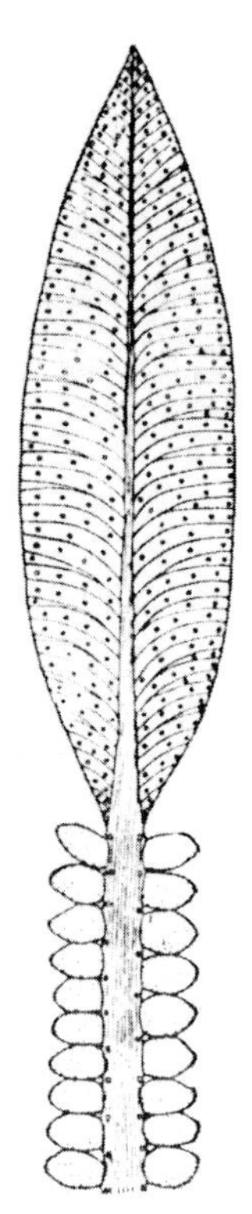

**FIGURE 12.6 A reconstruction of the megasporophyll of the Lower Permian cycad *Archaeocycas*.**

*SOURCE:* From S. H. Mamay, *Science*, vol. 164, pp. 295–296, 18 April 1969. Copyright 1969 by the AAAS.

no longer represented in the living flora. The Pinophyta first appear abruptly in the Upper Devonian, with the greatest diversity during the early Mesozoic, and then decline during the Cretaceous. All the major groups of pinophytes lack fossil ancestors. Ideas as to the origin of these groups remain unsupported.

## *Fossil Magnoliophyta (Flowering Plants)*

The sudden appearance and the abundance of flowering plant fossils in Cretaceous strata has been one of the great mysteries of biology. Charles Darwin wrote that to him the origin of the flowering plants (Magnoliophyta, or angiosperms) was "an abominable mystery," and over a century later it is still a puzzle. However, recent hypotheses have been proposed that shed some new light.

To begin with, one should recognize that a number of distinctive features set the flowering plants apart from the gymnosperms. These features are the enclosure of ovules within an ovary, the lack of free nuclear division in the development of the embryo, the process of endosperm formation, utilization of animal pollination, and double fertilization. True double fertilization, whereby one sperm fertilizes the egg and the other sperm unites with the

**FIGURE 12.7 Wax model of a fertile branch of Pinophyta *Cordaites*.**

*SOURCE:* Photo courtesy of the Field Museum of Natural History, Chicago.

polar nuclei (not just a neck canal cell, as recently reported in *Ephedra)* to ultimately become endosperm, is the only feature that is completely unique to the Magnoliophyta. Each of the other features has at least one or a few shared exceptions with other seed plants.

Unquestionable flowering plant fossils have not been found in pre-Cretaceous deposits. There are some reports of Triassic or Jurassic fossils of flowering plant leaves or palm trunks, but these remains are of questionable determination or of more recent age.

Flowering plant fossils are found as fruits, leaves, pollen, stems, (especially wood), and less commonly as flowers. Paleobotanists have typically classified flowering plant fossils in modern genera and families, especially when dealing with leaf compression fossils. However, current thinking is that the leaves and pollen of the past are not always attributable to extant angiosperm groups.

The sudden appearance and abundance of the flowering plants has led some botanists to postulate that the angiosperms must have evolved for a considerable time before the Cretaceous. To account for the lack of earlier Mesozoic fossils, two hypotheses have been proposed. One hypothesis, called the **upland theory,** suggests that the early flowering plants evolved in tropical upland habitats, where preservation as fossils would not be as likely as in lowland basins. It is reasoned that even if fossils were formed in the upland basins, erosion probably would have destroyed them. Certainly it could be reasoned that in upland regions, with their more diverse habitats, there would have been greater opportunity for evolution of new adaptive types. A second hypothesis, the **lowland theory,** proposes that the Magnoliophyta were not present before the Cretaceous. Support for this idea comes from the great diversity seen in today's tropical forests. The lowlands are thought to have been centers of origin, diversification, and adaptive radiation. Both of these ideas are not without critics and are mostly speculative.

What can be offered with confidence is that flowering plant fossils are present in Lower Cretaceous strata. Fossils found in Middle Cretaceous formations show great diversity in flowers and fruits, and some exhibit characters considered by botanists as "primitive" (Fig. 12.8). These so-called earliest angiosperms are not classified with any living taxa, even though they are rather complex. The fossil remains from which the reconstructed plant in Fig. 12.8 was constructed were remains of stems, leaves, petals, fruits, and seeds.

The precursor to the Magnoliophyta has received much speculation. During the late nineteenth and early part of the twentieth century, there were those who considered the inconspicuous, imperfect type flower as a "primitive" condition in the development of flowering plants. Proponents of this idea considered the gymnosperm alliance of *Ephedra-Gnetum-Welwitschia (Gnetales)* that lacks archegonia as support for this hypothesis. Other theories have considered a hypothetical group, the "Hemiangiospermae," some pteridophytes, the seeds of ferns, cycads, and extinct cycad-like plants, called cycadeoid, as archetypical. This latter group has been emphasized by those botanists who consider the perfect flower with both stamens and individual carpels present as the most ancestral. The presumed ancestor would have been a woody shrub, having large, evergreen, alternate, simple leaves; large perfect flowers with many individual sepals, petals, stamens, and carpels. The flower is considered to have been insect pollinated. This type of flower would most closely resemble contemporary magnolia-like flowers (Fig. 12.9, p. 422).

Recently a third hypothesis has been proposed, suggesting that the ancestral angiosperm was a herbaceous diminutive rhizomatous perennial with simple alternate leaves. In the axils of these leaves were found reduced cymose unisexual flowers subtended by bracts and small bracteoles. The fossils forming the basis for these hypothesized angiosperms *(Ascarina* and *Hed-*

**FIGURE 12.8 A suggested reconstruction of a leafy branch and flower of a "primitive" flowering plant.**

*SOURCE:* Based on Dilcher and Crane, 1984b; photo courtesy of D. Dilcher.

*yosmun)* have features fitting both Magnoliopsida (dicots) and Liliopsida (monocots), suggesting that these two groups separated early in geological history.

## *Origin of Characteristics*

A discussion on flowering plant origins would not be complete without something being said about the origin of the distinctive characteristics of the Magnoliophyta, mentioned earlier. In addition to the problem of a most incomplete fossil record, there is much difference of opinion among botanists as to the extent of so-called primitive characteristics.

The first feature to consider is the enclosing of the ovules within the ovary (**angiospermy**). Paleobotanical evidence suggests that the seed fern cupule may be a logical beginning point for a single carpel or pistil. Various isolated fossils from Carboniferous strata indicate various amounts of fusion of the appendages providing for some sac-like enclosing of the ovule. It must be remembered that through the human imagination a carpel can be derived, but many other floral features cannot be properly developed.

**FIGURE 12.9 Flower of *Liriodendron tulipifura,* showing characters thought by some botanists to be like the earliest flowering plant.**

The second hypothesis suggests that the carpel may have developed by the longitudinal folding of an ovule-bearing leaf, and then attachment to the two margins, thereby enclosing the ovules inside a walled structure. Because the fusion-evidence phenomenon is lacking from the fossil record, botanists have turned to developmental morphology of living plants for evidence, especially so-called primitive groups, such as the order Magnoliales. This order does have some of the features observed in various Cretaceous flowering plants. In some taxa the carpel is folded and is not completely sealed until maturation of the fruit. Observations of various genera *(Degeneria, Drimys,* etc.) show a progression in complexity and degree of fusion, as well as development of the stigma. This extant evidence gives credence to the hypothesis, but lacks fossil evidence to support the origin of carpels from gymnosperm leaves.

The origin of double fertilization and the resulting formation of endosperm following union of sperm and polar nuclei is a feature that is wholly distinctive to flowering plants. Unfortunately, the fossil record is of no value in unraveling this puzzle. However, recent research in the nonflowering plant *Ephedra* has demonstrated double fertilization, with the second sperm fusing with the ventral canal nucleus, thereby lending credence to an *Ephedra*–flowering plant homology.

The last distinctive Magnoliophyta characteristic is the lack of the free-nuclear stages in the development of the embryo. Within the conifers (i.e., Coniferales) there is a reduction in the number of free nuclei present. For example, *Pinus* has only four nuclei and *Sequoia sempervirens* has none. Among the flowering plants, members of the genus *Paeonia* have free nuclear development during embryological formation.

If, after reading the above brief accounts dealing with the origin of flowering plants, the reader is somewhat skeptical, this attitude is not without foundation. To be honest, we really do not know very much about origins of major plant groups, and what information is known is rather unsubstantial and nebulous. Perhaps the enigma of the origin of the angiosperms (and vascular plants in general) can best be summarized in the words of a contemporary student of the flowering plants, A. Cronquist (1988), who recently wrote:

> The origin of the angiosperms was an "abominable mystery" to Charles Darwin, and it remains scarcely less so to modern students of evolution. It is clear that they are vascular plants, related to other vascular plants, that they belong to the pteropsid rather than the lycopsid or sphenopsid phylad within the vascular plants, and that their immediate ancestors must have been, by definition, gymnosperms. Beyond that, much is debatable.

## *Evolution of Species*

Plant species are found in populations, which are not genetically static but dynamic and constantly changing. Variations observed within the populations are determined by various environmental factors (i.e., climate, soils, etc.), breeding systems (i.e., outcrossing, selfing, etc.), and internal variables (i.e., mutations, etc.).

From time to time, individuals and populations bearing distinctive characteristics will diverge from their original populations in such a way that gene exchange between the two will be prevented. The barriers to exchange of genetic material are called **isolating mechanisms** and include environmental, reproductive, and spatial factors.

**Environmental isolation** occurs when habitats suitable for supporting two separate taxa are limited, thereby restricting gene exchange between populations. Some examples of environmental factors are different soils, light intensity differences, and variation in moisture availability. The populations can live in the same region and be **sympatric** geographically but inhabit different habitats. In regions of low topographic relief and uniform climate, the same species may have a wide distribution. Species populations, in this case, form gradients of character variation (called **clines**) where different populations intergrade. Genetic variability between these joining populations may then become barely observable, while at the extremes of the range, significant differences may be noticed. In areas of extreme climatic and topographic variation, species become more restricted to specific zones and usually have a more restricted distribution. More species are generally found in diverse regions, with less species diversity in broad climatic regions.

Second, **reproductive isolation** occurs where gene exchange is inhibited by genotypically controlled differences in reproductive behavior between individuals of different populations. In certain interbreeding populations, the barrier to gene exchange is effective in blocking fertilization externally. This could be a structural feature, such as not having the reproductive organs of the flower coadapted structurally, so that interspecific pollination does not occur. It might also be physiological, where the period of flowering or shedding of pollen occurs at different times or seasons. For example, near Monterey Bay, California, the following pines *(Pinus)* occur sympatrically: *P. attenuata* (knobcone pine), *P. muricata* (Bishop pine), and *P. radiata* (Monterey or radiata pine). These sympatric pines are artificially fertile with vigorous hybrids produced, yet very little hybridization occurs between them, because each sheds its pollen at different times in the spring.

Reproductive barriers to gene exchange can also function after gametes or sex cells come together inside the bodies of organisms. The blocking may come from noncompatible interaction between chromosomes, gametes, or genes. Sometimes this will be observed quickly (i.e., flower shrivels up and aborts after pollination) or later when the $F_1$ hybrid grows to maturity but has aborted sex organs or an $F_2$ progeny has reduced reproductive potential or is sterile.

**FIGURE 12.10 Diagram illustrating how species can evolve through time.**

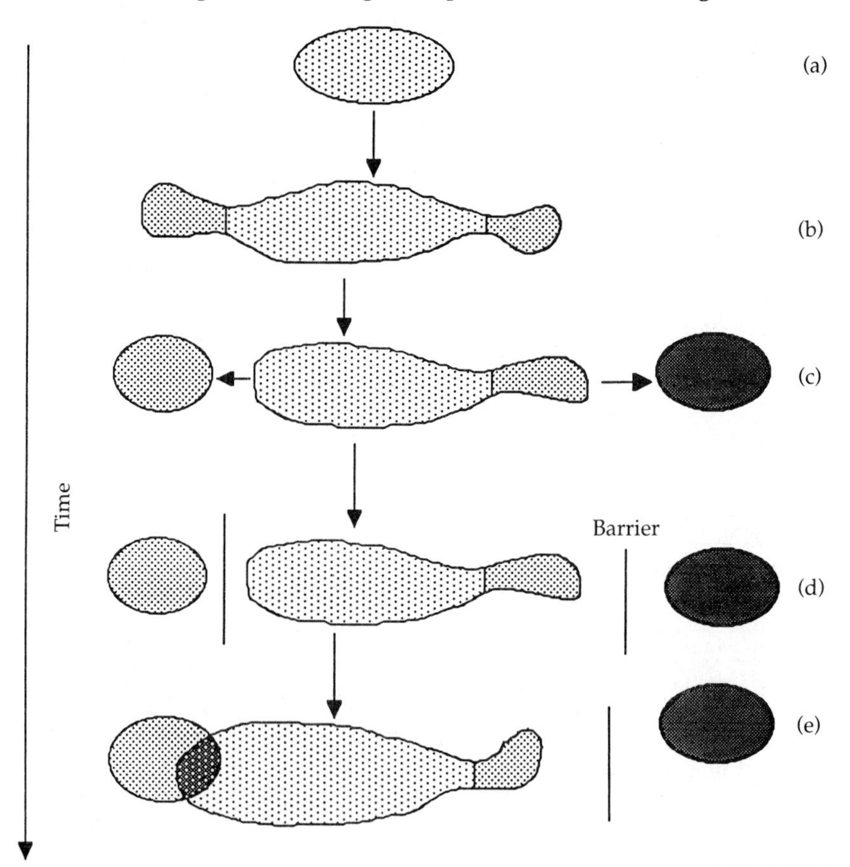

The final isolating mechanism is **spatial isolation**. This is where gene exchange is limited or prohibited because of the great distance between populations. An excellent example of this is the interfertility of *Platanus occidentalis* (North American sycamore) and *P. orientalis* (Oriental plane tree). The species are geographically separated by the Atlantic Ocean and both are found in different climates. A hybrid resulting from a cross between these two species resulted in the completely fertile *P. acerifolia*, the London plane tree.

When these isolating mechanisms become so strong that the populations become isolated reproductively, they now may be termed **distinct species.** This process is shown in sequence in Fig. 12.10: (A) A single population inhabits a uniform environment. Over time the environment may begin to change. (B) Differentiated habitats allow for migration of different population **biotypes** into these new environments. (C) Further migration differentiation and barrier formation produce isolated races and subspecies. (D) The isolated subspecies can begin through time to change genetically and chromosomally (i.e., become polyploids) and become more reproductively **allopatric** (geographically isolated). (E) Continued environmental change may allow previously isolated taxa to coexist sympatrically, but they now remain distinct because of the established reproductive barriers.

This discussion has been rather simplified, but should give a brief idea as to how species can and do originate.

## *SELECTED REFERENCES*

Banks, H. P. 1975. The Oldest Vascular Land Plants: A Note of Caution. *Rev. Palaeobot. Palynol.* 20:13–25.

Beck, C. B. 1971. On the Anatomy and Morphology of the Lateral Branch System of *Archaeopteris*. *Amer. J. Bot.* 58:758–784.

Beck, C. B. ed. 1976. *Origin and Early Evolution of Angiosperms.* Columbia Univ. Press, New York, 341 pp.

Bold, H. C., C. J. Alexopoulos, and T. Delevoryas. 1987. *Morphology of Plants and Fungi*, 5th ed. Harper & Row, New York, 912 pp.

Crepet, W. L. 1984. Ancient Flowers for the Faithful. *Nat. Hist.* 93:38–45.

Cronquist, A. 1988. *The Evolution and Classification of Flowering Plants*, 2nd ed. New York Botanical Garden, Bronx, NY, 555 pp.

Delevoryas, T. and R. C. Hope. 1971. *A New Triassic Cycad and Its Phyletic Implications.* Postilla 150, Peabody Museum, Yale Univ., New Haven, CT, pp. 1–21.

Dilcher, D. L. 1974. Approaches to the Identification of Angiosperm Leaf Remains. *Bot. Rev.* 40:1–157.

Dilcher, D. and P. R. Crane. 1984a. In Pursuit of the First Flower. *Nat. Hist.* 93:56–61.

Dilcher, D. L. and P. R. Crane. 1984b. *Archeanthus:* An Early Angiosperm from the Western Interior of North America. *Ann. Missouri Bot. Gard.* 71:380–388.

Doyle, J. A. and L. J. Hickey. 1976. Pollen and Leaves from the Mid-Cretaceous Potomic Group and their Bearing on Early Angiosperm Evolution. In C. B. Beck (ed.), *Origin and Early Evolution of Angiosperms.* Columbia Univ. Press, New York, pp. 139–206.

Edwards, D. S. 1980. Evidence for the Sporophytic Status of the Lower Devonian Plant *Rhynia gwynne-vaughanii* Kidston and Lang. *Rev. Palaeobot. Palynol.* 29:177–188.

Edwards, D. S. 1986. *Aglaophyton major,* a Non-Vascular Land Plant from the Devonian Rhynie Chert. *J. Linnean Soc. Bot.* 93:173–204.

Edwards, D. and J. Feehan. 1980. Records of Cooksonia-Type Sporangia from the Late Wenlock Strata in Ireland. *Nature* 287:41–42.

Friedman, W. E. 1990. Double Fertilization in *Ephedra,* a Non-Flowering Seed Plant: Its Bearing on the Origin of Angiosperms. *Science* 247:951–954.

Friis, E. M., W. G. Chaloner, and P. R. Crane (eds.). 1989. *The Origins of Angiosperms and their Biological Consequences.* Cambridge Univ. Press, Cambridge, 358 pp.

Gensel, P. G. and H. N. Andrews. 1987. The Evolution of Early Land Plants. *Am. Sci.* 75:478–489.

Gifford, E. M. and A. S. Foster. 1989. *Morphology and Evolution of Vascular Plants,* 3rd ed. W. H. Freeman, New York, 626 pp.

Hickey, L. J. and J. A. Wolfe. 1975. The Bases of Angiosperm Phylogeny: Vegetative Morphology. *Ann. Missouri Bot. Gard.* 62:538–589.

Hueber, F. M. and H. P. Banks. 1967. *Psilophyton princeps:* The Search for Organic Connection. *Taxon* 16:81–85.

Hughes, N. F. 1976. *Palaeobiology of Angiosperm Origins.* Cambridge Univ. Press, Cambridge, 242 pp.

Jacobs, J. 1986. Teleology and Reduction in Biology. *Biol. Philos.* 1(4):389–400.

Kidston, R. and W. H. Lang. 1917–1921. On Old Red Sandstone Plants Showing Structure, from the Rhynie Chert Bed, Aberdeenshire. Parts I–V. *Trans. Roy. Soc. Edinburgh* 52:831–854.

Kubitzki, K. ed. 1977. *Flowering Plants: Evolution and Classification of Higher Categories.* Plant Systematics and Evolution. Suppl. I. Springer-Verlag, Wien, 416 pp.

Lawrence, G. H. M. 1951. *Taxonomy of Vascular Plants.* Macmillan, New York, 823 pp.

Mamay, S. H. 1969. Cycads: Fossil Evidence of Late Paleozoic Origin. *Science* 164:295–296.

Niklas, K. J. 1979. An Assessment of Chemical Features for the Classification of Plant Fossils. *Taxon* 28:505–516.

Pettitt, J. M. and C. B. Beck. 1967. Seed from the Upper Devonian. *Science* 156:1727–1729.

Remy, W. and R. Remy. 1980. Devonian Gametophytes with Anatomically Preserved Gametangia. *Science* 208:295–296.

Stewart, W. N. 1983. *Paleobotany and the Evolution of Plants.* Cambridge Univ. Press, Cambridge.

Taylor, D. W. and L. J. Hickey. 1990. An Aptian Plant with Attached Leaves and Flowers: Implications for Angiosperm Origin. *Science* 247:702–704.

Wiley, E. O. 1981. *Phylogenetics.* John Wiley, New York, 439 pp.

# 13

# Contemporary Methods of Studying Plant Systematics

The individual who takes time to observe plants closely cannot help but be impressed with the variation within and between taxonomic groups. To grasp the reality of natural relationships and to provide a workable taxonomy, botanists must consider information from a wide spectra of sources. These sources of data come from the various disciplines of plant biology: anatomy and morphology, pollen analyses (palynology) where special equipment is used (electron microscopy), biochemistry, cytology, and hybridization.

Sometimes the information helpful in developing a taxonomy of a plant group comes from other workers. Most commonly though, the data must be gathered by the taxonomist him/herself. This can be a formidable task, because of the large number of species in some groups, the many characters exhibited by organisms, and the time involved. This can lead researchers to choose very carefully the characters to be studied. It can also lead to very specific methods that provide the necessary data needed to make sound taxonomic judgments. The following sections introduce some of these ways.

## *Plant Anatomy as Applied to Systematics**

Most of the observations on which taxonomists depend for identification and interpretation come from the naked eye, a hand lens, or the stereoscopic (dissecting) microscope. These devices cover the magnifications from 0–20×. Characteristics visible within this narrow range have been, and will continue to be, the ones used most.

Many clues to the natural groups of plants, and to the identification of unknown

*This section contributed by Nels R. Lersten, Iowa State University, Ames, Iowa.

specimens, are provided by cells and tissues as seen in the range of 20–1000 × magnification through the compound microscope. This instrument and its accessory equipment (slides, cover slips, dyes, reagents, microtomes and cutting blades, light sources), after gradual improvement over about 200 years, have been routinely used in most botanical laboratories since the mid-nineteenth century.

Until the mid-twentieth century, the compound light microscope represented the ultimate in magnification. The limit of magnification inherent in the properties of light waves required that the next advance must depend on shorter electron beams instead. This led to the invention of the **transmission electron microscope (TEM)**, which went through about a 30-year period of improvement before it became a routine instrument by the early 1960s. While the TEM expanded the range of magnification from about 1000 × to perhaps 100,000 ×, most of the exquisite subcellular details it reveals have not proven to be of great usefulness to systematics, except for groups of organisms among algae, fungi, and bacteria.

A modified electron microscope, called the **scanning electron microscope (SEM)**, has become very useful in the last 20 years for plant systematics. It has a practical range of magnification that is far less than that of the TEM (up to perhaps 10,000 ×), but is invaluable at low magnifications of less than 100 × because it bounces electrons off the surface of thick objects, then collects and integrates the reflected electrons and forms a three-dimensional image. This provides a different view of plant surfaces and structures, such as pollen grains and seeds, even though the SEM has a range of magnification that overlaps that of the light microscope.

Systematic anatomy is usually considered as the domain that includes surface and internal features seen in the compound microscope. This is admittedly arbitrary, and the boundary is made even fuzzier by considering scanning electron microscopy, which mostly "sees" what the light microscope is capable of seeing, but in a different way. With these limits in mind, the contributions of systematic anatomy can be outlined in the following paragraphs, with a few examples.

### *The Kinds of Information Systematic Anatomy Provides*

Thin epidermal peels or cross and longitudinal sections of fresh, preserved, or even dried plant organs can yield information when viewed either unstained or enhanced by certain general stains or by polarized light or phase-contrast optics. Thin, whole organs, especially leaves and flower parts, can be "cleared" (protoplasts removed chemically) and stained so that one can focus the microscope at different levels and see entire empty cells, particularly those of the vascular bundles. Most commonly, however, pieces of plant organs are processed into either wax or a harder resin of some kind and then thin slices (1–25 μm) are cut from them using steel or glass knives attached to a precision cutting device called a **microtome.** The slices are placed on a standard glass slide and stained in any one of many possible color combinations (red safranin and green fast-green is probably most common), dehydrated, and a cover slip cemented over them using a clear resin that hardens.

These methods, combined with magnification by the microscope, allow one to see details of the surface, such as trichomes and other secretory structures, stomata and adjacent subsidiary cells, or just the general shape and degree of bulging of ordinary epidermal cells. Internally, the arrangement of tissues, and the types of cells that occur within them, can be determined for roots, stems, leaves, and flowers, and their subsequent fruits and seeds. Looking at plants in this way for well over a century, botanists have described an enormous range of variation in microscopic features.

Some of these microscopic structures, and the patterns they form, are unevenly distributed among plants and therefore can be used as taxonomic characters, just as those seen with the naked eye or hand lens. However, because it takes much more effort to look at plants with the microscope, the generalizations made by using these methods are typically based on rather small samples. This means that systematic anatomy provides a very incomplete record as compared to gross characters, such as stamen and carpel number, which are known for every species. Nevertheless, "success stories" can be mentioned at almost all levels in the taxonomic hierarchy.

### *Anatomical Characters and Vascular Plant Evolution*

Vascular tissue itself, both xylem and phloem, provides a beautiful example to reinforce ideas of evolutionary relationships among the vascular plant groups. Xylem, in all groups up through gymnosperms, with a couple of minor exceptions, is rather "simple" in construction, consisting almost entirely of long, slender tracheids with intact end walls. Water moves between these cells solely through bordered pits in the wall. In angiosperms, however, the xylem story becomes more complicated. Some tropical woody Magnoliopsida (dicot) families, which are considered to be close to the base of the angiosperm phylogenetic tree, also have only tracheids. Some Liliopsida (monocot) families, which are thought to be at the base of this group, also have only tracheids. Most of the more "advanced" dicots have, in addition to tracheids, vessel elements (broader, shorter than tracheids, with various patterns of actual holes, called **perforation plates,** in their end walls) arranged in vertical series to form vessels. Among Liliopsida, some families have vessels only in roots, some only in roots and stems, while still others have them in all organs. Thus the gradual anatomical adaptation to more efficient water transport is thought to be an indicator of "evolutionary advancement."

Phloem reveals a more general pattern. Lower vascular plant groups up to the Pinophyta have phloem tissue consisting almost entirely of elongate, parenchyma-like cells, called **sieve cells,** which are mostly distinguishable because they have very large protoplasmic threads clustered together in sieve areas, which connect these cells to each other. In at least some Pinophyta groups, the sieve cells are accompanied by closely associated parenchyma cells, called **albuminous cells,** or **Strasburger cells.** This appears to be an intermediate stage, leading to the more "specialized" angiosperm phloem, in which photosynthates move through sieve tube members with a unique protoplasm lacking a nucleus. Conspicuous protoplasmic connections (**sieve plate**) perforate the end walls in addition to lateral sieve areas. Specialized parenchyma cells (**companion cells**) are appressed to each sieve tube member and are thought to provide nuclear communication and to act in a secretory capacity to pump sugars into them. No dicots or monocots have yet been described that lack such specialized phloem, thus no pattern corresponding to that in the xylem seems to occur.

In addition to the cells that comprise vascular tissue, xylem and phloem arrangement in the stem has long been of interest as a taxonomic feature (Fig. 13.1). The general trend starts with a solid column of xylem surrounded by phloem (**protostele**), with several variations in the most "primitive" groups. In ferns and fern allies, the vascular tissue forms a tube (**siphonostele**) with several variations. In the Pinophyta (gymnosperms) and Magnoliopsida (dicots), the bundles are interpreted as true and independent bundles, which are arranged in a cylindrical pattern (**eustele**). The Liliopsida (monocots) commonly have many more vascular bundles than dicots, especially those with solid stems, and the bundles appear to be more irregularly arranged (**atactostele**); they are also interconnected at each node by a complex network of horizontal bundles (**nodal plexus**). Thus it has been of interest that the Magnoliopsida family Piperaceae and members of a few other families also have atactosteles. It has been speculated that this indicates where the Magnoliopsida and Liliopsida may have been connected in an evolutionary sense.

Another aspect of vascular tissue that provides some insight into vascular plant evolution is leaf venation. The prevalent pattern in the ferns and fern allies and the Pinophyta is open dichotomous branching, without any minor veins or vein endings in between. At least in the ferns, this seems to be correlated with continued leaf growth at the margins. In many Pinophyta, a peculiar type of diffuse vascular tissue, called **transfusion tissue**, is variously developed in leaves, which seems to be an independent and perhaps crude analogy to minor veins.

**FIGURE 13.1 Some types of arrangements of vascular tissue (xylem and phloem) in stems. Phloem is white; parenchyma is stippled; xylem is cross-lined. (a), (b), (c) Variations of protosteles. (d) and (e) Variations of siphonosteles. (f) Eustele. (g) Atactostele.**

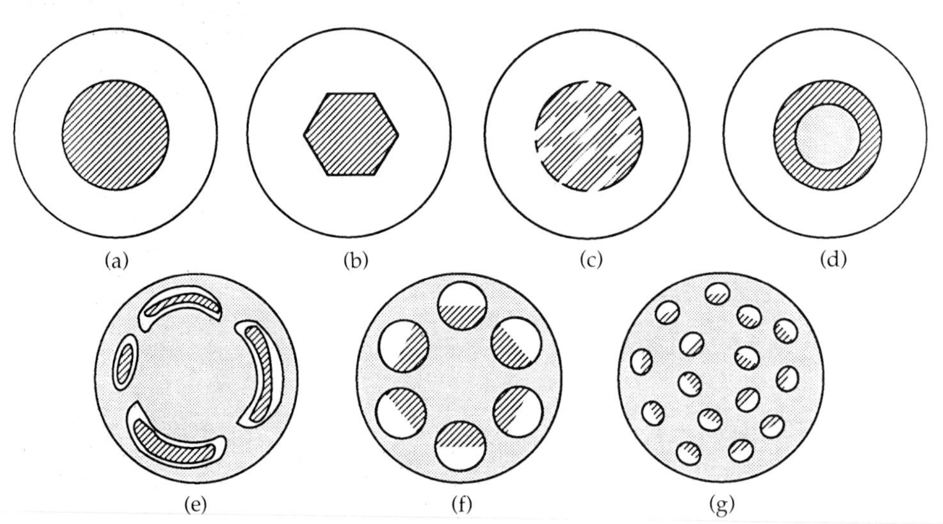

In flowering plants, in contrast to these groups, leaves are richly provided with both major and minor veins, and this may be correlated with the way these leaves develop, which does not involve prolonged marginal growth. Vein endings, which are the final expression of vascular development in leaves (and floral appendages also), are characteristic of Magnoliopsida but are rather uncommon in most Liliopsida. This has led to the idea that the common tubular or tongue-like (ligulate) leaf of monocots is fundamentally different from a Magnoliopsida leaf. Certainly the bizarre development of a palm leaf, which starts out as one structure folded like an accordion, and then splits into segments when the folds separate clearly, sets the Arecaceae and the related family Cyclanthaceae apart from other Liliophyte families.

### *Anatomical Characters Delimiting Families*

It is not easy to find airtight features at the family level. One of the best seems to be the peculiar stomatal apparatus characteristic of the grass family (Fig. 13.2). The elongate pair of dogbone- or dumbell-shaped guard cells, each flanked by a large hemispherical subsidiary cell, appears to be restricted to grasses.

If the legumes are considered to be an order (Fabales), then the familiar papilionoid legumes become the family Fabaceae, and they are distinguished from all others by an interesting seed character. Extending beneath the length of the **hilum,** which is the scar left where the seed was attached to the carpel wall, there is a horizontal cylinder of tracheid-like cells in which each cell is oriented perpendicular to the hilum surface. This "tracheid bar" is completely absent from the other two families in the order and, indeed, from seeds of all other angiosperms.

**Myrosin cells** are elongate parenchyma-like cells with very dark contents, consisting mostly of myrosin, an unusual protein involved in mustard oil production. These cells are associated with the phloem, and they were thought to be a unique characteristic of the mustard family, the Brassicaceae. Recent studies have also described them from some suspected related families, thus myrosin cells may be characteristic of an order rather than a family. This emphasizes the earlier statement that conclusions from anatomical characters are usually based on small samples, which means that new observations are

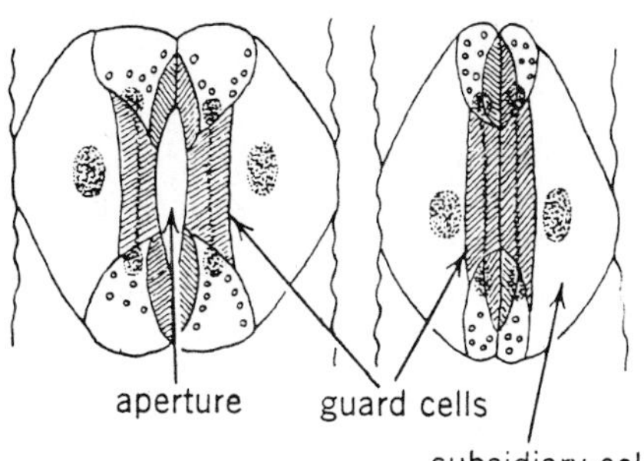


**FIGURE 13.2 The stomata of *Saccharum* (sugar cane), shown open and closed in relation to the adjacent subsidiary cells.**

*SOURCE:* From *Plant Anatomy,* 2nd ed., by K. Esau. Copyright © 1965 John Wiley and Sons. Reprinted by permission of Wiley-Liss, a division of John Wiley and Sons, Inc.

always possible that may change prevailing ideas of their distribution and, therefore, their significance.

### *Anatomical Characters Significant Within Families*

Families are often subdivided into subfamilies, sections, or other groupings of genera. At this level there are many examples of the use of anatomical characters, either as primary defining features or as additional characters that support a particular scheme of classification.

In the family Clusiaceae (Guttiferae), one subfamily (Hypericoideae) is distinctive because of its "gland-dotted" leaves. Anatomical study has shown that these are internal oil-containing structures of unusual, perhaps unique, anatomy and development.

Many groups of plants have what superficially seem to be similar "dots" scattered within their leaves, but more detailed anatomical study has shown that they vary considerably and thus present a variation that is useful for systematics. One of the most striking examples of this is provided by two tribes within the legume family Fabaceae. The Amorpheae and Psoraleeae had been combined by taxonomists earlier, in part because of the fact that they both had "gland-dotted foliage." Anatomical study of several members of each group, however, showed that the "dots," which are internal oil cavities, developed in radically different ways in each tribe, one being of internal cell origin (Amorpheae) and the other a peculiar inward thrusting of an epidermal cluster of cells (Psoraleeae). This character is of prime importance in showing that these tribes are separate entities and they they are not even closely related to each other within the family.

Of quite a different nature is the primary character used to delimit the tribe Guettardeae within the large family Rubiaceae. This tribe is defined solely on the presence of large crystals embedded in the cell wall of the trichomes. Crystals within epidermal cells can be found among various plant groups, but they are a defining characteristic, along with the unusual associated enlarged cells just below them, in leaves of the Aeschynomeneae (Fabaceae), the peanut tribe.

**Laticifers** are highly unusual cells that contain latex. They occur in several plant families, and are useful taxonomic characters. Worthy of special mention here is the tribe Lactuceae of the Asteraceae (Compositae), in which laticifers are among the defining characters that are so distinctive that this tribe has sometimes been set apart as a separate family. Recent research has shown that cells, generally called laticifers, in the family Convolvulaceae have distinctive subcellular features that may call for their redescription as an entirely separate kind of cell; thus it will take on new importance as a character in this family.

The grass family is one in which anatomical characters, especially of leaves, has been of primary importance in delimiting tribes. At an earlier time, when two subfamilies were recognized (excluding the bamboos), one (Panicoideae) was defined in part based on the presence of a single-layered swollen bundle sheath around the vascular bundles (e.g., as in corn leaves), while members of the Festucoideae had leaves with a small but double-layered bundle sheath (e.g., as in wheat leaves). These are still useful characters, but in recent grass classifications they are important at the tribal level. The bamboos and a couple of related tribes are different from other grasses internally in having a row of large but narrow, laterally elongate cells along each side of the leaf veins. These "fusoid cells" are distinctive and unique to these plants and are of unknown function.

### *Anatomical Characters at the Genus and Species Levels*

It is not easy to find distinctive characters at the genus or species levels. Since relationships are so close, it is mostly a matter of degree of variation; thus statistical methods are often applied to determine whether small variations are significant or not. Some exceptions have been observed by wood anatomists, who have found it possible to distinguish between species within some genera. It is of considerable economic importance to be able to make such distinctions here, and it may be that if it is important enough to do so, then appropriate characters may be found to distinguish species. A lively controversy in recent years has emerged over attempts to distinguish species of *Cannabis* (marijuana) by microscopic characters that can be detected in ground-up plant parts. Economic incentives are also behind attempts to identify species from bits of tissue in ground-up tobacco, tea, and other similar products in order to test for adulterants that might have been included.

Identifying species from partially digested bits of plant material is important in studying food habits of birds and other animals, and it has proved to be important in cases of suspected poisoning of both animals and people.

Much of paleobotany depends on identifying partial remains of plants from microscopic characters, which is usually all that remains after eons of time in the ground. Even more dependent on microscopy is the field of palynology, which seeks to identify plants to species or genera based on characters of one tiny part, the pollen or spore. Scanning electron microscopy has been particularly important in this field of study, as will be seen in a later section of this chapter.

## *Plant Morphology as Applied to Systematics**

Charles Darwin considered morphology the "very soul" of natural history. In modern biology, morphology may appear less important. However, to a careful observer it will become evident that it still plays a fundamental role, and that practically all biological disciplines ranging from molecular biology to systematic biology and ecology use morphological concepts and imply morphological theories. For example, a molecular geneticist may investigate the gene(s) that control leaf shape, or an ecologist may analyze nutrient distribution between metamers. In these investigations the morphological concepts "leaf" and "metamer" are used, and these concepts imply morphological theories that influence the formulation of questions and the direction of research.

What are the basic theories and concepts in morphology? There are too many to deal with in a brief introduction such as this; therefore, only three will be mentioned. (1) The **classical theory,** or **model,** has many different

*This section contributed by Rolf Sattler, McGill University, Montreal.

versions, distinguishing between two types of morphological organizations: thallus and cormus. The latter is defined by three types of characteristically positioned organs: root, **caulome** (i.e., stem or its homologs), and **phyllome** (i.e., leaf or its homologs). Roots may be lacking in some instances, and flowers are composed of an axis (i.e., caulome) and appendages (i.e., phyllomes). (2) The **telome theory** postulates one basic unit of plant construction, the telome. Morphological diversity observed is postulated to result from elementary processes acting on telomes and/or telome trusses. These processes are overtopping, planation, fusion, reduction, and recurvation, plus two processes for internal differentiation. For example, leaves of flowering plants are considered to be the result of the combination of overtopping, planation, and fusion. (3) According to a recent theory by Hagemann, the most fundamental unit of plant construction is not the axis-like telome, but a flattened thallus-like structure with roots and stems that evolved later.

With regard to the shoot of more complex vascular plants, such as flowering plants, most botanists have used the concepts of the classical theory. However, some authors have proposed additional theories or models (Fig. 13.3). Among these models, the **metameric model** has recently gained increased acceptance, especially among some ecologists. This model postulates the shoot being composed of **metamers,** each of which comprises a leaf with its node and an internode. This view emphasizes the morphological continuum of stem-node-leaf.

It would be inappropriate and destructive to quarrel about which of the models is correct. As there are many ways to cut a pie, there is more than one way to conceptually dissect a whole plant into units and subunits. It is therefore appropriate to consider the five models of Fig. 13.3 as complementary, each representing a different perspective.

In view of the differences resulting from the variety of perspectives that can be chosen, the systematist does not have an easy task. He or she cannot be content operating only within one conceptual framework, for to do this

**FIGURE 13.3 Five contrasting possibilities for conceptually dismembering the shoot of vascular plants: (a) Classical stem-and-leaf model (A = axillary branch, L = leaf). (b) Fertile leaf model (D = leaf-branch unit). (c) Leaf-skin model (L = leaf). (d) Phytonic model (Ph = phyton). (e) Metameric model (M = metamer).**

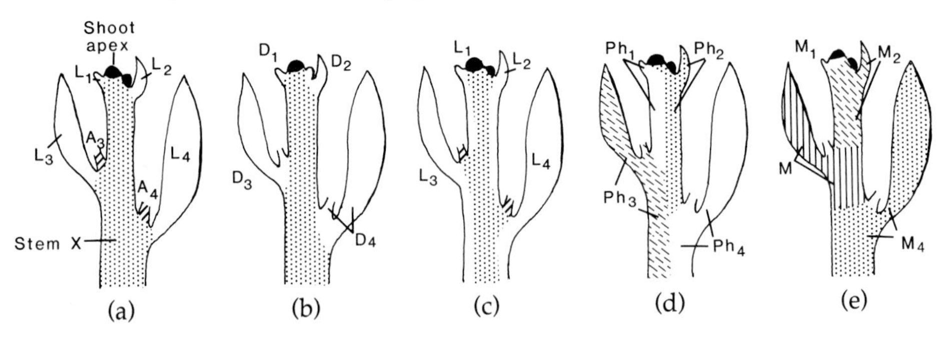


*SOURCE:* From Rutishauser and Sattler, 1985.

would lead to a one-sided and biased view of relationships. For example, the person who uses the leaf concept as a character will obtain character states with regard to this unit, whereas someone who refers to metamers will distinguish metamer character states. The question is to what extent classifications based on one or the other concepts will differ from each other. Perhaps there are cases where the difference would be negligible and others where it would be remarkable.

One assumption underlying the preceding views of plant form is the idea that a plant can and must be conceptually dismembered into discrete structural units and that the classes of these units are mutually exclusive; i.e., any one structure belongs to one class (category) or another. This means a structure is either a "stem" or a "leaf," and it cannot be intermediate between the two. In contrast to this view, various authors have stressed that plant structures cannot always be neatly pigeonholed into morphological categories. Intermediates do occur between different categories. To account for this complexity, the terms of the classical model have been redefined in a continuum model of the shoot (Fig. 13.4). Instead of being hierarchical, this model is based on a tetrahedron in whose corners we find the typical representatives of shoots, caulomes (i.e., stems and stem homologs), phyllomes (i.e., leaf and leaf homologs), and trichomes (hairs). The intermediates between any of the four typical forms are located along the surface and interior of the tetrahedron. For example, phylloclades of the Asparagaceae are intermediates between shoot and phyllome, or caulome and phyllome, and in the tetrahedron they are located accordingly.

This continuum view of plant form is fundamentally related to **homology,** one of the most central concepts of comparative morphology. Unfortunately, there are so many definitions of homology that the situation has become rather confusing. Mention will be made of only two concepts: **phenetic homology**, which is based on overall similarity, and **cladistic homology** (also called **synapomorphy** by cladists), which is "a similarity in structure in a group of organisms that delimits that group alone and that was found in their common ancestor" (Stevens, 1984) (discussed in Chapter 11). Both phenetic and cladistic homologies are usually seen as 1:1 correspondences. Thus, a spine of a cactus is considered to correspond to a leaf. However, in contrast to cladistic homologies that are always total (i.e., 1:1 correspondence), phenetic homologies may be partial. Therefore, the phylloclade of the Asparagaceae is considered partially homologous to a leaf and a branch or stem. This notion of partial homology changes fundamentally our approach to the

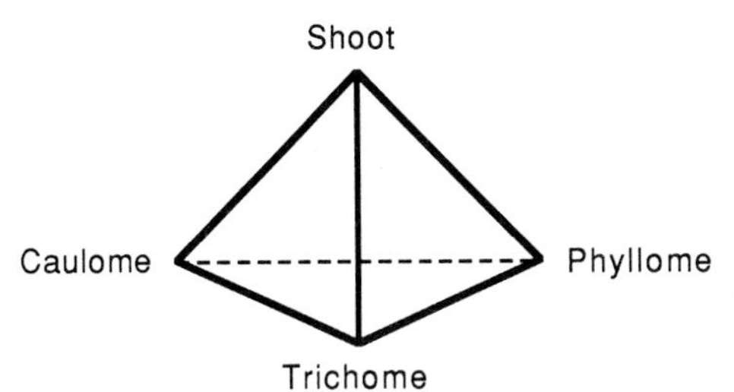


**FIGURE 13.4 Continuum model of the shoot.**

*SOURCE:* Courtesy of Rolf Sattler.

study of plant structure and systematics. We can therefore no longer take for granted the questions we used to ask as being appropriate. Instead of asking whether structure X is homologous to *either* A *or* B, we may have to ask totally different questions, such as the following: Is X more homologous to A or B? Or to what extent is X homologous to A and B? Such questions are appropriate for the continuum model mentioned above.

Cladistic homology is also based on similarity, but only on the similarity within a clade. Inasmuch as cladistic homologies imply phylogenetic hypotheses, they tell us more than phenetic homologies. However, since partial correspondences are ignored in cladistic homologies, they tell us less than partial homologies in a phenetic sense.

Both phenetic and cladistic homologies may be based only on mature structures, or they may take into account the development of structures. Needless to say, homologies based on development (which include the mature structures as a stage) are more comprehensive and provide more information. Fortunately, an increasing number of morphologists study development. This leads to a greater appreciation of the dynamics of plant form whose recognition is, however, limited, because of the use of structural categories that are static. The following example will be used to illustrate this idea. The statement "This leaf grows" is a dynamic statement because it implies the process of growth. However, the dynamics is limited due to the use of the structural category "leaf," which as a category is static. Thus, if this leaf grows in an indeterminate fashion, and therefore becomes shoot-like, this change from leaf-like to the shoot-like form cannot be conveyed by the statement "This leaf grows," because this statement implies that, although the structure grows, it remains a leaf.

The above example is not fictitious. In *Guarea* (Meliaceae), for example, so-called pinnate "leaves" show indeterminate growth. Many other examples where a transgression of morphological categories occurs could be cited. Evidently, nature is more dynamic than any conceptual framework based on mutually exclusive structural categories can convey. The question is how this dynamics can be represented. The answer is, in terms of process morphology.

What is **process morphology?** It is plant morphology that is not based on structural categories such as "stem" or "leaf," but on morphogenetic processes. Thus, the evolution of plant form is not seen as the modification of a few fundamental structural units, but as a change in process combinations. From this perspective, the question is not whether two structures are essentially the same, and hence should be grouped under the same category; the question is how process combinations are related such that a relationship may be formal or phylogenetic. With regard to evolution, the question is as follows: How and why have process combinations changed during ontogeny and phylogeny?

It is important to realize that during phylogeny process combinations may change in such a way that 1:1 correspondences can no longer be recovered and hence are inadequate to represent the dynamics of morphological change. Even from one generation to the next, developmental pathways may intermingle in such ways that their identities are lost. Certain homoetic mutations that involve partial homeosis illustrate this phenomenon. **Homeosis** is the partial or total replacement of a structure by another one of the same organism.

If, for example, a stamen is replaced by a hybrid structure of a stamen and petal, we have a case of partial homeosis. It does not make sense to ask whether this hybrid structure is a stamen or petal. It is neither, and yet it is partially one and the other. In terms of process morphology, it is simply stated how the process combination has changed. No identification with structural categories is needed. No 1:1 correspondence need be imposed. The goal of process morphology is to uncover the dynamics of ontogeny and phylogeny.

Although process morphology allows a more dynamic description and comparison of ontogenetic and phylogenetic change, morphology in terms of structural categories will still remain useful for many purposes. Which approach is preferable will depend on the situation and goal. In general, all or most approaches have a certain usefulness and therefore complement each other. Nonetheless, process morphology appears to be more adequate, inasmuch as it mirrors more comprehensively the natural dynamics of plant form.

## *Chemosystematics**

Chemosystematics (also called chemotaxonomy) uses the chemical constituents of plants to assess inter- and intraspecific relationships and to infer phylogeny. Although sometimes seen as a recent innovation, chemosystematics is as ancient as the field of plant taxonomy itself. The pervading interest of early humans in plants concerned their chemical contents, that is, which plants were edible or had important medicinal properties. These characteristics were the basis for simple classifications of plants into categories such as poisonous versus nonpoisonous, and herbs of medicinal value versus herbs with no known medical use.

As human society grew more complex, analyses of the chemical constituents of plants became more sophisticated. It was recognized that related plants generally had the same medicinal properties or chemical constituents, and by the seventeenth century a strong association had developed between botany and medicine. This association has continued, and numerous important drugs are still being derived from wild plants.

Fields other than medicine also had a strong influence on the early development of chemosystematics. For example, the fields of anatomy and morphology led to important discoveries about the chemical constituents of plant groups. This is because certain chemically based features of plants, such as color and different forms of crystals or starch grains, can also be considered as morphological or anatomical characteristics and are often most easily studied in the latter form.

Modern chemosystematics is a precise science that bears little resemblance to earlier forms of botanical chemistry. It has become a major field of study in only the past 30 years, and its emergence is largely due to three factors: (1) the development of new analytical techniques (e.g., chromatography and electrophoresis) for rapidly and accurately analyzing the chemical

*This section contributed by Loren H. Rieseberg, Rancho Santa Ana Botanic Garden, Claremont, California.

and molecular constituents of relatively small amounts of plant material, (2) the discovery of the tremendous variation in types of molecules produced by plants, and (3) the recognition of the informational content of certain molecular classes (i.e., proteins, RNA, and DNA).

The numerous types of molecules used in plant systematics can be divided into two major categories based on molecular weight: **micromolecules** and **macromolecules.** The former are sometimes referred to as secondary plant products and include a vast array of relatively low-molecular-weight compounds such as alkaloids, betalains, cyanogenic glucosides, flavonoids, glucosinolates (mustard oil glycosides), terpenoids, etc. Macromolecules are high-molecular-weight compounds, often composed of many repeating units, such as proteins and nucleic acids.

## *Micromolecular Data*

During the past 30 years, the contribution of micromolecular data to plant systematics has been enormous. Nevertheless, contrary to popular opinion, there is no reason to believe that micromolecular data are any more important or fundamental than other data sets, such as those derived from morphology, cytology, or anatomy. For example, the presence or absence of a certain compound is not, *a priori,* more important taxonomic evidence than the presence or absence of petals. This is because micromolecular characters, like morphological or anatomical characters, are controlled by several to many genes, with alleles at each locus contributing to the end product. Thus, the exact meaning of both morphological and micromolecular differences is often unclear from a genetic standpoint.

Micromolecular data are also similar to other data sets in that they have been found to be useful at all taxonomic levels; but as for other data sets, it is not possible to predict at what level they will be most helpful. Nevertheless, the majority of studies employing micromolecular evidence have been at the generic level or below.

Micromolecular data are most useful for assessing relationships when the biosynthetic pathways are known, because the pathways show the sequence in which various compounds are produced. This knowledge can be used for determining whether compounds are "primitive" or "advanced" and for assessing the taxonomic significance of various differences in compound structure.

Because of the large number of different kinds of micromolecules that have been used for taxonomic inference in plants, only two of the most commonly analyzed micromolecules will be discussed below: flavonoids and terpenoids. (For a more complete discussion see Harborne and Turner, 1984, or Giannassi and Crawford, 1982).

***Flavonoids*** Of all the micromolecules found in plants, flavonoids have been the most widely used taxonomically. Flavonoids have a relatively simple nucleus (Fig. 13.5), which is then elaborated with a variety of different side groups that characterize individual compounds. The flavonoids are highly variable in function, apparently playing important roles in pigmentation, cellular respiration, disease and herbivore resistance, and screening of important cellular constituents from ultraviolet light.

**FIGURE 13.5 Molecular structures of representative flavonoids, betalains, and terpenoids.**

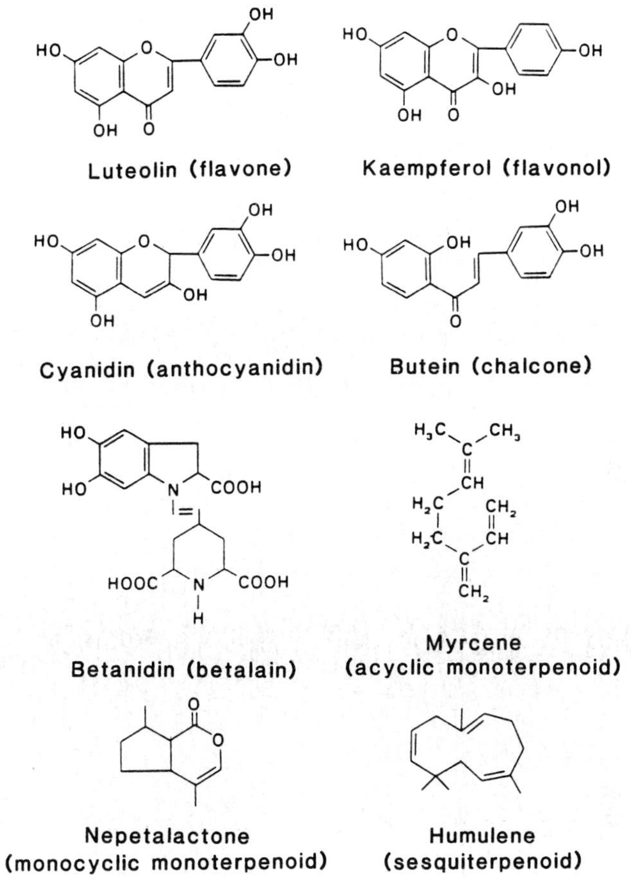

Luteolin (flavone)

Kaempferol (flavonol)

Cyanidin (anthocyanidin)

Butein (chalcone)

Betanidin (betalain)

Myrcene (acyclic monoterpenoid)

Nepetalactone (monocyclic monoterpenoid)

Humulene (sesquiterpenoid)

There are a number of reasons why flavonoids have been useful taxonomically; these include the following: (1) They occur in most vascular plants, as well as in mosses and liverworts (they do not occur in bacteria, algae, fungi, or animals). (2) They exhibit great structural diversity. (3) They are chemically stable for years after isolation.

Perhaps the best example of flavonoid data that provides significant taxonomic evidence involves the types of floral pigments found in various flowering plant families. Red- and blue-colored flowers, in most plant families, result from a particular class of flavonoids called anthocyanins. Anthocyanins are absent, however, from ten plant families that were considered members of, or closely related to, the order Centrospermae. In these families, the function of anthocyanins was replaced by an entirely unrelated class of micromolecules, betalains. Of the ten families characterized by betalains, only the Cactaceae were traditionally excluded from the Centrospermae. Based on the chemical evidence described herein, the Cactaceae are now classified as members of the Centrospermae.

Flavonoids have also been useful for analyzing evolutionary processes, such as hybridization and polyploidy. This is because chemical compounds, unlike many structural (morphological or anatomical) characters, are often inherited additively in hybrids. The best-documented study of interspecific hybridization is that by Alston and Turner in the legume genus *Baptisia*. They demonstrated that most species of *Baptisia* are characterized by a distinct set of flavonoids, and that hybrids are easily detected in mixed populations due to the additive patterns observed (Fig. 13.6). Furthermore, they were able to determine the parentage of individuals of subsequent generations that resulted from backcrossing of a hybrid to one or the other parental species.

**FIGURE 13.6 Diagrammatic representations of chromatograms of the leaf flavonoids of *Baptisia:* (a) pattern of leaf extract from *B. leucantha;* (b) pattern of leaf extract from *B. sphaerocarpa;* (c) extracts used to produce the patterns in (a) and (b) combined to produce this pattern (the letters S and L refer to the origin of a spot from either *B. leucantha* [L] or *B. sphaerocarpa* [S]); (d) pattern of natural hybrid of *B. leucantha* × *B. sphaerocarpa.***

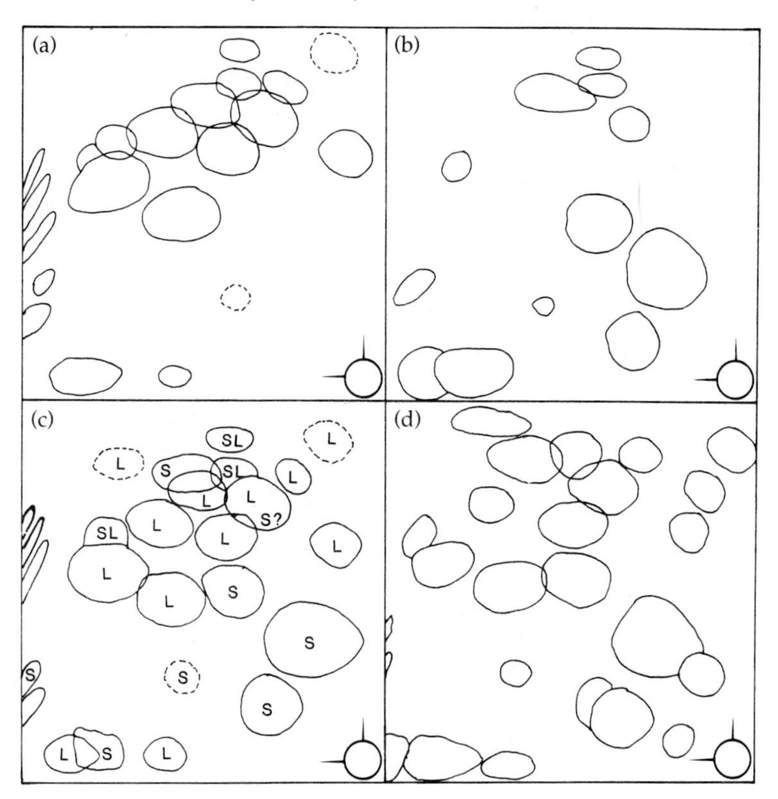

***Terpenoids*** Another set of micromolecules that has proven useful in plant systematics is the terpenoids. Terpenoids are a highly diverse group of compounds, united only by their common origin through a particular metabolic pathway. Natural plant products that are often terpenoid in origin include essential oils, resins, latex, cutins, and pigments.

Terpenoids have been used less widely in plant systematics than flavonoids, due to their occurrence in a limited number of plant groups and the necessity of sophisticated equipment and procedures for their analysis. Nevertheless, they have been very useful for systematic studies at the species level or below. Terpenoids have also been valuable for testing hypotheses of hybridization and introgression.

One example of the systematic value of these compounds involves studies of terpenoid variation in the widespread Douglas Fir *(Pseudotsuga menziesii)* of western North America. It was demonstrated that the coastal race, var. *menziesii,* and the inland race, var. *glauca,* are quite distinct chemically. The researchers also detected two well-defined chemical races within var. *glauca,* while another botanist was able to identify three chemical races within var. *menziesii.* These studies demonstrate that terpenoid evidence, like other micromolecular evidence, can be useful for detecting geographic variation that may not be detectable morphologically.

## *Macromolecular Data*

Macromolecular data are fundamentally different from micromolecular data in that macromolecules are direct gene products (proteins) or genetic material itself (nucleic acids), whereas micromolecules are the end products of enzyme-mediated biosynthetic pathways. Thus, macromolecular evidence allows the genetic relationships among organisms to be determined in a precise, objective manner that is not possible using micromolecular or structural evidence.

***Proteins*** A number of different methods have been used for analyzing protein variation, including serology, electrophoresis, and amino acid sequencing.

The technique of **serology** relies on the immunological reaction shown by animals when foreign proteins are injected into them; they in turn produce antibodies, each specific to a protein (or antigen) and capable of reacting with it. Serum containing these antibodies can then be removed from the injected mammal and allowed to react with protein extracts from other test plant species. The degree of reaction obtained is used as an indicator of the closeness of relationship. Research in this area has been limited, however, by the need to use animals for antibody production, debate over the best method for measuring the antigen/antibody reaction, and questions about the advisability of using crude protein extracts versus single proteins as antigens.

In contrast, **electrophoresis** has been used much more widely in plant systematics. This technique, possibly more than any other in plant chemosystematics, has the advantage of combining a simple methodology with maximum extractable information. Put simply, plant proteins are separated along a voltage gradient in starch, acrylamide, or agarose gels. After a suitable period of separation, proteins are located on the gels by use of general protein stains or enzyme-specific assays (Fig. 13.7). Each protein is a specific gene product.

By far the majority of electrophoretic work has involved the analysis of specific enzymes using the assays mentioned above (reviewed in Gottlieb, 1981; Crawford, 1983, 1985). By examining a number of different enzymes electrophoretically, a measure of overall genetic similarity (at enzyme loci) between populations, subspecies, and species can be obtained. It is now evident that populations of the same species are generally very similar at a selection of enzyme loci, but comparisons of populations of distinct species are usually characterized by much greater differentiation.

Electrophoretic evidence is particularly useful for examining putative cases of hybridization and polyploidy. This is because any two species are likely to be characterized by species-specific enzyme forms (allozymes) and hybrids will have the allozymes of both species.

An example of the taxonomic value of enzymatic evidence involves the origin of two tetraploid species of *Tragopogon*. *Tragopogon* is an Old World genus represented in North America by three introduced diploid species, *T. dubius, T. porrifolius,* and *T. pratensis.* The two tetraploid species, *T. mirus* and *T. miscellus,* originated in the early 1900s in eastern Washington state. The presumed diploid parental taxa were found to be divergent at close to 40% of the 20 enzyme loci examined, and the tetraploid taxa possessed completely additive enzyme patterns: *T dubius* × *T. porrifolius* gave rise to *T. mirus,* and *T. dubius* × *T. pratensis* gave rise to *T. miscellus.* These results corroborated the conclusions previously drawn from morphological and cytological evidence.

In contrast to electrophoretic data, which are most useful at the generic level or below, amino acid sequences of proteins have been used to study phylogenetic relationships at the level of families and orders. A major advantage of amino acid sequence data relative to morphological or micromo-

**FIGURE 13.7 Agarose gel stained for the enzyme phosphoglucoisomerase in *Helianthus.* Only the cytosolic isozyme is shown here. Single-banded phenotypes are from plants homozygous for one allele, a, b, or c, whereas the three-banded phenotype is from a heterozygous individual.**

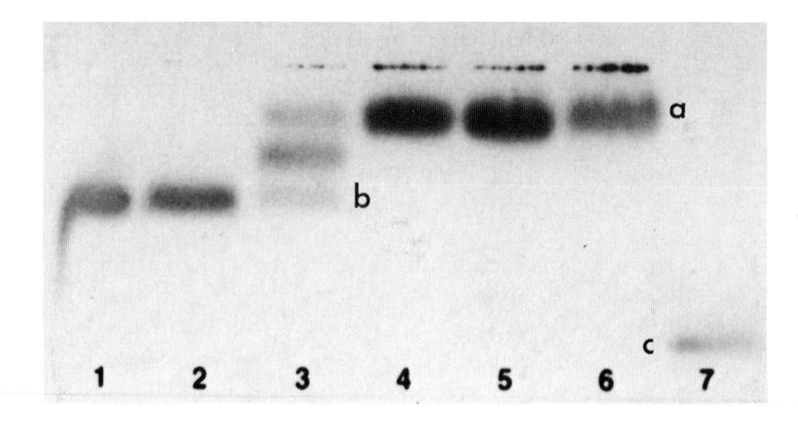

lecular data is that the proteins being compared are much more likely to represent homologous characters than are the morphological or micromolecular characters. Furthermore, because amino acid sequences are a transcribed copy of the DNA sequences that code them, quantitative estimates of genetic divergence can be obtained. Nevertheless, amino acid sequence data have had little impact on systematics because of the time and effort involved in purifying and sequencing proteins. In addition, given the greater ease of sequencing nucleic acids, the future usefulness of amino acid sequences for phylogenetic studies is limited.

*Nucleic Acids* Studies of nucleic acids (DNA, RNA) represent the latest and clearly the most powerful approach to chemosystematics of plants. It can be argued with some justification that direct study of the genetic material provides the most fundamental and objective assessment possible of the genetic similarities of plant taxa. Like most other taxonomic characters, nucleic acid data are useful at all taxonomic levels.

Early studies assessed nucleic acid variation by means of DNA-DNA and DNA-RNA hybridization. This method provides quantitative estimates of the sequence similarity of two DNA or DNA/RNA samples by measuring the overall pairing of hybrid molecules formed between them. The future use of this method is limited, however, by the difficulty of the technique, combined with its low resolving power. Instead, most studies have concentrated on obtaining primary sequence information using two different techniques: restriction fragment and map comparisons, and DNA sequencing. In the former method, used primarily at the generic level and below, restriction endonucleases are used to selectively sample specific subsets in a DNA molecule. Actual DNA sequencing is more expensive and time-consuming and requires more sophisticated protocols than does restriction analysis, but it represents a more straightforward approach to comparative systematics and is most valuable for higher-level phylogenetic comparisons. Both methods are presently being applied to numerous problems in plant systematics.

Studies of nucleic acid sequences in plants have concentrated on one of the three plant cellular genomes, the chloroplast genome (Fig. 13.8, p. 444), rather than the nuclear or mitochondrial genome. This is due to a number of reasons, including the following. (1) Chloroplast DNA (cpDNA) can be easily isolated and analyzed. (2) Chloroplast DNA is essentially unaltered by evolutionary processes, such as gene duplication and concerted evolution, which can distort the evolutionary history of DNA sequences. (3) The chloroplast genome is highly conserved in organization, size, and primary sequence, providing it with a distinct advantage in phylogenetic studies relative to the much more diverse nuclear and mitochondrial genomes.

Although plant nuclear DNA is analyzed less easily than cpDNA, it does possess two attributes that make it useful taxonomically. First, certain nuclear sequences appear to evolve more rapidly than any cpDNA sequences, allowing a finer level of discrimination at the population level than for cpDNA. Second, the nuclear genome is inherited biparentally, whereas the chloroplast genome is inherited maternally or at least clonally. Thus a hybrid plant will possess the nuclear complement of both parents but only the cpDNA of the maternal plant.

**FIGURE 13.8 Physical map of the chloroplast genome of the cinnamon fern, *Osmunda cinnamomea*. Each concentric circle shows the position (given by the short radial lines) of cleavage sites for the indicated restriction endonuclease (Pvu II, Sac I, or Bst EII) and the size in kilobases of the resulting restriction fragments. The positions of the following genes are shown: ribosomal RNA genes (16 S, 23 S), gene for the large subunit of ribulose-1,5-bisphosphate carboxylase (rbcL), genes for two subunits of the chloroplast ATPase (atpA and atpB), and the gene for a 32,000-dalton photosystem II polypeptide (psbA).**

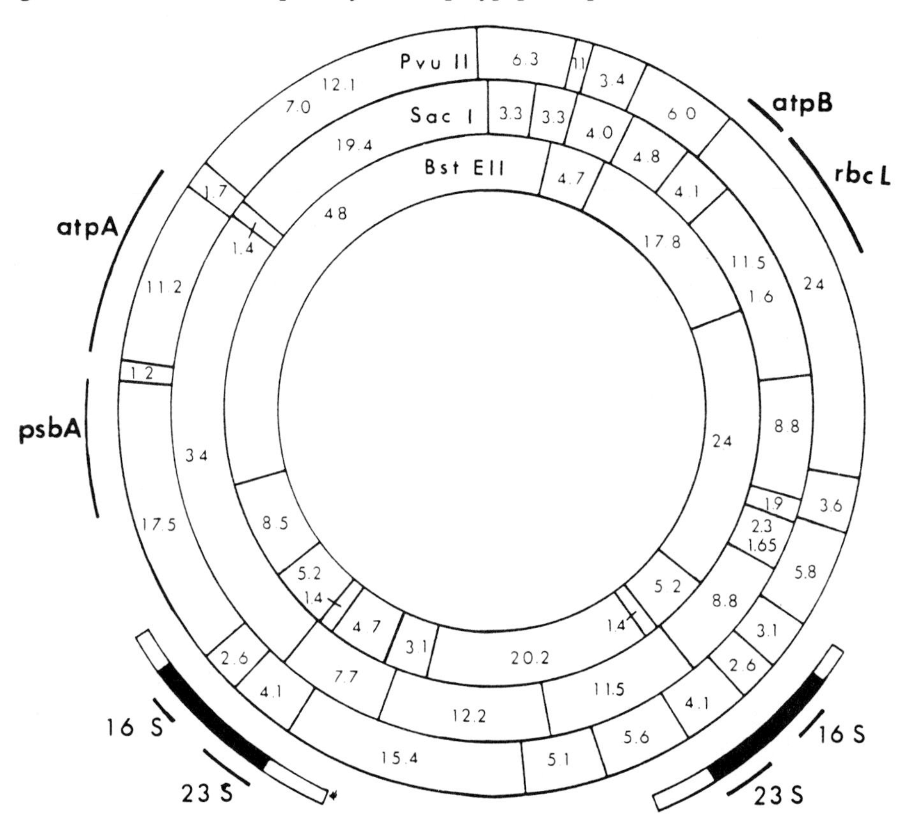


*SOURCE:* From Palmer, 1986.

*Brassica* provides an example of the unique power of chloroplast DNA for documenting the parentage of hybrids and allopolyploids. Two groups of investigators independently analyzed the cpDNA of 6 *Brassica* species in order to elucidate the maternal and paternal parentage of the 3 allopolyploid species whose putative nuclear origins are shown in Fig. 13.9. Both groups were able to assign the maternal progenitors for the allopolyploids, *B. carinata* and *B. juncea*, and *B. nigra* and *B. campestris*, respectively (and, by subtraction, *B. oleracea* and *B. nigra* must have served as the paternal parents in the cross).

Although there is a wealth of cpDNA data at the interspecific level,

**FIGURE 13.9 Classical phylogeny for cultivated and allopolyploid species of *Brassica*.**

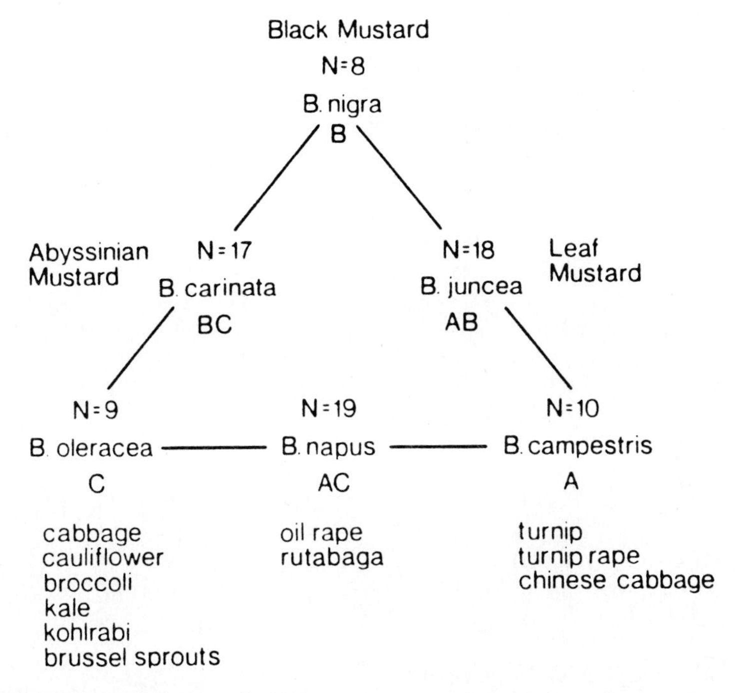


*SOURCE:* From Palmer et al., 1983.

fewer studies have applied cpDNA data to higher-level questions. One type of application involves the use of major structural rearrangements in the chloroplast genome to demarcate major plant groups. In the Asteraceae, for example, it was revealed that there are two chloroplast genome arrangements. Chloroplast DNAs from species in the subtribe Barnadesiinae (Mutisieae) have the same structure as the cpDNAs of almost all land plants. All other Asteraceae share a 22-kilobase inversion, suggesting that Barnadesiinae is the most primitive group in the family and that the inversion is a derived feature that groups all other Asteraceae together. Additional restriction fragment and map comparisons have confirmed this hypothesis.

## *Palynology as Applied to Systematics**

Palynology is a science that deals with the study of the walls of pollen grains and spores, but not with their live interior. More recently, however, much interest has been devoted to the study of the physiological processes

*This section contributed by Cliff Crompton, Biosystematics Research Center, Research Branch, Agriculture Canada, Ottawa.

within these fundamental agents related to sexual and often asexual reproduction in the plant kingdom.

The origin of the word *pollen* is from the Latin, meaning "fine flour," or "fine meal." Pollen grains, as discussed in Chapter 8, are correctly referred to as microspores in flowering plants.

Since the invention of the microscope, pollen grains have interested botanists. Between 1711 and 1832 various terms were applied to pollen grain morphology, such as exine and intime, nucleus and colpi. Indeed, Lindley in 1830 used pollinia morphology as the basis for his taxonomic treatment of the Orchidaceae. With the discovery that many plant taxa can be identified by their pollen grains, the science of pollen analysis became useful to geologists-botanists for explaining vascular plant succession in the postglacial Pleistocene of Europe and North America. This resulted from the fact that pollen, by virtue of their readily preserved walls, were discovered in strata of lake bottoms and bogs in north temperate regions.

One major limiting factor in the study of plant succession and climatology was the lack of an adequate pollen taxonomy. This in turn led to the development of many identification keys and descriptions by scientists for use at the regional level. These works usually had a diagnostic key and were illustrated by pen-and-ink drawings or with photomicrographs.

Another point of interest in the study of pollen was the very important discovery that certain taxa caused allergic reactions in some people. Therefore, knowledge of the kind and type of pollen or spore in the air or sample had medical implications.

### *Pollen Grain Morphology*

To understand the taxonomic implications of the morphology of pollen grain walls, a review of their physiological development, which is influenced by both external and internal processes, is necessary.

Selection pressure and evolution play an important role in the form and function of pollen grains. These factors, therefore, allow researchers to use morphological features of pollen grain walls to identify trends in the phylogeny of various taxa, or groups of related taxa, which subsequently aid us in understanding taxonomical relationships (Fig. 13.10).

### *Development of the Exine*

The angiosperm pollen grain is commonly comprised of a multilayered cell wall. Essentially, the process of exine formation is governed in the cytoplasms by the microtubule structures and their vesicles; subsequently, the vesicles and plasma membrane interact and organize to imprint the protoplast surface. The process is achieved by a layering of the vesicles or by protoplasmic projections. As the process proceeds, a secondary layer, comprised of deposited carbohydrates, displaces the plasma membrane to form and increase the depth of primordial **exine.** This early exine (called sexine by some botanists) is thickened by polymers that are similar to those found in spore walls. This may take place on "membrane lamellae, either walls of vesicles or convoluted membrane structures," or it "may be polymerized between cytoplasmic protrusions rather than into them." Therefore, this process has an effect like a negative photographic print of the protrusion arrangements.

**FIGURE 13.10 Examples of pollen exine morphology as observed by scanning electron microscopy (SEM) in the genus *Lotus* (Fabaceae): (a) *L. agrophyllus,* × 2650; (b) *L. argophyllus,* × 6600; (c) *L. glabra,* × 2000; (d) *L. glabra,* × 5000; (e) *L. junceus,* × 2000; (f) *L. junceus,* × 5000.**

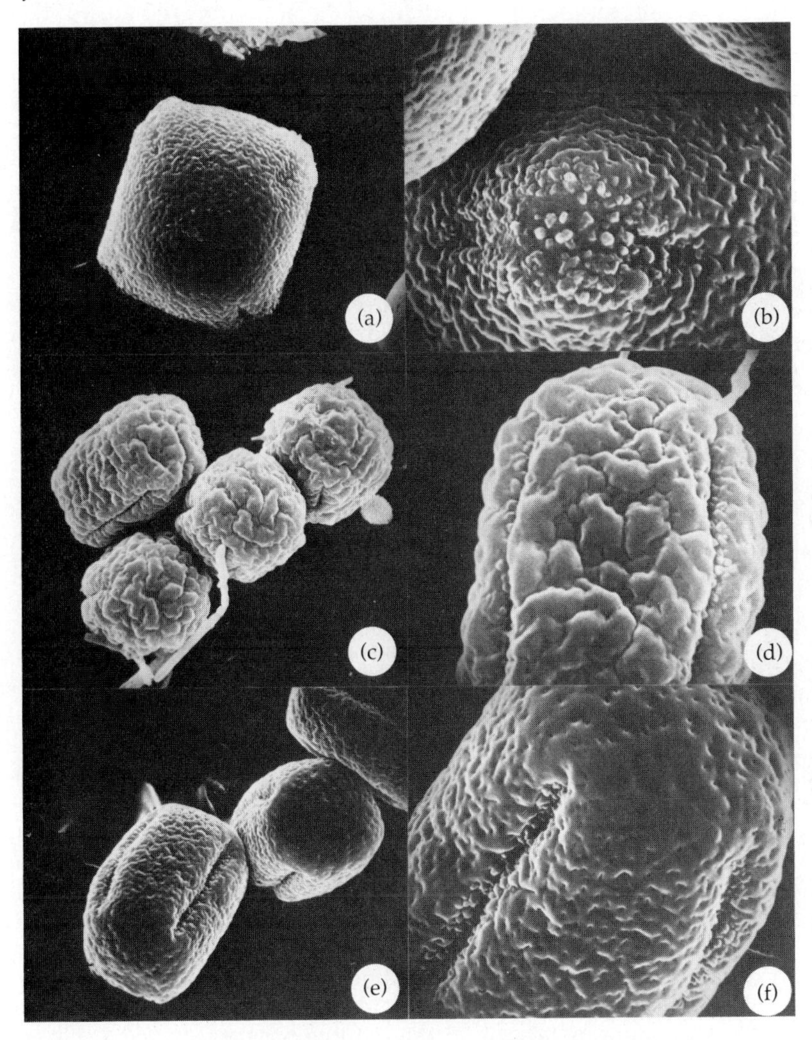


After the pollen grains are released from the pollen mother cell (PMC) tetrad callosic wall, the young exine is augmented from without by sporopollenin produced by the nurse cells of the tapetum. An excellent series of transmission electron microscope (TEM) photomicrographs of these processes in the Saxifragaceae and Euphorbiaceae has been shown.

Ontogenetic botanical studies have revealed the following steps in wall formation:

1. The callose pattern is determined in the cytoplasm by the generation of vesicles or microtubules in association with quantities of the endoplasmic reticulum; if highly formulated processes are to be developed, a massive protrusion is visible.
2. Microtubules and other cell constituents are beneath the protrusion, but the protrusions are too electron dense to be associated with any particular organelle or part of an organelle. Banked arrays of endoplasmic reticulum are formed, and are coated with ribosomes, producing other vesicles whose destinies are to become the developing pollen walls.
3. At the cell surface, outside the plasma membrane, larger bodies appear that stain with lipid stains. These bodies are quickly modified, and they permeate between the protrusions and form a layer over the whole surface, except for **colpus** areas.
4. Several layers are produced prior to pollen release from the PMC tetrad. The pollen wall then expands due to unequal morphogenesis between the two parts of the wall (ektexine) on the outside and the endexine toward the inside. Laminated areas are produced near the **colpi** and the process is completed.

### *Pollen Grain Evolution—Form and Function*

The principle given for the many different pollen morphological types was explained by Klacker (1965), who pointed out that at their surfaces, cells interact with their special environments. Pollen grains are cells developed at the anther interface, namely, the tapetum. This interface is comprised of sugars and proteins. Rowley (1971) informed us in a review of this subject that the composition of glycocalyx is very important in the synthesis of pollen wall architecture, heteropolar systems, dimorphisms, the cell membrane, colpi forms, etc. Rowley summarized the following development: (1) The exine is plastic to allow the alteration of shape; (2) exine modifications require transfer of ions, hence energy is required; (3) margins of germinal apertures expand during development and shrink at maturity; and (4) polybasic molecules form an agglutination bridge between negatively charged sites. Many botanists have linked pollen grain morphology to its form and function or evolution. Nowhere is this more evident than when a person examines certain discernible trends associated or correlated to the pollination mode of species, for example, the viscine threads of Onagraceae pollen, the bladders present on many Pinaceae pollen, the small size and buoyancy of Urticaceae pollen, and the large production of pollen of chasmogamous flowers and the low pollen production of cleistogamous flowers in the Plantaginaceae.

Some botanists feel that general features of pollen grains can now be interpreted as having adaptive significance and can be shown to discharge important biological functions. This means the exine is a repository for sporophytic material. This can be summarized as follows:

1. The pollen grain wall is adapted to carry physiologically active materials derived from the sporophytic parent (the exine domain) and the gametophyte (the intine domain).

2. The incompatibility system is mediated by the sporophytic proteins contained in the exine cavities. Material of similar origin may also play a role in the control of interspecific compatibility.
3. Intine enzymes function in the penetration of the stigma and in the early growth of the pollen tube, and intine proteins may be involved in intraspecific compatibility control.
4. Apertures due to the modified structures present in the exine function somewhat like the stomata of a leaf. They are potential outlets for the pollen tube and for mobile gametophytic chemicals from the intine domain.

It has been found that many anomalous monocotyledenous taxa, such as *Crocus* and *Canna,* do not have these proteins and that some families with reduced exines may not convey tapetal proteins. Therefore, the incompatibility function may be contained in the cytoplasm, the intine, or the ovary.

Indeed, selected species of some legume genera, such as *Trifolium, Lotus,* and *Vicia,* have nonornamented **sculpturing** (patterns on the pollen wall surface) and few cavities. Some of these species are well known to have self-incompatibility problems and sterility complications.

Some interesting comparisons can be made between selected taxa of various "hayfever" plants and using geometry to explain the various configurations, shapes, colpi, and **pore** arrangements of pollen grains. This led to the trichoclassic system of explaining their various morphologies. However, this system does not take most anomalous pollen shapes into account (see Lesins and Lesins, 1979; Small et al., 1981a), particularly in the genus *Medicago.* Some shapes may be explained as being fixed through speciation, or perhaps due to chromosome loss or gain. The pollen grain shapes may be attributed to aneuploidy, or to unreduced or reduced gametes, as shown by multiple spindle (microtubule) formation and irregular division of the pollen grains during and after telophase II of meiosis in the PMCs. These diverse forms may then become fixed characters.

Walker and Doyle (1975) showed a pollen wall-pattern evolution and the apparent divergence of exine trends by comparing pteridophytes, gymnosperms, and primitive angiosperm spores and pollen grains. These trends are summarized as follows: (1) atectate exine = pteridophytes, (2) acectate-granular-incipient-alveolaet-alveolate exine = gymnosperms, (3) acectate-granular-incipient columellate-tectate columellate exine = angiosperms (see Fig. 13.10).

### *Pollen Grain Classifications*

There are, in essence, two main systems of classifying and identifying pollen grains. One is a numerical classification in which pollen grains are categorized on the basis of their pore number, colpi number, and position of colpi or pores. The other system is based on comparing the pollen grains to 22 classes of characterizations (palynomorphs). Both systems are complementary and useful to the pollen taxonomist or pollen identifier.

Unfortunately, the keys are only useful for pollen of north temperate, South American, and Southeast Asian taxa. In addition, the pollen grains of many taxa are very similar in morphology and often are not keyed out to the species level or included in the treatment. Therefore, keys are often inadequate, emphasizing that pollen taxonomies and special keys to many groups are inadequate and incomplete. A world pollen flora is presently being produced, and it will include palynological accounts of various families at diverse taxonomic levels.

## *Cytology and Genetics as Applied to Systematics*

During plant cell division a number of rod- or thread-like bodies become visual in a contracted form and hence are known as **chromosomes.** These bodies carry the genetic units of heredity, called the **genes,** for the plant. The chromosomes have a strong affinity for certain stains, thereby allowing for various information to be obtained: number, shape and size, and behavior during different stages of mitosis or meiosis.

### *Chromosome Number and Morphology*

Study into the cytological nature of plants began early in the twentieth century. Early investigation revealed that much variation existed in the number and morphology of chromosomes present in species. Sometimes the variation was characteristic of different taxonomic subspecies or varieties. The development of specific, easy, quick-staining procedures made cytological information available for taxonomic purposes. For systematic studies today it is customary to use differential staining techniques of root tips or pollen mother cells (PMCs) from the developing anther for microscopic study. The resulting **karyotype** is the appearance of the somatic complement of chromosomes of a species at the metaphase of mitosis. The investigator can observe karyotype and number differences and then use the resulting characters in applied systematic conclusions. For example, Fig. 13.11 shows the karyotypes of 8 species of the genus *Crepis* in the Asteraceae (Babcock, 1947). Observed is the variation in number ($x$ = 3, 4, 5, and 6), size, and morphological differences of the chromosomes. (In the study of cytology, $x$ refers to the lowest or base chromosome number of a species, $n$ refers to the haploid number of chromosomes, $2n$ refers to the diploid number, $3n$ to the triploid, etc.)

### *Polyploidy*

Plants with cells having two or more complete sets of chromosomes are called **polyploids.** Polyploid cells arise as a result of difficulty during mitosis or meiosis in which the chromosomes divide but the cell does not. Therefore, a cell with twice the normal chromosome number is produced. Cells continuing through the various stages of division can give rise to new individuals with twice the chromosome number as the original parents (i.e., original, $2n = 13$; polyploid, $2n = 26$). Complete doubling of the entire haploid chromosome compliment of a species is termed **autopolyploidy** (also called **autoploidy**). Autopolyploid individuals can be produced artificially by treating plant cells, etc. with certain chemicals such as colchicine; cold treatment; etc. The chemical or cold treatment at the right time disrupts spindle formation

**FIGURE 13.11 Diagrammatic sketches of karyotypes of eight species of *Crepis* (Asteraceae) showing variations in basic number, shape, and size of chromosomes.**

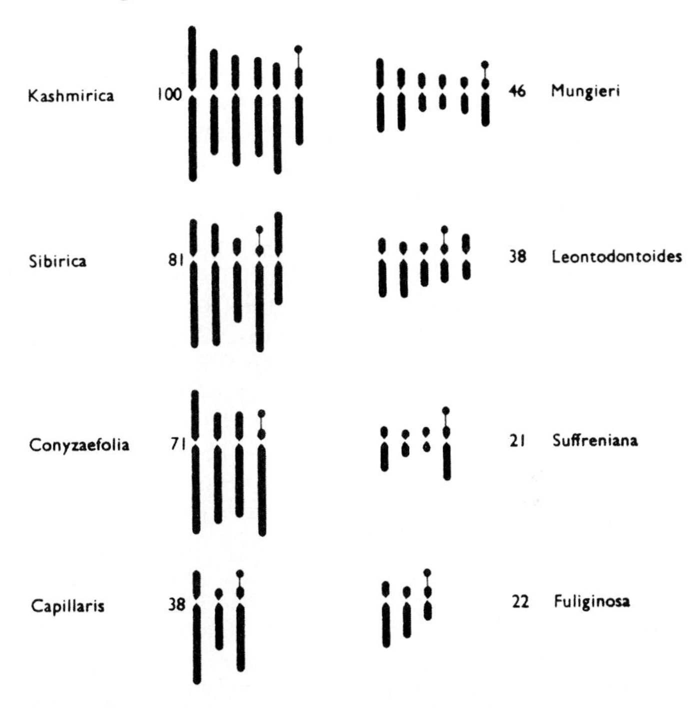


*SOURCE:* From Babcock, 1947.

during mitosis or meiosis, allowing for twice the number of chromosomes in the resulting cells.

The range of variation in polyploids is often less than that found in the parental diploids because the genes are present twice as often. Many polyploids, however, have larger flowers and phenotypic features, such as increased leaf size, and are often self-pollinated, even when the related diploids are generally cross-pollinated.

Many polyploids originate through hybridization between different diploid populations or taxa. The resulting hybrid polyploid is termed an **allopolyploid** (also called **alloploid**) in that one or more sets of chromosomes come from different species or different genetic strains. The resulting alloploid may be relatively fertile and may reproduce asexually only. Sometimes polyploids of hybrid origin have arisen by the doubling of chromosomes in sterile diploid or triploid hybrids. The resulting chromosome doubling may then restore the plant's fertility. An example of this is observed in the grass genus *Spartina,* which grows in salt marsh habitats. *Spartina maritima,* a native species in Great Britain, has 60 chromosomes (somatic number of $2n = 60$) and inhabits coastal salt marshes, while *S. alterniflora* is native to North America and has 62 chromosomes. The North American *S. alterniflora* was first collected in Great Britain in 1839, with a sterile hybrid *Spartina* × *townsendii* (that reproduced asexually by rhizomes) collected at the same locality in 1870. In the early 1890s, a fertile vigorous polyploid, *S. anglica,* was collected that has 122 chromosomes. This new polyploid taxon has now spread throughout other salt marshes in Great Britain and other European countries. Figure 13.12 illustrates how a polyploid complex could develop from four diploid taxa.

Another type of polyploidy occurs in which chromosome number differences vary in the presence or absence of a single chromosome or a few chromosomes instead of a whole chromosome set. This is called **aneuploidy.** For example, in the genus *Vicia* are found chromosomes revolving around base numbers of $x = 5$, 6, or 7, with diploid and tetraploid numbers of $2n = 10$, 12, 14, 24, and 28. The addition of or subtraction of a chromosome from a chromosome set of the diploid level is usually harmful to the plant, whereas if it occurs at a higher level this abnormality is more tolerated. The normal diploid fertile plant is called a **disomic,** while a plant having a diploid stage (that is one chromosome present three times) is called a **trisomic.** A plant lacking one chromosome is termed **monosomic.**

Polyploids in nature are selected by the environment, and therefore polyploidy is an important evolutionary mechanism and is of interest to students of systematics. Many important economic crops are polyploids, such as, bananas, cotton, peppers, potatoes, sugar cane, tobacco, and wheat, as well as lovely garden plants such as chrysanthemums, day lilies, and *Impatiens.*

### *Hybridization*

During our early understanding of plant species in nature, it was thought that crossing between two different species was a rare occurrence. However, it is realized today that species hybridize much more commonly than thought, and therefore hybrids are important for recombining parental characteristics.

Environments and habitats are constantly changing, and therefore individuals formed by hybrid origin may be better adapted to colonize these new environments than the original parents. It is known that the recombination of two species' shared genetic material has a greater potential for producing offspring than do gradual changes within single populations and taxa, and that these offspring are better able to flourish in new habitats. Two examples will be given. The North American sycamore tree, *Platanus occidentalis,* is naturally distributed throughout eastern North America in moist woodland habitats. It was introduced into the colder regions of Europe. The oriental plane tree, *P. orientalis,* is native to the eastern Mediterranean region, extending east to the Himalayas, and has been cultivated for many centuries throughout southern Europe in less cold climates. Toward the end of the seventeenth century (approximately 1670), hybrids were noted that were intermediate and fertile. The resulting London plane tree, *Platanus acerifolia,* was very rigorous in growth, more cold hardy than either parent, and has shown a greater tolerance as a street tree toward urban pollution. The second example is centered with the genus *Penstemon,* the beard tongues. In southern California, *P. grinnellii,* mountain penstemon, has pale blue, open two-lipped corollas, is pollinated by large carpenter bees, and is found in pine forests. A second species, *P. centranthifolius,* scarlet bugler, has scarlet, long tubular corollas, is pollinated by hummingbirds, and is widely distributed in drier habitats. A third species, *P. spectabilis,* showy penstemon, is intermediate

FIGURE 13.12 Flow chart showing various types of polyploidy for four hypothetical fertile diploid species: AA, BB, CC, and DD. Taxa surrounded by boxes are completely fertile. Taxa lacking boxes are at various stages of sterility. Pairings of chromosomes during meiosis are given in brackets. The chromosome level and number are given at the left. The base chromosome number ($x$) is 8.

Haploid: $n = x = 8$

Diploid: $2n = 2x = 16$

Triploids: $2n = 3x = 24$

Tetraploids: $2n = 4x = 32$

Pentaploids: $2n = 5x = 40$

Hexaploids: $2n = 6x = 48$

Heptaploids: $2n = 7x = 56$

Octaploids: $2n = 8x = 64$

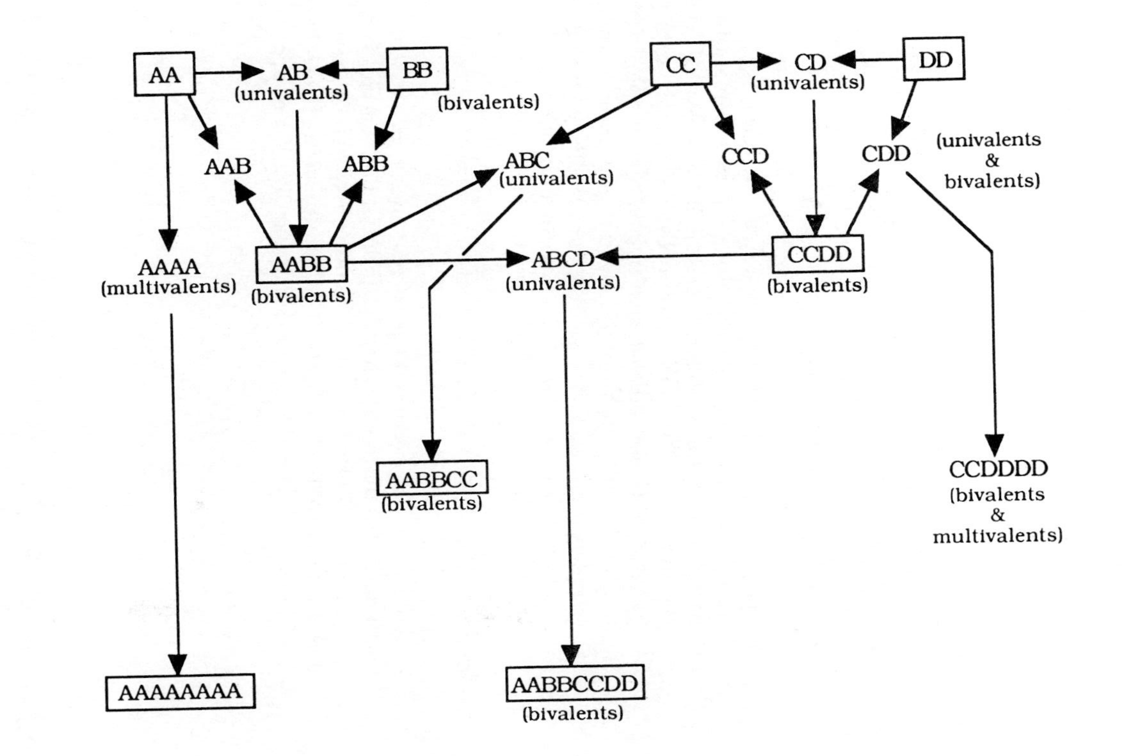

between the mountain penstemon and the scarlet bugler by having blue to purplish flowers with intermediate-shaped corollas. This putative hybrid species is found on brush-covered foothills, a recent habitat between the habitats of the mountain penstemon and scarlet bugler, and is pollinated by solitary wasps. A fourth recent hybrid species, *P. parishii,* has been observed to be a backcross hybrid between the scarlet bugler and the showy penstemon. It has bluish-purple flowers, intermediate in shape, is pollinated by a different insect, and inhabits dry road cuts (Fig. 13.13).

What is illustrated by *Penstemon* is how two environmentally isolated and morphologically distinct species have given rise to other species through hybridization in the past, when the environment in southern California began to change from more forest-covered mountains and drier slopes to include more brush-covered foothills. With the availability of a new brushy habitat or dry road cuts, some *Penstemon* hybrids were able to move into new environments that were frequented by different pollinators. **Natural selection,** therefore, brought about a stabilization of the gene combinations, adapting the phenotype of the plant to survive well in the new habitats. What is seen here is the evolution of new species through hybridization, radiation of new species into new habitats, and isolation of these new species by ecological and pollinator barriers.

**FIGURE 13.13 Interrelationships of four species of *Penstemon* (Scrophulariaceae), beard tongue, found in Southern California. Each species is found in different habitats and pollinated by specific pollinators. Two species, (c) and (d), were derived by hybridization between (a) and (b) and between (b) and (c), respectively, followed by isolation and stabilization of populations.**

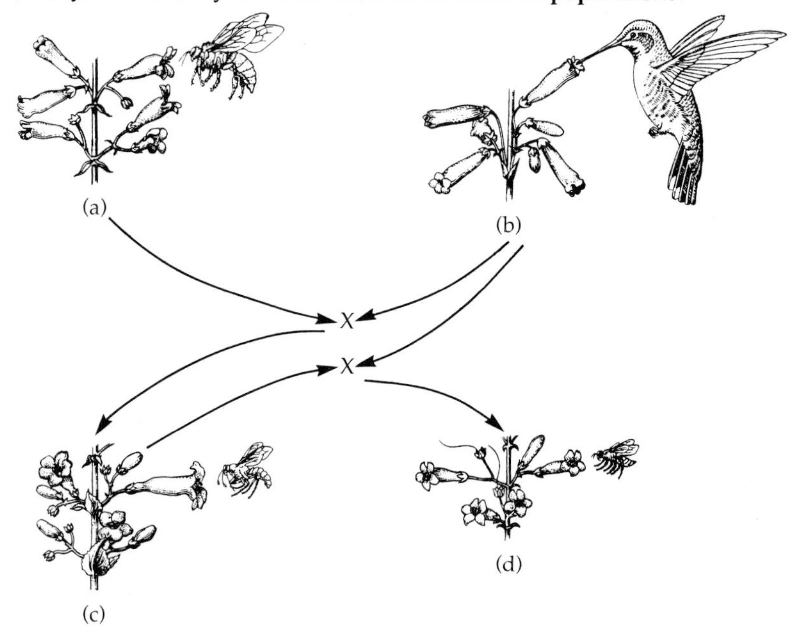


*SOURCE:* From G. Ledyard Stebbins, *Processes of Organic Evolution,* 3rd ed. © 1977, p. 121. Reprinted by permission of Prentice-Hall, Englewood Cliffs, New Jersey.

## SELECTED REFERENCES

### General

Stuessy, T. F. 1990. *Plant Taxonomy*. Columbia Univ. Press, New York, 514 pp.

### Plant Anatomy

Bailey, I. W. 1957. The Potentialities and Limitations of Wood Anatomy in the Study of the Phylogeny and Classification of Angiosperms. *J. Arnold Arbor.* 38:243–254.

Barthlot, W. 1981. Epidermal and Seed Surface Characters of Plants: Systematic Applicability and Some Evolutionary Aspects. *Nordic J. Bot.* 1:345–355.

Behnke, H. D. and W. Barthlot. 1983. New Evidence from the Ultrastructural and Micromorphological Fields in Angiosperm Classification. *Nordic J. Bot.* 3:43–66.

Brisson, J. D. and R. L. Peterson. 1977. The Scanning Electron Microscope and X-Ray Microanalysis in the Study of Seeds: A Bibliography Covering the Period of 1967–76. *Scann. Electron Microsc.* (IIT Research Inst., Chicago) 2:697–712.

Carlquist, S. 1961. *Comparative Plant Anatomy*. Holt, Rinehart & Winston, New York, 146 pp.

Corner, E. J. H. 1976. *The Seeds of Dicotyledons*. 2 vols. Cambridge Univ. Press, Cambridge.

Cotthem, W. R. J. van. 1970. A Classification of Stomatal Types. *J. Linn. Soc., Bot.* 63:235–246.

Cronquist, A. 1981. *An Integrated System of Classification of Flowering Plants*. Columbia Univ. Press, New York, 1262 pp.

Cutler, D. F., K. L. Alvin, and C. E. Price (eds.), 1982. *The Plant Cuticle*. Academic Press, London.

Esau, K. 1965. *Plant Anatomy*, 2nd ed. John Wiley & Sons, New York.

Heywood, V. H. ed. 1971. *Scanning Electron Microscopy. Systematic and Evolutionary Applications*. Academic Press, London.

Melville, R. 1976. The Terminology of Leaf Architecture. *Taxon* 25:549–561.

Metcalfe, C. R. 1960. *Anatomy of the Monocotyledons. I. Gramineae*. Oxford Univ. Press, Oxford.

Metcalfe, C. R. 1960. Current Developments in Systematic Plant Anatomy In: V. H. Heywood (ed.), *Modern Methods in Plant Taxonomy*, Academic Press, London, pp. 45–57.

Metcalfe, C. R. 1971. *Anatomy of the Monocotyledons. V. Cyperaceae*. Clarendon Press, Oxford, 597 pp.

Metcalfe, C. R. and L. Chalk. 1950. *Anatomy of the Dicotyledons*. Oxford Univ. Press, Oxford.

Metcalfe, C. R. and L. Chalk. 1979. *Anatomy of the Dicotyledons*, 2nd ed. Vol. I. Oxford Univ. Press, Oxford.

Palser, B. F. (1975, 1976). The Bases of Angiosperm Phylogeny: Embryology. *Ann. Missouri Bot. Gard.* 62:621–646.

Patel, J. D. 1979. A New Morphological Classification of Stomatal Complexes. *Phytomorphology* 29:218–229.

Payne, W. W. 1978. A Glossary of Plant Hair Terminology. *Brittonia* 30:239–255.

Payne, W. W. 1979. Stomatal Patterns in Embryophytes: Their Evolution, Ontogeny and Classification. *Taxon* 28:117–132.

Philipson, W. R. 1977. Ovular Morphology and the Classification of Dicotyledons. *Pl. Syst. Evol.* Suppl. 1:123–140.

Rasmussen, H. 1981. Terminology and Classification of Stomata and Stomatal Development—A Critical Survey. *J. Linn. Soc., Bot.* 83:199–212.

Roth, I. 1977. Fruits of Angiosperms. In: *Handbuch der Pflanzenanatomie,* Vol. IX. (1). Borntraeger, Berlin, Stuttgart, 675 pp.
Roth, I. 1981. Structural Patterns of Tropical Barks. In: *Handbuch der Pflanzenanatomie,* Vol. IX. (3). Borntraeger, Berlin, Stuttgart.
Singh, H. 1978. Embryology of Gymnosperms. In: *Handbuch der Pflanzenanatomie,* Vol. X. (2). Borntraeger, Berlin, Stuttgart, 302 pp.
Sprone, K. R. 1977. Some Problems Associated with Character Correlations. *Pl. Syst. Evol.* Suppl. 1:33–51.
Stace, C. A. 1966. The Use of Epidermal Characters in Phylogenetic Considerations. *New Phytol.* 65:304–318.
Tomlinson, P. B. 1969. *Anatomy of the Monocotyledons. III. Commelinales–Zingiberales.* Clarendon Press, Oxford, 446 pp.
Tomlinson, P. B. 1974. Development of the Stomatal Complex As a Taxonomic Character in the Monocotyledons. *Taxon* 23:109–128.
Tomlinson, P. B. 1982. *Anatomy of the Monocotyledons. VII. Helobiae.* Oxford Univ. Press, London.
Wagner, W. H. Jr. 1979. Reticulate Veins in the Systematics of Modern Ferns. *Taxon* 28:87–95.
Zimmermann, M. H. and P. B. Tomlinson. 1972. The Vascular System of Monocotyledonous Stems. *Bot Gaz.* 133:141–155.

## *Plant Morphology*

Cooney-Sovetts, C. and R. Sattler. 1987. Phylloclade Development in the Asparagaceae: An Example of Homeosis. *J. Linn. Soc. Bot.* 94:327–371.
Croizat, L. 1962. *Space, Time, Form: The Biological Synthesis.* Caracas, published by the author.
Cusset, G. 1982. The Conceptual Bases of Plant Morphology. *Acta Biotheor.* 31A:8–86. (Also in R. Sattler (ed.) *Axioms and Principles of Plant Construction,* Nijhoff/Junk Publ., The Hague, pp. 8–86.)
Eyde, R. H. 1975. The Foliar Theory of the Flower. *Amer. Sci.* 63:430–437.
Guédès, M. 1979. *Morphology of Seed-Plants.* Cramer, Vaduz.
Hagemann, W. 1976. Sind Farne Kormophyten? Eine Alternative zur Telomtheorie. *Pl. Syst. Evol.* 124:251–277.
Hagemann, W. 1984. Morphological Aspects of Leaf Development in Ferns and Angiosperms. In: R. A. White and W. D. Dickison (eds.), *Contemporary Problems in Plant Anatomy.* Academic Press, New York, pp. 301–350.
Howard, R. A. 1974. The Stem-Node-Leaf Continuum of the Dicotyledoneae. *J. Arnold Arbor.* 55:125–181.
Kaplan, D. R. 1970. Comparative Foliar Histogenesis in *Acorus calamus* and its Bearing on the Phyllode Theory of Monocotyledonous Leaves. *Amer. J. Bot.* 57:331–361.
Komaki, M. K., K. Okada, E. Nishino, and Y. Shimura. 1988. Isolation and Characterization of Novel Mutants of *Arabidopsis thaliana* Defective in Flower Development. *Development* 104:195–203.
Leins, P., S. C. Tucker, and P. K. Endress (eds.), 1988. *Aspects of Floral Development.* Cramer, Berlin, Stuttgart.
Meyen. S. V. 1987. *Fundamentals of Paleobotany.* Chapman and Hall, London.
Rutishauser, R. and R. Sattler, 1985. Complementarity and Heuristic Value of Contrasting Models in Structural Botany. I. General Consideration. *Bot. Jahrb. Syst.* 107:415–455.
Rutishauser, R. and R. Sattler. 1989. Complementarity and Heuristic Value of Contrasting Models in Structural Botany. III. Case Study of Shoot-Like "Leaves"

and Leaf-Like "Shoots" in *Utricularia macrorhiza* and *U. purpurea* (Lentibulariaceae). *Bot. Jahrb. Syst.* 111:121–137.
Sattler, R. 1973. *Organogenesis of Flowers.* A Photographic Text-Atlas. Univ. of Toronto Press, Toronto.
Sattler, R. 1984. Homology—A Continuing Challenge. *Syst. Bot.* 9:382–394.
Sattler, R. 1986. *Biophilosophy. Analytic and Holistic Perspectives.* Springer-Verlag, Berlin.
Sattler, R. 1988a. Homeosis in Plants. *Amer. J. Bot.* 75:1606–1617.
Sattler, R. 1988b. A Dynamic Multidimensional Approach to Floral Morphology. In: P. Leins, S. C. Tucker, and P. K. Endress (eds.). *Aspects of Floral Development.* Cramer, Berlin, Stuttgart, pp. 1–6.
Steingraeber, D. A. and J. F. Fisher. 1986. Indeterminate Growth of Leaves in *Guarea* (Meliaceae): A Twig Analogue. *Amer. J. Bot.* 73:852–862.
Stevens, P. F. 1984. Homology and Phylogeny: Morphology and Systematics. *Syst. Bot.* 9:395–409.
Stewart, W. N. 1964. An Upward Outlook in Plant Morphology. *Phytomorphology* 14:120–134.
Troll, W. 1937–1943. *Vergleichende Morphologie der Höheren Pflanzen.* 3 vols. Borntraeger, Berlin (Reprint: Koeltz, Koenigstein, 1967–1968; Register by I. and A. Siegert, Koeltz, Koenigstein, 1971).
White, J. 1984. Plant Metamerism. In: R. Dirzo and J. Sarukhan (eds.) *Perspectives on Plant Population Ecology.* Sinauer, Sunderland, MA, pp. 15–47.
Zimmermann, W. 1959. *Die Phylogenie der Pflanzen,* 2nd ed. Fischer, Verlag, Stuttgart.

## *Chemosystematics*

Alston, R. E. 1967. Biochemical Systematics. In: T. Dobzhansky, M. K. Heckt, and W. C. Steere (eds.). *Evolutionary Biology.* Vol. I. Appleton Century Crofts, New York, pp. 197–305.
Alston, R. E. and K. Hempel. 1964. Chemical Documentation of Interspecific Hybridization. *J. Heredity* 55:267–269.
Alston, R. E. and B. L. Turner. 1963. *Biochemical Systematics.* Prentice-Hall, Englewood Cliffs, NJ, 404 pp.
Alston, R. E. and B. L. Turner, 1963. Natural Hybridization Among Four Species of *Baptisia* (Leguminosae). *Amer. J. Bot.* 50:159–173.
Crawford, D. J. 1983. Phylogenetic and Systematic Inferences from Electrophoretic Studies. In: S. D. Tanksley and T. J. Orton (eds.). *Isozymes in Plant Genetics and Breeding.* Elsevier, Amsterdam, pp. 257–287.
Crawford, D. J. 1985. Electrophoretic Data and Plant Speciation. *Syst. Bot.* 10:405–416.
Crawford, D. J. and D. E. Giannasi. 1982. Plant Chemosystematics. *BioScience* 32:114–125.
Erikson, L. R., N. A. Straus, and W. D. Beversdorf. 1983. Restriction Patterns Reveal Origins of Chloroplast Genomes in *Brassica* Amphidiploids. *Theor. Appl. Genet.* 65:201–206.
Giannasi, D. E. and D. J. Crawford. 1982. *Chemosystematics of Plants.* John Wiley, New York.
Gibbs, R. D. 1974. *Chemotaxonomy of Flowering Plants.* Vols. I–IV. McGill-Queen's Univ. Press, Montreal, 2372 pp.
Gottlieb, L. D. 1981. Electrophoretic Evidence and Plant Populations. *Prog. Phytochem.* 7:1–46.
Harborne, J. B. and B. L. Turner. 1984. *Plant Chemosystematics.* Academic Press, London.
Hawkes, J. G. (ed.), 1968. *Chemotaxonomy and Serotaxonomy.* The Systematics Assoc. Spec. Vol. No. 2. Academic Press, London, 299 pp.

Hills, D. M. and C. Moritz (eds.). 1990. *Molecular Systematics.* Sinauer Associates, Sunderland, MA, 588 pp.

Jansen, R. K. and J. D. Palmer. 1987. A Chloroplast DNA Inversion Marks an Ancient Evolutionary Split in the Sunflower Family (Asteraceae). *Proc. Natl. Acad. Sci. USA* 84:5818–5822.

Jansen, R. K. and J. D. Palmer. 1988. Phylogenetic Implications of Chloroplast DNA Restriction Site Variation in the Mutisieae (Asteraceae). *Amer. J. Bot.* 75:753–766.

Mabry, T. J. 1977. The Order Centrospermae. *Ann. Missouri Bot. Gard.* 64:210–220.

Mabry, T. J., K. R. Markham, and M. B. Thomas. 1970. *The Systematic Identification of Flavonoides.* Springer-Verlag, New York, 354 pp.

McClure, J. W. 1975. Physiology and Functions of Flavonoids. In: J. B. Harborne, T. J. Mabry, and H. Mabry. (eds.). *The Flavonoids.* Academic Press, New York, pp. 970–1055.

Oxford, G. S. and D. Rollinson eds. 1983. *Protein Polymorphism: Adaptive and Taxonomic Significance.* The Systematics Assoc. Special Vol. 24. Academic Press, London, 405 pp.

Palmer, J. D. 1986. Chloroplast DNA and Phylogenetic Relationships. In S. K. Dutta (ed.). *DNA Systematics.* Vol. II. CRC Press, Boca Raton, FL, pp. 63–80.

Palmer, J. D., C. R. Shields, D. B. Cohen, and T. J. Orten. 1983. Chloroplast DNA Evolution and the Origin of Amphidiploid *Brassica* Species. *Theor. Appl. Genet.* 65:181–189.

Roose, M. L. and L. D. Gottlieb. 1976. Genetic and Biochemical Consequences of Polyploidy in *Tragopogon. Evolution* 30:818–830.

Smith, P. M. 1976. *The Chemotaxonomy of Plants.* Edward Arnold Publishers, London.

Snajberk, K. and E. Zavarin. 1976. Mono- and Sesquiterpenoid Differentiation of *Pseudotsuga* of the United States. *Biochem. Syst. Ecol.* 4:159–163.

Soltis, D. E. and P. S. Soltis (eds.). 1989. *Isozymes in Plant Biology.* Adv. Pl. Sciences Ser. Vol. 4. Dioscorides Press, Portland, OR.

von Rudloff, E. 1972. Chemosystematic Studies of the Genus *Pseudotsuga.* I. Leaf Oil Analysis of the Coastal and Rocky Mountain Varieties of Douglas Fir. *Canad. J. Bot.* 50:1025–1040.

von Rudloff, E. 1975. Volatile Oil Analysis in Chemosystematic Studies of North American Conifers. *Biochem Syst. Ecol.* 2:131–167.

Zuckerkandl, F. and L. Pauling. 1965. Molecules as Documents of Evolutionary History. *J. Theor. Biol.* 8:357–366.

## Palynology

Audran, J. C. and M. T. M. Willemse. 1982. Wall development and its Autofluorescence of Sterile and Fertile *Vicia faba* L. pollen. *Protoplasma* 110:106–111.

Bassett, I. J., C. W. Crompton, and J. A. Parmalee. 1978. *An Atlas of Airborne Pollen Grains and Common Fungus Spores of Canada.* Research Branch Can. Agric. Monogr. 18, 321 pp.

Brooks, J. et al. (eds.). 1971. *Sporopollenin.* Academic Press, London, 718 pp.

Dickinson, H. G. 1976. Common Factors in Exine Deposition. In: I. K. Ferguson and J. Muller (eds.). *The Evolutionary Significance of the Exine.* Linnaen Society Symposium Series No. 1. Academic Press, London, 591 pp.

Dobrofsky, S. and W. F. Grant, 1980. Electrophoretic Evidence Supporting Self-incompatibility in *Lotus cornicalatus. Canad. J. Bot.* 58:712–716.

Echlin, P. and H. Godwin. 1968. The Ultrastructure of Pollen in *Helleborus.* I. The Development of the Tapetum. *J. Cell Sci.* 3:161–174.

Erdtman, G. 1960. The Acetolysis Method. A Revised Description. *Svensk Bot. Tidskr.* 54:561–564.
Erdtman, G. 1964. Palynology. In: W. B. Turrill (ed.). *Vistas in Botany,* Vol. 4, Macmillan, New York, 314 pp.
Erdtman, G. 1966. *Pollen Morphology and Plant Taxonomy.* Hafner, New York, 553 pp.
Faegri, K. and J. Iverson. 1964. *Textbook of Pollen Analysis.* Blackwell Scientific, Oxford, 237 pp.
Ferguson, I. K. and J. Muller (eds.). 1976. *The Evolutionary Significance of the Exine.* Linnean Society Symposium Series No. 1 Academic Press, London, 591 pp.
Heslop-Harrison J. (ed.) 1971. *Pollen Development and Physiology.* Butterworth, London, 333 pp.
Hyde, J. 1944. Pollen Analysis and the Museums. *Museums J.* 44:145–149.
Ikuse, M. 1956. *Pollen Grains of Japan.* Hirokawa, Tokyo, 303 pp.
Kremp, G. O. W. 1956. *Morphologic Encyclopedia of Palynology.* Univ. Arizona Press, Tucson, AZ, 189 pp.
Linskins, H. F. 1963. *Pollen Physiology and Fertilization.* North-Holland, Amsterdam, 257 pp.
McAndrews, J. H., A. A. Berti, and G. Norris. 1973. *Key to the Guarternary Pollen and Spores of the Great Lakes Region.* Life Sci. Misc. Publ. Royal Ontario Museum.
Muller, J. 1979. Form and Function in Angiosperm Pollen. *Ann. Mo. Bot. Gard.* 66:593–632.
Nowicke, J. and J. J. Skvarla. 1979. Pollen Morphology: The Potential Influence in Higher Order Systematics. *Ann. Mo. Bot. Gard.* 66:633–700.
Skvarla, J. J. 1966. Fine Structural Studies of *Zea mays* pollen I. Cell Membranes and Exine Ontogeny. *Amer. J. Bot.* 53:1112–1125.
Skvarla, J. J. and B. L. Turner. 1971. Fine Structure of the Pollen of *Anthemis nobilis* L. (Anthemideae-Compositae). *Proc. Okla. Acad. Sci.* 51:61–62.
Small, E., I. J. Bassett, and C. W. Crompton. 1981. Pollen Variation in Tribe Trigonelleae (Leguminosae) with special reference to *Medicago. Pollen et Spores* 23:295–320.
Stanley, R. G. and H. F. Linskens. 1974. *Pollen Biology, Biochemistry, Management.* Springer-Verlag, Berlin, 307 pp.
Thanikaimoni, G. 1972. *Index bibliographies sur la morphologie des pollens d'angiospermes.* All India Press, Pondichery, 164 pp.
Thanikaimoni, G. 1973. *Index bibliographies sur la morphologie des pollens d'angiospermes.* Supplement 1. Sri Aurobindo Ashram Press, Pondichery, 386 pp.
Thanikaimoni, G. 1976. *Index bibliographies sur la morphologie des pollens d'angiospermes.* Supplement 2. All India Press, Pondichery, 337 pp.
Walker, J. W. and J. A. Doyle. 1975. The Bases of Angiosperm Phylogeny. *Ann. Missouri Bot. Gard.* 62:664–723.
Wodehouse, R. P. 1935. *Pollen Grains.* McGraw-Hill, New York, 574 pp.
Wodehouse, R. P. 1972. *Hay Fever Plants,* 2nd rev. ed. Hafner Press, New York, 280 pp.

## Cytology and Genetics

Brandham, P. E. and M. D. Bennett (eds.). 1983. *Kew Chromosome Conference II.* Allen & Unwin, London.
Darlington, C. D. 1963. *Chromosome Botany and the Origins of Cultivated Plants,* 2nd ed. George Allen & Unwin, London, 231 pp.
Darlington, C. D. and L. F. LaCour. 1962. *The Handling of Chromosomes,* 4th ed. George Allen & Unwin, London, 263 pp.

Davis, P. H. and V. H. Heywood. 1963. *Principles of Angiosperm Taxonomy*. Van Nostrand, Princeton, NJ, 556 pp.
Grant, V. 1981. *Plant Speciation*. Columbia Univ. Press, New York.
Grant, W. F. ed. 1984. *Plant Biosystematics*. Academic Press, Orlando, FL, 674 pp.
Jackson, R. C. 1971. The Karyotype in Systematics. *Ann. Rev. Ecol. Syst.* 2:327–368.
Jackson, R. C. and D. P. Hauber. 1983. *Polyploidy*. Hutchinson Ross, Stroudsburg, PA.
Lewis, W. H. 1970. Chromosomal Drift, A New Phenomenon in Plants. *Science* 168:1115–1116.
Lewis, W. H. (ed.). 1980. *Polyploidy: Biological Relevance*. Plenum Press, New York, 594 pp.
Löve, A. and D. Löve. 1975. *Plant Chromosomes*. J. Cramer, Vaduz.
King, R. C. 1968. *A Dictionary of Genetics*. Oxford Univ. Press, New York, 291 pp.
Sharma, A. K. and A. Sharma. 1980. *Chromosome Techniques*. Butterworths, London.
Sinha, U. and S. Sinha. 1980. *Cytogenetics, Plant Breeding and Evolution*, 2nd ed. Vikas, Sahibabad, India.
Solbrig, O. T. 1970. *Principles and Methods of Plant Biosystematics*. Macmillan, London, 226 pp.
Stace, C. A. 1989. *Plant Taxonomy and Biosystematics*, 2nd ed. Edward Arnold, London, 264 pp.
Stebbins, G. L. 1950. *Variation and Evolution in Plants*. Columbia Univ. Press, New York, 643 pp.
Stebbins, G. L. 1959. The Role of Hybridization in Evolution. *Proc. Amer. Phil. Soc.* 103:231–251.
Stebbins, G. L. 1971. *Chromosomal Evolution in Higher Plants*. Addison-Wesley, Reading, MA, 216 pp.
Stuessy, T. F. 1990. *Plant Taxonomy*. Columbia Univ. Press, New York, 514 pp.

# 14

# Endangered and Threatened Species

Less than 25 years ago, few botanists and still fewer members of the general public were concerned about whether some plant species were becoming less common. There were some well-known animals that were disappearing, such as the grizzly bear and timber wolf in the continental United States, the giant panda in China, the snow leopard in the Himalayas, the California condor, and many species of whales. But plants? "Who cares! As long as there is something green growing, that's all that matters, isn't it?" Should students of systematics be concerned with this problem?

In 1966, the U.S. Congress passed the first Endangered Species Act. This law was designed to determine the extent of disappearing plants and animals, not to just protect them. The law specified that a list be made of the endangered species, with ranges of each and an estimate of the number of individuals remaining. The new law defined an **endangered species** (Table 14.1, p. 462) as a taxon in danger of becoming extinct in the near future within all or part of its geographical range if the same causal factors continue to operate. An **extinct species** is a taxon that no longer exists in the wild within all or part of its geographical range. In 1973, the law was strengthened with various amendments. It defined what is a threatened species and added the endangered species category to the list. **Threatened** (or **vulnerable**) **species** are taxa that are thought to be likely to become endangered in the wild in the near future within all or part of their geographical distribution if the same causal factors continue to operate. This law went on to state that the actions of agencies of the federal government must not "jeopardize the continued existence of endangered and threatened species or result in the destruction or modification of habitat of such species which is determined to be critical."

When this act was debated by the politicians, many Congressional representatives were probably thinking thoughts of brown-eyed furry animals and soaring, winged creatures when they cast their vote. In the early 1970s, however, the halted construction of two hydro dam projects made many politicians, business persons,

**TABLE 14.1 Terms used in the study of endangered species.**

| *Term* | *Meaning* |
|---|---|
| **Extinct** | Taxa that are no longer known to exist in the wild. |
| **Extirpated** | A taxon extinct within a particular portion of its range but present within other segments of its distribution. |
| **Endangered** | A taxon in danger of completely disappearing in the near future within all or part of its geographical range. |
| **Threatened (or vulnerable)** | Taxa that are thought likely to become endangered in the wild in the near future within all or part of their geographical range. |
| **Rare** | Taxa with small populations, usually localized in distribution, that may be at risk but are not at present endangered or threatened. |
| **Out of Danger** | A taxon that is presently secure due to conservation measures and is no longer in one of the other categories. |
| **Indeterminate** | Taxa with so little information known about them that it is not known whether they are extinct, extirpated, endangered, threatened, or rare. |
| **Endemic** | A taxon found naturally only in a particular geographical location and in no other place in the world. |

biologists, and the general public in North America take note. A small 7.5-cm (3-in.) fish, the snail darter, and a 8.5-dm (2½-ft.) plant, Furbish's lousewort, helped delay construction of these projects in Tennessee and on the U.S.-Canadian border in Maine and New Brunswick, respectively.

As a result of the small darter controversy, the act was amended in 1978 to be more flexible when conservation conflicts arise at government projects. The act provided for two committees to study the impasse: one committee considers whether those in charge of the project have considered all alternatives and the other committee evaluates whether a project's benefit outweighs preserving an endangered species. This second committee is comprised of the Secretaries of Agriculture, Army, and Interior; the Chairman of the Council of Economic Advisors; and representatives from the state where the project is located.

In 1982 the United States Endangered Species Act was reapproved, but with some amendments. These amendments strengthened the protection of endangered plants. The act states that the reason for listing a plant should be biological (i.e., in danger of extinction or not) and not economic. In addition, the habitat associated with the plant is to be defined. In this area local citizen groups can play a helpful role. Various state or provincial groups, such as the Michigan Natural Features Inventory and the California Native Plant Society in the United States, or the National Trust in Great Britain, have been instrumental in determining the distribution and rarity of a species. When the U.S. Congress passed the 1973 amendments to the Endangered Species Act, it also ratified the Convention on International Trade in Endangered Species of Wild Fauna and Flora (CITES). This cooperative treaty controls permits and regulations for importing and exporting endangered and threatened species directly or products made from them. International trade in any almost extinct species is strictly forbidden. Many airports frequented by international travelers now have warning signs and displays in the waiting rooms explaining this act.

Very serious questions are now being asked. Of what value is a particular species? Should some "insignificant" species be allowed to hinder modern progress? Why should we try to save species from extinction? Are some species more worth saving than others? How do species become endangered? What role should taxonomists play in the controversy? These questions need some answers!

## *Value of Plant Species to Humans*

Various plant species have great practical value in agriculture and medicine. Plants provide us with food, drugs, and raw materials for home use and industry. In the United States, 25% of all prescription drugs include plant substances that cannot be made synthetically. Some of these drugs include various antibiotics, pain killers, and extracts used to treat heart disease, high blood pressure, and cancer. Of the estimated 250,000–500,000 plant species in the world today, only 5000 species have been researched thoroughly as to their pharmaceutical potential. Some botanists feel that at least 5000 usable medicinal plant species are out there waiting to be discovered; most of these will be from tropical rain forest regions of the world, where species diversity is often enormously high. Extinct species do society no good, and their potential is lost forever.

In modern-day Western agriculture, only about 18 species are used for human or domestic animal food. Most cultivated plant species have been grown for thousands of years (maize, oats, wheat, rye, potatoes, etc.). The pressure for food has forced agriculturalists to look at growing different plant species and to develop new cultivars of old cultigens. Some species have proved helpful in biological control, where one species may inhibit the growth of another.

The science of genetic engineering is in its infancy, but some botanists are optimistic that desirable genetically controlled characters, such as disease resistance, nutrient quality, salt tolerance, etc., from wild plants may be transferred from one plant to another. Reducing the number of species in the world reduces the size of the genetic gene pool. For example, maize or corn (*Zea mays*) is an annual that has to be planted each year. Recently, a wild kind of maize (*Zea diploperennis*) was discovered on a high mountain in the central highlands of Mexico (Fig. 14.1, p. 464). In addition to being a perennial, the species grows well in heavy wet soil and has nearly absolute disease resistance to maize viruses. *Zea diploperennis,* discovered in 1977, is now known to grow in only three sites. The great economic potential of this endangered, insig-

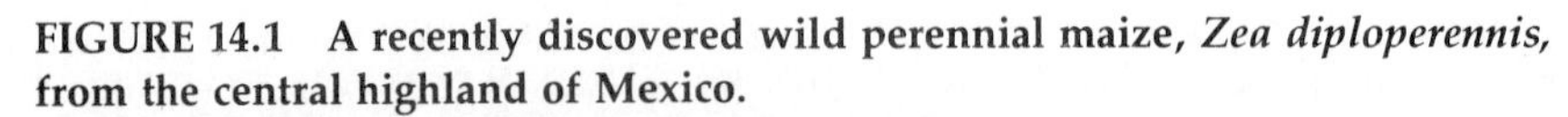

**FIGURE 14.1 A recently discovered wild perennial maize, *Zea diploperennis*, from the central highland of Mexico.**

*SOURCE:* Photo courtesy of H. Iltis.

nificant-looking grass is obvious for developing disease-resistant maize. This could save the world's farmers many billions of dollars annually in planting and pest control costs. Today, the natural populations have been integrated into the UNESCOMAB (Man and Biosphere) worldwide system of preserved areas.

Almost all animal life is dependent initially on plants for their energy. Plants are, therefore, the energy providers in most biological food chains and webs. The removal or loss of an important plant species or group of species could have far-reaching effects upon the plant community where the plant lives. Many birds, insects, snails, etc. live in a particular plant community or use particular plants for food. Destruction or loss of the plant species is also likely to make the animal(s) less common.

In addition to the applied practical reasons for preserving plant species, there are philosophical reasons. The extinction of any species means that the organism is gone forever. It means that future generations will not have the same opportunities for the enjoyment of diversity that we have enjoyed. Enjoyment is one aspect of humans maintaining their own mental health; a healthy mental state due to aesthetics created by unique plant communities or particular plants. Some cultures and subcultures in the world have particular philosophical and religious beliefs that emphasize the preservation of the natural world. In fact, some societies live very much in harmony with nature. Plants sometimes play an important part in culture. For example, in New Zealand, a rare plant, *Desmoschoenus spiralis* (pingao), has been extirpated in many areas by habitat destruction. It is a basic material for Maori crafts. It would be unfortunate to lose some of these cultural traditions that help make the world an interesting place in which to live.

## *How Species Become Endangered*

### *Habitat Destruction*

The majority of plant species are threatened because of a loss of their habitat—that area where they are best adapted to live, reproduce, and survive. With increasing population growth throughout the world, especially in some Third World countries, there has been an increasing need for more housing, roads, shopping areas, land for agriculture, and pressure for different types of land use. As a result, bays, estuaries, and marshes are filled; forests are cut down; land is overturned by mining operations; and surface vegetation is destroyed by overgrazing of domestic animals.

The problem is most critical in the tropical rain forests (Fig. 14.2, p. 466). An area the size of the British Isles is modified, destroyed, or clear-cut each year, or about 20 ha (50 acres) every minute. It is estimated that if this destruction rate continues at its present pace, most rain forests will not exist in 20–25 years. Some individuals want to blame "slash and burn" agriculture by ethnic peoples, but this is responsible for only one tenth to one fifth of the destruction. On the other hand, large-scale clearing of rain forests for cattle-raising operations in southern Mexico, Central America, and South America have destroyed many millions of hectares of forests, only to be planted in non-native grasses. The beef raised is not used locally but is sold to large, fast-food chains in North America and Europe. In North Queensland, Australia, over a million hectares of coastal rain forest have been cleared, drained, and planted in sugar cane. The demand for tropical woods has destroyed a large portion of the tropical forests in Africa, Southeast Asia, and Brazil. Also in Brazil, large stands of native rain forest have been cleared and burned, only to be planted with fast-growing conifers.

This massive habitat modification has also had great impact on the animal species that live in them. It has been estimated by some botanists that for every plant that becomes extinct, 10–30 other plants and animals face extinction.

A final point about tropical ecosystem destruction is that most of the major nutrients of the ecosystem are confined to the living plants themselves (biomass). When the forests are cleared, the low-nutrient soils remain without vegetation cover. The result is loss of productivity, erosion, flooding, and abandonment.

Many endangered plant species exist in unique ecological niches found in bogs, special soil types, or isolated islands, mountain ranges, or valleys.

FIGURE 14.2 **Tropical rain forest destruction in southern Veracruz, Mexico.**

They are rare because they are adapted precisely to their present location and will survive only if their habitat is preserved.

### *Competition with Introduced Species*

A number of species have become endangered or extinct because they were unable to compete with new ones introduced into their habitats. This applies more to animals than to plants, but botanical examples do exist. For example, in the Galapagos Islands, the remaining forests and shrub lands are being invaded by species of *Cinchona* and *Lantana*; the South African Cape province by various *Acacia*; and wetlands of eastern North America by *Lythrum salicaria*. The ecosystems found on isolated islands are especially vulnerable to competition. For example, 97% of the native species found in the Hawaiian Islands are **endemic**, that is, not found anywhere else. Many of these species have not been able to compete with introduced species. This can be seen most markedly on the island of Oahu, where habitat modification due to increased human pressures has resulted in most of the plants now being non-native. Of the native Hawaiian species, over 35% are endangered and over 10% may already be extinct.

### *Air Pollution*

Various air pollutants and acid rain have affected many previously undisturbed habitats. Poisonous heavy metals from smelters at Sudbury, Ontario, Canada; Anaconda, Montana; and Polk County, Tennessee have left eroded, moon-like landscapes, devoid of much vegetation. In the mountains east and north of Los Angeles, California, conifers have been damaged by the city smog. Many forest trees in the Black Forest of Germany and in the Appalachian Mountains in northeastern North America are showing acid rain damage or dying. The effect of acid rain on plants and ecosystems will continue to spread until there is a reduction in the pollutants emitted from massive burning of fossil fuels, especially coal.

### *Illicit Plant Collection*

Certain types of plants are highly prized by plant fanciers. These especially include cacti, carnivorous plants, orchids, ferns, palms, and bromeliads. Dealers in Arizona, Texas, and Mexico dig up, or pay young people to dig up, large numbers of desert plants and cacti. These are sold in markets and to collectors who are willing to pay a high price for a "rare" cactus. Unscrupulous landscapers sell unsuspecting homeowners cacti collected during the night in isolated desert localities. This "cactus rustling" has become a multimillion dollar business that is difficult to control. Recently an international orchid smuggler was apprehended and convicted of illegally smuggling and trading in rare orchids, some worth thousands of dollars (Fig. 14.3, p. 468).

It should be pointed out that this is much different from the professional botanist, who is collecting uncommon plants for proper study and understanding, but even the professional botanist needs to exercise caution in how many specimens are collected and in how the plants are cared for, as well as adhering to laws governing plant collecting.

## *Current Knowledge about Endangered Plants*

Until 1970 very little substantial knowledge about threatened plants had been published. Today there is a wealth of information. For example, in Europe all countries but five have produced endangered or threatened plant lists or so-called Red Data Books. These latter books are in-depth information sources, some prepared in association with the International Union for Conservation of Nature and Natural Resources (IUCN), about a country's or geographical region's endangered flora. Many states in the United States and provinces in Canada have produced lists of endangered plants. The states with the most varied climates, topography, and geology are the ones that have the greatest number of endangered or threatened species or that have a high number of **candidate species** (species that may have the potential of becoming endangered or threatened within a state) (Table 14.2, p. 469). Most of the candidate species have restricted distributions and are more vulnerable to extinction pressures.

In the United States, between 10% and 11% of the total native flora has been designated as rare or threatened. In Canada, 25–30% of the total native flora is rare. In Europe, the percentage is somewhat higher because of the older inhabited, more dense population, extensive industrialization of countries, and the high degree of endemic species in the Mediterranean region.

**FIGURE 14.3 Newspaper clippings dealing with unscrupulous cactus traders and the recent trial and conviction of an international orchid smuggler.**

*Cactus-rustlers riding the purple sage*

**Smuggler 'raped world of rare, wild orchids'**

AN international orchid smuggler "raped beauty spots around the world" of rare, wild orchids that may now be extinct, it was stated at the Old Bailey yesterday.

Armenian-born survived shipwrecks, tropical diseases and head-hunters but he met his nemesis in the Royal Botanic Gardens at Kew, home of the world's oldest and largest orchid collection.

There, botanists became suspicious of his motives in persistently asking about the rediscovery of a rare orchid called sanderianum, whose location was a closely-guarded secret.

a self-taught botanist, had discovered in 1972 on the border of China and Vietnam a species of slipper orchid so rare that no-one had known of its existence. The plant had been named after him: Paphiopedilum Henryanum.

**'Fed an obsession with orchids'**

**Orchid Smuggler Convicted**

on trial in England for smuggling, harboring, offering for sale and selling restricted orchids in violation of the Convention in International Trade in Endangered Species (CITES), pleaded guilty and has been sentenced to one year in prison (⅔ of it suspended), fined $15,700 and ordered to pay $15,700 toward the prosecution's costs.

Northern Europe has a lower number of world threatened species, because the total diversity of the flora is relatively low due to the cold climate and the more recent invasion of plants to the region following the last Ice Age.

In the Southern Hemisphere, there are lists of threatened plants for South Africa, Australia, New Zealand, and Chile. Preliminary lists are woefully lacking for most of Africa, South America, and especially for the countries that contain the diverse tropical rain forests.

New Zealand is a country that is probably the best documented in terms of understanding its threatened species. Since the first plant register appeared in 1976, four updated treatments have been produced, with the most recent in 1987. New Zealand's native flora comprises 2300 species, of which 82% are endemic, and over 325 taxa are threatened.

The Mediterranean-type climatic regions of the world have a high percentage of threatened or rare species: California, the Cape of South Africa,

**TABLE 14.2 Numbers of endangered and threatened taxa and candidate taxa in the United States by state.**

| *State* | *Endangered and Threatened Taxa* | *Candidate Taxa* | *State* | *Endangered and Threatened Taxa* | *Candidate Taxa* |
|---|---|---|---|---|---|
| Alabama | 3 | 84 | Nebraska | 0 | 2 |
| Alaska | 0 | 29 | Nevada | 7 | 75 |
| Arizona | 9 | 91 | New Hampshire | 3 | 6 |
| Arkansas | 1 | 34 | New Jersey | 1 | 24 |
| California | 27 | 647 | New Mexico | 10 | 31 |
| Colorado | 7 | 58 | New York | 2 | 27 |
| Connecticut | 1 | 1 | North Carolina | 5 | 76 |
| Delaware | 0 | 15 | North Dakota | 0 | 2 |
| Dist. of Columbia | 0 | 5 | Ohio | 2 | 17 |
| Florida | 24 | 174 | Oklahoma | 0 | 14 |
| Georgia | 5 | 80 | Oregon | 2 | 131 |
| Hawaii | 14 | 747 | Pennsylvania | 1 | 19 |
| Idaho | 1 | 37 | Rhode Island | 1 | 3 |
| Illinois | 3 | 29 | South Carolina | 5 | 49 |
| Indiana | 1 | 23 | South Dakota | 0 | 3 |
| Iowa | 2 | 9 | Tennessee | 4 | 66 |
| Kansas | 1 | 7 | Texas | 16 | 124 |
| Kentucky | 2 | 29 | Utah | 11 | 102 |
| Louisiana | 1 | 19 | Vermont | 3 | 5 |
| Maine | 2 | 11 | Virginia | 2 | 52 |
| Maryland | 1 | 26 | Washington | 0 | 47 |
| Massachusetts | 1 | 9 | West Virginia | 1 | 20 |
| Michigan | 1 | 14 | Wisconsin | 2 | 20 |
| Minnesota | 1 | 15 | Wyoming | 0 | 29 |
| Mississippi | 1 | 23 | Guam | 0 | 2 |
| Missouri | 4 | 26 | Puerto Rico | 2 | 94 |
| Montana | 1 | 12 | Virgin Islands | 0 | 14 |
| | | | Total | 344 | 3308 |

*SOURCE:* Information obtained from the U.S. Fish and Wildlife Service, 1988.

Western Australia, and the Mediterranean countries themselves. It has been estimated that this climatic region contains approximately 25,000 species, of which almost half this number are narrow endemic species, and a high percentage of these are threatened in some way. For example, in the Cape Province of South Africa, there are over 1620 threatened taxa, with 98 endangered and 36 extinct.

The isolated islands of the world are known for high levels of endemism, not just species endemism but endemic genera and families as well. For example, Cuba has a flora of between 6000 and 7000 species of which 3000–4000 are endemic species. Of these species, 832 taxa are endemic and 959 are threatened or extinct. Many oceanic islands have a high percentage of threatened and rare species (Table 14.3). For example, St. Helena Island, which had domestic goats introduced to the island in 1513, now has 96% of its endemic native flora threatened or rare. The Hawaiian Islands, with a total endemic flora of over 2300 species, has between one-third and one-half of its species as extinct or at some level of concern (Fig. 14.4).

Our understanding of **threatened species** (taxa that are thought likely to become endangered in the wild in the near future within all or part of its geographical distribution if the same causal factors continue to operate) is most complete in the northern countries of Europe and North America. It is in these regions that more botanists and preserved specimens are found, where adverse pressures on vegetation have been studied longer, and where there are fewer species of plants. Two-thirds of the world's estimated 250,000–500,000 different plants grow in the tropical regions. One half of these species are found in the New World tropics and one half are shared with the tropical regions of Africa and Asia (Davis et al., 1986). It is not surprising then, that a comparison of the information known about world floras showed a major lack of understanding of the status of plant species in the tropical world. This is made more difficult because plant species in the tropical rain forests tend to have scattered distributions, with only a few individuals of a taxon present per square kilometer. If only a small portion of a tropical rain forest is destroyed, the chances of recording the existence of a species is greatly enhanced.

The tropical regions of the world generally lack financial and botanical resources to study their floras. For example, a small country like Colombia has an estimated 45,000 different species, with only a few botanists to study the flora. Great Britain, on the other hand, has approximately 1370 taxa, with

**TABLE 14.3 Endemic vascular plant taxa from selected oceanic islands.**

| | *Category* | | | | | | | | *Categories* |
|---|---|---|---|---|---|---|---|---|---|
| | *A* | *B* | *C* | *D* | *E* | *F* | *G* | *Total* | *A–E* |
| Ascension Island | 1 | 5 | — | 4 | — | 1 | — | 11 | 10 (91%) |
| Azores | 1 | — | 5 | 18 | 6 | 11 | 14 | 55 | 30 (55%) |
| Bermuda | 3 | 4 | 1 | 6 | — | ? | ? | ? | 14 |
| Canary Islands | 1 | 126 | 119 | 132 | 5 | 26 | 160 | 569 | 383 (67%) |
| Galapagos | — | 9 | 15 | 111 | 15 | 2 | 77 | 229 | 150 (66%) |
| Juan Fernandez | 1 | 52 | 32 | 9 | 1 | 17 | 6 | 118 | 95 (81%) |
| Lord Howe Island | — | 2 | 10 | 58 | 3 | — | 2 | 75 | 73 (97%) |
| Madeira | — | 17 | 30 | 39 | — | 22 | 23 | 131 | 86 (66%) |
| Mauritius | 19 | 65 | 35 | 39 | 14 | 69 | 39 | 280 | 172 (61%) |
| Norfolk | 5 | 11 | 29 | — | 1 | 2 | — | 48 | 46 (96%) |
| Rodrigues | 10 | 20 | 8 | 8 | — | — | 2 | 48 | 46 (96%) |
| Seychelles | — | 21 | 35 | 15 | 2 | 17 | — | 90 | 73 (81%) |
| Socotra | 1 | 84 | 17 | 29 | 1 | 2 | 81 | 215 | 132 (61%) |
| St. Helena | 7 | 23 | — | 17 | — | 2 | — | 49 | 47 (96%) |

*SOURCE:* From Davis et al., 1986.

*NOTE:* A = extinct; B = endangered; C = threatened; D = rare; E = indeterminate; F = insufficiently known; G = nonthreatened.

FIGURE 14.4 The status of endemic taxa on some islands in the Pacific Ocean.

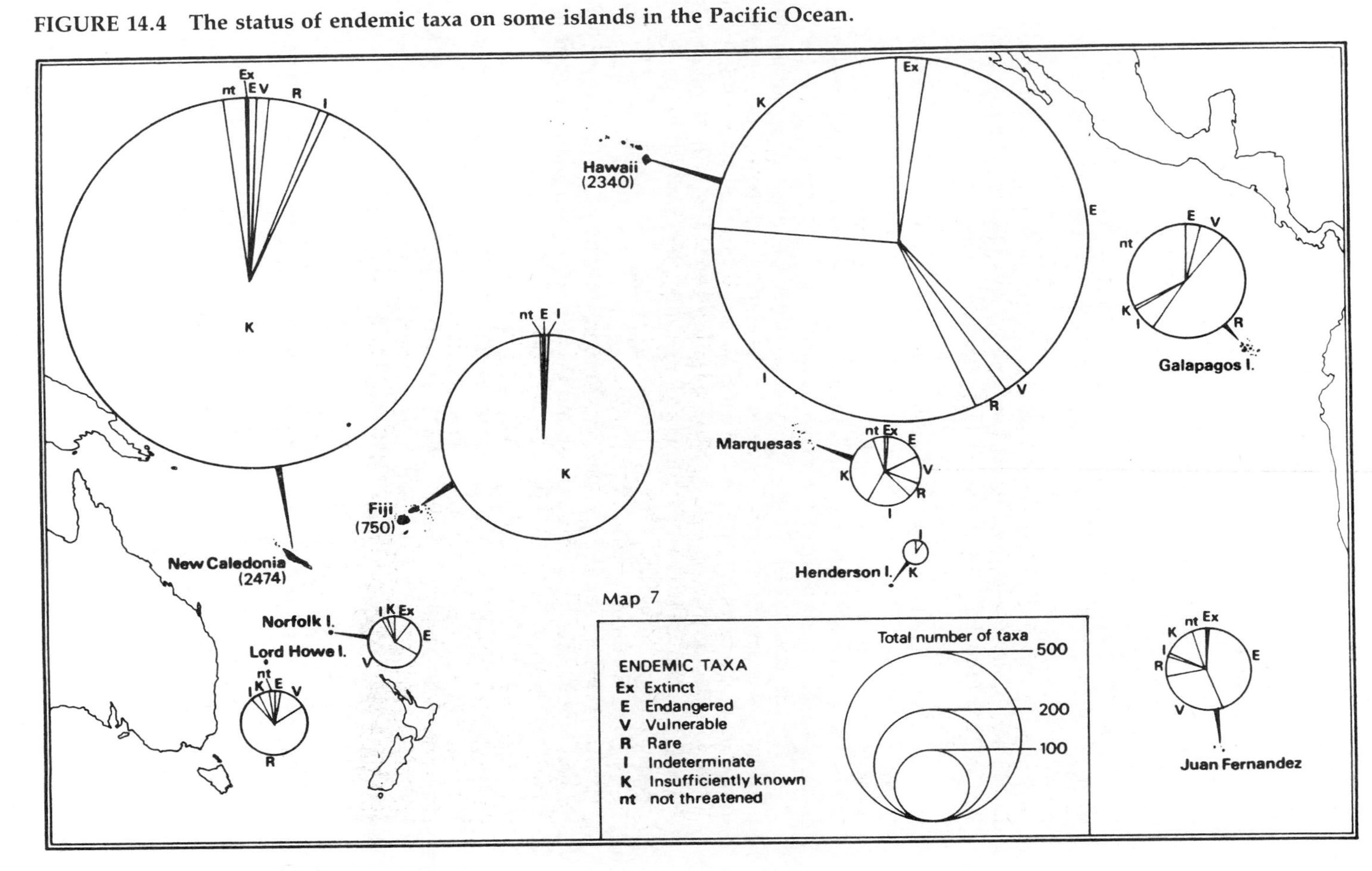


SOURCE: From Davis et al., 1986.

thousands of amateur and professional botanists available to document the status of species and their habitats. As a result of this discrepancy, parts of or whole forest areas may be cut down and destroyed, and the presence or absence of a species, or the potential value of a species, is destroyed with the removal of the forest.

## What about the Future?

Whenever one looks at the rapid destruction of the tropical rain forest regions of the world, views the dying forests of Europe and North America due to acid rain, witnesses the destruction of mountain habitats due to logging and mining, or watches coastal marshes and estuaries being drained or filled, it is easy to become discouraged about plant preservation. However, there are some things that can be done to help solve these problems.

**1.** *Train more professional and amateur botanists.* It is estimated that there are approximately 3000 persons trained throughout the world today in the field of taxonomy. It has been estimated that there needs to be at least six times this many well-trained taxonomists if the flora of the world are to be inventoried and studied before it is too late.

The ultimate goal is to inventory and survey the plant distribution and biodiversity of specific regions (especially the tropics) before they are severely modified or destroyed. Until we have a basic knowledge about a flora, it will be impossible to develop a sound conservation plan. This will involve well-trained persons in botany, ecology, and taxonomy, and not ill-prepared individuals (Fig. 14.5).

**2.** *More Red Data Books are needed.* The purpose of a Red Data Book is to provide the necessary information required to keep a particular plant species from becoming extinct. Except for North America, most of Europe, New Zealand, Australia, and some islands, the status of endangered and threatened species in the world is relatively uncertain. This is especially true for most of the tropics, including Latin America, most of Africa, and tropical Asia. An exerted effort to document the endangered flora of regions of high endemism should be of prime concern. This includes islands with many endemics (i.e., Cuba, Madagascar, New Caledonia), geographical regions with varied ecology and geology (i.e., California, Colombia), locations under various types of human pressures (i.e., Malta, Cape South Africa), and those regions where the most species could be saved (i.e., tropical rain forest areas).

**3.** *Detailed monitoring of species is needed.* To identify a species as endangered is one thing. What is also needed is identification of the distribution of the taxon, population size, critical parts to its life cycle, the community structure to which it belongs, and the proper management techniques needed to preserve the species from extinction.

**4.** *More action toward conservation is required.* The first park created to preserve species was Yellowstone National Park, Wyoming, in 1872. Since then many national parks, state parks, and small preserves have been created.

**FIGURE 14.5 Note found on the door of a professor illustrating how ill-prepared some are and how frivolously they perceive proper botanical training.**

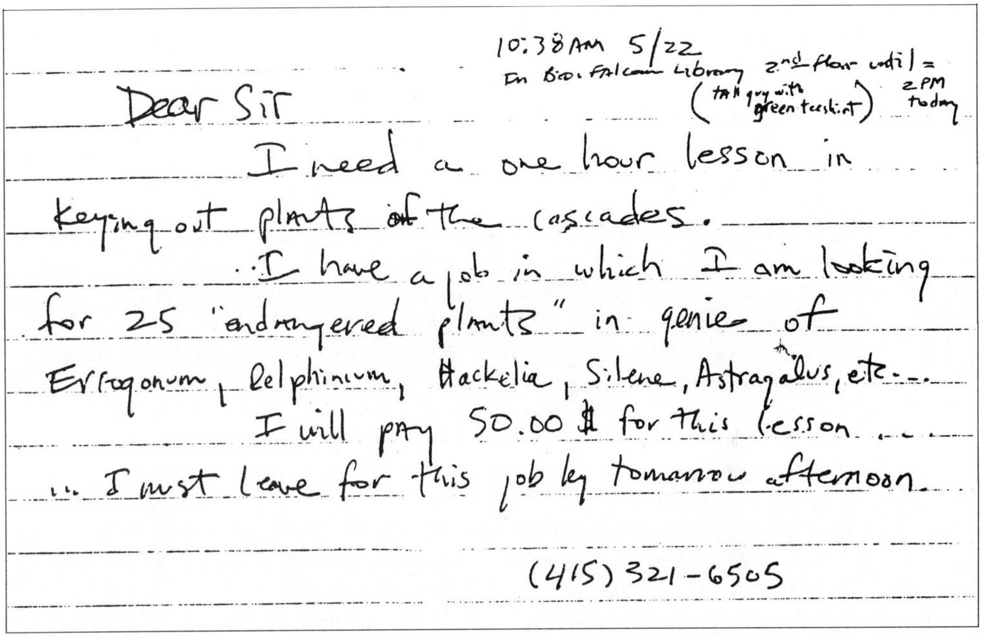

10:38 AM 5/22
In Bio. Falcon Library 2nd floor until 2PM today (tall guy with green teeshirt)

Dear Sir

I need a one hour lesson in keying out plants of the cascades.

I have a job in which I am looking for 25 "endangered plants" in genies of Eriogonum, Delphinium, Hackelia, Silene, Astragalus, etc...

I will pay 50.00 $ for this lesson...

... I must leave for this job by tomorrow afternoon.

(415) 321-6505

*SOURCE:* Courtesy of J. Thomas.

In spite of increasing the areas of preserved land from 174 million ha during 1972–1982 to 386 million ha throughout the world at present, only a small percentage of the earth's surface has been preserved, while much of the earth's surface is being modified.

Creative methods to encourage governments and private individuals to set aside native habitats need to be emphasized. For example, instead of "just setting aside areas," emphasis may be put on harvesting the excess plants and animals found in a preserve for food or utilization of the clear unpolluted water found there for drinking. Another solution is to find better ways to use the land productively while still preserving the wild species. Greater cooperation between conservationists, foresters, and agriculturists needs to take place, with all parties realizing that sustained yield and conservation can go together.

**5.** *There should be education of more professionals and of the public.* All of the above-mentioned activities will come to naught unless there are well-trained individuals to carry out the programs. We must overcome the shortage of managers, technicians, and scientists, especially in the tropics. This training must go along with general education of the public toward an awareness of the positive aspects of conservation. Conservation practices must be explained in the press, on radio, and on television. The survival of many plant species will depend on the success of this education effort. The future survival of the human race may also depend on it.

## *How Individuals Can Help Endangered Wild Plants*

It is easy to look at the magnitude of problems dealing with plant conservation and to become discouraged with the many difficulties. However, much progress has been made in the United States, Canada, Great Britain, and many other countries by individuals just like you, combined with responsible organizations. To help individuals to realize that it is not just "doom and gloom," the following points are suggestions to encourage individuals wishing to help protect endangered wild flowers.

1. Support conservation organizations that protect land needed for the survival of endangered plants. For example, a superb organization in North America is The Nature Conservancy, 1800 North Kent, Suite 800, Arlington, VA 22209 U.S.A.
2. Support and participate in local and regional botanical organizations, botanical gardens, garden clubs, and native plant societies that support native plant protection.
3. Educate yourself about the native wild plants of your area and about plants in general. The information you are learning from this book should help you in many ways.
4. Learn how your political system works on the various levels of government and how you can influence legislation that pertains to endangered species.
5. Purchase native plants or plants such as bromeliads, cacti, carnivorous plants, and orchids only from a dealer who does not sell wild-collected species. Encourage reputable dealers to indicate that their plants are all artificially propagated.
6. Do not pick or transplant wildflowers, especially from different habitats, except as a last resort to preserve them. This should only be done as a unified group conservation effort and with the advice of experts.
7. Do not give location information about endangered plant species to anyone except appropriate authorized individuals.
8. Educate your family and friends about your concern toward endangered wildflowers and share your understanding with everyone you can.

## *SELECTED REFERENCES*

Argus, G. and K. Pryer, 1990. *Rare Plants of Canada,* in press.

Barreno, E. et al. (eds.) 1984. *Listado de Plantas Endemicas, Raras o Amenazadas de Expana.* Informacion Ambiental Conservactionismo en Espana. no. 3, 7 pp.

Borhidi, A. and O. Muniz, 1983. *Catalogo de Plantas Cubanas Amenazadaso Extinguidas.* Edit. Academia, 85 pp.

Bramwell, D., O. Hamann, V. Heywood, and H. Synge (eds.). 1988. *Botanic Gardens and the World Conservation Strategy.* Academic Press, Orlando, FL, 420 pp.

Davis, S. D. et al. 1986. *Plants in Danger, What Do We Know?* IUCN, Gland, Switzerland, 461 pp.

FAO/UNEP. 1981. *Tropical Forest Resources Assessment Project* (in the Framework of the Global Environment Monitoring System—GEMS) UN 32/6.1301-78-04. Technical Reports No. 103, Food and Agricultural Organization of the United Nations, Rome.

Fisher, J. 1987. *Wild Flowers in Danger.* Victor Gollancz, London, 194 pp.

Given, D. R. 1981. *Rare and Endangered Plants of New Zealand.* Reed, Wellington, N.Z., 154 pp.

Given, D. R., W. R. Sykes, P. A. Williams, and C. M. Wilson. 1987. *Threatened and Local Plants of New Zealand. A Revised Checklist.* Botany Division Report. DSIR, Christchurch, N.Z., 16 pp.

Hall, A. V. et al. 1980. *Threatened Plants of Southern Africa.* South African National Scientific Programmes Report No. 45, Pretoria, 244 pp.

Hedberg, I. (ed.). 1979. *Systematic Botany, Plant Utilization and Biosphere Conservation.* Almquist & Wiksell, Stockholm, 157 pp.

IUCN Commission on National Parks and Protected Areas (CNPPA). 1982. *IUCN Directory of Neotropical Protected Areas.* Tycooly International, Dublin, 436 pp.

IUCN Threatened Plants Committee Secretariat. 1980. *First Preliminary Draft of the List of Rare, Threatened and Endemic Plants for the Countries of North Africa and the Middle East.* IUCN, Kew, U.K., 170 pp.

Jain, S. K. and K. L. Mehra (eds.) 1983. *Conservation of Tropical Plant Resources.* Bot. Survey India, Dept. Environ., Government of India, P. O. Botanic Garden, Howrah 711103, India, 253 pp.

Kral, R. 1983. *A Report on Some Rare, Threatened, or Endangered Forest-Related Vascular Plants of the South.* 2 Vols. USDA Forest Service, Southern Region, 1720 Peachtree Rd. NW. Atlanta, GA, 30367. Tech. Publ. R8-TP2, 1305 pp.

Leigh, J., R. Boden and J. Briggs. 1984. *Extinct and Endangered Plants of Australia,* MacMillan Co. of Australia, Melbourne, 369 pp.

Leigh, J., J. Briggs and W. Hartley. 1981. *Rare or Threatened Australian Plants.* Aust. Nat. Park Wildlife Sev. Sp. Publ. No. 7, Canberra, 178 pp.

Lucas, G. and H. Synge. 1978. *The IUCN Plant Red Data Book.* IUCN, Morges, Switzerland, 540 pp.

Morse, L. E. 1987. Rare Plant Protection, Conservance Style. *Nature Conservancy Mag.* 37(5):10–15.

Morse, L. E. and M. S Henifin. (eds.). 1981. *Rare Plant Conservation: Geographical Data Organization.* New York Botanical Garden, Bronx, NY.

Myers, N. 1983. *A Wealth of Wild Species.* Westview Press, Boulder, CO, 274 pp.

Myers, N. 1984. *The Primary Source: Tropical Forests and Our Future.* Norton, New York, 399 pp.

National Parks & Conservation Association. 1975. *Help Save Our Endangered Plants.* Washington, D.C., 24 pp.

Norton, B. G. (ed.). 1986. *The Presentation of Species: The Value of Biological Diversity.* Princeton Univ. Press, Princeton, NJ, 272 pp.

Prance, G. and T. S. Elias (eds.). 1977. *Extinction is Forever.* The New York Botanical Garden, Bronx, NY, 437 pp.

Saunders, D. A. et. al. (eds.). 1987. *Nature Conservation: The Role of Remnants of Native Vegetation.* Surrey Beatty & Sons, Chipping Norton, N.S.W., Australia.

Simmons, J. B. et al. (eds.). 1976. *Conservation of Threatened Plants.* Plenum Press, New York, 336 pp.

Soulé, M. E. and B. A. Wilcox. (eds.). 1980. *Conservation Biology.* Sinauer Association, Sunderland MA, 395 pp.

Specht, R. L., E. M. Roe, and V. H. Boughton. (eds.). 1974. Conservation of Major Plant Communities in Australia and Papua New Guinea. *Austral. J. Bot.* Suppl. Ser. No. 7. 667 pp.

Synge, H. (ed.). 1981. *The Biological Aspects of Rare Plant Conservation.* John Wiley & Sons, Churchester, 558 pp.

Synge, H. and H. Townsend (ed.) 1979. *Survival or Extinction.* The Bentham-Moxon Trust, Royal Botanic Gardens, Kew, 250 pp.

Threatened Plants Unit, IUCN Conservation Monitoring Centre. 1983. *List of Rare, Threatened and Endemic Plants in Europe* (1982 ed), 2nd ed. Nature and Environment Series No. 27, Council of Europe, Strasbourg, 357 pp.

Wilson, E. O. (ed.). 1987. *Biodiversity.* National Academy of Sciences Press, Washington, DC, 400 pp.

# 15

# *The Role of Botanical Gardens in Society*

Botanical gardens are marvelous places to spend time. They usually have lawns, trees, shrubs, and herbaceous plants in different types of plantings and arrangements. Many people have the misconception that they are parks or floral gardens to "go and see flowers in bloom." Botanical gardens are much more than that.

## *Objectives of Botanical Gardens*

Botanical gardens have in the past, and continue today, to play major roles and purposes in society and world cultures. There are at least six major objectives undertaken by botanical gardens:

**First,** botanical gardens were established for comparative research studies of many different kinds of plants from various parts of the world, either living or preserved as herbarium specimens. In this way the taxonomy of groups can be studied morphologically, experimentally, and with greater precision.

**Second,** applied or economic botany can be practiced. Plants that might have value as food crops, for medicinal uses, or be put to some human use can be grown, acclimatized, hybridized, and tested. Following extensive study the plants can then be released. Economically valuable plants today, such as chocolate, coffee, hemp, and vanilla, along with many hundreds of other plants from remote parts of the world, have been introduced for the benefit of society. For example, Philip Miller brought cotton seed in 1732 from the Chelsea Garden in London to Georgia, various rubber tree species from Brazil and the Americas were introduced to Singapore and then to Malaysia via the Singapore Botanic Gardens, and tea was introduced into India via the Calcutta Botanic Garden.

**Third,** botanical gardens have provided a horticultural service. By service is meant the collection, selection, hybridization, and introduction into the retail market of many new and improved kinds of ornamental plants. Three good examples of introduction of ornamental plants by botanical gardens are Longwood Gardens' cooperative program with the U.S. Department of Agriculture—the New Guinea *Impatiens wallerana* hybrids; Minnesota Landscape Arboretum of the University of Minnesota—"Northern Lights" hybrid series of azaleas and other hardy ornamental shrubs; the Plant Introduction Scheme of the Botanical Garden of the University of British Columbia, Vancouver, Canada—ornamental shrubs and perennials, including *Genista pilosa,* "Vancouver Gold," *Arctostaphylos uva-ursi,* "Vancouver Jade," and *Anagallis monellii* var. *linifolia,* "Pacific Blue." These three botanical gardens' introductions have all been made during the last 15 years.

**Fourth,** the botanical gardens have become centers of botanical education. Some of the great gardens, such as Brooklyn, Edinburgh, and Kew, have schools associated with them where landscape design, taxonomy, horticulture techniques, and management are taught. Inspection of the plants in the gardens reveals that all plants are named, recorded, and information on their distribution and history revealed. Recently in the United States, the development of the cooperative program between the University of Delaware and Longwood Gardens is an excellent example of an effective program for the stimulation and establishment of careers in public horticulture.

**Fifth** is the development of the aesthetic presentation of plants and exhibits. The complexity of modern twentieth-century urban society leads to many diverse stresses on the individual. The pleasure felt by individuals because of the care in selection of plant diversity, landscape design, color coordination, and proper interpretation and labeling of plants makes a calming, aesthetic experience of a garden of value.

**Sixth,** and possibly most important during this contemporary period, is the role botanical gardens play in conservation, preservation, and maintenance of endangered plants and gene pools. With greater pressure being exerted on oceanic islands, Mediterranean climates, and tropical rain forests by increased human population, habitat destruction, and massive plant collection for illicit trade, it is estimated by some that as many as 10% of the world's flora will be on the verge of extinction by the year 2000. Certainly the horticultural skills of the staff of botanical gardens that have enriched society with such an abundance of ornamental cultivars can now be directed toward preserving unique flora elements, many which undoubtedly will benefit humankind in the future. One such garden that is in the lead in conservation of plant species, especially in the Neotropics, is the Missouri Botanical Garden, St. Louis, Missouri, U.S.A.

How does a botanical garden differ from a "flower garden"? A flower garden is just that—an assemblage of flowers arranged for their aesthetic beauty. A good example of this type of garden would be the famous Butchart Gardens near Victoria, British Columbia, Canada. Here an old limestone quarry and surrounding area have been developed into a magnificent display of flowers.

## *History of Botanical Gardens*

A botanical garden is an area set aside and maintained by an organization for growing and studying various groups of plants for aesthetic, conservation, economic, educational, recreational, and scientific purposes.

Gardens have existed in China, Egypt, and India since the days of the first recorded records. These gardens were not botanical gardens in the true sense of the word, but they existed for growing food, herbs, ornamentals, and for aesthetic, religious, and status reasons. The famous "hanging gardens" of Babylon in Mesopotamia would be in this category. The Father of Botany, Theophrastus, had a garden attached to his school near Athens that had been given to him by his famous teacher Aristotle.

The first gardens in Europe were constructed during Roman times and were devoted to growing medicinal plants. During the medieval times of the eighth century, monastic gardens originated to enhance medical training in the monasteries. The Papal physic garden, begun in the 1400s, was one of these early famous gardens.

The Middle Ages in Europe found some form of religious philosophy dominating everyday thinking. Most Europeans believed that the human race had originated in a garden "paradise" filled with trees, shrubs, and flowers. The exact location of this paradise garden was not known but it was thought to have been located in some warm climate. When the great exploratory voyages failed to discover this garden in the New World and East Indies, European thought changed from trying to find the original "Garden of Eden" to reconstructing what the original garden might have been like. It was the attempt to answer this question that led to the development of the first botanic gardens in Europe (for a most interesting discussion relating to botanical garden development and the art of the same period, see Prest, 1981).

The origin of the modern botanical garden can be attributed to the Italians. The first garden was founded in Padua in 1545. This claim is not without question, for some have stated that the garden at Pisa is older by 2 years (for a discussion, see Hyames and MacQuitty, 1969). Other gardens soon followed: Florence (1545); Bologna (1567); Leiden, Netherlands and Heidelberg, Germany (1593). In the 1600s: Strasbourg, France (1619); Oxford, England (1621); Paris, France (1635); Groningen, Netherlands (1642); Berlin (West), Germany (1646); Uppsala, Sweden (1655); Edinburgh, Scotland (1670); Chelsea (London), England (1673); Amsterdam, Netherlands (1682); and Tokyo, Japan (1684). In the 1700s, botanic gardens were founded in Vienna, Austria (1754); Kew, England (1759); Cambridge, England (1762); Coimbra, Portugal (1773); Calcutta, India (1787); and Dublin, Ireland (1795).

The early botanical gardens were constructed along very formal plans. The gardens were relatively small, with close rectangular beds bordered by brick and stone edges or surrounded by iron railings or well-clipped hedges. Archways, arbors, fountains, and paths were positioned to portray some meaning that the designer held about the "original garden paradise."

The founding of the early British gardens in the seventeenth century brought a revolutionary change in design. For the first time the grounds were larger in size and were laid out in more natural designs: open and park-like, with curving flower beds, grassy vistas, areas devoted to various types of plants such as herbaceous or woody, and collections from specific geographical regions. The artistic design of these gardens so appealed to the European botanists of this period that almost all new gardens were founded using this natural, less restrictive concept.

## *Survey of Some Botanical Gardens*

The following will be a survey of some of the significant botanical gardens of the world. These were chosen from almost 800 gardens listed in Henderson's *International Directory of Botanical Gardens* IV (1983). The gardens chosen for discussion give a broad spectra of the contributions made by botanical gardens in conservation, aesthetics, education, horticulture, landscape design, research, and as living museums. They also serve to illustrate the historical roles these gardens have played in the past and the manner in which each contributes to our contemporary time period.

The lack of discussion of other fine gardens located throughout the world only reflects space considerations in a book like this and not the quality of the gardens or their contribution to botany.

### *Europe*

***Jardin des Plantes*** The Jardin des Plantes, which is associated with the Museum National d'Histoire Naturelle, Paris, is part of a world-class complex of institutions (Fig. 15.1). It was founded in 1635 in the heart of the city along the banks of the Seine River. During its long history, many famous botanists were involved with the garden, such as Vespasien Robin (mentor of Robert Morrison of Oxford), A.-L. de Jussieu, Bernard de Jussieu, Jean Baptiste Lamarck, and Georges Louis Leclercq de Buffon.

The garden was first developed as a physic garden, but in time expanded

**FIGURE 15.1 Front view of the main tropical plant conservatory, Jardin des Plantes, Paris.**

into more of a botanical garden. The size is 13 ha (32.5 acres), with closely placed flower beds. In this space and six greenhouses are found over 15,000 taxa. The collection is rich in tropical plants, including aroids, Bromeliaceae, ferns, and Orchidaceae. The alpine garden and school of botany, with over 6300 entries, and the Australian plant collection are noteworthy. One of the first trees planted in the garden is still flourishing, *Robinia pseudo-acacia*. The major collections of woody plants are found in the associated Arboretum de Chévreloup outside of Paris on the road to Versailles.

The extensive botanical and horticultural research conducted here at Paris during the nineteenth and early twentieth centuries, and its influence upon botanical thought during this period is only surpassed by that of Kew and Berlin. During contemporary times, the institutions have fallen upon "hard times," with reduced governmental support and staff. In spite of this, taxonomists have needed more than ever before to use the specimens in the 8,900,000-sheet worldwide herbarium, which is very rich in specimens from South America, Africa, and Southeast Asia. This is especially true as botanists attempt to preserve endangered and threatened plants from regions of the world, the species names of which are based on preserved specimens in the Paris Herbarium. It is heartening to know that this fine, old botanical center of the past will continue to play a major role in contemporary botanical issues.

***Berlin-Dahlem Botanical Museum*** The Berlin-Dahlem Botanical Museum had its beginning in 1679 (Fig. 15.2). The garden never amounted to

**FIGURE 15.2 Restored main palmhouse, Berlin-Dahlem Botanical Museum, West Berlin.**

very much until 1801, when the famous botanist Carl Ludwig Willdenow was made director of an old run-down royal garden. As director he was most active in cleaning up the garden, building new glasshouses for tropical plants, collecting and establishing an Alpine collection, and making pools of water for aquatics. The number of living specimens grew in 11 years to 7700 species. Many other famous botanists followed Willdenow as director, notably Adolf Engler, Ludwig Diels, and Robert Pilger (who had the task of rebuilding the garden after its destruction during World War II).

The policy of the botanical garden since its beginning has been to serve the scientific community first and the general public second. The result is that, except for the Royal Botanic Gardens, Kew, no other garden in the world has influenced botany and horticulture more than the Berlin garden.

The garden's 18,000 plantings on 42 ha (105 acres) reflect the phytogeographical interest of the staff. The flora of the five geographical regions of the world are well represented. There is a woodland garden, a superb alpine and rock garden, an Italian garden, a systematic collection, and an arboretum. The classic glasshouses have recently been restored and contain a fine collection of diverse tropical plants. There is also a good collection of insectivorous plants and an excellent presentation of cacti and succulents, one of the best under glass. The North American flora collection is very well represented.

The botanical museum, with its displays, diagrams, and exhibitions of economic botany, and the herbarium have been most important to the garden. The herbarium today has over 2,000,000 specimens from all over the world, but would no doubt be one of the world's largest if it had not been almost completely destroyed during World War II in 1943. New and remodeled space for research and for the herbarium has recently been completed.

A review of the botanical history and a walk through the garden make it very clear that the contributions from this garden have greatly enriched our knowledge of botany. The future looks bright indeed for this botanical center.

***Botanical Garden of the University of Leiden*** The Dutch people are world famous for their beautiful flowers, gardens, and ways of propagating plants. The Botanical Garden of the University of Leiden is small and only covers about 2.63 ha (5 acres) in the middle of the city. What is lacking in size is made up for in botanical history.

The Leiden garden was established in 1587, which makes this the sixth earliest botanic garden in Europe. The first planting of 1100 species in the garden was by two early botanists, Clusius and Clutius. Their *Hortus Botanicus* was important in the training of students in medicine. This garden has been reconstructed from discovered old notes. Paul Herman was the next person to have an effect on the garden. He added many more species to its holdings and rearranged the collection after the systematic collection at Oxford University. Herman Boerhaave was the person in charge when the young Swedish botanist Carl Linnaeus came to visit. At this time almost 8000 taxa were growing. Boerhaave became good friends with Linnaeus, but more importantly, greatly influenced Linnaeus' thinking. Certainly this personal influence and the garden's rich systematic collections at that time are felt even today through Linnaeus' early publications *Genera Plantarum* and *Systema Naturae*.

Today the garden is the same size, but with 19 small greenhouses housing tropical plants. The research of previous years has shifted from the garden to the university and the Rijksherbarium, which has over 2,500,000 preserved specimens.

It is hoped that this inner city garden will survive the urban pressures exerted upon it and will continue to enrich the field of botany through its rich history.

***Jardin botanique de la Ville de Genève*** Founded in 1817 and situated along the shores of Lac Genève is the Jardin botanique de la Ville de Genève (Fig. 15.3). Though rather small at 13 ha (about 32 acres), the garden has contributed greatly to our knowledge of botany and horticulture.

The Geneva botanical garden was made prominent in its early infancy by the research, collecting, and plantings conducted by the famous de Candolle family of botanists. The results of their efforts can be seen today, with many unique temperate plants; magnificent tree specimens such as *Sequoiadendron gigantea* (giant sequoia) and *Sequoia sempervirens* (California redwood) from California; and an outstanding alpine and rock garden.

From its early beginning, the garden grew rapidly in its living and her-

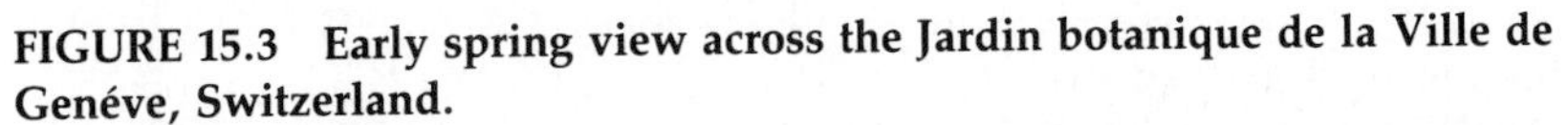

**FIGURE 15.3 Early spring view across the Jardin botanique de la Ville de Genéve, Switzerland.**

barium specimen holdings, until today there are 15,000 taxa in the garden and 5,000,000 specimens in the herbarium. The library is also one of the better botanical libraries in Europe.

The garden is very much in the forefront of botanical thought and research. There is a new building housing the research laboratories, library, and herbarium. The herbarium is housed underground in climate-controlled rooms in special compactors to save space. Each room is isolated from the others to facilitate fumigation and the control of specimen utilization. It is probably the largest herbarium in the world to have the families arranged alphabetically within major taxonomic groups (pteridophytes, etc.). Research today is in the area of plant biochemistry, biosystematics, morphology, and numerical taxonomy.

The Swiss people can be proud of this garden. A visit to this lovely city would not be complete without a visit to it.

***Cambridge University Botanic Garden*** The Cambridge University Botanic Garden was founded in 1762. It was later moved to its present location of 15.6 ha (39 acres). Being connected with a botany program of a major university, the garden has reflected some of the interests of that parent body.

The garden is artistically landscaped with systematic plantings, winter-hardy trails, an alpine garden, and plants grown in British gardens arranged according to their date of introduction (Fig. 15.4). There is a nice collection of temperate conifers; in particular, the collection of *Ephedra* is very good. Reflecting a more contemporary view, there is a collection of endangered and threatened plants of England and a garden for the handicapped.

The garden has 13,000–14,000 taxa of plants growing. The herbarium, together with the university collection, reaches over 550,000 specimens. The botanical staff of the garden and university have been most active in research dealing with British and European flora, biosystematics, and the taxonomy of ornamental plants, most recently *Geranium, Ruscus, Acaena,* and *Bergenia.*

The Cambridge Botanic Garden is an excellent example of how an academic institution, research in botany and horticulture, and public interest can be blended into common goals, with betterment for all involved.

**FIGURE 15.4 View of the palm house, Cambridge University Botanic Garden, Cambridge, England.**

***Royal Botanic Garden, Edinburgh*** The city of Edinburgh, Scotland, is one of those truly great classic cities that has retained much of its Old World charm. The skyline is dominated by the Castle and surrounding old town, with its cobblestone streets, old buildings, and "closes." Fitting perfectly against this backdrop is the Royal Botanic Garden (Fig. 15.5). It was founded in 1670 and was in different locations until it moved to its present site in 1820–1824. Today it covers 24.8 ha (62 acres) of beautifully maintained grounds and conservatories.

This botanic garden maintains over 12,000 living plants from all over the world. It pioneered the art of "natural rock gardens," and today is recognized as having the finest alpine rock garden in the world. It is also known for its fine collection of *Primula, Rhododendron,* and other Ericaceous plants and its artistic horticultural plantings. The new large greenhouse is constructed without internal supports, allowing for a wide range of plant-growing climates under one roof.

The research service to botany and horticulture by the garden's staff has also been significant. The herbarium contains almost 2,000,000 specimens, with many early collections from China and the Orient. The garden is joining with other major gardens in encouraging the preservation of endangered and threatened species.

It is hoped that the Royal Botanic Garden will continue to have an illustrious history.

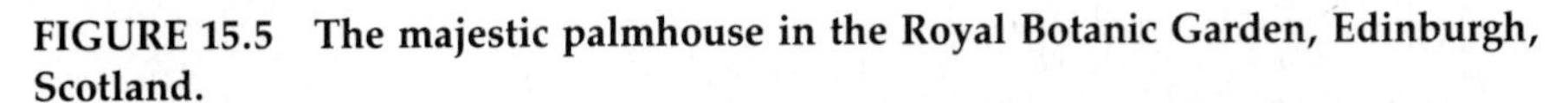
**FIGURE 15.5 The majestic palmhouse in the Royal Botanic Garden, Edinburgh, Scotland.**

***Royal Botanic Gardens, Kew*** "Kew Gardens," as the Royal Botanic Gardens is affectionately known to many a Britisher, is without question the finest botanical garden and botanical research and resource center in the world (Fig. 15.6). It originated in 1759, when Augusta, Dowager Princess of Wales, the widow of Frederick, Prince of Wales, commissioned a small 9-acre (3½-ha) garden to be constructed south of the present Orangery. George III came to the English throne in 1760, and during his reign the gardens expanded and became famous due to the efforts of the head gardener, William Aiton, and the unofficial director, Sir Joseph Banks. The first official director was Sir William J. Hooker, who, when the garden was obtained by the nation in 1841,

**FIGURE 15.6 Royal Botanic Gardens, Kew, England. (a) The Victorian-style Temperate House. (b) The Princess of Wales House, opened in 1987, has computer-controlled multi-climate sections. (c) Spiral stairs in C wing of the herbarium.**

**(a)**

**(b)**

**(c)**

expanded the size to over 120 ha (300 acres), founded the library and herbarium, and established the Museum of Economic Botany. Later in the 1800s, the grounds around Kew Palace and Queen Charlotte's Cottage were added.

Even though the extensive gardens, arboretum, and live plant collections in magnificent Victorian and modern conservatories are interesting and beautiful to view, it is the scientific resources present and the research and education that goes on at Kew that make it stand out even more from other gardens. The massive number of collections of living and preserved plant material form the basis for studies in plant anatomy, biochemistry, conservation, cytogenetics, horticulture, landscape planning, physiology, and taxonomy. Since 1984 the gardens have been governed by a Board of Trustees under the British National Heritage Act of 1983, with financial support provided by the British Ministry of Agriculture.

The living collections are most diverse, with 351 families, 5465 genera, and over 28,000 species growing successfully in the garden. The number of individually maintained entries (species, cultivars, etc.) is over 89,000. Every day plant material leaves Kew in the form of seeds, cuttings, and whole plants for many parts of the world.

The arboretum covers the greatest area of the garden. Here trees and shrubs are grouped according to systematic relationships whenever possible. Large mature trees of *Acer, Cedrus, Fagus, Juglans, Pinus,* and *Quercus* give the grounds and vistas a stately appeal.

Tropical plants are maintained indoors in several display houses. These include the Aroid House, Filmy Fern House, Palm House, the new Princess of Wales Conservatory (1987), the Temperate House, the Water Lily House, and the Australian House. Unusual plants found here include *Victoria amazonica* (giant water lily) from South America, *Alluaudia* from Madagascar, and *Welwitschia mirabilis* from Angola.

The rock garden, with its winding paths, sandstone outcrops, and trickling streams, complements well the new Alpine House (1981), with its refrigerated system that permits the cultivation of species from equatorial mountains and high latitudes. The Duke's Garden, grass collection, and the National Order Beds are invaluable to the students of taxonomy.

There is a wealth of botanical art of outstanding beauty in the collections of the Royal Botanic Gardens but the *tour de force* is in the Marianne North Gallery. Here is displayed the prodigious output of Miss North's talent and energy executed during her travels around the world in the nineteenth century.

The collections of the Museum of Economic Botany, first put on public display by Sir William Hooker in 1847, are now housed in several buildings in the Royal Botanic Gardens grounds. Only a selection from this large collection is now seen by the public, principally in the building designed by Decimus Burton and opened in 1857, opposite the Palm House, and in Cambridge Cottage, which has a display of timbers and their uses. The collections, totaling over 70,000 items, are being systematically curated and cataloged by computer, together with a modern display of the importance of plants to humankind in the new Sir Joseph Banks Building—Centre for Economic Botany, opened in 1989.

The herbarium and library date from 1852. In 1854, George Bentham presented his extensive herbarium and library to Kew and began his long botanical association with Joseph Hooker, later to succeed his father as director of Kew. Today the herbarium has almost 6,000,000 preserved specimens and is adding between 50,000 and 70,000 more each year to the collection from all over the world. There are over 275,000 type specimens as well. This vast collection forms the basis for the preparation of various floras, monographs, plant lists, and revisions to many regions of the world by the staff of Kew and by other researchers as well. These specimens, along with the living collections in the garden, have provided the major basis for more than 30,000 new species described by botanists working at Kew since its founding.

The library is the largest botanical collection of books, journals, pamphlets, illustrations, etc. in the world. It comprises approximately 750,000 items, made up mostly of over 120,000 volumes, 140,000 pamphlets, 10,000 microforms, over 200,000 botanical illustrations, 10,000 plus items in the map collection, and over 250,000 letters and manuscripts from the eighteenth century to the present.

Kew Gardens and its staff are responsible for publishing many journals, books, floras, articles, and monographs about plants. The *Kew Magazine* (which incorporated *Curtes's Botanical Magazine,* the oldest extant botanical journal, founded in 1787) is published quarterly, the *Kew Record of Taxonomic Literature* is published quarterly, and *Index Kewensis* (a reference source to all the world's original plant Latin names) is published annually. The Conservation Unit is now providing a computer data bank and information (Red Data Books) on the international status of plant conservation with regard to endangered and threatened species.

The Jodrell Laboratory, which opened in 1876 (rebuilt in 1965), has established itself as a world center in the study of plant anatomy, cytogenetics, and plant biochemistry. The plant physiological research section that was originally at Kew has moved to the most recent addition to the Royal Botanic Gardens system, Wakehurst Place, a former private estate with an Elizabethan mansion that became part of Kew in 1965. Its 202 ha (505 acres) of rich soils in milder and wetter Sussex in South England has expanded the potential for propagating many more types of plants. Here, research dealing with the physiology of seed germination is conducted. Here also is maintained an international seed bank of 5000 wild plant species of known history, which can be supplied to researchers throughout the world.

Recently a foreign botanist on leaving the Royal Botanic Gardens was heard to remark, "Where would our knowledge of plants be today if it were not for Kew Gardens?" The answer? Much, much poorer!

### North America

***Jardin botanique de Montréal*** The Jardin botanique de Montréal is located on 73 ha (180 acres) of du Maisonneuve Parc adjacent to the 1976 Olympic Games Park and Stadium (Fig. 15.7). It stands as a fitting tribute to one of Canada's greatest botanists, Fr. Marie-Victorin.

The garden was founded in 1931 and has since been under the financial control of the city of Montréal. As conceived and designed by Henry Teuscher and Marie-Victorin, the garden was to be a complex of smaller specialized gardens surrounding and radiating out from the main administrative complex of the library, herbarium, information center, and offices. The classrooms and

**FIGURE 15.7 Main conservatory entrance of the Jardin botanique de Montréal, Canada.**

research laboratories of the Institut botanique de l'Université de Montréal are also located here, and this provides one of the scientific research aspects of the garden.

The U-shaped public conservatories and associated supporting glasshouses are located behind the administrative complex. The large central house is used for educational purposes with a visitor's center organized along six themes: flora of Quebec, plants of the world, anatomy and physiology, the botanical garden, horticulture, and the utilization of plants by humans. Conservatory rooms branching off of this central core are devoted to tropical plants, economic plants, aroids and orchids, Begonias and Gesnarias, and two attractive cacti and desert plant rooms. On opposite ends of the conservatory complex are a house used for seasonal displays that are constructed around a particular theme and an excellent house of tropical ferns. There is also a house used only for bonsai and penjins that is one of the largest collections outside of Asia.

The garden grounds include a relatively young but diverse arboretum, an alpine garden illustrating plants from various alpine regions of the world, and a fine water garden, with pools of various water depths and quality to provide for a greater diversity of aquatic flora.

Several other gardens of unique quality are the annual garden, display garden, economic plant garden, fruit garden, monk's garden, physic garden, and children's garden. The children's garden is devoted to helping city children learn the joys of gardening and of working with plants.

At the present time, the garden has over 20,000 living plants and an herbarium comprised of over 600,000 specimens. The botanical library is one of the best in Canada.

A visit to the culturally rich, second largest French-speaking city outside of France would not be complete without spending a day at the Montreal Botanical Garden.

***University of California Botanical Garden*** The first major academic center for studying botany in North America developed at Harvard University in the mid-nineteenth century. During this same period in the San Francisco Bay area of California, a rival center originated around the famous western botanists W. R. Dudley at Stanford University and Edward L. Greene at the University of California, Berkeley.

The University of California Botanical Garden is probably the best university botanical garden in North America (Fig. 15.8). It was founded in 1890 and covers 12.8 ha (33 acres) of hilly terrain above the university campus. There are over 12,000 species (15,000 accessions), many of them large, mature trees. The plants are arranged into specialty gardens and into geographical regions. There are collections of cacti and succulents, especially from South America, ferns, South African plants, Orchidaceae, and rhododendrons. The rich, diverse flora of California is well represented. There are 11 greenhouses.

The research output of the staff and students of the university has been

**FIGURE 15.8 The New World Desert Collection at the University of California Botanical Garden, Berkeley. Many of the specimens displayed here were collected during a series of expeditions to the Andes in 1935–1958.**

*SOURCE:* Photo courtesy of J. Affolter.

considerable and has greatly influenced botany in North America. The large worldwide herbarium of over 1,500,000 specimens, mostly from the Americas and the Pacific Basin, reflects the different interests of the staff. The botanical book collection in the university library is one of the best in North America.

The botany program at the university and at the botanical garden has been built on a strong past. Today this is continuing and will hopefully continue into the future, making this a major center for studying botany.

***New York Botanical Garden*** A key center for studying botany and horticulture today in North America is the New York Botanical Garden (Fig. 15.9). The establishment of the garden was authorized by the state legislature in 1891. Today it is operated as a private corporation, with some support from the City of New York and the State of New York.

The botanical garden today is not just a garden in a large urban area, but a botanical resource center like Kew Gardens in England or the Berlin-Dahlem garden in West Berlin.

The garden covers 100 ha (250 acres) in the heart of the city along the Bronx River. There are 15,000 species planted in various collections and gardens. These include demonstration gardens, the large Montgomery conifer collection, Stout day lily garden, Havemeyer lilac collection, a large *Rhododendron* and *Azalea* collection, Everett rock garden, perennial garden, herb garden, rose garden, native plant garden, arboretum, and Enid A. Haupt Conservatory complex, which covers almost .4 ha (1 acre) of space. In the conservatory, there is a palm and tropical plant collection, aquatic plant displays, and a collection of cacti from the Americas and succulents from Southern Africa. The strong emphasis by garden staff on the Neotropics can be seen here.

The botanical research emphasis of the garden covers many areas. These include morphology, chemotaxonomy, economic botany, taxonomy, and phylogeny. Many famous botanists have been or are associated with the garden

**FIGURE 15.9 The recently restored Enid A. Haupt Conservatory complex at the New York Botanical Garden, The Bronx, covers almost 0.4 ha (1 acre) and is rich in plants from the Neotropics.**

*SOURCE:* Photo © Allen Rokach, courtesy of the New York Botanical Garden.

or its affiliated academic institutions of the past: N. L. Britton, A. Cronquist, H. A. Gleason, G. Prance, B. Maguire, P. A. Rydberg, J. K. Small, W. C. Steere, and John Torrey.

The herbarium is a worldwide collection of over 5 million specimens, with emphasis on the flora of the New World. The collection is one of the fastest growing collections, with over 90,000 specimens being added each year. The herbarium has the most active loan program in the world, lending more than 50,000 specimens a year for study.

The library and botanical resource center is world class. There are over 200,000 volumes housed and over 500,000 items, including pamphlets, photographs, botanical prints, letters, etc. It is also set up with major botanical data bases for doing extensive literature searches.

The education and publishing aspects of the garden have also been extensive. Many students and individuals of the general public have studied subjects in horticulture, plant propagation, gardening, economic botany, and taxonomy. In recent years the garden has become a major center for studying economic botany in North America. The research staff have been very active in collecting specimens, writing floras and books, and publishing monographs on various groups of plants. The garden also publishes a number of botanical journals, considered outstanding in the field, and its reference books on horticulture are considered the best.

In 1971, the 778-ha (1945 acres) Mary Flagler Cary Arboretum at Millbrook, NY was added to the jurisdiction of the garden. The Garden's Institute of Ecosystem Studies is located here, with an ecology education program, display gardens, and collections of woody plants from temperate North America and Asia. The understanding of ecosystem structure and function is the focus of the research program, and specific interests include nutrient cycling, landscape ecology, aquatic ecology, plant-animal interactions, microbial ecology, forest ecology, chemical ecology, and wildlife management.

The New York Botanical Garden is to North America what Kew Gardens is to England. A similar question could be asked about this garden: Where would our knowledge of North American botany be today if it were not for the New York Botanical Garden? Answer: much, much poorer!

***Rancho Santa Ana Botanic Garden*** The Rancho Santa Ana Botanic Garden in Claremont, California, had its beginnings in 1927 and is one of the more recent of the major botanical gardens to come on the scene in the continental United States (Fig. 15.10). The garden, a nonprofit institution, is supported by a private foundation, which has as its major objectives studying and growing plants from the deserts of the American Southwest, Mexico, and the flora of California.

In the garden's 34-ha (86-acre) space are planted over 2000 species of native wildflowers, yuccas, *Calachortus*, cacti, and cedars. Woody native trees and shrubs are also well represented. All these plantings are artistically arranged and landscaped into the garden's location at the base of the San Gabriel Mountains. Of special interest are the conifer collection, California Coastal Garden, and the Desert Garden.

Research at the garden has centered in the areas of plant anatomy, evolution, morphology, phylogeny, and taxonomy. During its short lifespan

**FIGURE 15.10 A secluded stream habitat near the administration building at the Rancho Santa Ana Botanic Garden, Claremont, California.**

some of its botanists have become famous: Richard Benjamin, Lyman Benson, Sherwin Carlquist, Philip Munz, and Robert Thorne. The herbarium, comprised of over 1,000,000 preserved specimens, was formed by the merging of the herbaria of several academic institutions with that of the garden. Academic affiliation with The Claremont Graduate School provides a graduate program in botany and an added research stimulus for the garden's staff. Its library is one of the better botanical libraries in North America, with over 35,000 volumes and a 15,000-volume periodical collection. Since 1984, a community education program has been most active in providing botanical instruction for the large urbanized area of Southern California.

In spite of the garden's youth, the future looks bright. The garden's present direction of preserving the diverse southwestern and California flora will make it greatly appreciated in years to come.

***Arnold Arboretum*** The Arnold Arboretum of Harvard University, located just outside Boston, Massachusetts, is recognized as having the finest collection of woody plants in North America (Fig. 15.11, p. 494). It was founded originally in 1872 on 50 ha (125 acres) of land willed to the university. A unique long-term agreement between the university and the City of Boston places the supervisory role with Harvard staff, while the city is responsible for road maintenance and security.

Its first chosen director was Charles Sprague Sargent, an independently wealthy former student of the famous Harvard botanist Asa Gray. Sargent and the famous plant hunter E. H. Wilson introduced over 500 plants into

**FIGURE 15.11 Spring in the lilac collection of the Arnold Arboretum of Harvard University, Jamaica Plain, Boston.**

*SOURCE:* Photo courtesy of the Arnold Arboretum.

North America. One of the more recent of these introductions was *Metasequoia glyptostroboides* (dawn redwood) from China in 1947. These plants, along with over 7000 other woody species, are planted artistically throughout the present-day 106-ha (265-acre) arboretum, which was designed by the famous landscape architect Frederick Law Olmsted.

The arboretum includes many special collections, such as the azalea-rhododendron collection, large maple *(Acer)* collection (one of the biggest in the world), the Larz Anderson bonsai collection, *Malus* (crab apple) collection, *Syringa* (lilac) collection, mock orange collection, and the honeysuckle family collection of *Lonicera* and *Viburnum*. Unique specimen trees are common and include bald cypress, Carolina hemlock, golden larch, and Sargent's weeping hemlock.

Since the arboretum was founded, research has been very important at the arboretum. The staff have been active in the areas of anatomy, morphology, ecology, and taxonomy of woody plants. A large herbarium of 900,000 specimens makes up part of the Harvard University Herbaria on the university campus in Cambridge. Most recently the arboretum is leading in North America in the preservation of endangered and threatened flora of the United States.

***Fairchild Tropical Garden*** The Fairchild Tropical Garden was founded on 33 ha (83 acres) in 1935 on the outskirts of Miami, Florida (Fig. 15.12). The garden was established by Colonel Robert Montgomery, a retired career officer, as a place to house the many tropical plants introduced to Florida by Mr. David Fairchild of the Plant Introduction Bureau of the United States Department of Agriculture. The title to the garden land was divided between the local government (Dade County), which is responsible for 23 ha (58 acres), and the Fairchild Tropical Garden Association, which is responsible for 10 ha (25 acres). The association has the responsibility for the administration of the garden.

The garden is renowned for its palm collection, which is thought to be the world's largest. Noteworthy are *Corypha* palms; *Nephroespermum*, oil palms—*Atalea* and other genera; *Ptychosperma*, bamboo palm; *Phoenix*, date palms; and *Mascarena*, bottle palms; and over 500 other species of palms.

A second renowned collection of the garden is the cycad collection, second only in number to the collection at Kirstenbosch, South Africa. The genera *Ceratozamia, Cycas, Dioon, Strangeria,* and *Zamia* are well represented.

Other groups, such as Bromeliadaceae, Orchidaceae, figs, and climbing vines, are also distinctive.

Botanical research using tropical plants has gone forward in the areas of morphology, anatomy, pollination, biology, and taxonomy. Education of

**FIGURE 15.12 View of some of the palms in the Fairchild Tropical Garden, Miami, Florida.**

*SOURCE:* Photo courtesy of Margarita Mattingly.

children and adults is also fostered. A visit to the Fairchild garden is truly a rewarding experience when visiting southern Florida.

***Missouri Botanical Garden*** The Missouri Botanical Garden is affectionately known to the people of St. Louis as "Shaw's garden" after its founder and benefactor Henry Shaw, who in 1859 founded the garden for the people of Missouri. The garden today covers 28 ha (70 acres) of landscaped grounds on the former Shaw estate. The most unique aspect is the Climatron, a geodesic-dome greenhouse where plants from tropical and subtropical climates are found under its 0.2-ha (0.5-acre) roof (Fig. 15.13). Over 1000 exotic plants are planted along a stream, pond, and waterfall. These include many *Hibiscus*, Orchidaceae, trees of Africa and the American tropics, and economic plants.

There are four other greenhouses, including the Temperate House, with a fine specialized collection, and the Linnean House, the oldest continually operating display greenhouse west of the Mississippi. It is filled with many camellias.

The outside plantings of over 4300 species include a superb water lily garden, English woodland garden, herb and scented garden for the handicapped, and the largest authentically designed Japanese garden in North America. Its collection of *Hemerocalles* (day lilies), *Iris,* roses, *Hosta,* and economic plants is very good. In addition, there are excellent collections of orchids, bromeliads, insectivorous plants, and the largest collection of living aroids (over 6000 plants) in the world.

The new Ridgeway Center has greatly enhanced the public awareness of the garden through its education, strong publication programs, lecture series, symposia, and flower shows.

The garden today, under its dynamic director Peter Raven, has become one of the top biosystematic-taxonomy, conservation research centers in the world. Each year over 45,000 specimens are loaned to botanists at other resource institutions. Its herbarium is strong in North, Central, and South American flora, and the flora of Africa. More specimens are being added to this collection each year (100,000) than any other world herbarium so that it will soon house over 4,000,000 specimens. The specimens and fine library are preserved in modern moveable-aisle compactors (the first herbarium in the United States to do so) in a climate-controlled building. Its library is rich in rare books (about 5000 volumes) and has over 90,000 bound volumes; 20,000 pamphlets, letters, paintings, photographs etc., and receives over 1300 botanical journals. Academically the garden and some research staff are affiliated with several universities in St. Louis that have graduate programs. The garden staff also work closely in helping taxonomists from Central and South America to develop proper curatoral programs that can be utilized in their own herbaria.

The garden has a 640-ha (1600-acre) arboretum that was acquired about 1926. It is located outside the city at Grey Summit, Missouri. It serves as an ecological preserve and is used for many educational classes, which adds another dimension to the programs of the garden.

After a period of neglect and decline, the Missouri Botanical Garden is now most active again as a botanical garden and research center. The great traditions laid down in the past by the likes of Edgar Anderson, George

**FIGURE 15.13 (a) The Climatron and water lily pond at the Missouri Botanical Garden, St. Louis, Missouri. (b) Area near the Japanese garden.**

(a)

(b)

Engelmann, and William Trelease have provided an able foundation for the present generation of botanists. May Shaw's garden "live long and prosper."

## *Africa*

***Kirstenbosch Botanic Garden*** The National Botanic Gardens of South Africa has its headquarters in the Cape Province at Kirstenbosch Botanic

Garden. This garden is recognized as one of the more beautiful botanical gardens in the world for two reasons—its setting at the foot of Table Mountain, Cape Town, and its unique native flora (Fig. 15.14).

The botanical garden was founded in 1913, when a young Englishman, Harold Welch Pearson, convinced the political leaders of the Cape to provide financial support from the government for the upkeep and staff. The ancestor to the Kirstenbosch Garden was the Company's Garden (now often simply called the Cape Town Gardens) in Cape Town, which had been developed for cereal, vegetable, and fruit tree cultivation. However, the soil was poor, and it was a wind swept area with little room for expansion. Two of the arguments used successfully by Pearson before the political leaders were that the Cape had a most unique flora, some of which was almost extinct, and the "Cape Colony was the only British possession without a botanic garden."

The Kirstenbosch Garden is a wild woodland garden with a modified formal garden at its nucleus. This landscaping approach fits in well with the slopes, hills, forests, streams, and bold mountain background. Its main theme is the cultivation, display, and study of the indigenous flora of Southern Africa. There is a large natural amphitheater devoted to the Proteaceae, where

**FIGURE 15.14 View through the *Erica* garden of the Kirstenbosch Botanic Garden at the base of the eastern slope of Table Mountain, Cape Province, South Africa.**

*SOURCE:* Photo courtesy of F. Getliffe Norris.

all the species can be viewed. Africa is noted for its unique cycads and in the Cycad Amphitheater the world's finest collection of species is displayed; many are endangered, threatened, or extinct in the wild. The shrub garden is remarkable for its *Protea* and heath members. A very special feature of the garden is the Braille Trail, where the blind and partially sighted can enjoy the scents, textures, and sounds of the garden.

The Compton Herbarium at Kirstenbosch has over 250,000 specimens, largely specializing in the flora of southern Africa, and a new research laboratory is dedicated to the investigation of problems of the culture of rare and endangered species. A plant utilization section promotes the use of indigenous flora in landscaping.

The National Botanic Garden includes six other regional gardens throughout South Africa. Their mission is to promote the knowledge and appreciation of Southern Africa flora and to undertake the *ex situ* conservation of threatened species. Together these gardens are becoming more and more important as conservation of the natural flora becomes more critical.

In spite of political unrest and financial problems, the gardens flourish. It can only be hoped that they will continue to benefit future generations of Africans and others.

### *Asia*

***Sun Yat-Sen Botanical Garden*** According to the early Chinese literature, the Chinese people have been introducing and acclimatizing plants for applied uses for at least 7000 years. Because plants were grown in their gardens for many centuries, a definite Chinese landscape style has developed. With the recent rebirth of interest in China today, this style appeals to many individuals throughout in the world.

The construction of scientific botanical gardens and arboreta began in China almost 60 years ago with the establishment of the Sun Yat-Sen Memorial Botanical Garden in Nanjing in 1929 (Fig. 15.15, p. 500). The garden was just beginning to develop when it was destroyed during World War II.

Following the founding of the People's Republic of China, the reconstruction of the botanical garden took place in 1954. Today the Sun Yat-Sen Botanical Garden covers 186 ha (465 acres) and is under the jurisdiction of the Jiangsu Institute of Botany. The number of taxa being grown numbers about 3000, with special emphasis on medicinal plants, shrubs, and trees of the Central and North Asian subtropics, and rare and endangered plants. There is an arboretum, conifer and ornamental plant garden, and a systematic garden. Its one public greenhouse houses over 800 different species. It has a herbarium of over 600,000 preserved specimens and a library housing about 50,000 books. The botanists on staff have been active in researching the flora of China and of Jiangsu province, and in more contemporary research work, such as developing plants to grow in polluted environments and to improve environmental quality. Discovering new economic or medicinal uses for plants is important to the Chinese. For example, the domestication of *Dioscorea* and the introduction of aromatic plants, forage plants, plants used for food additives, new crop plants, etc. have consumed much of the time of the garden's staff. Recently an introduced plant from Paraguay, *Stevia rebaudiana*, has been shown to contain a new natural sweetener.

**FIGURE 15.15 Main building of the Sun Yat-Sen Botanical Garden, Nanjing, People's Republic of China.**

*SOURCE:* Photo courtesy of the Sun Yat-Sen Botanical Garden.

The Chinese are to be commended for having accomplished so much in such a short time. Perhaps we in more western societies should learn from the Chinese instead of the other way around.

***Botanic Gardens of the University of Tokyo*** The people of Japan have taken the art of gardens to a level not seen in Western cultures. The early gardens of Japan were a blend of Chinese monochrome landscaping and abstract uses of stone plants and raked sand. Pure botanical gardening, as begun in Europe, is a recent introduction into Japan since the end of World War II. An exception to this statement was the Koishikawa Botanical Garden, the oldest true botanic garden in Japan, founded as a medicinal plant garden in 1684 during the Tokugawa Shozunate in the Yedo era. The Koishikawa is now called the Botanic Gardens of the University of Tokyo, and its affiliation with the university dates from 1877 (Fig. 15.16). The garden has 4000 species in various sections—the Arboretum, Fern Garden, Japanese-style garden, Medicinal Plant Reserve Garden, Montane Plant Garden, and Systematic Garden—spread over 16 ha (about 35 acres). The arboretum is well represented with species from eastern Asia, a good collection of *Prunus,* broad-leaved trees, evergreen *Quercus,* and conifers. There are excellent collections of Japanese *Primula, Rhododendron,* and ferns.

**FIGURE 15.16 A large Ginkgo tree *(Ginkgo biloba)* in the Botanic Gardens of the University of Tokyo, Japan, from which the first spermatozoid in the "spermatophytes" was observed.**

*SOURCE:* Photo courtesy of K. Iwatsuki.

Associated with this garden is the Nikko Botanic Garden in Tochigi Prefecture, which is an alpine garden. This specialized garden, founded in 1902, has over 2200 temperate and alpine plants of Japan and the surrounding regions of eastern Asia.

The botanical research for the garden is undertaken by botanists and staff of the University of Tokyo. It is here that a 1,500,000 worldwide specimen herbarium is situated.

Current research is in all of the various fields of botany and concentrates on plants from Japan and eastern Asia.

***Singapore Botanic Gardens*** Like other early tropical botanic gardens, the Singapore Botanic Gardens was founded (in 1822) primarily as an economic plant garden. This was only 2 years after the founding of the city of Singapore. This first garden only lasted for 7 years, but was restarted in its current site in 1859 under the Agri-Horticultural Society. The society received a grant and loan from the Indian government of 24 ha (60 acres) and convict labor to begin the new garden. The garden flourished for a time, and in 1866 another 10 ha (25 acres) was added to its space. This proved financially burdensome, and in 1874 the garden was taken over by the government in the public interest.

From 1866 to 1888 two botanists from Kew Gardens were asked to be directors. It was during this time that the garden began to add exotics from other parts of the world to its 47-ha (117-acre) present-day plant garden, including such plants as cocoa, coffee, *Eucalyptus*, rubber, maize, sugar, and tea. During this time various European vegetables were introduced via the garden into Malayan horticulture. The most important plant contribution was the development of the rubber tree from Brazil *(Hevea brasiliensis)* and methods to extract latex from it.

During the first part of the twentieth century the garden not only continued to expand its economic botany program, but also contributed greatly to the taxonomic understanding of this region of the world, with many papers on the Malayan flora and the description of thousands of new species.

The garden today is well-known for its varied-level terraced lawns; its bamboo, palm, and Orchidaceae collections; and its pergolas of climbing leguminous vines (Fig. 15.17). Over 2000 different tropical species are grown today on the garden grounds. The herbarium has over 700,000 specimens preserved, and its publications on the economic botany of tropical plants are recognized throughout the world. It still maintains its small native forest from its early days, which is unique in a densely populated high-rise island country.

Situated only 1°15′ north of the Equator, the Singapore Botanic Gardens

**FIGURE 15.17 View near the rose garden of the Singapore Botanic Gardens.**

*SOURCE:* Photo courtesy of J. Regalado.

is one of a mere handful of truly tropical botanic gardens in the world. Tropical moist forests are recognized as centers of the richest species diversity on Earth and are also the most threatened habitats due to population pressures and timber exploitation. The role of gardens such as the Singapore Botanic Gardens as repositories of germ plasm and technical information on tropical botany and horticulture has become increasingly relevant and important today.

### *Australia and New Zealand*

***Royal Botanic Gardens, Sydney*** The Royal Botanic Gardens, Sydney is considered by many botanists to have one of the most beautiful settings of any of the world's major gardens. Its 30 ha (75 acres) of lawns, flower beds, groups of native and exotic trees, and glasshouses are situated with the picturesque Sydney Harbor and Sydney Opera House on two sides, and old governmental buildings and grounds on the remaining sides. The main business district of Sydney is nearby.

The garden site has been continuously cultivated from the founding of Australia in 1788, when "a farm of nine acres in corn" was planted. The government farm was soon moved to better soils west of Sydney, and the foundation of the Botanic Gardens dates from 13 June 1816 when a supervisor of the garden was appointed and the gardens were enclosed.

Today the gardens are under the jurisdiction of a trust and are associated with the Department of Environment and Planning of the New South Wales Government. Approximately 5000 species of plants are grown throughout the grounds and in two large display glasshouses—one pyramid-shaped and the other in the form of an arc (Fig. 15.18). Inside them native and exotic tropical plants are tastefully arranged. The garden is strong in cycads, Australian trees, orchids, and a good palm collection. The herbarium is large, with over 1,000,000 preserved specimens. There is also a good botanical library. Research by the gardens' botanists has centered around Australian members of such families as Proteaceae, Myrtaceae, Fabaceae, Cyperaceae, Casuarinaceae, Poaceae, and Lamiaceae, as well as vegetation surveys, conservation, and the

**FIGURE 15.18 The Pyramid House in the Royal Botanic Gardens, Sydney, Australia, contains native and exotic tropical plants.**

preparation of an Australian flora. Recent research projects have included several that emphasize cladistic methods, biogeography, and the study of relationships at and above the generic level. An illustrated flora of New South Wales is another major project.

The Royal Botanic Gardens has developed two new satellite gardens. The Mount Tomah Garden of 28 ha (56 acres) opened on 1 November 1987 and is about 120 km northwest of Sydney, at an altitude of about 1000 m. It displays cool-climate plants with emphasis on conifers, rhododendrons, and Southern Hemisphere floras. The Mount Annan Garden for Australian native plants consists of 500 ha (1250 acres) located about 60 km southwest of Sydney and opened in October, 1988.

A tourist visit to the Sydney Opera House is not complete without spending time in the nearby botanic gardens.

***Christchurch Botanic Garden*** The small city of Christchurch on the South Island of New Zealand can trace its beginning to a group from England who wished to found an English colony in New Zealand. In the original surveys, a large 200-ha (500-acre) area was set aside as an open space, or "lung of the city." The actual beginning of the Christchurch Botanic Garden was in 1863–1864, when a group of citizens wanted a botanical garden for the acclimatization, propagation, and enjoyment of useful exotic plants. In true English fashion, the first tree planted was a commemorative tree, the "Albert Edward oak." This tree, which still stands, commemorated the marriage of Princess Alexandra of Denmark to the then Prince of Wales.

The real development and expansion of the garden came in the late 1800s and the early twentieth century. It was during this time, in spite of very little support money from the controlling municipal government, that the garden expanded to its present-day 26 ha (about 65 acres), introduced no less than 4000 species of ornamental and useful plants to New Zealand, provided trees for enhancing the growing city of Christchurch, and brought into cultivation many specimens of the unique New Zealand flora, especially alpine plants, *Hebe,* and *Leptospermum* taxa.

The garden is located in an oxbow bend of the Avon River. It has a fine collection of unique New Zealand flora, exotic cacti and succulents, tropical exotics, shrubs, and trees. Six greenhouses are open for display; most noteworthy is the Cunningham House, patterned after the Palm House in Glasnevin, Ireland (Fig. 15.19). There is no significant herbarium, and pure botanical research is lacking. There is some training of students in horticulture. It has become primarily a city park garden under the control of the municipality's parks department. This is what happens many times when the policies governing museums, botanical gardens, and zoos become subject to the changing ideas of local politicians.

Botanical research dealing with the New Zealand flora is being conducted in Christchurch, however, at the New Zealand government's Department of Scientific and Industrial Research (DSIR) and by the botany staff of the nearby University of Canterbury.

It can only be hoped that the controllers of the garden will allow for the staff to become more active in the fields of applied economic and pure botany

**FIGURE 15.19 The Cunningham Palm House and Rose Garden in the Christchurch Botanic Garden, Christchurch, New Zealand.**

*SOURCE:* Photo courtesy of F. Coats.

to build on the already strong points of the garden. The horticultural and botanical world would be greatly enhanced if this happened.

## *SELECTED REFERENCES*

Agarwal, V. S. 1983. *Perspectives in Botanical Museums.* Today and Tomorrow's Printers and Publ., New Delhi, 390 pp.

Anonymous. 1985. *Royal Botanic Gardens Kew and Wakehurst Place.* Royal Botanic Gardens, Kew, 16 pp.

Ashton, P. 1984. Botanic Gardens and Experimental Grounds. In: V. H. Heywood and D. M. Moore (eds.). *Current Concepts in Plant Taxonomy,* Academic Press, London, pp. 36–46.

Begley, S. and E. Jones. 1989. Research Amid the Carnellies. *Newsweek* 113(20):58–59.

Bramwell, D. et al. (eds.). 1987. *Botanic Gardens and the World Conservation Strategy.* Published for IUCN by Academic Press, London, 367 pp.

Britton, N. L. 1896. Botanical Gardens. *Bull. Torrey Bot. Club* 23:331–345.

Browne, P. 1986. Kew Safeguards Tomorrow's Plants. *Reader's Digest* 48(284):100–105.

Dejun, Y. 1983. *The Botanical Gardens of China.* Science Press, Beijing, 320 pp.

Gibbons, B. 1990. Missouri's Garden of Consequence. *National Geographic* 178(2):124–140.

Henderson, D. M. 1983. *International Directory of Botanical Gardens.* IV, 4th ed. Koeltz Scientific Books, Koenigstein, W. Germany, 288 pp.

Hill, A. W. 1915. The History and Functions of Botanic Gardens. *Ann. Missouri. Bot. Gard.* 2:185–240.

Hyams, E. and W. MacQuitty. 1969. *Great Botanical Gardens of the World.* Thomas Nelson & Sons, London, 288 pp.

Jacob, I. and W. Jacob. 1985. *Gardens of North America and Hawaii.* Timber Press, Portland, OR, 368 pp.

Moore, K. J. 1974. Botanic Gardens and Arboreta. In: A. E. Radford et al. (eds.). *Vascular Plant Systematics.* Harper & Row, New York, pp. 775–790.

Ornduff, R. 1978. Using Living Collections—The Problems. *Bull. Amer. Assoc. Bot. Gard.* 12:113–117.

Pepper, J. 1978. *Planning the Development of Living Collections.* Longwood Program Seminars 10:24–27.

Prest, J. 1981. *The Garden of Eden. The Botanic Garden and the Recreation of Paradise.* Yale Univ. Press, New Haven, CT, 122 pp.

Shukla, P. and S. P. Misra. 1982. *An Introduction to Taxonomy of Angiosperms,* 3rd ed. Vikas Publ. House, New Delhi, 556 pp.

Simmons, J. 1983. Tropical Worlds Under Glass. *New Sci.* 99 (1370):401–403.

Stafleu, F. A. 1969. Botanical Gardens before 1818. *Boissiera* 5(3):31–46.

Stern, W. T. 1971. Sources of Information About Botanic Gardens and Herbaria. *J. Linn. Soc., Biol.* 3(3):225–233.

Stone, B. C. ed. 1977. *The Role and Goals of Tropical Botanic Gardens.* Rimba Ilmu Universiti Malaya, Penerbit Universiti Malaya, Kuala Lumpur, Malyasia, 249 pp.

Teuscher, H. 1933. The Botanical Garden of the Future. *J. New York Bot. Gard.* 34:49–62.

Thompson, P. A. 1972. The Role of the Botanic Garden. *Taxon* 31(1):115–119.

Vos, F. de. 1978. The Nature of Present Day Collections. *Bull. Amer. Assoc. Bot. Gard.* 13:12–15.

Woodland, D. 1989. Recreating the Garden of Eden and More. *Focus* 25(1):13–15.

# *Epilogue*

Throughout the 15 chapters of this book the author has attempted to meet the original three objectives: (1) to teach basic botanical facts, (2) to relate these facts to systematic principles, and (3) to show how systematic principles relate to contemporary society. The first two comprise the bulk of the book; the last point will now be discussed.

## *The Relevance of Systematics to Society*

The human species *(Homo sapiens)* is the one organism in the biological world that can willfully change and modify the environment. To completely understand what impact humans have on our environment today, the limits of this impact and human influence on the future survival of species and communities should be understood.

On a less philosophical and more applied level, the systematic botanist may be involved in programs of biological control of plant diseases and pests. Detailed knowledge of wild plant species and their biological relationships is necessary for programs of this nature to succeed.

Some systematists study wild plant species so that relationships and variability seen in cultivated taxa can be better understood. Research of this type can lead to much improvement in our major food sources. For example, knowledge about disease resistance observed in native wild potatoes in the mountains of Peru is being applied to cultivars of the common "Irish potato" (not native to Ireland) to improve disease resistance, plant vigor, and overall yield. Botanists trained in systematics have been involved in discovering new plants that have potential as medicinal plants. Sometimes this involves working with chemists or medical personnel in identifying potential drug plants. On occasion, the knowledge of systematics can save the lives of persons who have ingested unknown plants by identifying potentially poisonous plants and mushrooms.

The contemporary systematist is being consulted more and more by many types of people. For example, helping drug enforcement agents and police identify unknown plants and plant parts correctly; helping environmental quality workers to identify the plant species contaminating water; advising real estate developers or civil engineers on the plants of a tract of land and how the plants may best be preserved and managed; and answering all types of questions that center around the identification of species.

Plant systematics was one of the very first disciplines of science to explore the world around us. As societies have evolved through time, so has the profession. Today the field is very much alive with data from classic observational information gathered from the field and dried herbarium specimens. It is also vibrant with information from chemistry, development, electron microscopy, genetics, and molecular biology.

## *Job Opportunities and Qualifications*

At first glance, the potential for employment and having a career using systematic principles may appear limited. Just the reverse is true! A large number of systematists work in the academic fields teaching students at all educational levels and doing research, discovering facts new to science. Others travel the world collecting, describing, and classifying new taxa. Still others become curators, caring for living and preserved plant collections in private and public botanical gardens, museums, and institutions. However, the great majority of persons use their knowledge directly or indirectly in many different ways. Some are employed by state, provincial, or federal agencies; or certain agricultural, pharmaceutical, petroleum, mining, or environmentally oriented corporations; and others in nature centers and parks. Table E.1 gives a much abridged list of professions that use systematic principles.

In general, students become interested in science for the intrigue and challenge of the subject, not to get rich. Systematists in North America, for example, earn between $15,000 and $35,000, with some individuals in prestigious universities earning between $40,000 and $75,000. Some museum or governmental workers may earn lower or higher salaries. The salary is usually commensurate with a person's training, experience, capabilities, and job responsibilities, and the region of the country or world.

Some positions lend themselves to combining field trips, research, and vacation time. Certain academic positions are 9–10 months in duration, instead of 12 months, and thereby allow for engaging in other employment or research.

The availability of jobs for systematic botanists has varied unpredictably. The demand has increased in recent years because of increased public environmental awareness. Environmental problems increase job opportunities for the systematist. Employment potential for women and minorities has never been better in many parts of the world.

The minimum background for a well-trained systematist is usually 4 years of college or university with a bachelor's degree in biology or botany. Elective courses in chemistry, computer science, and field and molecular biology are advantageous. Field experience during the summer or vacation

**TABLE E.1 An abridged list of professions that use the principles of plant systematics.**

| | |
|---|---|
| Allergist | Horticulturist |
| Aquatic biologist | Laboratory technician |
| Aquaculturist | Land reclamation |
| Biological control specialist | Landscape designer |
| Biological illustrator | Museum technician |
| Biological photographer | Natural resources manager |
| Biological writer | Nature education |
| Biotechnologist | Nursery management |
| Biostatistician | Park naturalist |
| Botanical consultant | Park ranger |
| Botanist | Parks management |
| Brewmaster | Peace Corps |
| Computer analyst | Pest control specialist |
| Conservation officer | Plant breeder |
| Cooperative Extension Service | Plant hunter |
| Customs | Plant production |
| Curator | Plant propagator |
| Disease control specialist | Plant protection and inspection |
| Ecologist | Range manager |
| Economic botanist | Resource management |
| Environmental consultant | Systematics research |
| Environmental law | Teacher |
| Environmentalist | Toxicologist |
| Farming advisor | Vegetable scientist |
| Forensic scientist | Viticulturist |
| Forester | Water quality manager |
| Genetic engineering research | Weed scientist |
| Greenhouse management | |

periods at biological stations, in research laboratories, or in an applicable job is highly recommended. Most individuals in systematics, however, have master's or doctorate (D. Sci.; Ph.D.) degrees, requiring 4–6 additional years of schooling beyond the bachelor's degree. This is not excessive when considering that most professions in medicine, education, and science require advanced training for their professionals.

Systematics is a field for almost anyone who has been fascinated with nature. It is the author's hope that this text has kept that flame of fascination burning brightly.

## SELECTED REFERENCES

Anderson, G. J. and J. A. Slater. 1986. *Careers in Biological Systematics,* 2nd ed., American Society of Plant Taxonomists, c/o Rancho Santa Ana Botanic Garden, Claremont, CA, 12 pp.

National Association of Biology Teachers. *Careers in Biology. A Challenge.* NABT, 11250 Roger Bacon Drive, #19, Reston, VA 22090.

Saigo, R. H. and B. W. Saigo. *Careers in Botany.* Botanical Society of America, c/o School of Biological Sciences, Univ. of Kentucky, Lexington, KY, 16 pp.

Saigo, R. H. and B. W. Saigo. 1985. *Careers in Biology–II.* Carolina Biological Supply, Burlington, NC, 4 pp.

Young, S. P. 1984. Careers in the Biological Sciences—Finding Your Niche. *Amer. Biology Teacher* 46(1):12–17, 64.

# *Appendix I: Floras of the World*

The following list of floras is greatly abridged. Emphasis is given to North America, recent treatments, and those covering a broader geographical region. Most "picture book" guides and "laundry lists" of plants have been omitted, unless keys, descriptions, etc. are lacking for a region. This bibliography should provide an initial introduction to the flora of a region for individuals who wish a beginning understanding.

## *Bibliographies*


Blake, S. F. 1961. *Geographical Guide to Floras of the World.* Part II. Misc. Publ. 797. U.S. Dept. Agr., Washington, D.C., 742 pp.

Blake, S. F. and A. C. Atwood. 1942. *Geographical Guide to Floras of the World.* Part I. Misc. Publ. 401. U.S. Dept. Agr., Washington, D.C., 336 pp.

Davis, S. D. et al. 1986. *Plants in Danger. What Do We Know?* IUCN Conservation Monitoring Centre. Cambridge, U.K., 461 pp.

Fish, E. 1974. *Wildflowers of North America: A Selected, Annotated Bibliography of Books in Print.* Library of the New York Botanical Garden, Bronx, 34 pp.

Frodin, D. G. 1984. *Guide to Standard Floras of the World.* Cambridge Univ. Press, Cambridge, 619 pp.

Gunn, C. R. 1956. An Annotated List of State Floras. *Trans. Kentucky Acad. Sci.* 17(2):114–120.

Kartesz, J. T. and R. Kartesz. 1980. *A Synonymized Checklist of the Vascular Flora of the United States, Canada, and Greenland.* Univ. North Carolina Press, Chapel Hill, NC, 498 pp.

Little, E. L. Jr. and B. H. Honkala. 1976. *Trees and Shrubs of the United States: A Bibliography for Identification.* U.S.D.A. Misc. Publ. 1336, U.S. Gov. Printing Office, Washington, D.C., 56 pp.

Sachet, M. H. and F. R. Fosberg. 1955. *Island Bibliographies.* Natl. Acad. Sci/Natl. Res. Council Publ. No 335. Washington, D.C., 427 pp.

Shetler, E. R. 1966. *Floras of the United States, Canada and Greenalnd.* Smithsonian Inst. Washington, D.C. (Mimeographed), 12 pp.

Shetler, E. R. 1966. *Floras of the United States, Canada and Greenland.* Smithsonian Inst. Washington, D.C. (Mimeographed), 12 pp.

Shetler, S. G. and L. E. Skog. 1978. *A Provisional Checklist of Species for Flora North America.* Monogr. Sys. Bot. Vol. 1. Missouri Bot. Gard., St. Louis, 199 pp.

Simpson, N. D. 1960. *A Bibliographical Index of the British Flora.* Bournemouth, 429 pp.

Stuckey, R. L. 1975. A Bibliography of Manuals and Checklists of Aquatic Vascular Plants for Regions and States in the Conterminous United States. *Sida* 6:24–9.

## General Keys

Cronquist, A. 1979. *How to Know the Seed Plants.* W. D. Brown, Dubuque, IA, 153 pp.
Cronquist, A. 1981. *An Integrated System of Classification of Flowering Plants.* Columbia Univ. Press, New York, 1262 pp.
Davis, P. H. and J. Cullen. 1978. *The Identification of Flowering Plant Families: Including a Key to Those Native and Cultivated in North Temperate Regions.* Cambridge Univ. Press, Cambridge, 113 pp.
Engler, A. 1964. *Syllabus der Pflanzenfamilien.* (H. Melchior ed.) 12th ed. Gebruder Borntraeger, Berlin.
Hutchinson, J. 1967. *Key to the Families of Flowering Plants of the World.* Oxford Univ. Press, Oxford, 117 pp.
Keng, H. 1978. *Orders and Families of Malayan Seed Plants,* rev. ed. Singapore Univ. Press, Singapore, 437 pp.
Thonner, F. 1981. *Thonner's Analytical Key to the Families of Flowering Plants.* (rev. by R. Geesink, et al.) Wageningen: PUDOC, Leiden Univ. Press, The Hague, 231 pp.
Willis, J. C. 1973. *A Dictionary of the Flowering Plants and Ferns,* 8th ed. (Revised by H. K. Airy-Shaw.) Cambridge Univ. Press, Cambridge, 1245 pp.

## Polar Regions

Bocher, T. W., K. Holmen, and K. Jakobsen. 1968. *The Flora of Greenland* (Transl. by T. T. Eklington and M. C. Lewis). Haase, Copenhagen, 312 pp.
Chastain, A. 1958. *La Flore et la Végétation des Îles Kerguelen.* Mem. Mus. Nat'l. Hist. Nat. II/B (Bot.) Vol. 11(1). Paris, 136 pp.
Green, S. W. 1964. *The Vascular Flora of South Georgia.* Scient. Rep. British Antarctic Survey No. 45. London, 58 pp.
Hultén, E. 1968. *Flora of Alaska and Neighboring Territories: A Manual of the Vascular Plants.* Stanford Univ. Press, Stanford, CA, 1008 pp.
Huntley, B. J. 1971. Vegetation. In: E. M. van Zinderen Bakker et al., (eds.) *Marion and Prince Edward Islands.* A. A. Balkema, Cape Town, pp. 98–160.
Longon, R. E. and M. W. Holdgate. 1979. *The South Sandwich Islands. IV.* Bot. Sci. Rep. British Antarctic Survey No. 94., Cambridge, 53 pp.
Löve, Á. 1970. *Islenzk Ferdaflóra.* Almenna Bókafélagid, Reykjavíc, 428 pp.
Moore, D. M. 1968. *The Vascular Flora of the Falkland Islands.* Scient. Rep. British Antarctic Survey, No. 60., London, 202 pp.
Polunin, N. 1959. *Circumpolar Arctic Flora.* Oxford Univ. Press, Oxford, 514 pp.
Porsild, A. E. 1955. *Vascular Plants of the Western Canadian Arctic Archipelago.* Bull. Nat'l. Mus. Canada 135, Ottawa, 266 pp.
Ronning, O. 1964. *Svalbards Flora.* (Polarhandbok 1) Norsk Polarinstitutt, Oslo, 123 pp.
Skottsberg, C. J. F. 1954. Antarctic Flowering Plants. *Svensk. Bot. Tidskr.* 51:330–338.
Tatewaki, M. et al. 1931–64. *The Phytogeography of the Islands of the North Pacific Ocean.* Hokkaido Univ., Sapporo, Japan.
Tolmatchev, A. I. 1960–1981. *Arkticveskaja Flora SSSR* (Flora Arctica URSS). Vyp. 1–8. AN SSSR Press, Moscow/Leningrad.
Tolmacher, A. I. ed. 1969. *Vascular Plants of the Siberian North and the Northern Far East* (trans. from Russian by L. Philips). Israel Program for Scientific Translations, Jerusalem, 340 pp.
Wiggins, I. and J. H. Thomas. 1962. *A Flora of the Alaskan Arctic Slope.* Univ. Toronto Press, Toronto, 425 pp.

## North America

### Canada

Boivin, B. 1967–79. *Flora of the Prairie Provinces. Parts 1–5* (Provancheria 2–5). L'Universite Laval, Quebec (Reprinted from *Phytologia* 15:121–159, 329–446; 16:1–47, 219–339; 17:58–112; 18:281–293; 22:315–398; 23:1–140; 42:1–24, 385–414; 43:1–106, 223–251).

Calder, J. A. and R. L. Taylor. 1968. *Flora of the Queen Charlotte Islands.* 2 Vols. Canada Dept. Agr. Res. Br. Monogr. No. 4. Queen's Printer, Ottawa. Vol. 1, 659 pp.; Vol 2, 148 pp.

Dore, W. G. and J. McNeill. 1980. *Grasses of Ontario.* Res. Br. Agr. Canada. Monogr. 26. Canadian Gov. Publ. Centre, Hull, Quebec, 566 pp.

Erskine, D. S. 1960. *The Plants of Prince Edward Island.* Canadian Dept. Agr. Res. Br. Publ. No 1088. Queen's Printer, Ottawa, 270 pp.

Fernald, M. L. 1950. *Gray's Manual of Botany,* 8th ed. American Book Co., New York, 1632 pp.

Gleason, H. A. 1963. *The New Britton and Brown Illustrated Flora.* 3 Vols. Hafner, New York.

Gleason, H. A. and A. Cronquist. 1963. *Manual of Vascular Plants of Northeastern United States and Adjacent Canada.* D. Van Nostrand, New York, 810 pp.

Hitchcock, C. L. and A. Cronquist. 1973. *Flora of the Pacific Northwest.* Univ. of Washington Press, Seattle, WA, 730 pp.

Hitchcock, C. L. et al. 1955–1969. *Vascular Plants of the Pacific Northwest.* 5 parts. Univ. of Washington Press, Seattle, WA.

Hosie, R. C. 1979. *Native Trees of Canada,* 8th ed. Fitzhenry & Whiteside, Don Mills, Ontario, 380 pp.

Hutlen, E. 1941–1950. *Flora of Alaska and Yukon.* 10 vols. Lunds Univ. Arsskrift. N.F. Avd. 2 Bd. 37–46, 1902 pp.

Hutlen, E. 1968. *Flora of Alaska and Neighboring Territories.* Stanford Univ. Press, Stanford, CA, 1008 pp.

Looman, J. and K. F. Best. 1979. *Budd's Flora of the Canadian Prairie Provinces.* Publ. Canada Dept. Agr. Res. Br. No. 1662. Canadian Govern. Publ. Centre, Hull, Quebec, 863 pp.

Marie-Victorin, F. 1964. *Flore Laurentienne,* 2nd ed. (revised by E. Rouleau) Les Presses de l'Université de Montréal, Montréal, 925 pp.

Morton, J. K. and J. M. Venn. 1984. *The Flora of Manitoulin Island,* 2nd ed. Dept. of Biology, University of Waterloo, Waterloo, Ontario, 106 pp.

Porsild, A. E. and W. J. Cody. 1980. *Vascular Plants of Continental Northwest Territories, Canada.* Natl. Mus. Canada, Ottawa, 667 pp.

Roland, A. E. and E. C. Smith. 1966–69. The Flora of Nova Scotia. Rev. ed. 2 parts. *Proc. Nova Scotian Inst. Sci.* 26(2):3–224; 26(4):277–743.

Rousseau, C. 1974. *Geographie Floristique du Québec-Labrador.* Distribution des Principales Espéces Vasculaires. Les Presses de l'Université Laval, Quebec, 799 pp.

Scoggan, H. J. 1950. *The Flora of Bic and the Gaspé Peninsula, Quebec.* Nat. Mus. Canada Bull. No. 115. Ottawa, 399 pp.

Scoggan, H. J. 1957. *Flora of Manitoba.* Bull. Natl. Mus. Canada, No. 140. Ottawa, 619 pp.

Scoggan, H. J. 1978–79. *Flora of Canada.* Parts 1–4. Natl. Mus. Nat. Sci. Canada Publ. Bot. No. 7. Ottawa, 1711 pp.

Soper, J. H. and M. L. Heimburger. 1982. *Shrubs of Ontario.* Royal Ontario Museum, Life Sciences Miscellaneous Publ. Toronto, 495 pp.

Taylor, R. L. and B. MacBryde. 1977. *Vascular Plants of British Columbia: A Descriptive*

*Resource Inventory*. Univ. B.C. Bot. Gard. Tech. Bull. 4. Univ. British Columbia Press, Vancouver, 752 pp.
Taylor, T. M C. 1970. *Pacific Northwest Ferns and their Allies*. Univ. Toronto Press, Toronto, 247 pp.

## United States

### General and regional floras

Abrams, L. 1923–1960. *An Illustrated Flora of the Pacific States*. 4 vols. (Vol. 4 by Roxana Ferris). Stanford Univ. Press, Stanford, CA.
Batson, W. T. 1972. *A Guide to the Genera of Native and Commonly Introduced Ferns and Seed Plants of the Southeastern United States Excluding Peninsular Florida*. Publ. by the author, Univ. of S. Carolina, Columbia.
Batson, W. T. 1977. *A Guide to the Genera of Eastern Plants*. Wiley, New York, 203 pp.
Benson, L. & R. A. Darrow. 1981. *The Trees and Shrubs of the Southwestern Deserts*, 3rd ed. Univ. of Arizona and Univ. of New Mexico Presses. Tucson and Albuquerque, 476 pp.
Blackburn, B. 1952. *Trees and Shrubs in Eastern North America*. Oxford Univ. Press, New York, 358 pp.
Brockman, C. F. 1968. *Trees of North America*. Golden Press, New York, 280 pp.
Brown, H. P. 1938. *Trees of Northeastern United States Native and Naturalized*, rev. ed. Christopher, Boston, 490 pp.
Case, F. W. 1964. *Orchids of the Western Great Lakes Region*. Cranbrook Inst. of Science. Bull, No. 48. Bloomfield Hills, MI.
Chapman, A. W. 1897. *Flora of the Southern United States*, 3rd ed. American Book Co., New York, 655 pp.
Clovis, J. F. et al. 1972. *Common Vascular Plants of the Mid-Appalacian Region*. West Virginia Book Exchange, Morgantown, WV, 306 pp.
Cobb, B. 1956. *A Field Guide to the Ferns*. Houghton-Mifflin, Boston, 281 pp.
Coker, W. C. & H. R. Totten. 1945. *Trees of the Southeastern States*, 3rd ed. Univ. of North Carolina Press, Chapel Hill, 419 pp.
Correll, D. S. 1950. *Native Orchids of North America (North of Mexico)*. Chronica Botanica Co., Waltham, MA, 399 pp.
Correll, D. S. and H. B. Correll. 1972. *Aquatic and Wetland Plants of the Southwestern United States*. U.S. Environmental Protection Agency. (reprinted by Stanford Univ. Press, Stanford, CA), 1777 pp.
Craighead, J. J., F. C. Craighead, and R. J. Davis. 1963. *A Field Guide to Rocky Mountain Wildflowers*. Houghton-Mifflin, Boston, 277 pp.
Cronquist, A. 1980. *Vascular Flora of the Southeastern United States. Vol. 1. Asteraceae*. Univ. North Carolina Press, Chapel Hill, NC, 261 pp.
Cronquist, A. et al. 1972–. *Intermountain Flora: Vascular Plants of the Intermountain West, U.S.A.* (To date volumes 1, 4, 7 of a projected six volumes have been published) Hafner Publ. Co. for the New York Botanical Garden, New York.
Duncan, W. H. 1975. *Woody Vines of the Southeastern United States*. Univ. of Georgia, Athens, GA, 76 pp.
Duncan, W. H. and M. B. Duncan. 1987. *The Smithsonian Guide to Seaside Plants of the Gulf and Atlantic Coasts from Louisiana to Massachusetts, Exclusive of Lower Peninsular Florida*. Smithsonian Inst. Press, Baltimore, MD, 409 pp.
Duncan, W. H. and M. B. Duncan. 1988. *Trees of the Southeastern United States*. Univ. of Georgia Press, Athens, GA, 322 pp.
Duncan, W. H. and L. E. Foote. 1975. *Wildflowers of the Southeastern United States*. Univ. of Georgia Press, Athens, GA, 296 pp.
Elias, T. S. 1987. *The Complete Trees of America*. Gramercy Publ., New York, 948 pp.

Enari, L. 1956. *Plants of the Pacific Northwest.* Binfords & Mort., Portland, OR, 315 pp.

Eyles, D. E. and J. L. Robertson, Jr. 1963. *A Guide and Key to the Aquatic Plants of the Southeastern United States.* U.S. Dept. of the Interior. Fish and Wildlife Service. Bureau of Sport Fisheries and Wildlife Circular 158, Washington, D.C.

Fassett, N. C. 1957. *A Manual of Aquatic Plnnts.* (rev. by E. C. Ogden) Univ. of Wisconsin Press, Madison, WI, 405 pp.

Fernald, M. L. 1950. *Gray's Manual of Botany,* 8th ed. American Book Co., New York, 1632 pp.

Gilkey, H. M. 1957. *Weeds of the Pacific Northwest.* Oregon State College. Corvallis, OR, 441 pp.

Gilkey, H. M. and L. R. Dennis. 1967. *Handbook of Northwestern Plants.* Oregon State Univ. Bookstores, Corvallis, OR, 505 pp.

Gleason, H. A. 1963. *New Britton and Brown Illustrated Flora of the Northeastern United States and Adjacent Canada.* 3 vols. Hafner, New York.

Gleason, H. A. and A. C. Cronquist. 1963. *Manual of the Vascular Plants of the Northeastern United States and Adjacent Canada.* Van Nostrand, New York. 810 pp.

Godfrey, R. K. and J. W. Wooten. 1979 & 1981. *Aquatic and Wetland Plants of Southeastern United States.* 2 vols. Univ. of Georgia Press, Athens, GA.

Gould, F. W. 1951. *Grasses of the Southwestern United States.* Univ. of Arizona Biol. Sci. Bull. No. 7. Tucson, AZ, 352 pp.

Great Plains Flora Association. 1986. *Flora of the Great Plains.* Univ. Press of Kansas, Lawrence, KS, 1392 pp.

Harrar, E. S. and J. G. Harrar. 1962. *Guide to Southern Trees,* 2nd ed. Dover, New York, 709 pp.

Haskins, L. L. 1967. *Wild Flowers of the Pacific Coast.* Binfords & Mort, Portland, OR, 406 pp.

Hermann, F. J. 1970. *Manual of the Carices of the Rocky Mountains and Colorado Basin.* Agr. Handbook No. 374. Forest Service. U.S.D.A. Supt. of Documents, Washington, D.C., 397 pp.

Hightshoe, G. L. 1988. *Native Trees, Shrubs, and Vines for Urban and Rural America. A Planting Design Manual for Environmental Designers.* Van Nostrand Reinhold, New York, 819 pp.

Hitchcock, A. S. 1951. *Manual of the Grasses of the United States,* 2nd ed. (rev. by Agnes Chase) Misc. Publ. No. 200 of the U.S.D.A., 1051 pp.

Hitchcock, C. L. and A. Cronquist. 1973. *Flora of the Pacific Northwest.* Univ. of Washington Press, Seattle, WA, 730 pp.

Hitchcock, C. L. et al. 1955–1969. *Vascular Plants of the Pacific Northwest.* 5 parts. Univ. of Washington Press, Seattle, WA.

Holmgren, A. H. and J. L. Reveal. 1966. *Checklist of the Vascular Plants of the Intermountain Region.* U.S. Forest Service Research Paper INT-32. Intermountain Forest and Range Exp. Sta. Ogden, UT, 160 pp.

Hotchkiss, N. 1967. *Underwater and Floating-Leaved Plants of the United States and Canada.* Resource Publ. 44 U.S. Dept. of the Interior. Bureau of Sport Fisheries and Wildlife, Washington, D.C.

Howell, J. T. and R. J. Long. 1970. The Ferns and Fern Allies of the Sierra Nevada in California and Nevada. *Four Seasons* 3(3):2–18.

Isely, D. 1973–. Leguminosae of the United States. 3 vols. (I. Subfamily Mimosoideae, *Mem. New York Bot. Gard.* 25(1):1–152; II. Subfamily Caesalpinioideae, *Mem. New York Bot. Gard.* 25(2):1–228).

Klimas, J. E. and J. A. Cunningham. 1974. *Wildflowers of Eastern America.* A. A. Knopf, New York, 273 pp.

Krüssmann, G. 1985. *Manual of Cultivated Conifers,* 2nd ed. (edited by H. D. Warda, transl. by M. E. Epp). Timber Press, Portland, OR, 361 pp.

Little, E. L., Jr. 1950. *Southwestern Trees. A Guide to the Native Species of New Mexico and Arizona.* U.S.D.A. Agric. Handbook No. 9. U.S. Govt. Printing Office, Washington, D.C., 109 pp.
Luer, C. A. 1975. *The Native Orchids of the United States and Canada, Excluding Florida.* New York Botanical Garden, New York, 361 pp.
McMinn, H. E. and E. Maino. 1946. *An Illustrated Manual of Pacific Coast Trees,* 2nd ed. Univ. of Calif. Press, Berkeley, CA, 409 pp.
Mitchell, A. 1987. *The Trees of North America.* Facts on File, Inc., New York, 208 pp.
Montgomery, F. H. 1964. *Weeds of Canada and the Northern United States.* Ryerson Press, Toronto, 226 pp.
Muenscher, W. C. 1950. *Keys to Woody Plants,* 6th ed. Comstock, Ithaca, New York, 108 pp.
Muenscher, W. C. 1955. *Weeds,* 2nd ed. Macmillan, New York, 560 pp.
New York Botanical Garden. 1905–1949; 1954–. *North American Flora.* Series I (Vols. 1–34), Series II in progress. New York Botanical Garden Bronx, NY.
Niehaus, T. F. 1976. *A Field Guide to Pacific States Wildflowers.* Houghton Mifflin, Boston, 432 pp.
Orr, R. T. and M. C. Orr. 1974. *Wildflowers of Western America.* A. A. Knopf, New York, 270 pp.
Peterson, R. T. and M. McKenny. 1968. *Field Guide to Wild Flowers of Northeastern and North Central North America.* Houghton-Mifflin, Boston, 420 pp.
Petrides, G. A. 1972. *A Field Guide to Trees and Shrubs,* 2nd ed. Houghton-Mifflin, Boston, 428 pp.
Pohl, R. W. 1968. *How to Know the Grasses,* 2nd ed. W. C. Brown Co., Dubuque, IA, 244 pp.
Preston, R. J. 1989. *North American Trees,* 4th ed. Iowa State Univ. Press, Ames, IA, 407 pp.
Ramseur, G. S. 1960. The Vascular Flora of High Mountain Communities of the Southern Appalachians. *J. Elisha Mitchell Sci. Soc.* 76:82–112.
Reed, C. F. 1970. *Selected Weeds of the United States.* U.S.D.A. Agr. Handbook No. 366, Washington, D.C., 463 pp.
Rehder, A. 1986. *Manual of Cultivated Trees and Shrubs Hardy in North America.* (reprint of 2nd ed.). Dioscorides Press, Portland OR, 996 pp.
Rickett, H. W. 1966–1973. *Wildflowers of the United States.* 6 vols. (1-Northeast, 2-Southeast, 3-Texas, 4-Southwest, 5-Northwest, 6-Central Mountains and Plains), McGraw-Hill, New York.
Rosendahl, C. O. 1955. *Trees and Shrubs of the Upper Midwest.* Univ. of Minnesota Press, Minneapolis, MN, 411 pp.
Ross, R. A. and H. L. Chambers. 1988. *Wildflowers of the Western Cascades.* Timber Press, Portland, OR, 140 pp.
Rydberg, P. A. 1922. *Flora of the Rocky Mountains and Adjacent Plains,* 2nd ed. New York Botanical Garden, Bronx, NY. (Reprinted 1954, 1961. Hafner, New York) 1143 pp.
Rydberg, P. A. 1932 *Flora of the Prairies and Plains of Central North America.* New York Botanical Garden, Bronx, NY. (Reprinted 1971 Dover, New York, 2 vols.), 969 pp.
Seymour, F. C. 1969. *Flora of New England.* C. E. Tuttle, Rutland, VT, 596 pp.
Shreve, F. and I. L. Wiggins. 1964. *Vegetation and Flora of the Sonoran Desert.* 2 vols. Stanford Univ. Press, Stanford, CA, 1740 pp.
Small, J. K. 1913. *Flora of the Southeastern United States,* 2nd ed. Publ. by the author, New York, 1394 pp.
Small, J. K. 1933. *Manual of the Southeastern Flora.* Univ. of North Carolina Press, Chapel Hill, NC, 1554 pp.

Small, J. K. 1938. *Ferns of the Southeastern United States.* Reprinted 1963. Hafner, New York, 517 pp.
Smith, J. P. 1975. *A Key to the Genera of Grasses of the Conterminous United States.* Mad River Press, Arcata, CA, 39 pp.
Stephens, H. A. 1973. *Woody Plants of the North Central Plains.* Univ. Press of Kansas, Lawrence, KS, 530 pp.
Stewart, A. N., L. R. Dennis, and H. M. Gilkey. 1963. *Aquatic Plants of the Pacific Northwest With Vegetative Keys,* 2nd ed. Oregon State Univ. Press, Corvallis, OR 261 pp.
Taylor, T. M. C. 1970. *Pacific Northwest Ferns and Their Allies.* Univ. of Toronto Press, Toronto, 247 pp.
Terrell, E. E. 1970. Spring Flora of the Chesapeake & Ohio Canal Area, Washington, D.C. to Seneca, Maryland. *Castanea* 35:1–26.
Tiner, R. W., Jr. 1987. *A Field Guide to Coastal Wetland Plants of the Northeastern United States.* Univ. Mass. Press, Amherst, MA, 285 pp.
Treshow, M., S. L. Welsh, and G. Moore. 1970. *Guide to the Woody Plants of the Mountain States.* Brigham Young Univ. Press, Provo, UT, 178 pp.
Vines, R. A. 1960. *Trees, Shrubs, and Woody Vines of the Southwest.* Univ. of Texas Press, Austin, TX 1104 pp.
Weber, W. A. 1976. *Rocky Mountain Flora,* 5th ed. Colorado Assoc. Univ. Press, Boulder, CO, 279 pp.
Weniger, D. 1969. *Cacti of the Southwest.* Univ. of Texas Press, Austin, TX, 249 pp.
Wherry, E. T. 1942. *Guide to Eastern Ferns,* 2nd ed. Univ. of Penn. Press, Philadelphia, PA, 252 pp.
Wherry, E. T. 1961. *The Fern Guide; Northeastern and Midland United States and Adjacent Canada.* Doubleday, Garden City, NY, 318 pp.
Wherry, E. T. 1964. *The Southern Fern Guide.* Doubleday, Garden City, NY, 349 pp.
Wood, C. E., Jr. et al. 1958–. Generic Flora of the Southeastern United States. *J. Arnold Arbor.*

### *Alabama*

Banks, D. J. 1965. A Checklist of the Grasses (Gramineae) of Alabama. *Castanea* 30:84–96.
Clark, R. C. 1972. The Woody Plants of Alabama. *Ann. Mo. Bot. Gard.* 58(2):99–242.
Dean, B. E. 1969. *Ferns of Alabama,* rev. ed. Southern Univ. Press, Birmingham, AL, 214 pp.
Dean, B. et al. 1973. *Wildflowers of Alabama and Adjoining States.* Univ. of Alabama Press, University, AL, 230 pp.
Martin, I. R. and W. B. Devall. 1949. *Forest Trees of Alabama.* Alabama Polytechnic Inst, Auburn, AL, 87 pp.

### *Alaska*

Hutlen, E. 1960. *Flora of the Aleutian Islands and Westernmost Alaska With Notes on the Flora of the Commander Islands.* J. Cramer, Weinheim, 376 pp.
Hutlen, E. 1968. *Flora of Alaska and Neighboring Territories.* Stanford Univ. Press, Stanford, CA, 1008 pp.
Potter, L. 1962. *Roadside Flowers of Alaska.* Roger Burt, Hanover, NH, 590 pp.
Viereck, L. A. and E. L. Little. 1972. *Alaska Trees and Shrubs.* U.S.D.A. Agr. Handbook No. 410, Washington, D.C., 265 pp.
Viereck, L. A. and E. L. Little, Jr. 1974. *Guide to Alaska Trees.* U.S.D.A. Agr. Handbook No. 472, Washginton, D.C., 98 pp.

Welsh, S. L. 1974. *Anderson's Flora of Alaska and Adjacent Parts of Canada.* Brigham Young Univ. Press, Provo, UT, 724 pp.
Wiggins, I. L. and J. H. Thomas. 1961. *Flora of the Alaska Slope.* Univ. of Toronto Press, Toronto, 425 pp.

### Arizona

Benson, L. 1969. *The Cacti of Arizona.* Univ. of Arizona Press, Tucson, AZ, 218 pp.
Kearney, T. H. and R. H. Peebles. 1960. *Arizona Flora,* 2nd ed. Univ. of California Press, Berkeley, CA 1085 pp.
Lehr, J. H. 1978. *A Catalogue of the Flora of Arizona.* Desert Bot. Gard. Phoenix, AZ, 203 pp.
McDougall, W. B. 1973. *Seed Plants of Northern Arizona.* Mus. N. Ariz. Flagstaff, AZ 594 pp.
Tidestrom, I. and T. Kittell. 1941. *A Flora of Arizona and New Mexico.* Catholic Univ. Press, Washington, D.C., 897 pp.

### Arkansas

Demaree, D. 1943. A Catalogue of Vascular Plants of Arkansas. *Taxodium* 1(1):1–88.
Lang, J. M. 1969. The Labiatae of Arkansas. *Proc. Arkansas Acad. Sci.* 20:75–84.
Moore, D. W. 1950. *Trees of Arkansas.* Ark. Division of Forestry and Parks, Little Rock, AR, 119 pp.
Smith, E. B. 1973. An Annotated List of the Compositae of Arkansas. *Castanea* 38:79–109.
Smith, E. B. 1978. *An Atlas and Annotated List of Vascular Plants of Arkansas.* Univ. Arkansas Bookstore, Fayetteville, AR, 592 pp.
Wilcox, W. H. 1973. A Survey of the Vascular Flora of Crittenden County, Arkansas. *Castanea* 38(3):286–297.

### California

Benson, L. 1969. *The Native Cacti of California.* Stanford Univ. Press, Stanford, CA, 243 pp.
Bowerman, M. 1944. *The Flowering Plants and Ferns of Mount Diablo, California.* Gillick Press, Berkeley, CA, 290 pp.
Collins, B. J. *Key to Coastal and Chaparral Flowering Plants of Southern California.* Calif. State Univ. Foundation, Northridge, CA, 249 pp.
Crampton, B. 1974. *Grasses in California.* Calif. Nat. Hist. Guides: 33. Univ. of Calif. Press, Berkeley, CA, 178 pp.
Dawson, E. Y. 1966. *Cacti of California.* Calif. Nat. Hist. Guides: 18. Univ. of Calif. Press, Berkeley, CA, 64 pp.
Ferlatte, W. J. 1974. *A Flora of the Trinity Alps of Northern California.* Univ. of Calif. Press, Berkeley, CA, 206 pp.
Ferris, R. S. 1968. *Native Shrubs of the San Francisco Bay Region.* California Nat. Hist. Guides: 24. Univ. of Calif. Press, Berkeley, CA, 82 pp.
Gillett, G. W., J. T. Howell, and H. Leschke. 1961. A Flora of Lassen Volcanic National Park, California. *Wasmann J. Biol.* 19(1):1–185.
Grillos, S. J. 1966. *Ferns and Fern Allies of California.* Calif. Nat. Hist. Guides: 16. Univ. of Calif. Press, Berkeley, CA, 104 pp.
Hall, H. M. and C. C. Hall. 1972. *A Yosemite Flora.* Paul Elder & Co., San Francisco, CA, 282 pp.
Holt, V. 1962. *Keys for Identification of Wild Flowers, Ferns, Trees, Shrubs, and Woody Vines of Northern California,* rev. ed. National Press Publ., Palo Alto, CA, 174 pp.

Hoover, R. F. 1970. *The Vascular Plants of San Luis Obispo County, California.* Univ. of Calif. Press, Berkeley, CA, 350 pp.

Howell, J. T. 1970. *Marin Flora: Manual of the Flowering Plants and Ferns of Marin County, California,* 2nd ed. Univ. of Calif. Press, Berkeley, CA, 295 pp.

Howell, J. T., P. H. Raven, and P. Rubtzoff. 1958. A Flora of San Francisco, California. *Wasmann J. Biol.* 16(1):1–157.

Howitt, B. F. and J. T. Howell. 1964. *The Vascular Plants of Monterey County, California.* Univ. of San Francisco Press, San Francisco, CA, 184 pp.

Howitt, B. F. and J. T. Howell. 1973. *Supplement to the Vascular Plants of Monterey County, California.* Pacific Grove Mus. of Nat. Hist. Assoc., 60 pp.

Jepson, W. L. 1909–1943. *A Flora of California,* 3 vols., incomplete. Assoc. Students Store, Univ. of Calif., Berkeley, CA.

Jepson, W. L. 1925. *A Manual of the Flowering Plants of California.* Univ. of Calif. Press, Berkeley, CA, 1238 pp.

Lloyd, R. and R. Mitchell. 1973. *A Flora of the White Mountains, California and Nevada.* Univ. of Calif. Press, Berkeley, CA, 202 pp.

Mason, H. L. 1957. *A Flora of the Marshes of California.* Univ. of Calif. Press, Berkeley, CA, 878 pp.

McClintock, E., W. Knight, and N. Fahy. 1968. A Flora of the San Bruno Mountains, San Mateo County, California, *Proc. Calif. Acad. Sci.,* 4th Ser. 32(2):587–677.

McMinn, H. E. 1951. *An Illustrated Manual of California Shrubs.* Univ. of Calif. Press, Berkeley, CA, 633 pp.

Metcalf, W. 1968. *Introduced Trees of Central California.* Calif. Nat. Hist. Guides: 27. Univ. of Calif. Press, Berkeley, CA, 159 pp.

Millspaugh, C. F. and L. W. Nuttall. 1923. *Flora of Santa Catalina Island.* Field Mus. Nat. Hist. Publ. 212. Bot. Series. V. 413 pp.

Muller, K. K., R. E. Broder, and W. Beittel. 1974. *Trees of Santa Barbara.* Santa Barbara Bot. Garden, Santa Barbara, CA, 248 pp.

Munz, P. A. 1959. *A California Flora.* Univ. of Calif. Press, Berkeley, CA, 1681 pp.

Munz, P. A. 1961. *California Spring Wildflowers: From the Base of the Sierra Nevada and Southern Mountains to the Sea.* Univ. of California Press, Berkeley, CA, 123 pp.

Munz, P. A. 1962. *California Desert Wildflowers.* Univ. of Calif. Press, Berkeley, CA, 122 pp.

Munz, P. A. 1963. *California Mountain Wildflowers.* Univ. of Calif. Press, Berkeley, CA, 122 pp.

Munz, P. A. 1965. *Shore Wild Flowers of California, Oregon, and Washington.* Univ. of Calif. Press, Berkeley, CA, 122 pp.

Munz, P. A. 1968. *Supplement to a California Flora.* Univ. of Calif. Press, Berkeley, CA, 224 pp.

Munz, P. A. 1974. *A Flora of Southern California.* Univ. of Calif. Press, Berkeley, CA, 1086 pp.

Niehaus, T. F. 1974. *Sierra Wild Flowers.* Calif. Nat. Hist. Guides: 32. Univ. of Calif. Press, Berkeley, CA, 223 pp.

Peterson, P. V. 1966. *Native Trees of Southern California.* Calif. Nat. Hist. Guides: 14. Univ. of Calif. Press, Berkeley, CA, 136 pp.

Peterson, P. V. and P. V. Peterson, Jr. 1975. *Native Trees of the Sierra Nevada.* Calif. Nat. Hist. Guides: 36. Univ. of Calif. Press, Berkeley, CA, 147 pp.

Philbrick, R. N. 1972. Plants of Santa Barbara Island. *Madrono* 21:329–393.

Pusateri, S. J. 1963. *Flora of Our Sierran National Parks.* Carl & Irving Printers, Tulare, CA, 170 pp.

Raven, P. H. 1966. *Native Shrubs of Southern California.* Calif. Nat. Hist. Guides: 15. Univ. of Calif. Press, Berkeley, CA, 132 pp.

Raven, P. H. and H. J. Thompson. 1966. *Flora of the Santa Monica Mountains, California.* Student Bookstore, Univ. of California, Los Angeles, CA, 189 pp.

Robbins, W. W., M. K. Bellue, and W. S. Ball. 1970. *Weeds of California*. State of Calif., Sacramento, CA, 547 pp.
Rodin, R. J. 1960. Ferns of the Sierra. Yosemite Nat. Hist. Assoc. Special issue of *Yosemite Nature Notes* 39(4), 79 pp.
Sharsmith, H. K. 1965. *Spring Wild Flowers of the San Francisco Bay Region*. Calif. Nat. Hist. Guides: 11. Univ. of Calif. Press, Berkeley, CA, 192 pp.
Smith, C. F. 1976. *A Flora of the Santa Barbara Region, California*. Santa Barbara Mus. of Nat. Hist., Santa Barbara, CA, 331 pp.
Smith, G. L. 1973. A Flora of Tahoe Basin and Neighboring Areas. *Wasmann J. Biol.* 31(1), 231 pp.
Thomas, J. H. 1961. *Flora of the Santa Cruz Mountains of California, A Manual of Vascular Plants*. Stanford Univ. Press. Stanford, CA, 434 pp.
Thomas, J. H. and D. R. Parnell. 1974. *Native Shrubs of the Sierra Nevada*. Calif. Nat. Hist. Guides: 34. Univ. of Calif. Press, Berkeley, CA, 127 pp.
Thorne, R. F. 1967. A Flora of Santa Catalina Island, California. *Aliso* 6(3):1–77.
True, G. H. 1973. *The Ferns and Seed Plants of Nevada County, California*. Cal. Acad. Sci., San Francisco, CA, 62 pp.
Twisselmann, E. C. 1956. A Flora of the Temblor Range and the Neighboring Part of the San Joaquin Valley. *Wasmann J. Biol.* 14:161–300
Twisselmann, E. C. 1967. *A Flora of Kern County, California*. Univ. of San Francisco, San Francisco, CA, 295 pp.
Witham, H. V. 1972. *Ferns of San Diego County*. San Diego Nat. Hist. Mus., 72 pp.

### Colorado

Barrell, J. 1969. *Flora of the Gunnison Basin*. Natural Land Inst., Rockford, IL, 494 pp.
Harrington, H. D. 1964. *Manual of the Plants of Colorado*, 2nd ed. Swallow Press, Denver, CO, 666 pp.
Matsumara, Y. and H. D. Harrington. 1955. *The True Aquatic Vascular Plants of Colorado*. Tech. Bull. 57. Colorado Agr. Exp. Sta.
Nelson, R. A. 1970. *Plants of Rocky Mountain National Park*. Rocky Mountain Nature Assoc., 168 pp.
Weber, W. A. 1976. *Colorado Flora: Western Slope*. Colorado Associated University Press, Boulder, CO, 530 pp.

### Connecticut

Graves, C. B. et al. 1910. Catalogue of the Flowering Plants and Ferns of Connecticut Growing Without Cultivation. *Bull. Conn. Geol. & Nat. Hist. Rev.* 14:1–569.
Harger, E. B. et al. 1931. Additions to the Flora of Connecticut. *Conn. State Geol. Surv. Bull.* 48:1–94.

### Delaware

Taber, W. S. 1937. *Delaware Trees*. Delaware State Forestry Dept., Dover, DE, 250 pp.
Tatnall, R. R. 1946. *Flora of Delaware and the Eastern Shore*. Soc. Nat. Hist. Delaware, Wilmington, DE, 313 pp.

### District of Columbia

Hitchock, A. S. and P. C. Standley. 1919. Flora of the District of Columbia and Vicinity. *Contr. U.S. Nat. Herb.* 21, 329 pp.

### Florida

Conrad, H. S. 1969. *Plants of Central Florida*. Ridge Audubon Soc., Lake Wales, FL, 143 pp.

Godfrey, R. K. 1988. *Trees, Shrubs & Woody Vines of Northern Florida & Adjacent Southern Georgia & Alabama.* University of Georgia Press, Athens, GA, 728 pp.
Kurz, H. and R. K. Godfrey. 1962. *Trees of Northern Florida.* Univ. of Florida Press, Gainesville, FL, 311 pp.
Lakela, O. and R. W. Long. 1976. *Ferns of Florida.* Banyan Books, Miami, FL 192 pp.
Lakela, O. et al. 1976. *Plants of the Tampa Bay Area.* Banyan Books, Miami, FL, 198 pp.
Long, R. W. and O. Lakela. 1971. *A Flora of Tropical Florida.* Univ. of Miami Press, Coral Gables, FL, 962 pp.
Luer, C. A. 1972. *The Native Orchids of Florida.* New York Botanical Garden, Bronx, NY, 293 pp.
Murrill, W. A. 1945. *A Guide to Florida Plants.* Publ. by the author, Gainesville, FL, 89 pp.
Small, J. K. 1913. *Flora of the Florida Keys.* Publ. by the author, New York, 162 pp.
Small, J. K. 1913. *Florida Trees.* Publ. by the author, New York, 107 pp.
Small, J. K. 1913. *Shrubs of Florida.* Publ. by the author, New York, 140 pp.
Tomlinson, P. B. 1980. *The Biology of Trees Native to Tropical Florida.* Harvard Univ. Press, Allston, MA, 480 pp.
Ward, D. B. et al. 1962–75. *Contribution to the Flora of Florida.* Agr. Exp. Sta. and Univ. of Florida, Gainesville, FL.
Ward, D. B. ed. 1977–. Keys to the Flora of Florida. 1–. *Phytologia.*

### *Georgia*

Bishop, G. N. 1948. *Native Trees of Georgia,* 2nd ed. Univ. of Georgia School of Forestry, Athens, GA, 96 pp.
Duncan, W. H. 1941. *Guide to Georgia Trees.* Univ. of Georgia Press, Athens, GA, 63 pp.
Duncan, W. H. and J. T. Kartesz. 1981. *Vascular Flora of Georgia.* Univ. of Georgia Press, Athens, GA, 143 pp.
Jones, S. B., Jr. 1974. The Flora and Phytogeography of the Pine Mtn. Region of Georgia. *Castanea* 39(2):113–149.
Jones, S. B., Jr. and N. C. Coile. 1988. *The Distribution of the Vascular Flora of Georgia.* Herbarium, Department of Botany, Univ. of Georgia, Athens, GA, 230 pp.
McVaugh, R. and J. H. Pyron. 1951. *Ferns of Georgia.* Univ. of Georgia Press, Athens, GA, 195 pp.
Thorne, R. F. 1954. The Vascular Plants of Southwestern Georgia. *Amer. Midl. Naturalist* 52:257–327.

### *Idaho*

Davis, R. J. 1952. *Flora of Idaho.* W. C. Brown Co., Dubuque, IA, 828 pp.
St. John, H. 1963. *Flora of Southeastern Washington and Adjacent Idaho,* 3rd ed. Outdoor Pictures, Escondido, CA, 583 pp.

### *Illinois*

Dobbs, R. J. 1963. *Flora of Henry County, Illinois.* Nat. Land Inst., Rockford, IL, 350 pp.
Fell, E. W. 1955. *Flora of Winnebago County, Illinois.* Nature Conservancy, Washington, D.C., 207 pp.
Fuller, G. D. 1955. *Forest Trees of Illinois.* Dept. of Conservation. Division of Forestry, Springfield, IL, 71 pp.
Jones, G. N. 1963. *Flora of Illinois,* 3rd ed. Amer. Midl. Naturalist Monog. No. 7. Univ. of Notre Dame, Notre Dame, IN, 401 pp.
Jones, G. N. and G. D. Fuller. 1955. *Vascular Plants of Illinois.* Univ. of Illinois Press, Urbana, IL, 593 pp.

Mohlenbrock, R. H. 1970–1982. *The Illustrated Flora of Illinois.* (Ongoing series, 10 vols. published thus far.) S. Illinois Univ. Press, Carbondale and Edwardsville, IL.
Mohlenbrock, R. H. and J. W. Voight. 1959. *A Flora of Southern Illinois.* S. Illinois Univ. Press, Carbondale, IL, 390 pp.
Swink, F. and G. Wilhelm. 1979. *Plants of the Chicago Region,* rev. ed. Morton Arboretum, Lisle, IL, 922 pp.
Winterringer, G. S. 1967. *Wild Orchids of Illinois.* Ill. State Mus. Popular Sci. Series. Vol. 6., Springfield, IL.
Winterringer, G. S. and A. C. Lopinot. 1966. *Aquatic Plants of Illinois.* Ill. State Mus. Popular Sci. Series, Vol. 6. Ill. State Mus. Div. and Dept. of Cons., Div. of Fisheries, Springfield, IL.

***Indiana***

Crovello, T. J., C. A. Keller, and J. T. Katesz. 1983. *The Vascular Plants of Indiana: A Computer Based Checklist.* Univ. of Notre Dame Press, Notre Dame, IN, 136 pp.
Deam, C. C. 1940. *Flora of Indiana.* Indiana Dept. of Conservation, Indianapolis, IN, 1236 pp.
Peattie, D. C. 1930. *Flora of the Indiana Dunes: A Handbook of the Flowering Plants and Ferns of the Lake Michigan Coast of Indiana and of the Calumet District.* Field Mus. Nat. Hist., Chicago, IL, 432 pp.

***Iowa***

Beal, E. O. 1953. Aquatic Monocotyledons of Iowa. *Proc. Iowa Acad. Sci.* 60:89–91.
Beal, E. O. and P. H. Monson. 1954. Marsh and Aquatic Angiosperms of Iowa. Monocotyledons. *Iowa Univ. Studies Nat. Hist.* 19(5), 95 pp.
Campbell, R. B. 1961. *Trees of Iowa.* Agr. and Home Econ. Exp. Sta, Cooperative Extension Service, Iowa State Univ., Ames, IA, 63 pp.
Conard, H. S. 1951. *Plants of Iowa, Being a Seventh Edition of the Grinnell Flora,* Publ. by the author, 90 pp.
Cratty, R. I. 1933. The Iowa Flora. *Iowa State Coll. J. Sci.* 7:177–252.
Crawford, D. J. 1970. The Umbelliferae of Iowa. *Univ. Iowa Stud. Nat. Hist.* 21(4). 36 pp.
Davidson, R. A. 1959. The Vascular Flora of Southeastern Iowa. *Univ. Iowa Stud. Nat. Hist.* 20(2). 102 pp.
Eilers, L. J. 1971. The Vascular Flora of the Iowan Area. *Univ. Iowa Stud. Nat. Hist.* 21(5). 137 pp.
Gilly, C. L. 1946. The Cyperaceae of Iowa. *Iowa State Coll. J. Sci.* 21(1):55–151.
Greene, W. 1907. *Plants of Iowa.* Bull. State Hort. Soc., Des Moines, IA, 264 pp.
Guldner, L. F. 1960. *The Vascular Plants of Scott and Muscatine Counties,* Davenport Publ. Mus., 228 pp.
Hayde, A. 1943. A Botanical Survey in the Iowa Lake Region of Clay and Palo Alto Counties. *Iowa State Coll. J. Sci.* 17(3):277–416.
Pohl, R. W. 1966. The Grasses of Iowa. *Iowa State Coll. J. Sci.* 40(4):341–566.
Pohl, R. W. 1975. *Keys to Iowa Vascular Plants.* Kendall-Hunt, Dubuque, IA, 198 pp.

***Kansas***

Barkley, T. H. 1978. *A Manual of the Flowering Plants of Kansas,* 2nd ed. Kansas State Univ. Endowment Assoc., Manhattan, KS, 402 pp.
McGregor, R. L., R. E. Brooks, and L. A. Hauser. 1976. *Checklist of Kansas Vascular Plants.* State. Biol. Surv. Kansas, Tech. Pub. 2., Lawrence, KS, 168 pp.
Owensby, C. E. 1980. *Kansas Prairie Wildflowers.* Iowa State Univ. Press, Ames, IA, 124 pp.

Stevens, W. C. 1961. *Kansas Wild Flowers,* 2nd ed. Univ. of Kansas Press, Lawrence, KS, 461 pp.
Stephens, H. A. 1969. *Trees, Shrubs, and Woody Vines in Kansas.* Univ. Press of Kansas, Lawrence, KS, 250 pp.
Stephens, H. A. 1973. *Woody Plants of the North Central Plains.* Univ. Press of Kansas, Lawrence, KS, 530 pp.

### Kentucky

Beal, E. O. and J. W. Thieret. 1987. *Aquatic and Wetland Plants of Kentucky.* Kentucky Nature Preserves Comm. Sci. and Tech. Series Number 5, 419 pp.
Braun, E. L. 1943. *An Annotated Catalog of Spermatophytes of Kentucky.* Publ. by the author, Cincinnati, OH, 161 pp.
Gunn, C. R. 1968. *The Floras of Jefferson and Seven Adjacent Counties, Kentucky.* Ann. Kentucky Soc. Nat. Hist. No. 2., Louisville, KY, 322 pp.
Meijer, W. 1972. *Tree Flora of Kentucky.* Univ. of Kentucky, Mimeographed, 144 pp.
Wharton, M. E. 1971. *A Guide to the Wild Flowers and Ferns of Kentucky.* Univ. Press of Kentucky, Lexington, KY, 344 pp.
Wharton, M. E. and R. W. Barbour. 1973. *Trees and Shrubs of Kentucky.* Univ. Press. of Kentucky, Lexington, KY, 582 pp.

### Louisiana

Brown, C. A. 1945. *Louisiana Trees and Shrubs.* Louisiana Forestry Comm. Bull. No. 1., Baton Rouge, LA, 262 pp.
Brown, C. A. 1972. *Wild Flowers of Louisiana and Adjoining States.* Louisiana State Univ. Press, Baton Rouge, LA, 247 pp.
Brown, C. A. and D. A. Correll. 1942. *Ferns and Fern Allies of Louisiana.* Louisiana State Univ. Press, Baton Route, LA, 186 pp.
Rafinesque, C. S. 1967. *Florula Ludoviciana; or A Flora of the State of Louisiana* (transl. from French by C. C. Robin). Hafner, New York, 178 pp.
Reese, W. D. and J. W. Thieret. 1966. Botanical Study of the Five Islands of Louisiana. *Castanea* 31:251–277.
Thieret, J. W. 1972. *Checklist of the Vascular Flora of Louisiana.* Part I. Lafayette Nat. Hist. Mus. Tech. Bull. 2., Lafayette, LA, 48 pp.

### Maine

Beam, R. C., C. D. Richard, and F. Hyland. 1966. *Revised Checklist of the Vascular Plants of Maine.* Bull. Joss. Bot. Soc. No. 8. Orono, ME, 71 pp.
Cambell, C. S. and F. Hyland. 1977. *Winter Keys to Woody Plants of Maine,* Univ. Maine at Orono Press, Orono, ME, 52 pp.
Ogden, E. B. 1948. *The Ferns of Maine.* Univ. of Maine, Orono, ME, 128 pp.
Wallace, J. E. 1951. *The Orchids of Maine.* Univ. of Maine Bull. Vol. 53. Univ. of Maine Press, Orono, ME.

### Maryland

Brown, R. G. and M. L. Brown. 1972. *Woody Plants of Maryland.* Maryland Univ. Press, College Park, MD, 347 pp.
Kaylor, J. F. 1946. *Trees of Maryland.* Solomons, Dept. of Research and Education, 23 pp.
Shreve, F. et al. 1910. *The Plant Life of Maryland.* Maryland Weather Serv. (reprint 1969). Sp. Publ. No. 3. Johns Hopkins Univ. Press, Baltimore, MD, 533 pp.

### *Massachusetts*

Bigelow, J. 1840. *A Collection of Plants of Boston and Its Vicinity*, 3rd ed. Chanlers C. Little and James Brown, Boston, MA, 468 pp.

Eaton, R. J. 1974. *A Flora of Concord.* Mus. of Comp. Zool., Cambridge, MA, 236 pp.

Harris, S. K. et al. 1975. *Flora of Essex County, Massachusetts.* Peabody Mus., Salem, MA.

MacKeever, F. C. 1968. *Native and Naturalized Plants of Nantucket.* Univ. of Mass. Press, Amherst, MA, 132 pp.

### *Michigan*

Barnes, B. V. and W. H. Wagner, Jr. 1981. *Michigan Trees.* Univ. of Michigan Press, Ann Arbor, MI, 343 pp.

Billington, C. 1949. *Shrubs of Michigan*, 2nd ed. Cranbrook Inst. Sci., Bloomfield Hills, MI, 339 pp.

Billington, C. 1952. *Ferns of Michigan.* Cranbrook Inst. Sci. Bloomfield Hills, MI, 240 pp.

Gleason, H. A. 1939. *The Plants of Michigan.* Wahr's Bookstore. Ann Arbor, MI, 204 pp.

Hall, M. T. 1959. *An Annotated List of the Plants of Oakland County, Michigan.* Cranbrook Inst. of Sci., Bloomfield Hills, MI, 93 pp.

Smith, H. V. 1966. *Michigan Wild Flowers.* Cranbrook Inst. of Sci. Bull. No. 42 (revised). Bloomfield Hills, MI, 468 pp.

Voss, E. G. 1967. A Vegetative Key to the Genera of Submerged and Floating Aquatic Vascular Plants of Michigan. *Mich. Bot.* 6:35–50.

Voss, E. G. 1972. *Michigan Flora. Part I: Gymnosperms and Monocots.* Cranbrook Inst. Sci. Bull. 55. Bloomfield Hills, MI, 488 pp.

Voss, E. G. 1985. *Michigan Flora, Part II: Dicots (Saururaceae–Cornaceae).* Cranbrook Inst. Sci. Bull. 59 and Univ. of Michigan Herbarium, Ann Arbor, MI, 724 pp.

### *Minnesota*

Clements, F. E., C. O. Rosendahl, and F. K. Butters. 1912. *Minnesota Trees and Shrubs.* Univ. Minnesota, Minneapolis, MN, 314 pp.

Lakela, O. 1965. *Flora of Northeastern Minnesota.* Univ. of Minnesota Press, Minneapolis, MN, 541 pp.

Monserud, W. and G. B. Ownbey. 1971. *Common Wildflowers of Minnesota.* Univ. of Minnesota Press, Minneapolis, MN, 331 pp.

Morley, T. 1969. *Spring Flora of Minnesota.* Univ. Minnesota Press, Minneapolis, MN, 283 pp.

Moyle, J. B. 1964. *Northern Non-woody Plants: A Field Key to the More Common Ferns and Flowering Plants of Minnesota and Adjacent Regions.* Burgess Publ. Co., Minneapolis, MN, 108 pp.

Rosendahl, C. O. and F. K. Butters. 1928. *Trees and Shrubs of Minnesota.* Univ. Minn. Press, Minneapolis, MN, 385 pp.

### *Mississippi*

Gunn, C. R. et al. 1980. *Vascular Flora of Washington County, Mississippi and Environs.* Sci. and Educ. Admin., USDA, New Orleans, LA.

Jones, S. B., Jr. 1974. Mississippi Flora. I. Monocotyledon Families with Aquatic or Wetland Species. *Gulf Res. Rep.* 4(3):357–379.

Jones, S. B., Jr. 1974. Mississippi Flora. II. Distribution and Identification of the Onagraceae. *Castanea* 39:370–379.

Jones, S. B., Jr. 1975. Mississippi Flora. III. Distribution and Identification of the Brassicaceae. *Castanea* 40:238–252.
Jones, S. B., Jr. 1975. Mississippi Flora. IV. Dicotyledon Families with Aquatic or Wetland Species. *Gulf Res. Rep.* 5(1):7–22.
Jones, S. B., Jr. 1976. Mississippi Flora. V. The Mint Family. *Castanea* 41:41–58.
Jones, S. B., Jr., T. M. Pullen, and J. R. Watson. 1969. The Pteridophytes of Mississippi. *Sida* 3:359–364.
Lowe, E. N. 1921. Plants of Mississippi. *Mississippi State Geol. Survey Bull.* 17:1–292.
Mattoon, W. R. and J. M. Beal. 1936. *Forest Trees of Mississippi.* Mississippi State College, State College, MS, 80 pp.

### *Missouri*

Eisendrath, E. R. 1978. *Missouri Wild Flowers of the St. Louis Area.* Monogr. Sys. Bot. Vol. 2. Missouri Bot. Gard., St. Louis, MO, 390 pp.
Handebrink, E. L. 1958. *The Flora of Southeast Missouri.* Publ. by the author, Kennett, MO, 78 pp.
Kucera, C. L. 1961. *The Grasses of Missouri.* Univ. of Missouri Press, Columbia, MO, 241 pp.
Settergren, C. and R. E. McDermott. 1962. *Trees of Missouri.* Agr. Exp. Sta., Univ. of Missouri, 123 pp.
Steyermark, J. A. 1940. *Spring Flora of Missouri.* Missouri Bot. Gard., St. Louis, MO, 582 pp.
Steyermark, J. A. 1963. *Flora of Missouri.* Iowa State Univ. Press, Ames, IA, 1725 pp.

### *Montana*

Booth, W. E. 1950. *Flora of Montana. Part I. Conifers and Monocots.* Montana State College Research Foundation. Bozeman, MT, 232 pp.
Booth, W. E. and J. C. Wright. 1962. *Flora of Montana. Part II. Dicotyledons.* Montana State College. Bozeman, MT, 280 pp.
Dorn, R. D. 1984. *Vascular Plants of Montana.* Mountain West Publ., Cheyenne, WY, 276 pp.
Drummond, J. 1949. *Native Trees of Montana.* Mont. State Col. Ext. Serv., Bozeman, MT, 44 pp.
Hahn, B. E. 1973. *Flora of Montana: Conifers annd Monocots.* Montana State Univ., Bozeman, MT, 143 pp.
Lackschewitz, K. 1986. *Plants of West-Central Montana—Identification and Ecology: Annotated Checklist.* U.S.D.A. Forest Service. Intermountain Res. Sta. Gen. Tech. Rep. INT-217, 128 pp.
Stanley, P. C. 1921. Flora of Glacier National Park, Montana. *Contr. U.S. Natl. Herb.* 22(5):235–438.

### *Nebraska*

Lommasson, R. C. 1973. *Nebraska Wild Flowers.* Univ. of Nebraska Press, Lincoln, NE, 185 pp.
Petersen, N. F. 1923. *Flora of Nebraska,* 3rd ed. Lincoln, NE, 220 pp.
Pool, R. J. 1951. *Handbook of Nebraska Trees,* rev. ed. Univ. of Nebraska Conserv. and Surv. Div., Lincoln, NE, 179 pp.
Winter, J. M. 1936. *An Analysis of the Flowering Plants of Nebraska.* Bull. Conserv. Dept. Nebr. No. 13, 203 pp.

### *Nevada*

Archer, W. A. et al. 1940–1965. *Contributions toward a Flora of Nevada.* Parts 1–50; incomplete. U.S. Natl. Arbor. and U.S.D.A., Washington, D.C.

Beatley, J. C. 1976. *Vascular Plants of the Nevada Test Site and Central-Southern Nevada: Ecologic and Geographic Distributions.* Technical Information Center, Energy Res. and Dev. Admin., 316 pp.

Clokey, I. W. 1951. *Flora of the Charleston Mountains, Clark County, Nevada.* Univ. of Calif. Press, Berkeley, CA, 274 pp.

Holmgren, A. H. 1942. *A Handbook of the Vascular Plants of Northeastern Nevada.* U.S. Dept. of Interior and Utah State Agr. College and Exp. Sta., Logan, UT, 214 pp.

Lloyd, R. and R. Mitchell. 1973. *A Flora of the White Mountains, California and Nevada.* Univ. of Calif. Press, Berkeley, CA, 202 pp.

Tidestrom, I. 1925. Flora of Utah and Nevada. *Contr. U.S. Natl. Herb.* 25:1–655.

### New Hampshire

Baldwin, H. I. 1974. The Flora of Mount Monadnock, New Hampshire. *Rhodora* 76:205–228.

Foster, J. H. 1941. *Trees and Shrubs of New Hampshire,* 2nd ed. Soc. for Protect. of N. H. Forests, Concord, NH, 112 pp.

Pease, A. S. 1964. *A Flora of Northern New Hampshire.* New England Bot. Club, Cambridge, MA, 278 pp.

Scamman, E. 1947. *Ferns and Fern Allies of New Hampshire.* New Hampshire Acad. Sci., Durham, NH, 96 pp.

### New Jersey

Britton, N. L. 1889. Catalogue of the Plants Found in New Jersey. *Final Rep. Geol. Surv. N. J.* 2:27–642.

Chrysler, M. A. and J. L. Edwards. 1947. *The Ferns of New Jersey Including the Fern Allies.* Rutgers. Univ. Press, New Brunswick, NJ, 201 pp.

Stone, W. 1911. *The Plants of Southern New Jersey.* (reprint 1973). MacCrellish and Quigley, Trenton, NJ.

### New Mexico

Martin, W. C. and C. R. Hutchins, 1980. *A Flora of New Mexico.* 2 vols. J. Cramer, Vaduz, 2591 pp.

Tidestrom, I. and T. Kittell. 1941. *A Flora of Arizona and New Mexico.* Catholic Univ. Press, Washington, D.C., 897 pp.

Wooten, E. O. and P. C. Standley. 1915. *Flora of New Mexico.* (Reprinted 1971.) Hafner, New York, 794 pp.

### New York

Brown, H. P. 1975. *Trees of New York State.* (Reprint of 1921 edition.) Dover, New York, 433 pp.

Gleason, H. A. 1962. *Plants of the Vicinity of New York,* 3rd ed. New York Botanical Gard., Hafner, New York, 307 pp.

House, H. D. and W. P. Alexander. 1927. *Flora of the Allegheny State Park Region.* N.Y. State Mus. Handbook No. 2. Univ. of State of New York, Albany, NY, 225 pp.

McVaugh, R. 1958. *Flora of the Columbia County Area, New York.* NY State Mus. and Sci. Serv. Bull. No. 360. Univ. of New York, Albany, NY, 400 pp.

Mitchell, R. J. and E. O. Beal. 1979. *Magnoliaceae through Ceratophyllaceae of New York State.* II. N.Y. State Mus. Bull. No. 435, 62 pp.

Torrey, J. 1843. *A Flora of New York,* 2 vols. Carroll and Cook, Albany, NY.

Wiegard, K. M. and A. J. Eames. 1926. *The Flora of the Cayuga Lake Basin, New York.* Cornell Univ. Agr. Exp. Sta. Mem. 92, 491 pp.

Zenkert, C. A. 1934. The Flora of the Niagara Frontier Region: Ferns and Flowering Plants of Buffalo, N.Y. and Vicinity. *Bull. Buffalo Soc. Nat. Sci.* No. 16, 328 pp.

### *North Carolina*

Blomquist, H. L. and H. J. Oosting. 1959. *A Guide to the Spring and Early Summer Flora of the Piedmont, North Carolina.* Publ. by the authors, Durham, NC, 181 pp.

Justice, W. S. and C. R. Bell. 1968. *Wild Flowers of North Carolina.* Univ. of North Carolina Press, Chapel Hill, NC, 217 pp.

Radford, A. E., H. E. Ahles, and C. R. Bell. 1968. *Manual of the Vascular Flora of the Carolinas.* Univ. of North Carolina Press, Chapel Hill, NC, 1183 pp.

Wilbur, R. L. 1963. *The Leguminous Plants of North Carolina.* N. Carolina Agr. Exp. Sta., Raleigh, NC, 294 pp.

### *North Dakota*

Stevens, O. A. 1963. *Handbook of North Dakota Plants.* North Dakota Inst. for Regional Stud., Fargo, ND, 324 pp.

### *Ohio*

Anliot, S. E. 1973. *The Vascular Flora of Glen Helen, Clifton Gorge, and John Bryan State Park.* Ohio Biol. Notes No. 5, 162 pp.

Braun, E. L. 1961. *The Woody Plants of Ohio. Trees, Shrubs, and Woody Climbers Native, Naturalized and Escaped.* Ohio State Univ. Press, Columbus, OH, 362 pp.

Braun, E. L. 1967. *The Vascular Flora of Ohio. The Monocotyledoneae.* Vol. 1. Ohio State Univ. Press. Columbus, OH, 464 pp.

Schaffner, J. H. 1928. *Field Manual of the Flora of Ohio.* R. G. Adams, Columbus, OH, 638 pp.

Weishaupt, C. G. 1960. *Vascular Plants of Ohio.* Publ. by the author, Columbus, OH, 309 pp.

### *Oklahoma*

Goodman, G. J. 1958. *Spring Flora of Central Oklahoma.* Univ. of Oklahoma Duplicating Service, Norman, OK, 126 pp.

McCoy, D. 1976. *Roadside Flowers of Oklahoma.* Publ. by the author, Lawton, OK, 115 pp.

Stemen, T. R. and W. S. Myers. 1937. *Oklahoma Flora.* Harlow, Oklahoma City, OK, 706 pp.

Waterfall, U. T. 1972. *Keys to the Flora of Oklahoma,* 5th ed. Oklahoma State Univ. Bookstore, Stillwater, OK, 246 pp.

### *Oregon*

Gilkey, H. M. and P. L. Packard. 1962. *Winter Twigs.* Oregon State Univ. Press, Corvallis, OR, 109 pp.

Hayes, D. W. and G. A. Garrison. 1960. *Key to Important Woody Plants of Eastern Oregon and Washington.* Agr. Handbook No. 148. U.S.D.A., 227 pp.

Horner, C. H. and E. S. Booth. 1953. Spring Flowers of Southeastern Washington and Northeastern Oregon. *Walla Walla Coll. Publ. Dept. Biol. Sci. and Biol. Sta.* 3(1):1–172.

Mason, G. 1975. *Guide to the Plants of the Wallowa Mountains of Northeastern Oregon.* Special Publ. Mus. of Nat. Hist. Univ. of Oregon, Eugene, OR, 411 pp.

Peck, M. E. 1961. *A Manual of the Higher Plants of Oregon,* 2nd ed. Oregon State Univ. Press, Corvallis, OR, 936 pp.

Weidemann, A. F., L. J. Dennis, and F. H. Smith. 1969. *Plants of the Oregon Coastal Dunes.* Oregon State Univ. Bookstores, Corvallis, OR.

### *Pennsylvania*

Cannan, E. D. 1946. *A Key to the Ferns of Pennsylvania.* Publ. by the author, Johnstown, PA, 112 pp.
Grimm, W. C. 1950. *The Trees of Pennsylvania.* Stackpole and Heck, Harrisburg, PA, 363 pp.
Grimm, W. C. 1952. *The Shrubs of Pennsylvania.* Stackpole, Harrisburg, PA, 522 pp.
Henry, L. K. et al. 1965. Western Pennsylvania Orchids. *Castanea* 40:93–168.
Jennings, O. E. 1953. *Wild Flowers of Western Pennsylvania and the Upper Ohio Basin.* 2 Vol. Univ. of Pittsburgh Press, Pittsburgh, PA 574 pp.
Kelly, J. P. 1937. *The Ferns and Flowering Plants of Central Pennsylvania.* Penn. State Coll., State College, PA, 120 pp.
Porter, T. C. 1903. *Flora of Pennsylvania.* Ginn and Co., Boston, 362 pp.
Small, J. K. and J. J. Carter. 1913. *Flora of Lancaster County.* Publ. by the authors, New York, 336 pp.
Wagner, P. R. 1943. *The Flora of Schuylkill County, Pennsylvania.* Univ. Pennsylvania, Philadelphia, PA, 230 pp.
Wherry, E. T., J. M. Fogg, Jr., and H. A. Wahl. 1979. *Atlas of the Flora of Pennsylvania.* Morris Arboretum, Philadelphia, PA, 390 pp.

### *Rhode Island*

Palmatier, E. A. 1952. *Flora of Rhode Island.* Dept. of Botany, Univ. of Rhode Island, Kingston, RI, 75 pp.

### *South Carolina*

Batson, W. T. 1970. *Wild Flowers in South Carolina.* Univ. of South Carolina Press, Columbia, SC, 146 pp.
Radford, A. E., H. E. Ahles, and C. R. Bell. 1968. *Manual of the Vascular Flora of the Carolinas.* Univ. of North Carolina Press, Chapel HIll, NC, 1183 pp.

### *South Dakota*

Van Bruggen, T. 1976. *Vascular Plants of South Dakota.* Iowa State Univ. Press, Ames, IA, 538 pp.

### *Tennessee*

Anderson, W. A., Jr. 1929. *The Ferns of Tennessee.* Univ. Tennessee, Knoxville, TN, 40 pp.
Gattinger, A. 1901. *The Flora of Tennessee and a Philosophy of Botany.* Publ. by the author, Nashville, TN, 296 pp.
Mahler, W. F. 1970. Manual of the Legumes of Tennessee. *J. Tenn. Acad. Sci.* 45:65–96.
Rogers, K. E. and F. D. Bowers. 1969–. Notes on Tennessee Plants. *Castanea* 34:294–397; 36:191–194; 38:335–339.
Shaver, J. M. 1954. *Ferns of Tennessee.* Geo. Peabody Coll. for Teachers, Nashville, TN, 502 pp.
Smith, C. R. and R. W. Pearman. 1971. A Survey of the Pteridophytes of Northeastern Tennessee. *Castanea* 36:181–191.
Wofford, B. E. and A. M. Evans. 1979–80. Atlas of the Vascular Plants of Tennessee. 1–3. *J. Tenn. Acad. Sci.* 54:32–8, 75–80; 55:110–114.

### Texas

Correll, D. S. 1956. *Ferns and Fern Allies of Texas.* Texas Research Foundation, Renner, TX, 188 pp.
Correll, D. S. and M. C. Johnston. 1970. *Manual of the Vascular Plants of Texas.* Texas Research Foundation, Renner, TX, 1881 pp.
Gould, F. W. 1975. *The Grasses of Texas.* Texas A & M Univ. Press, College Station, TX, 186 pp.
Johnston, E. G. 1972. *Texas Wild Flowers.* Shoal Creek Publ., Austin, TX, 205 pp.
Jones, F. B. 1975. *Flora of the Texas Coastal Bend.* Welder Wildlife Foundation, Sinton, TX, 262 pp.
Jones, F. B., C. M. Rowell, Jr., and M. C. Johnston. 1961. *Flowering Plants and Ferns of the Texas Coastal Bend Counties.* Welder Series. B-1. Welder Wildlife Foundation, Sinton, TX, 146 pp.
Lundell, C. L. et al. 1955–. *Flora of Texas* (projected 10-volume work). Vol. 1 (pt. 1–3) and Vol. 3 have been published. Texas Research Foundation, Renner, TX.
Mahler, W. F. 1971. *Keys to the Vascular Plants of the Black Gap Wildlife Management Area Brewster County, Texas,* 3rd ed. Publ. by the author.
Mahler, W. F. 1973. *Flora of Taylor County, Texas.* Southern Methodist Univ. Bookstore, Dallas, TX, 247 pp.
Reeves, R. G. 1972. *Flora of Central Texas.* Prestige Press, Ft. Worth, TX, 320 pp.
Reeves, R. G. and D. C. Bain. 1947. *Flora of South Central Texas.* Texas A & M College, College Station, TX, 298 pp.
Rickett, H. W. 1969. *Wild Flowers of the United States: Texas. Vol. 3* McGraw-Hill, New York, 553 pp.
Schulz, E. D. 1928. *Texas Wild Flowers.* Laidlaw Brothers, Chicago, 505 pp.
Shinners, L. H. 1972. *Spring Flora of the Dallas-Forth Worth Area, Texas,* 2nd ed. (Revised by W. F. Mahler.) Prestige Press, Fort Worth, TX, 514 pp.
Turner, B. L. 1959. *The Legumes of Texas.* Univ. of Texas Press, Austin, TX, 284 pp.
Vines, R. A. 1977. *Trees of East Texas.* Univ. of Texas Press, Austin, TX, 538 pp.
Wills, M. M. and H. S. Irwin. 1961. *Roadside Flowers of Texas.* Univ. of Texas, Austin, TX, 295 pp.

### Utah

Arnow, L. A. 1980. *Flora of the Central Wasatch Front, Utah.* Univ. Utah Printing Service, Salt Lake City, UT, 663 pp.
Flowers, S. 1944. Ferns of Utah. *Bull. Univ. Utah.* 35(7): *Biol. Sci.* 4(6):1–87.
Holmgren, A. H. 1948. *Handbook of the Vascular Plants of the Northern Wasatch.* Publ. by the author, 202 pp.
Tidestrom, I. 1925. Flora of Utah and Nevada. *Contr. U.S. Nat. Herb.* 25:1–655.
Welsh, S. L. and G. Moore. 1973. *Utah Plants: Tracheophyta,* 3rd ed. Brigham Young Univ. Press, Provo, UT, 474 pp.
Welsh, S. L., et al. 1987. *A Utah Flora.* Great Basin, Nat. No. 9, 894 pp.

### Vermont

Atwood, J. T., Jr., ed. 1973. *Checklist of Vermont Plants.* Vermont Bot. Bird Club, 90 pp.
Seymour, F. C. 1969. *The Flora of Vermont,* 4th Ed. Agr. Exp. Sta. Bull. 660. Univ. of Vermont, Burlington, VT, 393 pp.

### Virginia

Cooperrider, T. S. and R. F. Thorne. 1964. The Flora of Giles County, Virginia. II. *Castanea* 29:46–70.

Harvill, A. M., Jr. 1970. *Spring Flora of Virginia.* Publ. by the author, Farmville, VA, 240 pp.
Harvill, A. M., Jr. 1977. *Atlas of the Virginia Flora. I. Pteridophytes through Monocotyledons.* Virginia Bot. Assoc., Farmville, VA, 59 pp.
Hathaway, W. T. and G. W. Ramsey. 1973. The Flora of Pittsylvania County, Virginia. *Castanea* 38:38–78.
Massey, A. B. 1944. *The Ferns and Fern Allies of Virginia.* Virginia Polytech. Inst., Blacksburg, VA, 110 pp.
Massey, A. B. 1961. *Virginia Flora.* Virginia Agr. Exp. Sta. Tech. Bull. 152, Blacksburg, VA, 258 pp.
Mazzeo, P. M. 1972. The Gymnosperms of Virginia: A Contribution Towards a Proposed State Flora. *Castanea* 37:179–195.
Merriman, P. R. 1930. *Flora of Richmond and Vicinity (exclusive of grasses, sedges and trees).* Virginia Acad. of Sci., Richmond, VA, 353 pp.
Silberhorn, G. M. 1976. *Tidal Wetland Plants of Virginia.* Educ. Series No. 19, Virginia Inst. of Marine Science, Gloucester Point, VA, 86 pp.

### *Washington*

Brockman, C. F. 1947. *Flora of Mount Ranier National Park.* National Park Services. U.S. Govt. Printing OFfice, Washington D.C., 170 pp.
Horner, C. E. and E. S. Booth. 1953. Spring Flowers of Southeastern Washington and Northeastern Oregon. *Walla Walla College Publ. Dept. Biol. Sci. and Biol. Sta.* 3(1):1–172.
Jones, G. N. 1938. *The Flowering Plants and Ferns of Mt. Rainier.* Univ. of Wash. Publ. in Biol. 7, 192 pp.
Piper, C. V. 1906. Flora of the State of Washington. *Contr. U.S. Natl. Herb.* 11:1–632.
St. John, H. 1963. *Flora of Southeastern Washington and Adjacent Idaho,* 3rd ed. Outdoor Pictures, Escondido, CA, 583 pp.

### *West Virginia*

Clarkson, R. B. 1966. The Vascular Flora of the Monongahela National Forest, West Virginia. *Castanea* 31:1–120.
Core, E. L. and N. Ammons. 1946. *Woody Plants of West Virginia in Winter Condition.* Book Exchange. Morgantown, WV, 124 pp.
Core, E. L. et al. 1960. *Plant Life of West Virginia.* Scholar's Library, New York, 224 pp.
Strausbaugh, P. D. and E. L. Core. 1971. *Flora of West Virginia,* 2nd ed. Seneca Books, Grantsville, WV, 1079 pp.
Strausbaugh, P. D., E. L. Core, and N. Ammons. 1955. *Common Seed Plants of the Mid-Appalachian Region,* 2nd ed. Book Exchange, Morgantown, WV, 305 pp.

### *Wisconsin*

Fassett, N. C. 1939. *The Leguminous Plants of Wisconsin.* Univ. of Wisconsin Press, Madison, WI, 157 pp.
Fassett, N. C. 1951. *Grasses of Wisconsin.* Univ. of Wisconsin Press, Madison, WI, 173 pp.
Fassett, N. C. 1976. *Spring Flora of Wisconsin,* 4th ed. (Revised by O. S. Thomson.) Univ. of Wisconsin Press, Madison, WI, 413 pp.
Fassett, N. C. et al. 1929–1974. *Preliminary Reports on the Flora of Wisconsin. I–.* Trans. Wisconsin Acad. Sci., Madison, WI.
Freckmann, R. W. 1972. *Grasses of Central Wisconsin.* Reports on the Fauna and Flora of Wisconsin. Rep. No. 6. Mus. of Nat. Hist. Univ. of Wisc., Stevens Point, WI, 81 pp.

Hartley, T. G. 1966. The Flora of the Driftless Area. *Univ. Iowa Stud. Nat. Hist.* 21(1). 174 pp.
Tryon, R. M., Jr., et al. 1953. *The Ferns and Fern Allies of Wisconsin,* 2nd ed. Univ. of Wisconsin Press, Madison, WI, 158 pp.

### *Wyoming*

Dorn, R. G. 1977. *Manual of Vascular Plants of Wyoming.* 2 vols. Garland, New York, 1498 pp.
Dorn, R. D. 1988. *Vascular Plants of Wyoming.* Mountain West Publ., Cheyenne, WY, 340 pp.
Hallsten, G. P., Q. D. Skinner, and A. A. Beetle. 1987. *Grasses of Wyoming,* 3rd ed. Res. J. 202, Agr. Exp. Sta. Univ. of Wyoming, Laramie, WY, 432 pp.
Johnson, W. M. 1964. *Field Key to the Sedges of Wyoming.* Univ. Wyoming Agr. Exp. Sta., Laramie, WY, 239 pp.
McDougall, W. B. and H. A. Baggley. 1956. *Plants of Yellowstone National Park,* 2nd ed. Yellowstone Library and Mus. Assoc., Yellowstone Park, WY, 186 pp.
Porter, C. L. 1944–1961. *Contributions Toward a Flora of Wyoming.* No. 1–34. Rocky Mountain Herbarium. Univ. of Wyoming. Laramie, WY.
Porter, C. L. 1962–1972. *A Flora of Wyoming.* Issued in Parts by the Agr. Exp. Sta. Univ. of Wyoming, Laramie, WY.
Shaw, R. J. 1976. *Field Guide to the Vascular Plants of Grand Teton National Park and Teton County, Wyoming.* Utah State Univ. Press, Logan, UT, 301 pp.

### *Mexico and Central America*

Allen, P. H. 1977. *The Rain Forests of Golfo Dulce.* (reprint of 1956 ed.) Stanford Univ. Press, Stanford, CA, 417 pp.
Burger, W. ed. 1971–. *Flora Costaricensis.* In parts. *Fieldiana,* Bot. No. 35 etc., Chicago.
Conzatti, C. 1939–1947. *Flora Taxonómica Mexicana.* 2 vols. Sociedad Mexicana de Historia Natural, Mexico City, Mexico.
Coyle, J. and N. C. Roberts. 1975. *A Field Guide to the Common and Interesting Plants of Baja California.* Nat. Hist. Publ. Co., La Jolla, CA, 206 pp.
Croat, T. B. 1978. *Flora of Barro Colorado Island.* Stanford Univ. Press, Stanford, CA, 943 pp.
D'Arcy, W. G. 1987. *Flora of Panama Checklist and Index.* I. II. Monogr. Syst. Botany Missouri Bot. Garden. Vol. 17, 328 pp; Vol. 18, 672 pp.
Gentry, H. S. 1972. *Rio Mayo Plants.* Publ. Carnegie Inst. Wash. No. 527. Washington D.C., 328 pp.
Hemsley, W. B. 1879–88. *Biologia Centrali-Americana or Contributions to the Knowledge of the Fauna and Flora of Mexico and Central America.* (F. D. Godman and O. Salvin, eds.) Bot. 5 vols. R. H. Porter, London.
Holdridge, L. R. 1970. *Manual Dendrólogico Para 1000 Especies Arbóreas en la Repúblico de Panama.* FAO, United Nations, Panama, 325 pp.
Holdridge, L. R. and A. L. J. Poveda. 1975. *Arboles de Costa Rica.* Vol. 1. Centro Cientifico Tropical, San José, 646 pp.
Instituto de Investigaciones Sobre Recursos Bioticos. 1978–. *Flora de Veracruz* (A. Gomez and V. Sosa, eds.) Publ. in fascicles, 1–39 at present. INIREB, Xalapa, Mexico.
Johnson, I. M. 1949. *The Botany of San Jose Island (Gulf of Panamá).* Sargentia No. 8. Jamaica Plains, MA, 306 pp.
Knobloch, I. W. and D. S. Correll. 1962. *Ferns and Fern Allies of Chihuahua, Mexico.* Texas Research Foundation, Renner, TX, 198 pp.
Langman, I. K. 1964. *A Selected Guide to the Literature on the Flowering Plants of Mexico.* Univ. Pennsylvania Press, Philadelphia, PA, 1015 pp.

Martinez, M. and E. Matuda. 1979. *Flora del Estado de México.* I–III. (reprint ed.) Biblioteca Enciclopédica del Estado de México. México, D.F.
Mason, C. T., Jr. and P. B. Mason. 1987. *A Handbook of Mexican Roadside Flora.* Univ. Arizona Press, Tucson, AZ, 380 pp.
McVaugh, R. 1974–. Flora Novo-galiciana. (issued in fascicles) *Contr. Univ. Michigan Herb.*, Ann Arbor, MI.
Molina, R. A. 1975. Enumeración de las Plantas de Honduras. *Ceiba* 19(1):1–118.
Pennington, T. D. and J. Sarukhán. 1958. *Manual para la Identificación de Cempo de los Principales Arboles Tropicales de México.* Instituto Nacional de Investigaciones Forestales, México City, 413 pp.
Pesman, M. W. 1962. *Meet Flora Mexicana.* Dale S. King Publ., Globe, AZ, 278 pp.
Reiche, C. 1963. *Flora Excursoria en el Valle Central de México.* Instituto Politecnico Nacional, México D.F., 303 pp.
Rzedowski, J. and G. C. Rzedowski (eds.). 1979. *Flora Fanerogámica del Valle de México.* Vol. I. Compañia Editorial Continental, S.A. México, D.F., 403 pp.
Sánchez-Sánchez, O. 1969. *La Flora del Valle de México.* Herrera, Mexico City, 519 pp.
Shreve, F. and I. L. Wiggins (eds.). 1964. *Vegetation and Flora of the Sonoran Desert.* 2 vols. Stanford Univ. Press, Stanford, CA, 1740 pp.
Standley, P. C. 1920–26. Trees and Shrubs of Mexico. 5 parts. *Contr. U.S. Natl. Herb.* 23. Washington D.C., 1721 pp.
Standley, P. C. 1928. Flora of the Panama Canal Zone. *Contr. U.S. Natl. Herb.* 27. 416 pp.
Standley, P. C. 1930. Flora of Yucatán. *Field Columbian Mus. Bot. Ser.* 3(3):157-492, Chicago.
Standley, P. C. 1930. A Second List of the Trees of Honduras. *Trop. Woods* 21:9–41.
Standley, P. C. 1937–9. Flora of Costa Rica. 4 parts. *Field Mus. Nat. Hist. Bot. Ser.* No. 18, 1616 pp.
Standley, P. C. and S. Calderón. 1944 (1941). *Flora Salvadoreña.* Lista Preliminar de la Planta de El. Salvador, 2nd ed. Imprensa Nacional, San Salvador, 450 pp.
Standley, P. C. and S. J. Record. 1936. The Forests and Flora of British Honduras. *Field Mus. Nat. Hist. Bot. Ser.* No. 12, 432 pp.
Standley, P. C. et al. 1946–77. Flora of Guatemala. 13 parts. *Fieldiana Bot.* 24.
Stolze, R. G. 1976–81. Ferns and Fern Allies of Guatemala. I.II. *Fieldiana Bot.* No. 39, n.s. 6, Chicago.
Wiggins, I. L. 1980. *Flora of Baja California.* Stanford Univ. Press, Stanford, CA, 1075 pp.
Woodson, R. E., Jr. and R. W. Schery. 1943–81. *Flora of Panama.* Parts 2–9. Missouri Bot. Garden, St. Louis, MO.

## *West Indies*

Adams, C. D. 1972. *Flowering Plants of Jamaica.* Univ. West Indies, Mona, Jamaica, 848 pp.
Arnoldo, M. 1964. *Zakflora: Wat in Het Wild Groeit en Bloeit op Curacao, Aruba en Bonaire.* 2nd ed. Uitgaven Natuurw. Werkgroep Ned. Antillen, No. 16. Curacao, 232 pp.
Barker, H. D. and W. S. Dardeau. 1930. *Flore d'Haiti.* Dept. de l'Agriculture, Port-au-Prince, Haiti, 456 pp.
Britton, N. L. 1918. The Flora of the American Virgin Islands. *Brooklyn Bot. Gard. Mem.* 1:19–118.
Britton, N. L. and C. F. Millspaugh. 1920. *The Bahama Flora.* Publ. by the authors, New York (reprinted 1963), 695 pp.
Britton, N. L. and P. Wilson. 1923–30. Botany of Puerto Rico and the Virgin Islands. (in Sci. Survey of Puerto Rico). *New York Acad. Sci.* 5–6, 626, 663 pp.

Correll, D. S. and H. B. Correll. 1982. *Flora of the Bahama Archipelago.* J. Cramer, Vaduz, 1692 pp.
Fawcett, W. and A. B. Rendle. 1910–36. *Flora of Jamaica.* Vols. 1,3–5,7. British Mus. (Nat. Hist.), London.
Fournet, J. 1978. *Flore Illustrée de Phanérogames de Guadeloupe et de Martinique.* Inst. Nat. Res. Agrinomique, Paris, 1654 pp.
Gooding, E. G. B., A. R. Loveless, and G. R. Proctor. 1965. *Flora of Barbados.* HMSO, London, 486 pp.
Grisebach, A. H. R. 1859–64. *Flora of the British West Indian Islands.* Reeve, London (reprinted 1963, Cramer, Weinhein, W. Germany), 789 pp.
Hodge, W. H. 1954. Flora of Dominica (BWI). I. *Lloydia* 17:1–238.
Howard, R. A. ed. 1974–. *Flora of the Lesser Antilles: Leeward and Windward Islands.* Parts 1–3. Arnold Arboretum of Harvard Univ., Jamaica Plains, MA.
León, H. and H. Alain. 1946–62. *Flora de Cuba.* 5 vols. Havana.
Liogier, A. H. 1982, 1983. *La Flora de la Española.* I.II. Universidad Central del Este, Serie Cientifica XII and XV. San Pedro de Macoris, Santo Domingo, Dom. Republic.
Little, E. L., Jr. and F. H. Wadsworth. 1964. *Common Trees of Puerto Rico and the Virgin Islands.* Agr. Handbook U.S.D.A. No. 249. Gov. Printing Office. Washington D.C., 548 pp. (reprinted 1988).
Little, E. L., Jr. and R. O. Woodbury. 1974. *Trees of Puerto Rico and the Virgin Islands.* Vol. 2. Agric. Handbook U.S.D.A. No. 449. Govt. Printing Office, Washington D.C., 1024 pp.
Marshall, R. C. 1934. *Trees of Trinidad and Tobago.* Govt. Printing Office, Port of Spain, 101 pp.
Moscoso, R. M. 1943. *Catalogues Florae Domingensis.* I. Spermatophyta. Univ. Santo Domingo, New York, 732 pp.
Patterson, J. and G. Stevenson. 1977. *Native Trees of the Bahamas.* Patterson, Hope Town, Abaco, Bahamas, 128 pp.
Proctor, G. R. 1984. *Flora of the Cayman Islands.* Kew Bull. Additional Ser. XI. HMSO, London, 834 pp.
Proctor, G. R. 1985. *Ferns of Jamaica.* British Mus. (Nat. Hist.) London, 761 pp.
Questel, A. 1941. *The Flora of the Island of St. Bartholomew and Its Origin.* Imprimerie Catholique, Basseterre, Guadeloupe, 244 pp.
Stoffers, A. L. 1962–. *Flora of the Netherlands Antilles.* 3 vols. (incomplete). Uitgaven Natuurwet. Studieking Suriname Ned., Utrecht.
Williams, R. O. et al. 1928–. *Flora of Trinidad and Tobago.* Vols. 1(1–8), 2(1–10), 3(1–2). Ministry Agr., Lands and Fisheries, Port of Spain.

## *South America*

Augusto, I. 1946. *Flora do Rio Grande do Sul Brasil.* Vol. 1. Imprensa Oficial, Alegre, 639 pp.
Burkart, A. et al. (eds.) 1969–. *Flora Ilustrada de Entre Rios.* Parts 2, 5, 6. Colecc. Ci. INTA. No. 6. Librart, Buenos Aires.
Cabrera, A. L. ed. 1963–70. *Flora de la Provincia de Buenos Aires.* 6 parts. Colecc. Ci. INTA. No. 4. Librart, Buenos Aires.
Cabrera, A. L. ed. 1977–. *Flora de la Provincia de Jujuy.* Parts 2, 10. Colecc. Ci. INTA. No. 13. Librart, Buenos Aires.
Cabrera, A. L. and E. M. Zardini. 1978. *Manual de la Flora de los Alrededores de Buenos Aires.* Acme, Buenos Aires, 755 pp.
delValle, A. J. I. 1972. *Introducción a la Dendrología de Colombia.* Centro de Publicaciones, Universidad Nacional, Medellin, 351 pp.

Descole, H. et al. 1943–56. *Genera et Species Plantarum Argentinarum*. Vols. 1–5. Kraft, Buenos Aires.
Dodson, C. H. and A. H. Gentry. 1978. *Flora of the Rio Palenque Science Center, Los Rios Province, Ecuador*. Marie Selby Bot. Gard., Sarasota, FL, 628 pp.
Foster, R. C. 1958. *Catalogue of the Ferns and Flowering Plants of Bolivia*. Contr. Gray Herb., N.S. 184. Cambridge, MA, 223 pp.
Harling, G. and B. Sparre (eds.). 1973–. *Flora of Ecuador*. Parts 1–. Opera Bot., B. Lund, Sweden.
Herter, W. 1949–56. *Flora del Uruguay*. Vol. 1. 10 fasc. Publ. by the author. Montevideo, 280 pp.
Hoehne, F. C. et al. 1940–. *Flora Brasilica*. Fasc. 1–. Instituto de Botânica, São Paulo.
Huber, H. 1977. *Gehölzflora der Anden von Mérida*. I. Mitt. Bot. Staatssamm I. München, 13:1–127.
Kramer, K. U. 1978. *The Pteridophytes of Surinam*. Uitgaven Natuurw. Studie Kring Suriname Ned. Antillen. No. 93., Utrecht, 198 pp.
Lasser, T. ed. 1964–. *Flora de Venezuela*. Vol. 1–. Instit. Bot., Ministerio de Agr. Y Cria (MAC), Caracas.
LeCointe, P. 1947. *Arvores e Plantas uteis da Amazônia Brasileira,* São Paulo, 506 pp.
Lombardo, A. 1964. *Flora Arbórea y Arborescente del Uruguay, con Clave Para Determinar las Especies,* 2nd ed. Concejo Departmental de Montevideo, Montevideo, 151 pp.
Macbride, J. F. et al. 1936–71. Flora of Peru. 10 parts. 24 nos. Publ. *Field Mus. Nat. Hist. Bot.* Ser. 13 (1, 2, 3, 3A, 4, 5, 5A, 5B, 5C, 6), Chicago.
Macbride, J. F. et al. 1980–. Flora of Peru (N.S.) *Fieldiana Bot.* N.S. 5. Chicago.
Maguire, B. et al. 1953–. The Botany of the Guayana Highlands. Parts 1–. *Mem. New York Bot. Gard.*, Bronx, New York.
Martius, K. F. P. von, A. W. Eichler and I. Urban (eds.). 1940–1906. *Flora Brasiliensis.* 15 vols. Fleischer, Munich (Reprint 1966–67 Cramer, Lehre, Germany), 20,733 pp.
Moore, D. M. 1968. *The Vascular Flora of the Falkland Islands.* Sci. Rep. Antarctic Surv., London, 202 pp.
Moore, D. M. 1983. *Flora of Tierra de Fuego.* Anthony Nelson, Shropshire, England, 396 pp.
Muñoz-Pizarro, C. 1966. *Sinopsis de la Flora Chilena (Claves Para la Identificación de Familias y Generos),* 2nd ed. Ediciones de la Universidad de Chile, Santiago, 500 pp.
Muso Nacional de Historia Natural. Uruguay. 1958–. *Flora del Uruguay.* Fasc. 1–. Montevideo, Uruguay.
Navas-Bustamante, L. E. 1973–9. *Flora de la Cuenca de Santiago de Chile.* 3 vols. Ediciones de la Universidad de Chile, Santiago.
Organization for Flora Neotropica. 1968–. *Flora Neotropica: A Series of Monographs.* 1–. New York Bot. Gard., Bronx, NY.
Pinto-Escobar, P. et al. (eds.) 1966–. *Catalogo Ilustradoi de las Plantas de Cundinamarca.* Parts I–. Imprensa Nacional, Bogota.
Pittier, H. et al. 1945–47. *Catalogo de la Flora Venezolana.* 2 vols. Third Inter-American Conf. on Agr., Lit. y Tip. Vargas, Caracas.
Pulle, A. A., J. Lanjouw, and A. L. Stoffers (eds.). 1932–. *Flora of Suriname.* Vol. 1–. Brill, Leiden.
Reitz, R. P. (ed.). 1965–. *Flora Ilustrada Catarinense.* I. Fasc. 1–. Herbário Barbosa Rodrigues, Itajai, Brazil.
Roosmalen, M. G. M. van. 1976. *Surinaams Vruchtenboek.* 2 vols. Publ. by the author.
Schweinfurth, C. 1958–61. Orchids of Peru. *Fieldiana Bot.* 30(1–4):1–1026.
Steyermark, J. et al. 1951–57. Contributions to the Flora of Venezuela. *Fieldiana Bot.* 28(1–4):1–1225.
Steyermaark, J. and O. Huber. 1978. *Flora del Avila.* Ministerio del Ambiente y de los Recursos Naturales Removables, Caracas, 971 pp.
Tryon, R. M. 1964. *The Ferns of Peru.* Contr. Gray Herb., N.S. 194. Cambridge, MA.

Uriba, J. A. and L. Uriba-Uriba. 1940. *Flora de Antioquia.* Imprensa Departemental, Medellin, 383 pp.

Vareschi, V. 1970. *Flora de los Páramos de Venezuela.* Ediciones del Rectorado, Univ. de los Andes, Merida, 429 pp.

Wiggins, I. L. and D. M. Porter. 1971. *Flora of the Galápagos Islands.* Stanford Univ. Press, Stanford, CA, 998 pp.

## Australia, New Zealand, Southwest Pacific and East Indian Ocean Islands

Allan, H. H. 1961. *Flora of New Zealand* I: *Indigenous Tracheophyta.* Government Printer, Wellington, 1085 pp.

Anderson, R. H. 1968. *The Trees of New South Wales,* 4th ed. Government Printer, Sydney, 510 pp.

Aston, H. I. 1973. *Aquatic Plants of Australia.* Melbourne Univ. Press, Melbourne, 368 pp.

Beadle, N. C. W. and L. D. Beadle. 1971–80. *Student's Flora of North Eastern New South Wales.* Parts 1–4. Dept. of Bot., Univ. of New England, Armidale, 686 pp.

Beadle, N. C. W., O. D. Evans, and R. C. Carolin (with M. D. Tindale). 1972. *Flora of the Sydney Region.* Reed, Sydney, 724 pp.

Bentham, G. 1863–78. *Flora Australiensis.* 7 Vols. (reprinted 1967, Asher, Amsterdam) Reeve, London.

Blackall, W. E. and B. J. Grieve. 1974. *How to Know Western Australian Wild Flowers.* Parts 1–3. (reprint of 1954–65 edition) Univ. W. Australia Press, Perth, 595 pp.

Blackall, W. E. and B. J. Grieve. 1980–1. *How to Know Western Australian Wildflowers,* 2nd ed, Parts 3A, 3B, Univ. W. Australia Press, Perth.

Burbidge, N. T. and M. Gray. 1970. *Flora of the Australian Capital Territory.* Australian National Univ. Press, Canberra, 447 pp.

Cheeseman, T. F. 1925. *Manual of the New Zealand Flora,* 2nd ed. Government Printer, Wellington, 1163 pp.

Clifford, H. T. and J. Constantine. 1980. *Ferns, Fern Allies and Conifers of Australia.* Univ. Queensland Press, Brisbane, 150 pp.

Clifford, H. T. and G. Ludlow. 1978. *Keys to the Families and Genera of Queensland Flowering Plants.* Univ. of Queensland Press, Brisbane, 202 pp.

Cockayne, L. and E. P. Turner. 1967. *The Trees of New Zealand,* 5th ed. Government Printer, Wellington, 182 pp.

Costin, A. B., M. Gray, C. J. Totterdell, and D. J. Wimbush. 1979. *Kosiusko Alpine Flora.* C.S.I.R.O., Melbourne; Collins, Sydney, 408 pp.

Cunningham, G. M. et al. 1981. *Plants of Western New South Wales.* N.S.W. Government Printer, Sydney.

Curtis, W. M. 1956–. *The Students Flora of Tasmania.* Parts 1–3,4A. Tasmanian Government Printer, Hobart.

Ewart, A. J. 1930 (1931). *Flora of Victoria.* Government Printer (for Melbourne Univ. Press), Melbourne, 1257 pp.

Field, H. C. 1890. *The Ferns of New Zealand and Its Immediate Dependencies, with Directions for their Collection and Cultivation.* Griffith, Faren, Okeden and Welsh, London, 164 pp.

Francis, W. D. 1970. *Australian Rain-forest Trees,* 3rd ed. (revised by G. Chippendale). Australian Government Publishing Service, Canberra, 468 pp.

Gardner, C. A. 1985. *Wildflowers of Western Australia,* 16th ed. St. George Books, Perth, 159 pp.

Grieve, B. J. and W. E. Blackall. 1975. *How to Know Western Australian Wild Flowers.* Part 4. Univ. W. Australia Press, Perth, (1–149), 596–861 pp.

Healy, A. J. and E. Edgar. 1980. *Flora of New Zealand* 3: *Adventive Cyperaceous, Petalous, and Spathaceous Monocotyledons.* Government Printer, Wellington, 220 pp.
Hyland, B. P. M. 1971. *A Key to the Common Rain-forest Trees Between Townsville and Cooktown Based on Leaf and Bark Features.* Dept. of Forestry, Queensland, Brisbane, 103 pp.
Johnson, P. N. and D. J. Campbell. 1975. Vascular Plants of the Auckland Islands. *New Zealand J. Bot.* 13:665–720.
Jones, D. L. and S. C. Clemensha. 1981. *Australian Ferns and Fern Allies.* A. H. & A. W. Reed, Sydney, 294 pp.
Mark, A. F. and N. M. Adams. 1979. *New Zealand Alpine Plants,* 2nd ed. A. H. & A. W. Reed, Wellington, 262 pp.
Moore, L. B. and E. Edgar. 1971. *Flora of New Zealand. 2: Indigenous Tracheophyta.* Government Printer, Wellington, 354 pp.
National Herbarium of New South Wales. 1961–. *Flora of New South Wales.* National Herbarium of N.S.W., Sydney.
Poole, A. L. and N. M. Adams. 1963. *Trees and Shrubs of New Zealand.* Government Printer, Wellington, 250 pp.
Recher, H. F. and S. S. Clark (eds.). 1974. *Environmental Survey of Lord Howe Island.* New South Wales Government Printer, Sydney, 86 pp.
Robertson, R., et al. (eds.). 1981. *Flora of Australia.* vol. 1–. Bureau of Flora and Fauna, Australian Govt. Publ. Ser., Canberra.
Salmon, J. T. 1980. *The Native Trees of New Zealand.* A. H. & A. W. Reed, Wellington, 384 pp.
Salmon, J. T. 1986. *A Field Guide to the Native Trees of New Zealand.* Reed Methuen, Auckland, 228 pp.
Sykes, W. R. 1977. *Kermadec Islands Flora: An Annotated Checklist.* New Zealand D.S.I.R. Bull. No. 219, 216 pp.
Turner, J. S., C. N. Smithers, and R. D. Hoogland. 1968. *The Conservation of Norfolk Island.* Australian Conservation Foundation Spec. Publ. No. 1. Canberra, 41 pp.
Willis, J. H. 1970–72. *A Handbook to Plants in Victoria.* 2 vols. Melbourne Univ. Press, Melbourne.

## *Africa, Madagascar, and West Indian Ocean Islands*

Adamson, R. S. and T. M. Salter (eds.). 1950. *Flora of the Cape Peninsula.* Juta, Cape Town. 889 pp.
Ali, S., A. El-Gadi, and S. M. H. Jafri. 1976–. *Flora of Libya.* Parts 1–. Botany Dept., Al-Faateh Univ. of Tripoli, Tripoli.
Alston, A. H. G. 1959. *The Ferns and Fern Allies of West Tropical Africa.* Crown Agents (HMSO), London, 89 pp.
Aubréville, A. et al. (eds.). 1961–. *Flore du Gabon.* Fasc. 1–. Muséum National d'Histoire Naturelle, Laboratorie de Phanérogamie, Paris.
Aubréville, A. et al. (eds.). 1963–. *Flore du Cameroun.* Fasc. 1–. Muséum National d'Histoire Naturelle, Laboratoire de Phanérogamie, Paris.
Baker, J. G. 1877. *Flora of Mauritius and the Seychelles: A Description of the Flowering Plants and Ferns of Those Islands.* No. 19, 1. (reprinted 1970, Cramer, Lehre, W. Germany) Reeve, London, 557 p.
Berhaut, J. 1967. *Flore du Sénégal,* 2nd ed. Clairafrique, Dakar, 485 pp.
Blundell, M. 1982. *The Wild Flowers of Kenya.* Collins, London, 160 pp.
Bond, P. and P. Goldblatt. 1984. *Plants of the Cape Flora. A Descriptive Catalogue.* J. South African Bot. Suppl. No. 13., National Botanic Gard., Cape Town, 455 pp.
Bosser, J. et al. (eds.). 1976–. *Flore des Mascareignes* (La Reunion, Maurice, Rodriques).

Fasc. 1–. Sugar Industry Research Institute, Mauritius: ORSTOM, Paris; Royal Bot. Gard., Kew.

Burger, W. 1967. *Families of Flowering Plants in Ethiopia.* Okalahoma Agr. Exp. Sta. Bull. No. 45. Oklahoma State Univ. Press, Stillwater, OK, 236 pp.

Carrisso, L. et al. (eds.). 1937–. *Conspectus Florae Angolensis.* Vol. 1–. Junta de Investigacões do Ultramar, Lisbon.

Compton, R. H. 1976. *The Flora of Swaziland.* J. South African Bot. Supple. No. 11. Kirstenbosch, Mbabane, 684 pp.

Dale, I. R. and P. J. Greenway. 1961. *Kenya Trees and Shrubs.* Buchanan's Kenya Estates, Nairobi; Hatchards, London, 654 pp.

Dyer, R. A. 1975–6. *The Genera of Southern African Flowering Plants.* 2 Vols. S. African Government Printer, Pretoria, 1040 pp.

Dyer, R. A. 1977. *Flora of Southern Africa: Key to Families and Index to the Genera of Southern African Flowering Plants.* Bot. Res. Instit. Dept. Agr. Tech. Serv. Republic of S. Africa, Pretoria, 60 pp.

Engler, A. 1908–25. *Die Pflanzenwelt Afrikas Insbesondere seiner Tropischen Gebiete.* Grundzüge der Pflanzenverbreitung in Afrika und die Charakterpflanzen Afrikas. Parts 1–3,5(1). Die Vegetation der Erde No. 9. Englemann, Leipzig.

Fernandes, A. and E. J. Mendes (eds.) 1969–. *Flora de Mocambique.* Fasc. 1–. Junta de Investigacões Cientificas do Ultramar, Lisbon.

Fosberg, F. R. and S. A. Renvoize. 1980. *The Flora of Aldabra and Neighboring Islands.* Kew Bull. Addit. Ser. No. 7. HMSO, London, 358 pp.

Geerling, C. 1982. *Guide de Terrain des Ligneux Saheliens et Soudano Guineens.* Veenman & Zonen, Wagneningen, 340 pp.

Hedberg, O. 1957. *Afroalpine Plants: A Taxonomic Revision.* Symb. Bot. Uppsala, Uppsala, 411 pp.

Humbert, H. et al. (eds.) 1936–. *Flore de Madagascar et des Comores.* Fasc. 1–. Gouvernement Général de Madagascar; Tananarive: Laboratoire de Phanérogamie. Muséum National d'Histoire Naturelle, Paris.

Hutchinson, J. and J. M. Dalziel. 1953–72. *Flora of West Tropical Africa,* 2nd ed. (revised by R. W. J. Keay and F. N. Hepper) 3 vols. Crown Agents (HMSO), London.

Jardin Botanique National de Belgique. 1967–. *Flore du Congo, du Rwanda et du Burundi* (from 1971, *Flore d'Afrique Centrale.)* In fasc., Brussels.

Kornaś J. 1979. *Distribution and Ecology of the Pteridophytes in Zambia.* Pánstwowe Wydawnickwo Naukowe, Warsaw/Krakow, 207 pp.

Leistner, O. A. (ed.). 1988. *Flora of Southern Africa, Which Deals with the Territories of South Africa, Ciskei, Transkei, Swaziland, Bophathatswana, South West Africa/Namibia, Botswana, and Venda.* Botanical Research Inst., Pretoria. Vol. 16, Pt. 3, Fasc. 6, 436 pp.

Maíre, R. et al. 1952–. *Flore de l'Afrique du Nord.* Vol. 1–. Encyclopédie Biol. No. 33. Lechevalier, Paris.

Merxmuller, H. (ed.). 1966–72. *Prodromus einer Flora von Südwestafrika,* 35 fasc. Cramer, Lehre, Germany, 2188 pp.

Moll, E. 1981. *Trees of Natal.* Univ. Cape Town ECO-LAB Trust Fund, Cape Town, 567 pp.

Negre, R. 1962–3. *Petite Flore des Régions Arides du Maroc Occidental.* 2 Vols. Centre National de la Recherche Scientifique, Paris, 982 pp.

Oliver, D. et al. 1868–77. *Flora of Tropical Africa.* Vols. 1–3. Reeve, London.

Olovode, O. 1984. *Taxonomy of West African Flowering Plants.* Longman, London, 158 pp.

Ozenda, P. 1977. *Flore du Sahara,* 2nd ed. Centre National de la Recherche Scientifique (CNRS), Paris, 622 pp.

Palgrave, K. C. 1977. *Trees of Southern Africa.* C. Struik, Cape Town, 959 pp.

Pottier-Alapetite, G. 1979. *Flore de la Tunisie: Angiospermes–Dicotylédones.* Imprimerie Officielle, Tunis, 651 pp.

Quezel, P. and S. Santa. 1962–63. *Nouvelle Flore de l'Algerie et des Regions Désertiques Meridionales.* Vols. 1–2. Centre Nationale de la Recherche Scientifique, Paris, 1180 pp.

Renvoize, S. A. 1975. A Floristic Analysis of the Western Indian Ocean Coral Islands. *Kew Bull.* 30:133–152.

Robyns, W. 1958. *Flore du Congo Belge et du Ruanda–Urundi: Tableau Analytique des Familles.* Institut National pour l'Etude Agronomique du Congo, Brussels, 67 pp.

Ross, J. H. 1972(1973). *The Flora of Natal.* Mem. Bot. Survey S. Africa. No. 39, Pretoria, 418 pp.

Stanfield, D. P. and J. Lowe (eds). 1970–. *The Flora of Nigeria.* In fasc. Ibadan Univ. Press, Ibadan.

Stoddart, D. R. (ed.). 1970. *Coral Islands of the Western Indian Ocean.* Atoll Res. Bull. No. 136. Washington, D.C., 224 pp.

Tackholm, V. 1974. *Student's Flora of Egypt,* 2nd ed. Cairo Univ. Press, Cairo, 888 pp.

Thistleton-Dyer, W. T. et al. (eds.) 1897–1937. *Flora of Tropical Africa.* Vols. 4–9, 10(1). Reeve, London.

Thonner, F. 1915. *The Flowering Plants of Africa: An Analytical Key to the Genera of African Phanerogams.* (reprinted 1963, Cramer, Weinheim, Germany), Dulau, London, 647 pp.

Troupin, G. 1978, 1983. *Flore du Rwanda: Spermatophytes.* 2 vols. Ann. Mus. Royal Afrique Centrale, Serie in $8^0$, Sci. Econ. No. 9, 1, 3. Butane, Rwanda.

Troupin, G. 1982. *Flore des Plantes Ligneuses du Rwanda.* Instit. Nat. de Recherche Scientifique. Publ. No. 21. Butane, Rwanda, 747 pp.

Turrill, W. B. et al. (eds.) 1952–. *Flora of Tropical East Africa.* In fasc. Crown Agents, London.

Voorhoeve, A. G. 1965. *Liberian High Forest Trees.* PUDOC, Wageningen, 416 pp.

White, F. 1962. *Forest Flora of Northern Rhodesia.* Oxford Univ. Press, Oxford, 482 pp.

Wyk, B. van and S. Malan. 1988. *Field Guide to the Wild Flowers of the Witwatersrand & Pretoria Region, Including the Magadiesberg & Suikerbosrand.* Struik Publishers, Capetown, 352 pp.

## *Europe, Russia, and the Middle East*

Aloenius, O. 1953. *Finlands Kärlväxter: de Vilt Växande och Allmännast Odlade,* 12th ed. (revised by A. Nordström). Söderström, Helsinki, 428 pp.

Baroni, E. 1969. *Guida Botanica d'Italia,* 4th ed. (revised by S. Baroni Zanett). Cappelli, Bologna, 545 pp.

Beldie, A. 1972. *Plantele din Muntii Bucegi: Determinator.* Academia RSR. Bucharest, 409 pp.

Binz, A. 1986. Schul-und Exkursions flora für die Schweiz. (rev. by C. Heitz) Schwabe & Co. Ag, Basel, 624 pp.

Bonnier, G. and R. Douin. 1911–1935. *Flore complète illustrée en couleurs de France, Suisse et Belgique.* 13 vols. Various publishers, Paris, Brussels, Neuchâtel.

Bouchard, J. 1977 (1978). *Flore Pratique de la Corse,* 3rd ed. Societe des Sciences Historiques et Naturelles de la Corse, Bastia, 405 pp.

Clapham, A. R., T. G. Tutin, and E. F. Warburg. 1962. *Flora of the British Isles,* 2nd ed. Cambridge Univ. Press, Cambridge, 1269 pp.

Clapham, A. R., T. G. Tutin, and E. F. Warburg. 1981. *Excursion Flora of the British Isles,* 3rd ed. Cambridge Univ. Press, Cambridge, 499 pp.

Davis, P. H. ed. 1965–1988. *Flora of Turkey and East Aegean Islands.* Vols. 1–10. Edinburgh Univ. Press, Edinburgh.

DeLanghe, J.-E., et al. 1978. *Nouvelle Flore de la Belgique, du Grand-Duché de Luxembourg, du Nord de la France et des Regions Voisines,* 2nd ed. Patrimoine du Jardin Botanique National de Belgique, Meise, 899 pp.

Dostál, J. 1958. *Klic Kuplné Kvetene CSR,* 2nd ed. Ceskoslovenské Academie Ved., Prague, 982 pp.

Fedorov, A. A., et al. 1974–. *Flora Europejskoj casti SSSR.* Vol. 1–. Nauka, Leningrad.

Fenaroli, L. 1971. *Flora delle Alpi,* 2nd ed. Marrtello, Milan, 429 pp.

Fournier, P. 1977. *Les Quatre Flores de France, Corse Comprise,* 2nd ed. 2 vols. Lechevalier, Paris, 1106 pp.

Franco, J. do A. 1971. *Nova Flora de Portugal.* Vol. 1. Sociedade Astoria, Lisbon, 647 pp.

Garcke, A. 1972. *Illustrierte Flora von Deutschland und Angrenzende Gebiete,* 23rd ed. (revised by K. von Weihe ed.) Parey, Berlin. 1607 pp.

Govoruchin, V. S. 1937. *Flora Urala.* Oblast' Publ. Service. Sverdlovsk, 536 pp.

Grosshem, A. A., et al. (eds.). 1939. *Flora Káwkaza,* 2nd ed. Vol. 1. AN SSSR Press, Leningrad.

Guest, E., C. C. Townsend, and A. Al-Rawi (eds.). 1966–. *Flora of Iraq.* Vols. 1–. Ministry of Agriculture. Republic of Iraq, Baghdad.

Halácsy, E. von. 1900–8. *Conspectus Florae Graecae.* 3 Vols. (Reprinted in 1968, Cramer, Zehre, Germany.) Engelmann, Leipzig.

Haslam, S. M., P. D. Sell, and P. A. Wolseley. 1977. *A Flora of the Maltese Islands.* Malta Univ. Press. Msida, Malta, 560 pp.

Hayek, A. von. 1924–31. *Prodromus Florae Peninsulae Balcanicae.* In Fedde Rep. Beih, 30 (1 & 2). pp. 1–1193 & 1–1152.

Hegi, G. (ed.) 1966–. *Illustrierte Flora von Mitteleuropa,* 3rd ed. Band (Vol.) 1–. Hanser, Munich & Parey, Berlin.

Hermann, F. 1956. *Flora von Nord- und Mitteleuropa.* Fishcher, Stuttgart, 1134 pp.

Hess, H. E., E. Landolt, and R. Hirzel. 1967–73. *Flora der Schweiz und Angrenzender Bebiete.* 3 vols. Birkhauser, Basel.

Heukels, H. and S. J. van Ooststroom. 1977. *Flora van Nederland,* 19th ed. Wolters-Noordhoff, Groningen, 925 pp.

Horvatić, S. ed. 1967–. *Analiticka Flora Jugoslavije.* Vol. 1–. Institut za Botaniku Sveucilista, Zagreb.

Hylander, N. 1953–66. *Nordisk Kärlväxtflora.* Vols. 1–2. Almqvist & Wiksell, Stockholm.

Komarov, V. L. and E. N. Klobukova-Alisova. 1931–32. *Key for the Plants of the Far Eastern Region of the USSR.* 2 vols. Academy of Sciences of the USSR, Leningrad, 1175 pp.

Komarov, V. L. and B. K. Shishkin et al. (eds.); E. G. Bobrov et al. (comp.). 1934–64. *Flora SSSR.* 30 vols. AN SSSR Press, Moscow/Leningrad. (Eng. ed. 1960. *Flora of the USSR.* Trans. N. Landau and P. Lavoott. Vol. 1–21, 24. Israel Program for Scientific Translations, Jerusalem).

Krok, T. O. B. N. and S. Almqvist. 1960. *Svensk Flora. I: Fanerogamer och Ormbunkväxter,* 25th ed. (revised by E. Almqvist) Svensk Bokförlaget, Stockholm, 403 pp.

Krylov, P. N. et al. 1927–64. *Flora Zapadnoj Sibiri.* 12 vols. Russkoe Botaniceskoe Obscestro, Tomskoe Otdelenie, Tomsk.

Lázaro é Ibiza, B. 1920–1. *Botánica Descriptiva: Compendio de la Flora Español,* 3rd ed. 3 vols. Madrid.

Lid, J. 1974. *Norsk og Svensk Flora,* 2nd ed. Norske Samlaget, Oslo, 808 pp.

Martincic, A. and F. Susnik (eds.) 1969. *Mala Flora Slovenije.* Cankarjeva Zalozba, Ljubljana, 517 pp.

Meikle, R. D. 1977 & 1985. *Flora of Cyprus.* Vol. 1, 2. Bentham-Moxon Trustees. (Royal Botanical Gard. Kew). Vol. 1, 832 pp.; Vol. 2, 1131 pp.

Migahid, A. M. 1978. *Migahid and Hammouda's Flora of Saudi Arabiá,* 2nd ed. 2 Vols. Riyadh Univ., Riyadh, 940 pp.

Mouterde, P. 1966–. *Nouvelle flore du Liban et de la Syrie.* Vols. 1–. Dar el-Machreq, Beirut.

Nasir, E. and S. I. Ali (eds.) 1970–. *Flora of West Pakistan.* Fasc. 1–. Published by the editors, Dept. of Botany, Univ. of Karachi, Karachi.

Polunin, O. 1969. *Flowers of Europe. A Field Guide.* Oxford Univ. Press, Oxford, 662 pp.

Polunin, O. 1980. *Flowers of Greece and the Balkans.* Oxford Univ. Press, Oxford, 592 pp.

Polunin, O. and A. Huxley. 1978. *Flowers of the Mediterranean.* Chatto and Windus, London, 260 pp.

Post, G. E. 1932–3. *Flora of Syria, Palestine and Sinai,* 2nd ed. (Rev. by J. E. Dinsmore.) 2 vols. American Univ. of Beirut, Beirut.

Raciborski, M. et al. (eds.) 1919-80. *Flora Polska.* 14 vols. Polska Akademji Umiejetnosci, Cracow.

Rasmussen, R. 1952. *Foroya Flora,* 2nd ed. Jacobsens Bókhandils, Tórshavn, 232 pp.

Rechinger, K. H. (ed.) 1963. *Flora Iranica.* Fasc. 1–. Akademische Druck-und Verlagsanstalt, Graz, Austria.

Rollan, M. G. 1981 & 1983. *Claves de la Flora de España.* 2 Vols. Ediciones Mundi-Prensa, Madrid.

Rostrup, E. and C. A. Jorgensen. 1973. *Den Danske Flora,* 20th ed. (Rev. by A. Hansen). Gyldendal, Copenhagen. 664 pp.

Sabeti, H. 1976. *Forests, Trees and Shrubs of Iran.* National Agriculture and Natural Resources Research Organization, Iran, Teheran, 810 pp. (Persian text); 64 pp. (English text).

Soo, R. (ed.) 1964–80. *A Magyar Flóra és Vegetáció Rendszertani-növényföldrajzi Kézikönyve.* Vols. 1–6. Akadémiai Kiadó, Budapest.

Stojanov, N., B. Stefanov, and B. Kitanov. 1966–7. *Flora na Balgarija,* 4th ed. 2 vols. Naukai Izkustvo, Sofia, 1326 pp.

Strid, A. et al. 1986. *Mountain Flora of Greece, Vol. 1.* Cambridge University Press, Cambridge, 822 pp.

Täckholm, Y. 1974. *Student's Flora of Egypt,* 2nd ed. Cairo Univ. Press, Cairo, 888 pp.

Tutin, T. G., et al. (eds.) 1964–80. *Flora Europaea.* 5 Vols. Cambridge Univ. Press, Cambridge.

Webb, D. A. 1977. *An Irish Flora,* 6th ed. W. Tempest (Dundalgan Press), Dundalk, 277 pp.

Zangheri, P. 1976. *Flora Italica (Pteridophyta-Spermatophyta).* 2 Vols. A. Milani (CEDAM), Padua, 1157 pp.

Zohary, M. and N. Fembrun-Dothan. 1966–77. *Flora Palaestinia.* 3 Vols. Jerusalem.

Zohary, M. 1976. *A New Analytical Flora of Israel.* Am Oved, Tel Aviv, 540 pp.

## *Asia, Southeast Asia, Malaysia, and India*

Aubréville, A. et al. (eds.) 1960–. *Flore du Cambodge, du Laos, et du Viêt-Nam.* Fasc. 1–. Laboratoire de Phanérogamie, Muséum National d'Histoire Naturelle, Paris.

Backer, C. A. and R. C. Bakhuizen van den Brink. 1963–8. *Flora of Java.* 3 Vols. Noordhoff (later Wolters-Noordhoff), Groningen.

Balakrishnan, N. P. 1981–83. *Flora of Jowai.* 2 Vols. Botanical Survey of India, Howrah, 666 pp.

Bamber, C. J. 1916. *Plants of the Punjab.* Superintendent of Government Printing, Punjab, Lahore. (reprinted in 1976, Bishen Singh Mahendra Pal Singh, Dehra Dun), 652 pp.

Bhandari, M. M. 1978. *Flora of the Indian Desert.* Scientific Publishers (distributed by United Book Traders), Jodhpur, 472 pp.
Botanical Survey of India. 1978–. *Flora of India.* Fasc. Botanical Survey of India, Howrah, West Bengal.
Collett, H. 1921. *Flora Simlensis,* 2nd ed. Thacker, Spink, Calcutta. (Reprinted 1971, Bishen Singh, Dehra Dun), 652 pp.
Dassanayake, M. D. and F. R. Fosberg (eds.) 1980–. *A Revised Handbook to the Flora of Ceylon,* 2nd ed. Vol. 1. Amerind (for Smithsonian Inst. and Nat. Sci. Found., Wash. D.C.), New Delhi.
Forest Experimental Station, Korea. 1973. *Illustrated Woody Plants of Korea,* rev. ed. Seoul, 237 pp.
Grierson, A. J. C. and D. G. Long. 1983–. *Flora of Bhutan.* Vols. 1–. Royal Bot. Gard., Edinburgh, Scotland.
Hara, H., W. T. Stern, and L. H. J. Williams. 1978–9. *An Enumeration of the Flowering Plants of Nepal.* Vol. 1–2. British Museum (Nat. Hist.), London.
Holttum, R. E. et al. 1953–71. *A Revised Flora of Malaya.* Vols. 1–3. Government Printer, Singapore.
Holttum, R. E. (ed.) 1959–. *Flora Malesiana. Series II: Pteridophyta.* Vols. 1–. Noordhoff, Jakarta.
Hooker, J. D. 1872–97. *Flora of British India.* 7 Vols. Reeve, London. (Reprinted as *Flora of India* at various times to 1978).
Hundley, H. G. and U Chit Ko Ko. 1961. *List of Trees, Shrubs, Herbs and Principal Climbers, etc. Recorded from Burma.* 3rd ed. Supt. Govt. Printing and Stationery, Rangoon, 532 pp.
Institute of Botany, Academia Sinica. 1959–. *Chung Kuo Chung tzu Chih wu t'u Chih.* (Flora Reipublicae Popularis Sinicae) Vols. 1–. Ko h süeh chu pan she (Academia Sinica Press), Peking.
Johns, R. J. 1981. *The Ferns and Fern-allies of Papua, New Guinea.* Parts 6–12. Papua, New Guinea Univ. Technology Research Reports No. 48–81. Lea, Papua, New Guinea.
Johns, R. J. and A. Bellamy. 1979 (1980). *The Ferns and Fern Allies of Papua New Guinea.* Parts 1–5. Forestry College, Bulolo, Papua, New Guinea.
Keng, H. 1978. *Orders and Families of Malayan Seed Plants,* 2nd ed. Univ. Malaya Press, Lumpur, 437 pp.
Khan, M. S. and A. M. Huq. (eds.) 1972–. *Flora of Bangladesh.* Fasc. 1–. Agricultural Research Council, Bangladesh, Dacca.
Lee, S. C. 1935. *Forest Botany of China.* Commercial Pess. Shanghai, 991 pp.
Lee, T. B. 1979. *Illustrated Flora of Korea.* Hyangmunsa, Seoul, 990 pp.
Lee, Y. N. 1976. *Flowering Plants.* (In: *Illustrated Flora and Fauna of Korea.* Vol. 18. Samhwa (for Ministry of Education), Seoul, 893 pp.
Li, H. L. 1963. *Woody Flora of Taiwan.* Morris Arboretum, Philadelphia, PA, 992 pp.
Li, H. L. et al. (eds.) 1975–9. *Flora of Taiwan.* 6 vols. Epoch, Taipei.
Maheshwari, J. K. 1963. *The Flora of Delhi.* Council for Scientific and Industrial Research, New Delhi, 282 pp.
Matthew, K. M. 1983. *The Flora of the Tamilnadu Carnatic.* 3 vols. The Rapinat Herbarium, St. Joseph's Coll., Tiruchirapalli, India, 2154 pp.
Merrill, E. D. 1912. *A Flora of Manila.* Publ. Bur. Sci. Philippines. No. 5. Dept. of Agr. and Nat. Res. Philippine Islands, Manila. (Reprinted 1968, Cramer, Lehre, Germany), 490 pp.
Merrill, E. D. 1922–26. *An Enumeration of Philippine Flowering Plants.* 4 parts. Publ. Bur. Sci. Philippines. No. 18. Dept. of Agr. and Nat. Res. Philippine Islands, Manila. (Reprinted 1968, Asher, Amsterdam).
Nair, N. C. 1978. Flora of the Punjab Plains. *Rec. Bot. Surv. India* 21(1):1–326.
Noda, M. 1971. *Flora of the North-East Province (Manchuria) of China.* Kazama Bookshop, Tokyo, 1613 pp. (in Japanese).

Northwestern Institute of Botany, Academia Sinica. 1974. *Flora Tsinlingensis.* Vols. 1–2. Academia Sinica Press, Peking.
Oltwi, J. 1965. *Flora of Japan.* (F. G. Meyer and E. H. Walker. eds.) Smithsonian Institution, Washington D.C., 1067 pp.
Polunin, O. and A. Stainton. 1984. *Flowers of the Himalaya.* Oxford Univ. Press, Oxford, 580 pp.
Rau, M. A. 1975. *High Altitude Flowering Plants of West Himalaya.* Bot. Surv. of India. Howrah, West Bengal, 234 pp.
Ridley, H. N. 1922–5. *Flora of the Malay Peninsula.* 5 vols. Reeve, London. (Reprinted in 1968, Asher, Amsterdam).
Royen, P. van. 1979–83. *The Alpine Flora of New Guinea.* 4 vols. Cramer, Vaduz, Liechtenstein, 3160 pp.
Royen, P. van et al. 1964–70. *Manual of the Forest Trees to Papua and New Guinea.* 9 parts. Dept. of Forests, Territory of Papua and New Guinea, Port Moresby.
Saldanha, C. J. 1984. *Flora of Karnataka. Vol. 1. Magnol.–Fabac.* Mohan Primlani, Oxford & IBH Publ. Co., New Delhi, 535 pp.
Smitinand, T. and K. Larsen (eds.) 1970–. *Flora of Thailand.* In several vols. Applied Scientific Research Corporation of Thailand, Bangkok.
Somasundaram, T. R. Sri. 1967. *A Handbook on the Identification and Description of Trees, Shrubs, and Some Important Herbs of the Forest of the Southern States for the Use of the Southern Forest Rangers' College.* Manager of Publications, Government of India, Delhi, 563 pp.
Steenis, C. G. G. J. van (eds.) 1948. *Flora Malesiana.* Series I: Spermatophyta. Vols. 1–. Groningen (Later Noordhoff), Jakarta.
Steward, A. N. 1958. *Manual of Vascular Plants of the Lower Yangtze Valley, China.* Oregon State College, Corvallis, OR, 621 pp.
Wagner, W. H., Jr. and D. F. Grether. 1948. The Pteridophytes of the Admiralty Islands. *Univ. Calif. Publ. Bot.* 23(2):17–110.
Walker, E. H. 1976. *Flora of Okinawa and the Southern Ryukyu Islands.* Smithsonian Institution Press. Washington, D.C., 1159 pp.
Whitmore, T. C. 1966. *Guide to the Forests of the British Solomon Islands.* Oxford Univ. Press, Oxford, 226 pp.
Whitmore, T. C. and F. S. P. Ng (eds.) 1972–8. *Tree Flora of Malaya.* Vols. 1–3. Malayan Forest Rec. No. 22. Longman, Kuala Lumpur.
Womersley, J. S. (ed.) 1978–. *Handbooks of the Flora of Papua New Guinea.* Vols. 1–. Melbourne Univ. Press, Melbourne.
Wu, C. I. (ed.) 1977–9. *Flora Yunnanica.* (Kunming Institute of Botany, Academia Sinica), Academia Sinica Press, Peking.

## *Oceania*

Aubréville, A. et al. (eds.) 1967–. *Flore de la Nouvelle-Calédonie et Depéndances.* Fasc. 1–. Muséum National d'Histoire Naturelle, Paris.
Bramwell, D. and Z. Bramwell. 1974. *Wild Flowers of the Canary Islands.* Stanley Thornes, London, 261 pp.
Britton, N. L. 1918. *Flora of Bermuda.* Scribners, New York. (Reprinted 1965, Hafner, New York), 585 pp.
Brown, F. B. H. and E. W. Brown. 1931–5. *Flora of Southeastern Polynesia.* 3 vols. Bernice P. Bishop Mus. Bull. Nos. 84, 89, 130, Honolulu.
Brownlie, G. 1977. *The Pteridophyte Flora of Fiji.* Beih. Nova Hedwiga, No. 55. Cramer, Vaduz, Liechtenstein, 397 pp.
Copeland, E. B. 1938. Ferns of Southeastern Polynesia. *Occas. Pap. Bernice Pauahi Bishop Mus.* 14(5):45–101.

Degener, O. 1932–. *Flora Hawaiiensis; or the New Illustrated Flora of the Hawaiian Islands.* Vols. 1–. Published by the author, Honolulu (Vols. 1–4 reprinted 1946).
Corner, E. J. H. and K. E. Lee. 1975. Discussion on the Results of the 1971 Royal Society Expedition to the New Hebrides. *Philos. Trans. (London)* B. 272:267–486.
Drake del Castello, E. 1893. *Flore de la Polynésie Francaise.* Masson, Paris, 352 pp.
Fosberg, F. R. 1955. *Northern Marshalls Expedition 1951–1952: Land Biota; Vascular Plants.* Atoll Res. Bull. No. 30. Washington, D.C., 22 pp.
Fosberg, F. R., M. V. C. Falanruw, and M. H. Sachet. 1975. *Vascular Flora of the Northern Marianas Islands.* Smithsonian Contr. Bot. 22:1–45.
Fosberg, F. R. and M. H. Sachet. 1975–. *Flora of Micronesia.* Part 1–. Smithsonian Contr. Bot. No. 20., Washington, D.C.
Fosberg, F. R., M. H. Sachet, and R. Oliver, 1979. A Geographical Checklist of the Micronesian Dicotyledoneae. *Micronesia* 15:41–295.
Fosberg, F. R. et al. 1980. *Vascular Plants of Palau with Vernacular Names.* Dept. Bot. Smithsonian Institution, Washington, D.C., 43 pp.
Glassman, S. F. 1952. *The Flora of Ponape.* Bernice P. Bishop Mus. Bull. No. 209. Honolulu, 152 pp.
Guillaumin, A. 1948. *Flore Analytique et Synotique de la Nouvelle-Calédonie (Phanérogames).* Office de la Recherche Scientifique Coloniale, Paris, 369 pp.
Hillebrand, W. 1888. *Flora of the Hawaiian Islands.* Westermann, New York (Reprinted 1965, Hafner, New York), 673 pp.
Parham, J. W. 1972. *Plants of the Fiji Islands,* rev. ed. Govt. Printer, Suva, 462 pp.
Renvoize, S. A. 1975. A Floristic Analysis of the Western Indian Ocean Coral Islands. *Kew Bull.* 30:133–152.
Rock, J. F. 1974. *Indigenous Trees of Hawaii.* (Reprint of author's 1913 *The Indigenous Trees of the Hawaiian Islands,* published by the author, Honolulu.) Tuttle, Rutland, VT, 548 pp.
Skottsberg, C. J. F. 1922. The Phanerogams of Easter Island. In: C. J. F. Skottsberg. *Nat. Hist. of Juan Fernandez and Easter Island.* Part 2—Botany. Almqvist & Wiksell, Uppsala, pp. 61–84.
Smith, A. C. 1979. *Flora Vitiensis Nova: A New Flora of Fiji.* Vol. 1–. Pacific Trop. Bot. Gard. Lawai, Kauai, HI, 501 pp.
St. John, H. 1954. Ferns of Rotuma Island, A Descriptive Manual. *Occas. Pap. Bernice Pauahi Bishop Mus.* 21(9):161–208.
St. John, H. 1956. A Translation of the Keys in "Flora Micronesica" of Ryôzo Kanehira (1933). *Pacific Sci.* 10:96–102.
Stoddart, D. R. (ed.). 1970. Coral Islands of the Western Indian Ocean. *Atoll Res. Bull.* No. 136. Washington, D. C., 224 pp.
Stone, B. C. 1970. The Flora of Guam. *Micronesica* 6:1–629.
Wiggins, I. L. and D. M. Porter. 1970. *Flora of the Galapágos Islands.* Stanford Univ. Press, Stanford, CA, 998 pp.

# Appendix 2: Classification of the Division Magnoliophyta (Flowering Plants)

The following outline is the most recently updated classification system as proposed by Arthur Cronquist (*The Evolution and Classification of Flowering Plants*, 2nd ed. 1988. The New York Botanical Garden, Bronx, NY, 555 pp.). It has been used in this text because detailed information is readily available to anyone wishing further understanding beyond the scope of this book. The names in brackets are older, accepted, alternate names or family names considered by Cronquist to be included in the family listed.

CLASS MAGNOLIOPSIDA (DICOTYLEDONS)

Subclass I. Magnoliidae

Order 1. Magnoliales
Family 1. Winteraceae
2. Degeneriaceae
3. Himantandraceae
4. Eupomatiaceae
5. Austrobaileyaceae
6. Magnoliaceae
7. Lactoridaceae
8. Annonaceae
9. Myristicaceae
10. Canellaceae

Order 2. Laurales
Family 1. Amborellaceae
2. Trimeniaceae
3. Monimiaceae (Atherospermataceae, Hortoniaceae, Siparunaceae)
4. Gomortegaceae
5. Calycanthaceae
6. Idiospermaceae
7. Lauraceae (Cassythaceae)
8. Hernandiaceae (Gyrocarpaceae)

Order 3. Piperales
Family 1. Chloranthaceae
2. Saururaceae
3. Piperaceae (Peperomiaceae)

Order 4. Aristolochiales
Family 1. Aristolochiaceae

Order 5. Illiciales
Family 1. Illiciaceae
2. Schisandraceae

Order 6. Nymphaeales
Family 1. Nelumbonaceae
2. Nymphaeaceae (Euryalaceae)
3. Barclayaceae
4. Cabombaceae
5. Ceratophyllaceae

Order 7. Ranunculales
Family 1. Ranunculaceae (Glaucidiaceae, Helleboraceae, Hydrastidaceae)
2. Circaeasteraceae (Kingdoniaceae)
3. Berberidaceae (Leonticaceae, Nandinaceae, Podophyllaceae)
4. Sargentodoxaceae
5. Lardizabalaceae
6. Menispermaceae
7. Coriariaceae
8. Sabiaceae (Meliosmaceae)

Order 8. Papaverales
Family 1. Papaveraceae (Chelidoniaceae, Eschscholziaceae, Platystemonaceae)
2. Fumariaceae (Hypecoaceae, Pteridophyllaceae)

Subclass II. Hamamelidae

Order 1. Trochodendrales
Family 1. Tetracentraceae
2. Trochodendraceae

Order 2. Hamamelidales
Family 1. Cercidiphyllaceae
2. Eupteleaceae
3. Platanaceae
4. Hamamelidaceae (Altingiaceae, Rhodoleiaceae)
5. Myrothamnaceae

Order 3. Daphniphyllales
Family 1. Daphniphyllaceae

Order 4. Didymelales
Family 1. Didymelaceae

Order 5. Eucommiales
Family 1. Eucommiaceae

Order 6. Urticales
Family 1. Barbeyaceae
2. Ulmaceae (Celtidaceae)
3. Cannabaceae
4. Moraceae
5. Cecropiaceae
6. Urticaceae
* Physenaceae (temporarily placed here)

Order 7. Leitneriales
Family 1. Leitneriaceae

Order 8. Juglandales
Family 1. Rhoipteleaceae
2. Juglandaceae

Order 9. Myricales
Family 1. Myricaceae

Order 10. Fagales
Family 1. Balanopaceae
2. Fagaceae
3. Nothofagaceae
4. Betulaceae (Carpinaceae, Corylaceae)

Order 11. Casuarinales
Family 1. Casuarinaceae

Subclass III. Caryophyllidae

Order 1. Caryophyllales (Centrospermae)
Family 1. Phytolaccaceae (Agdestidaceae, Barbeuiaceae, Gisekiaceae, Petiveriaceae, Stegnospermaceae)
2. Achatocarpaceae
3. Nyctaginaceae
4. Aizoaceae (Ficoidaceae, Mesembryanthemaceae, Sesuviaceae, Tetragoniaceae)
5. Didiereaceae
6. Cactaceae
7. Chenopodiaceae (Dysphaniaceae, Halophytaceae, Salicorniaceae)
8. Amaranthaceae
9. Portulacaceae (Hectorellaceae)
10. Basellaceae
11. Molluginaceae
12. Caryophyllaceae (Alsinaceae, Illecebraceae, Silenaceae)

Order 2. Polygonales
Family 1. Polygonaceae

Order 3. Plumbaginales
Family 1. Plumbaginaceae (Limoniaceae)

Subclass IV. Dilleniidae

Order 1. Dilleniales
Family 1. Dilleniaceae
2. Paeoniaceae

Order 2. Theales
Family 1. Ochnaceae (Diegodendraceae, Lophiraceae, Luxemburgiaceae, Strasburgeriaceae, Sauvagesiaceae, Wallaceaceae)
2. Sphaerosepalaceae (Rhopalocarpaceae)
3. Sarcolaenaceae
4. Dipterocarpaceae
5. Caryocaraceae
6. Theaceae (Asteropeiaceae, Bonnetiaceae, Camelliaceae, Sladeniaceae, Ternstroemiaceae)
7. Actinidiaceae (Saurauiaceae)
8. Scytopetalaceae (Rhaptopetalaceae)
9. Pentaphylacaceae
10. Tetrameristaceae
11. Pellicieraceae
12. Oncothecaceae
13. Marcgraviaceae
14. Quiinaceae

15. Elatinaceae
16. Paracryphiaceae
17. Medusagynaceae
18. Clusiaceae (Guttiferae, an accepted alternate name; Garciniaceae, Hypericaceae)

Order 3. Malvales
Family 1. Elaeocarpaceae (Aristoteliaceae)
2. Tiliaceae
3. Sterculiaceae (Byttneriaceae)
4. Bombacaceae
5. Malvaceae

Order 4. Lecythidales
Family 1. Lecythidaceae (Asteranthaceae, Barringtoniaceae, Foetidiaceae, Napoleonaeaceae)

Order 5. Nepenthales
Family 1. Sarraceniaceae
2. Nepenthaceae
3. Droseraceae (Dionaeaceae)

Order 6. Violales
Family 1. Flacourtiaceae (Berberidopsidaceae, Neumanniaceae, Plagiopteridaceae, Soyauxiaceae)
2. Peridiscaceae
3. Bixaceae (Cochlospermaceae)
4. Cistaceae
5. Huaceae
6. Lacistemaceae
7. Scyphostegiaceae
8. Stachyuraceae
9. Violaceae (Leoniaceae)
10. Tamaricaceae
11. Frankeniaceae
12. Dioncophyllaceae
13. Ancistrocladaceae
14. Turneraceae
15. Malesherbiaceae
16. Passifloraceae
17. Achariaceae
18. Caricaceae
19. Fouquieriaceae
20. Hoplestigmataceae
21. Cucurbitaceae
22. Datiscaceae (Tetramelaceae)
23. Begoniaceae
24. Loasaceae (Gronoviaceae)

Order 7. Salicales
Family 1. Salicaceae

Order 8. Capparales
Family 1. Tovariaceae
2. Capparaceae (Cleomaceae, Koeberliniaceae, Pentadiplandraceae)
3. Brassicaceae (Cruciferae, an accepted alternate name)
4. Moringaceae
5. Resedaceae

Order 9. Batales
- Family 1. Gyrostemonaceae
  - 2. Bataceae

Order 10. Ericales
- Family 1. Cyrillaceae
  - 2. Clethraceae
  - 3. Grubbiaceae
  - 4. Empetraceae
  - 5. Epacridaceae (Prionotaceae, Stypheliaceae)
  - 6. Ericaceae (Vacciniaceae)
  - 7. Pyrolaceae
  - 8. Monotropaceae

Order 11. Diapensiales
- Family 1. Diapensiaceae

Order 12. Ebenales
- Family 1. Sapotaceae (Achraceae, Boerlagellaceae, Bumeliaceae, Sarcospermataceae)
  - 2. Ebenaceae
  - 3. Styracaceae
  - 4. Lissocarpaceae
  - 5. Symplocaceae

Order 13. Primulales
- Family 1. Theophrastaceae
  - 2. Myrsinaceae (Aegicerataceae)
  - 3. Primulaceae (Coridaceae)

Subclass V. Rosidae

Order 1. Rosales
- Family 1. Brunelliaceae
  - 2. Connaraceae
  - 3. Eucryphiaceae
  - 4. Cunoniaceae (Baueraceae)
  - 5. Davidsoniaceae
  - 6. Dialypetalanthaceae
  - 7. Pittosporaceae
  - 8. Byblidaceae (Roridulaceae)
  - 9. Hydrangeaceae (Kirengeshomaceae, Philadelphaceae, Pottingeriaceae)
  - 10. Columelliaceae
  - 11. Grossulariaceae (Argophyllaceae, Brexiaceae, Carpodetaceae, Dulongiaceae, Escalloniaceae, Iteaceae, Montiniaceae, Phyllonomaceae, Polyosmataceae, Pterostemonaceae, Rousseaceae, Tetracarpaeaceae, Tribelaceae)
  - 12. Greyiaceae
  - 13. Bruniaceae (Berzeliaceae)
  - 14. Anisophylleaceae (Polygonanthaceae)
  - 15. Alseuosmiaceae
  - 16. Crassulaceae
  - 17. Cephalotaceae
  - 18. Saxifragaceae (Eremosynaceae, Francoaceae, Lepuropetalaceae, Parnassiaceae, Penthoraceae, Vahliaceae)
  - 19. Rosaceae (Amygdalaceae, Drupaceae, Malaceae, Pomaceae)

20. Neuradaceae
21. Crossosomataceae
22. Chrysobalanaceae
23. Surianaceae (Stylobasiaceae)
24. Rhabdodendraceae

Order 2. Fabales (Leguminosae, an accepted alternate name)
Family 1. Mimosaceae
2. Caesalpiniaceae
3. Fabaceae (Papilionaceae, an accepted alternate name)

Order 3. Proteales
Family 1. Elaeagnaceae
2. Proteaceae

Order 4. Podostemales
Family 1. Podostemaceae (Tristichaceae)

Order 5. Haloragales
Family 1. Haloragaceae (Myriophyllaceae)
2. Gunneraceae

Order 6. Myrtales
Family 1. Sonneratiaceae (Duabangaceae)
2. Lythraceae
3. Rhynchocalycaceae
4. Alzateaceae
5. Penaeaceae
6. Crypteroniaceae
7. Thymelaeaceae
8. Trapaceae
9. Myrtaceae (Heteropyxidaceae, Kaniaceae, Psiloxylaceae)
10. Punicaceae
11. Onagraceae
12. Oliniaceae
13. Melastomataceae (Memecylaceae, Mouririaceae)
14. Combretaceae (Strephonemataceae)

Order 7. Rhizophorales
Family 1. Rhizophoraceae

Order 8. Cornales
Family 1. Alangiaceae
2. Cornaceae (Aralidiaceae, Aucubaceae, Curtisiaceae, Davidiaceae, Griseliniaceae, Helwingiaceae, Mastixiaceae, Melanophyllaceae, Nyssaceae, Toricelliaceae)
3. Garryaceae

Order 9. Santalales
Family 1. Medusandraceae
2. Dipentodontaceae
3. Olacaceae (Aptandraceae, Cathedraceae, Chaunochitonaceae, Coulaceae, Erythropalaceae, Heisteriaceae, Octoknemaceae, Schoepfiaceae, Scorodocarpaceae, Strombosiaceae, Tetrastylidaceae)
4. Opiliaceae (Cansjeraceae)
5. Santalaceae (Anthobolaceae, Canopodaceae, Exocarpaceae, Osyridaceae, Podospermaceae)
6. Misodendraceae
7. Loranthaceae

8. Viscaceae
9. Eremolepidaceae
10. Balanophoraceae (Cynomoriaceae, Dactylanthaceae, Sarcophytaceae)

Order 10. Rafflesiales
Family 1. Hydnoraceae
2. Mitrastemonaceae
3. Rafflesiaceae (Apodanthaceae, Cytinaceae)

Order 11. Celastrales
Family 1. Geissolomataceae
2. Celastraceae (Canotiaceae, Chingithamnaceae, Goupiaceae, Lophopyxidaceae, Siphonodontaceae)
3. Hippocrateaceae
4. Stackhousiaceae
5. Salvadoraceae
6. Tepuianthaceae
7. Aquifoliaceae (Phellinaceae, Sphenostemonaceae)
8. Icacinaceae (Phytocrenaceae)
9. Aextoxicaceae
10. Cardiopteridaceae
11. Corynocarpaceae
12. Dichapetalaceae

Order 12. Euphorbiales
Family 1. Buxaceae (Pachysandraceae, Stylocerataceae)
2. Simmondsiaceae
3. Pandaceae
4. Euphorbiaceae (Androstachydaceae, Hymenocardiaceae, Picrodendraceae, Putranjivaceae, Scepaceae, Stilaginaceae, Uapacaceae)

Order 13. Rhamnales
Family 1. Rhamnaceae (Camarandraceae, Frangulaceae, Phylicaceae)
2. Leeaceae
3. Vitaceae

Order 14. Linales
Family 1. Erythroxylaceae (Nectaropetalaceae)
2. Humiriaceae
3. Ixonanthaceae
4. Hugoniaceae (Ctenolophonaceae)
5. Linaceae

Order 15. Polygalales
Family 1. Malpighiaceae
2. Vochysiaceae
3. Trigoniaceae
4. Tremandraceae
5. Polygalaceae (Diclidantheraceae, Disantheraceae, Emblingiaceae, Moutabeaceae)
6. Xanthophyllaceae
7. Krameriaceae

Order 16. Sapindales
Family 1. Staphyleaceae (Tapisciaceae)
2. Melianthaceae
3. Bretschneideraceae
4. Akaniaceae
5. Sapindaceae (Ptaeroxylaceae)

6. Hippocastanaceae
7. Aceraceae
8. Burseraceae
9. Anacardiaceae (Blepharocaryaceae, Pistiaceae, Podoaceae)
10. Julianiaceae
11. Simaroubaceae (Irvingiaceae, Kirkiaceae)
12. Cneoraceae
13. Meliaceae (Aitoniaceae)
14. Rutaceae (Flindersiaceae)
15. Zygophyllaceae (Balanitaceae, Nitrariaceae, Penganaceae, Tetradiclidaceae, Tribulaceae)

Order 17. Geraniales

Family 1. Oxalidaceae (Averrhoaceae, Hypseocharitaceae, Lepidobotryaceae)
2. Geraniaceae (Biebersteiniaceae, Dirachmaceae, Ledocarpaceae, Rhynchothecaceae, Vivianiaceae)
3. Limnanthaceae
4. Tropaeolaceae
5. Balsaminaceae

Order 18. Apiales

Family 1. Araliaceae
2. Apiaceae (Umbelliferae, an accepted alternate name; Hydrocotylaceae, Saniculaceae)

Subclass VI. Asteridae

Order 1. Gentianales

Family 1. Loganiaceae (Antoniaceae, Desfontainiaceae, Potaliaceae, Spigeliaceae, Strychnaceae)
2. Gentianaceae
3. Saccifoliaceae
4. Apocynaceae (Plocospermataceae, Plumeriaceae)
5. Asclepiadaceae (Periplocaceae)

Order 2. Solanales

Family 1. Duckeodendraceae
2. Nolanaceae
3. Solanaceae (Goetziaceae, Salpiglossidaceae, Sclerophylacaceae)
4. Convolvulaceae (Dichondraceae, Humbertiaceae)
5. Cuscutaceae
6. Retziaceae
7. Menyanthaceae
8. Polemoniaceae (Cobaeaceae)
9. Hydrophyllaceae

Order 3. Lamiales

Family 1. Lennoaceae
2. Boraginaceae
3. Verbenaceae (Avicenniaceae, Chloanthaceae, Dicrastylidiaceae, Nyctanthaceae, Phrymaceae, Stilbaceae, Symphoremataceae)
4. Lamiaceae (Labiatae, an accepted alternate name; Menthaceae, Tetrachondraceae)

Order 4. Callitrichales

Family 1. Hippuridaceae
2. Callitrichaceae
3. Hydrostachyaceae

Order 5. Plantaginales
Family 1. Plantaginaceae

Order 6. Scrophulariales
Family 1. Buddlejaceae
2. Oleaceae (Fraxinaceae, Syringaceae)
3. Scrophulariaceae (Ellisiophyllaceae, Rhinanthaceae)
4. Globulariaceae (Selaginaceae)
5. Myoporaceae (Spielmanniaceae)
6. Orobanchaceae
7. Gesneriaceae (Cyrtandraceae)
8. Acanthaceae (Thunbergiaceae)
9. Pedaliaceae (Martyniaceae, Trapellaceae)
10. Bignoniaceae
11. Mendonciaceae
12. Lentibulariaceae (Pinguiculaceae, Utriculariaceae)

Order 7. Campanulales
Family 1. Pentaphragmataceae
2. Sphenocleaceae
3. Campanulaceae (Cyphiaceae, Cyphocarpaceae, Lobeliaceae, Nemacladaceae)
4. Stylidiaceae
5. Donatiaceae
6. Brunoniaceae
7. Goodeniaceae

Order 8. Rubiales
Family 1. Rubiaceae (Henriqueziaceae, Naucleaceae)
2. Theligonaceae (Cynocrambaceae)

Order 9. Dipsacales
Family 1. Caprifoliaceae (Carlemanniaceae, Sambucaceae, Viburnaceae)
2. Adoxaceae
3. Valerianaceae (Triplostegiaceae)
4. Dipsacaceae (Morinaceae)

Order 10. Calycerales
Family 1. Calyceraceae

Order 11. Asterales
Family 1. Asteraceae (Composite, an accepted alternate name; Ambrosiaceae, Carduaceae, Cichoriaceae)

## CLASS LILIOPSIDA (MONOCOTYLEDONS)

### Subclass I. Alismatidae

Order 1. Alismatales
Family 1. Butomaceae
2. Limnocharitaceae
3. Alismataceae

Order 2. Hydrocharitales
Family 1. Hydrocharitaceae (Haplophilaceae, Thalassiaceae)

Order 3. Najadales
Family 1. Aponogetonaceae
2. Scheuchzeriaceae

3. Juncaginaceae (Lilaeaceae, Maundiaceae, Triglochinaceae)
4. Potamogetonaceae
5. Ruppiaceae
6. Najadaceae
7. Zannichelliaceae
8. Posidoniaceae
9. Cymodoceaceae
10. Zosteraceae

Order 4. Triuridales
Family 1. Petrosaviaceae
2. Triuridaceae

Subclass II. Arecidae

Order 1. Arecales
Family 1. Arecaceae (Palmae, an accepted alternate name; Nypaceae, Phytelephasiaceae)

Order 2. Cyclanthales
Family 1. Cyclanthaceae

Order 3. Pandanales
Family 1. Pandanaceae

Order 4. Arales
Family 1. Acoraceae
2. Araceae
3. Lemnaceae

Subclass III. Commelinidae

Order 1. Commelinales
Family 1. Rapateaceae
2. Xyridaceae (Abolbodaceae)
3. Mayacaceae
4. Commelinaceae (Cartonemataceae)

Order 2. Eriocaulales
Family 1. Eriocaulaceae

Order 3. Restionales
Family 1. Flagellariaceae
2. Joinvilleaceae
3. Restionaceae (Anarthriaceae, Ecdeiocoleaceae)
4. Centrolepidaceae

Order 4. Juncales
Family 1. Juncaceae
2. Thurniaceae

Order 5. Cyperales
Family 1. Cyperaceae (Kobresiaceae)
2. Poaceae (Gramineae, an accepted alternate name; Anomochloaceae, Bambusaceae, Streptochaetaceae)

Order 6. Hydatellales
Family 1. Hydatellaceae

Order 7. Typhales
Family 1. Sparganiaceae
2. Typhaceae

Subclass IV. Zingiberidae

Order 1. Bromeliales

Family 1. Bromeliaceae (Tillandsiaceae)

Order 2. Zingiberales (Scitamineae)

Family 1. Strelitziaceae
2. Heliconiaceae
3. Musaceae
4. Lowiaceae (Orchidanthaceae)
5. Zingiberaceae
6. Costaceae
7. Cannaceae
8. Marantaceae

Subclass V. Liliidae

Order 1. Liliales

Family 1. Philydraceae
2. Pontederiaceae
3. Haemodoraceae (Conostylidaceae)
4. Cyanastraceae
5. Liliaceae (Agapanthaceae, Alliaceae, Alstroemeriaceae, Amaryllidaceae, Aphyllanthaceae, Anthericaceae, Asparagaceae, Asphodelaceae, Aspidistraceae, Asteliaceae, Blandfordiaceae, Calochortaceae, Campynemaceae, Colchicaceae, Convallariaceae, Dianellaceae, Eriospermaceae, Funkiaceae, Hemerocallidaceae, Herreriaceae, Hesperocallidaceae, Hyacinthaceae, Hypoxidaceae, Ixoliriaceae, Medeolaceae, Melanthiaceae, Nartheciaceae, Ruscaceae, Tecophilaeaceae, Uvulariaceae)
6. Iridaceae (Gladiolaceae, Hewardiaceae, Isophysidaceae, Ixiaceae)
7. Velloziaceae
8. Aloaceae
9. Agavaceae (Doryanthaceae, Dracaenaceae, Nolinaceae, Phormiaceae, Sansevieriaceae)
10. Xanthorrhoeaceae (Calectasiaceae, Dasypogonaceae)
11. Hanguanaceae
12. Taccaceae
13. Stemonaceae (Croomiaceae, Roxburghiaceae)
14. Smilacaceae (Lapageriaceae, Luzuriagaceae, Petermanniaceae, Philesiaceae, Rhipogonaceae)
15. Dioscoreaceae (Cladophyllaceae, Stenomeridaceae, Tamaceae, Trichopodaceae)

Order 2. Orchidales

Family 1. Geosiridaceae
2. Burmanniaceae (Tripterellaceae, Thismiaceae)
3. Corsiaceae
4. Orchidaceae (Apostasiaceae, Cypripediaceae, Limodoraceae, Neottiaceae, Thyridiaceae, Vanillaceae)

# *Glossary*

**Abaxial** Dorsal; on the side of an organ away from the main axis.
**Acaulescent** A naked stem, usually with leaves clustered at or near the base.
**Accessory** A fleshy fruit made up of a succulent receptacle covered with several to many individual pistils, each forming a dry achene-like fruit (i.e., strawberry).
**Achene** A dry one-seeded indehiscent fruit with the seed connected to the pericarp at only one point (i.e., sunflower seed).
**Acicular** *See* Needle-shaped.
**Acorn** A one-seeded dry indehiscent fruit with a hard coat and surrounded by a "cap" of dried bracts (i.e., oak).
**Actinomorphic** *See* Regular.
**Acuminate** When the leaf apex gradually tapers to a prolonged point, with the two leaf margins pinched slightly before reaching the tip. The tip may be short or long, or narrow or broad.
**Acute** Tapering to a point with the two margins having straight sides and forming an angle of less than 90°.
**Adaxial** Ventral; on the side of an organ toward the main axis.
**Adnate** The union of unlike parts (i.e., calyx and corolla).
**Adventitious** Term used for recently introduced plants.
**Adventitious bud** A bud that develops at places other than at the node.
**Adventitious root** Unusual roots that develop from other regions of the plant than usual—generally from stem or leaf tissue; also applies to other plant organs.
**Aerial roots** Roots originating and borne above the ground or water.
**Aggregate** A fleshy fruit with a receptacle that is not especially fleshy, and with several to many individual pistils, these each becoming fleshy drupes (i.e., blackberry, raspberry).
**Albuminous cells** Specialized parenchyma cells, sometimes found in gymnosperms closely associated with phloem sieve cells; also called Strasburger cells.
**Allelopathy** Inhibition of other plants from growing nearby.
**Allozyme** A different form of an enzyme where the polypeptides are determined by different alleles at one locus.
**Alpha taxonomy** Descriptive taxonomy.
**Alternate leaves** A type of phyllotaxy in which only one leaf is at a node.
**Alveolate** Appearing like the surface of a honeycomb.
**Ament** A deciduous, pendent, or erect spike-like inflorescence comprised of unisexual, apetalous flowers (syn. catkin).
**American Code** An early botanical code of nomenclature followed in North America during the late nineteenth and early twentieth centuries.
**Androecium** A term used collectively for all the stamens of the flower.
**Angeion** A Greek word meaning "enclosing vessel."
**Annotation label** A label added to a specimen when an authority verifies the proper identification of a plant.
**Annual** A plant that completes its life cycle in one growing season or one year.
**Annulus** A thickened band of cells in the sporongial wall that facilitates in releasing the spores.

**Anther** The enlarged terminal pollen-bearing portion of the stamen.
**Antheridium** The multicellular male reproductive organ of plants other than seed plants (pl. antheridia).
**Antipodals** Three nuclei of no known function in the embryo sac, furthest away from the micropyle.
**Antiseptic** Any substance that is nontoxic to cells but inhibits the growth of or kills disease-causing or harmful microorganisms.
**Apetalous** A flower that lacks a corolla.
**Apex** The end or tip.
**Apical placentation** Placentation type found in both simple and compound pistils in which one or more ovules are attached to the top of the locule.
**Apiculate** When the leaf apex ends with a slender but nonstiff tip.
**Apocarpous** Term applied to flowers that have two or more distinct carpels.
**Arachnoid** Having tangled cobweb-like hairs.
**Arborescent** Tree-like.
**Arboretum** A place where trees and shrubs are cultivated for educational and scientific purposes.
**Archegonium** The multicellular female reproductive organ of plants other than seed plants that contains the female gamete or egg (pl. archegonia).
**Areole** The spot on the stem of cacti where spines are present.
**Aril** A fleshy stalk or cup-like structure that subtends or partially encloses a seed.
**Aristate** When the leaf apex has a stiff bristle tip.
**Artificial classification** A classification system based on obvious or convenient characters for the purpose of categorizing irrespective of affinity.
**Ascending** Stem growing upward at about a 45–60° angle from the horizontal.
**Atactostele** The scattered arrangement of vascular bundles.
**Attenuate** Tapering gradually to a slender tip or point.
**Auricle** An appendage or ear-shaped lobe.
**Auriculate** Having ear-like lobed appendages.
**Awl-shaped** A small, sharp, flat, narrowly triangular evergreen leaf that tapers gradually to a point; resembles a leather punch (i.e., in some *Juniperus*).
**Awn** A bristle-like, sometimes stiff extension of a structure, usually from the apex or tip.
**Axil** The upper angle area between the leaf and the stem to which it is attached.
**Axile placentation** Placentation type when the ovules are attached where the septa of a compound pistil are united, usually in the middle but may be to the septa themselves.
**Axillary** Term used when branches, buds, or flowers are attached in the axil of a leaf.

**Banner** The upper, commonly largest petal of a "pea-like" flower; also called a standard.
**Barb** Short, sharp bristles, usually pointing backward or down.
**Barbellate** Bearing stiff hairs with barbs down the side.
**Basal** At the base.
**Basal cell** Enlarged base cell of the suspensor.
**Basal placentation** Placentation type found in both simple and compound pistils having one locule and one or more ovules attached to the floor of the locule.
**Basionym** The basic epithet name when two or more Latin names are being validly combined.
**Berry** A fleshy fruit formed from one compound ovary with few to many seeds (i.e., blueberry, grape).
**Beta taxonomy** Experimental systematics.

**Biennial** A plant that completes its life cycle in two growing seasons or years, with vegetative growth the first year, and blooming and fruiting the second.
**Binomial nomenclature** Two-word Latin names given to all species of known plants.
**Biomass** The amount of living plant material in a given area.
**Biosystematics** Term used to delimit the natural biotic units and to apply to these units a system of nomenclature that conveys precise information regarding relationships, variability, and dynamic structure; sometimes called experimental systematics.
**Bipinnately compound** A compound leaf with the leaflets divided again into secondary leaflets.
**Bipinnatifid** Twice branched.
**Bisexual** Having both male and female sex organs in the same flower, a perfect flower (syn. hermaphrodite).
**Blade** The expanded or flat portion of a leaf or petal (syn. lamina).
**Bladeless** Blade of the leaf is absent.
**Bloom** *See* Glaucous.
**Botanical name** The binomial Latin name of a plant comprised of the genus name and specific epithet (syn. Latin name).
**Bracket key** Type of key in which each pair of contrasting choices are placed on adjacent lines of the key and are not separated by intervening lines.
**Bract** A modified leaf.
**Bracteole** A small bract or secondary bract (syn. bractlet).
**Bud** A rudimentary undeveloped branch, flower, or leaf, usually enclosed by protective scales or sepals.
**Bud scale scar** Scar remaining when bud scales have dropped off.
**Bulb** A subterranean thickened stem bearing many fleshy or scale-like leaves surrounding a fleshy bud and with fibrous roots coming from the bottom.
**Bulblets** Small bulbs, usually borne above ground level.
**Bulbil** A small asexually formed bulb, usually found in the axils of leaves or inflorescences (i.e., *Allium canadensis*).
**Bundle scar** Scar left on a leaf scar where the vascular tissue was broken when the leaf dropped from the plant.

**Caespitose** Having stems growing in a clump or tuft.
**Calyptera** Forming a cap over another structure or organ.
**Calyx** A term used collectively for the outermost whorl of "floral leaves" of a flower.
**Canescent** Surface densely covered with short gray or white hairs that give color to the surface.
**Capitulum** *See* Head.
**Capsule** A dry, dehiscent multi-carpeled fruit with two or more placentae; there are various types depending on how they split.
**Carpel** The basic constructive unit of the pistil that has been interpreted as a modified "seed-bearing leaf."
**Caryopsis** An indehiscent, dry one-seeded fruit with a seed connected to the pericarp by all sides (i.e., maize, wheat).
**Catkin** *See* Ament.
**Caudix** A slow-growing, woody, upright underground base of a herbaceous perennial that each year gives rise to leaves and flowering stems.
**Caulescent** Stem being leafy above the ground.
**Centrifugally** When development begins at the middle and proceeds away from the center axis.
**Centripetally** Originating from the outside and proceeding toward the center of the axis.
**Chambered pith** Term applied to pith tissue that separates into solid tissue and many small air chambers.

**Character** A feature of an organism that can be counted, measured, or assessed in some way.
**Chemosystematics** Area of science that uses chemical constituents of plants to assess plant phylogenies.
**Chemotaxonomy** *See* Chemosystematics.
**Chromosome** Microscopic nucleoprotein bodies in the cell nucleus that carry genes.
**Chicle** Elastic plant substance used in chewing gum.
**Ciliate** With a marginal fringe of small hairs, these being finer than fimbriate.
**Circinate vernation** Coiled into a ring with the end or apex in the middle; a condition seen in most young fern leaves.
**Circumboreal** Being distributed or found around the top or bottom of the world in the boreal zone.
**Circumpolar** Being distributed or found around the North or South Pole.
**Circumscissile capsule** A capsule that opens by a lid forming along a horizontal circular suture (i.e., *Plantago*).
**Clade** A classification group defined by features exclusive to all its members and distinguishing the group from all other groups; a single phylogenetic lineage.
**Cladistics** A modern system of classification in which the only groups formally recognized are those that distinguish the group from other groups (called clades).
**Cladists** Proponents of cladistics.
**Cladogram** A branching diagram that shows how particular organisms are grouped into clades.
**Cladophyll** A branch, stem, or stem segment that is flattened and looks like a leaf (i.e., *Asparagus*).
**Class** A taxonomic unit between the higher rank of division and the lower rank of order.
**Classification** The orderly arrangement of plants into a hierarchal system.
**Cleft** Generally applies to leaf margin segments and sinuses, sharp and not cut over one-half way to the midrib.
**Cleistogamous** Flowers that do not open but produce viable seed by selfing.
**Climbing** Clinging or twining to other objects (syn. scandent).
**Clines** Gradients of character variation.
**Collection number** A number used only once by a plant collector and applied to a particular plant collection at a particular location and time; usually applied in a numerical sequence (i.e., 1, 2, 3, . . . ).
**Column** The union of stamens and style into a central flower structure, as seen in the orchid family; also a basal twisted part of an awn, as seen in grasses.
**Coma** The tuft of hairs attached at the end of a structure (i.e., apex of a seed).
**Common name** *See* Vernacular name.
**Comose** Having a clump or tuft of long hairs attached at the apex (i.e., apex of a seed).
**Companion cells** Specialized parenchyma cells appressed to each sieve tube cell and thought to provide nuclear communication to the sieve tube cell.
**Complete flower** A flower having all four of the floral parts—sepals, petals, stamens, and pistil(s).
**Compound inflorescence** Where the inflorescence is made up of two or more simple inflorescences (i.e., panicle, compound umbel).
**Compound leaf** A leaf where the blade is separated into two or more smaller segments or leaflets.
**Compound pistil** A pistil consisting of two or more united (connate) carpels.
**Connate** When like parts become united together in part or whole (syn. coalescent) (i.e., sepals united).
**Cordate** Broadly heart-shaped with the petiole attached to the broad end (syn. subulate).

**Coriaceous** Being thick, tough, and leather-like.
**Corm** A vertical, usually broader than tall, thickened fleshy underground stem covered with papery leaves.
**Corolla** A term used collectively for all the petals of the flower.
**Corona** An appendage of tissue situated between the corolla and the stamens; also called a crown.
**Corymb** A flat-topped, indeterminant inflorescence in which the lower pedicels become progressively elongate and the rachis shortened; may be simple or compound.
**Cosmopolitan** Having a worldwide distribution.
**Cotyledon** Seed leaves that provide nourishment to a germinating seedling.
**Couplet** A pair of contrasting choices, many of which make up a dichotomous key.
**Couplets** The contrasting dichotomous choices used in a key.
**Crenate** Leaf margin with low, broad, rounded teeth.
**Crenulate** Leaf margin having very small rounded teeth.
**Crown** *See* Corona.
**Crozier** A plant structure that is curled at the end resembling the narrow end of a violin; like the young developing leaves of most ferns (syn. fiddlehead).
**Culm** Name given to the stem of grasses, sedges, and rushes.
**Curator** A person responsible for the care of a collection or herbarium of specimens.
**Cuspidate** When the leaf apex has an abrupt, short, sharp, rigid tip.
**Cyathium** An inflorescence type in the family Euphorbiaceae consisting of a three-lobed stalked female flower surrounded by male flowers, both unisexual flower types within a cup-shaped involucre.
**Cyme** A determinate inflorescence where the terminal flower is older than the subtending two lateral flowers; may be simple or compound.
**Cystolith** A specialized crystal found in plant cells; especially present in some leaves (i.e., Urticaceae).

**Deciduous** Shed each year.
**Decumbent** Lying on the ground but with the ends curved upward.
**Decurrent** Extending down from the point of attachment.
**Decussate** Leaves opposite with the next leaf pair at right angles to the previous pair.
**Dehiscent fruit** A fruit that splits open as a result of drying.
**Deltoid** Triangular-shaped.
**Dendrogram** A branching diagram that shows levels of similarity between taxa.
**Dentate** Sharp marginal teeth of a leaf projecting at right angles to the margin.
**Denticulate** Very small, sharp marginal teeth of a leaf projecting at right angles to the margin.
**Determinate** A flower inflorescence that has the oldest flower terminating the rachis with the blooming pattern being outward and downward.
**Diaphragmed pith** Term applied when pith tissue is compartmentalized by thin membranes.
**Dichotomous** Continually forking twice into equal-sized branches.
**Diffuse** Spreading widely in all directions.
**Dimorphic** Two different forms on the same plant.
**Dioecious** Male and female reproductive structures on different plants of the same species (i.e., *Salix*, willow).
**Diploid** The production of two sets of chromosomes in each cell (abbr. $2n$).
**Discoid** Looking like a disk; a type of head inflorescence in the Asteraceae lacking ray flowers, having only disk flowers.
**Disk flowers** Regular tubular flowers in the head inflorescence of the family Asteraceae (Compositae) (syn. tubular flower).
**Dissected** Cut into many fine segments.

**Distichous** Leaves conspicuously in two rows (syn. two-ranked).
**Divided** Leaf margin segments cut to the leaf base or midrib.
**Division** A taxonomic unit between the higher rank of kingdom and the lower unit of class.
**Domatium** A depression or projection commonly on the undersurface of leaves that may house parasites (pl. domatia) (i.e., *Coprosma*).
**Dorsolateral** Relating to both the back and the sides.
**Double fertilization** The unique character confined to flowering plants, whereby fusion occurs between sperm and egg (giving a 2*n* zygote) and the simultaneous fusion of a second sperm with the two polar nuclei (giving a 3*n* primary endosperm nucleus).
**Double serrate** Larger sharp, forward-pointing, marginal teeth of a leaf, that in turn have small serrations.
**Drupe** A fleshy one-seeded indehiscent fruit with a stony endocarp (i.e., cherry, peach).

**Effectively published** When a new Latin name is published in a recognized botanical journal or book.
**Egg** The female gamete or sex cell.
**Elaters** Hygroscopic arm or band-like structures attached to the spores of *Equisetum* that aid in spore dispersal.
**Electrophoresis** A technique whereby certain chemical compounds, like plant proteins, are separated along an electrical gradient in various gels.
**Elliptic** Broadest at the middle with the length usually more than twice the width (syn. elliptical).
**Emarginate** When the leaf apex has a shallow notch at the broad apex (syn. retuse).
**Embryo** A young sporophytic plant before germination and growth in seed plants.
**Embryostegia** Marking or swelling on the seed coat that gives the location of the embryo inside.
**Emetic** A substance that causes vomiting.
**Endangered species** A taxon in danger of becoming extinct in the near future within all or part of its geographical range if the same causal factors continue to operate.
**Endemic species** A taxon found naturally only in a particular geographical location and in no other place in the world.
**Endocarp** The inner layer of the three layers making up the pericarp of the fruit.
**Endosperm** A nutritive tissue digested by the developing embryo in flowering plants formed from the fusion of male sperm and two polar nuclei (3*n*).
**Entire** A smooth leaf margin lacking any teeth or indentations.
**Environmental isolation** Where gene exchange is inhibited by environmentally controlled factors.
**Ephedrine** A decongestant alkaloid drug found in *Ephedra* plants.
**Epicalyx** An involucre or series of bracts looking like an additional calyx (i.e., *Malva*).
**Epigynous** A floral part arrangement where the parts are adnate to the ovary top or wall; epigynous flowers are always inferior.
**Epiphyte** A plant that lives attached to another plant but does not obtain nourishment from that plant.
**Equitant** Alternate-arranged leaves that have their bases overlappping and flattened lengthwise (i.e., *Iris*).
**Erect** Standing upright, perpendicular to the ground surface (syn. strict).
**Eusporangiate** Type of sporangium development where the sporangium forms from more than one cell and produces many spores.
**Eustele** The cylindrical arrangement of vascular bundles separated by parenchyma tissue.
**Even-pinnate** A pinnately compound leaf lacking a terminal leaflet at the end of the rachis.

**Evergreen** Applied to leaves that persist for longer than one growing season.
**Excurrent branching** With a main central axis or trunk and small lateral branches.
**Exine** The outer wall of the pollen grain.
**Exocarp** The outer layer of the three layers making up the pericarp of the fruit.
**Expectorant** A substance that promotes mucus removal from the nose and respiratory tract.
**Exstipulate** Lacking stipules.
**Extinct** A taxon that no longer exists in the wild within all or part of its geographical range.
**Extirpated species** A taxon extinct within a particular portion of its range but present within other segments of its distribution.

**Falcate** Curved sideways and tapering upward; asymmetric.
**False indusium** The covering of fern sporangia by the rolling of the leaf margin.
**False partition** A partition that does not originate from the carpel wall but extends between two parietal placentas (i.e., Brassicaceae)
**False septum** *See* False partition.
**Family** A taxonomic unit between the higher rank of an order and the lower rank of a genus.
**Fascicles** In clusters (i.e., needles of *Pinus*)
**Father of Modern Taxonomy** Carl Linnaeus, Swedish naturalist physician, 1707–1778.
**Female gametophyte** A multicellular phase in the plant reproductive cycle that produces the female sex organs and gametes.
**Fertile floret** A grass flower that can produce a grain or caryopsis.
**Fertilization** The fusion of two gametes or sex cells forming a diploid ($2n$) zygote.
**Fibrous roots** Several to many equal-sized roots that have replaced tap roots.
**Fiddlehead** A curled young developing fern leaf, resembling the narrow end of a violin (syn. crozier).
**Filament** The slender stalk of a stamen.
**Fimbriate** Fringed with hairs, these being coarser than ciliate.
**Flexostat** A small piece of stiff paper or cardboard with a slit in the middle used to hold bent specimens neatly while drying in a plant press.
**Floccose** Scattered patches of interwoven hairs.
**Flora** A list of the plants occurring in a particular geographic region; also the plants living in a designated area.
**Floret** The grass flower including the sexual parts and the lemma and palea; also applied to a small flower.
**Floristics** The collection and preparation of information for a flora.
**Flower** A specialized shoot that has 1–4 specialized "leaves," one of which will be involved in male or female reproduction; the reproductive organ of flowering plants (angiosperms).
**Flower bud** A bud that contains only embryonic flowers.
**Follicle** A dry, dehiscent one-celled, one-carpellate fruit splitting down only one side (i.e., milkweed).
**Forma** An intraspecific taxonomic name after variety, sometimes used for plants having one to few characters different from the typical species.
**Fragment packet** A small paper envelope attached to a herbarium mounting sheet to hold small plant parts.
**Free-central placentation** Placentation type found in one locule, compound pistils where the ovules are attached to a central column of tissue attached only at the base; developed from axile placentation.
**Frond** A leaf of a fern; also used when referring to palm leaves.
**Fruit** A mature ripe ovary and on occasion associated parts.

**Fruticose** Shrubby.
**Functional megaspore** Surviving haploid megaspore that gives rise to the female gametophyte or embryo sac in seed plants.

**Gamete** A sex cell that combines with another sex cell to form a zygote; haploid *(n)* in chromosome number.
**Gametophyte** The haploid *(n)* gamete-producing phase in alternation of generation in plants.
**Gamopetalous** *See* Sympetalous.
**Gene** The hereditary unit of inheritance located on the chromosome.
**Genera** The plural of genus.
**Generative cell** In flowering plants the cell that divides to ultimately form the two sperm.
**Genus** The first word of the species name of a plant or animal; used as a noun and capitalized.
**Germination** The beginning or resumption of growth.
**Glabrate** Becoming glabrous with age (syn. glabrescent).
**Glabrescent** *See* Glabrate.
**Glabrous** No hairs present; the surface smooth and free of characteristics.
**Gland** A secreting structure that produces nectar or some other liquid.
**Glandular** Bearing glands.
**Glandular hairs** Enlarged gland or secretory structure at the apex of the hair.
**Glaucous** Covered with a waxy covering giving a whitish appearance.
**Glochidiate** Stiff hairs barbed at the apex.
**Glomerule** A general term applied to a dense cluster of flowers.
**Glume** Usually a small bract; applied to the 1 or 2 empty, lower bracts at the base of a grass spikelet.
**Gymnos** A Greek word meaning naked.
**Gynobasic** A style that originates from the base of a lobed ovary.
**Gynoecium** Collectively the carpels in the flower.
**Gynostegium** A covering or sheath over the gynoecium regardless of its origin.

**Half-inferior ovary** Term applied to the stamens and perianth being attached to only part way up the ovary well.
**Halophyte** A plant adapted to growing in salty soil.
**Haploid** When each cell or tissue has only one set of chromosomes *(n)*.
**Hastate** Arrowhead-shaped with basal lobes flared outward.
**Head** An indeterminate dense cluster of sessile flowers (syn. capitulum).
**Herb** A plant made up of little woody tissue and dying back to the ground each year.
**Herbalists** Individuals who practiced the medicinal use of natural herbs and wrote books on the subject.
**Herbals** Name given to early European books on the medicinal use of plants.
**Herbarium** A collection of dried plants, mounted and arranged for reference; a place that houses a dried plant collection.
**Hermaphrodite** *See* Bisexual.
**Hesperidium** A berry-like fruit covered with a thick leathery rind and with the locules filled with fleshy hairs (i.e., orange).
**Heterosporous** The production of two distinct types of spores.
**Heterostyle** Styles and usually stamens being of two or more lengths in different flowers of the same species.
**Hip** A cluster of achenes surrounded by a hypanthium or cup-shaped receptacle (i.e., rose).
**Hirsute** Long, shaggy, stiff hairs.

**Hirsutulous** A less pronounced hirsute condition.
**Hispid** Having long, sharp, stiff hairs that may be capable of penetrating the skin.
**Hispidulous** A less pronounced hispid condition.
**Holdfast roots** Specialized aerial roots that are designed to adhere to porous surfaces.
**Holotype** The one specimen designated by the author of a species as the type; it is known and is available for others to study.
**Homonym** A Latin name rejected because the same designation was previously used for another taxon of the same rank.
**Homosporous** Producing spores that are all the same size.
**Hood** The concave portion of the corona in the family Asclepiadaceae.
**Horn** A tapering, sometimes stiff, appendage of tissue similar to a cow's horn.
**Hybrid** Offspring of homozygous parents.
**Hybridization** The process of forming a hybrid by cross-pollination of plants; interbreeding of varieties, races, species, etc. among plants.
**Hypanthium** A cup-like structure formed by the fusing of the base of the perianth and stamens.
**Hypha** A thread-like filament of a fungus (pl. hyphae).
**Hypocotyl** The elongating axis of an embryo or germinating seedling between the cotyledons and the radicle.
**Hypodermis** A special productive layer of cells beneath the epidermis of some gymnosperms' leaves.
**Hypogynous** A floral part arrangement where the perianth and stamens are inserted under the ovary directly to the receptacle; hypogynous flowers always have a superior ovary.

**Identification** The assigning of an existing name to an unknown plant.
**Illegitimate name** Invalid Latin name due to not following the guidelines of the nomenclature code.
**Imbricate** Overlapping like shingles on a roof.
**Immature male gametophyte** The pollen grain or microspore in the seed plant reproductive cycle.
**Imperfect flower** Having only male or female sex organs in the same flower, a unisexual flower.
**Incised** Sharp, irregularly cut leaf margin segments and sinuses, cut not more than one-third the way to the midrib.
**Incomplete flower** A flower that lacks one or more of its floral parts.
**Indehiscent fruit** A fruit that does not split open.
**Indented key** Type of key where each pair of contrasting choices are equally indented and given the same number, letter, or symbol.
**Indeterminate inflorescence** A flower inflorescence that has the youngest flower terminal and central, with the blooming pattern inward and upward.
**Indeterminate species** Taxa with so little information known about them that it is not known whether they are extinct, extirpated, endangered, threatened, or rare.
**Indusium** Scale-like outgrowth of the leaf epidermis that covers the sporangia of a fern sorus (pl. indusia).
**Inferior ovary** Term applied to the stamens and perianth being attached to the top wall of the ovary.
**Inflorescence** Any grouping or arrangement of flowers.
**Insertion of parts** How the floral parts are arranged or attached to one another.
**Integument** Layer or layers of tissue on the outside of the nucellus of the ovule that develops into the seed coat.
**Internode** The section of stem between two nodes.
**Irregular** In the broad sense, having bilateral symmetry (syn. zygomorphic).
**Isolating mechanisms** Barriers to the exchange of genetic material.

**Isotype** A duplicate specimen of a holotype.
**Isozyme** A different form of an enzyme where the polypeptides are determined by alleles at more than one locus.

**Keel** A dorsally, centrally extending rib, similar to a keel of a boat; also the fused lower petals of a "pea-like" flower.
**Key** A series of contrasting paired choices used to identify an unknown organism by the process of elimination.
**Keying** The process that is used to identify an unknown plant.

**Labellum** An odd petal or lip of an irregular corolla, especially that in the orchid family.
**Lamina** The blade or flattened part of a leaf or petal.
**Lanate** Long interwoven woolly hairs (syn. woolly).
**Lanceolate** Long tapering with the widest toward the base.
**Laticifer** A specialized plant cell that produces latex.
**Latin name** The binomial name of a plant written in Latin and comprised of the genus and the specific epithet (syn. botanical name).
**Leaf scar** Scar or mark left on a stem when the leaf falls from the plant.
**Leaflet** The smaller segments of a compound leaf (syn. pinna) (pl. pinnae).
**Lectotype** A type specimen chosen by a later author from the original specimens cited by the original author of a species name.
**Legitimate name** A validly published Latin name of a plant.
**Legume** A dry, one-celled, one-carpellate, dehiscent fruit splitting down two sides; loosely called a "pod" (i.e., bean, pea).
**Lemma** The outermost or lower of the two bracts that together surround the grass flower (floret).
**Lenticel** Spongy tissue area usually on the surface of woody stems that allows for gas exchange to and from the internal tissues of the plant.
**Leptosporangiate** Type of sporangium development where the sporangium forms from one initial cell and produces usually 64–128 spores.
**Ligulate flowers** *See* Ray flowers.
**Ligule** Small hair-like or scale-like outgrowth from a leaf; also, the flattened corolla of the ray flowers of the Asteraceae.
**Linear** Narrow with parallel sides and a length usually over four times the width.
**Lip cells** Thin-walled cells in fern sporangia that break apart with constriction of the annulus, facilitating the spores being released.
**Lobed** A loosely used term that technically applies to round segments and sinuses of leaf margins not cut over one-half the way to the midrib.
**Locule** A space; applied mostly to the space or spaces within the ovary of flowering plants.
**Loculicidal capsule** A capsule that splits open along the middle of the locule (i.e., Iris).
**Lodicule** The minute paired remains of the perianth of a grass flower that functions to push apart the lemma and palea.
**Loment** A dry legume constricted between the seeds so that the fruit falls apart in one-seeded segments (i.e., *Desmodium*).

**Macromolecules** Chemical components of a high molecular weight and often composed of many repeating units (i.e., proteins, nucleic acid).
**Male gametophyte** A multicellular phase in the plant reproductive cycle that produces the male sex organs and gametes.
**Malpighian hairs** Unicellular two-branched hairs found in the family Malpighiaceae.
**Manual** A publication with keys and descriptions that helps in identifying plants of a particular region.

**Margin** At the outside limit or edge (i.e., along the edge of the leaf blade).
**Marginal placentation** Placentation type found only in simple pistils where the ovules are attached to a placenta along the marginal wall of the carpel.
**Megagametophyte** *See* Female gametophyte.
**Megasporangium** The sporangium in a heterosporous plant that contains the larger of the two spore types, the megaspore.
**Megaspore** A spore produced in a megasporangium that is involved in female sex organ development.
**Megaspore mother cell** The diploid cell that enlarges and undergoes meiosis forming haploid megaspores.
**Megasporophyll** A bract or leaf-like structure that produces megasporangia.
**Mesocarp** The middle layer of the three layers making up the pericarp of the fruit.
**Microgametophyte** *See* Male gametophyte.
**Micromolecules** Chemical compounds of a relatively low molecular weight found as end products of biosynthetic pathways.
**Micropyle** The small opening in the integuments of the ovules of seed plants through which the pollen tube enters.
**Microsporangium** A sporangium in a heterosporous plant that produces microspores.
**Microspore** A small spore produced in a microsporangium that is involved in male sex organ development; an immature male gametophte.
**Microspore mother cell** A specific cell that undergoes meiosis and produces haploid pollen; synonymous with pollen mother cell.
**Microsporophyll** A bract or leaf-like structure that produces microsporangia.
**Midrib** The central or midvein of a leaf.
**Mixed buds** A bud that contains both flowers and leaves.
**Monoecious** Male and female reproductive structures on the same plant.
**Monograph** A comprehensive study of a group of plants including all aspects of the taxonomy and biology of the group to arrive at a workable classification based on this information.
**Monotypic** Having only one, as a genus with only one species.
**Monomorphic** Having one separate form.
**Mucronate** Having a leaf apex with an abrupt, short, soft tip.
**Multiple** A fleshy false fruit made up of more than one flower (i.e., mulberry, pineapple).
**Mycorrhiza** In fungi, the combination of the fungus and the root of a vascular plant.

**Natural classification** A classification system based mostly on large morphological features in an attempt to show affinity.
**Nectar gland** *See* Nectary.
**Nectary** A gland or group of glands that secrete nectar.
**Needle-shaped** Very long and narrow with parallel sides (i.e., *Abies, Pinus*) (syn. acicular).
**Neotype** A specimen chosen as a nomenclatural type when all plant material studied by the original author of a species name is lost.
**Netted** *See* Reticulate.
**Net veined** The small veins of a leaf branching and joining like a fish net.
**Nodal plexus** A complex network of horizontal vascular bundles.
**Node** The location on the stem where leaves or branches originate and occasionally other structures are attached (i.e., tendrils, flowers).
**Nommenclature** The application of technical names to plants according to an agreed set of rules.
**Nucellus** Tissue making up the bulk of the young plant ovule in which the embryo sac develops (syn. megasporangium).
**Nut** A loosely used term applied to a one-seeded, dry, indehiscent fruit with a hard coat (i.e., hickory, pecan, walnut).

**Ob-** A prefix meaning the reverse of the typical.
**Obcordate** Broadly heart-shaped with the petiole attached at the narrow end.
**Oblanceolate** Long and tapering with the widest part toward the apex and the petiole attached at the narrow end.
**Oblique** When both sides of a leaf base are unequal and asymmetrical.
**Oblong** The sides generally parallel with the ends rounded and two or three times longer than broad.
**Obovate** Egg-shaped or ovate but connected at the narrow end.
**Obtuse** Leaf apex being nonpointed but round.
**Ocrea** A sheathing stipule at the node formed by the fusion of two stipules (pl. ocreae) (i.e., Polygonaceae).
**Odd-pinnate** A pinnately compound leaf with a terminal leaflet at the end of the rachis.
**Opposite leaves** A type of phyllotaxy where two leaves are at a node and attached across from each other.
**Orbicular** Circular.
**Order** A taxonomic unit between the higher rank of a class and the lower rank of family.
**Out of danger species** A taxon that is presently secure due to conservation measures and no longer in one of the risk categories.
**Oval** A loose term meaning broadest at the middle but usually rounded at the ends and with the width over one-half the length.
**Ovary** The enlarged basal region of the carpel or pistil, housing the developing seeds and maturing into the fruit.
**Ovate** Egg-shaped with a stalk or petiole attached at the broad end.
**Ovule** Structure that develops into a seed and is not enclosed within another surrounding structure.

**Palea** The innermost scale, often chaffy, of the two bracts surrounding the grass flower.
**Palmately trifoliate** When the three leaflets are alike.
**Palmate veined** When the main veins of a leaf radiate from one point.
**Panicle** An indeterminate inflorescence comprised of two or more flowers on each pedicel; may be compound.
**Pantropical** Being distributed or found throughout the tropical regions of the world.
**Parallel veined** When the veins of a leaf are all about the same size and run equal distance from each other and with the interconnected veins obscure.
**Parasite** Living attached to another plant to the detriment of that plant.
**Parietal placentation** Placentation type found only in compound pistils where the ovules are attached to the placenta along the outer wall of the carpel, usually only one locule; sometimes includes marginal placentation.
**Parsimony** The shortest hypothetical route that explains the most likely evolutionary pathway.
**Parted** Leaf margin segments cut one-half to three-fourths of the way to the leaf base or midrib; the sinuses may be rounded or sharp.
**Pedicel** Supporting stalk of a flower.
**Peduncle** The supporting stalk of an entire inflorescence.
**Peltate** Umbrella-shaped leaf with the petiole attached to the lower surface of the blade usually away from the leaf margin; also applied to scales of some gymnosperm cones.
**Pepo** A fleshy fruit derived from a compound ovary where the outer wall (exocarp) becomes hard and tough (i.e., cucumber, watermelon).
**Perennial** A plant that lives for an indefinite number of years from the same root stock or underground system.

**Perfect flower** A flower with stamens and pistil functional.
**Perfoliate** When the stem apparently passes through a leaf blade or where opposite leaf bases seem to unite around the stem.
**Perianth** The petals and sepals of a flower taken collectively.
**Pericarp** The mature ovary wall of the fruit, usually comprised of three layers.
**Perigynium** A sac-like bract that surrounds the pistil in some members of the Cyperaceae (i.e., *Carex*).
**Perigynous** A floral part arrangement where the perianth segments are adnate into a "floral cup" or hypanthium from around the base of the ovary; perigynous flowers are always superior.
**Perispore** Wrinkled envelope covering the spore.
**Petal** The second set of "floral leaves" or perianth of a flower located between the sepals and stamens, usually colored or white (pl. corolla).
**Petaloid** Like petals in appearance.
**Petiole** The stalk of a simple leaf blade or compound leaf.
**Phylloclad** A flat branch looking like and functioning as a leaf (syn. cladophyll).
**Phyllode** A flattened petiole that takes the place of a blade functionally and morphologically (i.e., some *Acacia*).
**Phyllotaxy** How leaves are arranged on a stem.
**Phylogenetic classification** A classification system based on a wide variety of information in an attempt to construct phylogenetic relationships.
**Pilose** Scattered long, soft, nearly straight hairs.
**Pinna** The first primary division of a compound leaf, especially as applied to ferns.
**Pinnately trifoliate** When the terminal leaflet is different from the two lateral leaflets by having a longer stalk.
**Pinnatifid** A pinnately cleft, lobed, or parted leaf margin divided one-half to three-fourths of the way to the midrib.
**Pinnules** The pinnate (or second) segment of a compound leaf.
**Pistil** The female sex organ of the flower that produces seeds and is positioned in the center of the flower (pl. gynoecium); consists of stigma, style, and ovary; made up of one or more carpels.
**Pistillate** Flowers having pistils but no functional stamens.
**Pith** Large-celled, usually parenchyma tissue found in the central region of the root or stem.
**Placenta** The region of the ovary wall where ovules are attached.
**Placentation** The attachment pattern of ovules within the ovary of flowering plants.
**Plantlet** A small plant; also a small plant produced asexually in place of a flower, in an inflorescence or at the end of a runner.
**Plumule** The first bud of the embryo that develops into the future shoot.
**PMC** *See* Pollen mother cell.
**Pohlstöeffe** A name applied to a softening or wetting solution designed to soften dried plant parts for dissection and identification; named by graduate students after Dr. R. W. Pohl of Iowa State University, U.S.A., who developed the formula.
**Polar nuclei** Usually two nuclei located in the middle of the embryo sac that fuse with a male gamete to form the primary endosperm ($3n$) nucleus in flowering plants.
**Pollen** A general term applied to the haploid ($n$) pollen grains or microspores, which are the immature male gemetophytes in the seed plant reproductive cycle.
**Pollen mother cell** A specific cell that undergoes meiosis and produces haploid pollen; synonymous with microspore mother cell.
**Pollen sac** The chamber in the anther of the stamen that contains the pollen.
**Pollen tube** The tube formed after the pollen grain has germinated that carries the male gametes to the ovule during seed plant reproduction.
**Pollination** Transferring pollen from its point of origin (i.e., anther) to a receptive surface (i.e., stigma).

**Pollinium** A waxy mass of pollen; found in the families Asclepiadaceae and Orchidaceae (pl. pollinia).
**Polyclave keys** Keys that are multi-entry, any order, and use stacked cards with holes or punched edges corresponding to characters to be chosen.
**Polygamous** Having both perfect and unisexual flowers on the same plant.
**Polynomials** Long, cumbersome Latin names used by early European botanical writers; prior to the time of Carl Linnaeus.
**Polypetalous** Term applied when the petals are not united at all (distinct) to one another.
**Polyploid** An organism with two or more sets of chromosomes or genomes (i.e., triploid [$3n$], tetraploid [$4n$], hexaploid [$6n$], octaploid [$8n$]).
**Pome** A fleshy fruit formed from a compound, inferior ovary in which the receptacle (or calyx tube) becomes thick and fleshy (i.e., apple, pear).
**Pore** Opening in the exine wall of the pollen grain through which the pollen tube emerges.
**Poricidal capsule** A capsule that opens by pores near the top of the fruit (i.e., poppy).
**Position of ovary** The relative position of the ovary in relation to other parts.
**Prickle** Sharp outgrowth of the stem epidermis or cortex that lacks conductive tissue (i.e., rose).
**Primary endosperm nucleus** The resulting triploid ($3n$) nucleus following fusion of two polar nuclei and one male gamete nucleus (sperm) that gives rise to endosperm in flowering plants (syn. triple fusion nucleus).
**Primary root** The first root of a plant that develops as a continuation of the radicle of the embryo.
**Procumbent** Lying on the ground and not rooting at the nodes.
**Prop roots** Special adventitous roots that help support the lower regions of the plant.
**Prostrate** Lying flat on the ground and possibly rooting at the nodes.
**Prothallus** A multicellular, often flattened, thallus-like structure bearing the plant sexual organs, the antheridium and archegonium; the gametophyte stage in the fern reproductive cycle (pl. prothalli).
**Protostele** The simplest type of stele, made up of a solid column of vascular tissue.
**Pteridologist** A person who studies ferns.
**Pteridosperms** *See* Seed ferns.
**Puberulent** Very short hairs.
**Pubescent** A general term applied to hairs of any type; also applied to soft, short to medium length hairs.
**Pulvinus** Swollen tissue at the base of the blade that controls leaf orientation.
**Punctate** Having pits or dots formed by glands or waxy spots.

**Raceme** An indeterminate inflorescence with single flowers on pedicels arranged along the rachis.
**Rachilla** The small rachis or axis found in a grass spikelet; the secondary axis in sedges.
**Rachis** The elongate central stalk or axis in an inflorescence or compound leaf.
**Radicle** The first structure to emerge from a germinating seed to give rise to the primary root.
**Rare species** Taxa with small populations usually localized in distribution that may be at risk but are not at present endangered or threatened.
**Ray flowers** The marginal flowers of the head inflorescence in the Asteraceae (Compositae) that are different from the regular ones in the center of the head (syn. ligulate or strap-shaped).
**Receptacle** The apex end of a flower stalk that bears the parts of a flower.
**Regular** A flower with the perianth parts similar in color, form, shape and texture; having radial symmetry (syn. actinomorphic).

**Reniform** Kidney-shaped and broader than long.
**Repand** *See* Unulate.
**Reproductive isolation** When gene exchange is inhibited by geneotypically controlled differences in the reproductive behavior between individuals of different populations.
**Reticulate** Forming a network (syn. netted).
**Retuse** *See* Emarginate.
**Revolute** When the leaf margin is rolled under toward the underside of the leaf.
**Rhizoid** Root-like structures that lack the specialized structure of roots.
**Rhizome** An elongate modified stem growing usually below the ground surface and rooting at the nodes.
**Root** The basal axis of the plant that lacks nodes and internodes and provides anchorage and moisture absorption from the ground.
**Rootstock** A loosely used term applied to specialized underground stems (i.e., caudix).
**Rosette** A dense cluster of leaves radiating from a central point; may be at the base of the plant or at the end of a stem or branch.
**Rostellum** A slender appendage from the upper part of the stigma, as seen in orchids; a small beak.
**Rounded** With a broadly rounded leaf apex.
**Rugose** Wrinkled.
**Rule of priority** Rule of the nomenclature code stating that the oldest of conflicting Latin names is the correct name.
**Runner** A slender aboveground modified stem that roots only at the end.

**Sagittate** Arrowhead-shaped but with the basal lobes turned inward.
**Saline** Consisting of or containing salt; a solution or soil having much salt within.
**Samara** A dry, indehiscent, winged achene.
**Sapling** A young tree.
**Saprophyte** A plant lacking green color and depending upon decaying organic matter for necessary nutrients (i.e., *Monotropa*).
**Scabrous** A rough rasp-like surface to the touch.
**Scale-like** Overlapping as scales on a fish.
**Scandent** Same as climbing.
**Scape** A naked flowering stem without leaves.
**Scapose** Bearing a scape.
**Scar** The remains of a point of attachment.
**Scarious** Thin, dry, membranous and more or less translucent.
**Schizocarp** A dry, indehiscent fruit made up of two or more one-seeded carpels that separate from each other, leaving a connection between them (i.e., parsley family).
**Scientific name** *See* Latin name.
**Scorpioid cyme** A single axis inflorescence with one-sided flowers arranged along the rachis.
**Scurfy** Covered with overlapping scales; giving the appearance of cornmeal or "dandruff" on the surface.
**Seed** A mature ovule covered with a hard seed coat and bearing an embryo and sometimes leftover endosperm.
**Seed coat** Hard outer covering of a seed.
**Seed ferns** An extinct fossil group of seed plants that had fern-like or cycad-like leaves bearing seeds; also called pteridosperms.
**Seedling** A new, young plant.
**Semi-parasite** A plant that is green with chlorophyll but needs necessary substances from a host plant for particular biological processes (i.e., *Castilleja*).
**Sepal** The outermost whorl of "floral leaves" or perianth of a flower, usually green (pl. calyx).

**Septicidal capsule** A capsule that splits along the septa, separating it in half (i.e., century plant).
**Septum** The partition between locules of a fruit or ovary (pl. septa).
**Sericeous** Long, soft, appressed hairs, mostly all pointed in the same direction and giving the surface a silky appearance.
**Serology** A technique that relies on the comparison of an animal's immunological antibody-antigen reaction when invaded by foreign proteins.
**Serrate** Sharp teeth of the leaf margin directed forward toward the apex.
**Serrulate** Leaf margin with very small serrate teeth.
**Sessile** Lacking a stalk.
**Sheath** A thin tubular structure that surrounds partially or wholly another structure.
**Shrub** A woody perennial plant without a main stem but with several to many branches arising at ground level and usually less than 3–4 m tall.
**Silicle** A short (less than twice the width), two-locular dry fruit splitting with each half (valve) separating from the other, leaving a thin septum remaining (i.e., *Capsella, Lepidium*).
**Silique** A long (length more than twice the width) two-locular dry fruit, splitting, with each half (valve) separating from one another, leaving a thin septum remaining (i.e., *Brassica*).
**Simple inflorescence** Where flowers are attached directly to the main axis and the inflorescence is unbranched (i.e., raceme).
**Simple leaf** A leaf made up of a single blade.
**Simple pistil** A pistil consisting of one carpel and one placenta.
**Sinuate** Having a pronounced, wavy leaf margin.
**Sinus** The groove or notch between two adjacent structures (i.e., space between two lobes of a leaf).
**Siphonostele** A type of stele with a central core of pith surrounded by a cylinder of vascular tissue.
**Solid pith** Pith tissue that is uniform and solid throughout.
**Solitary** A single flower at the end of a peduncle.
**Sorophore** A gelatinous ring-like, hygroscopic structure inside the sporocarp to which sporangia are attached (i.e., aquatic fern *Marselia*).
**Sorus** A cluster of sporangia on a fern frond (pl. sori).
**Spacial isolation** When gene exchange is limited or prohibited because of the great distance between populations.
**Spadix** A fleshy, densely flowered spike with unisexual flowers and subtended by a spathe.
**Spathe** A large leafy-like bract subtending an inflorescence; sometimes showy and colored.
**Spathella** When a pair of small bracts enclose or subtend an inflorescence.
**Spatulate** Resembling a spatula and being rounded and broad at the apex and elongate tapering toward the base.
**Species** The binomial Latin or botanical name of a known plant or animal, made up of the genus name and the specific epithet.
**Specific epithet** The second name of the species name of a plant, used as an adjective or possessive noun and not normally capitalized.
**Specimen label** The permanent label applied to the lower right-hand corner of a herbarium sheet that gives all of the pertinent information about the specimen.
**Sperm** The motile, male reproductive or sex cell that is smaller than the female sex cell.
**Sperma** A Greek word meaning "seed."
**Spike** An unbranched inflorescence with sessile flowers.
**Spikelet** A small spike; the primary inflorescence in grasses and sedges having small, scale-like bractlets attached to the rachilla and subtending the small flowers.

**Spine** A sharp, stiff outgrowth of a stem, usually lacking vascular tissue; considered by some to be synonymous with thorn.
**Sporangiophore** A stalked structure bearing sporangia.
**Sporangium** A spore-producing structure (pl. sporangia).
**Spore** Usually a haploid *(n)* unicellular reproductive cell, the result of meiosis, with the ability to develop into another plant body by mitosis.
**Spore mother cell** A diploid (2*n*) cell that undergoes meiosis and usually produces four haploid *(n)* cells.
**Sporocarp** A hard or soft seed-like structure containing sporangia and spores.
**Sporophyll** Leaf or leaf-like organ that bears sporangia.
**Sporophyte** The phase in the plant reproductive cycle having alternation of generation and producing diploid (2*n*) spores.
**Stamen** The male sex organ of the flower that produces pollen (pl. androecium); made up of the anther and filament.
**Staminate** Flowers having stamens but no functional pistil.
**Staminode** A sterile, often modified stamen.
**Standard** The upper, commonly largest petal in a "pea-like" flower (syn. banner).
**Stele** The central vascular tissue cylinder inside the roots and stems of vascular plants.
**Stellate** Star-shaped hairs with the segments radiating from a central point.
**Stem** The normally aboveground supporting axis of vascular plants upon which leaves and flowers are attached at nodes.
**Sterile floret** A grass flower that lacks a pistil and is unable to produce a grain or caryopsis.
**Stigma** Area of the carpel with a surface receptive to pollen.
**Stipe** The stalk or petiole of a fern leaf; also the support stalk of the pistil (gynoecium).
**Stipitate** Having a stalk or stipe.
**Stipular spines** Stipules that are modified into sharp, pointed stiff outgrowths.
**Stipulate** Having stipules.
**Stipule** An appendage on either side of the base of the leaf or petiole at the attachment point to the stem.
**Stipule scar** Scar where stipules were attached.
**Stolen** A modified stem that trails along the ground and roots at the nodes.
**Strap-shaped flowers** *See* Ray flowers.
**Strict** *See* Erect.
**Strigose** Short, stiff, appressed hairs, all pointing in the same direction.
**Strobilus** A cone or cone-like structure (pl. strobili).
**Style** Elongated stalk between the ovary and stigma of the carpel through which the pollen tube grows.
**Subspecies** An infraspecific taxon name between species and variety names.
**Subulate** *See* Cordate.
**Superior ovary** When the ovary of the flower is distinct from all the other flower parts.
**Suspensor** An early formed part of many vascular plant embryos that pushes the embryo into the endosperm.
**Suture** The line or seam where a mature fruit splits or dehisces.
**Sympetalous** Term applied when the petals are fused (connate) together toward the base, forming a cup or tube (syn. gamopetalous).
**Symsepalous** Term applied when the sepals are fused (connate) together toward the base forming a cup or tube.
**Synangium** A group of fused sporangia.
**Syncarpous** Term applied to flowers with two or more fused (connate) carpels.
**Synconium** A fleshy fruit with a hollow receptacle or peduncle that houses many small achenes inside; a type of multiple fruit (i.e., *Ficus*).

**Synergids** Two short-lived cells next to the egg in the embryo sac of the ovule of flowering plants.
**Synonym** A Latin name designation rejected as being incorrectly applied to a taxon.
**Synoptical keys** *See* Polyclave keys.
**Syntype** One of two or more specimens mentioned by the author of a species when no type was mentioned.
**Systematics** The study of organisms and the diversity and relationships between them; treated by some botanists as a synonym for taxonomy.
**Systematist** Person trained in the principles of systematics.

**Tautonym** When the genus and specific epithet names are the same.
**Taproot** A tapering underground axis of the plant from which smaller lateral roots develop; may be thick or thin.
**Taxon** A taxonomic group or unit.
**Taxonomist** An individual involved in the identification, classification, and naming of plants and animals.
**Taxonomy** The study of classification, including its rules, theory, principles, and procedures.
**Temperate region** A region of varied-temperature climate that lies between the Tropic of Cancer and the Arctic Circle and between the Tropic of Capricorn and the Antarctic Circle.
**Tendril** A slender twisting appendage that attaches to other plants or structures, an organ for support.
**Tepal** Term applied to the perianth parts when the petals and sepals are indistinguishable from each other.
**Terminal bud scale scars** Clustered rings of scars remaining when the terminal bud scales have dropped.
**Ternate** *See* Trifoliate.
**Tetrad** In a group of four.
**Tetradynamous** Stamen arrangement with the inner filaments long and the two outer stamens with short filaments (i.e., Brassicaceae).
**Thorn** A sharp, stiff outgrowth of a stem usually with vascular tissue; considered by some to be synonymous with spine.
**Threatened species** Taxa that are thought likely to become endangered in the wild in the near future within all or part of their geographical distribution if the same causal factors continue to operate.
**Tomentose** Densely interwoven (woolly) soft hairs forming a covering that can conceal the true surface.
**Topotype** A specimen collected at the same locality as the type specimen but at a different time; not properly a type according to the code.
**Translator arm** A structure connecting pollinia from adjacent anthers to a gland on the side of the pistil or gynoecium (i.e., Asclepiadaceae).
**Translucent** Diffusing or transmitting light so that objects beyond are not clearly seen.
**Tree** A woody perennial plant with a single main stem or trunk and radiating branches on the upper portion of the plant, usually over 4 m tall at maturity.
**Trifoliate** A compound leaf when the leaflets are three in number (syn. ternate).
**Tripinnate** Three times branched.
**Tripinnately compound** A compound leaf with the original leaflets three times divided.
**Truncate** Being straight across or cut off abruptly.
**Tube cell** In the male gametophytes and pollen grains, the cell that gives rise to the pollen tube.
**Tuber** A fleshly underground modified stem covered with many "eyes" (buds) that grow into aboveground stems (i.e., potato).

**Tubular flower** Regular flower in the head inflorescence of the family Asteraceae (Compositae).
**Two-ranked** Leaves conspicuously in two rows (syn. distichous).
**Type** A specimen designated by the author of a species as the type or morphological example for that taxon (syn. holotype).

**Umbel** An indeterminate generally flat-topped or orbicular inflorescence with equal-length pedicels arising from a single point at the end of the rachis; may be compound.
**Uncinate** Stiff hairs or spines with a hook at the end.
**Unisexual flowers** Having only stamens or only pistils in the flower.
**Unulate** Having a slightly wavy leaf margin (syn. repand).

**Validly published** When a new Latin name is published according to the proper format given in the nomenclature code.
**Valve** A segment of the mature fruit wall that separates from the remaining wall segments.
**Variety** An intraspecific taxon name between subspecies and forma.
**Vascular** Refers to a region or tissue in a plant that conducts or gives rise to conducting tissues (i.e., phloem, xylem, vascular cambium).
**Vasculum** A cylindrical container made of metal or plastic used mostly in the past for collecting plants; has a hinged door on one side and is carried by a handle or strap.
**Vein** Slender threads of vascular tissue in a leaf or other organ.
**Velutinous** Velvet-like.
**Venation** The vein pattern found in leaves and other flattened organs.
**Vernacular name** Name given to a plant taken from the language of the culture or society where found (syn. common name).
**Verrucose** Covered with wart-like structures.
**Verticil** An inflorescence with the flowers in whorls at the nodes.
**Verticillate** *See* Whorled leaves.
**Vienna code** An early botanical code of nomenclature followed in Europe during the late nineteenth and early twentieth centuries.
**Villose** *See* Villous.
**Villous** Long, soft, wavy hairs that are not matted (syn. villose).
**Viscid** A sticky surface, as being covered with honey.
**Voucher label** A label attached to a specimen or herbarium sheet that establishes the authenticity of the specimen.

**Whorled leaves** A type of phyllotaxy where three or more leaves are attached to a node (syn. verticillate).
**Wing** A thin, flattened or membrane-like outgrowth of a structure or organ; also one of the two lateral petals in a "pea-like" flower.
**Woody** Hard textured with wood.
**Woolly** *See* Lanate.

**Xerophyte** A plant adapted to a dry or arid environment.

**Zygomorphic** *See* Irregular.
**Zygote** The resulting diploid ($2n$) cell following fusion of two haploid ($n$) male and female gametes or sex cells.

# Index

The index contains Latin names of species and higher groups, as well as English (common) names mentioned in this book. Names of individuals and certain selected topics are also included. Family names in current use are in bold type; other family names are not discussed in detail or are synonyms. When the Latin name is also the English name (e.g., *Aster*, aster; *Crocus*, crocus), the entry is the Latin form. Page numbers in bold indicate illustrations of the families.